Acarid Phylogeny and Evolution: Adaptation in Mites and Ticks

Acarid Phylogeny and Evolution: Adaptation in Mites and Ticks

Proceedings of the IV Symposium
of the European Association of Acarologists

Edited by

Fabio Bernini
Università di Siena

Roberto Nannelli
Istituto Sperimentale di Zoologia Agraria

Giorgio Nuzzaci
Università degli Studi di Bari

and

Enrico de Lillo
Università degli Studi di Bari

SIENA 2000
4th Symposium of the European
Association of Acarologists

KLUWER ACADEMIC PUBLISHERS
DORDRECHT / BOSTON / LONDON

A C.I.P. Catalogue record for this book is available from the Library of Congress.

ISBN 1-4020-0465-6

Published by Kluwer Academic Publishers,
P.O. Box 17, 3300 AA Dordrecht, The Netherlands.

Sold and distributed in North, Central and South America
by Kluwer Academic Publishers,
101 Philip Drive, Norwell, MA 02061, U.S.A.

In all other countries, sold and distributed
by Kluwer Academic Publishers,
P.O. Box 322, 3300 AH Dordrecht, The Netherlands.

Printed on acid-free paper

Cover Illustration:
Letter B with typical mites whose genus name starts with B. – Water-colour,
A. Berlese (from hand-written catologue "Acari della Collezione Berlese", 1917-1919,
Istituto Sperimentale per la Zoologia Agraria, Firenze, Italy).

Printed in the Netherlands.

This volume contains the Proceedings of the 4th Symposium of the European Association of Acarologists (Eur.A.Ac.) which was held at the Complesso Didattico del Laterino, Siena, Italy, from 24 to 28 July 2000.

Organizing Committee

Prof. F. Bernini (Chairman)
Dipartimento di Biologia Evolutiva, Università di Siena
Dr. E. de Lillo (Secretary)
Dipartimento di Biologia e Chimica Agro-forestale e Ambientale, Università di Bari
Dr. M. Migliorini (Secretary)
Dipartimento di Biologia Evolutiva, Università di Siena
Dr. M. Castagnoli
Istituto Sperimentale di Zoologia Agraria, Firenze
Dr. R. Nannelli
Istituto Sperimentale di Zoologia Agraria, Firenze
Prof. G. Nuzzaci
Dipartimento di Biologia e Chimica Agro-forestale e Ambientale, Università di Bari
Prof. S. Ragusa
Istituto di Entomologia Agraria, Università di Palermo
Servizio Congressi
Università di Siena

Editorial Committee of the Proceedings
Castagnoli Marisa, Di Palma Antonella, Fenton Brian, Ragusa Salvatore, Todisco Franca

Aknowledgements
The Organizing Committee acknowledges the support of the following institutions and organizations for their contribution, cooperation and support:
Monte dei Paschi di Siena
Università degli Studi di Siena
Università degli Studi di Bari
European Association of Acarologists

Contents

Preface

The words pronounced by Serge Kreiter during the meeting come to mind. They could record exactly the situation of Acarology in Europe and in the World: *"I think that in many European countries there are very few full time acarologists. It is very rare to have new positions available ... . And public money, from the European Community but also from national countries, is very hard to get when you want to work on mites. ... Could two acarological associations in Europe (Eur.A.Ac. and S.I.A.L.F.) work together or, better, get married?"*

So, the fourth symposium of the European Acarologists has not only been the occasion to have an idea on which direction the research is addressed today, but also it pointed out the difficulties of our *"scientific branch"*. On the basis of the presentations and invited papers we had evidences of a *"new"* Acarology based on modern techniques and methods of investigations but also the importance, often sheltered even if of relevant value, of the *"old"* Acarology made on the alpha taxonomy and basic studies. So, a *"new"* Acarology needs the *"old"* one.

In this context, the hope to put together the European acarologists has been coming up. This fact, of political meaning, can surely improve the *acarological movement* and the discussion on this point showed clearly the importance of several other activities and efforts in this direction.

We hope that the meeting in Siena will represent a significative stone for the progress of Acarology.

This fourth symposium was attended by 88 acarologists from 16 countries (12 European) with 77 contributions including invited papers, oral presentations and posters. All submitted manuscripts were critically reviewed by at least two competent referees with the acceptance of 46 original contributions included in this book.

We also thank all colleagues in Siena who helped in organizing the symposium and in particular Massimo Migliorini, the other members of the organizing committee, for their cheerful help in organization and registration of the partecipants, and for their general assistance at the registration desk, for assisting with slide and video projection. Finally, we wish to thank all the referees, who significantly improved the quality of this book by writing critical commentaries, Antonella Di Palma, who gave fundamental and essential contributions regarding the editorial work, and Brian Fenton for his English reviewing.

The editors

PRESIDENTIAL ADDRESS

Colleagues, ladies and gentleman,

On behalf of the Organizing Committee of the 4th Symposium of the European Association of the Acarologists, I have the pleasure and the honour of opening this meeting and welcoming all the participants.

First of all I intend to thank the University of Siena for its generous financial and logistic support.

After Graz (Austria), Krynica (Poland) and Amsterdam (The Netherlands), European acarologists meet in Siena (Italy).

What can we say about acarology and acarologists at the dawn of the new millennium? Let me to expose briefly three or four considerations.

1) Acarologists are more and more general biologists and zoologists, who choose mites and ticks as models. Turning the pages of the Proceedings of the past acarological congresses, we find it stressed that acarology *is a modern science*, that acarologists use *modern instruments and methods like other biologists*. This emphasis suggests that acarologists have an inferiority complex. However, this is no longer justified! For a long time Acarologists have been conditioned by the smallness of the creatures they study. Quite suddenly, the instruments available to us have become incredibly sophisticated: TEM and SEM microscopy, computer-based image analysis, immunochemical techniques, physiological and genetic experiments, molecular systematics and developmental biology (with sequencing and comparison of homeobox genes) are all available (and used!) for phylogenetic and applied studies on mites and ticks.

2) Acarology and other branches of science are increasingly expensive. Instruments, methods, scientific journals of high impact, and meetings have high costs. Good research is done almost exclusively in Universities and other public and private Institutions. Even the financial budgets available to these "professional" researchers are diminishing. Valid research has been interrupted or slowed down, as demonstrated by the repertories of acarological papers. The distance between professional and amateur researchers has therefore increased. This has a negative effect on classical systematics, the foundation of all other research.

3) Acarologists, and other scientists, tend to specialize, or even hyperspecialize, becoming systematists, ecologists, geneticists, stored products specialists, toxicologists, parasitologists and so on. Perhaps it is inevitable; in fact, as Krantz (1996) pointed out "the horizons of acarology have grown so vast that a generalist approach to the field no longer seems to be a realistic aspiration". However, Krantz continued, there is a strong risk that "the specialist will lose sight of the broader scientific, social and historic fabric into which his or her discipline is woven" (Krantz 1996).

This is why the Organizing Committee again thought it appropriate to propose the general theme of Phylogeny and Evolution for this Symposium. Evolution alone is a framework for many kinds of research and imparts biological significance.

4) Acarology and acarologists must not lose their cultural identity and their historical roots. Despite all these problems, acarologists should maintain the capacity to be touched and enlivened by images, such as these I am going to show you (figs 1-5). Joy, fantasy and irony are emotions that all scientists need!

I hope that the five sections concerning systematics, phylogeny, morphology, development and genetics of soil, parasitic, water and plant-dwelling Acari, and the ecological and applied aspects of these microarthropods bring about lively scientific discussion. The round table on the future of acarology is designed to promote another kind of discussion and proposals. Comparison of the results of different laboratories, exchange of views on materials, analyses and hypotheses will hopefully give impetus to future trends and solve standing questions.

I wish our Symposium much success! I wish all of you a very pleasant stay in Siena and the surrounding area.

References

Krantz G.W., 1996 - Specialization and systematics in Acarology: Reflections and Predictions. In: Mitchell R., Horn D.J., Needham G.R., Welbourn W.C. (eds.), Acarology IX. Proceedings, Columbus, Ohio, vol. 1: 1-4.

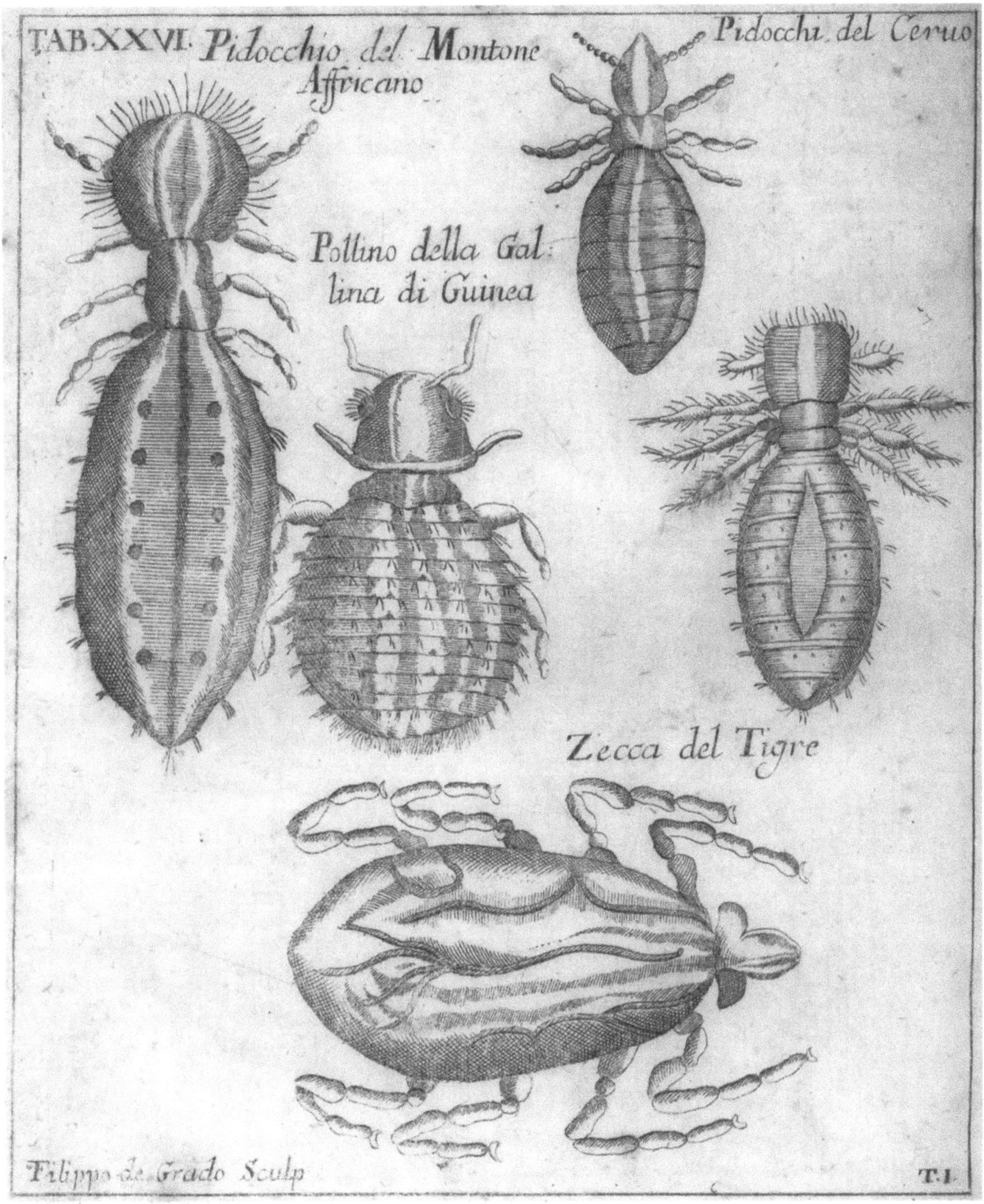

Fig. 1 - The tiger tick together with other parasites (louses) illustrated by Francesco Redi from an ancient edition published in Venice, 1762 (by courtesy of Prof. P. Omodeo).

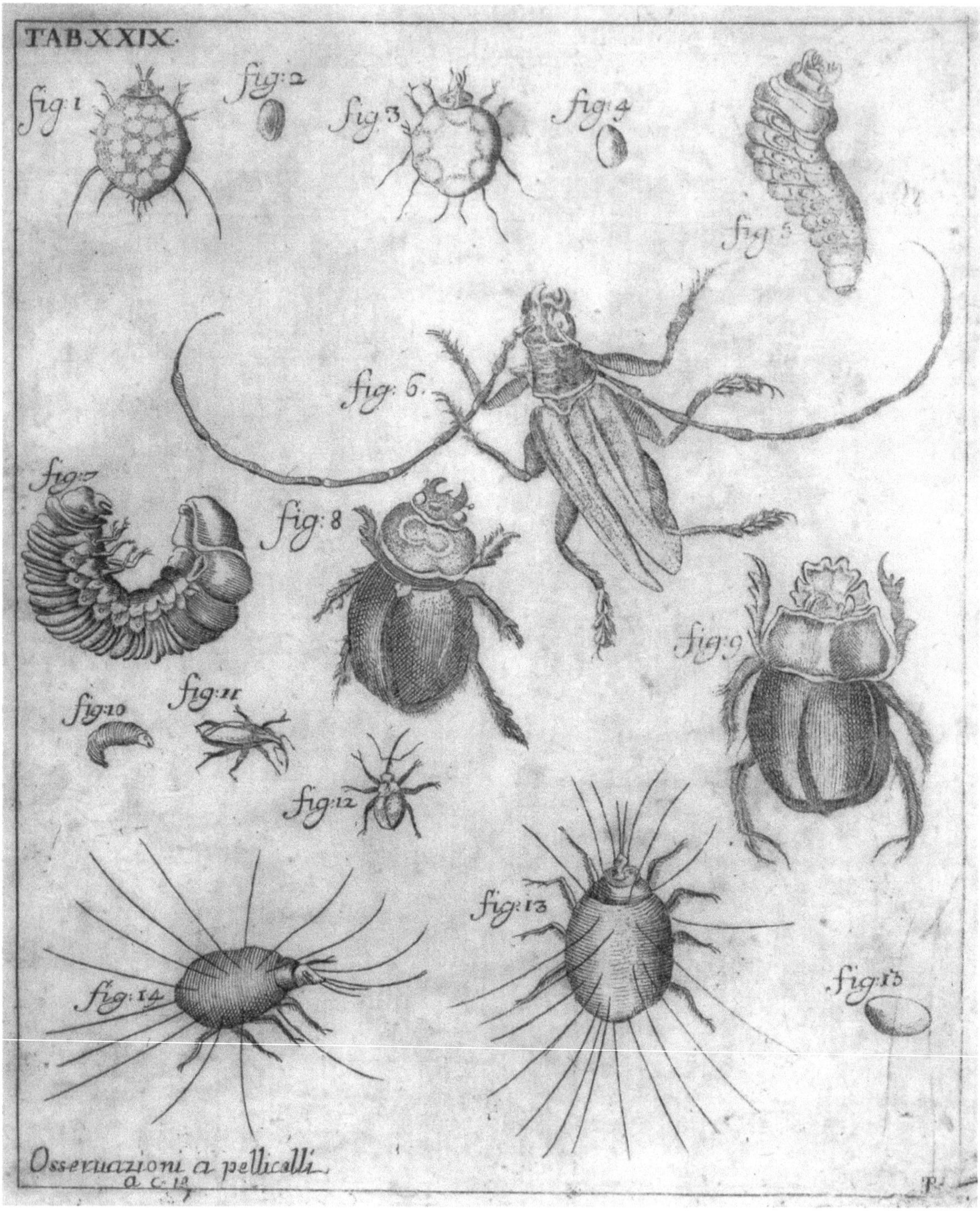

Fig. 2 - In the upper left side presumably two specimens of *Sarcoptes scabiei* (named by F. Redi, "baccherozzolino") are illustrated whilst downwards are two specimens of "vermi" from cheese (presumably *Tyrophagus* sp.) from the same book (by courtesy of Prof. P. Omodeo).

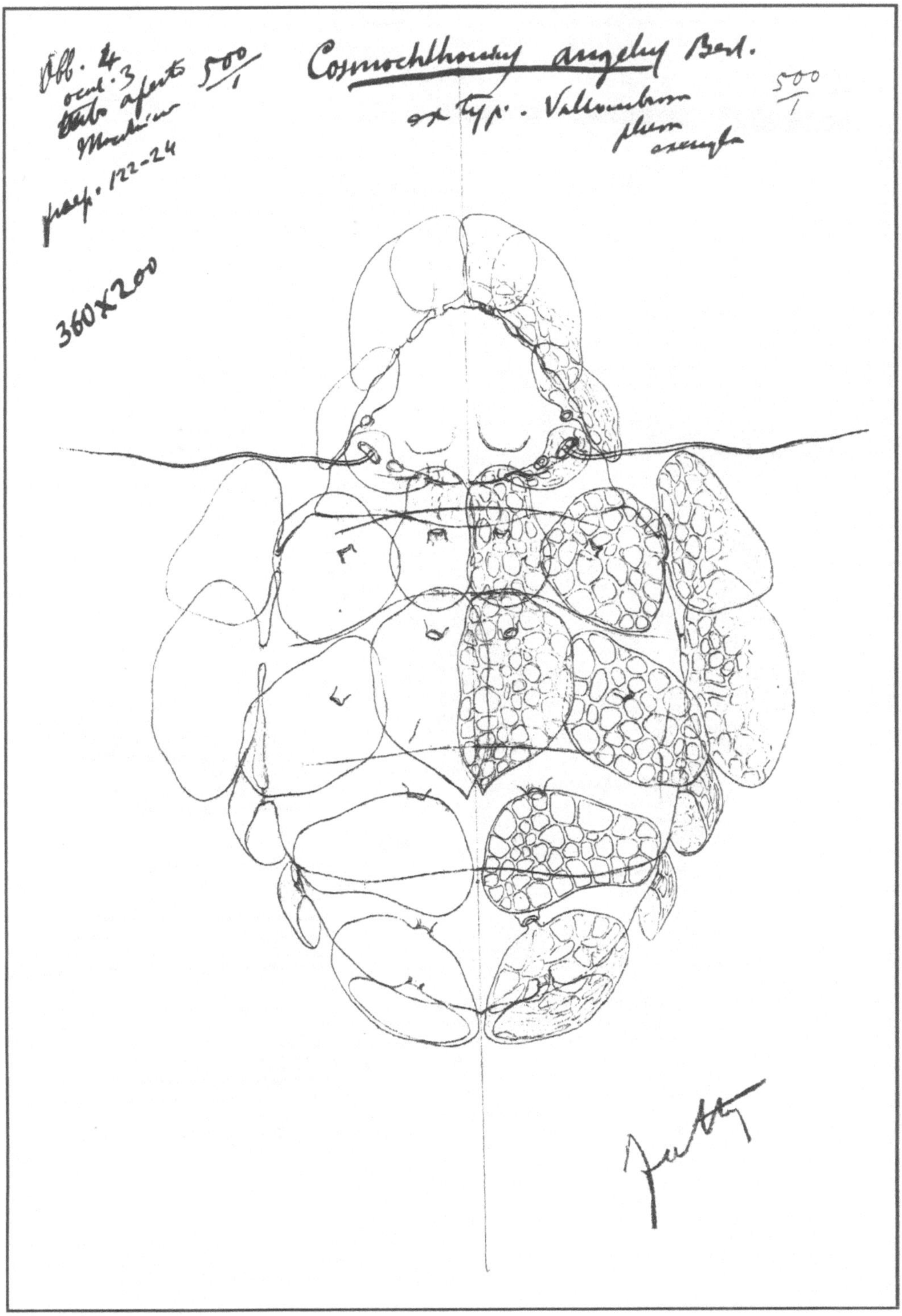

Fig. 3 - A page from Berlese's waste-book, representing the first drawing of the species *Cosmochthonius angelus* Berl. (now *Pterochthonius angelus*).

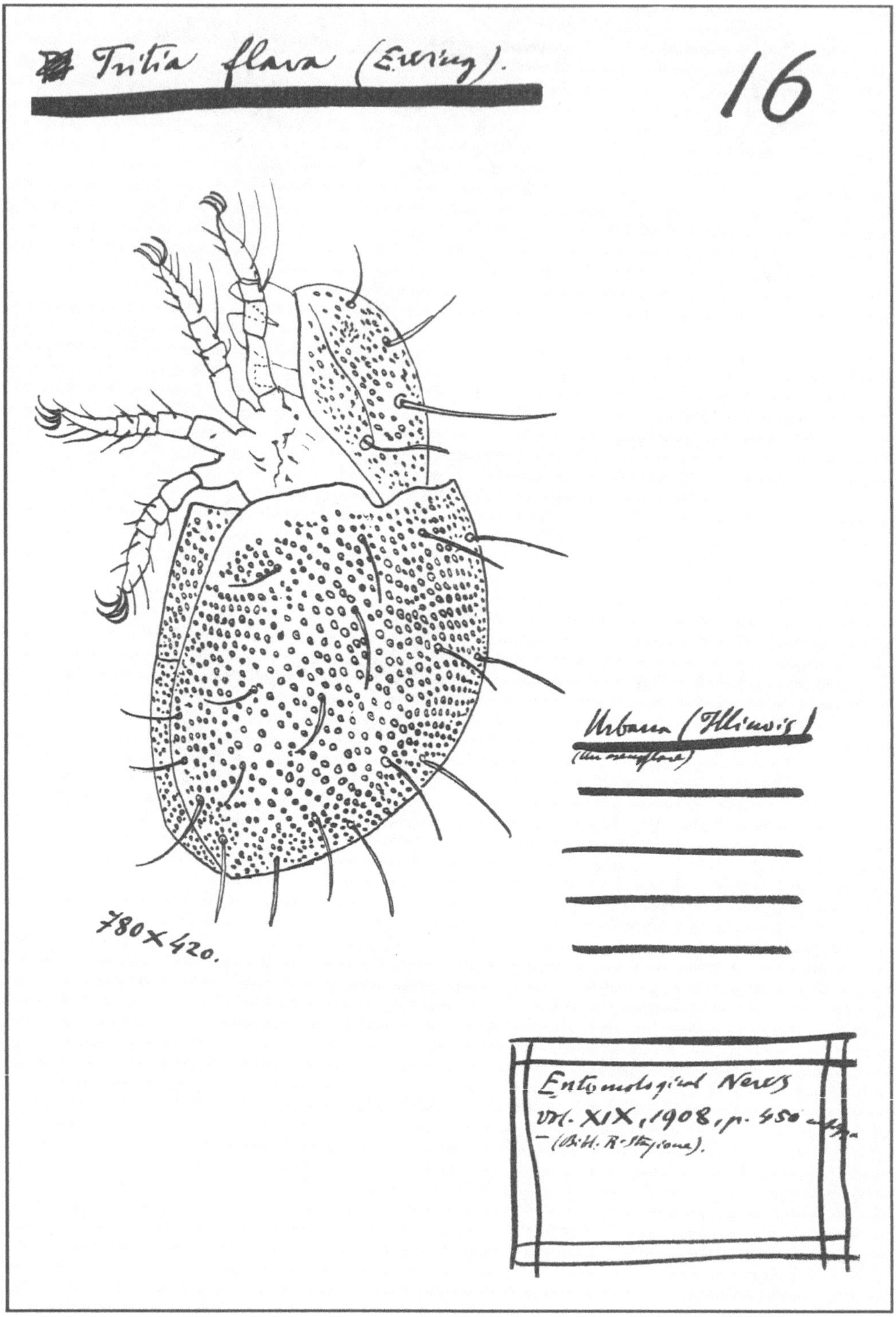

Fig. 4 - A page from the same paper, representing *Tritia flava* (Ewing) (now *Euphthiracarus flavus*).

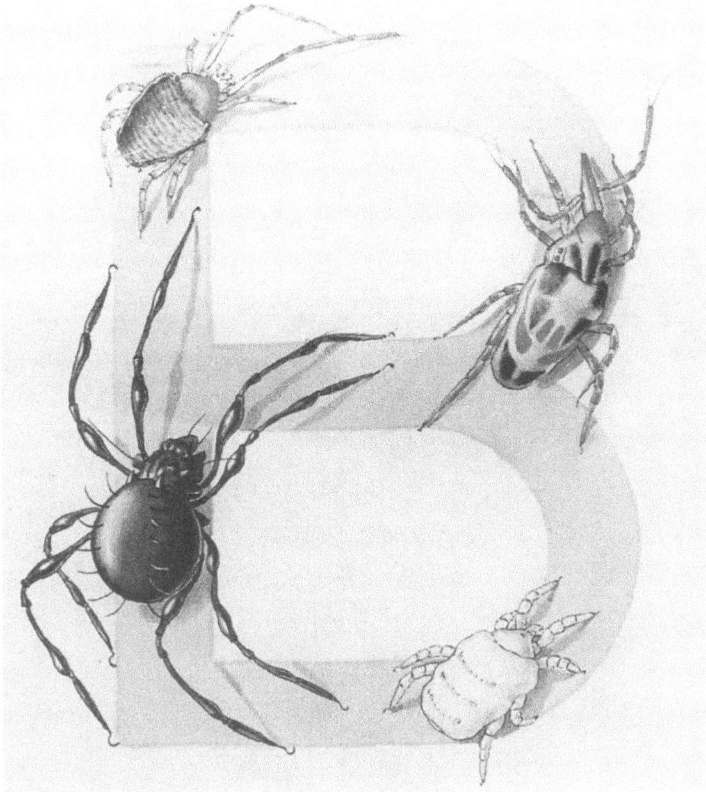

Fig. 5 - A page of Berlese's handwritten Catalogue of his Collection in Florence, representing the frontispiece of the part containing the genera starting with the word B, for instance *Belba* or *Bimichaelia* (by courtesy of the Istituto Sperimentale per la Zoologia Agraria of Florence).

As time goes by: a profile of Italian acarology

Salvatore Ragusa
Istituto di Entomologia agraria, Viale delle Scienze, I-90128 Palermo, Italy, ragusa@unipa.it

Key words: acarologists, research lines, Italy, historical review.

Abstract: A profile of the Italian acarology from the beginning up to now is reported. For each acarologist are mentioned news about the work and the field of interest.

There was scarce and fragmentary information on acarology in the past and scientists dealing with it were very few. At the beginning there were no scholars involved with mites "full time", as they usually were naturalists that seldom dealt with mites. The reasons why the development of acarology was delayed as a science, if compared to the development of entomology, were mainly two. The first one is doubtless the size of mites: too small to be detected at that time; the second and most important one was the existence of natural balance: no pesticides, no phospho-organic compounds were known and, of course, applied and, as a consequence, nature was able to balance everything without any intervention on the part of man. Mites existed, but they could not 'show off' as biological equilibrium did not allow them to damage plants and animals noticeably.

But as soon as mites began attacking people, causing them damage, scientists started to study them finally discovering such tiny individuals: the "general naturalist" at last encountered mites!

Acarology in the past

17th century

The first scientist who came across mites without having any knowledge of them, was **Battista Ferrari** (1584-1655). He was a botanist and in his book "Hesperides sive de malorum aureorum cultura et uso, Liber Quatuor", dated 1646 (fig. 1), described various new species of *Citrus*, drawing fruits, which seem to be attacked by the citrus bud mite (*Aceria sheldoni*). It should be said, however, that citrus plants exist in nature which naturally produce abnormal, digitate fruits, but if we consider Ferrari's description of one of them, i.e. Limon a Rio, Rivo (fig. 2), we can say that this lemon plant was not a new species, but a plant infested by citrus bud mite. We can say

F. Bernini, R. Nannelli, G. Nuzzaci, E. de Lillo (eds.), Acarid Phylogeny and Evolution. Adaptations in mites and ticks, 1-20.

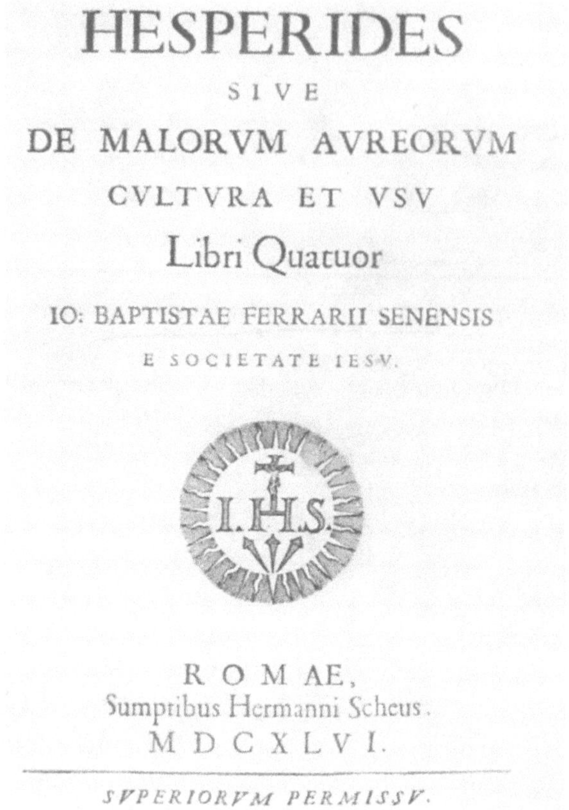
HESPERIDES
SIVE
DE MALORVM AVREORVM
CVLTVRA ET VSV
Libri Quatuor
IO: BAPTISTAE FERRARII SENENSIS
E SOCIETATE IESV.
I.H.S.
ROMAE,
Sumptibus Hermanni Scheus.
MDCXLVI.
SVPERIORVM PERMISSV.

Fig. 1 - Front page of Ferrari's book.

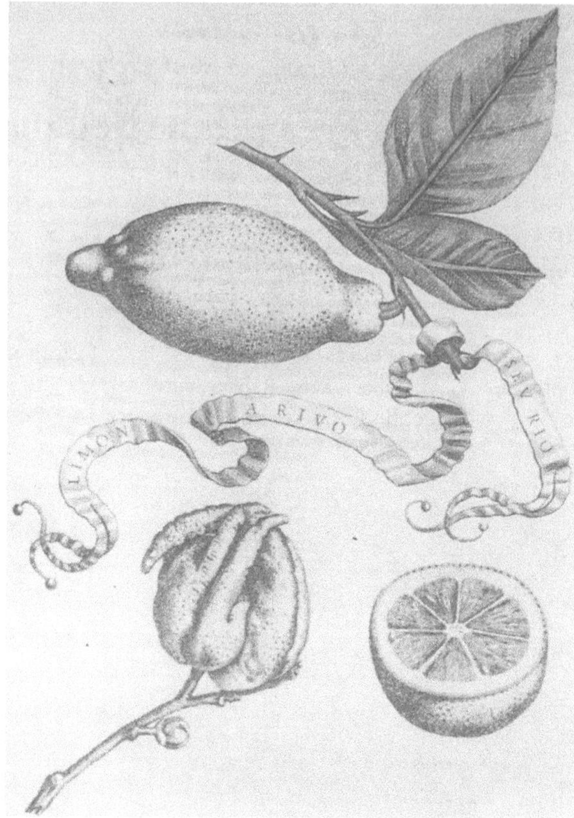

Fig. 2 - Ferrari's drawings of Limon a Rio, Rivo.

that because Ferrari affirms in his description "... Et huius limonis arbos, ... digitatos aliquando foetus enititur: & arboreo partu partes audet humanas ludere. [...] Pomum certe perlatum est Romam. [...] collectum, quod geminae manus, digitos pectinatim iungentis [...]". "[...] this lemon tree [...] sometimes gives birth to a digitate foetus: and with its arboreal birth dares to represent parts of a human body [...]. The fruit collected was brought to Rome [...] to show this aspect being like a double hand with connected fingers in the overall shape of a comb".

The first pieces of information on mites in Italy date to the 17th century. We owe them to very few scientists, the two most important being **Francesco Redi** (1626-1698) and **Marcello Malpighi** (1628-1694).

Francesco Redi (fig. 3) contributed to the discovery of the scabies mite, promoting the publication of a letter by Giovan Cosimo Bonomo on the "pellicelli" [small animals] of human body; in this letter the author described the "pellicello" asserting that the illness was caused by the "pellicello's" biting of the skin. As it is thought that Bonomo was a pseudonym of Redi, it seems that it was he who discovered this nasty species.

Marcello Malpighi (fig. 4), another important biologist and a teacher at the University of Bologna, had the privilege of founding cecidology, even if he did not know the agents which caused cecidia (galls). "De Anatome plantarum" is his most famous work, in whose second part he writes on "De seminum vegetationem, de gallis" (Atti 1847).

Besides Redi and Malpighi, other scientists have to be mentioned who, even if not as famous as the above, nevertheless contributed to the knowledge of mites, and in particular to the

Fig. 3 - Francesco Redi (1626-1698).

Fig. 4 - Marcello Malpighi (1628-1694).

knowledge of cecidology even before Malpighi. First we have to quote the abbot **F. Marucelli** (1625-1703), who listed the names of these scientists in his "Mare magnum", thus creating, according to Trotter, the oldest source of knowledge on cecidology. The list includes naturalists such as **P.A. Mattioli, U. Aldrovandi, P. Manunzio, F. Bonomo, C. Durante, P. Bellomo, P. Casulano,** and **B. Curzio.**

18th and 19th century

In 18th and 19th century, we have no "true" acarologists yet, but only scientists who sometimes published papers which dealt with mites. This is the case of **Nicola Contarini** (1780-1849), an ornithologist and entomologist, who published a catalogue of birds and insects in Padova and Venezia, listing 29 species of mites; of **Ciro Pollini** (1782-1833) who was interested in acarocecidia, without realizing what caused them; of **Camillo Rondani** (1808-1879), a very well-known dipterologist, who was also interested in mites, and described the new mite genus *Trichadenus* with the species *sericariae* (the silk-worm mite) and the mulberry mite, *Acarus mori*; of **Filippo De Filippi** (1814-1867) a famous zoologist, who described the new genus *Hypodectes* and *Lais*; of **Giovan Battista Ercolani** (1817-1883), a veterinary surgeon, the father of Italian veterinary science, who studied mites harmful to domestic animals; of **Adolfo Targioni-Tozzetti** (1823-1902) who was an outstanding professor of Zoology and Comparative Anatomy at the University of Firenze, and became the Head of the Stazione di Entomologia agraria (which was built under its initiative); he was mainly interested in insects, as his relevant studies on *Phylloxera* and coccids show, but also took mites into account as his description of some new genus (*Pediculoides*), and some morphological studies on the genus *Tetranychus* indicate; of **Giovan Antonio Bianconi** (1841-1875), a mathematician, who was mainly interested, during his short life in the tick, *Argas reflexus*.

Giovanni Briosi (1844-1907) a botanist at the University of Pavia, was the founder of the vine industry in Palermo and, as he was interested in vines, studied the eriophyoid mite *Colomerus vitis* also giving some ethological news which was first published in 1875 in the local newspaper "Il giornale di Sicilia".

Up to this point the knowledge of acarology in Italy was still fragmentary, not well defined, and mainly dealing with mites attacking people and animals. Agricultural mites were still looming on the horizon, like meteors: giving rare and brief glimpses, before they disappear again.

It is now, however, that two very important scientists, **Giovanni Canestrini** and **Antonio Berlese** appear on the acarological scene. They gave very important contributions to the knowledge of acarology for their accuracy and clear scientific methods which made them famous not only in Italy, but all over the world.

Giovanni Canestrini (1835-1900) attended the University of Vienna where he got his degree in 1861 and where he began to study ichthyology. After he left Vienna for Italy, he started working at the 'Museo di Storia naturale' in Genova where he continued his studies on salt-water fish. The following year he left Genova for Modena where he remained for a short while as in 1869 he was called by the University of Padova to hold the chair of Zoology and Comparative Anatomy. In this position he abandoned his studies of ichthyology showing a great interest towards arachnids and published the great work "Aracnidi italiani" [Italian Arachnids] in 1868 followed by a second edition in 1870 (Canestrini and Pavesi 1868 and 1870). From this time on Canestrini will publish 70 papers on arachnids especially on spiders and opilionids, but also on mites; on this latter subject he will leave extremely important traces writing many papers on taxonomy, and becoming one of the most important, internationally known taxonomists of mites. This great love for mites is evident in his own words: "[mites] have a great importance from the philosophical, zoological point of view and they give the evolutionist a rich material which has to be taken into great consideration". When he described the various species, he was very accurate, but showed his humble attitude at the same time. In 1890, for example, he wrote "Since 1884 I have been studying phytoptid mites several times, but only this year I got some success and this is because I learned two things: the way to observe such small animals and the way to distinguish one from the other" (Canestrini 1890). By 1890 he had become a very famous acarologist and received many collections to study sent to him from various parts of the world. From 1885 to 1899 he published his masterpiece "Prospetto dell'acarofauna italiana" [Prospect of Italian Acarofauna] (fig. 5) which was unfortunately left unfinished, and soon after his "Monografia dei Demodicidi e Sarcoptidi" [Monograph on Demodicidae and Sarcoptidae] which he published with Kramer. In 1891 he wrote a new "Sistema per la classe degli Acaroidei" [System for Acaroidei Class]. In the list compiled by Bernini, Castagnoli and Nannelli (1995), 146 species are reported as described by him alone or in cooperation with others and the descriptions seem to be still valid. Another important characteristic of Canestrini's was that he was a great taxonomist, who did not limit his activity to an arid description. He was also interested in the biology of these tiny creatures (Minelli 1998) whose way of living he explained in their great adaptations to different conditions of life. In so doing he introduced an evolutionist spirit to systematics. He also wrote his papers and books thinking of unspecialized readers as he wanted to bring science to the attention of common people. He was also the founder of different societies of natural sciences, such as the 'Società dei naturalisti' in Modena and the 'Società veneto-trentina di Scienze Naturali' in

PROSPETTO

DELL'ACAROFAUNA ITALIANA

PER

GIOVANNI CANESTRINI

PROF. NELLA R. UNIVERSITÀ DI PADOVA

Famiglie: **Oribatini, Gamasini, Hoplopini Trombidini, Rhyncholophini**

PADOVA
1885

Fig. 5 - Front page of G. Canestrini's "*Prospetto*".

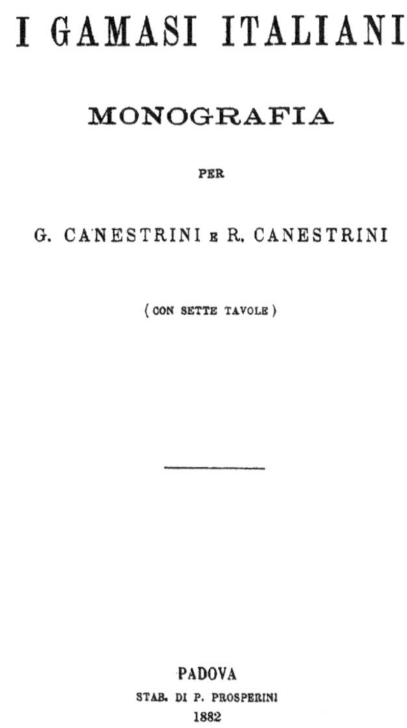

I GAMASI ITALIANI

MONOGRAFIA

PER

G. CANESTRINI E R. CANESTRINI

(CON SETTE TAVOLE)

PADOVA
STAB. DI P. PROSPERINI
1882

Fig. 6 - Front page of G. Canestrini and R. Canestrini work.

Padova. In both towns, Canestrini greatly improved the local zoological collections and in 1881 he started a mite collection which was really the first one in Italy. About this he wrote in 1883: "I am pursuing my acarological collection which at this point is one of the best known". His death was followed by an unfortunate, incredible oblivion: only his mite collection remained, even if it was 'frozen' in a Museum and was almost abandoned for a long time (Minelli 1994).

One of Canestrini's co-workers was his brother **Riccardo Canestrini** (1859-1891) with whom he published three papers dealing with gamasids "Nuove specie del genere *Gamasus*" [New Species of *Gamasus* Genus] and "I gamasi italiani" [Italian Gamasids] (fig. 6) (Canestrini G. and Canestrini R. 1881 and 1882a) and on some "Acari italiani nuovi o poco noti" [On Some New or Little Known Italian Mites] (Canestrini G. and Canestrini R. 1882b).

Canestrini was considered as a 'Maestro' and many young people followed in his footsteps. I will only mention those who worked on acarology such as **Filippo Fanzago** who published numerous notes on various Italian mites; **E. Sicher** (1865-1915) a nephew of Canestrini's who among other subjects such as anthropology, ethnology, ornithology, was also interested in mite biology and systematics, and gave an interesting contribution to mite embriology; **Gustavo Leonardi** (1869-1918) who started his career studying mites, especially from a systematic point of view, was later interested in scale insects and left us an important paper, the "Elenco dei

Fig. 7 - Francesco Minà-Palumbo (1814-1899).

Fig. 8 - Teodosio De Stefani-Perez (1853-1935).

Fitoptidi italiani" [List of Italian Phytoptids]; **Pietro Buffa** (1871-1941), who although interested in Thysanoptera, published some acarological notes introduced in Canestrini's "Prospetto dell'Acarofauna italiana"; **Felice Supino** (1871-1946) who studied some Hungarian and Birmanian mites and also the systematics and embryology of Ixodidae; **Alessandro Coggi** (1864-1917), whose main interest was in vertebrates but he also liked to further his knowledge of arthropods among them mites, with a preference towards the Italian Oribatidae fauna.

But what about acarology at that time in Sicily? We should remember the names of two scientists who were not really acarologists *sensu strictu*. One is **Francesco Minà-Palumbo** (1814-1899) (fig. 7) who started his studies in medicine at the University of Palermo, but went to the University of Napoli to complete them and returned to Sicily to start collecting various material: birds, insects, mites, and also plants. Even if only occasionally, he focused his attention on mites indicating those attacking vines and tobacco, his collections show what Sicilian acarological fauna was like at that time. The other scientist is **Teodosio De Stefani-Perez** (1853-1935) (fig. 8) a geologist and a school teacher, who was a great specialist of Hymenoptera. His masterpiece has to be considered the study on cecidology which was very limited and unexplored at that time (Mariani 1952). He described many agents which caused galls, mainly insects but also mites often producing a dichotomic key to identify the agents (De Stefani-Perez 1902), and giving a clear picture of the knowledge of Sicilian cecidology of the time. He also checked the collection of dried plants kept in the 'Royal Botanical Garden' in Palermo, showing the presence of cecidia and of their agents (De Stefani 1907), as well as cecidia coming from different parts of the world (Somalia, Eritrea, Libia, Tripolitania) (De Stefani-Perez 1908; 1909; 1910; 1912 and 1914).

Another scientist **Caro Benigno Massalongo** (1852-1928) was also working on cecidia in

Northern Italy. He made interesting contributions from both the taxonomical (as far as the producers of cecidia were concerned), and anatomical-pathological point of view (as far as the receivers were concerned). He also made good drawings which, up to now, are considered very clear and well done. His first studies date to 1881, when he noted some deformations on *Rumex* sp. When he met Trotter in 1896, they founded a circle of naturalists with the purpose of propagating the ideas they learned from other great naturalists such as Pollini and Seguier.

Alessandro Trotter (1874-1967) was another important cecidologist. When he was very young and still a student at secondary school, he showed a great interest in cecidology. For this reason he started visiting Massalongo and other naturalists (Goiran, Garbini, Forti, Stegagno) with whom he created a sort of scientific circle. His first work on cecidology date to 1897 (he was only 21 years old) where he mentions 124 galls, 21 of them caused by eriophyoid mites. Between 1899 and 1909, Trotter travelled all over Italy, collecting much material. In this decade he published 20 papers, describing 742 galls, 80 of them caused by mites, and also describing new species of eriophyoid mites. As the material collected was enormous, he decided to organize it, starting, at the beginning of 1900, first in Padova and then in Avellino, the important catalogue entitled "Cecidotheca italica" where almost one third (169) of 575 galls was caused by eriophyoids. In the meantime he travelled in Italy and abroad collecting and identifying new galls and new producers of galls. Morever, when still quite young (28), he decided to give a precise character to cecidology and in 1902 he founded in Padova the journal called "Marcellia" in honour of M. Malpighi "the undiscussed (acute, profound, absolute) founder of cecidology" (Trotter, Marcellia 1928, 25: IV, in Di Stefano 1967).

During his life Canestrini had given a great improvement to the acarological studies, but everything seemed almost lost at his death. As a matter of fact, the traces he left were quite deep and they appeared again in the active work of his pupil **Antonio Berlese** (1863-1927) (fig. 9), who never forgot his love for mites, which had developed while he had worked, as a young pupil, with Canestrini (Minelli 1994) and who collected his scientific acarological inheritance. In 1881 he dedicated his 'Maestro' the *Canestrinia* genus, which became the type of the family Canestrinidae and of the superfamily Canestrinioidea (Minelli 1998). Berlese was born in Padova, and still very young, displayed an aptitude towards natural studies. While still a student at the Liceo Foscarini in Venezia, he started focusing on insects and mites. Once his father went to visit G. Canestrini and showed him the collections and

Fig. 9 - Antonio Berlese (1863-1927).

ANTONII BERLESII

ACARI, MYRIOPODA ET SCORPIONES

HUCUSQUE IN ITALIA REPERTA

VOLUMEN I.
CUI TABULAE CENTUM
AB IPSO AUCTORE DELINEATAE, INCISAE ET DEPICTAE,
ADNECTUNTUR

PATAVII
Typis Fratr. Salmin.

Fig. 10 - Front page of Berlese work.

the researches the young boy had carried out. Canestrini was really surprised looking at the mite collection, but he was even more enthusiastic when he looked at the drawings of mites and insects the young scholar had made (Paoli 1927). Berlese became his pupil immediately and started publishing with him a short note on the "Streghia degli imenotteri" [The Toilet Organ of Hymenoptera] in 1880: Berlese was only 17. Afterwards, he graduated at the University of Padova in 1884: by that time he had already published a dozen papers on the biosystematics of mites and on the anatomy of insects. During the short period he was in Padova working with Canestrini, he published, besides some papers on insects, about 15 papers on mites. In 1885 he went to Firenze to the Istituto di Studi Superiori, Pratici e di Perfezionamento, called by Adolfo Targioni-Tozzetti and he remained as his assistant until he won a competition which brought him to the R. Alta Scuola di Agricoltura in Portici (Napoli), which he later transformed into the Laboratorio di Entomologia agraria where he got his academic chair and where he stayed until 1903 (Ragusa 1982). At that time, winning another competition, he succeeded to Targioni-Tozzetti at the R. Stazione di Entomologia agraria in Firenze where he remained as a director until his death. Under his direction the station became bigger and bigger and it was considered one of the most important scientific institutions in Italy.

Already in 1882, while still a student, he had started his most ambitious work, which was called "Acari, Myriopoda et Scorpiones hucusque in Italia reperta" whose first issue appeared on July 20th 1882 (fig. 10). The work continued to come out in a series of 101 issues, and the last one dated August 1st, 1903, included 988 tables drawn, etched and coloured by himself. In 1920 Berlese published the "Indexes", modifying the title of the work in "Acari, Myriopoda et Pseudoscorpiones hucusque in Italia reperta". By that time, using also Berlese's funnel, he had collected such a large amount of material that he was no longer able to draw each species on one table; therefore he started publishing only the diagnose with drawings and afterwards he no longer added drawings. He published the nine "Manipuli" and then the six "Centurie di acari nuovi" [Centuries of New Mites] and various other monographies on myrmecophilous mites, on gamasid mites, on trombidid mites, etc. Berlese also created a very important mite collection which is still kept nowadays in the containers he had selected. The collection is composed of slides kept in 235 boxes totalling today 11,195 slides, and by 2,348 flame-sealed vials containing specimens in alcohol (Pegazzano and Castagnoli 1984). If we analyze the collection, we see that Berlese described 1,671 species, 1,641 of which he considered as valid species; 1,523 valid species are still present in the collection while 118 are missing.

Most of the species studied by Berlese were terrestrial mites, only 3% being water mites. He mostly studied soil mites (45% of the species present in the collection); then moss mites (15%), insects (13%), vertebrates (13%, mostly birds), 5% of the species of the collection concerns plant mites and the remaining 8% are mites present in stored food, human surroundings, and other arthropods.

Berlese was used to working on different subjects of acarology and entomology at the same time. Returning to Firenze he started writing his magnificent book "Gli insetti, loro organizzazione, sviluppo, abitudini e rapporti coll'uomo" (fig. 11) in two volumes (1904-1923) (but a third volume was also in his mind) which is still considered the masterpiece of Italian entomology. Berlese was a very hard worker: he was in his laboratory before sunrise and for many years his working life in Firenze was as follows: in spring he worked at the diffusion of *Prospaltella*; in summer at the control of olive fly; in the remaining part of the year he was busy with technical reports, the organization of the next work, and studies on mites which remained his great love.

Berlese is considered one of the greatest entomologists and a milestone in his time as far as acarology and entomology are concerned. A profound researcher, he investigated nature, trying to understand its messages and trying to discover and describe what others before him had never seen. His personality was quite strong and was a combination of the different aspects of his character. He was an eclectic: a brilliant writer, a refined drawer and painter, a lover of classics, a philosopher, a sportsman, an excellent humorist and a very strong and invincible controversialist. He always found time for women too as he was also a great lover (Melis 1928).

He wounded himself with a rifle bullett, while he was looking in a field for special lizards

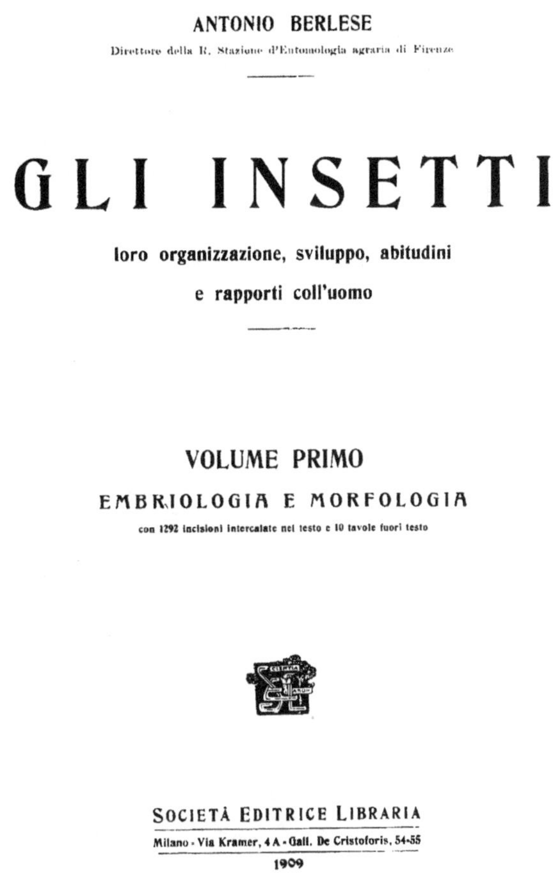
ANTONIO BERLESE

Direttore della R. Stazione d'Entomologia agraria di Firenze

GLI INSETTI

loro organizzazione, sviluppo, abitudini e rapporti coll'uomo

VOLUME PRIMO

EMBRIOLOGIA E MORFOLOGIA

con 1292 incisioni intercalate nel testo e 10 tavole fuori testo

SOCIETÀ EDITRICE LIBRARIA

Milano - Via Kramer, 4 A - Gall. De Cristoforis, 54-55

1909

Fig. 11 - "*Gli Insetti*" by A. Berlese.

to study. After his wounds healed, he decided to continue to work at the Stazione; but he was still weak and when the wound reopened, he was operated, but was not able to recover and died of a heart attack on 24th October 1927. He will be always present with people involved in acarology, as no one can study mites without knowing his fundamental studies and his large collections (Paoli 1927).

The first scientist who studied, among others, phytoseiid mites was **Costantino Ribaga** (1870-1945) (fig. 12) who worked under the guidance of Berlese. "Prof. Berlese advised me to study gamasid mites infesting plants as they were not well known", Ribaga wrote in the introduction to his

Fig. 12 - Costantino Ribaga (1870-1945).

work "Gamasidi planticoli" [Gamasids Living on Plants] published in 1904. In this paper the Author describes the new genus *'Phytoseius'* with the species *Ph. horridus, Ph. finitimus,* and another famous species, *Paraseiulus soleiger.* His descriptions are in Latin and in a very concise (but understandable) way; they analyze different characters, some of which (number of setae and their ratio, shape of ventral shields, chetotaxy of legs) are still used for the identification of phytoseiids. Berlese appreciated the work by Ribaga very much and following his paper made some adjustments to the taxonomic status of some species of phytoseiid mites. Unfortunately, as it is impossible to find Ribaga's collection, it has to be considered totally lost (Fontana and Duso 1995).

Fig. 13 - Guido Paoli (1881-1947).

At Berlese's death Italian acarology faced a pause again, as scientists after him abandoned the study of mites preferring insects. As a matter of fact we should say that a few scientists were still dealing with mites, in their "spare" time. Some of them were involved with Ixodidae like **Andrea Batelli** (1845-1917) who left us some interesting papers on their anatomy and physiology, and **Guido Paoli** (1881-1947) (fig. 13) who was a favourite pupil of Berlese's; he started his career working on olive fly, but also showed an interest in botany, anthropology, going around the world to collect materials. He was also interested in the control of locusts in various parts of Italy and published few papers on Ixodidae especially those in Somalia. He also took other mites into account, i.e. the *Dameosoma* genus and tarsonemids, on which he published a monograph (Binaghi 1947). **Maria Tonelli Rondelli** (1899- ?) was interested in the Ixodidae kept in the collection of the Museums of Milano and Torino. She also checked collections gathered during some scientific expeditions, such as Beccari's in the English Guyana and Zavattari's in Fezzan and in Somalia, and as a result of her work she described new species. She was also interested in South American Ixodidofauna, which led her to describe additional new species. Another pillar of Italian entomology, **Filippo Silvestri** (1873-1949) had some time for mites, even if his interest cannot be compared to Berlese's. He was one of the greatest Italian scientists, and was well known all over the world. He was a leader in biological control and made many trips to collect parasites which were introduced in Italy to control the newly arrived pests. But even if he is mainly mentioned as an entomologist, we should also remember him as a scientist who also paid attention to our small mites. As a matter of fact, during his life he published more than 470 papers and this shows his great preparation and interest in various subjects and the great knowledge he had on arthropods in general.

Another group of mites studied in this period were hydracnids. We should remember **Vittorio Largaiolli** (1868-1951) who got his degree at the University of Padova and was an assistant of G. Canestrini for two years; he had started his studies on arachnids and then on hydracnids of the Trentino region even if he was more interested in ichthyology (Dalla Fior 1951); **Rina Monti** (1871-1937) was also interested in Hydrachnidae from a taxonomic point of view, and related their presence to their environment; **Venanzio Marucci** (1875-1913) and **Carlo Maglio** (1878- ?) who drew up a critical list of the known species of Italian Hydrachnidae; **Giuseppe Ramazzotti** (1898-1984) who was working at the Institute of Hydrobiology in Pallanza, published few papers on mites as he was mainly interested in Tardigrada. We should remember his paper on hydracnids of Lago Maggiore, but especially that he was among the first scientists who described fresh water Porohalacarida in Italy. At Berlese's death **Giocondo Lombardini** (1886-1965) became the keeper of his collection. He was of course a follower of Berlese's and in 1936 prepared a list mentioning 1300 of the total species existing in the collection. But apart from this he was active as an acarologist mainly from the taxonomic point of view and published many papers on new genera and new species of mites belonging to different families, giving also information on their anatomy and histology.

Among other people working on mites, we should also remember **Leopoldo Chinaglia** (1890-1916). Unfortunately, he passed away when he was very young as he was killed during the first world war. Nevertheless he impressed Berlese, who wanted him as an assistant. He left us a note on the genus *Hydrozetes*, an almost unknown genus, describing new species and including an analytical key.

20th Century Acarology

In this section we report acarologists who were active in the first half of the 20th century.

Emilia Stella (1909-1994) contributed to the knowledge of Italian and African Ixodidae and of Trombididae and was also interested in deepening the knowledge on oogenesis and spermatogenesis of *Rhipicephalus sanguineus*. **Alessandro Filipponi** (1915-1974) started his career working at the Istituto di Zoologia in Roma University, where he had obtained his degree in 1940. He worked at the Istituto Superiore della Sanità first as a researcher and then as its Head from 1973. He was a keen and a passionate researcher and, besides a few studies on gregarines and on *Musca domestica*, he mainly worked on mites. The family on which he focused his interest was that of Macrochelidae, and he described new species, redescribed known ones, relegating about fifteen species in synonimy, trying to deepen the knowledge of their biology, fecundity, longevity, reproduction and postembryonic development with the aim of using them as possible antagonists of the domestic fly. He bred 22 species in the laboratory for the first time achieving the proof of arrhenotoky for 20 of them and the proof of thelytoky for the remaining two; he proved the

Fig. 14 - Paola Ivancich-Gambaro (1914-1996).

existance of a mechanism which regulates the sex ratio, linked to the population dynamics (Ricci 1974). Another scientist interested in Parasitiformes was **Paola Ivancich-Gambaro** (1914-1996) (fig. 14) who got her degree in Natural Sciences and Medicine at the University of Padova. Since the '40s she was interested in entomology, particularly pests of fruit crops. From the '60s she became interested in mites and started studying the causes of outbreaks of tetranychid mites, showing that they were due to the use (and abuse) of non selective chemicals which caused a great mortality of phytoseiid mites. While studying phytoseiid mites associated with vines, she highlighted the role of *Kampimodromus aberrans* in controlling the phytophagous mites and the role of *Amblyseius andersoni* as a key-predator on fruit trees, mentioning the presence of some resistant strains for the first time. She also tried to reintroduce on various crops those species of phytoseiids which had been killed by chemical sprays. Her research gave very important contributions to the development of biological and integrated control not only from a scientific point of view, but also from a practical one as she arranged technical meetings with farmers where she exposed her results trying to

Fig. 15 - Pierantonio Rota (1923-1996).

convince them to limit the continuous use of chemicals to control pests in favour of biological control. **Pierantonio Rota** (1923-1996) (fig. 15) worked at the Osservatorio per le Malattie delle Piante hosted in the Istituto di Entomologia agraria of the University of Milano. His scientific activity was linked to arthropods noxious to crops, both insects and mites, and he was more interested in the latter, especially tetranychids and eriophyoids. In particular he studied those species harmful to citrus trees, to ornamental plants in the field and in glasshouses, mites of stored products (cheese and sausage) and those associated with mushrooms and hay. His main important paper is "Osservazioni sugli acari tetranichidi dannosi alle piante coltivate ed ornamentali in Italia" [Observations on Tetranychid Mites Harmful to Cultivated and Ornamental Plants in Italy] in which he drew the most important species present in Italy giving details on their corology, host plants, and habitats (Martelli and Domenichini 1996). **Antonio Valle** (1925-1979) worked at the Museum of Bergamo. His research studies mainly dealt with ecology and mites, especially Parasitiformes Macrochelidae of which he described new subfamilies, genera and species. We would also like to remember him for the revision he made of various genera and subgenera described by Berlese and the revision of species described in Canestrini's Acarotheca. We close the section concerning the past (more or less recent) of Italian acarology mentioning **Giulio Manilla** (1929-1997) who started studying at the Faculty of Medicine in the University of Roma, but soon abandoned his studies for family reasons, dedicated himself to journalism and started his studies again only when the Faculty of Biological Sciences was established in L'Aquila, where he obtained his degree in 1971. Besides some research on various subjects (ichthyology, birds, insects, etc.), his main interests for about 20 years were ticks: he studied their pathogenic role and their distribution in different Italian regions. He specifically studied ticks of dogs, horses and birds, but took also into account those living inside caves and in areas only recently urbanized. In late spring 1997, he gave Prof. Minelli a bulky typescript, his last zoological work regarding the monograph of Italian ticks, which was published after his death as the 36th volume of the "Fauna d'Italia" [Italian Fauna]. His regret was not to have the opportunity to see the book through the press as his illness impeded it.

Modern Acarology

When we talk of the state of the art of Italian acarology today, one of the main reasons why we are compelled to start from Firenze, is that Berlese really built Italian acarology at the Stazione di Entomologia agraria (now Istituto Sperimentale per la Zoologia agraria in Cascine del Riccio). The second main reason is that we still maintain the hystorical memory of our acarology in Firenze, as here we find **Fausta Pegazzano** (1920), who has carried out research on entomology for a long time especially on olive fly and was the Head of the Acarology Section and the keeper of Berlese's collection until her retirement. She identified various species belonging to various families of Mesostigmata, particularly Macrochelidae. She revised some groups of the latter in cooperation with Filipponi, and described new species. But she was also interested in the Tenuipalpidae family and described new species too, giving some biological information. She had also the merit to revise the catalogue of Berlese's collection, a work she carried out with M. Castagnoli.

The Acarological laboratory in Firenze is composed of **Marisa Castagnoli** (1946) present Head of the laboratory, and two researchers, **Marialivia Liguori** (1950) and **Roberto Nannelli** (1948). Their work deals with systematics, biology, ecology and ethology of mites, especially pests of various crops, forest and house mites. Their research addresses different techniques for breeding mites, especially predaceous phytoseiid mites, which could be practically used in biological and integrated control programs. Their research lines are related to a) bio-ecology of some phytoseiid mites (*Cydnodromus californicus, Amblyseius cucumeris, Typhlodromus exhilaratus*) important for biological control in vineyards, strawberries and tomatoes, b) surveys of mites associated with different agroecosystems, their distribution and the population dynamics of the prey and predator species, c) relationships between corticicolous mite and chestnut cancer (*Cryphonectria parasitica*), d) surveys of eriophyoid mites on coniferous trees and of their related natural enemies, e) breeding techniques for phytoseiids, fungivorous species, and house dust and storage mites. Their specific competences are perfectly integrated in their common research lines. Castagnoli is mainly interested in the systematics of Prostigmata, especially eriophyoids, tenuipalpids and tydeids, and in bioecological studies of the phytoseiid mites and of their interrelationships with prey. She is also trying different breeding techniques. Liguori is interested in agricultural mites, particularly in the biology and ecology of phytoseiids and on their systematics. Nannelli focuses on oribatid mites and is particularly concerned with the biology and the role of this group in the transformation of the organic material in the forest soil. He also studied some aspects of *Varroa jacobsoni* from both the morphological and biological point of view. At present his research deals with the interrelationships existing between mites associated with chestnut cancer and the fungus responsible of this alteration. **Sauro Simoni** (1961) should also be added to the above group, as he started working on acarology cooperating with the above scientists on different aspects of their research.

Another scientist who works on *Varroa jacobsoni* at the Dipartimento di Biologia Applicata alla Protezione delle Piante at the Udine University is **Norberto Milani** (1950). He got his degree in physics and natural sciences at Padova University where he has been working as a physicist. Since he moved to Udine, he started studying *Varroa jacobsoni*. He is mainly interested in its biology and in the relationships with its host (semiochemicals implicated in the host/parasite relationship), but mainly he was, and still is, involved with the control of the above said mite and with its resistance to various acaricides and aims at solving this problem to protect bees.

We should remember the group of acarologists working on oribatid mites, but also on other subjects, at the Dipartimento di Biologia Evolutiva in Siena. They follow different research lines: a) functional morphology of mites of economic importance. This research tries to acquire the morpho-functional base data on two different taxa of mites, which are extremely important from a practical point of view, such as *Sarcoptes scabiei* and oribatid mites present in Siena and Southern Toscana. Oribatids are very important in recycling organic matter, as they have a positive influence on soil fertility; b) insular biogeography to represent an insular system still using oribatid mites living in continental and oceanic islands not far from the peninsula; c) appenninic fauna. In such a context the oribatid mite fauna gives clear examples of a population that is very differentiated from a chronological point of view and is useful to understand not only the historical components of the population, but also the running of biocenosis. The studies on oribatids regard taxonomical reviews based on morphological as well as on biochemical basis of species of *Carabodes* of the *marginatus* group. The above mentioned working group is composed by **Fabio Bernini** (1944), who is full professor of systematics and phylogenesis at the University of Siena. He is a specialist of oribatid mites, one of the most important groups of soil mites, which can be used as biological indicators. Bernini has focused his interest on bio- and paleogeographical problems of the Western Mediterranean area, working on the systematics of oribatids not only in a traditional way, but with the most modern techniques of population genetics. His studies are fundamental for any further research in ecology, physiology, anatomy, biogeography, etc. and underline how the Mediterranean Italian fauna is very different from the fauna of Central and North Europe. To summarize his interests we could say that they focus on taxonomy and systematics of oribatid mites using biochemical methods, and on the role of oribatid mites for the evaluation of the quality of the environment, and finally on functional morphology of the most important mites from an economical point of view. Together with Bernini we have to remember Avanzati and Migliorini. **Anna Maria Avanzati** (1958) is mainly interested in faunistical aspects as well as on biogeographical, ecological and systematics of families, genera and species of oribatid mites; she deepens their paleontology and evolution. **Massimo Migliorini** (1960) got his degree discussing a thesis on oribatid mites living in the North-Eastern parts of Sicilia and still continues to deepen the subject of his thesis. He is included in Bernini's research programs.

Only a few people used to work on Ixodidae and some of them have retired. We would like to mention **Lorenzo Sobrero** (1923), who was the Head of the Istituto Sperimentale Zooprofilattico of Puglia and Basilicata and was fundamentally a helminthologist. Since 1975 he was, however, interested in Ixodidae and gave us interesting contributions on their taxonomy. At present he is retired and is no longer working on mites.

Enrico Stella (1934) worked at the Parasitology laboratory of the Istituto Superiore di Sanità. His studies have contributed to the knowledge of Ixodidae surveyed in the parks of Circeo, Gargano and Castelporziano and of the seasonal distribution of some species. He has also studied the biology of *Hyalomma marginatus*, under laboratory conditions to understand the causes which determine the dispersion of this species in natural conditions. He has also the merit to have pointed out the existence of *Haemophysalis concinnae* in Italy. He is retired too. Also **Maria Grazia Petrelli** (1935) worked at the Istituto Superiore di Sanità and was interested in both Ixodidae and Macrochelidae. She studied various biological parameters of some species and analyzed the distribution of ticks, drawing a map of the territorial areas in relation to host/animals. At present she has retired.

Another scholar who has retired but is still working on Hydracnidae is **Anna Maria Nocentini** (1934) who worked at the Istituto di Idrobiologia of Pallanza (Novara). She has produced interesting contributions on fresh water mites, mainly Porohalacaridae and Hydracnidae, even if she is not a true acarologist.

Michele Maroli (1943) is the Head of the Sezione di Biologia, Sistematica ed Ecologia dei Vettori at the Istituto Superiore di Sanità (Roma). His research lines focus on arthropods of medical importance and mainly on their role as vectors of pathogens in Italy and in other Mediterranean countries. As far as mites are concerned he has enlarged our knowledge on dust mites (Pyroglyphidae), whose biology and distribution he has ascertained, and ticks (Ixodidae) and their distribution; he has studied the infection caused by *Ixodes ricinus* in a natural park in Central Italy by means of PCR, demonstrating co-infection by *Borrelia* spp. and *Ehrlichia* spp.

Giorgio Canestri-Trotti (1945), a professor of Parasitology, works at the Dipartimento di Biologia Animale e Umana of University of Torino. He has published papers on *Neotrombicula autumnalis, Cheyletiella yasguri* and *Argas reflexus* and has collected abundant material in Northern Italy (Emilia-Romagna, Veneto, Trentino Alto Adige and Friuli Venezia Giulia) as far as the fauna of Ixodid mites is concerned, and has built a list of those mites. He also studied the ixodid fauna in Boscone della Mesola in Ferrara. His future plans focus on surveying and studying the Ixodid fauna of Ferrara where previous verifications for the presence of *Borrelia* (Lyme disease) have been carried out. He is studying the above mentioned mites mainly from a parasitological point of view.

Salvatore Giannetto (1950) is professor of Veterinary parasitology in the University of Messina. His activity is addressed towards parasitology and zoology especially from a veterinary-medical point of view. In particular he has studied the following families: Ixodidae, Trombiculidae and Pyroglyphidae.

Another scientist interested in mites, also from a veterinary-medical point of view, is **Mario Principato** (1955) who works at the Faculty of Medicine at the Dipartimento di Scienze Biopatologiche Veterinarie of the University of Perugia. He started his activity working on ticks and focused then on various mites which cause damage to animals and people, particularly mites such as *Epidermoptes bilobatus*, *Knemidokoptes* *pilae*, *Notoedres muris* and, *Psoroptes* sp., which cause various types of itches in many animal species. More recently he studied *Demodex* sp., and *Glyciphagus* sp., which cause people a particular dermatitis. He is also involved in studying house mites and has demonstrated a method to verify their presence. He has recently established (in 1999) the link between *Glyciphagus domesticus* and the presence of structural anomalies which is so evident that he considers such a species as an indicator of the health status of a particular environment.

There are some scientists who used to work with mites, but are no longer dealing with them. This is the case of **Alfio Nucifora** (1932), an entomologist at the Dipartimento di Scienze e Tecnologie Fitosanitarie in the University of Catania, who was also interested in agricultural acarology especially in the control of some eriophyoids and tetranychids associated with citrus and fruit trees. His most important result was the revision of tarsonemid mites carried out together with R.E. Beer from the USA, where they described and illustrated many species in a very clear way. Since then he has no longer produced papers on mites. The same may be said about **Rita Arcidiacono** (1934) who works at the Istituto Policattedra di Biologia Animale in the University of Catania, where, besides other subjects, she used to study the distribution of oribatid mites and their seasonal fluctuations. She also studied oribatids from the fossil dunes along the coast of Gela (Sicilia). Unfortunately she has abandoned acarology.

In Firenze at the Istituto di Zoologia of Firenze University, **Giuseppe Messana** (1944) started studying fresh water Hydracarina collecting some of these mites in Somalia and Zaire and later on he moved his interest towards interstitial mites (Hydracaridae). Unfortunately at present he is no longer interested in mites as his main interest is now biospeleology.

In the list of those who have abandoned acarology, we have to mention also **Vincenzo Vomero** (1947) who now works on other topics which do not involve mites; but deep in his heart, he is still in love with mites as he still keeps in contact with other acarologists even if he has no time to be deeply involved with our tiny creatures. It is really a pity as his research lines were numerous and interesting: evolutionary systematics in parasitic mites (Myobiidae and Canestriniidae); ecological researches in soil mites of tropical and temperate forests; research on host/parasite relationships, on the parasitic specificity and on mites living in caves in Mexico and Guatemala. However, a metal model of *Sarcoptes scabiei* at the entrance of the Museo civico in Roma, where he is working, still reminds us of his love for mites.

Bruno Cicolani (1947) began his activity in 1972 as a Professor of Ecology at the Laboratorio di Acarologia ed Ecologia in the University of L'Aquila. He worked on a research program called "Ecology and ethology of Macrochelidae". His research now follows two lines: the first one is his first love, Macrochelidae analyzed from a morphological, biological and ecological point of view, and the other one concerns water mites (Hydrachnidia) whose systematics, biology and ecology he studies too; moreover he also deals with water mites and the biomonitoring of aquatic ecosystems. Cicolani is partly cooperated by **Antonio Di Sabatino** (1960), a researcher in ecology at the same university; he is also interested in water mites (Hydrachnidia) as far as biology, ecology and systematics are concerned, but also in their biodiversity and ecological importance in spring habitats and their role as indicators of environmental quality. He also studies taxonomy, distribution and ecology of mites belonging to the Torrenticolidae family in the Mediterranean area.

Other scientists who work on water mites are Morselli and Mari, both operating at the University of Modena. **Ivano Morselli** (1940) is mainly interested in the taxonomy and ecology of Halacaridae and Hydracnidae. He identified more than

50 species of halacarids, 13 of which were new, and about 40 of hydracnids, 2 of which were also new. As far as halacarids are concerned, surveys have been carried out along the coasts of Toscana and Lazio, at Ischia, along the coasts of Puglia and of Southern Veneto as well as in the fresh waters of some lakes located in Trentino and in the Appennines and in an artificial lake in the surroundings of Modena. As far as Hydracnidae are concerned surveys in sea water were done at Ischia and in brackish waters surrounding Ferrara and Ravenna, and in the fresh waters of the lakes of Calamona (Reggio Emilia) and in the surroundings of Mantova. Moreover, in cooperation with C.N.R in Firenze, he has collected three species of halacarids (two of which were new) on the sea shores in Somalia. At present he deals with hydracnids living in fresh waters in some parts of Northern Italy. **Marisa Mari** (1940), was the curator of the Museum of the Istituto di Zoologia in Modena University. She now teaches entomology at the Faculty of Sciences. Her main research lines concern taxonomy and ecology of halacarids and hydracnids living in sea and fresh waters. She works together with Morselli.

As regards stored products acarology, we would like to mention the work of **Marco Pagani** (1949) who works at the University of Piacenza. He attended various acarological courses in France and in England and has carried out research on the control of mites infesting stored products with particular attention to non chemical methods (infra-red, ultra-violet, microwaves, high temperatures, high pressure water steam), inactive powder (bentonite), antibiotics, etc. Moreover he studied the relationships between mites and mould of cheese at short seasoning. He has described, together with Fain, the new species of *Schwiebea codognoensis* (Acari, Acaridae) living in aqueducts, studying also its relationships with other organisms living in the same habitat. Finally he has also taken into account the Ixodidae living in the urban areas and in the farms ascertaining the possibility of diffusion of such arthropods.

There are quite a number of scientists working on mites which are important from an agricultural point of view. Their aim is to get information on different species, both noxious and useful to crops, from different points of view, to deepen our knowledge on their biodiversity, their biology and behaviour, on the control of phytophagous species using biological methods and on the morphology and ultrastructure of species of some groups. A team of scientists in Padova University works in this field. Unfortunately, one of them is no longer interested in mites as her main job is now scale insects. She is **Giuseppina Pellizzari** (1948), who in the recent past gave interesting contributions regarding tetranychids and phytoseiids in apple orchards and vineyards. Moreover she carried out trials to ascertain the side effects of pesticides on phytoseiid mites. **Vincenzo Girolami** (1940) teaches Agricultural Entomology and Biological and Integrated control. He is mainly an entomologist but has also worked on mites (and still does). His scientific activity focused on some aspects of biological equilibrium, side effects on useful arthropodofauna, mite/host plant relationships, biological and integrated control of some harmful species. **Carlo Duso** (1957) has worked both on entomology and acarology. In the 80's he carried out many laboratory and field trials to check the impact of pesticides on useful phytoseiid mites. At the same time he tried to evaluate damages caused by tetranychids associated with vines to define the intervention threshold. He started up laboratory breedings of some species of phytoseiid mites to check their biological parameters and has carried out long term trials to check the effects of the introduction of phytoseiid mites in vines to control tetranychid mites. In the 90's he took into account the interactions between spontaneous and cultivated vegetation, including the influence that pollen can have on the mite population dynamics. He has been recently interested in population genetics especially as far as *Typhlodromus pyri* is concerned.

Mariano Nicotina (1938) works at the Dipartimento di Entomologia e Zoologia agraria at the University of Napoli. He began his activity on viruses, rodents, gastropods, insects and started working on mites ten years ago focusing his attention on phytophagous and predaceous species associated with vines. In particular he has taken into account phytoseiid mites associated with vines located in Central and Southern Italy.

There is a group of people working at the University of Bari at the Dipartimento di Biologia e Chimica Agroforestale e Ambientale. **Giorgio Nuzzaci** (1943) at the beginning of his university career, was involved with insects injurious to vines, olive trees, citrus trees, etc. Later on, his activity was concerned with mites. He started deepening the systematics of Eriophyoidea, describing numerous new species, cooperating also with Boczek, and studying their internal morphology. At the end of 80's he abandoned such a research line, which was continued by de Lillo, and widened the morphological researches to various family such as Phytoseiidae, Penthaleidae, Tenuipalpidae, Tetranychidae, Varroidae, and Tarsonemidae with particular attention to the structure and function of their mouthparts, advancing hypotheses on their functionality, locating the salivary and alimentary ducts. In particular he found the glands present in the prosoma as well as the genital ones in the females in the Phytoseiidae family and described them. Moreover he has studied the ultrastructure and functional morphology of the nervous receptors enclosed in the mouthparth of the *Varroa jacobsoni* and of *Typhlodromus exhilaratus*. On these mites he located chemoreceptors, probably gustatory ones, present on the fixed and movable digits of chelicerae, on cornicula and on subcapitulum. Similar researches were carried out on *Tetranychus urticae* and *Cenopalpus pulcher*. **Enrico de Lillo** (1959) is mainly interested in eriophyoid mites, has surveyed them and has described new species especially since Nuzzaci left this research line. He is cooperating with various French, Greek and American scientists and has also started an archive on database of all systematic data concerning the above said mites, in cooperation with Amrine. At the moment his studies concern the salivary composition in eriophyoid mites. He was also interested in the sensorial structure of legs in *Phytoptus avellanae* and has also started studying the relationships between mites and mushrooms. Moreover he is cooperating with Nuzzaci as far as other research lines are concerned. **Antonella Di Palma** (1969) started her activity on mites studying the morphology of mouthparts of *Penthaleus major*. She cooperates with Nuzzaci and de Lillo on the description of various structures present in phytoseiid mites. At present she is interested in the genital structures of phytoseiids and in mouthparts of tarsonemids.

Vincenzo Vacante (1948) works at the University of Reggio Calabria. He is interested in mites injurious to economic plants, mainly eriophyoids and tetranychids, and in the taxonomy of the latter ones. At present he surveys mites associated with citrus trees and is making a revision of the Tarsonemidae family.

In Sicilia there are few people dealing with acarology in the University of Catania and Palermo. In the first one we have to mention **Giovanna Tropea Garzia** (1960), who works at the section of Entomologia agraria where she also carries out research on insects. As far as acarology is concerned her attention is addressed towards the epidemiology and the quali-quantitative knowledge of the acarofauna associated with Mediterranean crops and with stored products. In particular she is carrying out a) trials applying biological and integrated control against the two spotted spider mite associated with flowers and vegetables using *Phytoseiulus persimilis*; b) integrated control strategies against mites associated with vegetable in glasshouses using various *Amblyseius*; c) surveys of mites living on the main Mediterranean crops (vine, olive, citrus, apple, etc.). As far as stored product mites are concerned, she surveys mite species associated with pistachio nuts, with stored local cheeses, also studying the influence that the environmental climatic factors can have towards some species (*Tyrophagus putrescentiae*, *Lepidoglyphus destructor* and *Cheyletus eruditus*).

At the Istituto di Entomologia agraria in Palermo University, **Salvatore Ragusa** (1941) started his activity working on flowers and olive tree insects, taking part in a biological control project against the olive fly in Sicilia and in Puglia, cooperating with the Station de Lutte biologique in Antibes and with the Osservatorio per le Malattie delle Piante in Bari. In 1971 his interests moved to mites, mainly predaceous phytoseiid mites, and he visited the Agricultural Research Center of Bet-Dagan (Israel), where under the guide of E. Swirski acquired a thorough knowledge on phytoseiids. Since that period a nice and fruitful cooperation started between them and they have published many papers together on taxonomy, biology, and the side effects of the above said useful mites. In 1983 he visited M.me C. Athias-Henriot at the Station biologiques of Les Eyzies (France) and with her he deepened his knowledge on the taxonomy of phytoseiids. As a consequence they published a paper on the revision of *Neoseiulus* genus. His research lines are mainly concerned with phytoseiid mites, which are studied from different points of view, but always as possible natural control agents of phytophagous mites. He has made several surveys to check the presence of phytoseiid mites associated with cultivated and wild plants to deepen the knowledge on the distribution and the biodiversity of this family, whose general knowledge dates back to Berlese's time. His interests have been particularly dedicated to phytoseiids living in Sicilia where he tries to apply techniques of biological or integrated control against outbreaks of phytophagous mites using native phytoseiid mites, and trying to preserve and to increase them in the field. Attention has also been addressed to harmful species and to the relationships between them and phytoseiids, and also to their population fluctuations. At present he is involved in a common project with Chile on phytoseiids living on various plants with the aim of finding species which could be practically used. In particular a species was found associated with citrus trees in the desert located in Northern Chile (Pica). Some researches have started to ascertain its biological parameters which seem to be particularly interesting as this mite lives in a special place with particular climatic conditions, as besides this green spot of citrus trees, nothing else exists but the desert. Together with Ragusa, **Haralabos Tsolakis** (1964) works as a researcher, following the same research lines. In particular he has carried out surveys on phytoseiid mites associated with forests at various altitudes in Greece, and on the role of the spontaneous wild flora surrounding agro-ecosystems as far as phytoseiids are concerned. He is also interested in using ecocompatible pesticides to control harmful mites and to preserve the useful ones. At present he studies the effects of some natural pesticides, i.e. neem oil, on *T. urticae* and on phytoseiids associated with it, obtaining quite interesting results as, besides killing *T. urticae*, it also causes a diminished oviposition of that species on the surviving females.

This is the profile of the Italian acarology up to now. Unfortunately not many people and especially not many "full time" acarologists, are involved in it yet. Many groups of mites have not been taken into account yet. We hope, however, that as "time goes by", the seeds sown by past and present acarologists will germinate copiously and that an 'outbreak' of new acarologists will appear in the near future to face and finally balance our tiny creatures.

Acknowledgements

I am deeply indebted to all my colleagues who supplied information on acarologists as well as pictures of past acarologists and front pages of ancient books; I also wish to thank my wife Eleonora, who patiently assisted me in the revision of the manuscript, and Dr. Lucia Aliffi (Dipartimento di Scienze Filologiche e Linguistiche, Facoltà di Lettere, Università di Palermo) for her translation into Italian of the original Latin description of limon a Rio, Rivo by B. Ferrari.

References

Atti G., 1847 - Notizie edite ed inedite della vita e delle opere di Marcello Malpighi e di Lorenzo Bellini. *Tipografia Governativa Volpe*, Bologna 1847: 540 pp.

Bernini F., Castagnoli M., Nannelli R., 1995 - Arachnida Acari. In: Minelli A., Ruffo S., La Posta S. (eds), Checklist delle specie della fauna italiana, Calderini, Bologna, 22: 131 pp.

Binaghi G., 1947 - Guido Paoli. *Memorie della Società enomologica italiana*, 26: 3-19.

Canestrini G., 1890 - Ricerche intorno ai Fitoptidi. *Atti della Società veneto trentina di scienze naturali*, 12: 26 pp.

Canestrini G., Canestrini R., 1881 - Nuove specie del genere *Gamasus*. *Atti del Regio Istituto veneto di scienze, lettere ed arte*, ser. 5: 7.

Canestrini G., Canestrini R., 1882a - I gamasi italiani. *Atti della Società veneto trentina di scienze naturali*, 7: 8-32.

Canestrini G., Canestrini R., 1882b - Acari italiani nuovi o poco noti. *Atti del Regio Istituto veneto di Scienze lettere ed arte*, ser. 5(8): 913-930.

Canestrini G., Pavesi P., 1868 - Araneidi italiani. *Atti della Società italiana di scienze naturali,* 11: 738-872.

Canestrini G., Pavesi P., 1870 - Catalogo sistematico degli Araneidi italiani. *Archivio zoologico anatomia e fisiologia*, 2: 1-44.

Dalla Fior, 1951 - La scomparsa del Prof. Vittorio Largaiolli. *Rivista Studi Trentini di scienze naturali*, 28(1-2-3): 183-185.

De Stefani-Perez T., 1902 - I zoocecidi sulle piante del genere *Pistacia*. *Nuovi Annali dell'agricoltura Siciliana*, 13: 1-37.

De Stefani-Perez T., 1907 - A proposito di alcune galle dell'erbario secco del R. Orto botanico di Palermo. *Marcellia*, 6: 8-11.

De Stefani-Perez T., 1908 - I primi zoocecidi della Somalia italiana. *Marcellia*, 7: 142-149.

De Stefani-Perez T., 1909 - Altri zoocecidi dell'Eritrea. *Marcellia*, 8: 7-18.

De Stefani-Perez T., 1910 - I zoocecidi sin'ora noti della Eritrea e della Somalia italiana. *Bollettino del Regio Orto botanico e Giardino Coloniale di Palermo*, 9(1-2-3): 1-8.

De Stefani-Perez T., 1912 - Notizie su alcuni zoocecidi della Libia. *Ibidem*, 11(1-3): 1-8.

De Stefani-Perez T., 1914 - Altre notizie su zoocecidi della Tripolitania. *Bollettino di studi e informazioni. R. Giardino botanico (coloniale). Palermo*, 1(2): 1-4.

Di Stefano M., 1967 - Lineamenti cecidologici di un maestro (a quattro mesi dalla morte del Prof. Alessandro Trotter). *Marcellia*, 34(3-4): 119-133.

Fontana P., Duso C., 1995 - Costantino Ribaga e i "Gamasidi planticoli". *Atti dell'Accademia Roveretana degli Agiati,* ser. VII, 4, B: 85-99.

Mariani M., 1952 - Un Entomologo siciliano: Teodosio De Stefani-Perez (1853-1935). *Plinia*, 3 (1950-51) nota VIII: 1-13.

Martelli M., Domenichini G., 1996 - Pierantonio Rota (1923-1996). *Bollettino di zoologia agraria e bachicoltura*, Serie II, 28, 2: 233-239.

Melis A, 1928 - Antonio Berlese. *Rivista di Biologia*, 10(5-6): 7-14.

Minelli A., 1994 - L'opera zoologica di Giovanni Canestrini. In: G. Riccadonna (ed.), Giovanni Canestrini. Scienza ed evoluzionismo. *Atti Convegno*, Trento 16-marzo/Revo' 14-luglio-1991: 33-44.

Minelli A., 1998 - L'opera zoologica di Giovanni Canestrini e della sua Scuola. *Atti sesto seminario di Storia delle Scienze e delle tecniche nell'ottocento veneto*: 119-135.

Paoli G., 1927 - Antonio Berlese. *Memorie della Società Enomologica italiana*, 6: 55-84.

Pegazzano F., Castagnoli M., 1984 - The preparation of a modern catalogue of the Berlese acaroteca. *Acarology VI*, 1: 249-252.

Ragusa S., 1982 - Italy. In: Prasad V. (ed.), History of Acarology: 279-303.

Ricci M., 1974 - Alessandro Filipponi (17-11-1915 - 4-6-1974). *Rivista di Parassitologia*, 34(4): 234-236.

Mites as models in development and genetics

Richard H. Thomas
Department of Zoology, The Natural History Museum, Cromwell Road, London, SW7 5BD, UK, R.Thomas@nhm.ac.uk

Keywords: Mites, Acari, evolutionary development, genetic systems.

Abstract: The diversity of forms, habits and genetic systems found in mites make them excellent candidates for studies of fundamental evolutionary problems. In this review of the status and prospects for research on development and genetics in mites, the utility of certain mite groups in addressing problems ranging from the evolution of arthropod body form to the factors shaping genetic systems are considered. Illustrations of some of these issues are drawn primarily from work in my laboratory. I contend that such work on problems of wide interest in biology is necessary but not sufficient to raise the profile of systematic acarology and its need for increased resources. Cooperation between systematic acarologists and experimental workers is needed to effect change in funding patterns.

Introduction

Mites are by far the most diverse and abundant group of arachnids and they are clearly important in the functioning of many ecosystems. Yet, aside from the comparatively small number of species causing significant agricultural or human health problems, they are very poorly known. To those of us who find these animals endlessly fascinating this situation is difficult to understand. I contend in this review that one path to wider interest in the Acari lies in their use as model organisms to answer questions of broad relevance in biology. It should go without saying that the importance of mites in many ecosystems should make them good candidates to receive much more attention in biodiversity studies.

In this overview I will concentrate on the use of mites in development and genetics, largely using examples from my own laboratory. Others have demonstrated the utility of mites in areas such as population genetics, phylogeography and strain identification. For example, Fenton (this volume) uses eriophyids to look at host races while investigating problems of immediate economic importance. Navajas and colleagues (Navajas 1998; Navajas *et al.* 1998) combine molecular and morphological approaches to studies of the phylogeography of spider mites (Tetranychidae). The clonal structure of a parthenogenetic mite, *Penthaleus major*, has been investigated using allozyme electrophoresis (Weeks and Hoffmann 1998). More recently, analyses of amplified fragment length polymorphisms (AFLPs) have been used to study genetic diversity and population structure in several species (Weeks *et al.* 2000). AFLPs are now the technique of choice for genomic mapping and searching for quantitative trait loci (QTLs) and could certainly be applied for these purposes in any mite species in which laboratory crosses are possible. Several groups are using molecular sequence data to estimate phylogenies for several groups of mites, including ticks (Black *et al.* 1997; Klompen *et al.* 1996; Norris *et al.* 1996), and prostigmate mites (Soller *et al.* 2001).

F. Bernini, R. Nannelli, G. Nuzzaci, E. de Lillo (eds.), Acarid Phylogeny and Evolution. Adaptations in mites and ticks, 21-26.

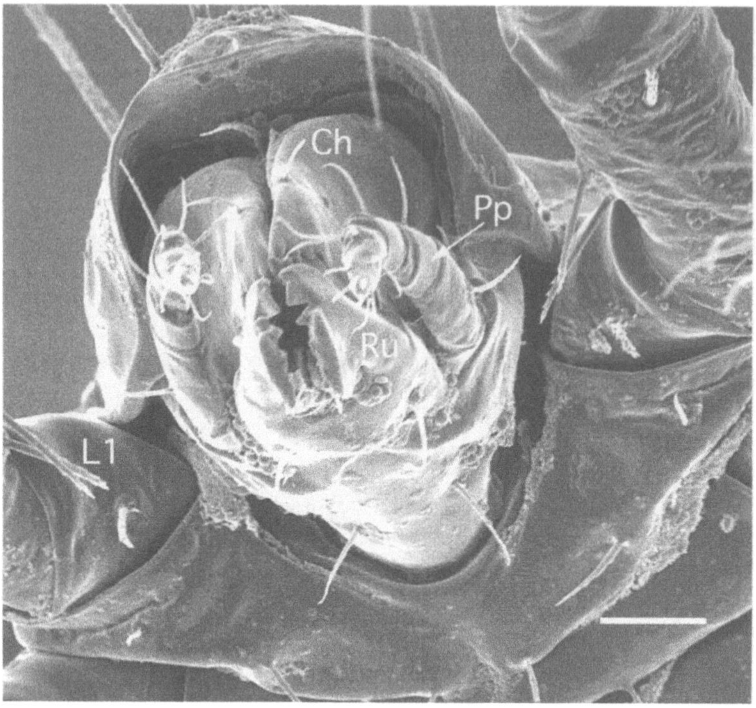

Fig. 1 - A model mite. Anterior view of an adult of the oribatid mite *Archegozetes longisetosus*. Abbr.: Pp, pedipalp; Ch, chelicera; Ru, rutellum; L1, leg I. Scale bar=20 µm.

Development

Mites are generally thought to be derived arachnids, whatever one's view of arachnid relationships (Shultz 1990; Weygoldt 1980; Wheeler and Hayashi 1998), but their early embryonic development is comparable to that of other chelicerate arthropods (Telford and Thomas 1998a). The relative ease of culturing many mite species when compared to most other arachnids and xiphosurans makes them good candidates for model chelicerate arthropods. We have been developing the trhypochthoniid oribatid mite, *Archegozetes longisetosus*, as a model organism for developmental and genetic studies (fig. 1).

Most of the literature on the embryonic development of mites is at least 40 years old, and much of it is approaching 100 years old or more. This work is generally of good quality and has provided us with the essential basis upon which to build using techniques developed in the past 25 years. Advances in imaging methods and preparatory techniques allow us to see morphological detail and localize molecules of all sorts. The huge effort expended on *Drosophila* and the nematode *Caenorhabditis* over recent decades has given us many techniques that can be transferred to projects on mites with little or no alteration. Our own work on the development of limbs in *Archegozetes* made use of a slightly modified standard embryo preparation protocol from *Drosophila* (Thomas and Telford 1999). This method allows the preparation of large numbers of embryos free of the tough vitelline membranes that hamper the penetration of stains and fixatives and also interfere with microscopy.

Surface morphology of developing mite embryos can be examined with scanning electron microscopy (SEM) (Thomas and Telford 1999; Walzl 1988). Our SEM study of *Archegozetes* (Thomas and Telford 1999) has solved an old problem in acariform mite morphology - the origin of the Claparède's organs found on the larval prosoma between legs I and II (fig. 2). Figure 3 shows clearly that the Claparède's organs ori-

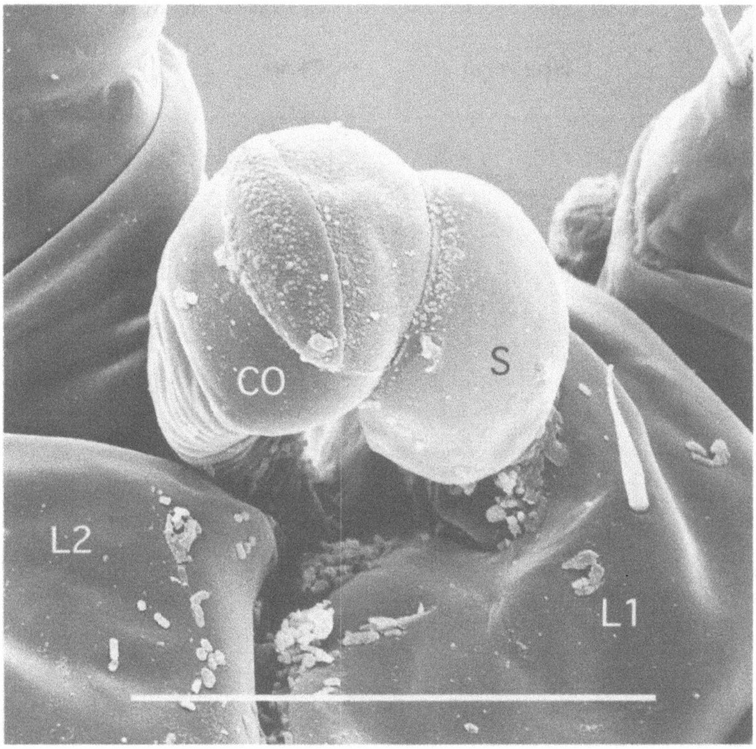

Fig. 2 - Close up view of the Claparède's organ (CO) on the right side of an *Archegozetes* larva. The cup shaped seta S1 (S) on the coxa of leg I (L1) covers the Claparède's organ when it is retracted. Scale bar=20 µm.

ginate at the base of legs II. This observation together with Alberti's (1979) beautiful ultrastructural work on acariform mite genital papillae and Claparède's organs suggests strongly that these organs are serially homologous. Thus, van der Hammen's (1989) hypothesis that genital papillae are limb derivatives becomes more plausible. Together, these data provide motivation for molecular developmental studies of opisthosomal limb derivatives across the Chelicerata.

Larvae of acariform mites typically have three pairs of legs, the fourth pair being suppressed until the nymphal and adult moults. SEMs of *Archegozetes* embryos (fig. 3) clearly show the rudiment of a fourth pair of limbs in apparent contradiction of previous work (reviewed in Evans 1992). This interpretation is also supported by molecular results mentioned below.

The development of laser scanning confocal microscopy (LSCM) provides us with a tool for examining the details of internal structures of mites and for the localisation of gene action. This can be achieved with considerably less effort than the more traditional approach of using thin sections and transmission electron microscopy,

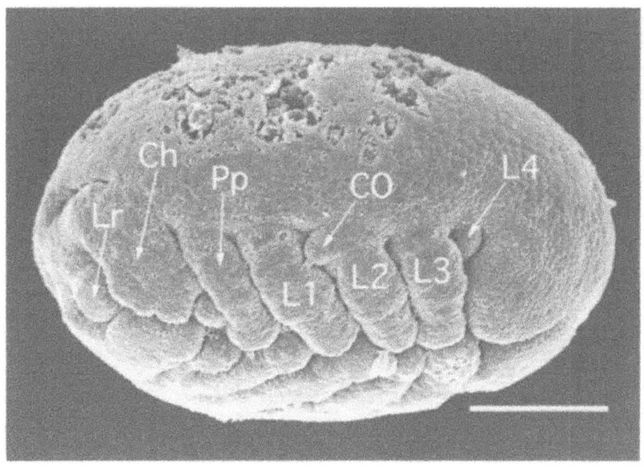

Fig. 3 - Lateral view of an *Archegozetes* embryo. The developing Claparède's organ is visible at the base of the second legs and the rudiment of the fourth leg is seen immediately behind the third legs. Abbr.: Lr, labrum; Ch, chelicera; Pp, pedipalp; CO, Claparède's organ; L1-4, legs I-IV. Scale bar=20 µm.

though in some cases with less resolution. LSCM is now straightforward to use and is widely available in universities.

Advances in molecular biology such as the polymerase chain reaction (PCR), efficient DNA sequencing and RNA interference now make it possible to take techniques developed on model organisms such as *Drosophila* and apply them to virtually any other taxon of interest. Technical advances and decreasing costs will allow molecular data to be collected in unprecedented amounts from a wide range of organisms. The field of evolutionary developmental biology is growing at a great rate as a result of these technological advances and this is allowing us to investigate questions that have intrigued biologists for many years. In this context model chelicerate arthropods are needed to throw light on the generality of results obtained from *Drosophila*, as well as to answer questions more specifically about mites.

We are using *Archegozetes* as a model chelicerate to study the molecular bases of body form evolution. The homeotic genes of the Hox cluster play a central role in patterning the anterior-posterior axis of the developing animal embryo (Wolpert *et al.* 1998). As such their expression patterns have proved useful in addressing an old problem in arthropod biology - the nature of the chelicerate head, and specifically the question of the apparent loss of the deutocerebral segment in the chelicerate lineage. We have examined the patterns of expression of several genes from the anterior Hox cluster of *Archegozetes* and compared them with data available from insects, and more recently, crustaceans. The anterior expression boundaries of these align exactly with those seen in insects, if we assume the chelicerate deutocerebral segment is not lost (Telford and Thomas 1998a). Comparable results have been obtained from two species of spiders (Damen *et al.* 1998; Popadic *et al.* 1998), thus increasing our confidence in our interpretation. These experiments are only a first step in examining the structure and functioning of the genes responsible for moulding the early events in chelicerate embryogenesis. Our research is now directed at understanding more about the mite Hox genes and the genes they interact with to produce the body forms we see. This work is facilitated by *Archegozetes* having a single, compact Hox cluster (*unpub. data*). Falciani *et al.* (1996) proposed that the *Drosophila* gene *zen*, which is located between the class 2 and class 4 genes of the Hox cluster would be shown to be a derived Hox gene based on amino acid sequence similarities across the insects. We have confirmed this suggestion by showing that the mite homolog of *zen* is expressed like a typical Hox gene (Telford and Thomas 1998b). This result has also been confirmed in a spider (Damen and Tautz 1998). The question now is to find out where in arthropod evolution the change in function of this Hox gene occurred.

The ability to prepare large numbers of embryos without tedious or very difficult dissections should allow the examination of early development in many more mite taxa with relatively little effort. I expect that wider use of these methods will yield characters important for higher level systematic questions in the mites and in arachnids as a whole.

Genetics

Mites present us with an astounding array of genetic systems (Norton *et al.* 1993). For example, it is clear that haplodiploid genetic systems (those in which males are haploid and females are diploid) have arisen several times within the mites. Haplodiploidy can be achieved by several means. The two best known mechanisms are arrhenotoky, where males are produced from unfertilized eggs and females are produced from fertilized eggs, and pseudo-arrhenotoky, where males become haploid following fertilization by the heterochromatization and elimination of their paternal chromosomes (Bull 1983). Something on the order of 10% of animal species are haplodiploid and these genetic systems have been implicated in the evolution of sociality in insects (Hamilton 1996). Sabelis and

Nagelkerke have considered the evolution of pseudo-arrhenotoky both in terms of theory and experiments on the ability of female mites to control the sex of their offspring (Sabelis and Nagelkerke 1988; Nagelkerke and Sabelis 1998). Perrot-Minnot and colleagues (Perrot-Minnot and Navajas 1995; Perrot-Minnot *et al.* 2000) have used molecular techniques to trace the fate of the paternal genome in male phytoseid mites. They detect the presence of paternal DNA in male offspring and show that it is selectively eliminated from male tissues. Paternal DNA was never found in the sperm of male offspring and there was no evidence for recombination between parental genomes prior to elimination of the paternal genome in the germ line. These studies provide an important context to guide future work on the mechanisms underlying pseudo-arrhenotoky.

The variety of genetic systems found in mites and their apparent multiple evolutionary origins are not encountered in other taxa and as such mites provide us with unparalleled opportunities to understand the evolution of these genetic systems. We have made a small start on this with our study of the Dermanyssina (Mesostigmata) (Cruickshank and Thomas 1999), an immense group that contains members with diplodiploid, pseudo-arrhenotokous and arrhenotokous genetic systems. A molecular phylogeny produced from a part of the 28S ribosomal RNA gene served as a framework upon which to map the genetic systems where known. Surprisingly, at least to us, our results support an hypothesis put forward in 1931 by Schrader and Hughes-Schrader (Schrader and Hughes-Schrader 1931). They suggested that arrhenotoky is derived from pseudo-arrhenotoky. Though based on the limited taxonomic sampling and very limited knowledge of the genetic systems of different taxa our results support a single origin of arrhenotoky from pseudo-arrhenotokous ancestors that in turn were derived from diplodiploid ancestors. This work can serve to focus future work on specific taxa to gain insight into the forces shaping these genetic systems. Studies of the cytogenetics of both pseudo-arrhenetokous and arrhenotokous taxa are badly needed. The scope for phylogenetic and experimental work in this area is immense.

Conclusions

Advances in microscopy and molecular biology over the last quarter century have given us powerful tools to advance evolutionary and developmental biology in general and acarology in particular. I hope that increased use of mites as experimental systems will drive renewed interest in their systematics and increased support for maintaining and enhancing museum collections. Accomplishing this will require close cooperation between systematists and experimental workers to obtain the funding for collections and to justify permanent positions for acarologists in universities and museums. Funding agencies and institutional administrators must be convinced of the necessity to support basic taxonomic work if they wish to see the more high-profile molecular work. Molecular systematics and population genetic studies in the service of biodiversity surveys and for agricultural and health-related problems likewise have a very important role to play in promoting the study of acarology. Much more effort will be required to ensure the long-term health of systematic acarology.

Acknowledgements

I am grateful to the organizers of the EURAAC 2000 meeting for their warm hospitality and patience. Dr Roy Norton's help over the years has been essential and I appreciate it very much. The Biotechnology and Biological Sciences Research Council has funded much of the work discussed here.

References

Alberti G., 1979 - Fine structure and probable function of genital papillae and Claparede organs of actinotrichida. In: Rodriguez J.G. (ed.), Recent Advances in Acarology. Academic Press, New York. Vol. 2: 501-507.

Black W.C., Klompen J.S.H., Keirans J.E., 1997 - Phylogenetic relationships among tick subfamilies (Ixodida: Ixodidae: Argasidae) based on the 18S nuclear rDNA gene. *Molecular Phylogenetics and Evolution*, 7: 129-144.

Bull J.J., 1983 - Evolution of Sex Determining Mechanisms. Benjamin/Cummings, Menlo Park: 316 pp.

Cruickshank R.H., Thomas R.H., 1999 - Evolution of haplodiploidy in dermanyssine mites (Acari: Mesostigmata). *Evolution*, 533: 1796-1803.

Damen W.G.M., Hausdorf M., Seyfarth E.A., Tautz D., 1998 - A conserved mode of head segmentation in arthropods revealed by the expression pattern of Hox genes in a spider. *Proceedings of the National Academy of Sciences of the United States of America*, 95: 10665-10670.

Damen W.G.M., Tautz D., 1998 - A Hox class 3 orthologue from the spider *Cupiennius salei* is expressed in a Hox-gene like fashion. *Development Genes and Evolution*, 208: 586.

Evans G.O., 1992 - Principles of Acarology. C.A.B. International, Wallingford: xviii + 563 pp.

Falciani F., Hausdorf B., Schröder R., Akam M., Tautz D., Denell R., Brown S., 1996 - Class 3 Hox genes in insects and the origin of *zen*. *Proceedings of the National Academy of Sciences of the United States of America*, 93: 8479-8484.

Hamilton W.D., 1996 - Narrow Roads of Gene Land. Vol. I: Evolution of Social Behaviour. W.H. Freeman, Oxford: 564 pp.

Klompen J.S.H., Black W.C., Keirans J.E., Oliver J.H., 1996 - Evolution of ticks. *Annual Review of Entomology*, 41: 141-161.

Nagelkerke C.J., Sabelis M.W., 1998 - Precise control of sex allocation in pseudo-arrhenotokous phytoseiid mites. *Journal of Evolutionary Biology*, 11:649-684.

Navajas M., 1998 - Host plant associations in the spider mite *Tetranychus urticae* (Acari: Tetranychidae): insights from molecular phylogeography. *Experimental and Applied Acarology*, 22: 201-214.

Navajas M., Lagnel J., Gutierrez J., Boursot P., 1998 - Species-wide homogeneity of nuclear ribosomal ITS2 sequences in the spider mite *Tetranychus urticae* contrasts with extensive mitochondrial COI polymorphism. *Heredity*, 80: 742-752.

Norris D.E., Klompen J.S.H., Keirans J.E., Black W.C., 1996 - Population genetics of *Ixodes scapularis* (Acari, Ixodidae) based on mitochondrial 16S and 12S genes. *Journal of Medical Entomology*, 33: 78-89.

Norton R.A., Kethley J.B., Johnston D.E., O'Connor B.M., 1993 - Phylogenetic perspectives on genetic systems and reproductive modes of mites. In: Wrensch D., Ebbert M. (eds.), Evolution and Diversity of Sex Ratio in Insects and Mites. Chapman & Hall, New York: 8-99.

Perrot-Minnot M.-J., Navajas M., 1995 - Biparental inheritance of RAPD markers in males of the pseudo-arrhenotokous mite *Typhlodromus pyri*. *Genome*, 38: 838-844.

Perrot-Minnot M.-J., Lagnel J., Migeon A., Navajas M., 2000 - Tracking paternal genes with DALP markers in a pseudoarrhenotokous reproductive system: biparental transmission but haplodiploid-like inheritance in the mite *Neoseiulus californicus*. *Heredity*, 84: 702-709.

Popadic A., Abzhanov A., Rusch D., Kaufman T.C., 1998 - Understanding the genetic basis of morphological evolution: the role of homeotic genes in the diversification of the arthropod bauplan. *International Journal of Developmental Biology*, 42: 453-461.

Sabelis M.W., Nagelkerke C.J., 1988 - Evolution of Pseudo-Arrhenotoky. *Experimental and Applied Acarology*, 4: 301-318.

Schrader F., Hughes-Schrader S., 1931 - Haploidy in Metazoa. *Quarterly Review of Biology*, 6: 411-438.

Shultz J.W., 1990 - Evolutionary morphology and phylogeny of Arachnida. *Cladistics*, 6: 1-38.

Soller R., Wohltmann A., Witte H., Blohm D., 2001 - Phylogenetic relationships within terrestrial mites (Acari: Prostigmata, Parasitengona) inferred from comparative DNA sequence analysis of the mitochondrial cytochrome oxidase subunit I gene. *Molecular Phylogenetics and Evolution*, 18: 47-53.

Telford M.J., Thomas R.H., 1998a - Expression of homeobox genes show chelicerate arthropods retain their deutocerebral segment. *Proceedings of the National Academy of Sciences of the United States of America*, 95: 10671-10675.

Telford M.J., Thomas R.H., 1998b - Of mites and zen: expression studies in a chelicerate arthropod confirm zen is a divergent Hox gene. *Development Genes and Evolution*, 208: 591-594.

Thomas R.H., Telford M.J., 1999 - Appendage development in embryos of the oribatid mite *Archegozetes longisetosus* (Acari, Oribatei, Trhypochthoniidae). *Acta Zoologica*, 80: 193-200.

van der Hammen L., 1989 - An Introduction to Comparative Arachnology. SPB Academic Publishing, The Hague: x + 576 pp.

Walzl M.G., 1988 - A scanning electron microscopic study of the embryonic development of the house dust mite, *Dermatophagoides farinae* (Pyroglyphidae, Actinotrichida) from blastula to the hatching of the larva. *Institute of Physical Conference*, Ser. 93: Vol. 3, Chapter 6: 159-160.

Weeks A.R., Hoffmann A.A., 1998 - Intense selection of mite clones in a heterogeneous environment. *Evolution*, 52: 1325-1333.

Weeks A.R., van Opijnen T., Breeuwer J.A.J., 2000 - AFLP fingerprinting for assessing intraspecific variation and genome mapping in mites. *Experimental and Applied Acarology*, 24: 775-793.

Weygoldt P., 1980 - Towards a cladistic classification of the Chelicerata. In: Gruber J. (ed.), VIII Internationaler Arachnologen-Kongress. H. Egermann, Wien: 331-334.

Wheeler W.C., Hayashi C.Y., 1998 - The phylogeny of the extant chelicerate orders. *Cladistics*, 14: 173-192.

Wolpert L., Beddington R., Brockes J., Jessell T., Lawrence P., Meyerowitz E., 1998 - Principles of Development. Current Biology Ltd., London: 484 pp.

Speciation and biogeography in eriophyid mites: a review

Brian Fenton
Scottish Crop Research Institute, Dundee, DD2 5DA, Scotland

Key words: eriophyid mite, rDNA, phylogeny, speciation, *Cecidophyopsis*.

Abstract: Eriophyid mites are amongst the smallest of all mites. They are exclusively associated with plants and have been for an extended period of time. The early forms colonized conifers. As they evolved, flowering plants were also colonized and now most flowering plants are home to at least one mite species. Each species is very restricted in its host range, usually colonizing only one species of plant. Clearly, such relationships, over extended periods of time, are an opportunity for evolutionary interactions to occur. One genus of eriophyid mites that has received more attention than most is *Cecidophyopsis*. One species, *C. ribis*, the black currant gall mite, is the most damaging pest of black currant plants. Many of the other *Cecidophyopsis* mites colonize commonly occurring plants. It has therefore been possible to use this group to investigate the evolutionary relationships between them and their host plants. Both mite and plant genomes were subjected to molecular biological analysis. DNA sequences were generated from their ribosomal genes. It was then possible to construct phylogenetic relationships between the different mite species and, in a similar fashion, between the host species. This information was then combined with information on the mite host range, induced plant alterations and geographical distribution, to build up a model of the likely speciation processes which occurred in the evolution of these *Cecidophyopsis* mites. The molecular phylogeny found three clear mite groupings, with one group being supported by an independent character. It was not clear what the basis of the other two groups were. The phylogenetic relationships of the plants did not mirror those of the mites. This overall dissimilarity of the relationships clearly indicated that, for these mites, there is little evidence of co-speciation between the mites and their plant hosts. The results also suggest that, for these eriophyid mites, there is no difference in the range of host specificity between galling and non-galling groups as there are monophagous and biphagous mites in both groups.

Introduction

The concepts underlying speciation are extremely complex in any group of organisms. This review will not enter into detail about the definition of a species.

Instead, it will compare and contrast conventional and molecular methods which examine species boundaries in eriophyid mites. These microscopic mites are important plant pests. Apart from causing direct damage to leaves or malformed plant structures, such as big buds (fig. 1), they can also transmit serious plant pathogens such as those associated with black currant reversion disease (BRDV), sterility mosaic disease (SMV) of pigeon pea, wheat streak mosaic virus (WSMV), etc. (Ghanekar *et al.* 1992; Oldfield and Proeseler 1996). In addition to their morphological specializations for phytophagy, most eriophyids are also adapted to colonize either one, or a small number of closely related plants (Oldfield 1996). Clearly, this specialization could lead to strong selection for coevolution between the

F. Bernini, R. Nannelli, G. Nuzzaci, E. de Lillo (eds.), Acarid Phylogeny and Evolution. Adaptations in mites and ticks, 27-34.

Fig. 1 - Various eriophyid galls. Different types of plant alterations caused by: A. *C. ribis* galled buds on black currant. B. *Eriophyes laevis* pimple galls on alder *Alnus glutinosa*. C. *Aculus acericola* red pimple galls on sycamore *Acer pseudoplatanus*.

mites and their host plant species. Using molecular techniques this hypothesis can now be examined. However, before doing so this review will address some of the issues which face the study of eriophyids and in particular the group of *Cecidophyopsis* mites which inhabit the commercially important currant plants (*Ribes* species).

Morphological analysis

One of the most reliable and simple methods for distinguishing species is to measure many different structures on individuals drawn from different populations (morphometrics). After a mathematical/statistical analysis it may then be possible to discover characters which clearly and consistently separate the populations, leading to a description of two species. This approach has been applied to *Cecidophyopsis* mites on *Ribes* and this has now defined five species (Amrine *et al.* 1994). However, for two of these species the first suggestion that they differed came from molecular evidence (see below). Morphometrics requires considerable time and skill as well as a measurements from a large number of individuals. Morphological characters often exhibit variation (e.g. lengths, widths, etc.) and the extreme ranges of a character in two species may overlap. It is also not possible to key out different stages of a mite. For the eriophyids as a whole, most authorities accept that it is very difficult to examine the deep phylogenies within this group. This is complicated by a long evolutionary history (leading to a huge array of current species), convergent evolution in terms of distribution, host-alteration, vagrancy, morphology, distribu tion, host specificity, etc., lack of comparative knowledge of internal organs and differences in over-wintering forms (Lindquist and Oldfield 1996). In figure 2 is an example of the genital flap

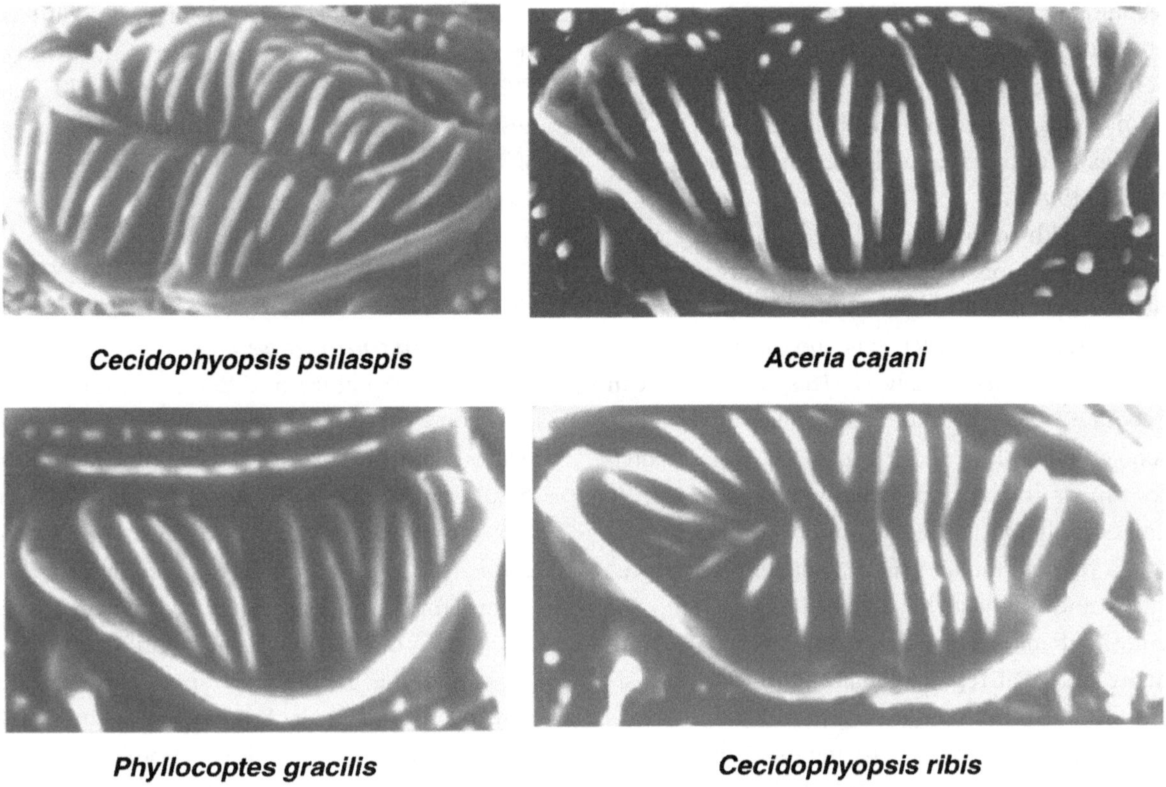

Fig. 2 - SEM Analysis of eriophyid mite female genital flaps. Scanning electron micrographs of the genital ridge patterns from different eriophyid species. Note how different the two *Cecidophyopsis* species are.

(epigynum) from four females of different eriophyid mite species: *C. psilaspis*, *Aceria cajani*, *Phyllocoptes gracilis* and *C. ribis*. In these images we see that the two *Cecidophyopsis* mite species show divergent characters in terms of the numbers and completeness of the ridges, whereas the genital flap from *A. cajani* closely resembles that of both *P. gracilis* and to a lesser extent *C. ribis*.

Molecular analysis

As outlined above the simplest method of determining a species is to use visual characters. Morphological techniques are therefore an indirect way of measuring whether two gene pools are distinct by measuring phenotypic differences which result from mutations in the two populations. When this works well there is no need to examine alternatives. However, there is an inherent assumption that, when two populations are morphologically indistinguishable, then they are the same species. In the first stages of reproductive isolation there are likely to be many mutations that do not alter shape or structure. There are examples where examination of the genetic structure of populations has revealed the presence of morphologically identical species. In the mosquito - *Anopheles gambiae* complex - the differences between species are only defined by chromosome inversions and these species are referred to as cryptic species (Collins and Paskewitz 1996). Another example can be found in aphids where an examination of different populations of the cereal aphid *Sitobion* species on different grass and wheat plants in the same area of England found a hybrid species (*S.* near *fragariae*) (Sunnucks *et al.* 1997). Clearly then, techniques which examine the gene pool (DNA) of populations will be far more sensitive at determining whether they are continuous (i.e. a from a single species), or distinct (i.e. from two species). For a review of concepts of species see Mallet (1995).

As with morphological methods, the use of molecular tools also requires careful consideration and skill. The advent of the polymerase chain reaction (PCR) has revolutionized all aspects of genetic analysis. This technique can generate DNA marker data and even gene sequences in days and it can work on individual specimens of any developmental stage. It is limited only by the need to have a defined target (i.e. a predetermined primer annealing sequence) and some of these have now been defined for eriophyid mites (Fenton *et al.* 1997). There are many potential regions suitable for PCR analysis and each can fulfil a different population analysis function. The ribosomal genes continue to provide many benefits for examining the species status of different individuals. These genes have been characterized from many examples of the major groups of eukaryotes and prokaryotes. They provide regions of rapid evolution (e.g. internal transcribed spacers, ITS1 and 2, useful in comparing closely related organisms) as well as slowly evolving genes (useful for comparing genera, families, etc.). Their use has been reviewed by Hillis and Dixon (1991). One of the major complications when using these genes is that they are a multigene family and as such they are subject to a process known as concerted evolution (Dover 1982). Concerted evolution causes the different copies of rDNA sequences to remain constant, despite each being subject to its own evolutionary forces i.e. the genes evolve in concert. The exact mode of operation of concerted evolution is still under study (Schlotterer and Tautz 1994). In some organisms incomplete homogenization can lead to heterogeneous sequences both within the population and within an individual (Vogler and DeSalle 1994), some of which may give different answers under different conditions (Fenton *et al.* 1998). However, in the case of eriophyid mites these sequences are uniform within a species, but different between species conforming to the ideal (see below).

Analysis of Cecidophyopsis mite species using ribosomal genes

The exact nature of different *Cecidophyopsis* specimens on different *Ribes* plants and their species status has been subject to a degree of confusion (Amrine *et al.* 1994). This is despite their importance as serious pests and disease vectors. Clearly, the application of DNA markers to this problem was likely to be of benefit, facilitating *Cecidophyopsis* identification and elucidating the number of species present on each plant and their host ranges. These parameters have implications both for plant resistance programmes and virus epidemiology. However, in addition to this, it was also possible to examine the longer term evolutionary histories of these mites in relation to their host plant specificity. PCR techniques were used to amplify the rDNA ITS regions from a wide variety of *Cecidophyopsis* samples over a period of years. These amplification products were either examined by restriction mapping or by determining their DNA sequences. This led to the identification of two new rDNA types from mites originating from *R. alpinum* and *R. aureum*. These mites were then described as two new species (Amrine *et al.* 1994). The molecular analysis continued to provide new insights and in 1995 *Cecidophyopsis* samples from Finnish black currants were found to have distinct rDNA sequences (Fenton *et al.* 1996). Interestingly, the same sequences were found in mites collected from buds of wild red currants. Alongside these mites on wild red currant was the previously recognized red currant gall mite *C. selachodon*. This meant that in Finland the two species were

Tab. 1 - The *Cecidophyopsis* mites in this study. The properties of the known *Ribes* infesting *Cecidophyopsis* species in terms of their hosts, descriptions and damage.

Species	Host	Molecular ID	Morphological ID	Damage
C. aurea	Golden currant	Yes	Amrine	Galling
C. alpina	Alpine currant	Yes	Amrine	Galling
C. grossulariae	Gooseberry/Black currant	Yes	Collinge	Leaf browning
C. ribis	Black currant	Yes	Westwood	Galling
C. selachodon	Red currant	Yes	van Eyndhoven	Galling
C. wc	Red currant	Yes	No	Leaf surface
C. spicata	Red/Black currant	Yes	No	Galling

sympatric and on the same plant species, yet their gene pools remained separated. Thus, the host range (galling on black currant and red currant), sympatric overlap with *C. selachodon* and new molecular sequence strongly suggest that this Finnish *Cecidophyopsis* mite belongs to an as yet undescribed species. In all of these mite species the DNA sequences from individuals and populations within the groupings were found to be remarkably uniform, indicating that, despite their propensity to live within galls where inbreeding is highly probable, homogenization of the rDNA locus is still very efficient.

A similar situation of sympatric *Cecidophyopsis* species was found as recently as 1998 (Kumar *et al.* 1999). Samples of *Cecidophyopsis* mites at two sites from red currant, white currant and gooseberry plants were examined using rDNA analysis. At both sites plants were in close proximity. At the first site different mite samples from white currant and red currant produced exactly the same rDNA ITS sequences. This sequence did not correspond to any of the known *Cecidophyopsis* sequences; rDNA from mites from gooseberry were found to have a different sequence. The latter sequence corresponded to that which had previously been identified from known specimens of *C. grossulariae*. Thus, at this site there were two species of mite, separated by their host range. At a second site over 1,000 km away there were also gooseberry and red currant plants and these were sometimes in contact. Here mites were only found on gooseberry plants. Their rDNA profile was the same as previously identified from *C. grossulariae* and from the mites on gooseberry at the first site. This work can be summarised as follows: 1) there are two non-galling populations of *Cecidophyopsis* mites on currant plants; 2) their gene pools remain separate even in close proximity; 3) one of these populations appears to be an undescribed species which is host-specific to red currant; 4) the other population is the previously identified *C. grossulariae*; 5) *C. grossulariae* appears to be unable to colonise red currant, even in the absence of competition and with direct contact between these plant species. Other work has shown that *C. grossulariae* can also colonise black currant and in some cases this is associated with severe damage to the plants (Fenton *et al.* 1996). Table 1 summarizes some of the current information known for these *Cecidophyopsis* mites.

Phylogenetic information

In addition to the useful information provided for the detailed analysis of the ecology and distribution of these mites the molecular sequences can also be used to establish which mites are more related to each other. This has revealed some interesting observations about their inter-relationships and the relationships that

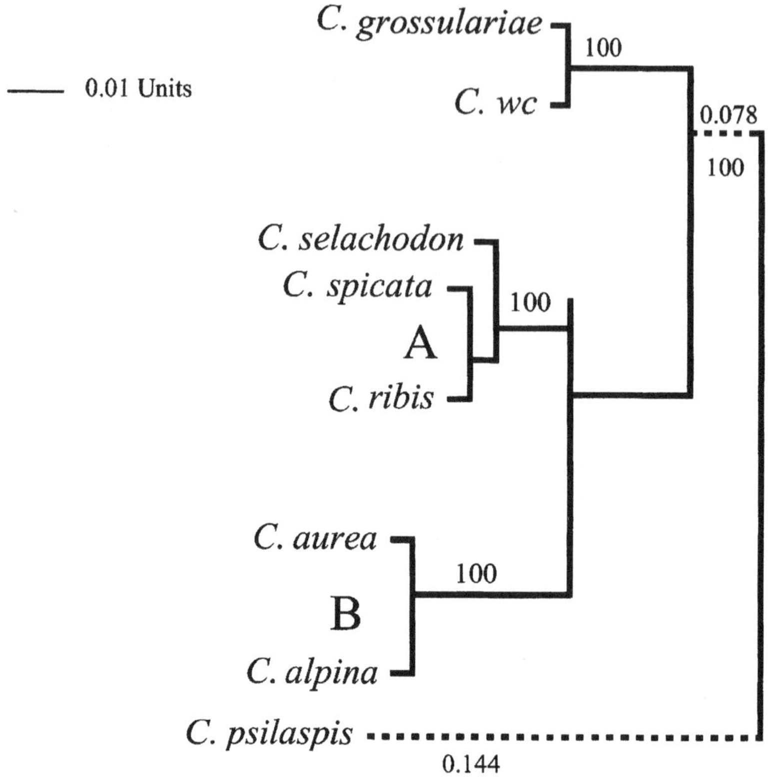

Fig. 3 - Phylogenetic tree of *Cecidophyopsis* mites. The phylogenetic tree constructed from a maximum likelihood analysis of the rDNA sequences of *Cecidophyopsis* mites. The filled-in circles represent those species which cause galls on the plant. The branch to *C. psilaspis* has been shortened so that it can fit into the diagram. The numbers at the junctions indicated how many times out of 100 the group to the left of the branch are found together. The branches are to scale apart from the dashed lines where the branch lengths are indicated below the lines.

they hold with their host plants (Fenton *et al.* 2001). Figure 3 is a phylogenetic tree constructed from mite rDNA sequences. In this tree there are three clear and highly significant groupings. The first is formed by the two non-gall-forming mites. The ancestral *Cecidophyopsis* mites were likely to be galling hence the loss of galling in this group is a derived character. This is also interesting as *C. grossulariae* has two hosts and *C. wc* has one. Clearly then, host specialization can differ in closely related mites. This is also true for a second group consisting of gall forming mites on red currant and black currant. Here the mite species *C. spicata* is capable of colonizing two plant species. Thus, even the interactions involved in gall forming do not necessarily restrict the mite to a single plant species. The final group of mites consists of *C. alpina* and *C. aurea*. The host plant *R. alpinum* is one of the most distinctive currant plants as it is dioecious. Clearly, this marks this plant well apart from the others [confirmed using molecular analysis (Messinger *et al.* 1999; Fenton *et al.* 2001)]. Yet the *Cecidophyopsis* mite which inhabits it is very closely related to the mite from *R. aureum*. There is a paradox here as *R. aureum* (the golden or buffalo currant, the host plant for *C. aurea*) was only recently introduced into Europe as an ornamental. The mite *C. aurea* has only been reported in Europe. There appears to be a restricted number of possibilities for this observation: 1) the mite *C. aurea* has recently evolved from *C. alpina*, or a common relative, to specialize onto *R. aureum* in Europe; 2) the mite *C. aurea* derives from common ancestry with *C. alpina*, but has another as yet unidentified European currant species as its host and has switched to *R. aureum*.

Conclusions

It appears that the *Cecidophyopsis* mites specialize onto one or two plant species. The limits to colonization do not appear to be based on plant ancestry, but some other character/s (e.g. volatile recognition, plant defence reactions). Specialization onto a single host is not necessary for either plant gall formation or the alternate free living life strategy. The widespread, ancestral, ability to form galls in *Cecidophyopsis* has been lost in the lineage to two closely related mites. These observations suggest that the high degree of host specialization of extant *Ribes*-colonizing *Cecidophyopsis* species disguises an otherwise adaptable group of plant pests which can alter their colonization and life strategies to exploit new resources. In one instance a *Cecidophyopsis* species has quickly colonized a plant host which it could not have previously encountered. These observations are in agreement with others made on the eriophyid mite *Eriophyes insidiosus* which existed on native plums in Mexico and the USA. It has expanded its host range and now also exists on introduced commercial peach trees, but only in those countries (Oldfield *et al.* 1995).

Acknowledgements

Thanks are due to Dr Lava Kumar and George Duncan for the images in figure 2. Thanks also go to Drs Gaynor Malloch, Nick Birch, Teifion Jones, Stuart Gordon, Pat Lanham and Rex Brennan. The Scottish Crop Research Institute is funded by grants from the Scottish Executive Rural Affairs Department and LK was funded by the Department for International Development.

References

Amrine J.W., Duncan G.H., Jones A.T., Gordon S.C., Roberts I.M., 1994 - *Cecidophyopsis* mites (Acari: Eriophyidae) on *Ribes* spp (Grossulariaceae). *International Journal of Acarology*, 20: 139-168.

Collins F.H., Paskewitz S.M., 1996 - A review of the use of ribosomal DNA (rDNA) to differentiate among cryptic *Anopheles* species. *Insect Molecular Biology*, 5: 1-9.

Dover G., 1982 - Molecular drive: a cohesive mode of species evolution. *Nature* (London), 299: 111-117.

Fenton B., Birch A.N.E., Malloch G.M., Lanham P.G., Brennan R.M., 2001 - Gall mite molecular phylogeny and its relationship to the evolution of plant host specificity. *Experimental and Applied Acarology*, 24: 831-836.

Fenton B., Malloch G., Jones A.T., Thomas W., 1996 - Molecular ecology of some *Cecidophyopsis* mites (Acari: Eriophyidae) on *Ribes* species and evidence for their natural cross colonisation of blackcurrant (*R. nigrum*). *Annals of Applied Biology*, 128: 405-414.

Fenton B., Malloch G., Moxey E., 1997 - Analysis of Eriophyid mite rDNA Internal Transcribed Spacer sequences reveals multiple simple sequence repeats. *Insect Molecular Biology*, 6: 23-32.

Fenton B., Malloch G.M., Germa F., 1998 - A study of variation in rDNA ITS regions shows that two haplotypes coexist within a single aphid genome. *Genome*, 41: 337-345.

Ghanekar A.M., Sheila V.K., Beniwal S.P.S., Reddy M.V., Nene Y.L., 1992 - Sterility mosaic of pigeonpea. In: Singh U.S, Mukhopadhyay A.N., Kumar J., Chaube H.S. (eds.), Diseases of Cereals and Pulses. Plant Diseases of International Importance. New Jersey: Prentice Hall. Vol. 1: 415-428.

Hillis D.M., Dixon M.T., 1991 - Ribosomal DNA: Molecular evolution and phylogenetic interference. *Quarterly Review of Biology*, 66: 411-429.

Kumar P.L., Fenton B., Jones A.T., 1999 - Identification of *Cecidophyopsis* mites (Acari: Eriophyidae) based on variable simple sequence repeats of ribosomal DNA internal transcribed spacer-1 sequences via multiplex PCR. *Insect Molecular Biology*, 8: 347-357.

Lindquist E.E., Oldfield G.N., 1996 - Evolution of eriophyoid mites in relation to their host plants. In: Lindquist E.E., Sabelis M.W., Bruin J. (eds.), Eriophyoid mites their biology, natural enemies and control. Elsevier, Amsterdam. World Crop Pests, 6: 277-300.

Mallet J., 1995 - A species definition for the modern synthesis. *Trends in Ecology and Evolution*, 10: 294-299.

Messinger W., Hummer K., Liston A., 1999 - *Ribes* (Grossulariaceae) phylogeny as indicated by restriction-site polymorphism of PCR-amplified chloroplast DNA. *Plant Systematics and Evolution*, 217: 185-195.

Oldfield G.N., Creamer R., Gispert C., Osorio F., Rodriguez T., Perring T.M., 1995 - Incidence and distribution of peach mosaic and its vector, *Eriophyes insidiosus* (Acari: Eriophyidae) in Mexico. *Plant Disease*, 79: 186-189.

Oldfield G.N., 1996 - Diversity and host plant specificity. In: Lindquist E.E., Sabelis M.W., Bruin J. (eds.), Eriophyoid mites their biology, natural enemies and control. Elsevier, Amsterdam. World Crop Pests, 6: 199-216.

Oldfield G.N., Proeseler G., 1996 - Eriophyid mites as vectors of plant pathogens. In: Lindquist E.E., Sabelis M.W., Bruin J. (eds.), Eriophyoid mites their biology, natural enemies and control. Elsevier, Amsterdam. World Crop Pests, 6: 259-275.

Schlotterer C., Tautz D., 1994 - Chromosomal homogeneity of *Drosophila* ribosomal DNA arrays suggests intra-chromosomal exchanges drive concerted evolution. *Current Biology*, 4: 777-783.

Sunnucks P., De Barro P.J., Lushai G., Maclean N., Hales D., 1997 - Genetic structure of an aphid studied using microsatellites: cyclic parthenogenesis, differentiated lineages and host specialization. *Molecular Ecology*, 6: 1059-1073.

Vogler A.P., DeSalle R., 1994 - Evolution and phylogenetic information content of the ITS-1 region in the Tiger Beetle *Cicindela dorsalis*. *Molecular Biology and Evolution*, 11: 393-405.

Mitochondrial DNA variation and phylogeography of *Steganacarus* on Tenerife (Canary Islands)

Nicola Salomone, Fabio Bernini
Department of Evolutionary Biology, University of Siena, via P.A. Mattioli 4, I-53100, Siena, Italy

Key words: Oribatida, phylogeography, genetic differentiation, mitochondrial DNA.

Abstract: Sequences from a fragment of the mitochondrial cytochrome oxidase I gene were used to analyze the population structure and phylogeography of the genus *Steganacarus* from Tenerife. The island has an oceanic origin; it became a single entity only in the Pleistocene (~2.0 Ma) as a result of the volcanic evolution of the Teide-Pico Viejo central complex which joined the earlier three massifs of Anaga, Roque del Conde and Teno. The mtDNA haplotype distribution and the estimates of sequence divergence among populations of *S. carlosi* reveal the presence of different lineages with strong phylogeographic structures which reflect the disjunct volcanic evolution of the island. The high levels of intra-specific differentiation suggest that biogeographic boundaries have an impact on the genetic structure of this low dispersal genus by limiting gene-flow among populations and determining isolated demes in which mutation and genetic drift may facilitate differentiation.

Introduction

The Canary Islands are located in the Eastern Atlantic Ocean, off the Northwest coast of Africa. They are a volcanic island chain comprising seven main islands and several islets. Tenerife, the largest and highest island of the Archipelago (fig. 1), occupies a central position; it is one of the most complex from a geological point of view (Watt and Masson 1995; Schmincke 1976). Since its emergence Tenerife was formed from two to three independent islands, corresponding to the massifs of Anaga in the Northeast, Teno in the Northwest and Roque del Conde to the South. The emergence of these three oldest rocks was followed by a period of volcanic quiescence. Only in more recent times, beginning approximately 2 Ma, the extensive volcanic activity that determined the emergence of the central volcano of Las Cañadas, later replaced by the Teide-Pico Viejo complex, joined the three massifs, giving the island its present configuration (fig. 2) (Ancochea *et al.* 1990).

These geological characteristics, together with remarkable within-island differences in ecology, make Tenerife an attractive environment for the study of speciation and genetic population divergence. The factors that limit the distribution of species may also give rise to genetically distinct lineages, modifying the genetic structure of conspecific populations via natural selection or regulation of inter population gene flow. This is particularly true for animals such as oribatid mites, with a poor ability to disperse (*vagility*) and limited range.

Classical taxonomy based on morphological differentiation distinguishes two species of the genus *Steganacarus* in the island of Tenerife: *S. (S.) carlosi* Niedbala, 1984, present in almost all

F. Bernini, R. Nannelli, G. Nuzzaci, E. de Lillo (eds.), Acarid Phylogeny and Evolution. Adaptations in mites and ticks, 35-39.

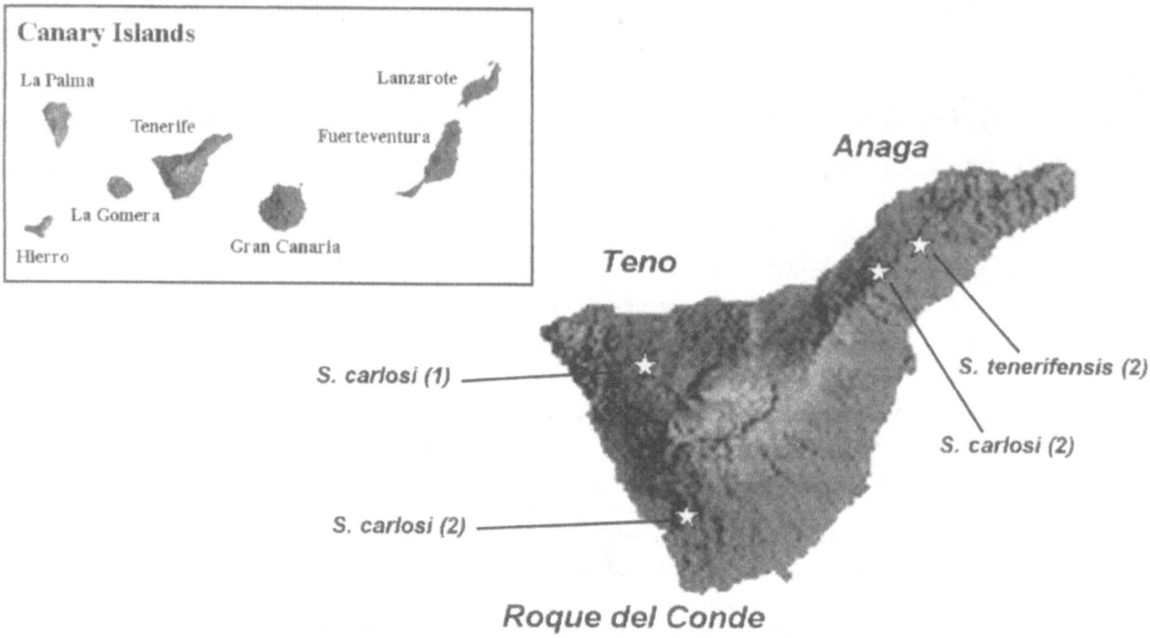

Fig. 1 - Map of the island of Tenerife showing the collection sites. The number of collected specimens is indicated in brackets.

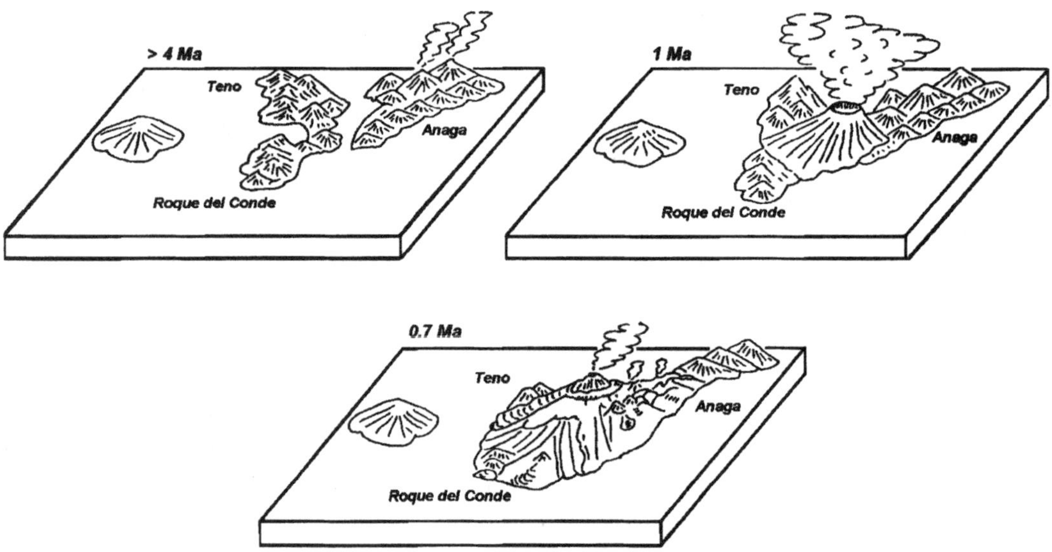

Fig. 2 - Evolution of the volcanic activity in Tenerife (adapted from Ancochea *et al.* 1990).

the islands of the Archipelago except Fuerteventura and Lanzarote, and *S. (S.) tenerifensis* Perez-Iñigo, 1972, endemic of Tenerife and restricted to small populations on the Anaga Massif.

The aim of this study was to provide a preliminary evaluation of the genetic structure and levels of divergence among populations belonging to the genus *Steganacarus* sampled on the three Massifs of the island of Tenerife, using sequence variations in a fragment of the mitochondrial cytochrome oxidase I gene.

Materials and methods

Steganacarid specimens were sampled in different sites as shown in figure 1. Whole genomic DNA was extracted from single individuals after digestion in 500 µl of C-TAB buffer (0.1 M Tris-HCl pH 8.0, 1.4 M NaCl, 0.02 M EDTA, 2% CTAB, 0.2% 2-mercaptoethanol) followed by a standard phenol-chloroform extraction and ethanol precipitation. Amplification of a 480 bp mitochondrial cytochrome oxidase I gene fragment was accomplished with primers C1-J-1751 and C1-N-2191 described in Simon *et al.* (1994). DNA amplification and sequencing was performed as described in Salomone *et al.* (1996).

Data analysis

Sequences were aligned by eye using the computer program ESEE (Cabot and Beckenbach 1989). The DnaSP version 3.14 (Rozas and Rozas 1999) was used to estimate DNA polymorphism, nucleotide diversity and genetic differentiation between populations. The proportion of nucleotide diversity attributable to variation between populations was calculated as N_{st} (Lynch and Crease 1990). This estimator of population structure is similar to the well-known F_{st} but incorporates a Jukes-Cantor model for multiple changes at a nucleotide site. Gene flow levels between populations, as measured by Nm, were calculated using N_{st}. Phylogenetic analysis was performed with PAUP* (vers. 4.0, Swofford 1998) and using a specimen of the same genus, *S. (S.) hirsutus* (Sila, Italy), as outgroup. Phylogenetic relationships were assessed using the neighbor-joining algorithm (Saitou and Nei 1987) based on a distance matrix obtained via Kimura's 2-parameter model (Kimura 1980). To assess the robustness of relationships, 500 bootstrap replications were performed.

Results

A total of 413 bp were aligned between sequences and many variations were found in analyzed samples. All the specimens revealed different haplotypes (gene diversity=1.00). As expected for protein coding genes, most substitutions occurred in third base codon. No insertions or deletions were observed. Within the *carlosi* group of populations, haplotypes from different localities are relatively divergent from each other with a mean number of pairwise differences of 22.0 and a nucleotide diversity of 0.05. Including the sequences of *S. tenerifensis*, the average number of nucleotide differences was 36.72, with a nucleotide diversity of 0.09. The maintaining of such high levels of differentiation requires that migratory events be restricted. Comparisons among the three populations of *S. carlosi* show minimal values of gene flow (average Nm 0.09; N_{st}=0.740), indicating very low levels of migration and a high degree of geographical isolation for this taxon. Phylogenetic relationships based on neighbor-joining method are shown in figure 3. Samples from Tenerife grouped into two major clades, one for *S. tenerifensis* and the other for all *S. carlosi* specimens. In the latter lineage a clear phylogeographic structure is evidenced, with samples from Anaga Massif in basal position.

Discussion and conclusions

Although the sample size is limited, sequence analyses show that *S. carlosi* on Tenerife has a high degree of genetic differentiation in the three examined populations and very low

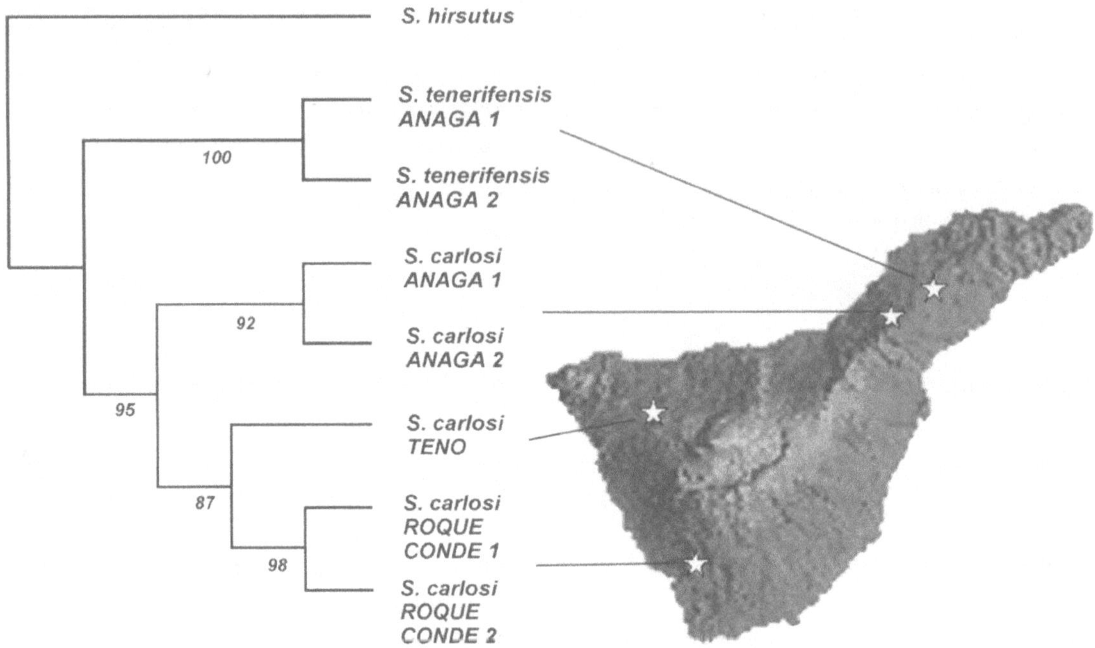

Fig. 3 - Neighbor-joining tree based on Kimura's (1980) genetic distances. The bootstrap values are shown at the nodes (500 reps.).

levels of interdeme contact. This pattern of haplotype diversification is probably the result of geographic isolation determined by the disjunct volcanic evolution of the island. This evidence suggests that biogeographic boundaries may have a great impact on speciation processes in *Steganacarus*. Even on a small geographic scale, physical barriers can limit gene flow and determine the isolation necessary to evolve intraspecific divergence among populations; this is the case in the populations of Teno and Roque del Conde which lie less than 40 Km apart.

Similar evidence is provided by phylogenetic analyses: all the *carlosi* haplotypes are divided into three different sublineages with a clear phylogeographic structure. Even in the hypothesis of frequent population bottlenecks, observed data suggest that these populations evolved independently after colonization of the three massifs; mutation and genetic drift has likely been the main agent shaping the pattern of genetic divergence.

The phylogenetic analysis supports the monophyletic origin of *Steganacarus* on Tenerife and the existence of two genetically distinct lineages, one for *S. tenerifensis* and the other for the three populations of *S. carlosi*. This result is consistent with the present taxonomic classification, even though earlier allozymatic studies suggested a different status (Avanzati *et al.* 1994). The question remains as to how *S. tenerifensis* and *S. carlosi* achieved distinct mtDNA lineages. Both taxa now live in the Laurisilva forest of the Anaga Massif from which colonization seems to have originated. This evidence may seem to contrast the proposed peripheral model of diversification. However, if the geographic barriers that initiate speciation are relatively short-lived, then inferences based on the current distribution of species would lead to erroneous conclusions. Even if selective factors cannot be excluded, the presence of the two species in the same natural environment makes this explanation unlikely. We believe that, although there is no obvious geographic feature that might act as a barrier to *Steganacarus* dispersal in Northern Tenerife, the volcanic history of the massif may have created the necessary conditions for divergence in allopatry followed by secondary intergradation of the two species. The violent volcanic eruptions and lava flows that characterized the evolution of the Anaga Massif (Ancochea *et al.* 1990) isolated peripheral populations favouring the diversification process that led to their speciation.

In conclusion, genetic data indicate that geographic isolation played an important role in the diversification of *Steganacarus* populations. Geographic boundaries may represent an effective barrier to gene flow even across small distances, allowing populations to diverge. Mutation, genetic drift and bottleneck events are expected to play a relevant role in this process.

More data is clearly needed to better understand the phylogeny of *Steganacarus* on Tenerife, an island which provides a useful model to investigate processes of colonization and genetic differentiation.

Acknowledgements

We would like to thank G.M. Hewitt, B.C. Emerson, H. Schatz and R.H. Thomas for helpful comments and suggestions on the manuscript.

References

Ancochea E., Fuster A.M., Ibarrola E., Cendrero A., Coello J., Hernan F., Cantagrel J.M., Jamond C., 1990 - Volcanic evolution of the island of Tenerife (Canary Islands) in the light of new K-Ar data. *Journal of Volcanology and Geothermal Research*, 44: 231-249.

Avanzati A.M., Baratti M., Bernini F., 1994 - Molecular and morphological differentiation between steganacarid mites (Acari: Oribatida) from the Canary islands. *Biological Journal of the Linnean Society*, 52: 325-340.

Cabot E.L., Beckenbach A.T., 1989 - Simultaneous editing of multiple nucleic acid and protein sequences with ESEE. Computer *Applications in the Biosciences*, 5: 233-234.

Kimura M., 1980 - A simple method for estimating evolutionary rate of base substitution through comparative studies of nucleotide sequences. *Journal of Molecular Evolution*, 16: 111-120.

Lynch M., Crease T.J., 1990 - The analysis of population survey data on DNA sequence variation. *Molecular Biology and Evolution*, 7: 377-394.

Rozas J., Rozas R., 1999 - DNAsp version 3: an integrated program for molecular population genetics and molecular evolution analysis. *Bioinformatics*, 15: 174-175.

Saitou N., Nei M., 1987 - The Neighbor-joining method: a new method for reconstructing phylogenetic trees. *Molecular Biology and Evolution*, 4: 406-425.

Salomone N., Frati F., Bernini F., 1996 - Investigation on the taxonomic status of *Steganacarus magnus* and *Steganacarus anomalus* (Acari: Oribatida) using mitochondrial DNA sequences. *Experimental and Applied Acarology*, 20: 607-615.

Schmincke H.U., 1976 - The geology of Canary Islands. In: Kunkel J. (ed.), Biogeography and ecology in the Canary Islands. The Haugue: Dr. W. Junk b.v. Publisher: 67-184

Simon C., Frati F., Beckenbach A.T., Crespi C., Liu H., Flook P., 1994 - Evolution, weighting and phylogenetic utility of mitochondrial gene sequences and a compilation of conserved polymerase chain reaction primers. *Annals of the Entomological Society of America*, 87: 651-701.

Swofford D.L., 1998 - PAUP*: Phylogenetic Analysis Using Parsimony (* and other methods), vers. 4.0. Sinauer, Sunderland, MA.: 128 pp

Watt A.B., Masson D.G., 1995 - A giant landslide on the North flank of Tenerife, Canary Islands. *Journal of Geophysical Research*, 100: 487-498.

Preliminary report on fossil oribatid mites from the Florisbad Quaternary Research Station, South Africa

Louise Coetzee
Department of Acarology, National Museum, P.O. Box 266, Bloemfontein 9300, South Africa, acarol@nasmus.co.za

Key words: fossil oribatid mites, Florisbad, South Africa, Quaternary.
Abstract: Fossil oribatid mites were extracted from the Florisbad Quaternary sediments in central South Africa. Samples were taken from a Middle Stone Age Horizon (125,000 years BP), a Holocene Horizon of aeolian sediments (<10,000 years BP) and an organic rich section (<10,000 years BP). Seventeen species of oribatid mites were recovered.

Introduction

Florisbad is a thermal spring situated 45 km NNW of Bloemfontein in central South Africa (28°46' S, 26°04' E) (fig. 1). Deposits of considerable archaeological and paleontological importance are associated with the spring. These include an archaic human skull fragment, Middle Stone Age (MSA) artifacts and a large collection of mammal bones (Brink 1987). In order to place the Quaternary deposits in the correct perspective, intensive investigations were carried out regarding the geology and dates of the deposits (Rubidge and Brink 1985; Loock and Grobler 1988; Grün *et al.* 1996).

A pit was made to establish the stratigraphy of the deposits and to obtain samples for dating (fig. 2). It was excavated down to the solid rock, 8 m deep. The top 2 meters consist of aeolian sediments, and are less than 10,000 years old (Holocene), underlain by sediments deposited during the last glacial event between 20,000-70,000 years ago. The layers below represent the last interglacial (warm cycle) and are 100,000-125,000 years old, followed by another 3 meters of deposit of which the base has been dated at 270,000 years BP. These age estimates were obtained by electron spin resonance (ESR) and optically stimulated luminescence (OSL) (Grün *et al.* 1996).

The main archaeological excavation is situated 30 m South of the pit, near the spring eye. Here numerous Middle Stone Age artifacts were found, indicating human activity 125,000 years BP when this area was a marsh due to water being discharged by the spring. In the cross section of the excavation, an organic layer was exposed which intersects the older layers (fig. 2). The estimated age of this organic layer was established by radiocarbon dating at 8,170±90 (1.4 m below surface) to 1,700±90 (0.05 m below surface) years BP (Nyakale 1999).

F. Bernini, R. Nannelli, G. Nuzzaci, E. de Lillo (eds.), Acarid Phylogeny and Evolution. Adaptations in mites and ticks, 41-44.

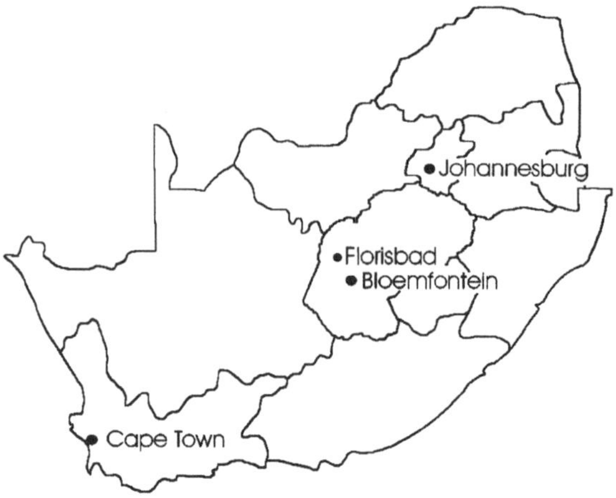

Fig. 1 - South Africa.

Material and methods

Thirty-five samples (400-500 gram each) were taken from the northern face of the pit, at 20-cm intervals starting from 20 cm below the modern soil. Nine samples (100 gram each) were taken from the organic layer in the main archaeological excavation at 20 cm intervals. Care was taken to remove the front face of the sections and collect samples from deeper inside the sediment.

Mites were extracted by means of the hyper saline flotation method described by Fain and Hart (1986). The sediments were disaggregated in 80% ethanol to which aqueous saturated sodium chloride (NaCl) was added. The difference in density between the 80% ethanol and saturated sodium chloride solution causes the ethanol impregnated mites to float to the surface.

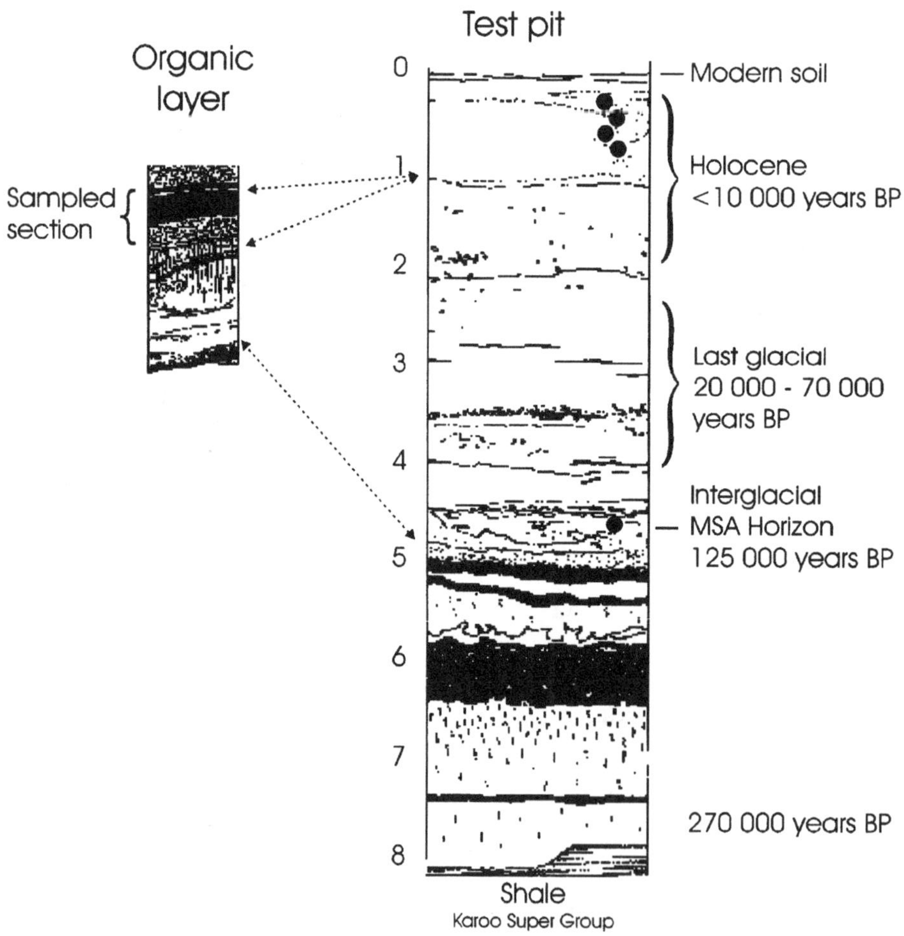

Fig. 2 - Section of the test pit and organic layer. Samples from the test pit in which fossil oribatids were encountered are indicated by black dots.

Results

Seventeen species of oribatid mites were retrieved from the organic layer (0.2-1.2 m), the Holocene layers of the test pit (upper 1 m) and from the Middle Stone Age Horizon (5 m below surface) (tab. 1). Two species of Uropodina (Mesostigmata) were collected from the organic layer. Insect fragments and pollen were also collected from all the samples.

Discussion

Humidity is one of the main factors influencing the distribution of oribatid mites. Rainfall and vegetation type go hand in hand, from high rainfall and lush vegetation in the eastern parts of the country, to arid conditions in the west. Florisbad falls in the grassland biome, which covers a large part of central South Africa. At Florisbad, the paleoclimate has been well established. Global temperature trends as reflected by Marine Oxygen Isotope Stages show that during the Middle Stone Age, 125,000 years ago, temperatures were high and the occurrence of hygrophilous mammals like Lechwe and Hippopotamus indicate more aquatic conditions in this region during the early Pleistocene (Brink 1987).

Based on distribution data from the National Museum oribatid reference collection, only *Epilohmannia pallida* subsp. nov., *Berlesezetes auxiliaris orkneyensis* and *Scheloribates elsi* have been recorded from the extant fauna of the central parts of South Africa.

Protoribates capucinus was recorded from the organic layer and the MSA Horizon, indicating

Tab. 1 - Fossil oribatid mites collected at Florisbad Quaternary Research Station, South Africa.

Species		Organic Layer	Test Pit Holocene	Test Pit MSA
		< 10,000 years BP	<10,000 years BP	125,000 years BP
		Number of specimens		
Epilohmanniidae	*Epilohmannia* sp.	2	3	
	Epilohmannia pallida subsp. nov.	3	3	
	Epilohmannia cylindrica cylindrica (Berlese, 1904)	12	25	
Lohmanniidae	*Papillacarus* sp. nov.	24	64	
	Cryptacarus promecus Grandjean, 1950	1		
	Torpacarus sp.		2	
Oppiidae	*Discoppia (Cylindroppia) cylindrica* (Pérez-Iñigo, 1965)	192	35	2
	Javieroppia cervus Minguez & Subias, 1986	3	7	
	Microppia minus (Paoli, 1908)	9		
	Multioppia wilsoni Aoki, 1964	39	1	
	Oppiidae sp.	14		
	Ramusella (Insculptoppia) sp.			1
Microzetidae	*Berlesezetes auxilliaris orkneyensis* Engelbrecht, 1972	4		
Haplozetidae	*Protoribates capucinus* Berlese, 1908	1		5
Scheloribatidae	*Scheloribates elsi* van Pletzen, 1965	2		
	Scheloribates (Hemileius) sp.	1		
Galumnidae	*Pergalumna* sp.	2		

similar soil conditions in those deposits, despite the difference in time. The absence of this species from the Holocene layers of the test pit, which is an aeolian deposit, stresses the dissimilar sedimentary conditions. *Epilohmannia c. cylindrica* and *Papillacarus* sp. nov. are well represented in the Holocene sediments of the test pit, but less so in the organic sediments, indicating the xerophilous nature of these species. *Discoppia* (*C.*) *cylindrica* is listed by Subias & Rodriquez (1986) as prefering mineral soil, but nevertheless was well represented in the organic layer. *D.* (*C.*) *cylindrica* therefore seems to have been a generalist regarding ecological requirements. These species have so far not been encountered in the extant fauna of Florisbad.

Acknowledgements

I thank fellow researchers at Florisbad, James Brink, (Quaternary Paleontologist) and Zoë Henderson (Middle Stone Age Archaeologist) for their helpful contributions, and Michelle Barlow for helping with the extractions.

References

Brink J.S., 1987 - The archaeozoology of Florisbad, Orange Free State. *Memoirs van die Nasionale Museum*, 24: 1-151.

Fain A., Hart B.J., 1986 - A new, simple technique for extraction of mites, using the difference in density between ethanol and saturated NaCl (Preliminary note). *Acarologia*, 27: 255-256.

Grün R., Brink J.S., Spooner N.A., Taylor L., Stringer C.B., Franciscus R.G., Murray A.S., 1996 - Direct dating of Florisbad homonid. *Nature*, 382: 500-501.

Loock J.C., Grobler N.J., 1988 - The regional geology of Florisbad. *Navorsinge van die Nasionale Museum*, Bloemfontein, 5: 489-497.

Nyakale M., 1999 - Palynology of Late Quaternary deposits from the central plateau, South Africa. Unpubl. M. Sc. Thesis, Department of Botany and Genetics, University of the Orange Free State.

Rubidge B.S., Brink J.S., 1985 - Preliminary survey of the extent and nature of the Pleistocene sedimentary deposits at Florisbad, South Africa. *Navorsinge van die Nasionale Museum*, Bloemfontein, 5: 69-76.

Subias L.S, Rodriquez P., 1986 - Oppiidae (Acari, Oribatida) de los sabinares (*Juniperus thurifera*) de España IX. *Subiasella* (*Lalmoppia*) n. subgen. y *Discoppia* (*Cylindroppia*) n. subgen. *Revista de Biologia de la Universidad de Oviedo*, 4: 111-121.

An Early Ordovician oribatid mite (Acari: Oribatida) from the island of Öland, Sweden

F. Bernini[1], G. Carnevale[2], G. Bagnoli[2] and S. Stouge[3]
[1] *Dipartimento di Biologia Evolutiva, Università di Siena, Italy*
[2] *Dipartimento di Scienze della Terra, Università di Pisa, Italy*
[3] *Geological Survey of Denmark and Greenland, Copenhagen, Denmark*

Key words: fossil mite, Ordovician period, oribatid phylogeny.

Abstract: The origin of the arachnids (spider, mites and so on), a typical terrestrial animal group dating back to the Early Devonian, from other chelicerate aquatic ancestors are complicated by the complete absence of fossil record.

Here we report the first mite from the Early Ordovician of the island of Öland, Sweden. It may contribute to the suggestion of a more ancient and diverse scenario for arachnid evolution.

A new exceptionally well preserved oribatid mite (fig. 1) has been found in the Lower Ordovician sediments of the island of Öland. The specimen was recorded from the residue of a sample of the Orthoceras Limestone (OTT 8) collected near Ottenby (fig. 2), that was treated with acetic acid and sieved for conodont collection. No contamination from younger rocks is possible because the outcrops in the island are no younger than Middle Ordovician. A laboratory contamination can be excluded as well, because all the laboratory equipment has been used only for Cambrian and Ordovician rocks.

The age of the sample is earliest Arenig on the basis of the associated fauna which is referred to the *Paroistodus proteus* conodont Zone (Lindström 1971).

The original chitinous cuticle is substituted by more resistant organic molecules of probably kerogen-like composition (see also Baas *et al.* 1995; Briggs *et al.* 1995; Stankiewicz *et al.* 2000).

The secondary body division into proterosoma (bearing the first two pairs of legs) and hysterosoma (with the last two pairs of legs), the pattern of the ano-genital region and the body outline allow us to recognize this specimen as an adult form of oribatid mite.

Moreover, the holoid form of the body, the gnathosomal diarthry, the presence of a genu clearly different from the tibia and the absence of pteromorphae, suggest a taxonomic assignement of this specimen to the Brachypylina Hull, 1918, and, more precisely, to the group of apterogasterina families (Krantz 1978).

On the basis of the shared exoscheletal characters the taxonomic placement of this specimen appears to be intermediate between the cymbaeremeid and the ameronothroid families

F. Bernini, R. Nannelli, G. Nuzzaci, E. de Lillo (eds.), Acarid Phylogeny and Evolution. Adaptations in mites and ticks, 45-47.

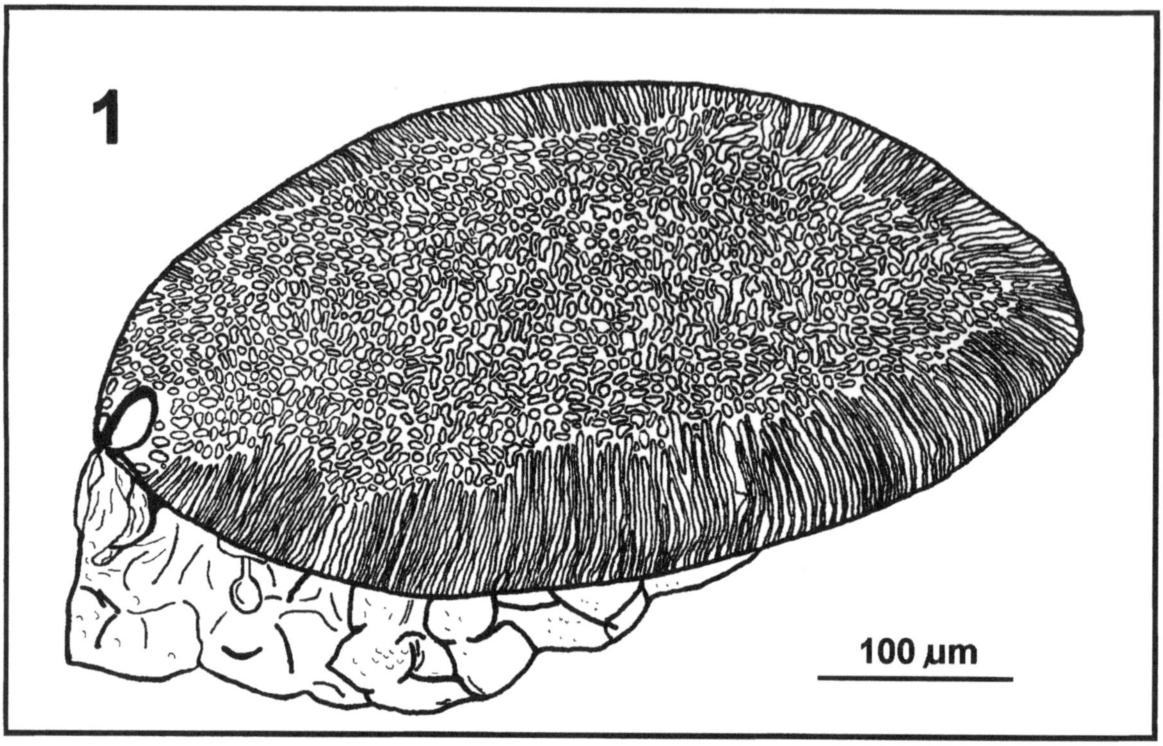

Fig. 1 - Schematic drawn of the oribatid mite from Öland showing the dorsal microsculpture.

Fig. 2 - A classic locality of the island of Öland, Sweden.

and it might represent an ancestral taxon to both groups (R. Schuster, *pers. comm.*).

Oribatid mites have been reported continously from Paleozoic sediments both with trace-fossil (coprolites and tunnels within plant tissues) and with early representatives of whole body fossils dating back to the Middle Devonian of Gilboa, USA, and the extant genera from the Jurassic of Sweden and the Burea Basin, Russia. All the remnants of oribatid mites as well as their modern representatives testify to the persistence of their feeding habit (detritivory) (Labandeira *et al.* 1997).

Modern representatives of Cymbaeremaeidae are arboricolous; in contrast, Ameronothridae include largely halophilous and intertidal species feeding almost exclusively on algae and fungi (Schulte 1976).

The Ordovician Period was characterized by a great biodiversification and by the adaptive radiation of many animal groups in the marine biota (Sepkoski 1995). Also plants attempted, to invade land, colonizing the moist terrestrial environments (Strother *et al.* 1996; Wellman 1999). Probably, this mite was not yet fully terrestrial in habits, but lived among the rich organic substance (algae and fungi) on the coastline. Perhaps, its life was not too much different from that of the present-day ameronothrids (Schulte and Weigmann 1977).

The age and the morphology of this specimen implies a reconsideration of the oribatid phylogeny, challenging the common opinion that the higher Oribatida are no more ancient than the Jurassic Period (Labandeira *et al.* 1997).

This specimen represent the oldest record of the arachnids. The Ordovician presence of arachnids confirms the results of recent morphological and molecular phylogenetic studies suggesting that the arachnid clade arose very early in the evolutionary history of the arthropods, and particularly of the "Arachnomorpha" (Wills *et al.* 1994; Averof 1998).

References

Averof M., 1998 - Origin of spider's head. *Nature*, 395: 436-437.

Baas M., Briggs D.E.G., van Heemst J.D.H., Kear A.J., De Leeuw J.W., 1995 - Selective preservation of chitin during the decay of shrimp. *Geochimica et Cosmochimica Acta*, 59: 945-951.

Briggs D.E.G., Kear A.J., Baas M., De Leeuw J.W., Rigby S., 1995 - Decay and composition of the hemichordate *Rhabdopleura*: implications for the taphonomy of graptolites. *Lethaia*, 28: 15-23.

Krantz G.W., 1978 - A manual of Acarology. Oregon State University Book Stores, Inc. Corvallis: 509 pp.

Labandeira C.C., Phillips W.G., Norton R.A., 1997 - Oribatid Mites and the Decomposition of Plant Tissues in Paleozoic Coal-Swamp Forest. *Palaios*, 12: 319-353.

Lindström M., 1971 - Lower Ordovician Conodonts of Europe. *Bulletin of Geological Society of America*, 127: 21-61.

Schulte G., 1976 - Zur Nahrungsbiologie der terrestrischen und marinen Milbenfamilie Ameronothridae (Acari, Oribatei). *Pedobiologia*, 16: 332-352.

Schulte G., Weigmann G., 1977 - The evolution of the family Ameronothridae (Acari: Oribatei) II. Ecological Aspects. *Acarologia (Paris)*, 19: 167-173.

Sepkoski J.J., 1995 - The Ordovician Radiations: Diversification and Extintion shown by Global Genus-Level Taxonomic Data. In: Cooper J.D., Droser M.L., Finney S.C. (eds.), Ordovician Odyssey: Short papers for the Seventh International Symposium on the Ordovician System: 393-396.

Stankievicz B.A., Briggs D.E.G., Michels R., Collinson M.E., Flannery M.B., Evershed R.P., 2000 - Alternative origin of aliphatic polymer in kerogen. *Geology*, 28: 559-562.

Strother P.K., Al-Hajri S., Traverse A., 1996 - New evidence for land plants from the lower Middle Ordovician of Saudi Arabia. *Geology*, 24: 55-58.

Wellman C.H., 1999 - Ordovician land plants: evidence and interpretation. *Acta Universitatis Carolinae-Geologica*, 43(1/2): 275-277.

Wills M.A., Briggs D.E.G., Fortey R.A., 1994 - Disparity as an evolutionary index: a comparison of Cambrian and Recent arthropods. *Paleobiology*, 20: 93-130.

Genetic polymorphism in prey preference at a small spatial scale: a case study of soil predatory mites (*Hypoaspis aculeifer*) and two species of astigmatic mites as prey

Izabela Lesna, Maurice W. Sabelis
Institute for Biodiversity and Ecosystem Dynamics, Section Population Biology, Kruislaan 320, 1098 SM Amsterdam, The Netherlands

Key words: predators, prey, preference, polymorphism, inheritance, mites, *Hypoaspis aculeifer*, *Tyrophagus putrescentiae*, *Rhizoglyphus robini*.

Abstract: Arthropod predators usually feed on a variety of different prey types. At one extreme populations consist of an ensemble of pure specialists for each prey type and at the other they are composed of generalists that indiscriminately seize any prey encountered. In between these extremes populations may be composed of generalist predators with learned or genetically fixed prey preferences. Here, we report on experimental evidence for genetic variation in prey preference in a local population of the haplo-diploid, soil-inhabiting predatory mite, *Hypoaspis aculeifer*. A sample of *c*. 150 predatory mites obtained from *c*. 0.25 m^2 sandy soil in a lily field (North-Holland), was collected and their offspring was subjected to two-way selection for prey choice, using two species of astigmatic mites as prey (*Rhizoglyphus robini* and *Tyrophagus putrescentiae*, further referred to as prey **R** and prey **T**). Individual female predators choosing exclusively prey **R** or exclusively prey **T** in three 10-minute dual-choice tests carried out at three-day intervals were set apart and reared (on prey **T**) as isofemale lines. Four generations of selection yielded three lines showing a high probability of choosing prey **R** and two lines choosing prey **T**. Cross-breeding in both directions resulted in hybrid females with intermediate preferences and the choices found in females from F1xParent backcrosses were best explained by assuming monogenic inheritance without dominance. Our selection and cross-breeding experiments demonstrate that polymorphism in prey preference can occur at a very small spatial scale within a local population. We discuss how such variability will alter the dynamics of otherwise monomorphic predator-prey models, and how the simple mode of inheritance of a preference trait can aid the evolution of ecotypes.

Introduction

Polyphagy of predatory arthropods is often inferred from observations at the species or population level. To what extent these observations also apply to individuals is not easy to answer because of the many possibilities and the amount of experimental work involved in discriminating one possibility from the other. At one extreme all individuals of a species feed indiscriminately on a range of food types. At the other, all individuals are specialists, but on different food types, while the species as a whole is still polyphagous. In between these extremes, there may be individuals that develop food preferences by learning, others that have innate preferences which can be altered by learning, and yet others that have genetically fixed preferences. Further complexities arise from the spatial scale at which these types of individ-

F. Bernini, R. Nannelli, G. Nuzzaci, E. de Lillo (eds.), Acarid Phylogeny and Evolution. Adaptations in mites and ticks, 49-64.

uals manifest themselves. For example, specialization may prevail in local populations, even though observations taken over the entire geographical range suggest the species to be a generalist (Fox and Morrow 1981; Futuyma and Moreno 1988; Jaenike 1990; Thompson and Pellmyr 1991). Similarly, individuals within a local population may all be specialists, but for different food types, so that polyphagy is observed when averaging over individuals population-wide. This is the question addressed in this paper: do local populations harbour genetic variation in food selection behaviour?

Whether mono- or polymorphic populations of fixed and/or flexible behaviour are expected at a local spatial scale, is the subject of theoretical explorations. One approach is based on individual optimality, thus ignoring competition among foragers. Suppose there is *one* habitat with just two food types that differ only in quantity (not quality) and in the time it takes the forager to find and feed on these food types. Then, a time-limited predator is expected to specialize above a critical density of the food type with the largest energy returns per time spent handling (provided food recognition is instantaneous) and feed on both food types below this threshold (Stephens and Krebs 1986). When food density fluctuates around this critical density, foragers should update their information on food availability and switch from being specialist to generalist (or *vice versa*) near the threshold. Food selection will then involve some form of learning behaviour. However, if the density of the two food types is constantly below or constantly above the threshold, foragers should be either generalist or specialist on the most profitable food type, and there is a tendency for genetic fixation for one or the other type of food selection behaviour. The theory of optimal food selection therefore predicts monomorphic local populations of either specialists (on the superior food), generalists, or flexible foragers that switch from generalist to specialist.

Polymorphism of genetically determined food-selection behaviour may evolve when foraging decreases food availability and predators compete for food. Assuming there are no costs involved in travelling and flexibility of food choice, predators should distribute themselves over the two food types so as to obtain equal net benefits, being a compromise between potential food gain and food loss due to interference (Sutherland 1996). Such an ideal free distribution may be achieved not only by flexible generalists, but also by a polymorphic population of specialists for either food type, as long as their frequencies are tuned to the net benefits of each food type. These two possible outcomes are equivalent when food supply occurs in simultaneous pulses of the same amounts of each food type. Under natural conditions, however, food availability tends to be variable in time, quantity and composition, making a generalist that switches to the currently more profitable food item a more likely evolutionary outcome. Ideal free foragers are therefore unlikely to show genetically determined polymorphism in food selection behaviour.

At a small spatial scale searching for food items may not be very costly and may approximate the ideal free situation. However, flexibility in food selection may well entail a cost, *e.g.* due to the sensory equipment, the time and energy needed to estimate the densities of the food types by sampling, or when food types differ in quality, due to the time and energy needed to tune the digestive enzymes to food composition. Specialists profit from not paying the cost of flexibility, and they coexist because they live on different food types, thereby avoiding interference with each other. Generalists have a larger food source at their disposal and profit from investing in flexibility, especially when the availability of the food types varies in time and space and the costs to survive until better times or to reach other locations are sufficiently high (Ward 1992). This leads to the prediction that local populations of foragers may consist of two specialists (on either of the two food types), one generalist or any mixture of specialists and generalists, depending on the degree of temporal and spatial variation in quality and availability of the food types (Wilson and Yoshimura 1994).

All of the above models are based on the assumption that specialists and generalists are genetically isolated from each other. Whether polymorphism is possible within populations that are homogeneously mixed with respect to mating is largely an open question. Population genetic models for one-locus two-allele traits predict that spatial and temporal variation in the environment can maintain polymorphism of specialists (homozygotes) and generalists (heterozygotes), but only under a very restrictive set of conditions with respect to relative fitness of the heterozygotes (Hoekstra *et al.* 1985).

In this paper, we analyze prey selection behaviour in a local population of predatory arthropods under conditions where prey composition is relatively constant. We test whether this local population harbours genetic variation for prey choice. Predatory arthropods are chosen because they are usually polyphagous (Sabelis 1992). A particularly striking example is the soil-inhabiting, predatory mite, *Hypoaspis aculeifer* (Canestrini) (Acari: Laelapidae). It can feed and reproduce on a diet consisting exclusively of astigmatic mites, tetranychid mites, soft-bodied immatures of cryptostigmatic mites, *Tribolium* eggs and larvae, fly eggs and larvae, thrips larvae and pupae, springtails, enchytraeid worms, nematodes or even pollen (Barker 1969; Lobbes and Schotten 1980; Usher and Davis 1983; Ragusa *et al.* 1986; Sardar and Murphy 1987; Ragusa and Zedan 1988; Murphy and Sardar 1991; Ruf 1993; Lesna *et al.* 1995 and 1996). This predator was chosen in the present study for the following reasons: (1) widespread occurrence in many different habitats, (2) ease of collection and availability in the field throughout the year, (3) short generation time (*c.* 12 days at 23°C; Lobbes and Schotten 1980), (4) ease of rearing in large numbers on astigmatic mites which in turn are fed on yeast flakes (Lesna *et al.* 1995), (5) reproduction by arrhenotokous parthenogenesis, which facilitates the creation of isofemale lines via mother-son mating (Lobbes and Schotten 1980; Usher and Davis 1983; Sardar and Murphy 1987; Ruf 1993).

In this paper we present experimental evidence that a local population of this predatory mite harbours genetic variation in choice with respect to two prey species from different genera of astigmatic mites: (1) the bulb mite *Rhizoglyphus robini* Claparède (prey **R**), and (2) the copra mite *Tyrophagus putrescentiae* (Schrank) (prey **T**). These predator and prey species co-occur in many different habitats and were collected from the soil in a lily field, *i.e.* an environment where the two prey species are present every year although bulb mites are much more abundant than copra mites. Prey preference was assessed by subjecting individual females to repeated two-choice tests, and by selection of those consistently attacking either of the two prey for creating isofemale lines. This procedure was continued for several generations until lines with opposite preferences were selected. Assessment of prey choice in F1-hybrids and F2-backcrosses were then carried out to elucidate the mode of inheritance.

Materials and Methods

Prey mites

Two species of astigmatic mites were offered as prey during the selection experiments: the bulb mite, *R. robini*, and the copra mite, *T. putrescentiae*. The first species was collected from lily bulbs cv. Pollyanna, which were harvested in December 1991 from a field in Anna Paulowna (Province North-Holland, The Netherlands). The bulb mites were reared in 8 cm ø Petri dishes provided with a bottom layer of moistened "Plaster of Paris" mixed with charcoal. The copra mites were obtained from Koppert BV, Berkel en Rodenrijs (The Netherlands) and reared in open glass jars placed in a closed plastic box filled up to *c.* 1 cm with water (to maintain a humid environment and prevent escape of mites from the open jars). Both species of prey mites were fed with pure bakers' yeast and kept in a dark climate box at *c.* 15°C.

Predatory mites

The predatory mite, *H. aculeifer*, was collected in September 1991 from lily bulbs grown in a field near Breezand (North-Holland). A sample of 20 bulbs infested by *R. robini* was taken from *c.* 0.25 m^2 of sandy soil. Approximately 150 individual predatory mites were collected from the bulbs. The population reared from these individuals in the lab has been referred to as strain "Breezand" in earlier publications (Lesna *et al.* 1995, 1996), and serves as the so called "Base population" for the selection experiments that were started in January 1995 and are described below. Predatory mites were reared in a climate room at 22°C and *c.* 70% RH in closed plastic vials (7 cm ø and *c.* 7 cm high) with a bottom layer of moist "Plaster of Paris" mixed with charcoal to maintain a humid environment. The vials were closed by a lid that was provided with pin holes to allow air exchange. A cover of fine mesh gauze sealed under the lid prevented escape via the pin holes. In all cultures only *T. putrescentiae* was used as prey to rear the predatory mites.

Prey offered in periods between experimental tests

The prey used for rearing may have had an impact on the motivational state of the predatory mites and the genetic composition of their population. This problem is unavoidable whatever our choice of prey. In this article, we report on experiments where *T. putrescentiae* (prey **T**) was used as a standard prey (1) for rearing of the Base population, (2) in between the prey choice tests during the artificial selection procedure and (3) for maintenance of the isofemale lines.

These experiments should be repeated with **R** (instead of **T**) as a standard prey, but as argued in Lesna and Sabelis (1999) and as confirmed in independent experiments (Lesna, *unpubl. data*) polymorphism is maintained on prey **T** (due to hybrid advantage) and fades away on prey **R** (due to hybrid inferiority). Hence, artificial selection appeared to be only successful with **T** as standard prey, which is the reason why only these experiments are presented in this article.

Selection experiments

A total of 81 female deutonymphs (*i.e.* either just before or undergoing their final moult) was taken from the lab-reared Base population and placed individually in small plastic vials (3 cm ø and *c.* 4 cm high), provided with a bottom layer of moist "Plaster of Paris" mixed with charcoal. Each vial was provided with five females of prey **T**, which approximates the daily food consumption of a female predator. After one day of feeding to satiation, predators were deprived of food (but not of water since the plaster was kept moist) for a period of three days. Adult females of *H. aculeifer* always survived this treatment as they can withstand starvation for several weeks, provided they have access to water. The starved predators were transferred by brush to the center of the bottom of a new, identical vial. Two individuals (an adult female and a protonymph) of each prey species were present at opposite sides on the bottom. Predators were observed twice per minute until the predator caught and fed upon an individual of either of the two prey species. In the few cases (*c.* 5%) where prey was not captured within 5 min., the predator was starved for an extra day and then subjected to the same test. When no feeding occurred during this second trial, the predator was discarded from further testing, but this was a rare event (1-2% of all trials).

The prey species selected in the two-choice experiment was recorded and the predator was allowed to complete consumption of it, but all other prey in the test vial were removed. Subsequently, five additional females of prey **T** were offered and these were consumed within a day following the choice test. In this way the predators reached satiation before they were subjected to food deprivation. After three days of starvation, they were again exposed to a two-choice test, as described above. In total, this procedure of starvation and choice-tests was repeated three times for each individual predator.

Since the females were collected as late deutonymphs, kept isolated for the *c.* 10 days involved in the prey choice tests, they had no opportunity to mate, not even with their own sons. Only those female predators that selected either only prey **R** or only prey **T** in three consecutive tests were kept isolated in small vials, whereas all other females were discarded. Since the females were virgin, they deposit haploid eggs that develop into males (arrhenotokous parthenogenesis). When the males were mature, mating occurred. Thus, after allowing the female to mate with her own son (oedipal mating), a subsequent generation of females was obtained, which were then isolated in the deutonymph stage and experienced the same experimental procedure as their mothers. In this way, isofemale lines were obtained by selection and maintained by mother-son mating. All these iso-female lines were kept at 22°C and *c.* 70% RH and reared on prey **T** (thus, irrespective of the choice made by the female predator).

This selection procedure was repeated for a total of four consecutive generations. The iso-female lines obtained from females that selected prey **R** in the first generation were subsequently selected for consistent preference for this prey species (*i.e.* three times choice for prey **R** out of three prey choice tests) and *vice versa* for the isofemale lines obtained from females that selected prey **T** in the first generation. After four generations this resulted in three isofemale lines, two lines preferring prey **R** (one of which was maintained and further referred to as the *R*-line) and one line preferring prey **T** (further referred to as the *T*-line). These lines were maintained in isolation and their preferences were assessed in samples of 70-80 young females in a series of three consecutive two-choice experiments. These tests were performed twice: directly after the selection experiments and, half a year later, just before the crossbreeding experiments.

Cross-breeding experiments

The *R*- and the *T*-line were subject to cross-breeding in both directions in an attempt to assess whether (1) the two lines hybridize and are thus not reproductively isolated, and whether (2) the observed difference in prey preference has a genetic basis.

Ten young, unmated females from each of the selected lines were placed in a vial together with ten males from the other line, *i.e.* (1) females from the *R*-line with males from the *T*-line, and (2) *T*-line females with *R*-line males. From the hybrid offspring of each of the two combinations of parental mates *c.* 80 young, unmated females were subjected individually to the series of three two-choice tests as described above.

In addition, F2-hybrids were obtained from backcrosses of F1-hybrid females with males from each of the parental lines: *RT*(F1)-hybrid females with either parental *R* or *T* males, and *TR*(F1)-hybrid females with either parental *R* or *T* males. From the offspring of each of the four types of females (*RTxR*, *RTxT*, *TRxR*, *TRxT*) *c.* 60 young females were subject individually to a series of three consecutive two-choice experiments as described above.

Statistical procedures

As explained in the section on Selection Experiments, there are two sets of data on preference assessments in the Base population as well as the two selected lines. As these two data sets were obtained half a year apart, one may ask whether they can be used in conjunction, thereby allowing full use of the information gained. Alternatively only the data set obtained just before assessment of preferences in the F1- and F1xParent-hybrids can be used. Hence, to test the null hypothesis that the two "frequency of choice" distributions (*i.e.* the numbers of female predators that selected prey **R** 3, 2, 1 or 0 times) stem from the same distribution, we first applied Mann-Whitney U-tests with correction for ties (Sokal and Rohlf 1995, Box 13.7, pp. 428-430). This test was also carried out to compare simultaneously obtained frequency distributions (the two types of F1-hybrids; the four types of F1xP-hybrids), as well as those obtained in different time periods (*e.g.* Base-line *vs.* hybrids).

In addition, we tested for independence of the three prey choices that each female made at intervals of 3-4 days. Under the null hypothesis, the observed frequencies are expected to be binomially distributed. Estimates for p, the probability that a female predator chooses **R** as prey, were obtained from the observed distributions. These estimates of p were used to calculate binomial distributions of 3-, 2-, 1- and 0-choices and these predicted binomial distributions were compared with the observed frequency distributions of prey choices, using a G-test for goodness of fit with the application of Williams' correction (Sokal and Rohlf 1995; Box 17.2, p. 706).

For the Base population and the two selected lines the predicted binomial distribution was construed from an estimate of p from these distributions, which goes at the expense of one degree of freedom for evaluating the G-test (df=c-1-e=2; df=degrees of freedom, c=number of classes (=4), e=number of parameters estimated from the observed distribution, *e.g.* e=1 for a binomial distribution). Assuming monogenic inheritance and intermediate inheritance of two alleles (no dominance), the estimates of p for each of the two selected lines were used to calculate the arithmetic mean, *i.e.* $(p_R+p_T)/2$, (or alternatively the geometric or harmonic mean). In this way, the probability of choosing prey **R** by the F1-hybrid females ($=p_{RT}=p_{TR}$) can be predicted, as well as the (binomial) distribution of prey choices, assuming independence of choices in subsequent trials. G-tests for both types of hybrids (*RT* and *TR*) were carried out to test for goodness of fit of the binomial distribution with predicted p to the observed distribution of choices (df=4-1=3). If these predictions hold, the estimates of p and $p_{RT}=p_{TR}$ can be used to predict the overall distribution, which is composed of two binomial distributions, one for homozygotes and one for heterozygotes in each of the four F1xParent backcrosses (again assuming monogenic inheritance and independent assortment without dominance). As before, G-tests for goodness of fit of predicted and observed distributions were carried out (df=4-1=3). As shown below, there was no need to test other more complex models than monogenic inheritance.

Results

A flow diagram of the selection and cross-breeding procedure is shown in figure 1, together with information on the number of isofemale lines selected and the number of replicates obtained at each phase in this procedure. The results of the three two-choice tests per female carried out with females of the Base population, the selected *R*- and *T*-lines, the F1-hybrids and the F1xP back-crosses are shown in table 1.

Selection

From the Base population 81 females were exposed to three two-choice tests carried out at intervals of 3-4 days. For various reasons (escape, death, lack of response to prey) 18 females had to be discarded, so that 63 females successfully completed this test phase. Females selecting prey **R** in all three tests were used to create isofemale lines and *vice versa* for prey **T**. This resulted in 14 isofemale lines preferring prey **R** and 14 isofemale lines preferring prey **T**. Of their progeny (first generation) 8-15 daughters were subject to the three two-choice tests. For various reasons (lack of consistent responses to the same prey species

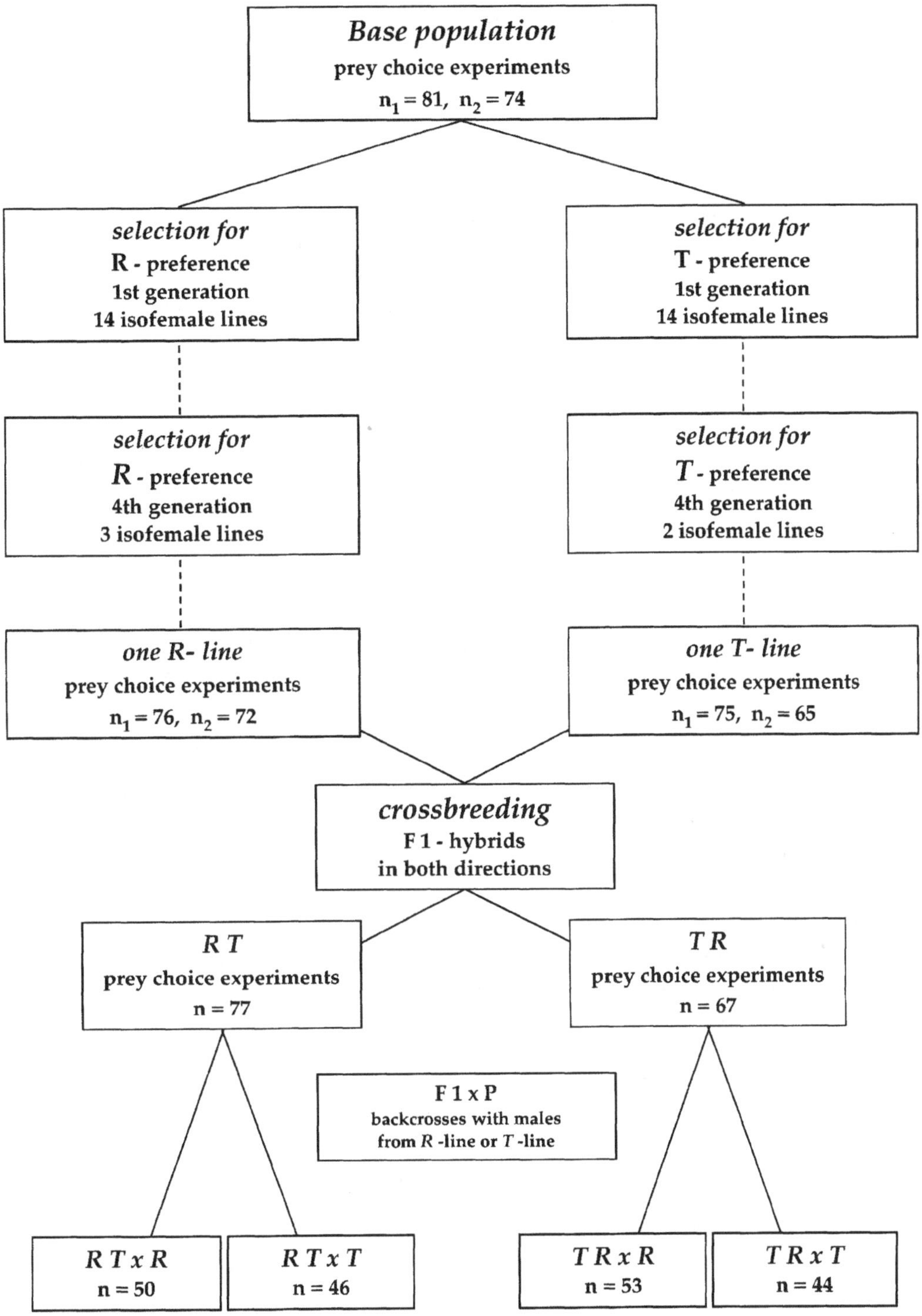

Fig. 1 - A flow diagram of the experimental procedure involving 4 generations of selection for preference for **R** (*Rhizoglyphus robini*) or **T** (*Tyrophagus putrescentiae*) prey, cross-breeding to obtain F1-hybrids, and F1xParent backcrosses. The crosses are presented under the assumption of monogenic inheritance of the preference trait and for the case of arrhenotokous parthenogenesis (diploid females – *RR*, *TR*, *RT*, *TT* – and haploid males – *R*, *T*). Shown are the number of females tested (n with indices 1 and 2 to indicate replicate experiments) in each phase of the experimental procedure and the number of isofemale lines obtained in the selection phase.

Tab. 1 - Frequencies of 3, 2, 1 and 0 successful attacks on *Rhizoglyphus robini* (**R**) among three two-choice tests per female of the Base population, selected lines (*R*-line and *T*-line), F1-hybrid females and F1xParent backcrossed females. The alternative prey was *Tyrophagus putrescentiae* (**T**).

Origin of females	Number tested	Number (and %) of **R**-choices among 3 two-choice tests 3 **R**	2 **R**	1 **R**	0 **R**
Base population	63	14 (22.2)	21 (33.3)	14 (22.2)	14 (22.2)
(>6 months)	69	18 (26.1)	23 (33.3)	12 (17.4)	16 (23.2)
R-line	76	36 (47.4)	27 (35.5)	11 (14.5)	2 (2.6)
(6 months)	72	29 (40.3)	25 (34.7)	16 (22.2)	2 (2.8)
T-line	75	5 (6.7)	21 (28.0)	19 (25.3)	30 (40.0)
(6 months)	65	4 (6.1)	20 (30.8)	17 (26.1)	24 (36.9)
hybrids					
R T	77	14 (18.2)	24 (31.2)	25 (32.5)	14 (18.2)
T R	67	14 (20.9)	18 (26.9)	22 (32.8)	13 (19.4)
(R T) x R	50	12 (24.0)	22 (44.0)	13 (26.0)	3 (6.0)
(R T) x T	46	9 (19.6)	15 (32.6)	15 (32.6)	7 (15.2)
(T R) x R	53	13 (24.5)	27 (50.9)	9 (17.0)	4 (7.5)
(T R) x T	44	4 (9.1)	11 (25.0)	16 (36.4)	13 (29.5)

in the three two-choice tests, lack of females among progeny, lack of response to prey) isofemale lines were discarded in the course of four generations of selection that followed. This resulted in three **R**-preferring isofemale lines and two **T**-preferring isofemale lines.

One **R**- and one **T**-preferring isofemale line were reared in large numbers and, together with the Base population, their offspring were subjected to three two-choice tests after three weeks and again after six months, just before the start of the cross-breeding experiments. The results of these two tests carried out half a year apart were not significantly different within each selected line and within the Base population (Mann-Whitney-U tests; P=0.266 for the *R*-line, P=0.744 for the *T*-line and P=0.691 for the Base population). Tests on pooled results of the two experiments half a year apart showed significant differences between the two selected lines ($P=5.28x10^{-20}$), between the *R*-line and the Base population ($P=2.29x10^{-6}$) and between the *T*-line and the Base population ($P=2.3x10^{-5}$).

In conclusion, selection led to two lines that differed from the original population and from each other. The selected lines showed opposite prey choices; on average *"R"* females chose prey **R** in 2.2 out of 3 two-choice tests and *"T"* females in 1.04 out of three tests, whereas "Base" females chose prey **R** in 1.6 out of three tests. In addition, since the prey choices of the *R*-line did not change over a 6 month period despite being reared on prey **T**, it can also be concluded that the *R*-line is genetically fixed.

Cross-breeding

To prove that there is a genetic basis for the differences between the two selected lines, reciprocal crossings were carried out between the

R- and *T*-lines. The two types of hybrids (*RT* and *TR*) were viable, in that there was no extra mortality compared to the parental lines. The hybrids showed prey choices that were virtually equal to each other (P=0.975), very similar to the Base population (P=0.459 for *RT vs.* Base; P=0.495 for *TR vs.* Base) and intermediate between those of the parents (P=3.44×10^{-7} for *RT vs. R*-line and P=2.13×10^{-6} for *TR vs. R*-line; P=0.0016 for *RT vs. T*-line and P=0.0034 for *TR vs. T*-line); *RT* females chose prey **R** in 1.51 out of three tests and *TR* females in 1.49 out of three tests.

The F1xParent backcrosses also led to viable offspring and very little mortality in all four cases. Thus, the selected lines are not reproductively isolated. The daughters from crossings of *RT* and *TR* females with *R*-males showed a preference for prey **R** that was significantly different from the *RT* and *TR* females in the F1, respectively (P=0.0378 for *RTxR vs. RT*; P=0.0478 for *RTxR vs. TR*; P=0.0108 for *TRxR vs. RT*; P=0.0164 for *TRxR vs. TR*). Those from *TR* females crossed with *T*-males showed a preference for prey **T** that was close to being significantly different from *RT* and *TR* females in the F1 (P=0.0579 for *TRxT vs. RT*; P=0.0776 for *TRxT vs. TR*). However, those from *RT* females crossed with *T*-males were not significantly different from *RT* females (P=0.7011) nor the *TR* females in the F1 (P=0.695). Thus, relative to the F1-hybrids, all F1xParent backcrosses except one (*RTxT*) showed a significant shift in prey preference in the direction of the father's preference.

Taken together, the results from cross-breeding studies show strong evidence for a genetic basis for prey preference in the predatory mites under study.

Mode of inheritance

To unravel the mode of inheritance we first tested the simplest case of monogenic determination without dominance. This was done by calculating the probability (***p***) of choosing **R** from the observed choices for each of the selected lines and by taking the arithmetic average of these two estimates (so-called mid-parent value) as a prediction of the choice in the F1-hybrids (NB: other modes of computing the mean, such as the geometric mean, did not affect the conclusions). The predicted value (***p***=0.540) was close to the observed values of the two F1-hybrids (***p***=0.49784 and 0.49751 for *RT* and *TR*, respectively). Next, the predicted values for the *R*- and *T*-lines and the F1-hybrids were used to predict the mean probabilities in the four F1xP backcrosses. For backcrosses with *R*-males, the prediction was equal to 0.636, whereas the observed values were 0.620 and 0.641 (*RTxR* and *TRxR* respectively). For the backcross of *TR* females with *T* males the predicted probability was somewhat higher than observed (***p***=0.443 *vs.* 0.379), whereas for the *RT* females backcrossed with *T* males the predicted probability was somewhat lower than observed (***p***=0.443 *vs.* 0.522). Thus, based on mean probabilities of choosing prey **R**, there is very good support for monogenic inheritance for the hybrids and the backcrosses with *R*-males, but less good support if it concerns the backcrosses with *T*-males.

A more critical test for monogenic inheritance is obtained when not only the mean but also the full distribution is used. If the three two-choice tests per female were made at sufficiently large time intervals to make the subsequent choices independent of each other, one would expect a binomial distribution with parameter ***p***. Using the estimates of for the selected lines, one can predict the distribution of the hybrids and that of hybrids and homozygotes in the offspring from the backcrosses. Using a G-test for goodness of fit, these predicted distributions can be tested against the observed distributions for the selected lines, the F1-hybrids and the F1xP backcrosses. The test results are presented in table 2 and these show that the binomial distribution is not rejected as a model to describe the frequencies of choices of the females in all four backrosses, in one of the F1-hybrids (*RT*) and in the *R*-line. The binomial model was rejected for the *TR*-hybrids (0.01<P<0.025) and for each of the two tests with the *T*-line (0.001<P<0.005; 0.005<P<0.01). Especially the latter tests with the *T*-line strongly indicate that the subsequent choices were not independent. This is probably caused by the fact that **T**-prey were used for rearing inbetween experiments. Most likely, the *T*-line harboured individuals that are more susceptible to conditioning. However, note that such effects of conditioning seem to be less important in the *TR*-hybrids and are undetectable in the F1xParent backcrosses.

Given the good fit of predicted distributions to those observed especially in the F1xParent backcrosses, prey preference of the predatory mite, *H. aculeifer* can be explained by assuming that it inherits as if determined by a single gene.

Discussion

This study demonstrates genetic variability in prey choice in a local population of soil predatory mites. It was possible to select lines with contrasting prey preferences, to hybridize the

Tab. 2 - Predicted and measured probabilities (***p***) of choosing *Rhizoglyphus robini* (**R**) by the selected lines, F1-hybrid females and F1xParent backcrossed females. The measured probabilities are calculated from the results in Table 1 as: p=[3 p(3R) + 2 p(2R) + p(1R)]/3. Parental (=measured) probabilities are used to predict those of F1-hybrid females whereas these two probabilities together were used to predict the probabilities in the F1xParent backcrosses (where offspring homozygotes and heterozygotes are expected to occur in 1:1 ratios). When two probability estimates are available (as in the selected *R*- and *T*-line), their arithmetic mean was used to calculate predicted values. The value of the test statistic and the critical significance levels for the G-test for goodness of fit to a binomial distribution (Sokal and Rohlf 1995; Box 17.2, p. 706) or the sum of two binomial distributions (one for homo- and another for heterozygotes in F1xParent backcrosses) are presented in the last two columns. When the expected distribution was calculated from ***p*** estimated from the measured distribution itself (Base population, *T*-line, *R*-line), the degrees of freedom (df) are equal to *c*-2, and when calculated from ***p*** estimated from other distributions than the one under test (all other cases), df equals *c*-1 (*c*=number of classes; here *c*=4).

Origin	Measured ***p***	Predicted ***p***	G-statistic	Critical level, P (df)
Base population	(1)0.518	-	12.363	0.001<P<0.005 (2)
	(2)0.541	-	21.002	0.0001<P<0.001 (2)
	mean: 0.529			
R-line	(1)0.759	-	1.685	0.500<P<0.900 (2)
	(2)0.708	-	2.588	0.100<P<0.500 (2)
	mean: 0.735			
T-line	(1)0.338	-	12.210	0.001<P<0.005 (2)
	(2)0.354	-	9.248	0.005<P<0.010 (2)
	mean: 0.346			
hybrids				
R T	0.498	0.540	6.520	P≈0.090 (3)
T R	0.498	0.540	9.398	P≈0.025 (3)
(R T) x R	0.620	0.636	0.366	0.900<P<0.975 (3)
(R T) x T	0.522	0.443	4.146	0.100<P<0.500 (3)
(T R) x R	0.641	0.636	2.947	0.100<P<0.500 (3)
(T R) x T	0.379	0.443	3.056	0.100<P<0.500 (3)

lines in both directions and to infer a genetic basis for the preferences. These results are remarkable for three reasons. First, they reveal genetic variability in a population collected from 0.25 m^2 of soil which, given the mobility of predatory mites, is a spatial scale small enough for the population to be homogeneously mixed. Second, they provide evidence for a simple Mendelian inheritance of food preference. Third, they show genetically determined preferences in a polyphagous, predatory arthropod. Thus, the notion of polyphagy does not necessarily imply that genetically determined preferences are absent, nor that it is the sum of the narrow dietary range of different geographic strains. In fact, genetic variability in prey preferences exists at a small spatial scale, *i.e.* small relative to the mobility of the predators.

Mode of inheritance

A genetic basis for food preferences has been shown for a variety of groups of organisms, such as rats and mice (Ramirez and Sprott 1978), garter snakes (Arnold 1981), cave fish (Schemmel 1980), moths (Phillips and Barnes 1975; Sheck and Gould 1995), butterflies (Tabashnik *et al.* 1981; Thompson 1988), bruchid beetles (Wasserman and Futuyma 1981), milkweed bugs (Leslie and Dingle 1983) and fruit flies (Andersson 1986; Jaenike 1985 and 1987; Wallin 1988). In most cases the inheritance was shown to be determined by multiple genes (moths, Phillips and Barnes 1975; bruchid beetles, Wasserman and Futuyma 1981; milkweed bugs, Leslie and Dingle 1983; fruitflies, Jaenike 1985 and 1987), by multiple autosomal loci (swallowtail butterflies, Hagen 1990), by one or more autosomal loci (moths, Sheck and Gould 1995), or primarily by one or more loci on sex chromosomes and additionally by one or more loci on autosomal chromosomes (swallowtail butterflies, Thompson 1988). There are only three cases where there is strictly or primarily one gene determining food preference: (1) diprionid sawflies with strict monogenic control (Knerer and Atwood 1973), (2) tephritid flies with primarily monogenic control of oviposition preference based on a recessive allele, the effect of which is modified by polygenes that affect susceptibility to conditioning (Huettel and Bush 1972), (3) fruit fly larvae with one or perhaps very few genes with large effect, but this was inferred only from the fast saturation in response to selection (Wallin 1988). However, the first two studies were carried out with individuals from different genetically isolated populations [either different species (Knerer and Atwood 1973) or different host races (Huettel and Bush 1972) and the third with a mixture of populations collected from three locations in south-western Sweden. In conclusion, these studies do not necessarily reveal within-population genetic variability, on which micro-evolutionary changes depend (Jaenike 1987). Monogenic inheritance of food preference is the most parsimonious explanation of the results presented in this article, and it is reported for the first time for a single natural population sampled at a single site.

Alternative genetic models have not been tested, however, and, although more complex, they may predict the same results. For example, monogenic inheritance can be mimicked by polygenic systems without linkage in the F1 generation (Falconer 1989, p. 18) and with tight linkage also in the F1xP backcrosses (*e.g.* Futuyma and Mayer 1980). Schemmel (1980) studied the inheritance of feeding behaviour of Mexican cave fish by crossing individuals from different river-dwelling and cave-dwelling populations. He inferred monogenic inheritance from F1, F2 and F1xParent backcrosses, but F2xParent crossings showed the presence of at least three genes. Thus except for the latter crosses, the expression of polygenes mimicked monogenic inheritance. We cannot exclude the possibility of tight linkage of several genes. In fact, it is quite likely that more genes are involved, *e.g.* because smell and taste receptors differ and both types are involved in expressing food preference.

Inheritance and the possibilities for sympatric speciation

Whether food preference inherits as if monogenically controlled or not, is of crucial importance to the possibilities for host race formation and subsequent sympatric speciation. The more genes involved in preference, the more fine-graded the distribution of intermediates will be. If intermediates have inferior reproductive success, selection to avoid producing hybrids will be weaker in the case of polygenic preference than in the monogenic case. Moreover, if food preference is determined by several genes and the preference genes are initially not tightly linked, then the probability of establishing close linkage with the gene complex coding for increased reproductive performance on the preferred food will be remote. Thus, monogenic inheritance of food preference will be conducive to sympatric speciation. However, for a preference gene to become linked to fitness-associated genes strongly depends on whether the fitness-associated genes are tightly linked, on the number of alleles per locus and on recombination between loci (Felsenstein 1981). Only, when there are biological reasons to assume that prey preference and mate choice are linked to some extent (*i.e.* not error-free), then the possibilities for sympatric speciation are broadened even when there are several loci involved in reproductive isolation (Kondrashov 1983a, 1983b and 1986; Johnson *et al.* 1996). We are not convinced that the predatory mites under study tend to mate more assortatively where they find their preferred prey. For one thing the two prey show considerable overlap in their distributions (Lesna, *pers. obs.*). For another, mate choice experiments presented in Lesna and Sabelis (1999) show disassortative mating for the *T*-line, and for the *R*-line when the two prey species **R** and **T** are offered together.

Hence, it is important to ask why food preference is polygenically controlled in some organisms and monogenically in others. Possibly, they represent different phases in sympatric speciation with preference initially associated with one gene and later with more. The number of genes will roughly correlate with the number of traits involved in food selection (Tabashnik *et al.* 1981), such as the ability to perceive and respond to auditory, visual, olfactory and chemotactile stimuli. Predatory mites lack auditory and visual perception and exclusively rely on taste and olfaction for selecting prey. It is tempting to suggest that this limitation is the underlying cause for the monogenic mode of inheritance found in our study, but this line of reasoning is overly simplistic. One could just as well argue that these constraints on perception impose stronger selection to invest in different specialized sensors for olfaction and taste. While this will increase the number of genes involved, the total investment in sensory ability will be economized by selection.

Caution should be exercised in extrapolating our finding of monogenic inheritance of prey preference to the preferences that become manifest under natural conditions. It is possible that we increased the chance of finding simple modes of inheritance due to the simplified environment that predatory mites experienced in our experimental tests. For example, our experimental design may only allow the predators to detect differences in taste whereas preference may additionally be based on smelling of chemical cues that originate directly from the prey or indirectly from the environment associated with prey. Testing the predictions from simple preference tests in more realistic environments is therefore the way forward.

Whereas the existence of just one locus for preference may be conducive to sympatric speciation, the existence of two alleles at this locus is not. This was pointed out by Felsenstein (1981) who studied the effect of between-population migration on population divergence. He compared models based on one allele at a modifier locus causing reduced migration (Maynard Smith 1966; Balkau and Feldman 1973) with other models based on two alleles at a locus for assortative mating (Maynard Smith 1966; Felsenstein 1981). It was found that increased migration had little effect on population divergence in one-allele models, but reduced divergence considerably in two-allele models. Hence, it is important to ask whether there are modifier genes that condition the forager to prefer the food type experienced early in life (imprinting), repeatedly eaten (conditioning) or encountered before (learning). We did not explicitly investigate this with respect to the predator-prey system studied, but a role of conditioning, learning or imprinting was indicated in prey choice tests where the choice distributions differed from binomiality. These deviations were only observed in the *T*-line and the *TR*-hybrids, probably caused by the fact that (1) all predators were reared on prey **T** in between the choice-tests and (2) the selection regimes did not differentiate between genetic and conditioned preferences for prey **T** (as opposed to selection for preference of prey **R**). These cases show that modifiers probably exist (*T*-line), but lines free of such modifiers exist as well (*R*-line), as indicated by the absence of a deviation between the observed data and the binomial distribution. Clearly, the role of modifier genes needs further consideration to establish a better connection with the theory of sympatric speciation (Felsenstein 1981).

Spatial scale of genetic variation

Genetic variation in food preference has been demonstrated among species and/or strains at various spatial scales, as illustrated by the following examples: (1) among related, interfertile species in the same habitat, as in moths (Sheck and Gould 1995) and butterflies (Thompson 1988), (2) among strains from the same species but different geographic origins, as in garter snakes (Arnold 1981b) and milkweed bugs (Leslie and Dingle 1983), (3) within a mixture of strains from different localities in a large geographic area, as in fruitflies in southwestern Sweden (Wallin 1988), (4) between populations living in a well-defined, 30x100 km region, as in butterflies on the eastern slope of the Sierra Nevada (Thomas *et al.* 1987) and (5) among isofemale lines from locations 10 km (and 3 years) apart, as in fruitflies (Jaenike 1985 and 1987). In some studies variation in polygenically determined oviposition preference has been found at a spatial scale small enough to suggest that it may occur within a single mating population. These studies involved pierid butterflies (Tabashnik *et al.* 1981; but note that the heritabilities were non-significant, possibly due to small sample sizes), swallowtail butterflies (Thompson 1988) and fruitflies (Jaenike and Grimaldi 1983). The results presented in this article add further evidence for genetic variation in food preference within populations and show for the first time that it can be found at the smallest spatial scale for preferences with a simple genetic (possibly monogenic) basis. Moreover, the genotypes are most likely part of a well-mixed population with respect to mating, because hybrids and their offspring are viable, mating barriers are absent, and there is *no* tendency to avoid mating with the alternative food preference genotype (Lesna and Sabelis 1999).

It should be emphasized, however, that the selection experiments started more than 3 years after collecting the predatory mites from the field. Hence, the possibility that prey preference for **T** is deleterious in nature, existing at a low frequency due to a balance between recurrent mutation and selection cannot be ruled out *a priori*. Rearing in the laboratory on prey **T** could increase the frequency of this phenotype in the Base population permitting its detection in the selection experiments. However, assessment of prey preferences in the female offspring reared from predators collected in September 1998 near the location sampled in 1991, showed no marked differences with the base strain reared for 3.5 years in the laboratory (Lesna and Sabelis 1999).

Conditions for polymorphism

The existence of genetic variation in food preferences at a small spatial scale within a homogeneously mixed population represents a challenge to evolutionary theory. It cannot be explained by individual-centered optimal foraging theory, which below a certain prey density threshold predicts only generalists and above the threshold only specialists on the most profitable of the two prey. Local genetic variation is also not a robust outcome from game theory on distributions of ideal-free foragers competing for food in patches of different quality. Whereas the mixed ESS prediction of ideal-free theory is compatible with genetic variation, temporal changes in the distribution of patch profitabilities would selectively favour flexible food patch selection. However, flexibility is unlikely to be cost-free, so that some variability in prey supply is necessary to make it worth. Conditions for within-population coexistence of two specialists, one on each of the two prey, plus a generalist have not been systematically explored for the case that there are costs to flexibility. Van Tienderen (1991) analysed a quantitative genetic model for specialists and generalists differing in costs and reproductive benefits on two constant food sources. He assumed random dispersal over the two food sources so that the genotypes best adapted to one food source were also forced to make the best of a bad job on the other. This assumption holds for plants and sessile animals, but not for mobile organisms. Hence, the predictions of van Tienderen (1991) do not apply to the predator-prey system studied by us. Hoekstra *et al.* (1985) explored the conditions for polymorphism in one-locus, diploid models, based on the assumption that adults form a single random mating population, but after mating females deposit offspring in their native food niche. They showed that polymorphism of homozygote specialists *and* their hybrids (the generalists) arises under rather restrictive conditions: heterozygote fitness on each food source should be close in fitness to the favoured homozygote specialist on that food source, which can only be achieved by a form of dominance-switching upon transfer of heterozygotes from one food source to the other. Conditions for genetic polymorphism become considerably less stringent for relative fitness of the heterozygotes if food choice is positively associated with a selective advantage in the preferred food niche (Jones 1980; Rausher 1984; Garcia-Dorado 1986) or alternatively if there is a locus controlling preference for the food type on which highest reproductive success will be attained (epistatically determined food preferences) (de Meeus *et al.* 1993). These requirements might be met in the predator-prey system studied here. However, we do not know to what extent genetical control of preference can be overruled by effects of conditioning, and we also have no *a priori* knowledge on the positive relation between food choice and relative fitness (nor on epistatically determined preferences). Hence, the assumptions and predictions of the model of Hoekstra *et al.* (1985) provide a more conservative starting point for predicting polymorphism.

Hoekstra *et al.* (1985) also showed that temporal (but spatially synchronous) variation in the supply of the food sources relaxes the conditions for polymorphism, but the impact is small and the conclusions remain qualitatively similar. They did not consider the possibility of asynchronous temporal variability in the supply of the two food sources that arises from independent stochastic variability among the two food sources (*e.g.* Wilson and Yoshimura 1994), or may arise from dynamic coupling between predators and its two prey. Although the latter processes might relax conditions for within-population variation, current theory predicts polymorphism only when heterozygotes perform almost as good as the homozygotes. This condition is shown to hold for the system under study (Lesna and Sabelis 1999).

Acknowlegdments

We are grateful for discussions with Yannis Michalakis and thank him, Hans Breeuwer, Sam

Elliot, Arne Janssen and Nico van Straalen for useful comments on the manuscript. Rob Lingeman provided statistical advice. Our research was funded as project ABI.003-2187 by the Technology Foundation (STW), affiliated with the Netherlands Organization of Scientific Research (NWO).

References

Andersson A., 1986 - Genetic influence on foraging behaviour in *Drosophila melanogaster* larvae. *Hereditas*, 105: 229-231.

Arnold S.J., 1981 - Behavioral variation in natural populations. II. Inheritance of a feeding response in crosses between geographic races of the garter snake, *Thamnophis elegans*. *Evolution*, 35: 489-509.

Balkau B.J., Feldman M.W., 1973 - Selection of migration modification. *Genetics*, 74: 171-174.

Barker P.S., 1969 - The response of a predator, *Hypoaspis aculeifer* (Canestrini) (Acarina, Laelapidae) to two species of prey. *Canadian Journal of Zoology*, 47: 343-345.

de Meeus T., Michalakis Y., Renaud F., Olivieri I., 1993 - Polymorphism in heterogeneous environments, evolution of habitat selection and sympatric speciation: soft and hard selection models. *Evolutionary Ecology*, 7: 175-198.

Falconer D.S., 1989 - Introduction to Quantitative Genetics. Longman Scientific and Technical, Harlow, England, 438 pp.

Felsenstein J., 1981 - Skepticism towards Santa Rosalia, or why there are so few kinds of animals. *Evolution*, 35: 124-138.

Fox L.R., Morrow P.A., 1981 - Specialization: Species property or local phenomenon. *Science*, 211: 887-893.

Futuyma D.J., Mayer G.C., 1980 - Non-allopatric speciation in animals. *Systematic Zoology*, 29: 254-271.

Futuyma, D.J., Moreno G., 1988 - The evolution of ecological specialization. *Annual Review of Ecology and Systematics*, 19: 207-233.

Garcia-Dorado A., 1986 - The effect of niche preference on polymorphism protection in a heterogeneous environment. *Evolution*, 40: 936-945.

Hagen R.H., 1990 - Population structure and host use in hybridizing subspecies of *Papilio glaucus* (Lepidoptera: Papilionidae). *Evolution*, 44: 1914-1930.

Hoekstra R.F., Bijlsma R., Dolman A.J., 1985 - Polymorphism from environmental heterogeneity: Models are only robust if the heterozygote is close in fitness to the favoured homozygote in each environment. *Genetical Research*, 45: 299-314.

Huettel M.D., Bush J.L., 1972 - The genetics of host selection and its bearing on sympatric speciation in *Procecidochares* (Diptera: Tephritidae). *Entomologia Experimentalis et Applicata*, 15: 465-480.

Jaenike J., 1985 - Genetic and environmental determinants of food preference in *Drosophila tripunctata*. *Evolution*, 39: 362-369.

Jaenike J., 1987 - Genetics of oviposition-site preference in *Drosophila tripunctata*. *Heredity*, 59: 363-369.

Jaenike J., 1990 - Host specialization in phytophagous insects. *Annual Review of Ecology and Systematics*, 21: 243-273.

Jaenike J., D. Grimaldi, 1983 - Genetic variation in for host preference within and among populations of *Drosophila tripunctata*. *Evolution*, 37: 1023-1033.

Johnson P.E., Hoppensteadt F.C., Smith J.J., Bush G.L., 1996 - Conditions for sympatric speciation: a diploid model incorporating habitat fidelity and non-habitat assortative mating. *Evolutionary Ecology*, 10: 187-205.

Jones J.S., 1980 - Can genes choose habitats? *Nature*, 286: 757-758.

Knerer G., Atwood C.E., 1973 - Diprionid sawflies: Polymorphism and speciation. *Science*, 179: 1090-1099.

Kondrashov A.S., 1983a - Multilocus model of sympatric speciation. I. One character. *Theoretical Population Biology*, 24: 121-135.

Kondrashov A.S., 1983b - Multilocus model of sympatric speciation. II. Two characters. *Theoretical Population Biology*, 24: 136-144.

Kondrashov A.S., 1986 - Multilocus model of sympatric speciation. III. Computer simulations. *Theoretical Population Biology*, 29: 1-15.

Leslie J.F., Dingle H., 1983 - A genetic basis of oviposition preference in the large milkweed bug, *Oncopeltus fasciatus*. *Entomologia Experimentalis et Applicata*, 34: 215-220.

Lesna I., Sabelis M.W., 1999 - Diet-dependent female choice for males with 'good genes' in a soil predatory mite. *Nature*, 401: 581-584.

Lesna I., Sabelis M.W., Conijn C.G.M., 1996 - Biological control of the bulb mite, *Rhizoglyphus robini*, by the predatory mite, *Hypoaspis aculeifer*, on lilies: predator-prey interactions at various spatial scales. *Journal of Applied Ecology*, 33: 369-376.

Lesna I., Sabelis M.W., Bolland H.R., Conijn C.G.M., 1995 - Candidate natural enemies for control of *Rhizoglyphus robini*, Claparède (Acari: Astigmata) in lily bulbs: Exploration in the field and preselection in the laboratory. *Experimental and Applied Acarology*, 19: 655-669.

Lobbes P., Schotten C., 1980 - Capacities of increase of the soil mite *Hypoaspis aculeifer* Canestrini (Mesostigmata: Laelapidae). *Zeitschrift für angewandte Entomologie*, 90: 9-22.

Maynard Smith J., 1966 - Sympatric speciation. *The American Naturalist*, 100: 637-650.

Murphy P.W., Sardar M.A., 1991 - Resource allocation and utilization contrasts in *Hypoaspis aculeifer* (Can.) and *Alliphis halleri* (G. & R. Can.) (Mesostigmata) with emphasis on food source. In: Schuster R., Murphy P.W. (eds.), The Acari, Reproduction, Development and Life-History Strategies. Chapman and Hall, London: 301-312.

Phillips, P.A., Barnes M.M., 1975 - Host race formation among sympatric apple, walnut, and plum populations of the codling moth, *Laspeyresia pomonella*. *Annals of the Entomological Society of America*, 68: 1053-1060.

Ragusa S., Zedan M.A., 1988 - Biology and predation of *Hypoaspis aculeifer* (Canestrini) (Parasitiformes, Dermanyssidae) on *Rhizoglyphus echinopus* (Fum. & Rob.) (Acariformes, Acaridae). *Redia*, 71: 213-225.

Ragusa, S., Zedan M.A., Sciacchitano M.A., 1986 - The effects of food from plant and animal sources on the development and egg production of the predacious mite *Hypoaspis aculeifer* (Canestrini) (Parasitiformes, Dermanyssidae). *Redia*, 69: 481-488.

Ramirez I., Sprott R.L., 1978 - Genetic mechanisms of drinking and feeding. *Neuroscience and Biobehavioral Reviews*, 2: 15-26.

Rausher M.D., 1984 - The evolution of habitat preference in subdivided populations. *Evolution*, 38: 596-608.

Ruf A., 1993 - Die Morphologische Variabilität und Fortpflanzungsbiologie der Raubmilbe *Hypoaspis aculeifer* (Canestrini, 1883) (Mesostigmata: Laelapidae). Ph.D. Dissertation, Univ. Bremen.

Sabelis M.W., 1992 - Arthropod predators. In: Crawley M.J. (ed.), Natural Enemies, the Population Biology of Predators, Parasites and Diseases, Blackwell, Oxford: 225-264.

Sardar M.A., Murphy P.W., 1987 - Feeding tests of grassland soil-inhabiting gamasine predators. *Acarologia*, 28: 117-121.

Schemmel C., 1980 - Studies on the genetics of feeding behaviour in the cave fish *Astyanax mexicanus* f. *anoptichthys*. An example of apparent monofactorial inheritance by polygens. *Zeitschrift für Tierpsychologie*, 53: 9-22.

Sheck A.L., Gould F., 1995 - A genetic analysis of oviposition behavior of *Heliothis virescens* and *Heliothis subflexa* (Lepidoptera: Noctuidae). *Environmental Entomology*, 24: 341-347.

Sokal, R.R., Rohlf F.J., 1995 - Biometry. 3rd Edition. Freeman and Company, New York, 887 pp.

Stephens, D.W., Krebs J.R., 1986 - Foraging Theory. Princeton University Press, Princeton, 247 pp.

Sutherland W.J., 1996 - From Individual Behaviour to Population Ecology. Oxford University Press, New York, 213 pp.

Tabashnik B.E., Wheelock H., Rainbolt J.D., Watt W.B., 1981 - Individual variation in oviposition preference in the butterfly, *Colias eurytheme*. *Oecologia*, 50: 225-230.

Thomas C.D., Ng D., Singer M.C., Mallet J.L.B., Parmesan C., Billington H.L., 1987 - Incorporation of a European weed into the diet of a North-American herbivore. *Evolution*, 41: 892-901.

Thompson J.N., 1988 - Evolutionary genetics of oviposition preferences in swallowtail butterflies. *Evolution*, 42: 1223-1234.

Thompson J.N, Pellmyr O., 1991 - Evolution of oviposition behavior and host preference in Lepidoptera. *Annual Review of Entomology*, 36: 65-89

Usher M.B., Davis P.R., 1983 - The biology of *Hypoaspis aculeifer* (Canestrini) (Mesostigmata): Is there a tendency towards social behaviour? *Acarologia*, 24: 243-250.

van Tienderen P.H., 1991 - Evolution of generalists and specialists in spatially heterogeneous environments. *Evolution*, 45: 1317-1331.

Wallin A., 1988 - The genetics of foraging behaviour: artificial selection for food choice in larvae of the fruitfly, *Drosophila melanogaster*. *Animal Behaviour*, 36: 106-114.

Wasserman S.S., Futuyma D.J., 1981 - Evolution of host plant utilization in laboratory populations of the southern cowpea weevil, *Callosobruchus maculatus* Fabricius (Coleoptera: Bruchidae). *Evolution*, 35: 605-617.

Ward S.A., 1992 - Environmental uncertainty and polyphagy in herbivorous insects. *Oikos*, 63: 506-512.

Wilson D.S., Yoshimura J., 1994 - On the coexistence of specialists and generalists. *The American Naturalist*, 144: 692-707.

Experimental hybridization of *Rhizoglyphus robini* and *R. echinopus* (Acari: Acaridae)

Juliusz Unrug, Jacek Radwan
Department of Zoopsychology and Animal Ethology, Institute of Environmental Sciences, Jagiellonian University, Cracow, Poland

Key words: hybridization, reproductive isolation, *Rhizoglyphus*.

Abstract: *Rhizoglyphus robini* Claparède and *R. echinopus* Fumouze et Robin are the most common species of the genus. They are distinguished on the basis of the *Sci* setae length (short in *R. robini*, long in *R. echinopus*) and the shape of the penal base, but these diagnostic features may vary within species, and both species were reported from the same habitats. We carried out experimental hybridization of *R. robini* and *R. echinopus* in reciprocal crosses to see if they are reproductively isolated. The results showed strong genetic incompatibility in the parental generation, more pronounced in *R. robini* females mated to *R. echinopus* males (7% fecund) than in *R. echinopus* females mated to *R. robini* males (20% fecund). We recorded complete sterility in F_1 hybrids, confirming the reproductive isolation of these species.

Introduction

Mites of the genus *Rhizoglyphus* (Acari: Astigmata: Acaridae) are important acarine pests, damaging a variety of crops and vegetables. We studied two species of this genus, which we identified according to Manson (1972) as *Rhizoglyphus robini* Claparède and *R. echinopus* Fumouze et Robin, and we will adhere to Manson's terminology throughout this paper. Van Eyndhoven (1961) and Manson (1972) distinguished these species on the basis of the length of *Sci* setae (short in *R. robini*, long in *R. echinopus*) and shape of the penal base, which is narrower and more conical in *R. robini*. Hughes (1976) regarded the species with long *Sci* setae as *R. callae*, reserving the name *R. echinopus* for species entirely lacking *Sci* setae. However, Gerson and Capua (1982) reported that the length of *Sci* setae in *R. robini* shows allometric variation, i.e. they are relatively longer in larger individuals. Also penis base may appear differently when subjected to pressure of the cover slip (Hughes 1976). Thus, these diagnostic characters may vary within species. Moreover, both species are found in similar habitats, e.g. winter grains, onion and garlic in the field (Rawlins 1955; Sömermaa 1972; Gerson *et al.* 1981), bulbs, corms and tubers in storage (Garman 1937; de Munk 1972; Müller and Hollinger 1980), so they may have many opportunities to hybridize. Our experiment was designed to determine if *R. robini* and *R. echinopus* are reproductively isolated.

Methods

Rhizoglyphus robini mites used in this study were derived from a colony of about 50 individuals found on onions in a garden near Cracow, Poland, 2 years before the experiment. *Rhizogly-*

F. Bernini, R. Nannelli, G. Nuzzaci, E. de Lillo (eds.), Acarid Phylogeny and Evolution. Adaptations in mites and ticks, 65-68.

phus echinopus came from a culture maintained at the Institute of Apiculture, Pulawy, Poland and was kindly supported by W. Chmielewski. Both species were identified independently by us and by W. Chmielewski.

The mites were cultured in desiccators (20-25°C) at about 90% relative humidity maintained by potassium hydroxide solution (153 g KOH l^{-1}) and fed on a mixture of powered yeast and wheat germ (3:1). The stock cultures were kept in 2.5 cm diameter plastic tubes and most of the time consisted of 100-200 adults and several hundred immatures.

Virgin males and females of both species used in cross-mating experiments were obtained by isolation of protonymphs (the first nymphal instar). All matings and egg counting were carried out outside the desiccators. Thirty males and thirty females of each reciprocal crossing groups and ten pairs for both parents' control generation were mated. Each pair was kept separately in 0.5 cm diameter glass cells with plaster and charcoal bases (see Radwan 1991, for details). At the start of the experiment we observed mating frequency in the parental generation pairs and their reciprocal crosses for 12 hours, checking the cells every two hours. Mating frequency observations were repeated after three days. Parents' control generation (RR - pairs of *R. robini*, EE - pairs of *R. echinopus*) eggs were counted twice a week during the first 2 weeks. Eggs laid by the reciprocal crosses were counted twice a week throughout the experiment. Hatched larvae of the control generation (RR and EE) and their reciprocal crosses were collected and each was placed in a separate cell to develop. The reciprocal crosses of *R. robini* and *R. echinopus* were used to demonstrate mating ability and fertility in all experimental combinations.

The text and tables use the symbols describing progeny of different crosses, where the first letter always indicates the female species (e.g., RE - progeny of female *R. robini* and male *R. echinopus*).

Results

The reciprocal crosses of *R. robini* and *R. echinopus* demonstrated some degree of mating ability and fertility in all experimental combinations. These groups (EE, RR, RE and ER) did not differ in mating frequency at 0.05 level (Kruskal-Wallis H=7.29, df=3, p=0.063).

The mean numbers of eggs differed between the experimental groups (Kruskal-Wallis H=83.46, df=3, p<0.001): in the control generations (RR and EE) it was significantly higher than for the reciprocal RE and ER crosses (tab. 1; U=0, n=100, p<0.001). There were no differences in the number of eggs laid between EE and RR females (tab. 1; U=167.5, n=40, p=0.379) and between ER and RE females (tab. 1; U=417.5, n=60, p=0.630).

The number of ovipositing females was significantly higher in the parents' control generation than in their reciprocal crosses (tab. 1; χ^2=488, p<0.001). The genetic incompatibility of the parental generation was stronger in *R. robini* females mated to *R. echinopus* males (7% fecund) than in *R. echinopus* females mated to *R. robini* males (20% fecund). Reproduction of reciprocal crosses (RE and ER) started at least two weeks later than in parental pairs.

The hatchability of eggs in ER crosses was significantly higher than that in RE crosses (tab. 1; χ^2=115.1, p<0.001).

The reproductive parameters of the hybrid generation indicated the complete sterility of both RE brother-sister crosses and ER brother-sister crosses. Experimental backcrosses were not fertile either (tab. 2).

Discussion and conclusions

Previous interspecific crossbreeding experiments between different species of Astigmata produced bisexual progeny after reciprocal crossing or only after crossing in one direction e.g. crossing of *Acarus farris* Oud. with *A. immobilis* Griffiths (Chmielewski 1975). There is some evidence for stronger incompatibility than

Tab. 1 - Reproductive parameters of the control generation of *R. robini*, *R. echinopus* and their reciprocal crosses (mean value ± SD).

	RR	RE	ER	EE
Mean number of matings	1.3 ± 1.17	0.7 ± 0.88	0.83 ± 1.05	1.0 ± 0.8
Mean numbers of eggs laid by females within first two weeks	31.7 ± 10.7	0.14 ± 0.28	0.72 ± 0.16	36.28 ± 13.94
% of ovipositing females	100	7	20	100
% of eggs hatched	99	29	44	97

Tab. 2 - Reproductive parameters of the control generations of *R. robini*, *R. echinopus* and their hybrids (mean value ± SD).

	RR	RE	ER	EE
Mean numbers of eggs laid by females within first two weeks	32.93 ± 14.94	0.0	0.0	32.48 ± 15.63
% of ovipositing females	100	0	0	100

recorded in the present study: no oviposition, no egg hatching, or the inability of progeny to finish ontogenesis e.g. crossing of *Suidasia medanesis* Oud. with *S. nesbitti* Hughes (Hughes 1955); *Tyrophagus palmarum* Oud. with *T. longior* Gerv. (Hughes 1961); *Acarus siro* L. with *A. farris* Oud. (Chmielewski 1975). Chmielewski (1975) noted that matings between different acarid species occurred very often, especially when they live in the same habitats (e.g. *Acarus* spp.). Moreover, matings also take place between species of different genera (e.g., *Acarus* and *Tyrophagus*), but only species of the same genus produced progeny. On the other hand, morphologically similar species (e.g., *A. siro* and *A. farris*) are reproductively isolated (Chmielewski 1975), as shown for *Rhizoglyphus robini* and *R. echinopus* in this study.

Experimental reciprocal hybridization of *R. robini* and *R. echinopus* proved the ability of males to mate with females of the other species. However, during the first day, *R. echinopus* males were seen to act very aggressively against *R. robini* females, which caused 30% mortality among them. Aggressive behaviour of heteromorphic males of dimorphic acarid species is typically directed towards other males and not towards conspecific females (Woodring 1969; Timms *et al.* 1981; Radwan *et al.* 2000). It seems that *R. echinopus* males might have difficulty in recognizing the sex of the congeneric species.

Only a few females reproduced successfully with males of the other species. The females from both reciprocal matings laid significantly lower numbers of eggs. The bisexual progeny obtained from these females proved that hybridization occurred, but the lower hatchability of eggs in both reciprocal crosses indicates strong genetic incompatibility between these species. This was supported by the complete sterility of the F_1 hybrid generation, both in male and female progeny.

It needs to be emphasized that the onset of oviposition after reciprocal crosses occurred much later than in homo-specific pairs. That relaxation of barriers against incompatible fertilization with age may be explained by the decrease in female residual reproductive value. Low expected future reproduction can favour more risky reproductive strategies (Williams 1966).

In summary, the results of the experiments demonstrated reproductive isolation between the studied populations of *R. robini* and *R. echinopus*. Although they mate heterospecifically and to some degree reproduce, hybridization produces infertile progeny. It would be interesting to see if stronger barriers against heterospecific mating, which may result in a substantial waste of reproductive effort to females, occurs between sympatric populations of these two species. Future research should also investigate if any reproductive barriers occur between different populations of the same species. The lack of such isolation would further corroborate the current taxonomic status of *R. echinopus* and *R. robini*.

Acknowledgments

We thank W. Chmielewski for providing the mites and for independently identifying the species, and the anonymous referee for his comments on the previous version of our manuscript.

References

Chmielewski W., 1975 - Interspecific hybridisation of mites belonging to the superfamily Acaroidea and some biological features of hybrids. *Zeszyty Problemowe Postepów Nauk Rolniczych*, 171:189-211.

de Munk W.J., 1972 - Bud necrosis, a storage disease of tulips. III. The influence of ethylene and mites. *Netherlands Journal of Plant Pathology*, 78: 168-178.

Garman P., 1937 - A study of the bulb mite (*Rhizoglyphus hyacinthi* Banks). *Bulletin of Connecticut Agricultural Experimental Station*, 402: 889-907.

Gerson U., Capua S., 1982 - Allometric variation in *Rhizoglyphus robini* Claparède (Acari: Astigmata: Acaridae). *Israel Journal of Entomology*, 16: 69-72.

Gerson U., Yathom S., Katan J., 1981 - A demonstration of bulb mite control by solar heating of the soil. *Phytoparasitica*, 9: 153-155.

Hughes A.M., 1955 - On *Suidasia medanensis* Oudemans (Acaridae) - a mite new to Great Britain. *Proceeding of the Zoological Society of London*, 125: 41-47.

Hughes A.M., 1961 - The mites of stored food. Her Majesty's Stationery Office, 1st Ed., London, 287 pp.

Hughes A.M., 1976 - The mites of stored food and houses. Her Majesty's Stationery Office, 2nd Ed., London: 116-123.

Manson D.C.M., 1972 - A contribution to the study of the genus *Rhizoglyphus* Claparede, 1869 (Acarina: Acaridae). *Acarologia*, 13(4): 621-650.

Müller P.J., Hollinger T.C., 1980 - Damage to *Rhizoglyphus* mites in some ornamental bulbous crops. *Acta Horticulturalica*, 109: 449-456.

Radwan J., 1991 - Sperm competition in the mite *Caloglyphus berlesei*. *Behavioural Ecology and Sociobiology*, 29: 291-296.

Radwan J., Czyz M., Konior M., Kolodziejczyk M., 2000 - Aggressiveness in two male morphs of the bulb mite *Rhizoglyphus robini*. *Ethology*, 106: 53-62.

Rawlins W.A., 1955 - *Rhizoglyphus solani*, a pest of onions. *Journal of Economic Entomology*, 48: 334.

Sömermaa K., 1972 - Die Verbreitung der Wurzelmilbe *Rhizoglyphus echinopus* (Fumouse et Robin, 1868) (Acarina, Tyroglyphidae) in Hälsingland. Studien der Biologie des Befalls und der Empfindlichkeit gegenuber einigen Pflanzenschutzmitteln im laboratorium. *Acta Agriculturalica Scandinavica*, 22: 65-89.

Timms S., Ferro D.N., Embetson R.M., 1981 - Andropolymorphism and its heritability in *Sancassania berlesei* (Michael) (Acari: Acaridae). *Acarologia*, 22: 391-398.

Van Eyndhoven G.L., 1961 - Artunterschiede beim Genus *Rhizoglyphus* (Acar.). XI International Kongress für Entomologie, I: 247-276.

Williams G.C., 1966 - Natural selection, the cost of reproduction and a refinement of the Lack's principle. *American Naturalist*, 100: 687-690.

Woodring J.P., 1969 - Observations on the biology of six species of acarid mites. *Annals of the Entomological Society of America*, 62: 102-108.

The water mites (Acari, Hydrachnidia) of a little disturbed forest stream in southwest Germany - a study on seasonality and habitat preference, with remarks on diversity patterns in different geographical areas

Reinhard Gerecke

Biesingerstr. 11, D-72070 Tübingen, Germany, reinhard.gerecke@uni-tuebingen.de

Key words: Acari, Hydrachnidia, diversity, habitat preference, phenology, European fauna, zoogeography.

Abstract: The water mite fauna was studied in the Goldersbach, a stream draining triassic sediments in the Schönbuch Nature Park near Tübingen. Mites were collected on ten dates between June 1995 and June 1996. Semiquantitative samples were obtained from a section 200 m in length, by hand-netting separately three different habitat types (gravels, moss carpets, pools). Additional samples, including also qualitative collecting in the interstitial habitat, were taken in three further stations in the catchment. A total of 2,742 specimens were collected (2,541 of which came from the seasonal investigation) representing a total of 47 species. Habitat preference, as well as seasonality of deutonymphs, males and females, are discussed for the more frequent species. All lenitophiles and most rheobionts display a strong seasonality and overwinter as eggs or nymphal resting stages, but adult *Atractides spinipes* and *Lebertia fimbriata* are found year-round. *L. fimbriata* lives mainly on gravel substrata in summer, but invades pool reaches during winter.

The results are compared with studies from other parts of Europe, and the diversity patterns of water mites and other lotic invertebrates are discussed. As regards species richness, water mites regularly turn out to be in second place (behind Diptera). A total number of 40-60 species is normally found in undisturbed mountain streams of central Europe.

A general tendency of reduction in species richness of rheobionts is observed with increasing geographical latitude in Europe. In most genera and families, the water mite fauna of Central and Northern Europe can be described as a 'diluted South European fauna'.

Introduction

For a long time, Central and Northern Europe were the focus centres of systematic and ecological research on fresh water mites. On the positive side, the intensity of this early research produced a particularly dense net of faunistic information. However, the number of taxonomic problems increased, hampering the exploitation of the available data base. For example, in 2000, not even a check list is available for the well-studied faunas of Germany and Switzerland.

This is the main reason for ignoring water mites in most limnofaunistic inventories. Indeed, our information on the general composition of the water mite fauna in individual streams or catchments in Europe derives mostly from studies carried out by water mite specialists. These studies include little, or no, information on the taxonomic composition of the crustacean and insect fauna of the streams in question (Walter 1944; Viets 1957; Schwoerbel 1959 and 1964; Láska 1963; Biesiadka 1970 and 1974; Böttger and

F. Bernini, R. Nannelli, G. Nuzzaci, E. de Lillo (eds.), Acarid Phylogeny and Evolution. Adaptations in mites and ticks, 69-89.

Ullrich 1974; Kowalik and Biesiadka 1981; Cicolani and Di Sabatino 1985; Angelier *et al.* 1985; Böttger and Hoerschelmann 1991; Gerecke and Schwoerbel 1991; Meyer 1994; Böttger and Martin 1995; Martin 1996 and 1997; Martin and Speth 1996). General limnofaunistic surveys dealing with detailed information on water mites have only been published from central Germany (Dittmar 1955), the French Pyreneans (Vinçon 1987) and Sicily (Ferrito 1994). From investigations in various parts of Europe (for a revision see Dohet *et al.* submitted, and the bibliography cited therein), an outstanding sensitivity of water mites to many types of disturbance has been observed.

Published monographic studies on the water mite fauna of individual running waters in Central Europe deal with streams draining crystalline bedrock (Harz: Viets 1957 and 1966; Schwarzwald: Schwoerbel 1959 and 1964, Gerecke and Schwoerbel 1991; Meyer 1994), limestone (Alps: Walter 1944), a mosaique of crystalline bedrock and limestone (Tatra: Láska 1963), tertiary and cretacean sediments (Polish carpathians: Biesiadka 1974) or holocene glacial deposits (region of Poznan: Biesiadka 1970; Schleswing-Holstein: Böttger and Hoerschelmann 1991; Böttger and Martin 1995; Martin 1996 and 1997; Martin and Speth 1996). Here, for the first time a synopsis is given of the water mite fauna colonizing a catchment draining upper Trias (Keuper) sediments. This geological formation, rich in clay sediments, is found outcropping along the northern margin of the Southwestern German Jura mountain chain, and in some patchy areas in Central Germany and in the Lorraine (France).

The study of the Goldersbach mites was carried out to obtain valuable documentation of the seasonal morphological variability for some widespread, but littlefewly known species (Gerecke 1999 and *in prep.*). The aim of this paper is to consider the faunistic results under phenological, ecological and zoogeographical aspects.

Study area and collecting sites

The catchment of the Goldersbach (fig. 1) covers 72 km^2 and has been the subject of an intense interdisciplinary research on geology, pedology, hydrology, vegetation history and some biological aspects between 1978 and 1982 (Einsele 1986).

The Goldersbach and most of its tributaries have their sources in weakly flowing rheohelocrenes at the boundary between sandstone and silty substrata, or from linear springs with temporarily flowing upper reaches. Additional groundwater is supplied from local upwellings in the stream bed, often marked by a particularly luxuriant moss vegetation. Only some of the first order reaches are rather straight and steep, while the remaining parts of the stream system are characterized by meandering flow in gently sloped, open valleys. Only locally, near bridges and tourist areas (e.g. downstream sites A, C and D, and upstream site B), is the water course slightly changed by stone embankments and wood cross-bars.

The mean annual precipitation in the study area is 745 mm (Einsele and Agster 1986). The mean runoff in the Goldersbach middle course at site B (see fig. 1) in 1979/1980 was 187 l/s, but in the period 1977-1981 extreme values from 20 to 1208 l/s were measured; the relative values for the Arenbach upstream from its confluence are 29 (4.5-319) l/s (Kracht and Klee 1986). A slight turbidity of the stream at medium and low water levels testifies for the prevalence of easily eroded sediments. In pools, extended deep layers of weakly permeable silt alternate with banks of well-oxygenated sand.

The Goldersbach is summer-cool, with annual mean temperatures about 7°C, and summer maximum below 18°C. Water chemistry was found to be rather homogenous in different sections of the stream and its affluents (Kracht and Klee 1986). Conductivity values fluctuate between 300 µS/cm (during spates, mainly in winter and spring) and 900 µS/cm (in autumn),

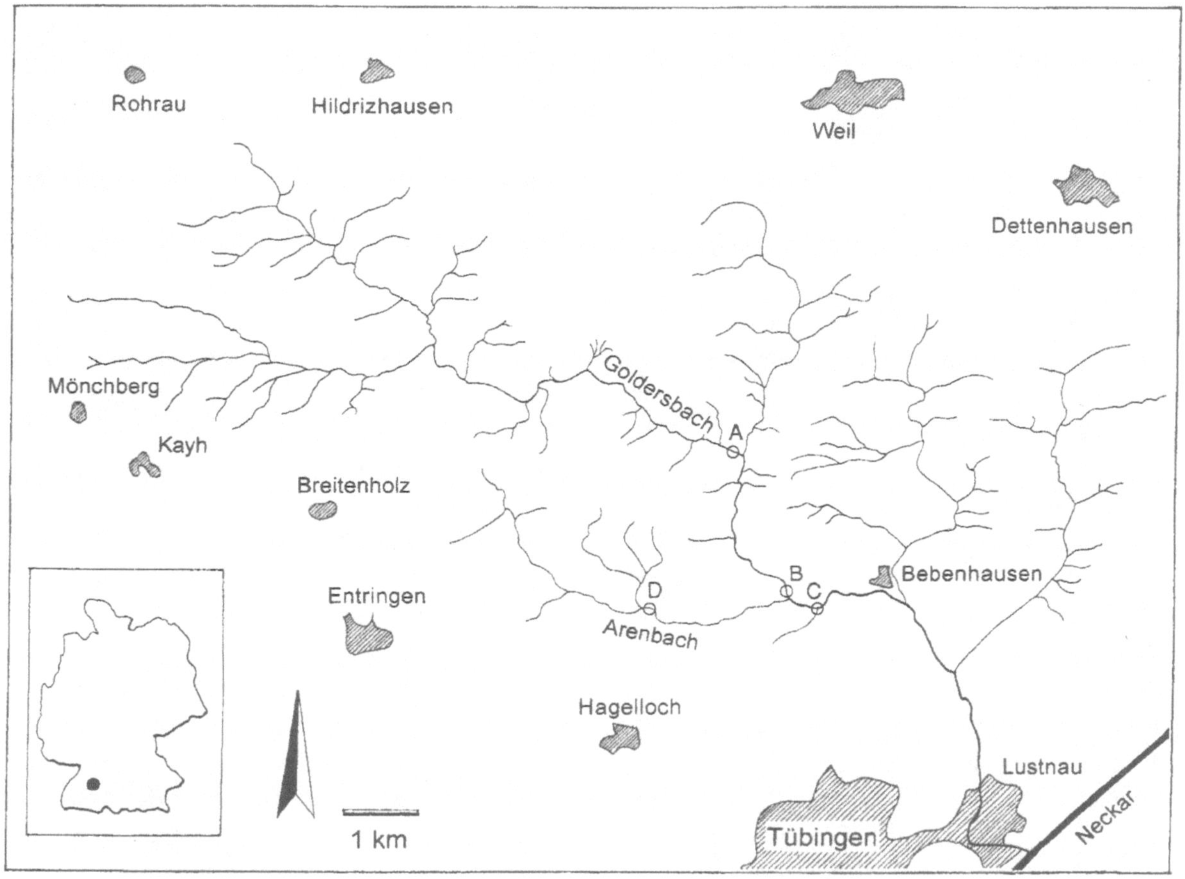

Fig. 1 - Goldersbach catchment: a map of the study area. Inlet: position of the study area in Germany.

generally with a continuous increase from May to October (Agster 1986). A year-round increased concentration of Cl^- in the eastern part of the study area is due to road-deicing, and a slight waste-water influence downstream from the village of Bebenhausen are the only traces of direct anthropogenic impact (Kracht and Klee 1986), in the stream downstream from the study sites presented here. From 1963 to 1983, an increase in NO_3 was observed in springs originating from sediment layers with a short retention time, probably reflecting atmospheric N input (Agster and Einsele 1986).

About 200 years ago, the vegetation consisted of degraded remnants of forest and shrubland due to overexploitation. Reforestation began in 1807, starting with angiosperms (*Fagus, Quercus*) that form the natural climax vegetation of the area. Later, forest administration favoured gymnosperms (*Pinus, Picea*), and the percentage of angiosperm coverage diminished consequently from 96% at the start of the nineteenth century to 44% in 1986 when 86% of the area was covered by forest vegetation (Arnold 1986). This change probably had important consequences for water supply and water chemistry in the area.

Published inventories of the limnofauna in the study area are only available for a few taxonomic groups (Oligochaeta: Hofmann and Klee 1986; Ostracoda: Janz 1983; fauna of stagnant pools in the surroundings: Nickel 1987; Janz 1988). Three unpublished state examination theses finished in 1970 (Buddeberg, Manz, Weissert, Zoological Institute University of Tübingen) have been used for a comparison with the data published here.

The submerged vegetation of the Goldersbach is characterized by a nearly complete absence of macrophytes, and dominance of mosses *(Fontinalis antipyretica)*. During winter, hard substrata become covered by mats of diatoms. Downstream from Bebenhausen, green algae (*Cladophora*) grow densely during summer. In riffles, *Alnus* roots make up an important component of the substrata.

The sampling sites are as follows:

A: Goldersbach upstream from Teufelsbrücke, 390 m a.s.l., in riffle areas 2-3 m in width and 10 cm deep, regular sequence of pools and riffles; right bank with forest, left bank separated from grassland by a line of *Alnus* and *Salix* trees with luxuriant herbaceous undergrowth.

B: Goldersbach near confluence of the Arenbach (Geschlossener Brunnen), 365 m a.s.l., in riffle areas 2-3 m in width and 10 cm deep, extended pool areas, below the Arenbach mouth riffles with large stones; right bank mainly steep slopes with forest vegetation, locally with outcroping and eroding silty sediments; left bank flat, with young *Picea abies* plantations and a confining forest road; riparian vegetation on both sides made up of *Alnus* and *Salix* trees.

C: Goldersbach upstream from the Bettelbach mouth, 360 m a.s.l., in riffle areas 3-4 m in width and 10-20 cm deep; extended pool areas with small riffles; at right bank steep slopes with forest vegetation, left bank separated from confining grassland by an embankment with *Alnus* and *Salix*.

D: Arenbach near Becklesgarten, 405 m a.s.l., steeper than A-C, extended sections with very slow flow instead of stagnant pools, in riffle areas about 1 m in width and 5-10 cm deep; banks with brush wood vegetation, mainly *Salix* and *Alnus*, separating the stream from surrounding grassland.

Material and methods

Semiquantitative samples for the seasonal investigation of site A were taken on ten occasions: 27.06.1995, 29.07.1995, 05.09.1995, 05.10.1995, 11.11.1995, 17.12.1995, 03.02.1996, 17.03.1996, 19.04.1996, 20.05.1996. Thus, the time interval was 4-5 weeks during the vegetation period, but 6-7 weeks between December and March.

On each date, for a four hour duration, mites were collected with a hand net and sorted on the spot in a white pan. Three habitat types (pools, gravels and stones, moss carpets) were sampled separately, but with equal intensity. Compared with standardized sampling of defined surface areas, this method is more flexible and allows the careful checking of different microhabitat types (e.g. leaf litter accumulations, decaying wood, and submerged roots within pools). All studies of interstitial habitats were of qualitative character only, in order to give a preliminary survey of the differences between the species composition in surface and subsurface habitats of the catchment; for a description of methods see Schwoerbel (1989).

Additional material was collected at the following sites and dates:

A: 27.03.1989, interstitial, Karaman-Chappuis; 25.06.1989 dito.

B: 16.06.1997, benthos, semiquantitative sampling as described for site A, results used in order to corroborate the seasonal study (in view of the very low species numbers collected in June 1995); 30.08.1998 interstitial, Karaman-Chappuis; 05.07.1999 interstitial, Karaman-Chappuis and Bou-Rouch.

C: 27.03.1989, interstitial, Karaman-Chappuis.

D: 06.07.1995, benthos; 29.03.2000, interstitial, Karaman-Chappuis and Bou-Rouch (ground water course Zoological Institute University of Tübingen).

Mites were conserved in Koenike's fluid and determined in the laboratory. A large part, e.g. all representatives of the genera *Atractides*, *Lebertia*

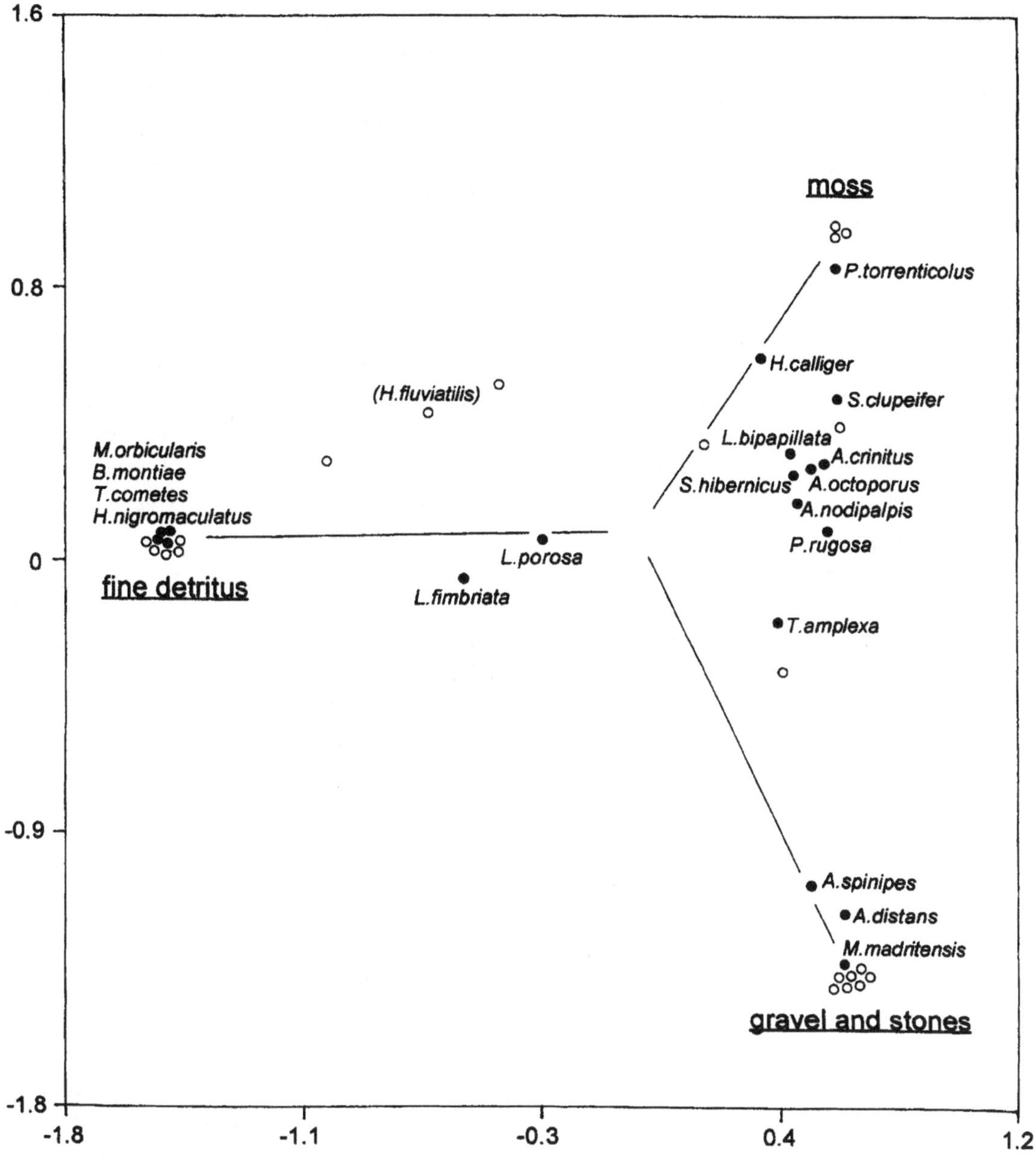

Fig. 2 - Goldersbach, site A (Teufelsbrücke); a CCA diagram illustrating the habitat preference of the most frequent water mite species; dark circles: species with >1% total abundance; open circles: species with <1% total abundance.

and *Protzia*, and all Aturidae, were dissected and slide-mounted in Hoyer's fluid for determination and measurements. The complete material is labelled and available in the collection Gerecke, Tübingen.

The plot of figure 2 was obtained from a matrix of numerical abundance in three microhabitat types, applying the method of correspondence analysis (Rohlf 1990). Due to the reduced number of variables, this method was used only in order to give an explicative illustration of the cumulative abundance of the species.

For the comparison of water mite diversity in different geographical areas (figs 6-10), species lists obtained from bibliography were critically revised, excluding synonyms and obviously erroneous records.

Results and discussion

The water mite taxocoenosis

The quantitative composition of the material on which this study is based, is compiled in Table 1 (surface water), and table 2 (interstitial habitats). A detailed list, differentiating the data for substrata, developmental stages, and collecting dates, is available from the author.

The total number of specimens amounts to 2,742. Some 2,363 come from the seasonal study at site A, but for the analysis of phenology, an additional 178 taken at site B were included in the June column of table 1. The interstitial samples produced a total number of 81 specimens. Due to the noteworthy different conditions at site D, a further 118 specimens collected here were not used in the ecological analysis. This material represents 12 species: *Atractides distans*, *A. nodipalpis* (amounting to 66% of the total number of specimens), *A. octoporus*, *A. spinipes*, *A. tener*, *Aturus crinitus, Barbaxonella angulata* (the only male specimen collected in the study area), *Brachypoda montiae, Hygrobates calliger*, *Lebertia fimbriata*, *L. porosa*, *Ljania bipapillata*, *Monatractides madritensis*, *Protzia rugosa* and *Torrenticola amplexa*.

Forty seven species were found, 38 in the benthos samples, 17 in interstitial habitats, and 8 in both kinds of habitat. The 200 m segment from where samples for the year-round study were taken (site A) produced a total of 41 species.

The material collected from subsurface waters is poor in specimens, probably due to the patchy distribution of sediments suitable for colonization. However, the community is obviously highly diverse, with 10 strictly hyporheobiont species: *Albaxona lundbladi*, *Atractides denticulatus*, *Axonopsis vietsi*, *Azugofeltria motasi*, *Kongsbergia pectinigera*, *Neoacarus hibernicus*, *Stygomomonia latipes*, *Wandesia racovitzai*, *W. stygophila*, and *Lobohalacarus weberi*. *Wandesia racovitzai* was previously only known from the type locality in Great Britain (Gledhill 1970; Gerecke 1999). *Ljania macilenta* is a hyporheophile, typically replacing *L. bipapillata* in deeper sediment layers, while the remaining species in table 2 are hyporheoxenes, accidentally taken in interstitial samples when the Bou Rouch technique was applied. Four hyporheobionts (*Neoacarus hibernicus*, *Wandesia racovitzai*, *Axonopsis inferorum* and *Barbaxonella angulata)*, were collected as single specimens in benthos gravel samples. The latter two species in this study were taken exclusively from surface, but not from interstitial habitats. This observation, as well as the high number of species so far found only as single specimens, suggest that a more detailed study of the interstitial fauna of the study area could produce a considerable further increase in water mite species.

A comparison of the results of the surface samples shows that the composition of the fauna is rather homogenous between sites A-D, in agreement with the similar geomorphology in the middle course of the Goldersbach and its affluents. However, the following quantitative considerations pertain to the fauna of site A.

Six species are dominant (>5% total abundance, in decreasing order): *Atractides nodipalpis, Hygrobates calliger, Torrenticola amplexa, Mideopsis orbicularis, Brachypoda montiae* and *Lebertia fimbriata*. Five of them are widely distributed in Europe, but *Brachypoda montiae* has only been recorded very rarely, mostly as single specimens, from Northern Italy (Lombardia), Slovenia, Austria, Southeastern Germany and Sweden. The presence among the dominant species of two strict lenitobionts (*B. montiae, M. orbicularis*) characterizes the Goldersbach as a rather slowly flowing stream with extended pool reaches.

Ten species are subdominant (abundance 2-5%, in order of decreasing abundance): *Hygrobates nigromaculatus, Monatractides madritensis, Teu-*

Tab. 1 - A list of the water mites collected in the Goldersbach and the Arenbach, benthos samples (A-C: collecting sites). Species are sorted following their relative abundance.

year	1995/ 1997	1995	1995	1995	1995	1995	1996	1996	1996	1996	TOT.	%
date	16.06. 27.06.	29.07.	05.09.	05.10.	11.11.	17.12.	03.02.	17.03.	19.04.	20.05.		
Sperchonopsis verrucosa (Protz, 1896)										1	1	0.04
Atractides fonticolus (K.Viets, 1920)								1			1	0.04
Axonopsis inferorum Motas & Tanasachi, 1947			1								1	0.04
Barbaxonella angulata (K.Viets, 1955)		1									1	0.04
Neumania deltoides (Piersig, 1894)	1										1	0.04
Wandesia racovitzai Gledhill, 1970		1									1	0.04
Lebertia glabra Thor, 1897					1					1	2	0.08
Monatractides stadleri (Walter, 1922)			2								2	0.08
Neoacarus hibernicus Halbert, 1911		1								1	2	0.08
Sperchon glandulosus Koenike, 1886		1		1							2	0.08
Sperchon thienemanni Koenike, 1907									1	1	2	0.08
Atractides tener (Thor, 1899)			2			1	1	1			5	0.20
Torrenticola andrei E.Angelier, 1949	1	1	1						1	1	5	0.20
Nautarachna crassa (Koenike, 1908)	1	1	3	1							6	0.24
Aturus scaber Kramer, 1875		2			1					5	8	0.32
Hydrodroma torrenticola (Walter, 1908)			1	2						5	8	0.32
Kongsbergia largaiollii (Maglio, 1909)		1	2	1			1		2	1	8	0.32
Mideopsis crassipes Soar, 1904		1		6						2	9	0.36
Sperchon setiger Thor, 1898			1	1	2	1		1	4	1	11	0.44
Hygrobates fluviatilis (Ström, 1768)	4			3	1					7	15	0.59
Panisus torrenticolus Piersig, 1898	1	2	8	2	1	2	1	2	2	2	23	0.91
Sperchon clupeifer Piersig, 1896		9	10	3					3	2	27	1.07
Atractides distans (K.Viets, 1914)	4		6	7	1	1			16	2	37	1.47
Hygrobates nigromaculatus Lebert, 1879	3	1	3	9	4	1	1	27	8	6	63	2.50
Monatractides madritensis (K.Viets, 1930)	24	11	14	2					2	12	65	2.58
Teutonia cometes (Koch, 1837)	6	7	5	19	3	1			11	17	69	2.74
Ljania bipapillata Thor, 1898	3	8	8	15	7	8	8	2	5	10	74	2.93
Sperchon hibernicus Halbert, 1944	1	27	6	11	2		1	2	10	16	76	3.01
Atractides octoporus Piersig, 1904	4	12	12	13	6		3	2	13	17	82	3.25
Atractides spinipes Koch, 1837			7	15	7	14	12	18	7	2	82	3.25
Aturus crinitus Thor, 1902	7	12	8	9	7	13	12	4	13	16	101	4.00
Lebertia porosa Thor, 1900	16	24	2	3	4	1		4	9	37	102	4.04
Protzia rugosa Walter, 1918	15	18	12	8	2	3	2	2	12	28	102	4.04
Lebertia fimbriata Thor, 1899	9	16	9	16	12	20	5	27	17	10	141	5.59
Brachypoda montiae Maglio, 1924	74	24	8	11					1	53	171	6.78
Mideopsis orbicularis (Müller, 1776)	16	22	43	35	3	1		11	24	68	223	8.84
Torrenticola amplexa (Koenike, 1908)	41	72	55	44	20	6	5	6	17	48	314	12.45
Hygrobates calliger Piersig, 1896	34	18	95	84	12	3	1	7	48	17	319	12.65
Atractides nodipalpis (Thor, 1899)	48	152	27	17	7	13	2	6	11	78	361	14.31
Total number	**313**	**445**	**351**	**340**	**103**	**89**	**55**	**123**	**237**	**466**	**2523**	**100**

tonia cometes, Ljania bipapillata, Sperchon hispidus, Atractides spinipes, Atractides octoporus, Aturus crinitus, Lebertia porosa and *Protzia rugosa*. Also most of these species are widely distributed in Europe. *M. madritensis* is a hyporheophile, and probably due to this habitat preference documented from scattered sites only. *A. spinipes* was for a long time a taxonomic problem, and all published records are doubtful. After the redefinition of this species on the basis of the Goldersbach population (Gerecke, *in prep.*), future faunistic research can explore its actual distribution area. Also among the subdominants, three frequently collected lenitobionts testify for the presence of extended pool areas: *H. nigromaculatus, T. cometes* and *L. porosa*.

The remaining 24 species (abundance <2%) include 9 accidental records of hyporheobionts (*Axonopsis inferorum, Barbaxonella angulata, Wandesia racovitzai, Neoacarus hibernicus*), crenobionts (*Atractides fonticolus, Lebertia glabra, Sperchon thienemanni*), or still water species (*Limnesia undulata, Neumania deltoides*). As the catchment of the Goldersbach is regularly subject to heavy spates, the presence of crenobionts is best explained by drift, while still water adapted species occasionally may be introduced by parasitized insect hosts. Among the rare rheobionts, *Torrenticola andrei* is a hyporheophile, and *Hydrodroma torrenticola, Mideopsis crassipes* and *Nautarachna crassa* are further typical elements of the lenitophilous pool fauna. *Nautarachna crassa* has so far been found only on rare occasions, and the collections from the Goldersbach have produced important records of the little known male.

Habitat preference

The habitat preference of the water mite species in the Goldersbach, resulting from the interpretation of separate collections in pool, gravel and moss choriotopes, is illustrated in figure 2. The lenitobiont pool-dwellers, with the dominant species *Mideopsis orbicularis, Brachypoda montiae, Hygrobates nigromaculatus* and *Teutonia cometes*, form a well-defined cluster at the left margin of the diagram, together with the less abundant *Mideopsis crassipes, Nautarachna crassa* and three accidental species found as single specimens. An obvious, but less distinct preference for lenitic habitats is also observed in *Lebertia porosa* and *L. fimbriata* (see below), as well as in *Hygrobates fluviatilis*, a species in naturally conserved running waters mostly present in low numbers, but capable of building up to large populations in organically polluted streams (Gerecke and Schwoerbel 1991, Dohet *et al.*, submitted).

The remaining, strictly rheobiont species, are arranged in the right part of the diagram along a vertical gradient between gravel substrata (at the bottom) and mosses (at the top). Obviously, *Protzia rugosa, Atractides nodipalpis, Sperchon hibernicus* and *A. octoporus* equally colonize both choriotope types. *Panisus torrenticolus, Hygrobates calliger, Sperchon clupeifer, Kongsbergia largaiollii, Ljania bipapillata* and *Aturus crinitus* were collected more frequently in mosses than on hard substrata, while *Torrenticola amplexa* has a slight preference for the gravel choriotope.

A further distinct cluster is formed in the inferior right angle of the diagram by species obviously bound to gravel and stone substrata. Characteristic species of this ecological group are *Monatractides madritensis, Atractides spinipes* and *Atractides distans*, all three species also have a tendency to invade interstitial habitats. They are accompanied by 7 further, mostly hyporheobiontic species, recorded as single specimens.

A comparison of tables 1 and 2 shows net differences between the fauna of surface and subsurface habitats. The species which are most frequent in the hyporheic interstitial were completely absent or taken only as single specimens in surface samples. On the other hand, surface species appeared in interstitial samples mainly when the Bou-Rouch technique was

Tab. 2 - A list of the water mites collected in the Goldersbach and the Arenbach, interstitial samples (A-C: collecting sites). Species are sorted in alphabetic order.

site	A	A	B	B	C	D	**Total**
date	23/03/1989	25/06/1989	05/07/1999	30/08/1989	27/03/1989	29/02/2000	
Hydrachnidia							
Albaxona lundbladi Motas & Tanasachi, 1947			2		4	3	**9**
Atractides denticulatus (Walter, 1947)	1				2		**3**
Atractides spinipes Koch, 1837					1		**1**
Aturus crinitus Thor, 1902	2		1		1	1	**5**
Axonopsis vietsi Motas & Tanasachi, 1947			1				**1**
Azugofeltria motasi Schwoerbel, 1961						1	**1**
Kongsbergia pectinigera Motas & Tanasachi, 1946			1				**1**
Ljania bipapillata Thor, 1898			1				**1**
Ljania macilenta Koenike, 1908			1				**1**
Neoacarus hibernicus Halbert, 1944	1	2	4	2		1	**10**
Panisus torrenticolus Piersig, 1898			1				**1**
Protzia rugosa Walter, 1918			1				**1**
Sperchonopsis verrucosa (Protz, 1896)						3	**3**
Stygomomonia latipes Szalay, 1943	2	1	3	3	1	2	**12**
Wandesia racovitzai Gledhill, 1970	4	3			2		**9**
Wandesia stygophila Szalay, 1944		1					**1**
Halacaridae							
Lobohalacarus weberi (Romijn & Viets, 1924)			1				**1**
Total	**10**	**7**	**17**	**5**	**11**	**11**	**61**

applied. Interstitial pumping in the Goldersbach was successful only in reaches with coarse gravel substrata near to the surface, where habitats of surface and subsurface species interchange.

Life cycles

For a detailed analysis of the life cycle of single species (periods of egg deposition, presence and behaviour of larvae, nymphal stages, males and females), an extended set of information is needed: collection of significant numbers of adults and nymphs over shorter time intervals in the field; catching of emerging larvae, at best with emergence traps that yield information on host preference; and laboratory cultures in order to relate larvae to adults in species with unknown larval morphology. Thus, the results presented here are rather incomplete, with a complete lack of information on larval stages, which only allows a more detailed discussion and comparison for the most frequent species. Descriptions of larval morphology are available only for less than 50% (20) of the Goldersbach water mite species, and in four cases, information on larval morphology is not even available at genus level: *Albaxona, Azugofeltria, Barbaxonella* and *Kongsbergia*.

The whole water mite taxocoenosis shows a strong seasonal fluctuation with nearly 120 specimens collected per hour between May and October, but a very low density during winter, reaching a minimum of 14 specimens/hour in February (fig. 3A). However, for many species, overwintering of small populations of adults could be observed.

Atractides nodipalpis represents the most abundant species in the Goldersbach, with nearly 15% of the total individuals collected. It is widely distributed in Europe, but poorly understood mostly due to taxonomic problems. Twenty-five subspecies have been introduced in the past century, most of which synonyms or separate species without particular relation to the nominate taxon (Gerecke, *in prep.*). As can be deduced from figure 3B, a low number of (mainly female) adults is found during winter, and population density increases slowly from March onwards. During the vegetation period, no particular bias in sex ratio is observed. Deutonymphs appear in large numbers in May, and then they are found continuously, but in decreasing density, until October.

On the basis of more detailed information (including data on larval emergence and parasitism) Martin (1998) hypothesized for *A. nodipalpis* a life cycle with overwintering in egg clutches, and, during springtime, passing the sequence of preadult stages, until the appearance of adults from June onwards. The data from the Goldersbach agree with this scenario as regards deutonymphs (presence from May to October, with an abundance peak in May), but the detection of considerable numbers of adults only from March onwards is contradictory. The existence of a bivoltine developmental cycle (as could be suggested by the observation of two peaks in larval emergence, Ullrich 1976; Martin 1998) is improbable as no deutonymphs could be observed neither in late autumn, nor in early spring. If *A. nodipalpis* in the Goldersbach does display bivoltinism, we should hypothesize a very short (or bypassed) deutonymphal stage that should result in size differences between spring and summer adults. As no such differences could be observed in the Goldersbach population (Gerecke, *in prep.*), the phenology is best explained with a 'double-track strategy'. At least a part of adults and/or tritonymphs could be capable of overwintering in refuges not detected with surface substratum sampling. However, overwintering at the tritonymphal stage is only possible if some of the deutonymphs from each generation can enter diapause without completing their life cycle within one vegetation period. Such a developmental cycle with parts of the population overwintering in egg clutches, other parts as tritonymphs or adults, would represent an interesting adaptation to seasonally unstable life conditions in running waters. A further interesting aspect that merits future attention are possible geographical differences in life cycle, as is suggested by overwintering adults observed in Southern, but not in Northern Germany.

A completely different phenology was published by Efford (1966) for a population of *A. nodipalpis* from Great Britain. The fact that adults

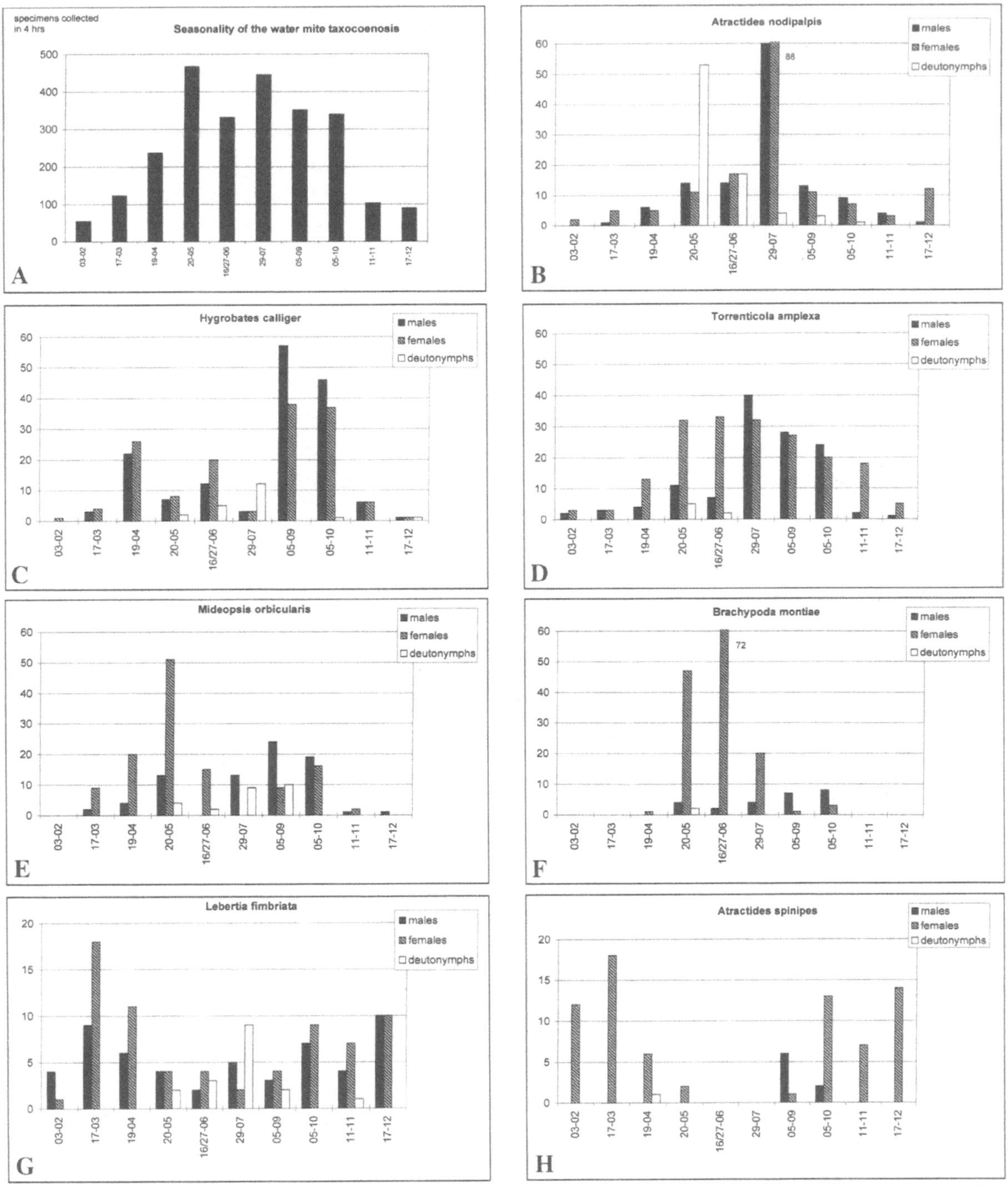

Fig. 3 - Goldersbach, site A (Teufelsbrücke): seasonality of the water mite taxocoenosis, of the six most frequent water mite species, and of *Atractides spinipes*. A) the whole water mite taxocenosis, B) *Atractides nodipalpis*, C) *Hygrobates calliger*, D) *Torrenticola amplexa*, E) *Mideopsis orbicularis*, F) *Brachypoda montiae*, G) *Lebertia fimbriata*, H) *Atractides spinipes*.

appeared in high individual numbers during winter, but were absent in summer, and males were found only in autumn, suggests that these data refer to a species different from *A. nodipalpis* (Martin 1998). In the Goldersbach, exactly this type of seasonality was observed in *A. spinipes* (see below), a species often confused with *A. nodipalpis*, but strongly different in the male sex (Gerecke, *in prep.*).

Hygrobates calliger, the second most abundant species in the Goldersbach, is almost completely absent from winter samples (fig. 3C). Individual numbers increase from March to April, but adults disappear nearly completely in July. Then, in late summer and spring, adults are found in very high numbers, but disappear during November; sex ratio is not biased. Deutonymphs were found from May to July, and as single specimens in October and December. Ullrich (1976) observed in Mittelgebirge Germany a continuous hatching of *H. calliger* larvae from May to October, with a peak in August. Obviously, the first larvae in May are offspring of adults emerging from March onwards, and begin to develop further to deutonymphs in May. Adults of this new generation appear in late summer, and probably reproduce and contribute to the emergence of larvae in autumn. It remains unclear, at which stage the winter period is passed. As in *A. nodipalpis*, the appearance of adults in spring could indicate overwintering as tritonymphs or refuged adults, absence of deutonymphs in early spring makes overwintering in egg clutches improbable. The presence of deutonymphs in autumn could indicate a bivoltine life cycle in *H. calliger*.

For *Torrenticola amplexa*, found with nearly the same abundance as *H. calliger*, a description of preadult stages and information on details of the life cycle were published by Tuzovskij (1981). In the Goldersbach, adults overwinter in low individual numbers (fig. 3D). During springtime, a strongly female-biased population is built up that obviously begins to reproduce in April. Deutonymphs are found in May and June. Suddenly, in July the sex ratio is nearly equal, indicating that the early summer deutonymphs have now developed to the following adult generation. The further course of the life cycle and the principal way of overwintering remain unclear. Probably, as observed by Tuzovskij (1981), there is a second generation of deutonymphs in September/October. The sudden decrease of male abundance in November, and the extended longevity of females suggest that there is egg production until early winter, and, that the population possibly survives the winter partly in egg clutches, partly as adults.

Adults of *Mideopsis orbicularis* are absent during winter, but emerge as a strongly female-biased population from March onwards (fig. 3E). In June, males have disappeared completely, but from May to July deutonymphs are found as the offspring of this springtime generation. They moult to a second adult generation from July onwards. The deutonymphs found in September are probably the offspring of this second generation. The autumn adults disappear almost completely between October and November. Overwintering could take place at the tritonymphal stage. Information on the life cycle of *M. orbicularis* was, until now, published from standing waters (Nocentini 1960; Meyer and Schwoerbel 1981). In lake Mindelsee, Meyer and Schwoerbel observed continuous egg production of females from April to November, but with a significantly higher share of ovigerous females in the first half of this period (April-July). They observed the appearance of deutonymphs from July to December, but at different times in different parts of the lake. Therefore, it is possible that also in Lake Mindelsee, deutonymphs begin emerging just earlier, but at sites that were not documented by the authors. Also the phenology of the adults was different in various parts of the lake, but locally with considerable numbers of both males and females until December. Thus, in stream dwelling populations of *M. orbicularis*, adults appear obviously for shorter periods compared with lake populations. In both types of habitat overwintering at the tritonymphal stage is probable, but the coexisting of dark- (older) and light-coloured (newly hatched) adults in springtime samples in lake Mindelsee indicates that overwintering at the adult stage is possible, at least in standing water populations.

For *Brachypoda montiae* (fig. 3F), no life cycle information was available so far. Our data show an extremely female-biased population in spring and summer, but males obviously do not disappear earlier, and are even more frequent than females in autumn. In view of their small size, the finding of a very low number of deutonymphs, exclusively in May, is probably a collecting artifact. However, we cannot exclude that they are also short-lived and hidden in particular microhabitats, and therefore relatively rarely caught. The phenological pattern of *B. montiae* is typical for a pool-dwelling species, with a strong summer peak and complete absence over an extended winter period.

Lebertia fimbriata (fig. 3G) differs from all so far discussed species in the year-round presence of adults, with maximum numbers in December and March. Deutonymphs appear from May to September, and in November. If we integrate these observations with data published by Gledhill (1973, deutonymphs in April and October/November), Lundblad (1968, young adults found in July and October), and Martin (1998, ovigerous females March-May, deutonymphs in July, two distinct peaks of larval emergence in May/June and October), there is much evidence that *L. fimbriata* has a bivoltine reproduction cycle, with two principal oviposition periods, in early spring, and autumn. This scenario agrees largely with the hypothesis of Martin (1998). In the Goldersbach, *L. fimbriata* inhabits lotic moss and gravel habitats during summer, but emigrates into pools during winter (fig. 4): from November to April, 80% of the specimens (n=81) were collected in pools, from May to October only 16% (n=51).

Regarding the more rare species (abundance <5%), only preliminary considerations are possible. Most of them, in particular the hyporheophilous *Monatractides madritensis* and the lenitobiont *Teutonia cometes*, are rare in, or completely absent from, samples taken in winter. Significant exceptions are made by *Aturus crinitus, Atractides spinipes* and *Ljania bipapillata*. For *A. crinitus*, no data concerning the life cycle have been published so far. In the Goldersbach, populations are generally female-biased, only in October more males than females were found. Throughout winter, males as well as females are present. *Atractides spinipes*, a species whose biology was so far unknown due to taxonomic problems, has an interesting phenology (fig. 3H): in June and July, no specimens were found at all; in September appears a generation of adults followed by the disappearance of males in November; during the whole winter season a stable population exclusively of females is maintained. The appearance of deutonymphs in April indicates that oviposition takes place in early spring; the summer is possibly bridged at the stage of a diapausing tritonymph. This interesting scenario of a water mite with an 'anticyclic phenology', needs confirmation by further investigations, based on larger individual numbers. For *L. bipapillata*, also the data published by Martin (1998), indicate a year-round presence of adults, with two distinct oviposition periods in spring and autumn. In contrast, Meyer (1994) found in the Southern Black Forest strong changes in phenology, with absence of adults in winter and in June. This contradiction might be explained by the use of different collecting methods: The standard surber technique restricts sampling to selected microhabitats, and seasonal changes in habitat preference of certain species (see below) could therefore be misinterpreted as seasonal changes in population density.

Stream-dwelling populations of *Hygrobates nigromaculatus* differ strongly from lake inhabiting *H. nigromaculatus* with regard to their

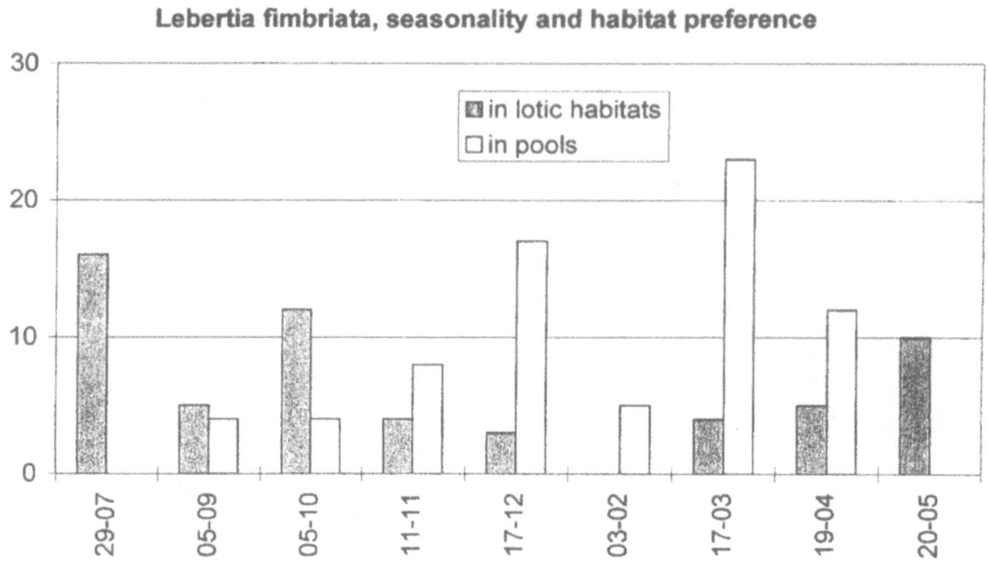

Fig. 4 - Goldersbach, site A (Teufelsbrücke): seasonal change of habitat preference in *Lebertia fimbriata*.

life cycle and possibly represent a separate, still undescribed species (Martin and Davids, this volume). Martin (1998) gives details of their life cycle: oviposition takes place from May to August, and the offspring develops to the adult stage during late summer and autumn. Martin hypothesizes overwintering of adult females, but in the Goldersbach nearly complete absence of adults from winter samples, and appearance of considerable numbers of males and females in March suggest that most specimens survive the winter as tritonymphs.

Seasonal changes in habitat preference

These data reveal an increased relative abundance of the water mite taxocoenosis in pools during winter compared with the summer season. If we calculate the winter/summer ratio of relative abundance [%] of the water mite taxocoenosis in pool habitats, this abundance increased by a factor of 1.3. This observation merits particular attention as the true pool-dwelling species are completely absent in winter. Thus, the change is caused mainly by the emigration of lotic-adapted species into pools, as has been demonstrated above for *Lebertia fimbriata* (fig. 4). Further species in pools represented with a remarkably increased relative density during winter are: *Atractides nodipalpis*, *Lebertia porosa* and *Sperchon hibernicus*.

Diversity of mites and other invertebrate groups

Diversity in single catchments

Our present knowledge of the invertebrate diversity in the Goldersbach is illustrated in figure 5. As in most cases, this list is incomplete due to differences in collecting technique, and collecting intensity: data for both Oligochaeta and Ostracoda is based on field work during a whole vegetation period, while Ephemeroptera, Plecoptera and Trichoptera were studied during winter only, and exclusively at the larval stage.

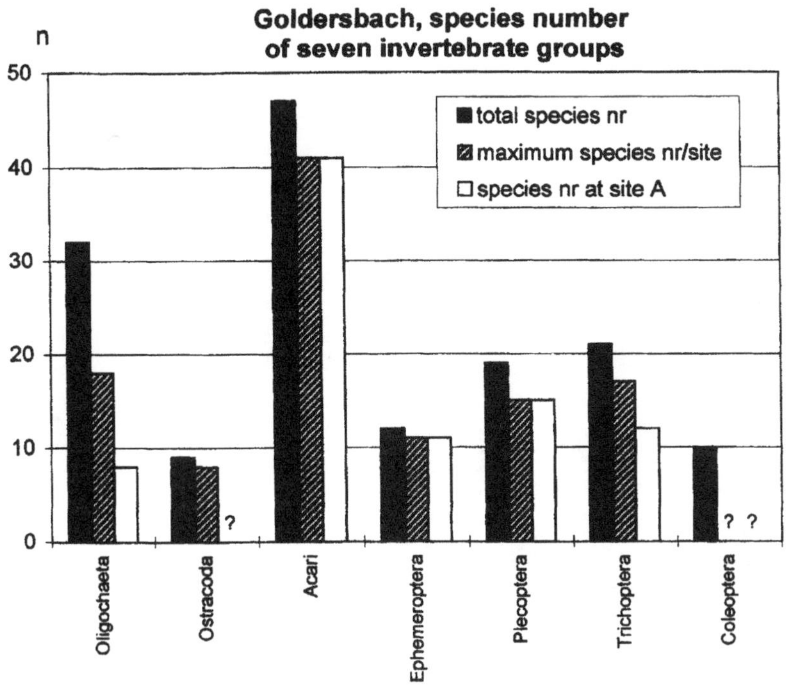

Fig. 5 - Goldersbach catchment; a comparison of the species number of important invertebrate groups (origin of the data: see text); question marks: no data available.

Information on Coleoptera is available only from occasional collections of the author and unpublished lists of entomological field courses of the university of Tübingen. Some additional information is included in two theses (Nickel 1987; Janz 1988) dealing with small standing waters where some further species were found that probably also occur in the Goldersbach (e.g. four species of the stonefly genus *Nemoura*). Not withstanding these constraints, figure 5 gives a representative illustration of the importance of water mites as a particularly diversified group of invertebrates in a naturally conserved stream habitat. Furthermore, mite diversity reflects the importance of another insect order too often neglected in limnofaunistic studies: all water mite species from this study for which information on host preference is available, are parasites of Diptera at the larval stage, generally midges of the family Chironomidae. The rich water mite fauna of the Goldersbach indicates the presence of an interesting and species-rich chironomid fauna.

Diversity in more extended geographical areas

From a comparison of monographic studies on the fauna of three stream catchments in different parts of Europe (fig. 6), Diptera were by far the most species rich group. In all three cases, water mites species richness was similar to that of Plecoptera and Trichoptera. More extended studies, investigating all stream basins of selected mountain areas, regularly produced more than 80 running-water dwelling species (and by far higher numbers, if springs and spring brooks are included, as in Biesiadka 1979).

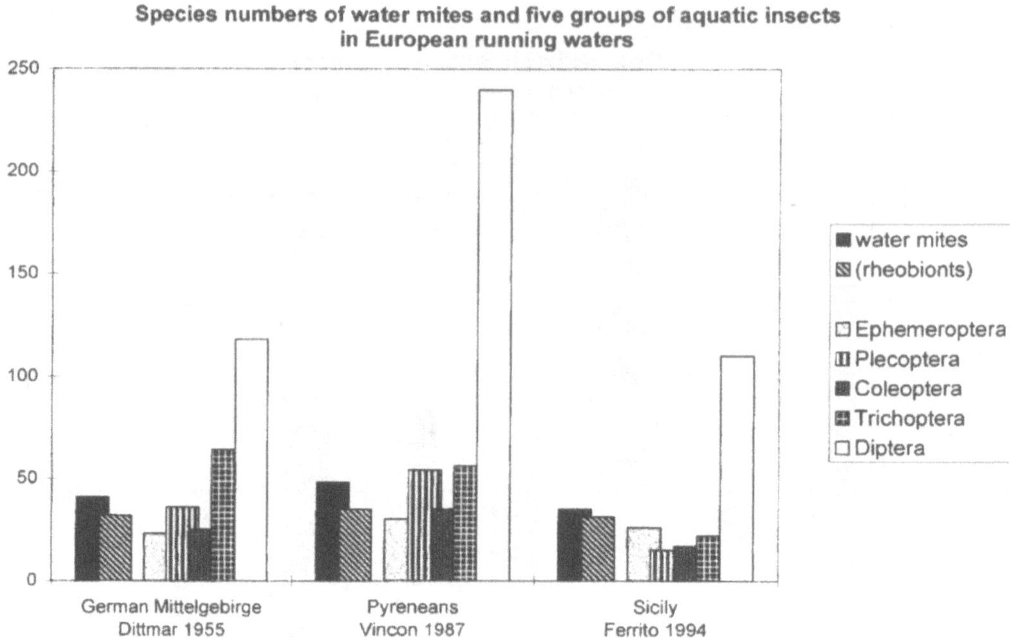

Fig. 6 - Species numbers of water mites, rheobiont water mites, and five insect orders. A comparison of three monographic studies on invertebrate diversity in different European areas.

Trends in distribution patterns of lotic water mites

Investigations on water mite communities in Western, Central and Eastern European mountain ranges demonstrate the presence of rheobiont species in rather similar numbers of 60-70 (fig. 7). However, a comparison of selected stream catchments in different areas of Europe (fig. 8) indicates a tendency for impoverishment of rheobionts with increasing geographical latititude. This trend is surely reinforced by a generally reduced diversity of geological formations in Central-Northern Europe with the presence of glacial deposits unsuitable for colonization over extended areas. Furthermore, anthropogenic interference obviously caused the extinction of many rheobiont water mite species from the Central-North European lowlands as has been demonstrated by Van der Hammen and Smit (1996) for The Netherlands.

However, an analysis of data from a larger geographical scale shows that the northerly decrease of rheobiont species numbers is not sufficiently explained by local factors and the influence of our civilization. At present state of knowledge (with many taxa not yet sufficiently studied and the detection of numerous additional species to be expected in Italy), also a comparison of the national faunas of Italy, Britain plus Ireland, and Sweden (fig. 9) indicate for the same tendency. An interpretation is complicated by the fact that the ecological requirements of numerous species are different at various latitudes. Lundblad's catalogue (1968) gives numerous examples of species that in South and Central Europe are confined to springs and/or streams, but in Fennoscandia colonize standing waters. Therefore, figure 10 concentrates on two genera that throughout Europe are represented by strict rheobionts only. In both cases, species richness is strongly reduced with increasing latitude, and in Fennoscandia only single species restricted to the southernmost parts of that area are found. The

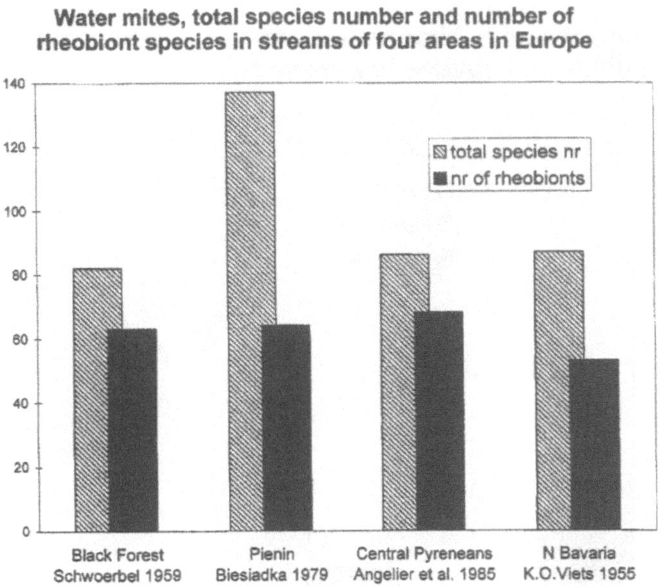

Fig. 7 - Total species number and number of rheobiont species in the streams of four areas in Europe.

Central and North European fauna of *Protzia* and *Torrenticola* consists exclusively of taxa that are found in Southern Europe as well. These areas are not characterized by distinct zoogeographic elements, but colonized by a 'diluted mediterranean fauna'. A similar scenario can be observed in numerous other rheobiont water mite genera and families, and is at most slightly balanced by opposite tendencies in groups with different ecological requirements. Out of the more than 900 water mite species listed in the last inventory of the European fauna (Viets 1978), only some 19 (3%) are restricted to Fennoscandia, all lake- and pool-dwellers, and more than half of them of uncertain taxonomic value. At present, lentic species, e.g. most species of the family Arrenu-

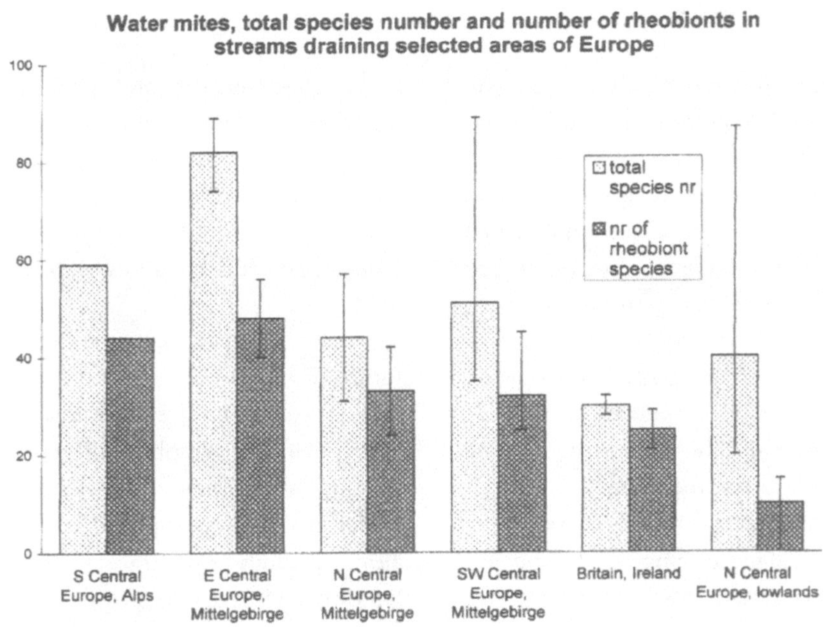

Fig. 8 - Mean values (and variability range) of total water mite species number and number of rheobionts in running water catchments in selected areas of Europe. Data are critically selected and compiled from the following publications: Southern Central Europe, Alps: Walter 1944; Viets 1955; Eastern Central Europe: Láska 1963; Biesiadka 1974; Southernwestern Central Europe: Schwoerbel 1959 and 1964; Meyer 1986 and 1994, this study; Northern Central Europe, Mittelgebirge: Viets 1957; Viets 1966; Britain, Ireland: Gledhill 1973; Conroy 1978; Northern Central Europe, lowlands: Biesiadka 1970; Böttger and Ullrich 1974; Böttger and Hoerschelmann 1991; Böttger and Martin 1995; Martin 1996 and 1997; Martin and Speth 1996.

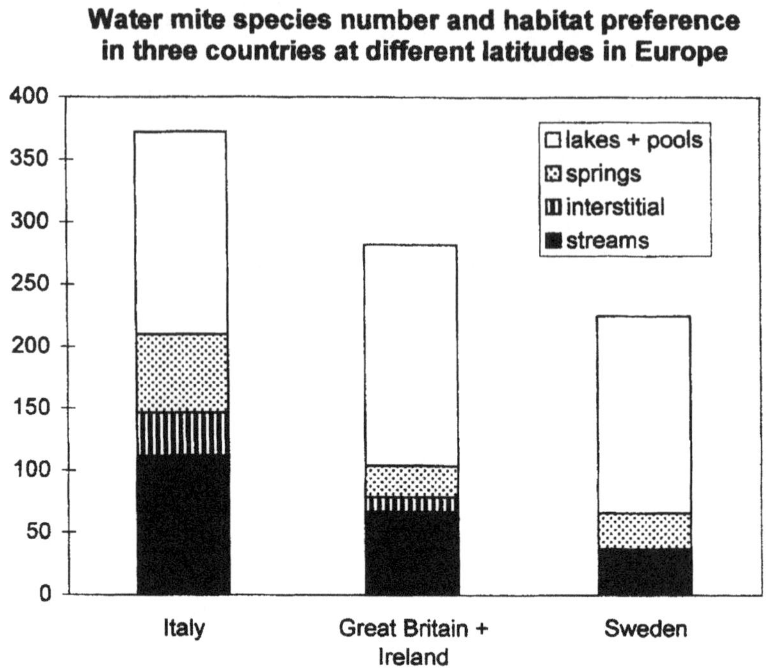

Fig. 9 - Total number and ecological composition of the national faunas of Italy (Di Sabatino and Gerecke, *unpubl.*), Great Britain and Ireland (Gledhill and Viets 1976), and Sweden (Lundblad 1968).

ridae, are recorded with nearly equal frequency in Southern, Central and Northern Europe (fig. 9). However, this is a preliminary observation as habitats of this type are poorly investigated in Southern Europe, and faunistic studies in this area continuously add new records of species that so far appeared to be restricted to more northern parts (Smit *et al.* 2000; Gerecke 2000).

Former authors (e.g. Thor 1926; Viets 1940) tried to explain the reduced species richness of the Northern European lotic water mite fauna with a restricted postglacial recolonization capacity due to a hypothesized absence of parasitic-phoretic larval stages in running water species. New evidence on parasitism in stream water mites does not support this theory. Obviously, the distribution of this fauna is controlled by the present climatic situation rather than by historical factors. Conditions for successfully completing the life cycle are reduced with increasing duration of snow and ice cover: the summer season might become too short for the production of sufficient numbers of offspring, the winter season too harsh for survival of adults. Consequently, with increasing latitude, more and more species would reach their distribution borderline when the individual survival probability falls below a certain limit.

Two aspects merit particular attention for future research.

1) we should get more insight into intra- and interspecific geographic differences of life cycles in lotic water mites. The differences in habitat preference between populations of one species along a south-north transect as documented by Lundblad (1968), are probably accompanied by differences in duration of life cycle stages and host preference. As is suggested by the new observations on overwintering adults presented here, former reconstructions of water mite life cycles are probably more schematic than the natural reality. More consistant information on

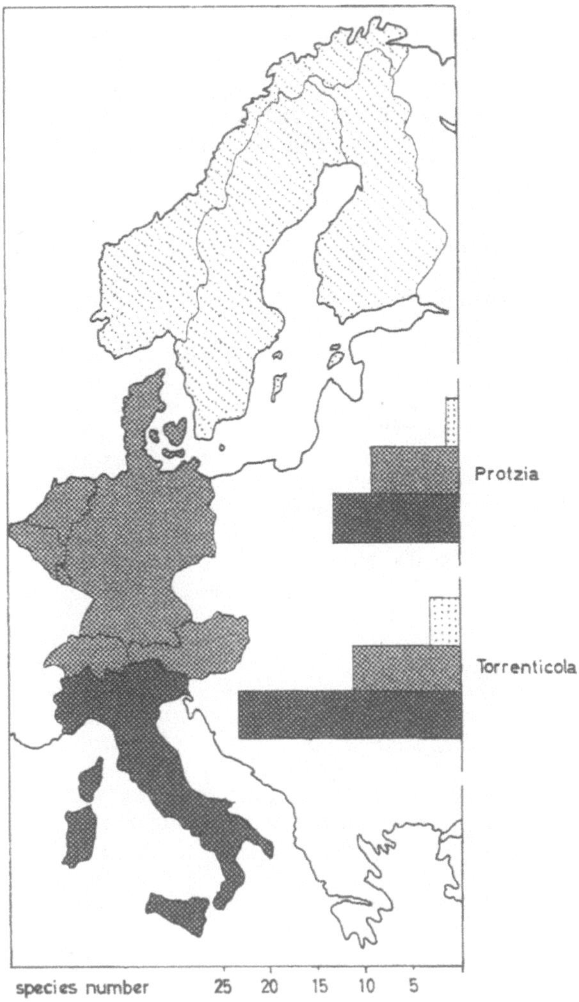

Fig. 10 - The species number of the genera *Protzia* Piersig, 1896 and *Torrenticola* Piersig, 1896 in Southern, Central and Northern Europe (revised data from Di Sabatino and Gerecke *unpubl.*; Viets 1978; and Lundblad 1962).

this question could contribute to our understanding of both geographic variability in water mites, and of their significance as indicators of environmental quality. If taxonomic research were paralleled by studies of this type, we could expect a noteworthy improvement of our understanding of speciation in this group of mites.

2) As illustrated in figure 9, no interstitial water mites are recorded from Scandinavia. A generally accepted scenario interpretes the absence of interstitial and groundwater species from selected areas as a consequence of delayed recolonization of formerly glaciated areas (see Strayer 1994). However, due to the flight radius of their hosts, interstitial water mites should be pioneer colonizers of such areas. Faunistic inventories of the interstitial of selected areas in Scandinavia are therefore desireable for a better understanding of the distribution and colonization capacity of groundwater-dwelling mites.

Acknowledgements

The field work and parts of the basic taxonomic studies for this investigation were financed by the Deutsche Forschungsgemeinschaft (DFG) in a research project on the water mite genus *Atractides*. Furthermore, I am very thankful to Terry Gledhill for critically reading a former draft of this paper and improving the language, and to Antonio Di Sabatino, Peter Martin and Jürgen Schwoerbel for helpful discussions. Antonio Di Sabatino assisted also with the preparation of figure 2, and, together with Reinhart Schuster, as watchful and constructive referees of the submitted manuscript. Martin Buddeberg, Andreas Manz and Klaus Weissert allowed me kindly to publish data from their state examen theses.

References

Agster G., 1986 - Wasser- und Grundwasserhaushalt der Einzugsgebiete des Schönbuchs in Abhängigkeit von Waldbestand und Untergrund. In: Einsele G. (ed.), Das landschaftsökologische Forschungsprojekt Naturpark Schönbuch. - VCH Verlagsgesellschaft Weinheim: 85-112.

Agster G., Einsele G., 1986 - Geogene und atmogene Einflüsse auf die Beschaffenheit der Grund- und Bachwässer im Schönbuch. In: Einsele G. (ed.), Das landschaftsökologische Forschungsprojekt Naturpark Schönbuch. - VCH Verlagsgesellschaft Weinheim: 325-342.

Angelier E., Angelier M.-L., Lauga, J., 1985 - Recherches sur l'ecologie des Hydracariens (Hydrachnellae, Acari) dans les eaux courantes. *Annales de Limnologie*, 21(1): 25-64.

Arnold W., 1986 - Der Wald im Naturpark Schönbuch. In: Einsele G. (ed.), Das landschaftsökologische Forschungsprojekt Naturpark Schönbuch. VCH Verlagsgesellschaft Weinheim: 67-74.

Biesiadka E., 1970 - Wodopójki (Hydracarina) dolnego biegu rzeki Welny. *Fragmenta Faunistica*, 16(5): 43-55.

Biesiadka E., 1974 - Wodopójki (Hydracarina) rzeki Raby i niektórych jej doplywów. *Acta Hydrobiologica*, 16(1): 31-50.

Biesiadka E. 1979 - Wodopójki (Hydracarina) Pienin. *Fragmenta Faunistica*, 24(4): 97-173.

Böttger K., Hoerschelmann U., 1991 - Zur Faunistik und Ökologie der Wassermilben (Hydrachnidia, Actinedida, Actinotrichida, Acari) des norddeutschen Tieflandsbaches Kossau. Limnologische Studien im Naturschutzgebiet Kossautal (Schleswig-Holstein) III. *Faunistisch-ökologische Mitteilungen*, 6(5/6): 219-228.

Böttger K., Martin P., 1995 - Faunistisch-ökologische Untersuchungen an den Wassermilben (Hydrachnidia, Acari) dreier kleiner Fließgewässer des Norddeutschen Tieflandes, unter besonderer Berücksichtigung der rheobionten Arten. *Limnologica*, 25(1): 61-72.

Böttger K., Ullrich F., 1974 - Wassermilben (Hydrachnellae, Acari) der Eider. Faunistische und biologisch-ökologische Angaben. *Faunistisch-Ökologische Mitteilungen*, 4: 419-435.

Buddeberg M., 1970 - Die Trichopterenlarven des Goldersbachs bei Tübingen. Unpublished state examination thesis University Tübingen: 1-93.

Cicolani B., Di Sabatino A., 1985 - Le idracnelle del Vomano (Gran Sasso): Considerazioni sul loro impiego per una caratterizzazione biologica del corso d'acqua. Atti XIV Congresso nazionale italiano di Entomologia: 169-176.

Conroy J.C., 1978 - The Acari (Hydrachnellae) of the River Finn system and the Oweneagh River, Co., Donegal. *Proceedings of the Royal Irish Academy*, Section B, 78(7): 113-123.

Dittmar H., 1955 - Ein Sauerlandbach. *Archiv für Hydrobiologie*, 50(3/4): 305-552.

Dohet A., Hoffmann L., Gerecke R., submitted - Water mites (Acari, Actinedida, Hydrachnidia) in the rhithron of streams in Luxembourg: community composition and potential use as water quality indicators. *Freshwater Biology*.

Efford I.E., 1966 - Observations on the life-history of three stream-dwelling watermites. *Acarologia*, 8(1): 86-93.

Einsele G. (ed.), 1986 - Das landschaftsökologische Forschungsprojekt Naturpark Schönbuch. Wasser- und Stoffhaushalt, Gio-, Geo- und Forstwirtschaftliche Studien in Südwestdeutschland. VCH Verlagsgesellschaft Weinheim: 636 pp.

Einsele G., Agster G., 1986 - Überblick zur Geologie und Morphologie des Schönbuchs. In: Einsele G. (ed.), Das landschaftsökologische Forschungsprojekt Naturpark Schönbuch - VCH Verlagsgesellschaft Weinheim: 1-14.

Ferrito V., 1994 - Les macroinvertébrés benthiques de la rivière Simeto (Sicile) et de quelques-uns de ses affluents. *Annales de Limnologie,* 30(1): 33-56.

Gerecke R., 1999 - Further studies on hydryphantoid water mites in the W Palaearctic region (Acari, Actinedida). *Archiv für Hydrobiologie,* Supplement 121/2, Monographic Studies: 119-158.

Gerecke R., 2000 - New Data on the Morphology and Distribution of *Mesobates forcipatus* Thor, 1901, and Observations on the Systematics of *Atractides*-like Water Mites (Acari, Actinedida, Hygrobatidae). *Aquatic Insects,* 22(3): 209-217.

Gerecke R., *in prep.* - The water mites of the genus *Atractides* (Acari: Parasitengona: Hygrobatidae) in the W palaearctic region. A review.

Gerecke R., Schwoerbel J., 1991 - Water Quality and Water-Mites in the Upper Danube Region, 1959-1984. In: Dusbabek F., Bukva V. (eds.), Modern Acarology, 1 (Academia) Prague (SPB Academic) The Hague: 483-491.

Gledhill T., 1970 - The genus *Wandesia* Schechtel (Hydrachnellae, Acari) and a description of a new species *Wandesia (Wandesia) racovitzai* from England. In: Orghidan T. (ed.), Livre du centenaire Emile G. Racovitza. Éditions de l'Académie de la République Socialiste de Roumanie: 231-238.

Gledhill T., 1973 - The water-mites (Hydrachnellae, Acari) of a stony stream. Proceedings of the III International Congress of Acarology, Prague, 1971. Academia, Publishing House of the Czechoslovak Academy of Sciences: 159-167.

Gledhill T., Viets K.O., 1976 - A synonymic and bibliographic check-list of the freshwater mites (Hydrachnellae and Limnohalacaridae, Acari) recorded from Great Britain and Ireland. *Freshwater Biological Association, Occasional Publications* 1: 59 pp.

Hofmann K., Klee O., 1986 - Ökologie der Oligochaeten unter verschiedenen Milieubedingungen eines oligosaproben Gewässersystems. In: Einsele G. (ed.), Das landschaftsökologische Forschungsprojekt Naturpark Schönbuch. VCH Verlagsgesellschaft Weinheim: 567-612.

Janz H., 1983 - Die Ostracoden (Crustacea) des Schönbuchs bei Tübingen. *Jahreshefte des Vereins für Naturkunde in Württemberg*, 138: 245-259.

Janz H., 1988 - Untersuchungen über die Zusammensetzung der Bodenfauna von Fallaubtümpeln im Naturpark Schönbuch in Abhängigkeit vom Zersetzungsgrad des Rotbuchenlaubs. PhD thesis, University of Tübingen: 201 pp.

Kowalik W., Biesiadka E., 1981 - Occurence of water mites (Hydracarina) in the River Wieprz polluted with domestic-industry sewage. *Acta Hydrobiologica*, 23(4): 331-348.

Kracht V., Klee O., 1986 - Der Stoffhaushalt in der fließenden Welle eines oligosaproben Bachsystems im jahreszeitlichen Wechsel. In: Einsele G. (ed.), Das landschaftsökologische Forschungsprojekt Naturpark Schönbuch. VCH Verlagsgesellschaft Weinheim: 549-566.

Láska F., 1963 - Der Einfluß der Orava-Talsperre auf die Wassermilbenfauna der Orava und ihrer Zuflüsse. *Sborník Vysoké ´skoly chemicko-technologické v Praze, Technologie vody*, 7(2): 139-170.

Lundblad O., 1962 - Die Hydracarinen Schwedens. II. *Arkiv för Zoologi*, (2) 14(1): 635 pp.

Lundblad O., 1968 - Die Hydracarinen Schwedens. III. *Arkiv för Zoologi*, (2) 21(1): 633 pp.

Manz A., 1970 - Die Plecopterenlarven des Goldersbachs bei Tübingen. Unpuplished state examination thesis University Tübingen: 50 pp.

Martin P., 1996 - Faunistisch-ökologische Benthosstudien an den Wassermilben (Hydrachnidia, Acari) zweier Bäche des Norddeutschen Tieflandes (Ostholsteinisches Hügelland, Schleswig-Holstein). *Faunistisch-Ökologische Mitteilungen,* 7: 153-167.

Martin P., 1997 - Zur Faunistik und Substratpräferenz der Wassermilben (Hydrachnidia, Acari) zweier durch feine mineralische Substrate geprägten Bäche in Schleswig-Holstein - *Faunistisch-Ökologische Mitteilungen,* 7: 221-237.

Martin P., 1998 - Zur Autökologie der Wassermilben (Hydrachnidia, Acari) zweier norddeutscher Tieflandbäche. PhD thesis Univ. Kiel: 269 pp.

Martin P., Speth S., 1996 - Die substratspezifische Verteilung der Hydrachnidia (Acari) zweier Bäche in jungglazialen Sandergebieten Schleswig-Holsteins (Osterau und Rodenbek). *Limnologica,* 26(1): 93-103.

Meyer E., 1986 - Die Wassermilben (Hydrachnellae, Acari) in den Fließgewässern um Freiburg i. Br. *Mitteilungen des Badischen Landesvereins für Naturkunde und Naturschutz,* N. F., 14(1): 147-206.

Meyer E., 1994 - Species composition and seasonal dynamics of water mites (Hydracarina) in a mountain stream (Steina, Black Forest, southern Germany). *Hydrobiologia,* 288: 107-117.

Meyer E., Schwoerbel J., 1981 - Untersuchungen zur Phänologie der Wassermilben des Mindelsees. *Archiv für Hydrobiologie,* Supplement, 59: 192-251.

Nickel E. 1987 - Besiedlung und Abbau von Buchenlaub durch die Bodenfauna - synökologische Untersuchungen an feuchten und trockenen Standorten des Naturparks Schönbuch. PhD thesis University of Tübingen: 330 pp.

Nocentini A.M., 1960 - Hydrachnellae del Lago di Mergozzo. *Memorie dell'Istituto Italiano di Idrobiologia,* 12: 245-287.

Rohlf J.F., 1990 - NTSYS-pc, numerical taxonomy and multivariate analysis, version 1.6. Applied Biostatistics Inc., Setauket, New York: 60 pp.

Schwoerbel J., 1959 - Ökologische und tiergeographische Untersuchungen über die Milben (Acari, Hydrachnellae) der Quellen und Bäche des südlichen Schwarzwaldes und seiner Randgebiete. *Archiv für Hydrobiologie,* Supplement 24, 3(3-4): 385-546.

Schwoerbel J., 1964 - Die Wassermilben (Hydrachnellae und Limnohalacaridae) als Indikatoren einer biozönotischen Gliederung von Breg und Brigach sowie der obersten Donau. *Archiv für Hydrobiologie,* Supplement 27, 1(4): 386-417.

Schwoerbel J. 1989 - Methoden der Hydrobiologie. Süßwasserbiologie. Gustav Fischer Verlag, Stuttgart, Jena: 261 pp.

Smit H., Gerecke R., Di Sabatino, A., 2000 - A catalogue of water mites of the superfamily Arrenuroidea (Acari: Hydrachnidia) from the Mediterranean countries. *Archiv für Hydrobiologie,* Supplement 121/3-4: 201-267.

Strayer D.L., 1994 - Limits to Biological Distributions in Groundwater. In: Gibert J., Danielopol D.L., Stanford J. (eds.), Groundwater ecology. Academic Press, London: 287-313.

Thor S., 1926 - Über das glaziale Relikt *Hygrobates albinus* Sig Thor und die Zeit der Verbreitung dieses Tieres. *Archiv für Hydrobiologie,* 16(3): 399-416.

Tuzovskij P.V., 1981 - Postembrionalnye stadii vodjanogo klesca *Torrenticola amplexa* (Koenike, 1908) (Acariformes, Torrenticolidae). *Entomologiceskoe Obozrenie,* 60(4): 205-212.

Ullrich F., 1976 - Biologisch-ökologische Studien an rheophilen Wassermilben (Hydrachnellae, Acari), unter besonderer Berücksichtigung von *Sperchon setiger* (Thor 1898). PhD thesis University of Kiel: 241 pp.

Van der Hammen H., Smit H., 1996 - The water mites (Acari, Hydrachnidia) of streams in The Netherlands: distribution and ecological aspects on a regional scale. *Netherlands Journal of Aquatic Ecology,* 30(2-3): 175-185.

Viets K., 1940 - Ausbreitungswege und nacheiszeitliche Verbreitung der Kaltwasser und Strömung liebenden Wassermilben in Europa. *Archiv für Hydrobiologie,* 37(2): 278-319.

Viets K.O., 1955 - Wassermilben aus Nordbayern (Hydrachnellae und Porohalacaridae, Acari). *Abhandlungen der Bayerischen Akademie der Wissenschaften, Mathematisch-Naturwissenschaftliche Klasse,* N.F., 73: 106 pp.

Viets K.O., 1957 - Wassermilben aus der Salmoniden-Region von Harzbächen. *Abhandlungen des naturwissenschaftlichen Vereins zu Bremen,* 35(1): 135-161.

Viets K.O., 1966 - Zur Analyse der Wassermilben-Fauna eines Harz-Baches. *Internationale Revue der gesamten Hydrobiologie,* 51(3): 451-460.

Viets K.O., 1978 - Hydracarina. In: Illies J. (ed.), Limnofauna Europaea. Fischer Verlag, Stuttgart: 154-181.

Vinçon G., 1987 - Etude hydrobiologique de la vallée d'Ossau (Pyrénées-Atlantiques) II. Le milieu et la structure du peuplement benthique. *Annales de Limnologie,* 23(3): 225-243.

Walter C., 1944 - Die Hydracarinen der Ybbs. 1. Teil. *Internationale Revue für Hydrobiologie und Hydrographie,* 43(4-6): 281-367.

Weissert K., 1970 - Die Ephemeropterenlarven des Goldersbachs bei Tübingen, mit Bemerkungen zur Biologie, Ökologie und Entwicklung. Unpuplished state examination thesis University Tübingen: 1-70.

The biodiversity of Neotropical water mites

Tom Goldschmidt
Department of Limnology, Institute of Zoology, University of Münster, Hüfferstraße 1, D-48149 Münster, Germany

Key words: diversity, species richness, water mites, neotropics, Costa Rica, freshwater, global patterns, extrapolation, database.

Abstract: The first studies on water mites from Central and South America date back to the 1880s, and although 1,360 species in 173 genera have been recorded so far, our knowledge of the diversity of this group in the Neotropic is rather incomplete. In order to estimate the total number of water mite species in this faunal region, I surveyed the state of the art in the exploration of neotropical water mites. The survey was based on a database evaluating 217 publications on water mites from 1,198 sample sites from Mexico to Patagonia. A graph of cumulative species number shows a continuous increase without reaching a plateau. Moreover, the analysis proved great differences in the knowledge of the fauna of different Latin American countries and few faunistic similarities between countries. A frequency analysis demonstrates that 45% of the known neotropical species have been found only at one locality. The analyses indicate that our knowledge of the neotropical water mite fauna is still very fragmentary. In addition, I examined the comparably well known water mite fauna of Costa Rica. Within eleven subfamilies surveyed in detail, 75% of the species were new to science. Extrapolating the Costa Rican situation to the entire Neotropic, I suggest a species number fourfold higher than currently known.

Introduction

The rapid and massive destruction of natural habitats, especially in the tropics, has drawn much attention to the vanishing diversity. Likewise, the consciousness developed, that a comprehensive knowledge of the present biodiversity is indispensable for the planning of their conservation and sustainable use (May 1988 and 1990; Linsenmair 1990; Wilson 1992; Colwell and Coddington 1994; Pimm *et al.* 1995; Peterson and Slade 1998). Moreover, biodiversity is an issue of growing public interest and received much recognition when the 'Convention on Biological Diversity' was passed at the environmental conference in Rio de Janeiro in 1992. Nevertheless, knowledge of basic data on biodiversity is still surprisingly low for most habitats and taxa (May 1992).

Apart from being a measure of species richness, the term biodiversity also comprises criteria like evenness, structure of dominance and abundance, range size, distribution patterns, degree of endemism, genetic diversity, character diversity, population ecology and ecological function (Begon *et al.* 1996, Barthlott *et al.* 1999). In the following I will only refer to the biodiversity component of species numbers.

The neotropical faunal region, following Illies (1971), comprises South America, the Westindies and Central America including the Southern part of Mexico. For practical reasons I will include Mexico as a whole.

F. Bernini, R. Nannelli, G. Nuzzaci, E. de Lillo (eds.), Acarid Phylogeny and Evolution. Adaptations in mites and ticks, 91-99.

General state of knowledge

How many species are there?

Systematic biology 'officially' started in 1758 with the 10th edition of Carl von Linné's 'Systema naturae'; some 9,000 species of animals and plants are described there. The systematic description of water mites originates about 20 years later, when Otto Friedrich Müller described 50 species from Denmark. Two and a half centuries later approximately 1.8 million plant and animal species have been described (Stork 1988; May 1988 and 1990; Barthlott *et al.* 1999). Within these there are about 1 million insects, 45,000 terrestrial mites and about 5,000 water mites (Smith and Cook 1991).

Within the last 15 years several estimates of the global total species number have been published, varying from 2 million to 100 million species (Stork 1988 and 1993; May 1988 and 1990; Linsenmair 1990; Hodkinson and Casson 1991; Gaston 1992; Rödder *et al.* 1993; Götting 1997; Pimm and Raven, 2000). Especially the extrapolations of canopy fogging data from the tropics resulted in very high total estimates (Erwin 1982 and 1988; Stork 1988; Morse *et al.* 1988). However there are no data on global species richness available for freshwater ecosystems. Although 'biodiversity' is also a frequently used and fashionable expression in freshwater studies, it is difficult to find species number estimates at a global scale. In the majority of studies diversity is strictly used in the sense of α-diversity, representing species numbers at single locations (Stout and Vandermeer 1975; Flowers 1991; Dumont and Segers 1996; Townsend *et al.* 1997; Ward *et al.* 1998; Toham and Teugels 1999).

Global patterns of biodiversity

In terrestrial ecosystems there is a strong latitudinal gradient in species richness, with a diversity maximum in the humid tropical region (Gaston 1992). Fifty percent (maybe 90%) of all species live in tropical lowland rainforest, and α-diversity is 10- to 100-fold higher in tropical terrestrial communities than in temperate ecosystems (Linsenmair 1990). These general patterns of diversity cannot be confirmed for aquatic habitats.

Several studies of stream insect communities showed a very heterogeneous picture of diversity patterns (Stout and Vandermeer 1975). Whereas species richness was higher for insects in tropical streams (Costa Rica) than in mid-latitude streams (US) (Stout and Vandermeer 1975), Stanford and Ward (1983) found tropical aquatic insects less diverse than temperate ones. Flowers (1991) showed that results of tropical-temperate diversity comparison largely depend on the group of insects investigated. For example, Riffle beetles are more diverse in the tropics (Brown 1981), whereas Plecoptera are more diverse in temperate streams (Zwick 1986), and Panamanian mayfly samples fall in the mid-range of diversity values from temperate areas (Wolda and Flowers 1985). The overall pattern for streams is that α-diversity of invertebrates varies widely with broad overlap between temperate and tropical regions. Hence, there is no 'diversity gap' between temperate and tropical stream ecosystems (Flowers 1991). Similar results were reported for pelagic Cladocera (Dumont and Segers 1996) and benthic Copepoda (Rundle *et al.* 2000), and are mainly explained by the lack of latitudinal microhabitat variation. 'Latitudinal trends in the diversity of freshwater meiofauna are virtually unknown and an investigation of such gradients is significantly hampered by lack of taxonomic information for the tropics' (Rundle *et al.* 2000, p. 126). Very little is known about species diversity in groundwater (Ward *et al.* 1998), springregions and interstitial habitats (Götting 1997). Therefore a lot of new species may be expected in these habitats. No tropical-temperate diversity comparisons exist for water mites, but the few α-diversity data for neotropical water mites (Böttger 1980) range a little below those from North America (Smith and Cook 1991).

Diversity estimates for Neotropical water mites

Water mites depend on aquatic insects as prey and hosts, therefore the above considerations on species richness of aquatic insects should also apply to water mites. One would expect a large number of undescribed neotropical water mite species, although perhaps not as large as for canopy-dwelling insects. The estimates I will try to make in the following are based on two fundamentals: the analysis of a database containing 1,198 datasets from 217 publications on water mites from Latin America and my own investigations on Costa Rican water mites.

Cumulative species number

The history of water mite research in Latin America can be subdivided into several phases (fig. 1). The first neotropical water mites were described in the 1880s from Mexico (Alfredo Dugés 1884), Guatemala (Otto Stoll 1886-1893) and Argentina (Antonio Berlese 1888). The first steep increase is marked by the description of 340 species from Southern Brazil and Paraguay published by Olov Lundblad (1941-1944). Publications by Lundblad (1953) on water mites from Colombia and Karl Viets (1954) on material from the Amazon region mark two further steps. During the following 25 years the knowledge of the neotropical fauna grew very slowly until 1980, when David Cook published 236 new species from Mexico, Costa Rica and Argentina, followed by 102 new species from Chile in 1988. Since then the number of known Latin American species rose slowly but continously to a total of 1,360 species. The graph is not showing a plateau yet, which should be the case if the description of the fauna was close to being complete (fig. 1).

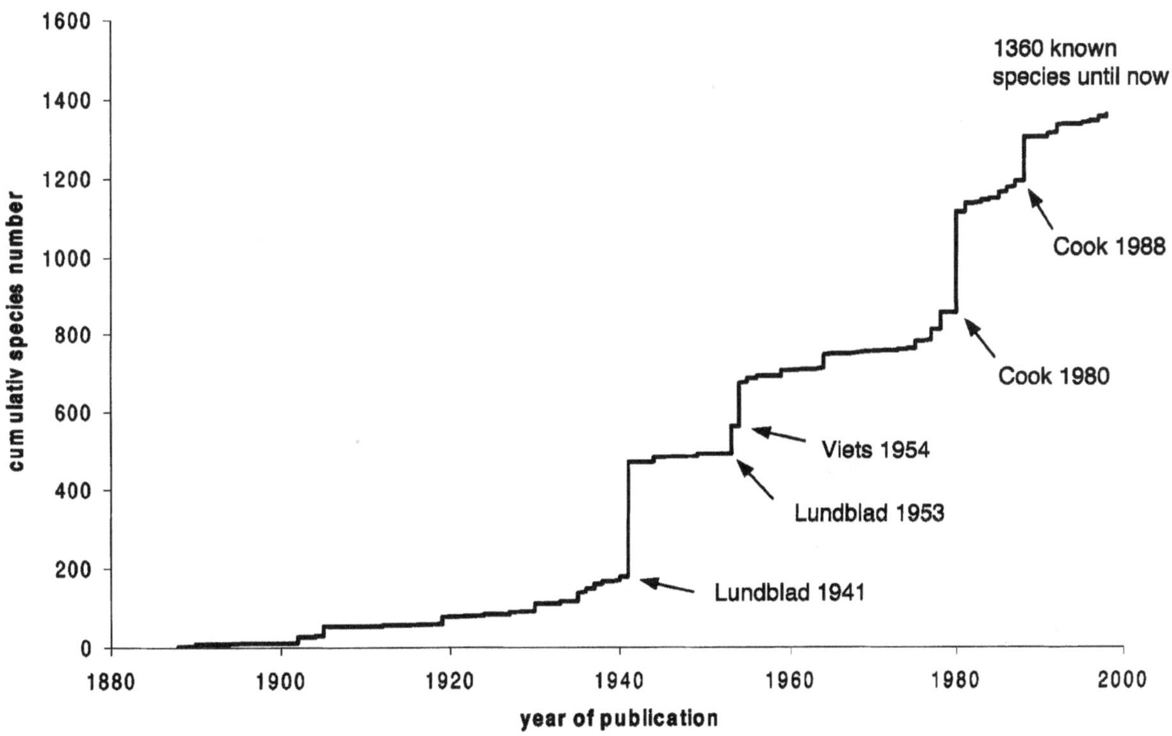

Fig. 1 - Graph of cumulative species number. The annotations in the graph refer to the most extensive publications on neotropical water mites.

Data availability in different regions

The knowledge of the water mite fauna is very heterogenous for the different regions of Latin America (fig. 2). The regions best investigated are Southern Brazil and Southern Mexico (tab. 1). Information on the water mite fauna is completely lacking for Northern Mexico, Belize, Nicaragua, French Guiana, East-Central Brazil, Jamaica and most of the Lesser Antilles.

Similarity of the better known faunas

For the comparably 'well-known' faunas (15 regions and countries with more than 50 known species) I calculated a similarity matrix, showing

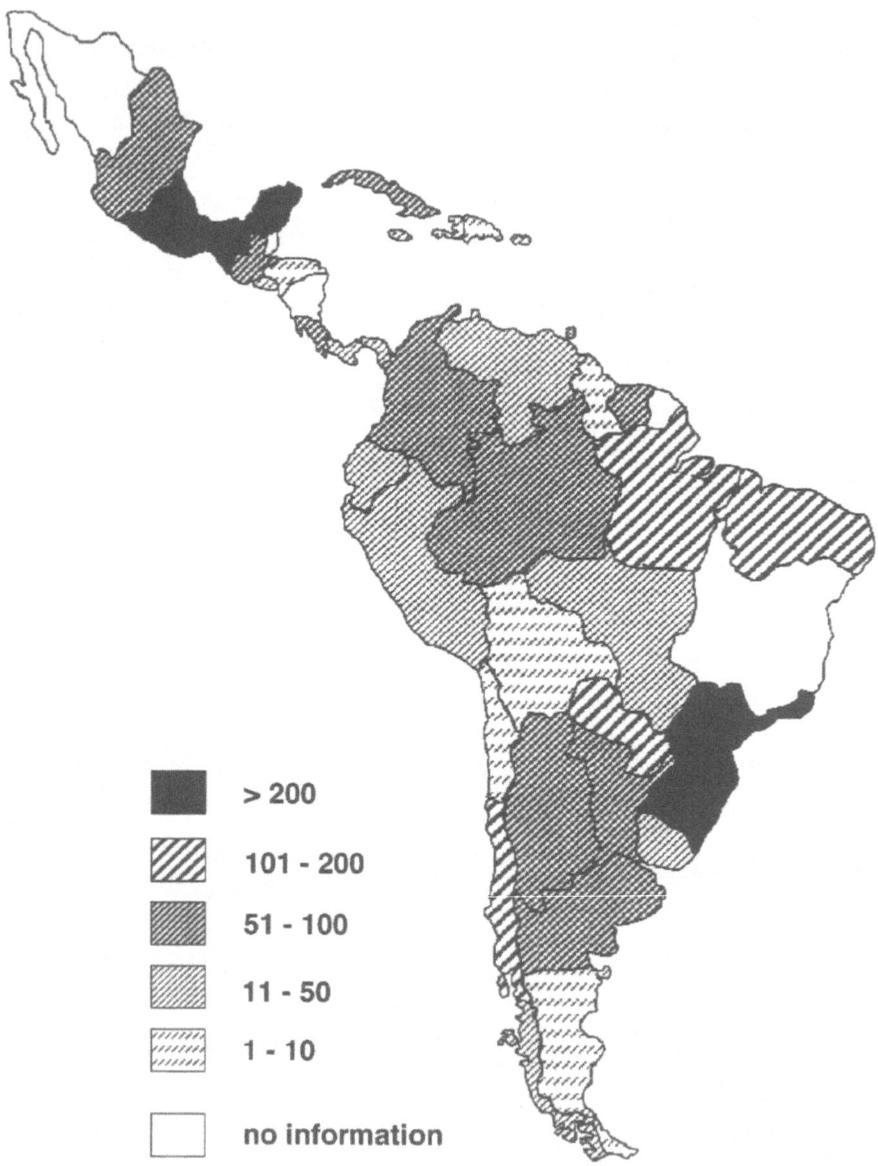

Fig. 2 - State of knowledge. Five different shades mark the different numbers of known species. The darkest grey represent more than 200 species, the lighter shades of grey 101-200, 51-100, 11-50 and 1-10. The white areas represent regions or countries without any information on the water mite fauna.

Tab. 1 - The number of known water mite species from the countries and regions of Latin America. The Dutch Caribbean Islands comprise Aruba, Bonaire and Curacao.

Latin American countries and regions	water mite species
Central Mexico	81
Southern Mexico	221
Guatemala	63
El Salvador	3
Honduras	1
Costa Rica	91
Panama	8
Cuba	39
Domenican Republic	6
Haiti	8
St. Vincent	4
Dutch Caribbean Islands	4
Trinidad	4
Colombia	91
Venezuela	31
Surinam	51
Guyana	3
Ecuador	15
Peru	37
Northwestern Brazil	56
North-Central Brazil	148
Northeastern Brazil	153
Centralwestern Brazil	28
Southern Brazil	267
Bolivia	5
Paraguay	190
Uruguay	20
Northern Chile	7
Central Chile	178
Southern Chile	16
Northwestern Argentina	91
Northeastern Argentina	53
Central Argentina	76
Southern Argentina	7
total	1360

the percentage of common species within the complete species list of two regions (tab. 2). Even the better known faunas show few similarities. There are three groups of regions with more than 10% similarity: (1) Central Mexico, Southern Mexico and Costa Rica, (2) Paraguay, Surinam, Northwestern, North-Central, Northeastern and Southern Brazil and (3) Northeastern, Northwestern and Central Argentina and Central Chile (tab. 2). The highest percentage of similar species is 24% between Paraguay and Southern Brazil followed by 19% between Central and Southern Mexico. Especially between Central and South America the similarity is very low and ranges between 1 and 2% between most areas. The two best known regions, Southern Mexico and Southern Brazil, have just 1% of their water mite species in common. The similarity of the Costa Rican fauna with the Southern and Central Mexican is 15 and 12% respectively. On the other hand, the similarity to Colombia, which is not farther away than Mexico, is only 4%, and only 1% of species are shared with Northern Brazil and Surinam.

Frequency analysis

A possibility to assess the completeness of an inventory is to check, whether individual species have been sampled repeatedly during a particular investigation. A census may be regarded as complete when all species have been observed 'multiple' times (Colwell and Coddington 1994). Therefore, I analyzed the frequency of the known species within the Latin American database. Most of the 1,360 species known from Latin America were rarely found; 611 (45%) were single findings, 258 (19%) were found only twice, just 5 species were quite abundant with more than 50 localities (fig. 3).

Extrapolation from the ratio of unknown species of Costa Rican water mites

The conclusion from the above analyses is that the neotropical water mite fauna is far from being completely described. Within my PhD I collected approximately 20,000 water mites from different freshwater habitats all over Costa Rica (Goldschmidt 2001). In order to calculate a probable total species number, I selected 11 subfamilies from Costa Rica for detailed taxonomic analysis. Within these I identified 44 species of which 33 species (75%) are new to science. Assuming that this ratio is representative for all neotropical water mites, an extrapolation of the 1,360 known species yields a possible total species number of 5,440.

Tab. 2 - Similarity matrix. Similarities between the Latin American water mite faunas with more than 50 known species. The similarity is calculated by the formula C/(A+B)-C; C=number of common species, A=number of known species in country A, B=number of known species in country B.

similarity common species	C Mexico	S Mexico	Guatemala	Costa Rica	Colombia	Surinam	NW Brazil	NC Brazil	NE Brazil	S Brazil	Paraguay	NW Argentina	NE Argentina	C Argentina	C Chile
C Mexico		**19%**	3%	**12%**	2%	-	1%	1%	1%	1%	1%	2%	3%	3%	<1%
S Mexico	48		4%	**15%**	2%	1%	1%	1%	2%	1%	2%	2%	1%	3%	1%
Guatemala	5	11		5%	6%	3%	-	1%	2%	2%	2%	1%	-	2%	<1%
Costa Rica	18	41	8		4%	1%	1%	1%	1%	1%	1%	2%	1%	2%	<1%
Colombia	4	7	9	7		1%	2%	2%	4%	4%	4%	6%	4%	6%	3%
Surinam	-	4	3	2	2		6%	6%	**13%**	5%	8%	1%	8%	5%	2%
NW Brazil	2	4	-	1	3	6		**14%**	8%	4%	6%	3%	2%	3%	1%
NC Brazil	3	7	3	3	5	12	25		**13%**	7%	7%	2%	5%	4%	1%
NE Brazil	3	8	4	2	9	23	15	34		**15%**	**15%**	3%	8%	5%	2%
S Brazil	5	7	6	4	14	16	11	26	54		**24%**	4%	7%	4%	2%
Paraguay	4	9	6	2	11	17	15	21	46	89		4%	9%	6%	2%
NW Argentina	3	7	2	3	11	1	4	5	7	13	11		8%	**12%**	2%
NE Argentina	4	4	-	1	5	8	2	9	16	21	20	11		**11%**	5%
C Argentina	5	8	3	4	9	6	4	9	11	14	15	18	13		**15%**
C Chile	1	4	1	1	8	4	2	3	7	8	8	6	11	34	

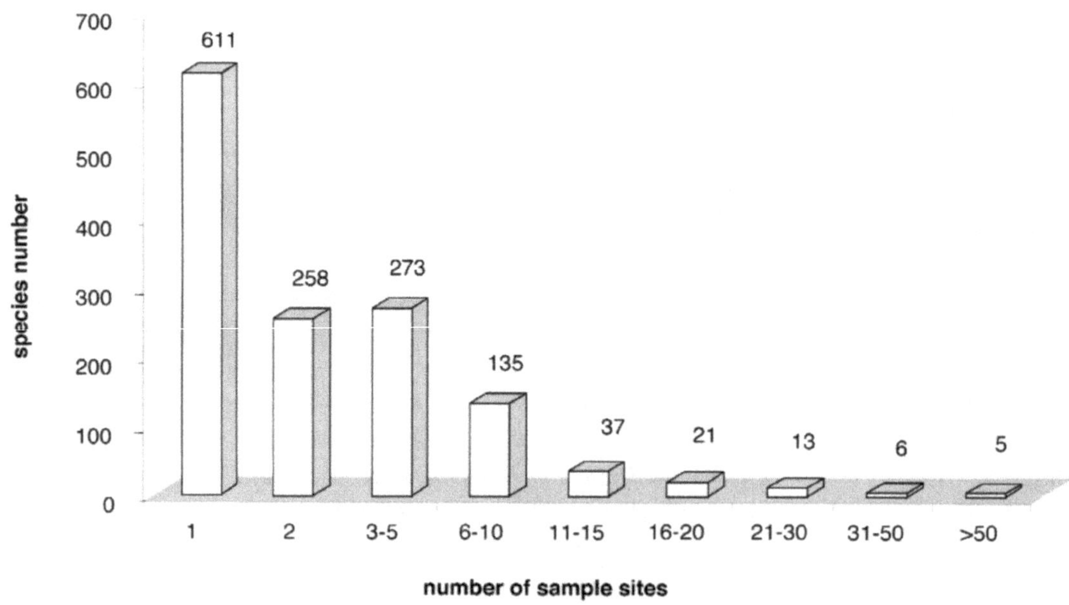

Fig. 3 - Frequency of the known neotropical water mites. The columns show the number of species reported from a certain number of localities.

Discussion

The analyses of a Latin American water mite database demonstrates that the total species number is probably much higher than currently known. Large differences, even between better explored faunas, could be caused by a high degree of endemism, but more likely are mainly artificial, showing the poor knowledge of the fauna. In both cases the heterogeneity suggests that a large proportion of species is still unknown in Latin America. The Neotropical region (especially South America itself) shows a strong particularity, as it has been isolated since the mesozoic (65 million years ago), until the Central American landbridge arose 2 million years ago (Rich and Rich 1983). The clear difference between the Central and South American fauna is probably also due to historical reasons.

Several publications deal with calculation methods to extrapolate inventory data (Heltshe and Forrester 1983; Chao 1984; Colwell and Coddington 1994). However, most of these methods produce reasonable results only when the inventory is quite comprehensive, their main aim is to calculate rather than count the last remaining species of an inventory. Therefore, these procedures will result in underestimates of the true richness when the inventory is too fragmentary (Colwell and Coddington 1994) as is probably the case for neotropical water mites. Most suitable for such situations are extrapolations that are based on the number of rare species (single and double findings) in the sample (Chao 2, second-order jackknife in Colwell and Coddington 1994). Applied to the present data set of neotropical water mites, these calculations lead to a total estimate of about 2,100 to 2,300 species. Other estimates use ratios of unknown species of a certain group in a certain area to extrapolate the global number of species in this group (Hodkinson and Casson 1991). Starting from my own investigation on Costa Rican water mites, this extrapolation model yields about 5,500 Neotropical Water Mites. I believe that this higher estimate is more reliable, because the Costa Rican water mite fauna is relatively well known in comparison to other areas. Therefore the percentage of new species could be much higher in other regions resulting in even higher total numbers.

Outlook

The enormous species richness of the tropics has been an issue of much debate. The two principal explanations discussed are: Equilibrium models, which explain high species numbers by highly specialized niche separation, and non-equilibrium models explaining the possibility of niche overlaps, allowing high species richness, due to stochastic local disturbance (Connell 1978; Linsenmair 1990). Data to test these models (mobility, dispersal, fertility, etc.) are not available for the vast majority of species (Linsenmair 1990). The fact that freshwater communities seem to be an exception to the general rule of high species richness in the tropics may give them a key position in the understanding of the fundamental mechanisms. Besides further taxonomic investigations, future research on neotropical water mites should focus on heterogeneity of water mite diversity, possible hotspots of diversity, degrees of local endemism, and the life history of certain species.

References

Barthlott W., Biedinger N., Braun G., Feig F., Kier G., Mutke J., 1999 - Terminological and methodological aspects of the mapping and analysis of the global biodiversity. *Acta Botanica Fennica*, 162: 103-110.

Begon M.E., Harper J.L., Townsend C.R., 1996 - Ecology, 3rd Edition. Blackwell Science Limited, Oxford: 750 pp.

Berlese A., 1888 - Acari austro-americani quos collegit Aloysius Balzan et illustravit Antonius Berlese. *Bollettino della Società entomologica italiana,* 20, 1-4: 171–222.

Böttger K., 1980 - Qualitative und quantitative Benthon-studien an Bergbächen in Guatemala (Zentral-amerika) unter besonderer Berücksichtigung der Hydrachnellae (Acari). *Archiv für Hydrobiologie*, 88: 96-119.

Brown H.P., 1981 - A distributional survey of the world genera of aquatic dryopoid beetles (Coleoptera: Dryopidae, Elmidae and Psephenidae *sens. lat.*). *Pan-Pacific Entomologist*, 57: 133-148.

Chao A., 1984 - Non-parametric estimation of the number of classes in a population. *Scandinavian Journal of Statistics*, 11: 265-270.

Colwell R.K., Coddington J.A., 1994 - Estimating terrestrial biodiversity through extrapolation. *Philosophical Transactions of the Royal Society of London*, Series B, 345: 101-118.

Connell J.H., 1978 - Diversity in tropical rain forests and coral reefs. *Science*, 199: 1302-1310.

Cook D.R., 1980 - Studies on neotropical water mites. *Memoirs of the American Entomological Institute*, 31: 1-645.

Cook D.R., 1988 - Water mites from Chile. *Memoirs of the American Entomological Institute*, 42: 1-356.

Dugés A., 1884 - Atax alzatei. *La Naturaleza. Periódico cientifico de la Sociedad Méxicana de historia natural*, 6(22): 344-347.

Dumont H.J., Segers H., 1996 - Estimating lacustrine zooplankton species richness and complementarity. *Hydrobiologia*, 341: 125-132.

Erwin T.L., 1982 - Tropical forests: their richness in Coleoptera and other arthropod species. *Coleopterists' Bulletin*, 36: 74-75.

Erwin T.L., 1988 - The tropical forest canopy. The heart of biotic diversity. In: Wilson E.O. (ed.), Biodiversity. National Academy Press, Washington: 123-129.

Flowers R.W., 1991 - Diversity of stream-living insects in northwestern Panamá. *Journal of the North American Benthological Society*, 10: 322-334.

Gaston K.J., 1992 - Regional numbers of insect and plant species. *Functional Ecology*, 6: 243–247.

Goldschmidt T., 2001 - Die Wassermilbenfauna Costa Ricas (Hydrachnidia: Acari) - Systematik, Ökologie, Diversität und Zoogeographie. Dissertation, Universität Münster.

Götting K.-J., 1997 - Über die bekannte und unbekannte Mannigfaltigkeit des Tierreiches. Wieviele Tierarten gibt es eigentlich? *Tierärztliche Umschau*, 52: 470-472.

Heltshe J.F., Forrester N.E., 1983 - Estimating species richness using the jackknife procedure. *Biometrics*, 39: 1-11.

Hodkinson I.D., Casson D., 1991 - A lesser predilection for bugs: Hemiptera (Insecta) diversity in tropical forests. *Biological Journal of the Linnean Society*, 43: 101-109.

Illies J., 1971 - Einführung in die Tiergeographie. G. Fischer Verlag, Stuttgart: 91 pp.

Linsenmair K.E., 1990 - Tropische Biodiversität: Befunde und offene Probleme. *Verhandlungen der Deutschen Zoologischen Gesellschaft*, 83: 245-261.

Lundblad O., 1941-1944 - Die Hydracarinenfauna Südbrasiliens und Paraguays, Erster-Fünfter Teil. *Kungliga Svenska Vetenskapsakademiens Handlingar*, Stockholm, 19/7: 1-183, 20/2: 1-175, 20/5: 1-148, 20/8: 1-171, 20/13: 1-182.

Lundblad O., 1953 - Die Hydracarinenfauna von Colombia. *Arkiv för Zoologi*, Serie 2, 5(8): 435-585.

May R.M., 1988 - How many species are there on earth? *Science*, 241: 1441-1449.

May R.M., 1990 - How many species? *Philosophical Transactions of the Royal Society of London*, Series B, 330: 293-304.

May R.M., 1992 - How many species inhabit the earth? *Scientific American*, 268(10): 18-24.

Morse D.R., Stork N.E., Lawton, J.H., 1988 - Species number, species abundance and body-length relationships of arboreal beetles in Bornean lowland rain forest trees. *Ecological Entomology*, 13: 25-37.

Peterson A.T., Slade N.A., 1998 - Extrapolating inventory results into biodiversity estimates and the importance of stopping rules. *Diversity and Distributions*, 4: 95-105.

Pimm S.L., Raven P., 2000 - Extinction by numbers. *Nature*, 403: 843-845.

Pimm S.L., Russell G.J., Gittleman J.L., Brooks T.M., 1995 - The future of biodiversity. *Science*, 269: 347-350.

Rich P.V., Rich T.H., 1983 - The Central American dispersal route. In: Janzen D.H. (ed.), Costa Rican Natural History. The University of Chicago press, Chicago, London: 12-34.

Rödder G., Ziegler F.-K., Falk E., 1993 - Wie viele Arten? Der Stand der Forschung gegen Ende des Jahrhunderts. *Paläontologische Zeitschrift*, 67/1-2: 215-222.

Rundle S.D., Bilton D.T., Shiozawa D.K., 2000 - Global and regional patterns in lotic meiofauna. *Freshwater Biology*, 44: 123-134.

Smith I.M., Cook D.R., 1991 - Water mites. In: Thorp J.H., Covich A.P. (eds.) Ecology and classification of North American freshwater invertebrates. Academic Press, San Diego: 523-592.

Stanford J.A., Ward J.V., 1983 - Insect species diversity as a function of environmental variability and disturbance in stream ecosystems. In: Barnes J.R., Minshall G.W. (eds.), Stream ecology: application and testing of general ecological theory. Plenum Press, New York: 265-278.

Stoll O., 1886-1893 - Arachnida Acaridea. In: Biologia Centrali - Americana: 1-53.

Stork N.E., 1988 - Insect diversity: facts, fiction and speculation. *Biological Journal of the Linnean Society*, 35: 321-337.

Stork N.E., 1993 - How many species are there? *Biodiversity and Conservation*, 35: 321-337.

Stout J., Vandermeer J., 1975 - Comparison of species richness for stream-inhabiting insects in tropical and mid-latitude streams. *The American Naturalist*, 109: 263-280.

Toham A.K., Teugels G.G., 1999 - First data on an Index of Biotic Integrity (IBI) based on fish assemblages for the assessment of the impact of deforstation in a tropical West African river system. *Hydrobiologia*, 397: 29-38.

Townsend C.R., Scarsbrook M.R., Dolédec S., 1997 - The intermediate disturbance hypothesis, refugia, and biodiversity in streams. *Limnology and Oceanography*, 42: 938-949.

Viets K., 1954 - Wassermilben aus dem Amazonasgebiet. *Schweizerische Zeitschrift für Hydrologie* 16/1, 2: 78-151, 161-247.

Ward J.V., Bretschko G., Brunke M., Danielopol D., Gibert J., Gonser T., Hildrew A.G., 1998 - The boundaries of river systems: the metazoan perspective. *Freshwater Biology*, 40: 531-569.

Wilson E.O., 1992 - Ende der biologischen Vielfalt? Spektrum Akademischer Verlag, Heidelberg, Berlin, New York: 557 pp.

Wolda H., Flowers R.W., 1985 - Seasonality and diversity of mayfly adults (Ephemeroptera) in a 'nonseasonal' tropical environment. *Biotropica*, 17: 330-335.

Zwick P., 1986 - The Bornean species of the stonefly genus *Neoperla* (Plecoptera: Perlidae). *Aquatic Insects*, 8: 1-53.

Life history strategies of *Hygrobates nigromaculatus*, a widespread palaearctic water mite (Acari, Hydrachnidia, Hygrobatidae)

Peter Martin[1], Cornelius Davids[2]
[1] *Zoologisches Institut, Christian-Albrechts-Universität zu Kiel, Olshausenstr. 40, D-24098 Kiel, Germany*
pmartin@zoologie.uni-kiel.de
[2] *Department of Aquatic Ecotoxicology, University of Amsterdam, Kruislaan 320, NL-1098 SM Amsterdam, The Netherlands*
davids@bio.uva.nl

Key words: larval morphology, life cycle differences, parasitic and non-parasitic larvae, stream pool and lake littoral populations, speciation, *Hygrobates* sp. nov. near *nigromaculatus*.

Abstract: Details of the life-cycle and morphology of *Hygrobates nigromaculatus* larvae are presented for two populations which originated from different habitats. One population was derived from pools of a lowland stream and the other was isolated from the littoral zone of a eutrophic Baltic lake. In the population originating from the stream, the larvae live by parasitising chironomids, attaching themselves to single abdominal segments of the host. The larvae of the population inhabiting a Baltic lake are not parasitic, and after hatching, they generally remain in the clutches where they develop into nymphs. Furthermore, the larvae of the two populations also differ significantly with regards to their morphology. So far, *H. nigromaculatus* is the only known water mite species that has been found in both standing and flowing waters and has changed its life-cycle. Due to the differences in their larval morphology and life cycles, *H. nigromaculatus* is considered a lake-living species and its stream-living relative as a still undescribed sibling species preliminary named *H.* sp. nov. near *nigromaculatus*.

Introduction

The water mite *Hygrobates (Hygrobates) nigromaculatus nigromaculatus* Lebert, 1879, is widespread in the Palaearctic, from Western Europe to Siberia (Viets 1978). In contrast to most water mites which are bound to a specific habitat, the occurrence of this species is unusual. *Hygrobates nigromaculatus* is found in stable populations in standing waters (mainly lakes) as well as in pools of running waters. There are some other species whose wide ecological adaptations allow them to live not only in lakes but also in streams (e.g. *Lebertia inaequalis*, *H. longipalpis*). However, these species are limnobiotic and colonize only larger streams, rivers or degraded smaller streams. *Hygrobates nigromaculatus* however, occurs both in the littoral zone of eutrophic lakes and in small or larger running waters which are not subject to human impact. For this reason, among others, the conspecificity of ecologically distinct populations of *H. nigromaculatus* had previously been called into question (Viets 1960). Although marked morphometrical differences (e.g. in length and width) between adult specimens originating from standing and running waters were found, it was impossible to find exact characters for separating the species into different taxa. Van Hezewijk and Davids (1985) studied the life cycle of a population of *H. nigromaculatus* occurring in a Dutch lake and compared it with two other species of the same genus. *Hygrobates nigromaculatus* was found to produce larvae

F. Bernini, R. Nannelli, G. Nuzzaci, E. de Lillo (eds.), Acarid Phylogeny and Evolution. Adaptations in mites and ticks, 101-110.

which, in contrast to most other water mite larvae, do not parasitize. The larvae transform without feeding into the next free developmental stage: the nymph. During a study on water mites from small streams in Northern Germany (Martin 1998), the life cycle of *H. nigromaculatus* was investigated for a population originating from running waters. To compare these results with those of an additional population of *H. nigromaculatus* from standing waters, specimens from a Baltic lake were cultured.

Thus, the first aim of the present study was to compare the morphology of larvae derived from populations of running and standing waters. In addition, results on the life cycles of the two populations are presented and compared. Based on these results, the taxonomic status of *H. nigromaculatus* is again called into question.

Collecting sites, material and methods

Two populations of *H. nigromaculatus* from running and standing waters were investigated separately.

The stream-living population originated from the Farver Au, a small stream in the Northern German lowland (for more information see Martin 1996). Besides single specimens of *Lebertia inaequalis*, *H. nigromaculatus* was the only species living on fine detritus in pools of a near-natural stream section. Some individuals of *H. nigromaculatus* were cultivated in the laboratory at temperatures close to those measured in the field (see Martin 1998), others at constant temperatures of about 7°C. For sampling of potential hosts, emergence traps which covered one square metre of stream surface were set up at the Farver Au. The collections were carried out weekly in 1995, from March until November. For details of sampling and cultivating methods see Martin (1998).

Specimens of *H. nigromaculatus* from a standing water population were collected in the littoral zone of the Great Pohlsee, a eutrophic Baltic lake. The collecting site was described in detail by Böttger and Mierwald (1990). Eighteen females were sampled on May 27th, 1999 at the North-Eastern edge of the lake and transported to the laboratory. Reproduction took place in tissue culture plates at 13°C (12 h day). Each female was treated separately. Chironomid larvae, as well as different ostracods and cladoceras, were offered as food.

The nomenclature and abbreviations of larval morphology follow Prasad and Cook (1972) and Smith and Cook (1991).

Results

Hygrobates nigromaculatus *from the stream population*

Laboratory results

One male and one female of *H. nigromaculatus* were placed together in the laboratory from the 12th of August 1994 on. The first eggs were laid in June 1995 and several clutches were produced, each with 5-40 eggs. After three weeks larvae hatched and died after a few weeks without feeding. Altogether about 100 eggs were produced by the female. Six other females were mated in the year 2000. They produced between 1-50 eggs in different clutches. One female produced up to 120 larvae. The eggs were an average of 178 µm in diameter (n=20, min. 172 µm, max. 184 µm). The larva is elongated in shape (fig. 1) and as in most *Hygrobates* larvae, with a large dorsal plate covering nearly the entire dorsal idiosoma. Similarly, the ventral idiosoma is almost completely covered by the coxal plates and the large excretory pore plate. The length of the dorsal plate (for all following measurements n=6, see tab. 1) is 297 µm (292-300) and the width 172 µm (168-176). The coxal plates CXI-CXIII are 198 µm long (193-203) and 104 µm wide (100-108). The excretory pore plate is slightly indented in its distal margin and the setae V4 are very long (201 µm, 193-210). The leg length increases from leg I to leg III; the length of leg I is 234 µm (229-240), leg II 257 µm (251-262) and leg III 298 µm (290-307). Other morphometrical data are listed in table 1.

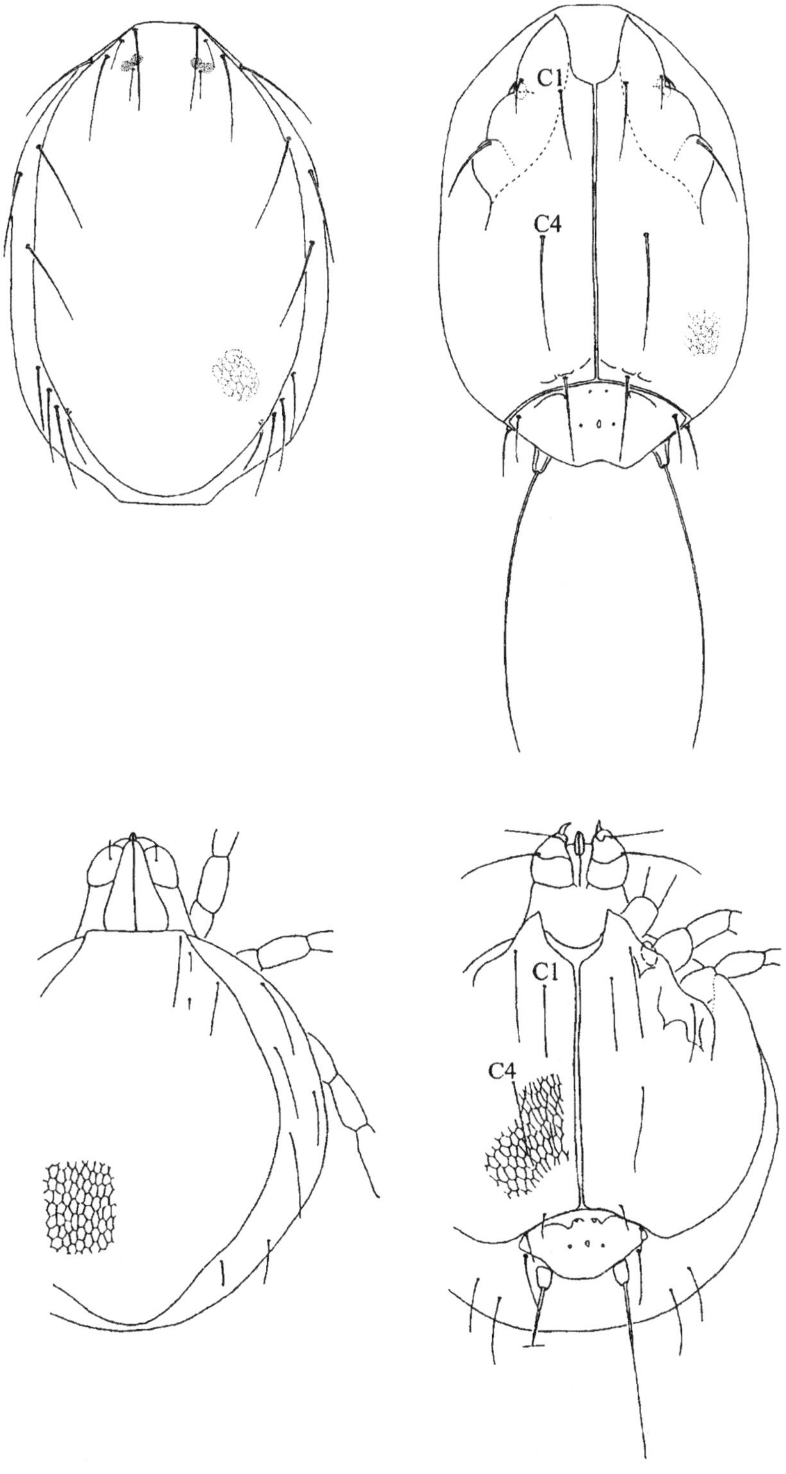

Fig. 1 - Larvae of *Hygrobates nigromaculatus* from a stream population (above, Farver Au) and from a standing water population (below, Lake Maarsseveen I, after Van Hezewijk and Davids, 1985). On the left: dorsal idiosoma, on the right: ventral idiosoma.

Field results

Altogether 213 larvae were found in the emergence traps of the Farver Au, 178 attached to their hosts and 35 detached larvae. The smallest larva was 276 µm long and 232 µm wide, the largest was 344 µm long and 264 µm wide. The average larval length in the emergence traps was 305 µm, the width 228 µm. *Hygrobates nigro-*

Tab. 1 - Morphometric data of larval *Hygrobates nigromaculatus.* On the right: data for larvae from the small stream Farver Au, North Germany, on the left: data from Van Hezewijk and Davids (1985) for larvae from the breeding population of lake Maarsseveen I, The Netherlands. Abbr.: n, number of specimens measured; x, mean.

		stream-living population				lake-living population			
				range				range	
		n	x	min	max	n	x	min	max
dorsal plate	length	6	297	292	300	10	297	274	321
	width	6	172	168	176	6	201	195	209
coxal plates CXI-III	length	6	198	193	203	18	176	165	187
	width	6	104	100	108	17	131	122	141
gnathosomal bay	depth	6	50	48	52	8	40	36	46
	width	6	35	34	35	8	46	41	52
excretory pore plate	length	6	50	48	53	8	46	41	52
	width	6	105	103	108	8	87	80	93
cheliceres	length	6	81	78	84	10	69	64	73
	width	6	21	20	21	11	21	19	23
leg I	total	6	234	229	240	10	184	173	197
	L1	6	35	34	36	-	-	-	-
	L2	6	44	42	45	10	39	37	40
	L3	6	42	41	43	10	33	31	36
	L4	6	51	50	51	10	42	38	46
	L5	6	63	62	65	10	52	45	56
leg II	total	6	257	251	262	10	204	192	214
	L1	6	37	36	38	-	-	-	-
	L2	6	44	43	45	10	38	35	40
	L3	6	44	43	45	10	36	33	38
	L4	6	60	59	60	10	50	45	52
	L5	6	72	70	74	10	59	56	66
leg III	total	6	298	290	307	10	237	216	248
	L1	6	47	45	49	-	-	-	-
	L2	6	49	48	50	10	41	37	43
	L3	6	51	50	53	10	41	38	44
	L4	6	70	68	71	10	58	50	61
	L5	6	82	79	84	10	69	61	73
setae on palp	Ge1	6	96	88	99	14	66	62	70
	Ge2	6	54	53	55	9	36	34	39
setae on idiosoma	Mh1	6	67	64	68	9	49	45	54
	Mh2	6	67	64	68	7	58	48	73
	C1	6	59	57	60	9	44	38	48
	C4	6	82	78	85	10	56	51	66
	V2	6	39	38	40	7	23	19	27
	V3	6	54	51	60	9	47	38	60
	V4	6	201	193	210	13	166	157	177
projection of V4	length	6	23	21	27	8	16	15	18
	width	6	13	10	17	8	11	9	13

maculatus larvae were exclusively parasites of chironomid adults (see Martin, 2000). The 178 larvae were attached to 119 chironomids from the subfamilies Chironominae (both tribes) and Orthocladiinae (Tab. 2). Among the Chironominae-tribe Tanytarsini, the species *Cladotanytarsus vandervulpi* (12.4% of the larvae on 10.9% of the hosts) as well as the genera *Micropsectra* (33.7% of the larvae on 29.4% of the hosts) and *Tanytarsus* (28.7% of the larvae on 28.6% of the hosts) were parasitised most frequently. Among the Orthocladiinae, *Paratrichocladius skirwithensis* (12.6% of hosts) was the most frequent host species (10.1% of larvae). The mean intensity for all host-parasite associations was 1.5 larva per host, the highest mean intensity for a host taxon was found for *Synorthocladius semivirens* (3.0). Several species of the Chironominae-tribe Tanytarsini had a maximum of four larvae parasitising a single host. There was no marked difference in parasitism of the host sexes; on 61 male hosts, 88 larvae were attached (1.44 larvae/host), whereas 90 larvae (1.55) were found on 58 female hosts. All *H. nigromaculatus* larvae were attached to the hosts' abdomen. Most larvae were attached ventromedially on the 1st abdominal segment,

Tab. 2 - Distribution of the *Hygrobates nigromaculatus* larvae on the host taxa of the Farver Au. Presented are numbers of the individuals and the percentages of the chironomid taxa (left) and the distribution of the water mite larvae on their hosts, the percentage of larvae, the mean intensity and the derivation of the intensity (right). Abbr.: ind, individuals; per, percentage; der, derivation of intensity.

	chironomids		water mite larvae			
ind	per [%]		ind	per [%]	mean intensity	der
		Chironominae				
		tribe Chironominae				
5	4.2	*Microtendipes pedellus*	5	2.8	1.00	-
1	0.8	*Polypedilum convictum*	1	0.6	1.00	-
		tribe Tanytarsini				
13	10.9	*Cladotanytarsus vandervulpi*	22	12.4	1.69	1-4
6	5.0	*Micropsectra atrofasciata*	7	3.9	1.17	1-2
3	2.5	*Micropsectra junci*	5	2.8	1.67	1-3
26	21.8	*Micropsectra* sp.	48	27.0	1.85	1-4
6	5.0	*Tanytarsus ejuncidus*	7	3.9	1.17	1-2
14	11.8	*Tanytarsus pallidicornis*	20	11.2	1.43	1-4
14	11.8	*Tanytarsus* sp.	24	13.5	1.71	1-4
2	1.7	Tanytarsini indet.	4	2.2	2.00	1-2
		Orthocladiinae				
15	12.6	*Paratrichocladius skirwithensis*	18	10.1	1.20	1-3
2	1.7	*Paratrichocladius* sp.	2	1.1	1.00	-
6	5.0	*Rheocricotopus fuscipes*	6	3.4	1.00	-
1	0.8	*Synorthocladius semivirens*	3	1.7	3.00	-
5	4.2	Chironomidae indet.	6	3.4	1.00	1-2
-----			-----			
119			178		1.50	
-		detached larvae	35			
119			213			

near the metathorax. Some larvae were found further posteriorly on the abdomen, but never further than the segment border between the 3rd and 4th segment.

All 213 larvae of *H. nigromaculatus* were found in the emergence traps between the 14th of June and the 20th of Sept. 1995 (Fig. 2). The maximum number were found on the 12th of July 1995 with 51 individuals/m^2 (including the free larvae which were detached from their hosts).

Based on the results above and on reference data (see Martin 1998), the following simplified life cycle for the Farver Au is presented. Oviposition takes place from May until August. After about three weeks, the larvae hatch and parasitise the chironomid subfamilies Chironominae and Orthocladiinae. Parasitic larvae were found on hosts from June until September. The last larvae return to the stream at the end of September. The nymphs developing from the larvae moult into adults after a few months. The females from this generation overwinter and lay eggs the following year. The spermatophore release of the males can occur from the autumn of their appearance until the summer of the next year. Phenological data (cf. Martin 1998) and laboratory observations leads to the conclusion that the life span of both sexes is at least one year. The fact that the sex ratio is female biased points to a longer life span of females.

Hygrobates nigromaculatus from the lake population

Only laboratory observations were carried out for the *H. nigromaculatus* population from the lake Great Pohlsee. The females of *H. nigromaculatus* were cultured starting on the 27th of May 1999. The females were observed accepting only chironomid larvae as food. The first eggs

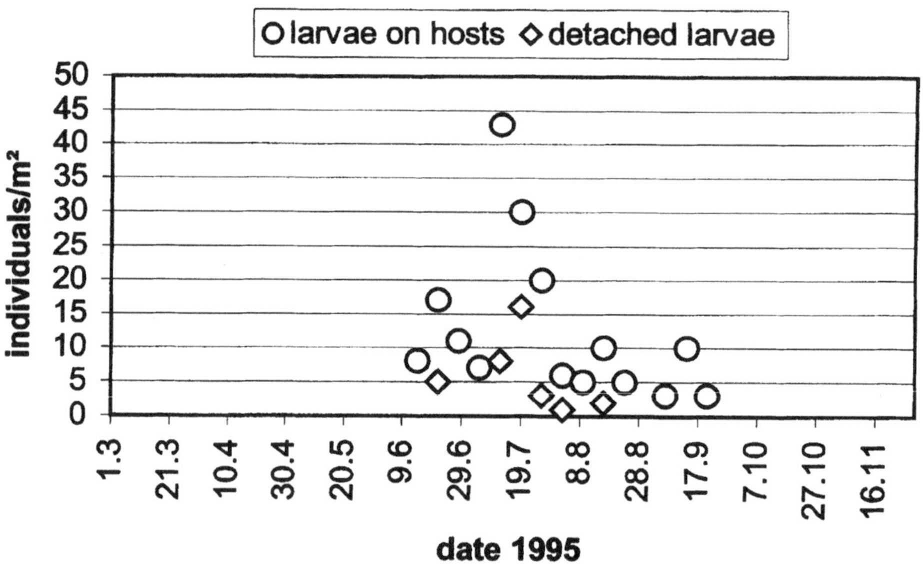

Fig. 2 - Seasonal occurrence of the larvae of *Hygrobates nigromaculatus* in the emergence traps of the Farver Au in 1995 (n=213). Samples were taken from the 1st of March until the 29th of November, larvae of *H. nigromaculatus* were found from the 14th of June until the 20th of September. Apart from the larvae on hosts, those larvae are presented which were detached from their hosts.

were found in the plates on the 31st of May 1999. Eggs were laid individually and in small clutches of 2-6 eggs. The females produce between 0-13 offspring. The first active larva was seen on the 28th of June 1999, thus after about four weeks after ovipositioning. Only a few larvae left their clutches and swam around for a short period; the others never left their clutch and developed into the nymphal stage therein. The first free nymphs were found about six weeks after ovipositioning. The last female was found lifeless on the 12th of July 1999. The larvae reared from these experiments are nearly round, their dorsal plates are approximately 1.5 times longer than wide and their coxal plates are wide. The larvae are morphologically identical to the larvae described by Van Hezewijk and Davids (1985). There were no obvious differences in the shape and the morphometric data of their larvae from the Lake Maarsseveen, comparatively presented in figure 1 and table 1.

Thus, the life cycle of *H. nigromaculatus* from the standing water populations differs from the population originating in the running water by producing a larva which shows marked morphological differences and which develops into the nymphal stage without the parasitic phase.

Discussion and conclusion

Before the differences in larval morphology and life cycle strategies in the distinct *Hygrobates nigromaculatus* populations are discussed, some aspects of the parasitology of stream-living larvae should be taken into account. The host taxa of *H. nigromaculatus* were reported by Martin (2000). Chironomids were the hosts for all examined parasitic *Hygrobates* larvae with the exception of the lake-living populations of *H. nigromaculatus* (Smith and Oliver 1986; Davids *et al.* 1994). The host`s abdomen is believed to be the attachment site for the whole hygrobatid family (Smith and Oliver 1986). The preference of larval *H. nigromaculatus* for attachment sites near the host`s thorax may be an adaptation which allows the host organism practically unhindered freedom of movement or the reason could lie in a reduction of concurrence with other parasites, but there are only very few studies dealing with small scale spatial distribution on hosts.

The distinct morphological differences in larval morphology of representatives from the two different populations concerns e.g. the shape of the dorsal plate, the shape of the excretory pore plate, the legs and the chaetotaxy (see fig. 1, tab. 1). The more elongated dorsal idiosoma of the stream-living larva bears an oblong dorsal plate with nearly parallel lateral margins (length 297 µm, width 172 µm). The larval idiosoma of the lake-living population is more rounded and the dorsal plate is also rounded by non-parallel lateral margins (length 297 µm, width 201 µm). The excretory plate of the stream-living population is wider (105 µm) than that of the lake-living relative (87 µm) and is indented on its caudal margin. All legs of the larvae from the stream are longer (leg I 234 µm, leg II 257 µm, leg III 298 µm) than those of the lake-living larvae (leg I 184 µm, leg II 204 µm, leg III 237 µm). The setae of the stream-living population are generally longer than their lake-living relatives, a feature particularly distinct e.g. in palpal seta Ge1 or the coxal setae C1 and C4. The morphological differences in larvae listed for *H. nigromaculatus* are typical for parasitic and non-parasitic sibling species (Smith 1998 and literature reviewed there).

The larvae reared from females of the Farver Au are nearly identical to the *H. nigromaculatus* larvae as described by Wainstein (1980). Wainstein's larvae were derived from a female which was taken from a small stream (P. Tuzovskij, *pers. com.*). On the other hand the larvae described in detail by Van Hezewijk and Davids (1985) look like that of the non-parasitic population from the lake Great Pohlsee. We therefore believe that the differences in larval morphology and the occurrence of parasitic larvae in streams and non-parasitic larvae in lakes is the normal condition for *H. nigromaculatus* and not an accidental observation.

Another indication of a stable occurrence of the different populations is the conspicuous phenomenon that adult *H. nigromaculatus* in lakes are generally smaller than those in streams (Viets 1960). This can be explained by the hypothesis that the non-parasitic populations in lakes can afford to produce fewer but bigger eggs eliminating the risky parasitic stage. In contrast, the bigger females of the parasitic population in streams must produce more but smaller eggs to ensure enough offspring for the risky parasitic larval life. This connection between life cycle and egg size and number is typical for sibling species (Davids 1997; Smith 1998) and also true for *H. nigromaculatus*. For lake-living *H. nigromaculatus*, Van Hezewijk and Davids (1986) found 4-5 eggs (maximally 12) per clutch. The average egg diameter was 194 µm, a size only slightly more than the diameter of eggs from the stream-living population found in the present study (178 µm, min. 172, max. 184, n=20). However, there is a high number of offspring per female in stream-living *H. nigromaculatus* (at least about 120 larvae) in contrast to the lake-living population (about 13 offspring). The fact that larger eggs represent a non-parasite lineage also applies to single species which produce different eggs in different seasons (e.g. *Piona nodata*, Böttger 1962). The laboratory observations of *H. nigromaculatus* let us exclude the possibility that there may be a seasonal change of production of parasitic and non-parasitic larvae.

The loss of parasitism in stream-living *H. nigromaculatus* is not the only reported case for the genus; *H. neooctoporus* and *H. longiporus* also lack larval feeding and host association (Wainstein 1980; Proctor in Smith 1998). In all reported cases of loss in parasitism in water mites the two lineages of sibling taxa are found in comparable habitats, mainly in standing waters (Smith 1998). *Hygrobates nigromaculatus* is the only known taxon whose parasitic and non-parasitic lineages live in streams and standing waters respectively; this is another reason to doubt the species status of *H. nigromaculatus*. Because this species also occurs in degraded streams, it will be very interesting to discover whether such populations belong to the parasitic or to the non-parasitic lineages. It will also be interesting to study the life cycle of other species with a similar ecological distribution in different habitats.

There are several possibilities to explain the loss of parasitism in general and for *H. nigromaculatus* in standing waters in particular. Davids (1997) pointed out that females which do not need to find a host are able to lay eggs earlier in the year than their parasitic relatives. This point can not be definitively determined for *H. nigromaculatus* because there is no comprehensive phenological data for a standing water population, but it is possible. In laboratory studies, Van Hezewijk and Davids (1985) found that in their lake-living population all eggs were transformed into nymphs by the end of June, whereas in the present study, the last larvae of the running water population only returned to the water at the end of September.

Another benefit for non-parasitic lineages may be an avoidance of competition. This applies to the nymphs which appear earlier in the year and are thus able to feed earlier, but particularly to the larvae. For example, for arrenurid larvae which live parasitically on Odonata, Mitchell (1964 and 1969) found that the main factor permitting high levels of sympatry in water mites is the behavioural flexibility of the larvae in selecting and exploiting their odonate hosts. Whereas in the littoral zone of the Great Pohlsee there are a lot of species besides *H. nigromaculatus* (see Böttger and Mierwald 1990), in the lentic biotopes of the Farver Au, *H. nigromaculatus* lives virtually alone. Thus, one reason of loss of parasitic larvae for lake-living *H. nigromaculatus* may be the increased competition for hosts in standing water populations.

The existence of sibling species in *H. nigromaculatus* could be tested by investigating its sexual behaviour (cf. interbreeding experiments in Smith 1998), e.g. by checking whether the spermatophores of the non-parasitic lineage are taken up by females from the other lineage and *vice versa*. It would be necessary to see if the descendants of these experiments were also capable of reproduction, and if so, to which lineage they would belong. However, there are additional possibilities to prove the status of the two lineages of *H. nigromaculatus*. Ten Winkel *et al.* (1989) reported that *H. nigromaculatus* from a lake also feeds on cladocera as well as on chironomid larvae. If both prey are offered, cladocera are preferred. In the absence of cladocera, the individuals show a burying behaviour during the search for chironomid hosts, which are then fed on heavily (Ten Winkel 1985). In the presence of cladocera they stop this behaviour. Thus, it is possible to compare feeding behaviour of the two lineages, e.g. by offering stream-living *H. nigromaculatus* cladocera as food. In the case of *H. nigromaculatus*, a study using molecular biological methods (genetic or allozyme equipment) may also be useful for clarifying the taxonomic status. As a result, *H. nigromaculatus* could serve as a wide-spread indigenous model for experiments in loss of parasitism in water mites.

Smith (1998) reviewed the loss of larval parasitism in Parasitengona in general, which occurs obviously relatively rarely, but in nearly every limnic habitat. Often there are either two forms of the same species or two closely related species, one with parasitic and one with non-parasitic larvae. The taxonomic and life cycle differences of *H. nigromaculatus* populations from running and standing waters presented here indicate the existence of two taxa in the former *H. nigromaculatus*, probably sibling species. In addition to our arguments Prof. E. Biesiadka (Teacher Training College, Olsztyn, Poland) thinks that there are morphological differences in adult *H. nigromaculatus* that are sufficient to separate the former taxon into two species (Dr. R. Stryecki, University of Lublin, Poland, *pers. comm.*). The first description of the species (Lebert 1879) is based on specimens from Lake Léman (Switzerland) meaning that *H. nigromaculatus* is the taxon from standing waters. Until an ultimate taxonomical study is carried out, the stream-living taxa should named *Hygrobates* sp. nov. near *nigromaculatus*.

References

Böttger K., 1962 - Zur Biologie und Ethologie der einheimischen Wassermilben *Arrenurus (Megaluracarus) globator* (Muell.), 1776, *Piona nodata nodata* (Muell.), 1776 und *Eylais infundibulifera meridionalis* (Thon), 1899 (Hydrachnellae, Acari). *Zoologische Jahrbücher. Abteilung für Systematik, Ökologie und Geographie der Tiere*, 89: 501-584.

Böttger K., Mierwald U., 1990 - Vergleichend faunistisch-ökologische Studien an den Wassermilben (Hydrachnellae, Acari) dreier norddeutscher Seen. *Faunistisch-Ökologische Mitteilungen*, 6(3/4): 107-126.

Davids C., 1997 - The influence of larval parasitism on the life history strategies in water mites (Acari, Hydrachnidia). *Archiv für Hydrobiologie*, 141(1): 35-43.

Davids C., Ten Winkel E.H., De Groot C.J., 1994 - Temporal and spatial patterns of water mites in Lake Maarsseveen I. *Netherlands Journal of Aquatic Ecology*, 28: 11-17.

Lebert H., 1879 - Description de quelques espèces nouvelles d'Hydrachnides du Lac Léman. - Mémoire posthume. *Bulletin de la Société Vaudoise des Sciences Naturelles*, 16(82): 327-377.

Martin P., 1996 - Faunistisch-ökologische Benthosstudien an den Wassermilben (Hydrachnidia, Acari) zweier Bäche des Norddeutschen Tieflandes (Ostholsteinisches Hügelland, Schleswig-Holstein). *Faunistisch-Ökologische Mitteilungen*, 7(3/4): 153-167.

Martin P., 1998 - Zur Autökologie der Wassermilben (Hydrachnidia, Acari) zweier norddeutscher Tieflandbäche. PhD Thesis, University of Kiel: 269 pp.

Martin P., 2000 - Larval morphology and host-parasite associations of some stream-living water mites (Hydrachnidia, Acari). *Archiv für Hydrobiologie/Supplement*, 121/3-4, Monogr. Stud.: 269-320.

Mitchell R., 1964 - A study of sympatry in the water mite genus *Arrenurus* (Family Arrenuridae). *Ecology*, 45(3): 546-558.

Mitchell R., 1969 - A model accounting for sympatry in water mites. *The American Naturalist*, 193(932): 331-346.

Prasad V., Cook D.R., 1972 - The taxonomy of water mite larvae. *Memoirs of the American Entomological Institute*, 18: 1-326.

Smith B.P., 1998 - Loss of larval parasitism in parasitengonine mites. *Experimental and Applied Acarology*, 22(4): 187-200.

Smith I.M., Oliver D.R., 1986 - Review of parasitic associations of larval water mites (Acari: Parasitengona: Hydrachnidia) with insect hosts. *The Canadian Entomologist*, 118: 407-472.

Smith I.M., Cook D.R., 1991 - Water Mites. In: Thorp J.H., Covich A.P. (eds.), Ecology and Classification of North American Freshwater Invertebrates. Academic Press San Diego: 523-592.

Ten Winkel E.H., 1985 - The influence of predation by the water mite *Hygrobates nigromaculatus* on a population of chironomid larvae. *Verhandlungen der Internationalen Vereinigung für Limnologie*, 22: 3230-3232.

Ten Winkel E.H., Davids C., De Nobel J.G., 1989 - Food and feeding strategies of water mites of the genus *Hygrobates* and the impact of their predation on the larval population of the chironomid *Cladotanytarsus mancus* (Walker) in lake Maarsseveen. *Netherlands Journal of Zoology*, 39(3-4): 246-263.

Van Hezewijk M.J., Davids C., 1985 - The larvae of three water mite species of the genus *Hygrobates* and their development (Acari, Hydrachnellae). *Bulletin Zoologisch Museum*, Universiteit van Amsterdam, 10(14): 97-105.

Viets K.O., 1960 - Über *Hygrobates nigromaculatus* Lebert 1879 (Hydrachnellae, Acari). *Mitteilungen aus dem Zoologischen Museum*, Berlin, 36(2): 445-461.

Viets K.O., 1978 - Hydracarina. In: Illies J. (ed.), Limnofauna europaea. Eine Zusammenstellung aller die europäischen Binnengewässer bewohnenden mehrzelligen Tierarten mit Angaben über ihre Verbreitung und Ökologie. Gustav Fischer, Stuttgart: 154-181.

Wainstein B.A., 1980 - Opredelitel lichinok vodjanych kleshchei. Institut Biologii Vnutrennich Vod, Nauka: 238 pp.

The impact of human activity on the water mite fauna (Acari, Hydrachnidia) of the 'Lasy Janowskie' Landscape Park (South-Eastern Poland)

Robert Stryjecki
University of Agriculture, Department of Zoology, Akademicka 13, PL-20-950 Lublin, Poland, Robstry@agros.ar.lublin.pl

Key words: water mites, Hydrachnidia, human impact, water pollution, species diversity.

Abstract: Water mites of the 'Lasy Janowskie' Landscape Park (South-Eastern Poland) were investigated, with the aim of documenting the qualitative and quantitative structure of coenoses in various habitats in the Park. Particular attention was paid to the impact of human activity on the *Hydrachnidia* fauna in the surface waters of the study area. The main types of human impact in running waters of the Park were: 1) river-bed regulation (causing a decrease in the species number, mainly in rheobionts and rheophiles), 2) discharge of fish pond waters into the river (causing an increase in the number of species and specimens, a domination of stagnobionts and a reduction of rheobionts and rheophiles, and a disturbance of population dynamics), 3) water pollution (causing the reduction of numbers of species and specimens). In the fish ponds exposed to human activity (regular draining, pond bottom ploughing, fertilization, water pollution by fish food), the water mite fauna was sparser and less diversified than in big shallow lake type ponds. The water mite fauna of transitional waters (canals) had a stagno-rheophilous character and was under a strong anthropogenic impact which, among others, resulted in a disturbance of population dynamics.

Introduction

The rapid civilisation development in the last few decades has caused increasing human impact on the natural environment. All elements of the environment, including water ecosystems, are exposed to various kinds of anthropogenic pressure. Most frequently, human impact causes unfavourable changes in the environment, resulting in the transformation and/or depletion of biocenoses. In the water mite fauna, such changes were found among others by: Schwoerbel (1964), Kowalik and Biesiadka (1981), Biesiadka and Kowalik (1991), Pöpperl and Martin (1995), Böttger and Martin (1995), Martin (1996), Cicolani and Di Sabatino (1991). A frequent result of strong human impact is the extinction of species from a given area (Gerecke and Schwoerbel 1991; Van der Hammen and Smit 1996). Therefore, it is essential to register the current fauna status as well as the forms of human impact and its influence on the biocenoses.

This paper presents examples of the reactions of water mite communities to various forms of human activity.

Study area and methods

The material was collected from the lowland 'Lasy Janowskie' Landscape Park (South-Eastern Poland) (fig. 1). The Park is mainly covered by forest vegetation, but large areas within the

F. Bernini, R. Nannelli, G. Nuzzaci, E. de Lillo (eds.), Acarid Phylogeny and Evolution. Adaptations in mites and ticks, 111-117.

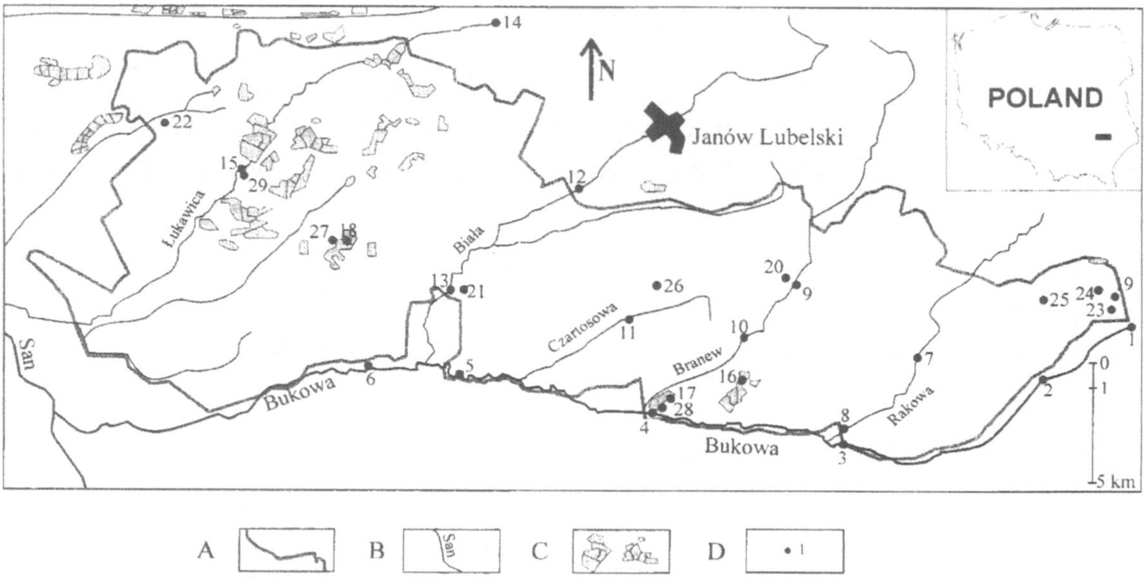

Fig. 1 - The 'Lasy Janowskie' Landscape Park, map of the study area. A - boundary of the Park, B - rivers, C - ponds, D - sampling sites (1-29); 1-15 - sites of running waters: 1, 14 - springs, 2-13, 15 - rivers; 16-27 - sites of stagnant waters: 16-18 - fish ponds, 19-22 - astatic reservoirs, 23-27 - peatbogs reservoirs; 28-29 - sites of transitional waters (canals).

boundaries of the Park are occupied by peatbogs and fish ponds, and a well-developed system of running waters is present. Most rivers in the Park are still in their natural state, but some of them, or at least some river sections, are exposed to a strong human impact (water pollution by domestic and industrial sewage, river-bed regulation, discharge of fish pond waters, hydrotechnical facilities, changes in the structure of sediments). Also artificial fish ponds undergo various anthropogenic pressure, such as drainage, fertilization, water pollution by fish food, agro-technical treatment. In transitional waters of canals human impact is related mainly to strong water flow fluctuations.

Samples were collected from 29 sites (fig. 1), taking into account all types of surface waters existing in the area of the Park: - running waters: springs (2 sites), rivers (13 sites); - stagnant waters: fish ponds (3 sites), astatic reservoirs (4 sites), peatbog reservoirs (5 sites); - transitional waters: canals (2 sites). Samples were taken in 1996-1997, at monthly intervals, from March to November with a hand net (from all sites), a dredge (in fish ponds) and with the Kajak sediment sampler (rivers and fish ponds). Species diversity of the water mite fauna was calculated via the application of the Shannon-Weaver index.

Results

A total number of 148 water mite species was found, representing around 35% of the *Hydrachnidia* species known from Poland. A considerable number of species have been found in most habitat types (tab. 1).

The best represented ecological group in the running waters were rheophiles (59.6% of the specimens, 24 species - data excluding the Lukawica River). As regards the individual numbers, in second place were rheobionts (27.0% of the specimens, 16 species). The largest number of species and specimens was caught from river sites that had remained in their natural state (different sediments, rich water vegetation, well-developed lotic and lentic habitats) (sites 8, 9 and 10). The

Tab. 1 - Numbers of water mite species collected in different types of surface waters in the 'Lasy Janowskie' Landscape Park.

Type of waters	Running waters			Stagnant waters				Transitional waters
Kind of waters	Springs	Rivers	**Total**	Fish ponds	Astatic reservoirs	Peatbog reservoirs	**Total**	Canals
Number of species	18	97	**110**	57	45	42	**89**	55

lowest number of species and/or specimens was found in springs (site 1) and in river sites polluted by sewage (sites 12 and 13) (tab. 2).

Only sampling sites exposed to human impact were thoroughly analyzed.

The Bukowa river at site 2 was transformed into a straight ditch, and during the vegetation period its surface was almost completely covered by macrophytes. A relatively low number of specimens (248) and species (12) was caught here. Most numerous were the lenitobionts: *Mideopsis crassipes*, *Hygrobates longipalpis* and *Lebertia inaequalis*, together amounting to 89.1%. The abundance and species number of rheobionts was very low.

At site 4, fish pond waters were discharged into the Bukowa river periodically. The autumn water inflow caused a sudden quantitative (from 10.1% to 60.9%) and qualitative (from 3 to 10 species) increase of stagnobionts, mainly deutonymphs of *Piona* and the lake species *Arrenurus crassicaudatus* (tab. 2).

In the Biala river (site 12 and 13), extremely low numbers of both species and specimens were observed (tab. 2). Here, municipal and industrial waste waters from Janów Lubelski were

Tab. 2 - Numbers of the most abundant water mite species at the particular sites in the running waters of the 'Lasy Janowskie' Landscape Park.

Species / Sampling site	1	2	3	4	5	6	7	8	9	10	11	12	13	14	15
Hydryphantes planus Thon	21					+				+	+			+	
Sperchon clupeifer Piers.			+	73	40	46	+	+	324	+	+				
Sperchon setiger Thor			+	+	58	+	+	+	+	+			+		
Lebertia fimbriata Thor			+	+	+	+	+	80	67	249	+				
Lebertia inaequalis (Koch)		62	96	91	49	46	43	430	246	105					
Lebertia rivulorum Viets			+	+	+	24	+	+	408	+	+		+		
Lebertia shadini Sok.			64	59	+	27	29	195	+	257					
Torrenticola amplexa (Koen.)			184	+	+	+	+	+		+					
Hygrobates calliger Piers.			113	60	38	42	+	181		+		+			
Hygrobates fluviatilis (Ström)		+	33		+	+	+	40	+	+	+	4	+		
Hygrobates longipalpis (Herm.)		69	+		+	+	36	+		+	+	+	33		+
Hygrobates nigromaculatus Leb.		+	+	33	38	29	292	97	346	168	73	4	4		
Piona (deutonymphs)				134	+	+		+		+			+		148
Forelia variegator (Koch)		+	+	+	34	39	27	157	+	103	+		6		+
Mideopsis crassipes Soar		90	89	+	+	+	46	78		76	37				+
Mideopsis orbicularis (Müll.)				+		+									1163
Arrenurus crassicaudatus Kram.				147	+	+									1172
Arrenurus mediorotundatus Thor						+								122	
Numbers of other species	8	27	294	245	185	133	186	275	163	273	31	5	14	120	1325
Specimens together	29	248	873	842	442	386	659	1533	1554	1231	141	13	57	242	3808
Species together	5	12	31	33	31	31	30	36	16	33	21	7	12	16	47

note: + - species occurred but was not abundant at a given site

discharged into the river. At site 12, during two years of research, only 13 specimens belonging to 7 species were caught. *Hygrobates fluviatilis* and *Hygrobates nigromaculatus* both occurred in 4 specimens, the remaining species *(Lebertia dubia* Thor, *L. inaequalis, Hygrobates calliger, H. longipalpis* and *Mideopsis roztoczensis* Bies. et Kow.) were caught as single individuals only. At site 13, 12 km downstream from the discharge of the sewage into the river, the water mite fauna was richer (57 specimens, 12 species), but still poor in comparison with other, unpolluted river sites (tab. 2).

Water from fish ponds was constantly flowing into the Lukawica river (site 15). In comparison with other Park rivers, the inflow and the character of the habitat (low water flow, muddy river bottom), resulted in a very atypical species composition and abundance. Large numbers of water mites were caught - 47 species and 3808 specimens, with maximum densities up to 2,450 specimens/m^2. The fauna was dominated by two stagnobionts, *Arrenurus crassicaudatus* (30.8%) and *Mideopsis orbicularis* (30.5%). Also other species typical of stagnant waters were rather abundant, such as *Limnesia maculata* (Müll.), *Neumania limosa* (Koch) and *Arrenurus sinuator* (Müll.). On the other hand, rheobionts and rheophiles of the genera *Sperchon, Lebertia* and *Hygrobates*, otherwise commonly and numerously occurring in running waters of the Park, were completely absent (tab. 2).

In the fish ponds (site 16, 17, 18) a total of 57 water mite species were caught. The ponds differed from each other in the composition of their water mite coenoses. In more eutrophicated reservoirs, littoral and pond species were dominant: in the Momoty Duze pond (site 16) *Arrenurus crassicaudatus* (13.7%), *Piona pusilla* (Neum.) (9.6%), *Limnesia maculata* (8.7%) and *L. undulata* (Müll.) (8.7%), in the Momoty Male pond (site 17) *Limnesia maculata* (18.2%), *L. undulata* (13.1%), *Neumania limosa* (6.9%) and *Piona variabilis* (Koch) (6.2%). Instead, fauna of the larger dystrophic Imielty Lug pond (site 18) had a different, more lake-like character. The most numerous species were: *Piona paucipora* (Thor) (13.5%), *Piona pusilla* (10.0%), *Piona coccinea* (Koch) (8.8%) and *Arrenurus crassicaudatus* (8.2%). A series of further species with a habitat preference for lakes were caught in the Imielty Lug pond in higher densities as compared with the other ponds.

The water mite fauna of transitional waters had a mixed (stagno-rheophilous) character. However, both qualitatively and quantitatively, the stagnophiles were distinctly predominating. In the canals, most numerous species were *Arrenurus crassicaudatus* (15.2%), *Limnesia maculata* (14.5%), *Midea orbiculata* (Müll.) (13.4%), and *Limnochares aquatica* (L.) (10.2%). In the more fastly flowing sections, single specimens of rheophiles occurred, representing the genera *Sperchon, Lebertia* and *Hygrobates*. The two highest individual numbers were observed in May and in August, obviously resulting from the inflow of water mites from the fish ponds.

In the running waters the greatest species diversity was found in the naturally conserved mid stream sites with different habitat types (site 5 and 6, H=4.10 and H=4.11, respectively) (Stryjecki 1999). Environmental stress, both natural and antrophogenic, caused a decrease in diversity (e.g. the polluted Biala river - sites 12 and 13) (fig. 2). In the fish ponds the highest diversity (H=4.03) was observed in the Imielty Lug pond. In the canals (site 28 and 29) the diversity was moderate.

Discussion and conclusions

In the area of the 'Lasy Janowskie' Landscape Park, human impact most frequently caused a qualitative change or a decrease of both species and individual numbers in the water mite fauna in comparison with naturally conserved sites. For example, in the regulated Bukowa river (site 2), a scarce number of species (12) and specimens (248) was found, and non-rheobiontic species pre-

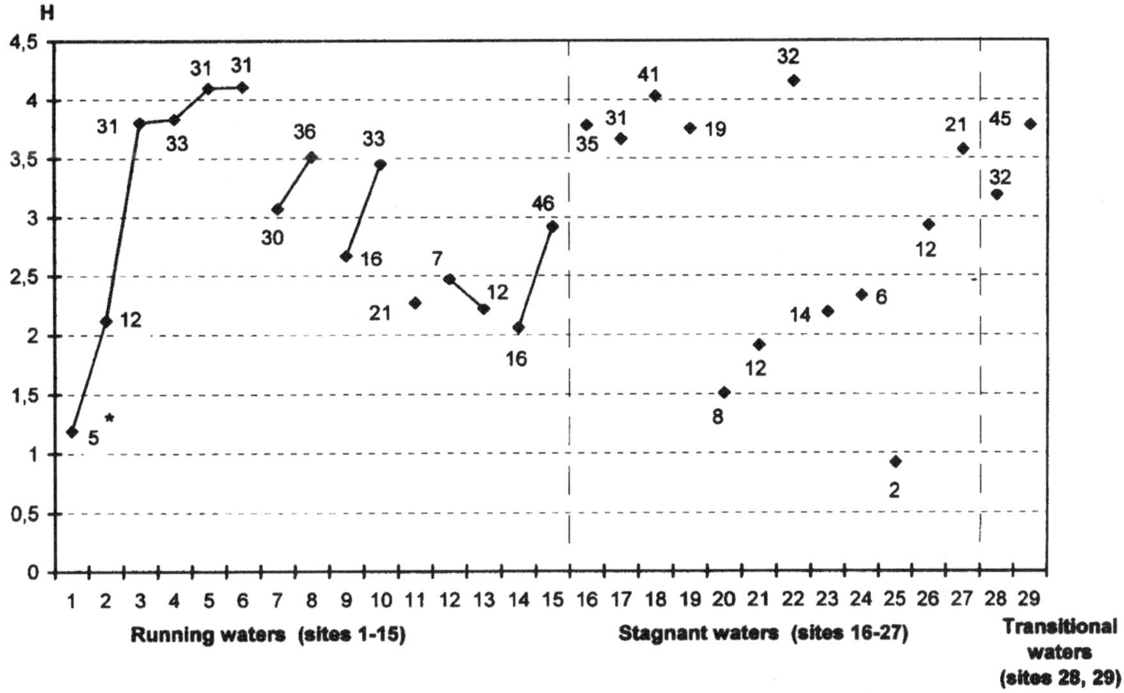

Fig. 2 - Values of the Shannon-Weaver index at sampling sites.* - number of species at a given site. The continuous lines connect sampling sites located in the same river.

dominated. According to many authors (Biesiadka 1972; Böttger and Martin 1995; Martin 1996; Van der Hammen and Smit 1996) the disappearance of rheobionts is a primary effect of river and stream regulation. At site 2, the scarce number of rheobionts was related mainly to the lack of shadowing riparian vegetation, causing a total cover by macrophytes, leading to a slowing down of the current. A similar phenomenon was found by Böttger and Martin (1995) and Martin (1996). Moreover, the river regulation caused a decay of stagnant habitats and river meanders and, as a result, a decreased habitat diversity.

The most common human impact on aquatic ecosystems is water pollution. The strong impact of water pollution on water mites was demonstrated by many authors (Schwoerbel 1959 and 1964; Kowalik and Biesiadka 1981; Cicolani and Di Sabatino 1991; Gerecke and Schwoerbel 1991; Van der Hammen and Smit 1996). The best conditions for the development of Hydrachnidia are clean and well oxygenated waters. An increase in water pollution causes a drastic fall in the number of species and specimens caught, and water mites are completely absent from strongly polluted waters. Research results from the Biala river (site 12 and 13), polluted by municipal and industrial sewage confirm these observations. Only the two species *Hygrobates fluviatilis* and *Hygrobates nigromaculatus* were found in higher individual numbers. The presence of these species at this site confirms their tolerance against water pollution, which was observed also by Schwoerbel (1959), Biesiadka (1979), Kowalik and Biesiadka (1981), Cicolani and Di Sabatino (1991) and Gerecke and Schwoerbel (1991).

Human impact relating to the discharge of pond waters into rivers seems to be quite interesting. As a result of this inflow, the number of species and specimens on river sites increased, but species characteristic of stagnant waters were predominating, and the seasonal population dynamics was disturbed. In naturally conserved river sections (site 4) subject to periodical fish pond water discharge only, the changes in water mite groups were reversible. Stagnant waters species floated down the river and only single specimens were caught in the lentic habitats of the river. Instead, into the Lukawica river (site 15) fish pond waters were discharged constantly. Here, the water current was slow and the bottom was muddy, and stagnobionts occurred year-round in large numbers.

The faunistic differences observed between eutrophicated (sites 16, 17) and dystrophic (site 18) ponds, undoubtedly resulted from differences in water chemistry due to a different trophy of these reservoirs. However, human impact may also influence these differences. Smaller ponds like Momoty Duze (site 16) and Momoty Male (site 17) were regularly drained, and their bottoms were ploughed. They were also fertilized and polluted by fish food. Instead, human impact on the larger Imielty Lug pond was relatively low, which in connection with its larger area and natural character resulted in a larger occurrence of lake species.

The analysis of the species diversity of the water mite fauna in the investigated waters showed that it is influenced by human impact. A low diversity was found on site 2 (river regulation) and on sites 12 and 13 (water pollution) (fig. 2). On the sites where fish pond waters were discharged into the rivers, species diversity was lower than on natural sites, although there was an increase in the number of species. An example is the Lukawica river at Bania village (site 15), where a large number of species was found (47), but the Shannon-Weaver index amounted only to 2.92 (fig. 2). The observed differences in the diversity of the Hydrachnidia fauna in the Park agree with the hypothesis that mite species diversity reflects the level of disturbance in the environment and may be a measure of the environment transformation (Cicolani and Di Sabatino 1991; Gerecke and Schwoerbel 1991).

In lotic ecosystems the effect of disturbance is, first of all, the disappearance of rheobionts, which have a narrow environmental tolerance. Instead, more tolerant species appear that frequently occur also in stagnant waters (Kowalik and Biesiadka 1981; Gerecke and Schwoerbel 1991; Van der Hammen and Smit 1996). In stagnant waters human impact (e.g. increase of trophy) causes an increase in the share of small water reservoir and phytophilous species which have a broader tolerance to environmental factors, while the share of species typical for lakes decreases (Biesiadka and Kowalik 1991). It should be stressed that each kind of human impact causes a fall in species diversity.

References

Biesiadka E., 1972 - Wodopójki (Hydracarina) Wielkopolskiego Parku Narodowego (in Polish). Water mites (Hydracarina) of the Wielkopolski National Park. Prace monograficzne nad przyroda Wielkopolskiego Parku Narodowego pod Poznaniem. Tom 5, zeszyt 3: 1-102.

Biesiadka E., 1979 - Wodopójki (Hydracarina) Pienin. Water mites *(Hydracarina)* of the Pieniny Mountains. *Fragmenta faunistica,* 24: 97-173 (in Polish).

Biesiadka E., Kowalik W., 1991 - Water mites (Hydracarina) as indicators of trophy and pollution in lakes. In: Dusbábek F., Bukva V. (eds.), Academia Prague, Modern Acarology, 1: 475-481.

Böttger K., Martin P., 1995 - Faunistisch-ökologische Untersuchungen an den Wassermilben (Hydrachnidia, Acari) dreier kleiner Fließgewässer des Norddeutschen Tieflandes, unter besonderer Berücksichtigung der rheobionten Arten. *Limnologica,* 25(1): 61-72.

Cicolani B., Di Sabatino A., 1991 - Sensitivity of water mites to water pollution. In: Dusbábek F., Bukva V. (eds.), Academia Prague. Modern Acarology, 1: 465-482.

Gerecke R., Schwoerbel J., 1991 - Water quality and water mites (Acari, Actinedida) in the upper Danube region. In: Dusbábek F., Bukva V. (eds.), Academia Prague. Modern Acarology, 1: 483-491.

Kowalik W., Biesiadka E., 1981 - Occurrence of water mites in the River Wieprz polluted with domestic-industry sewage. *Acta Hydrobiologica,* 23: 331-348.

Martin P., 1996 - Faunistisch-ökologische Benthosstudien an den Wassermilben (Hydrachnidia, Acari) zweier Bäche des Norddeutschen Tieflandes (Ostholsteinisches Hügelland, Schleswig-Holstein). *Faunistisch-Ökologische Mitteilungen,* Band 7, Heft 3/4: 153-157.

Pöpperl R., Martin P., 1995 - Die Wassermilben (Hydrachnidia, Acari) eines norddeutschen Seeabflusses (Belauer See, Schleswig-Holstein). *Faunistisch-Ökologische Mitteilungen*, Band 7, Heft 1/2: 61-72.

Schwoerbel J., 1959 - Ökologische und tiergeographische Untersuchungen über die Milben (Acari, Hydrachnellae) der Quellen und Bäche des südlischen Schwarzwaldes und seiner Randgebiete. *Arch. Hydrobiol.*, Suppl. 24: 385-546.

Schwoerbel J., 1964 - Die Wassermilben (Hydrachnellae und Limnohalacaridae) als Indikatoren einer biocönotischen Gliederung von Breg und Brigach sowie der obersten Donau. *Arch. Hydrobiol.*, Suppl. 27: 386-417.

Stryjecki R., 1999 - Róznorodnosc gatunkowa fauny wodopójek (Hydracarina) Parku Krajobrazowego „Lasy Janowskie". Species diversity of water mite fauna (Hydracarina) in the „Lasy Janowskie" Landscape Park. [w:] Puszkar T. (red.) Bioróznorodnosc obszarów stykowych Kotliny Sandomierskiej, Wyzyny Kielecko-Sandomierskiej oraz Wyzyny Lubelskiej. Wydawn. Towarzystwo Naukowe Sandomierskie, Wyzsza Szkola Humanistyczno-Przyrodnicza, Sandomierz: 125-133 (in Polish).

Van der Hammen H., Smit H., 1996 - The water mites (Acari: Hydrachnidia) of streams in the Netherlands: distribution and ecological aspects on a regional scale. *Netherlands Journal of Aquatic Ecology*, 30(203): 175-185.

The occurrence of water mites (Hydrachnidia, Acari) in saline waters from a stone coal-mine in Bogdanka (South-Eastern Poland)

Witold Kowalik
University of Agriculture, Department of Zoology, Akademicka 13, PL-20-950 Lublin, Poland

Key words: surface water reservoirs, coal mine saline waters, water chemistry, water mites, Hydrachnidia, influence of salinity.

Abstract: The occurrence of fresh-water hydrachnid mites in a sedimentation reservoir, a discharge canal and the Swinka River, all influenced by saline waters from a stone coal-mine in Bogdanka (Poland, Lublin voivodeship), was investigated. All habitats are characterized by mixooligohalinic chloride and sodium waters with an average salinity from 0.12 to 1.61% and with an electrolytic conductivity from 594 to 3,420 µS/cm. Highest values were measured in the coal mine water sedimentation reservoir and in the canal carrying waters from the reservoir to the river. In the study area, water mites showed a high sensitivity to water salinity. With increasing salt contents, a drastic decrease in both species number and quantity of specimens was observed. A considerable tolerance to increased water salinity (from 1.11 to 1.61%) was found in some stagnant waters species *Hydrodroma pilosa* (Bess.), *Limnesia maculata* (Müll.), *L. undulata* (Müll.), *Mideopsis orbicularis* (Müll.), and in species colonizing astatic waters (of the genera: *Hydrachna, Eylais, Hydryphantes*). Instead, most rheobionts and rheophiles seem to be more sensitive to increased water salinity.

Introduction

The occurrence of fresh-water water mites in different kinds of inland saline waters and laboratory tests relating to this problem is infrequently written about and what has been written is hard to find. A review is presented by Gerecke (1991) in a broad paper concerning the occurrence and ecology of water mites in springs and rivers (including saline ones) in Sicily. I know of no papers, except for a paper by Heuss (1966), analysing the occurrence of water mites in inland reservoirs polluted by a coal mine saline water discharge.

Hydrobiological research (Radwan *et al.* 1987; Kowalczyk *et al.* 1990) in saline waters draining a stone coal-mine in Bogdanka (Lublin voivodeship) was done in Poland in the years 1981-1987, and in 1999. Physical and chemical properties of water, and the composition of zooplankton, zoobenthos and necton in the coal-mine waters sedimentation reservoir, the discharge canal and the river were investigated. The results of the investigation on the water mite fauna of these waters are presented in this paper.

Study area and method

Saline waters draining away from the coal-mine in Bogdanka are collected in a sedimentation reservoir dammed up by barren rock (fig. 1). The reservoir was built in 1979 and has a surface of 8 ha, and a maximum depth 3.5 m; the bottom is stony, covered with fine grained stock sediment and very poor vegetation *(Potamogeton)*. The canal discharging the saline water from the reservoir is 5 km in length, 1 m in width, and has a maximum depth of 0.4 m. Its sandy

F. Bernini, R. Nannelli, G. Nuzzaci, E. de Lillo (eds.), Acarid Phylogeny and Evolution. Adaptations in mites and ticks, 119-124.

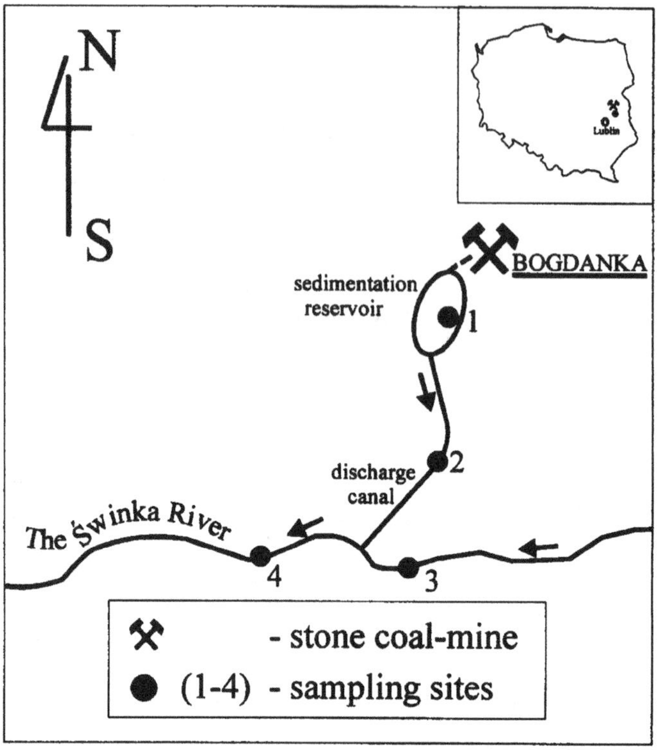

Fig. 1 - Schematic plan of the study area.

bottom is sparsely covered with *Potamogeton* vegetation. The width of the Swinka river is 3-5 m, the depth of 0.3-2 m. It has a sandy bottom which is peaty in certain river sections. Vegetation consists of *Elodea canadensis* Rich., *Nuphar luteum* (L.) Sm., and *Potamogeton* in the centre of the stream, and of *Glyceria aquatica* (L.) Wahlb. at the river bank. In the course of a year, salinity in the river downstream from the canal varied, depending on water flow fluctuations both in the canal and in the river.

Water mites were caught in the years 1981-1987 and in 1999, mainly once a month (between March and November) with a hand net and a dredge. From each of the 4 collecting sites, two samples were taken: in the sedimentation reservoir (site 1), in the discharge canal (site 2), in the Swinka River above the canal outlet (site 3) and below the canal outlet (site 4) (fig. 1).

Results

Selected chemical properties of water

Coal mine waters from Carboniferous and Jurassic strata at a depth of ca. 900 m below the surface are of chloride - sodium type (tab. 1), characterized mainly by a high average concentration of sodium chlorides, and in a lower degree by the concentration of sulphates (Radwan *et al.* 1987).

The annual average of Na^+, Cl^- and SO_4^{-2} concentration ranged from 1,112.1 to 1,578.8 mg/l (tab. 1). The annual average salinity of coal-mine waters from 1.15 to 1.61%, and their conductivity from 3,240 to 3,420 µS/cm indicate that they are mixoligohalinic according to the "Venice system". Moreover, they are weakly alkaline (pH 8.43-8.58) and do not freeze up in winter (temperature minimum about 4°C). The average calcium and potassium content is given in table 1, other chemical factors (O_2, BOD_5, ammonia, nitrites, nitrates, phosphates) correspond to the first class of the Polish water purity standard.

Water in the discharge canal had a slightly lower average salinity (0.57-1.48%, total of Na^+, Cl^- and SO_4^{-2} 521.8-144.7 mg/l). It was alkaline (pH 8.30) with a considerable calcium content (20.6-49.6 mg/l), and did not freeze up in winter.

It should be stressed that the salinity of coal-mine waters in the sedimentation reservoir and the canal was very low (tab. 1) in comparison with waters draining from coal mines in Upper Silesia (Southern Poland). There, Cl^- concentrations of 9,500-17,000 mg/l, Na^+ concentrations of 3,450-6,000 mg/l, and conductivity values of 20.7-30.1 µS/cm were measured (Rózycka 1967).

In the Swinka river upstream from the inflow of saline waters (site 3) a good quality of water was found and physical and chemical parameters corresponded to the 1st class of water purity. The calcium concentration was clearly higher than in the mine draining waters (tab. 1).

Downstream from the inflow of saline waters (fig. 1), water salinity (total of Na^+, Cl^- and SO_4^{-2} 384.3-532.2 mg/l), electrolytic conductivity (1,819-2,160 µS/cm), and potassium concentration (6.0-9.2 mg/l), were found drastically increased.

In 1999, the average water salinity level both in the sedimentation reservoir and in the discharge canal was found to have increased. This change resulted from increased coal production as compared with the period 1981–1984 and the following increase in the output of saline waters from the coal-mine.

The water mite fauna

Thirty two water mites species were found in the investigated waters (see tab. 2): 9 in the sedimentation reservoir (site 1) and in the canal (site 2), 16 in the river downstream from the inflow of saline waters (site 4), and 24 upstream from the inflow of these waters (site 3). In the course of this study (1981-1999), except for the canal, the number of species decreased as the water salinity increased (tab. 2). In 1999 samples were taken from the sedimentation reservoir, but no water mites were caught, probably due to an increase in salinity (tab. 1) and an increase in water pollution (*unpubl. data*).

In the sedimentation reservoir, only stagnophilous species were found: *Hydrodroma pilosa*,

Tab. 1 - Annual average values and ranges of the selected chemical factors of the investigated waters (Bogdanka, Poland).

Sampling sites / Factors	Years	Sedimentation reservoir (site 1)		Discharge canal (site 2)		The Świnka River upstream from the discharge of saline waters (site 3)		The Świnka River downstream from the discharge of saline waters (site 4)	
		1981-1984	1999	1981-1984	1999	1981-1984	1999	1981-1984	1999
pH		8.43	8.58	8.30	8.35	7.40	7.84	7.55	7.90
Conductivity µS/cm	range	-	3240 - 3420	-	2320-3140	-	532 - 644	-	1819 - 2160
	average		3333		2670		594		1948
Σ Na^+, Cl^- and SO_4^{-2}	mg/l	1112.1-1488.4	1578.8	521.8 - 1314.6	1446.7	131.8 - 169.7	122.3	384.3 - 532.2	488.1
K^+	mg/l	17.3 - 26.6	15.2	10.0 - 23.9	8.5	1.8 - 5.7	1.6	8.8 - 9.2	6.0
Ca^{++}	mg/l	15.7 - 34.7	19.5	39.8 - 49.6	20.6	67.5 - 89.9	72.1	57.6 - 76.2	30.0
Salinity	‰	1.15 - 1.55	1.61	0.57 -1.38	1.48	0.20-0.26	0.20	0.45 - 0.62	0.52

Tab. 2 - Occurrence of water mites (Hydrachnidia, Acari) in investigated waters in years 1981-87 and 1999 (Bogdanka, Poland).

Sampling sites	Sedimentation reservoir (site 1)					Discharge canal (site 2)					The Świnka River upstream from the discharge of saline waters (site 3)					The Świnka River downstream from the discharge of saline waters (site 4)				
Salinity ‰	from 1.15 to 1.61					from 0.57 to 1.48					from 0.20 to 0.26					from 0.45 to 0.62				
Species / Years	81-82	83-84	85-87	**Total 81-87**	1999	81-82	83-84	85-87	**Total 81-87**	1999	81-82	83-84	85-87	**Total 81-87**	1999	81-82	83-84	85-87	**Total 81-87**	1999
H. globosa (Geer)	+			**3**									+	**1**						
Eylais extendens (Müll.)		+		**1**								+		**3**			++		**6**	
Hydryphantes octoporus Koen.										+										
H. planus Thon																	+		**1**	
Hydrodroma despiciens (Müll.)												+		**2**			+		**2**	
H. pilosa (Bess.)	+++	++	++	**31**								+	+	**5**		+	+		**3**	
Lebertia inaequalis (Koch)							+		**1**		++	+++	+	**27**	+	+	++	+	**11**	
L. insignis Neum.										+	+	+	++	**12**	+++	+		+	**5**	++
L. pilosa Maglio												+		**5**						
L. porosa Thor							+		**2**		++	+	+	**15**		+	+		**8**	
Lebertia sp.												+++		**14**						
Torrenticola amplexa (Koen.)												+	+	**5**						
Limnesia maculata (Müll.)	++++	+++		**60**							+	++++	+++	**62**			++	+	**11**	
L. undulata (Müll.)	++	+		**9**								+		**4**						
Hygrobates calliger Piers.										+	+	+	+	**9**	+	+		+	**3**	+
H. fluviatilis (Ström)										++	++	++++	++	**41**	+++	+	+		**5**	+
H. longipalpis (Herm.)												+++		**13**			+		**2**	
H. longiporus Thor										+	+			**1**	+					
H. nigromaculatus Leb.												+	+	**2**			++		**8**	
H. trigonicus Koen.											++		+	**10**		+	+		**3**	
Atractides nodipalpis (Thor)												+		**2**	++		+		**1**	+
A. ovalis Koen.													+	**4**	+					
Unionicola crassipes (Müll.)		+		**1**																
Piona coccinea (Koch)																	+		**1**	
Piona sp. (nymphs)	+++			**11**																
Forelia variegator (Koch)											+	+		**5**			+		**1**	
Brachypoda versicolor (Müll.)	+			**1**								+		**1**						
Mideopsis crassipes Soar							+		**1**			+		**3**						
M.orbicularis (Müll.)						+			**1**		++	+++	+	**33**	++	+	++	+	**13**	
Arrenurus batillifer Koen.											+			**2**						
A. globator (Müll.)	+			**5**																
Number of specimens	89	25	8	**122**	0	1	4	0	**5**	16	52	165	64	**281**	50	17	56	11	**84**	13
Number of species	6	5	1	**9**	0	1	3	0	**4**	5	11	20	13	**25**	8	8	15	5	**17**	4

note: + - species occurring rarely (1-5 specimens), ++ - not numerous (6-10), +++ - moderately numerous (11-20), ++++ - numerous (> 20 specimens)

Limnesia maculata and nymphs of *Piona* sp. occurred quite numerously, while *Limnesia undulata, Hydrachna cruenta, Eylais extendens, Unionicola crassipes, Brachypoda versicolor* and *Arrenurus globator* were caught only sporadically (tab. 2).

In the discharge canal, water mite fauna was very scarce. Only single specimens of 9 species, mainly rheophiles were caught: *Lebertia inaequalis, L. insignis, L. porosa, Hygrobates longiporus, H. calliger, Mideopsis orbicularis, M. crassipes* and *Hydryphantes octoporus*. In 1999, 9 specimens of the rheophile *Hygrobates fluviatilis* were caught.

Both the low individual numbers and the complete change in species composition during the investigation period, suggest that, at least in the canal, water mites are not able to complete their life cycle. The presence of water mites in these waters is probably due to their introduction by flying aquatic insects (Diptera, Coleoptera, Heteroptera) on which water mite larvae parasitize.

Upstream from the inflow of saline waters, the Swinka River had a quite rich water mite fauna (24 species), characteristic for clean and well oxygenated waters. The most frequent species were the stagnophiles: *Limnesia maculata* and *Mideopsis orbicularis,* and the rheophile *Hygrobates fluviatilis* (tab. 2). Species found frequently, but in lower numbers of specimens, were the stagnophiles *Hydrodroma pilosa, Limnesia undulata, Hygrobates longiporus* and *H. trigonicus* and the rheophiles: *Lebertia inaequalis, L. porosa, L. insignis, Hygrobates calliger* and *Torrenticola amplexa*. The quantity of specimens was highest at this site.

Downstream from the outlet of the canal to the river (site 4), the number of both species (16) and specimens (97) was distinctly lower. *Lebertia inaequalis, L. porosa, Limnesia maculata* and *Mideopsis orbicularis* were caught in higher numbers; other species occurred as single specimens.

Conclusions

The statement of Viets (1925), Remane and Schlieper (1958), Heuss (1966), Gerecke (1991) and others that most inland water invertebrates are oligohalinic, and only a few species are euryhalinic, finds confirmation from studies carried out in Poland (Rózycka 1967; Radwan *et al.* 1983; Kowalczyk *et al.* 1990). Also most fresh water mites, are haloxenes, and thus sensitive to increased electrolyte concentration (Viets 1925; Gerecke 1991; Di Sabatino *et al.* 1992). Obviously, communities of water mites in the sedimentation reservoir, the discharge canal and in the river below the saline water discharge with high salt concentrations are degraded ecologically and not stable. Water mite species show a differentiated sensitivity to salinity: certain stagnant waters species (e.g. *Hydrodroma pilosa, Limnesia maculata, L. undulata, Mideopsis orbicularis*) and astatic waters species from the genera: *Hydrachna, Eylais* and *Hydryphantes* are relatively tolerant to increased water salinity (up to 1.61‰), whereas rheobionts and rheophiles (except for *Hygrobates fluviatilis, H. calliger, Lebertia inaequalis* and *L. porosa*) seem to be more sensitive to higher water salinity (Table 2). Similar observations were made by Gerecke (1991) and Di Sabatino *et al.* (1992) during investigations on water mites of Sicilian springs and rivers. Further studies are necessary in order to understand the importance of ionic composition to the sensitivity of different water mite species.

References

Di Sabatino A., Gerecke R., Cicolani B., 1992 - The water mites of the family Torrenticolidae Piersig, 1902 (Acari, Actinedida) in springs and running waters of Sicily (South Italy). *Archiv für Hydrobiologie (Monograpische Beiträge)*, suppl. 90, 2: 253-282.

Gerecke R., 1991 - Taxonomische, faunistische und ökologische Untersuchungen an Wassermilben (Acari, Actinedida) aus Sizilien unter Berücksichtigung anderer aquatischer Invertebraten. *Lauterbornia*, 7: 1-303.

Heuss K., 1966 - Beitrag zur Fauna der Werra, einem salinaren Binnengwässer. *Gewässer und Abwässer*, 43: 48-64.

Kowalczyk C., Radwan S., Zwolski W., Paleolog A., Kowalik W., 1990 - Wplyw zasolonych wód kopalnianych na ekologiczne wlasciwosci wód powierzchniowych. The influence of coal-mine saline waters on ecological properties of surface waters. In: Kajak Z. (ed.), Funkcjonowanie ekosystemów wodnych ich ochrona i rekultywacja, czesc II. Wydawnictwo SGGW-AR, Warszawa: 183-193 (in Polish).

Radwan S., Zwolski W., Kowalczyk C., Kowalik W., 1987 - Wplyw górnictwa na chemizm i biocenozy wód powierzchniowych na terenie LZW. The influence of the mining industry on the chemistry and biocenoses of surface waters in the Lublin Coal Basin). Ekologiczno-gospodarcze problemy rozwoju górnictwa Lubelskiego Zaglebia Weglowego. Lublin: 75-94 (in Polish).

Remane A., Schlieper C., 1958 - Biologie des Brackwassers. Die Binnengwässer 22, Stuttgart: 348 pp.

Rózycka T., 1967 - Zagadnienia zasolenia rzek Slaska w aspekcie biologicznym (in Polish). The problem of salinity of Silesian rivers in the biological aspect. Instytut Gospodarki Wodnej. Materialy badawcze 3, 2. Warszawa: 105-116.

Viets K., 1925 - Süßwasser - Hydracarinen aus salzhaltigem Wasser. *Mitt. geogr. Ges. naturh. Mus. Lübeck* (2 Reihe), 30. Lübeck: 137-142.

Reproductive systems of gamasid mites (Acari, Anactinotrichida) reconsidered

Gerd Alberti

Zoological Institute and Museum, J. S. Bach Str. 11/12, University of Greifswald, D-17489 Greifswald, Germany

Key words: behaviour, biodiversity, evolution, morphology, open questions, terminology.

Abstract: Gamasida (=Mesostigmata) are by far the most species rich and ecologically diverse anactinotrichids. Phylogenetic relationships between gamasid taxa are not solved and even highly controversial. In parallel with this discussion, which mainly refers to characters from external morphology, arguments from anatomy, fine structure, and behaviour in reproductive systems have accumulated that are herein updated and likely may help to solve at least part of this controversy. In general, two types of insemination systems are distinguished: podospermy and tocospermy. Podospermy is found in those taxa in which the females have paired sperm induction pores (=solenostomes) close to the bases of legs III/IV in addition to an unpaired genital opening that is used only for oviposition. The males have chelicerae with a modified movable digit that bears a process termed spermatodactyl and is used to manipulate the spermatophore into the sperm induction pore. Tocospermy is found in mites in which no solenostomes are found and the unpaired genital opening is used for both purposes: acceptance of a spermatophore and egg laying. However, tocospermy in this definition obscures profound anatomical and spermatological differences between gamasid taxa like Sejina, Uropodina, Epicriina, Zerconina on the one hand and Parasitina on the other hand. The latter show similarities with podospermous taxa (Dermanyssina) that likely represent synapomorphies: e.g., unpaired oviduct, ovary with nutritive tissue, ribbon-type spermatozoa, males with modified chelicerae. Outgroup comparison indicates that taxa like Sejina, Uropodina, Epicriina, Zerconina present plesiomorphic characters in their reproductive systems and most likely have to be regarded as early derivative groups. Instead Parasitina and Dermanyssina, considered to represent sister groups, are more terminal. To indicate these differences in reproductive systems more distinctly the following terminology is suggested. The plesiotypic mode of reproduction is referred to as architocospermy. The apotypic reproductive system of Parasitina and Dermanyssina is termed neospermy. Podospermy is regarded a subtype of neospermy. Hence the reproductive system of Dermanyssina should be classified more specifically as neopodospermy and that of Parasitina as neotocospermy accordingly. The evolution of these characters is hypothetisized as a consequence of sperm competition and selection for higher efficiency in offspring production.

Introduction

Gamasida (=Mesostigmata) comprise according to an estimation by Lindquist (*in litt.*) about 6,000 species known to science. Thus, though it is not a too large number, Gamasida represent the most species-rich taxon among Anactinotrichida. Members of this taxon occur world wide, and have adapted to many different habitats and life styles (e.g., carnivores, fungivores, pollinovores, ecto- and endoparasites regarding only nutritional biology) sometimes of economic importance for man, either as parasites (e.g., Dermanyssidae,

F. Bernini, R. Nannelli, G. Nuzzaci, E. de Lillo (eds.), Acarid Phylogeny and Evolution. Adaptations in mites and ticks, 125-139.

Varroidae) or as antagonists of crop pests (e.g., Phytoseiidae) (see Evans 1992). They also have been considered valuable bioindicators in soil ecology (e.g., Karg 1989a). Consequently, Gamasida have been investigated for a long time quite thoroughly, in particular regarding taxonomy.

Nevertheless, there exists still controversy about phylogenetic relationships among the various taxa. There is even no consensus about definition and arrangement of higher categories (e.g. compare Karg 1965, 1989b, 1993 and 1998; Krantz 1978; Evans and Till 1979; Johnston 1982; Krantz and Ainscough 1990; Evans 1992; Norton *et al.* 1993; Moraza and Lindquist 1998). It appears, that reasons for this controversy are based mainly on a lack of general knowledge of this diverse group. Most investigations have only focussed on the larger and abundant taxa such as Uropodina and so-called Gamasina more or less neglecting several small and rare taxa that are almost not known to science. Furthermore - as a general tendency - anatomical and behavioural studies have only been performed in a very few representatives, also regarding the outgroups. However, information from such studies could help to define higher categories and understand phylogenetic-systematical relationships among them.

In the present paper, a review of observations on reproductive systems (including morphological and behavioural data) is given and new results are added. These are discussed in context with what is known from the outgroups of Gamasida, i.e. Opilioacarida (=Notostigmata), Holothyrida (=Tetrastigmata) and Ixodida (=Metastigmata). Regarding the ranking of these major groups, the interpretation of Lindquist (1984) is adopted, in so far as all these groups are treated as subtaxa of the Anactinotrichida, i.e., Opilioacarida are not considered as representing a third group of Acari. Furthermore, I will follow the systematic arrangement given by Norton *et al.* (1993) that is not explained in the referred paper, but that is at least partly supported by the results presented here.

Results and discussion

The compilation of this contribution was done over many years already and struggled with a number of "problems" which the reader herewith is invited to share.

1. Probably the main problem:

Systematical concepts of Gamasida are based in most cases on characters of external morphology, e.g.:

*arrangement of genital plates in females (Trägardh 1946; Camin and Gorirossi 1955: one plate =Monogynaspida or three plates: =Trigynaspida,

*position of genital opening in males: midsternal and presternal position (e.g., Trägardh 1950; Karg 1965),

*males with modified chelicerae: spermatodactyl, spermatotreme or no modification (e.g., Karg 1965; Evans and Till 1979; Johnston 1982),

*arrangement of setae, slit sense organs (poroids), gland pores and sigilla (chaeto- poroido-, adeno-, sigillotaxy (e.g., Sellnick 1944 and 1958; Hirschmann 1957; Evans 1963; Karg 1965, 1989 and 1993; Lindquist and Evans 1965; Athias-Henriot 1975a; Lindquist and Moraza 1998),

*presence of ambulacra on the first legs or not, etc. (see, e.g., Karg 1965 and 1993; Krantz 1978; Evans and Till 1979; Johnston 1982; Evans 1992).

Further characters are derived from behaviour, e.g.:

*mode of sperm intromission (Michael 1892), either by using the genital opening that is also used for oviposition: tocospermy or via insemination pores (solenostomes): podospermy (Athias-Henriot 1975b).

Since more than 20 years also spermatological characters have been described and are used to define taxa: vacuolated spermatozoa, ribbon spermatozoa (e.g., Witaliňski 1975, 1976 and 1979; Witaliňski and Dallai 1991; Alberti

1980a, b, 1984, 1988, 1991 and 2000a; see also Weygoldt 1999).

However, the interpretations of these characters given by the various authors do not conform and result in controversy. What is apomorphic, what plesiomorphic? For example, podospermy originally has been regarded a plesiomorphic feature in Gamasida by Athias-Henriot (1968) and further authors. This was, presumably, induced by an observation reported already by Michael (1892) in his unrivalled paper "On the variation in the internal anatomy of the Gamasinae, especially in that of the genital organs, and their mode of coition" where he writes that the empty structure shows "endosmosis" when water is applied and compared it with coxal glands. But he himself felt this comparison unrealistic since these organs only are present in females. However, later authors concluded that the sperm access system of these mites may have evolved from nephridia (Athias–Henriot 1968 and 1975b; Young 1968; Karg 1998) and hence this character would represent a primitive condition inherited from annelid ancestors (Karg 1998). Consequently, spermatodactyls, strange modifications of male chelicerae that are used for spermatophore transfer in podospermic gamasid mites (Michael 1892; Krantz and Wernz 1979), had to be regarded as more primitive as the spermatotreme found in Parasitidae that was regarded as being a product of secondary anterior fusion of the spermatodactyl with the movable digit (Karg 1965). Also, the location of the male genital opening, presternal in the podospermic taxa, had to be regarded a plesiomorphic state (Karg 1965).

However, as was previously stated (Alberti and Hänel 1986), this nephridia-hypothesis is extremely unlikely, since no mite is known with nephridia in this position (i.e. close to legs III and IV) and there is no fine structural indication of such a function. The presumably closest related group of Acari (or Anactinotrichida), Ricinulei, does not have nephridia (frequently called coxal glands in Chelicerata; see Alberti *et al.* 1996) in this region, instead they are located close to legs I (see, e.g., Moritz 1993). Also Palpigradi -- regarded by van der Hammen (1989) as representing the sister group of Actinotrichida -- have nephridia connected to the first pair of legs. Furthermore, outgroup comparison reveals easily that the plesiomorphic way of transferring sperm to the female is via the genital opening that is used for oviposition and only in rare cases doubtless secondary insemination pores are used for copulation (e.g., in entelegyne spiders or some Acari such as spider mites and their relatives or Acaridida; see Alberti and Coons 1999 with further references). These pores obviously have evolved convergently.

Evidently, this mentioned alternative (podospermy or tocospermy plesiomorphic) is the problem of finding the correct direction of interpretation of character state.

2. The problem of sperm morphology: a very exciting, beautiful and fertile contribution to Acarology.

The investigations on sperm morphology with the use of the electron microscope brought probably - in the author's opinion - a major contribution for solving this controversy since it was possible to recognize a direction of sperm evolution within Gamasida without any doubt (Alberti 1980a, b, 1984 and 1991). It was evident from these studies (see also Witaliñski 1975, 1976 and 1979) that a sperm type called ribbon type evolved *within* the Gamasida, hence this sperm type has to be regarded as an apomorphy. This sperm type, which is in some taxa very complex, in other rather simple (see below), seems -- according to present knowledge -- to be correlated with further characteristics of the reproductive system (see Alberti 1988, 1991 and 2000a; Alberti and Coons 1999 for further ref.):

*there is an unpaired oviduct that is differentiated into a vaginal duct (oviduct II) and an uterus (oviduct I),

*within the uterus embryonic development starts; only one egg at a time is in the uterus,
*consequently, eggs containing an embryo are deposited meaning that fewer eggs are laid with shorter "dangerous" immobile period of the offspring; this is certainly a more economic and effective way of reproduction,
*the female gonad is differentiated into a germinative layer and a nutritive layer (nutrimentary egg development) - this may be a prerequisite to enhance egg development; the gonad is solid,
*extensive vaginal glands are present that are used to cover the eggs upon oviposition with a more or less complex layer,
*males have a presternal genital opening,
*males have modified chelicerae (the movable digit is modified: spermatotreme, spermatodactyl),
*sperm cells migrate towards the oocytes via the haemocoel, through a peculiar sperm access system (laelapid type), or their route in the female is obscure (phytoseiid type).

These character states support, as synapomorphies, a monophylum -- not yet termed (e.g., Neospermatida? - see below or Alberti 1988; Evans 1992) -- that consists of the sister groups Parasitina (autapomorphies: e.g., nutritive part of female gonad without extensive arms, spermatotreme, regular type of ribbon sperm) and Dermanyssina (=Dermanyssides; Evans and Till 1979; autapomorphies: differentiation of nutritive part of gonad into conspicuous arms =lyrate organ, spermatodactyl, irregular type of ribbon sperm, sperm access system present).

On the other hand, the vacuolated sperm cells are always very complex, develop always during a very complicated spermatogenesis (note, e.g., peripheral cisternae and their transformation into a large vacuole, so-called cellular processes, peculiar acrosomal complex, etc.). They undergo a very unusual capacitation after sperm transfer (so-called prospermia turn inside out to become so-called spermiophores; process known only in ticks in detail) (Breucker and Horstmann 1968 and 1972; Feldman-Muhsam and Filshie 1979; Alberti 1980a; Alberti and Coons 1999 with further ref.). This vacuolated type occurs in Opilioacarida, Holothyrida, Ixodida and several Gamasida (see below). Hence it is an autapomorphy of Anactinotrichida (it does neither occur in Actinotrichida nor in other Arachnida; see Alberti 2000a) and is a plesiomorphic character in Gamasida (Alberti 1980a, 1984, 1988 and 2000a; see also Alberti and Coons 1999; Alberti *et al.* 2000).

Vacuolated sperm cells are apparently correlated with the following features of genital systems:

*paired oviducts, no differentiation into a "true" uterus (i.e. an unpaired part that serves for developing an embryo),
*no embryonized eggs are laid, several or many eggs are developed simultaneously (there might be exceptions; Michael, 1889),
*female gonad not differentiated into a germinative and nutritive part (solitary egg development); gonad is tubular or hollow,
*no extensive vaginal glands (the 'vaginal glands' of Uropodina are probably not homologous with the glands in Parasitina and Dermanyssina, rather they might correspond with the lateral gland complexes in Parasitina),
*males have a midsternal genital opening (but see below: Celaenopsoidea),
*males have no modified chelicerae (but see below: Celaenopsoidea),
*sperm cells move after complex capacitation through oviducts towards the ovary (known from ticks and Sejina; indications from Opilioacarida; Michael 1892; Wagner-Jevseenko 1958; Alberti and Coons 1999).

The evident definition of these sperm cell types as being plesiomorphic (vacuolated type) or apomorphic (ribbon type) in Gamasida should not neglect that *both* types are highly derivative comparing sperm structures in the whole animal kingdom (see Alberti 1991 and 2000a). Alberti

(1988) suggested the terms "archispermy" for those Anactinotrichida using vacuolated sperm cells and transferring them via the primary genital opening in contrast to "neospermy" for those taxa using ribbon sperm and presenting the modified genital system as listed above.

3. Problems with the young: why so difficult?

If we accept the above lists as either apomorphic or plesiomorphic characters of high complexity, a further problem arises. There is -- obviously -- no correlation between the two sperm types and the mode of sperm transfer, i.e. tocospermous taxa have vacuolated sperm (most taxa) and ribbon sperm (Parasitina) and ribbon sperm possessing taxa represent tocospermy (Parasitina) and podospermy (Dermanyssina). The solution is quite evident from the above-mentioned considerations: since all taxa with vacuolated sperm present tocospermy and the arguments for podospermy as being the primitive type is rejectable, Parasitina almost necessarily have to be regarded as representing an intermediate step in evolution of gamasid reproductive systems: they show plesiomorphic transfer but apomorphic internal anatomy and sperm morphology. (*Nota bene*: Parasitina show diplodiploidy, the plesiotypic genetic system, in contrast to Dermanyssina that represent the apotypic haplodiploidy; Norton *et al.* 1993). Hence the simple term tocospermy that describes the mode of sperm transfer via the primary genital opening may be used only in this descriptive pure sense and is, of course, of value. However, it is necessary to stress that it obscures the complexity of character differences between Parasitina on the one hand and of the other tocospermous taxa on the other hand (that partly need still to be explored: see below). Hence it may be more indicative to use a more differentiated nomenclature to make these differences evident: the mode of reproduction (sperm structure, mode of sperm transfer, genital tract) of Parasitina should be termed *neotocospermy*. The mode of reproduction of those Anactinotrichida with vacuolated sperm and showing the plesiomorphic features listed above should be termed *architocospermy* and that of those mites with ribbon sperm and sperm access system *neopodospermy* (based on a suggestion by Witaliñski, *in litt.*). To define these systems, electron microscopy is needed, and this may be a disadvantage compared to the simple tocospermy - podospermy terminology that still makes sense, since electron microscopy is not always easily available. But the new terminology should help to keep in mind that more delicate results are needed in peculiar cases to be more precise. Further, this knowledge should induce attempts to get the whole set of information from this complex system (see also below).

It needs to be stressed that to make these conclusions reliable the definition of Parasitina and Dermanyssina as sister groups is an important prerequesite (see above).

4. A delicate problem: diversity of sperm access systems.

There is the problem with the remarkable diversity found in Dermanyssina,which is by far the most species-rich group of Gamasida. Hence the question once more may be considered: are Dermanyssina monophyletic? (Monophyly of Parasitina apparently is not doubted). With respect to the character states considered here, it may seem that there are profound differences in sperm morphology and in the sperm access systems.

The irregular ribbon sperm type is much more variable than the regular type known from Parasitina (*Parasitus*, *Pergamasus*, *Holoparasitus* and *Porrhostaspis* investigated: Witaliñski 1975, 1976 and 1979; Alberti 1980a; Witaliñski and Dallai 1991 and 1993). Sperm cells of *Varroa*, *Rhodacarus*, *Macrocheles*, *Pachylaelaps* and *Periseius* show evidently structures which are related to the regular rows of chambers in Parasitina. In *Phytoseiulus* the mature sperm almost completely consist of parallel membranes or cisternae, only few cytoplasm is remaining (Alberti 1988). However, since early stages in spermatogenesis show peripheral structures similar to those observed in spermatogenesis of the mentioned taxa and Parasitina in particular, it seems reasonable to consider the phytoseiid sperm as a highly derived and simplified ribbon type (correlated with a different sperm access system; see below).

Probably even more difficult is the question whether the laelapid sperm access system and the phytoseiid sperm access system are homologous. There seem to be profound differences (see Evans and Till 1979; Alberti 1988; Alberti and Di Palma, this volume). Though the function of the phytoseiid system remains largely enigmatic it seems remarkable that there is continuity between the system of the right and the left body sides by the recently detected electron lucent tissue. This makes the phytoseiid system to be considered more similar to the laelapid type than previous knowledge would have allowed.

The insemination pores of the sperm access systems may have different positions (see Athias-Henriot 1968 and 1975b). Thus, one might ask whether the sperm access systems are homologous. It seems very likely that this is the case, since the internal structures -- except of the sperm access systems -- are so similar (see above). The insemination pores most frequently are found in the region between coxae III and IV. Hence, this may be regarded the plesiomorphic location and pores located elsewhere may be derivative.

Nevertheless, it is very important to continue and to initiate study of sperm access systems of further taxa to clarify the function and evolution of this peculiarity of Dermanyssina.

Presently, the monophyly of Parasitina and Dermanyssina seems to be well supported by the characteristics offered by the reproductive system.

5. Problems with the old: the Non-Parasitina/ Dermanyssina - why so extravagant?

Monogynaspida such as Sejina, Uropodina, Zerconina and Epicriina conform with the above mentioned lists (in fact these lists are based on results taken from them) and seem to be early derivative taxa. Regarding sperm morphology there are several peculiarities that have been described earlier (Alberti 1980a). An autapomorphy of at least some Uropodina seems to be the longitudinal cleft in the flat peripheral cisterna (acrosomal cisterna). It would be of interest to study taxa such as Thinozerconidae, Protodinychidae, considered to represent early derivative Uropodina (Krantz 1978; Evans and Till 1979) or Diarthrophallidae thought to be a derived group (Norton *et al.* 1993), in this respect. Since only one species yet has been investigated in Epicriidae and Zerconidae, one can only say that their sperm represent some peculiarities within the vacuolated type. Recently it was possible to investigate a further species of Sejina, e.g. *Asternolaelaps* sp. (Alberti and Klompen, *in prep.*) a member of the Ichthyostomatogasteridae. In contrast to the sperm cells of *Sejus togatus* (Sejidae) that are only fragmentarily known (Alberti and Blaszak 1995) and are of the vacuolated type showing a folded exterior (similar to *Epicrius*), the spermatozoa of *Asternolaelaps* are quite similar to those of certain uropodids (see Alberti 1980a) but apparently lack the longitudinal cleft.

In the list shown above (see point *2. The problem of sperm morphology*) it seems evident that a presternal genital opening of the male is correlated with ribbon sperm and spermatodactyl and hence with neopodospermy, i.e., characteristics relevant for (derivative) taxa belonging to the presumed monophylum Dermanyssina. However, there are species that apparently are quite distant from these taxa, i.e. Celaenopsoidea that also show a presternal male genital opening. There is also sexual dimorphism in cheliceral morphology (Trägardh 1950). These mites belong to the Trigynaspida, a branch of gamasid mites regarded to be systematically distant from the Monogynaspida, discussed above, by most authors. Furthermore, a putative sperm access system and spermatodactyls have been described from the (monogynaspidine) Heterozerconidae (Athias-Henriot 1975b; Flechtman and Johnston 1990), a taxon (Heterozerconina) also not regarded closely related with Dermanyssina (Athias-Henriot 1975b; Johnston 1982; Norton *et al.* 1993). Thus, these observations need careful consideration. If these characteristics are homologous, they could contradict the interpretation just given, since it would be perhaps more likely to accept a plesiomorphic state of the presternal opening, modified chelicerae, and sperm access system. Or, alternatively, e.g., Trigynaspida would represent a paraphyletic group, which includes a sister group of the Monogynaspida (the Celaenopsoidea if this would turn out to be true). The same could be the case with Heterozerconina. Another alternative could be that these similarities have evolved independently in Trigynaspida, Heterozerconina and the other Monogynaspida.

As sperm morphology seems helpful in at least getting an idea, it was a strong wish to get males from a representative of Trigynaspida. With the help of Prof. Blaszak and Prof. Rafalski (University of Poznan) it was possible to study *Celaenopsis badius*. First results were published on occasion of the 2nd EURAAC meeting in Krynica (Alberti and Blaszak 1995). Unfortunately, we were not able to understand the sperm morphology completely, since we had not enough mites to study spermatogenesis in sufficient detail. However, there were some strange features evident: a central region showing meandering structures that was hence called tentatively "meandering body". Even more peculiar was that there -- apparently -- were cellular processes exposed -- in the sperm found in the male -- to the exterior, not toward the lumen of a vacuole as in the typical prospermium. Very exciting was a region which appeared similar to the "chambers" characteristic of most of the ribbon spermatozoa. Hence it was stated that *Celaenopsis* sperm (or the sperm of Antennophorina to which Celaenopsidae belong) might be intermediate between the vacuolated type and the ribbon type (Alberti and Blaszak 1995; Alberti 1995 and 2000a). Recently -- through the kindness of Dr. Klompen (Ohio State University, Columbus) -- we were able to study another representative of Antennophorina, *Megisthanus floridanus* (Megisthanoidea), in more detail. It was possible to describe the sperm of this species quite precisely as a strange sperm that obviously represents the vacuolated type, though with many peculiarities (Alberti and Klompen, *in prep.*). These are mainly: a very extensive operculum (this is the very thin part of the peripheral cytoplasmic mantle at its anterior end which is shed during capacitation establishing an opening of the vacuole through which the inner cytoplasmic column or inner core emerges, thus turning the cell inside-out), and a deep invagination from behind that makes the cytoplasmic column a hollow structure and a peculiar thin walled hollow appendage. This appendage and the invagination contain a material consisting of many strands. Reexamining with this experience the *Celaenopsis* sperm it was evident that the meandering body is homologous to the material found in the invagination and appendage of *Megisthanus*. Furthermore, it became clear that the cellular processes are also in *Celaenopsis* not exposed to the exterior but are -- as in *Megisthanus* -- covered by a very thin external "opercular" membrane that often disrupts artificially. Hence, the *Celaenopsis* sperm are certainly of the vacuolated type though also quite peculiar. *Megisthanus* does not show the chamber-like structures found in *Celaenopsis*. In the latter these structures surround the insertion sites of the cellular processes. When comparing transverse sections between, e.g., *Parasitus* sperm and *Celaenopsis* sperm the cellular processes and the longitudinal crests of the ribbons look very much alike. But in the sperm of Parasitidae these crests contain a small cisterna that is lacking in the cellular processes. Hence this may be a superficial similarity. Another feature of much interest is the extensive "opercular" area. It much reduces the external cytoplasmic mantle. If this mantle would be completely reduced in an evolutionary process, the sperm no longer would be vacuolated. The former vacuole membrane (in particular the membrane of the cytoplasmic column) would become -- secondarily -- the definite membrane as it occurs during capacitation of, e.g., the tick sperm. Interestingly, during spermatogenesis of ribbon sperm the original external cell membrane undergoes considerable transformations and later is replaced (also) by a secondary one (Witaliñski 1976; Alberti 1980a).

Thus, there is some further evidence to consider these sperm intermediate between the vacuolated type and the ribbon type. Whether this *Celaenopsis*-type represents really a step in the evolution of the Parasitina/Dermanyssina sperm or not, it may help to understand how this transformation that necessarily has occurred during the evolution of these taxa could have been performed.

Perhaps, further studies of other representatives of Trigynaspida may bring more clarity. Comparing the *Celaenopsis* sperm and the *Megisthanus* sperm, they differ also with respect to the peculiar appendage that is lacking in *Celaenopsis*, but are very similar with respect to the invagination. Such an invagination occurs quite frequently in the vacuolated type, sometimes it appears not earlier than during capacitation as in ticks, and is called acrosomal canal (Breucker and Horstmann 1972; Feldman-Muhsam and Filshie 1979). In other species, it is already present - to different degrees -- in the so-called prospermium. In *Neocarus texanus* (Opilioacarida), it is quite extensive and hence this early presence may be a plesiomorphic feature. The peculiar contents have only been observed in *Celaenopsis* and *Megisthanus* and thus may be a synapomorphy of these taxa (or autapomorphy of all Antennophorina?).

Heterozercon is peculiar in our context because of its sperm access system. Athias-Henriot (1975b) already distinguished the Heterozerconina from the Dermanyssina because of the different modifications of the male chelicerae. To present knowledge this taxon is the only one outside the Dermanyssina possessing a sperm access system. In the related *Discozercon* it was not yet seen (Evans, *in litt.*), though Johnston (1982) mentioned the presence of a slender spermatodactyl. Hence it was much important to study the sperm of these mites. Fortunately, after a long time of various attempts to find *Heterozercon* it was available again through the kindness of Dr. Klompen and his student B. S. Gerdeman. According to first results the *Heterozercon* sperm seem to be much different from all sperm observed hitherto in Anactinotrichida. They do neither contain a typical vacuole nor chambers. They are quite long containing many mitochondria. However, in early spermatogenesis (unfortunately only few stages were yet observable) cisternae appear that are typical for vacuolated spermatozoa. Thus it may be conclusive to interprete this sperm-type as autapomorphy and the sperm access system likely to have evolved convergently from that of Dermanyssina. Of course, it would be very important to study also *Discozercon*. Are there vacuolated sperm?

Taking this material together, I would suggest or support the idea that most of these taxa are early derivative and Parasitina/Dermanyssina more terminal. The former, of course, have also evolved peculiarites. Among these, interestingly, a sperm access system was established in one taxon (Heterozerconina). This obviously has effects on sperm morphology, thus demonstrating again that sperm morphology is dependent on fertilization conditions (Franzén 1956; Alberti 1991). Since the sperm type differs from that of Dermanyssina and shows indications of the vacuolated type during spermatogenesis, it seems likely that it evolved independently of Parasitina/Dermanyssina from the vacuolated type. It is very much necessary to study internal anatomy/fine structure of these mites to get further material for detailed comparisons with Parasitina/Dermanyssina (Alberti *et al.* 1999; Alberti 2000b).

Accepting this convergence-hypothesis would be in accordance with most of the more recently published systematic concepts of Gamasida (Krantz 1978; Evans and Till 1979; Evans 1992; Norton *et al.* 1993) showing a basal dichotomy of Gamasida (though differing in details), one branch leading to the Trigynaspida (plus Heterozerconina; Norton *et al.* 1993), the other to the (remaining) Monogynaspida. Which type of female genital plates arrangement is the more plesiomorphic? Perhaps one should consider the genital region of Opilioacarida in which three movable plates occur in both sexes in the genital region: the unpaired, median genital sclerite behind the genital opening and a pair of lids covering the genital sacs (Alberti and Coons 1999; genital verrucae: van der Hammen 1989). A small region in front of the genital opening is also separated from the sternal area in *Neocarus texanus* (see Alberti and Coons 1999; fig. 324). In Holothyrida, four plates (one anterior, one posterior, two lateral plates) are found in the genital region of the females (van der Hammen 1989). Have the genital sacs of Opilioacarida been amalgamated into the vulval region leaving the plates as laterogynials? Has the anterior sclerite indicated in *Neocarus* become a more distinct plate? There are even four plates in certain trigynaspid mites, i.e. in Klinckowstroemiidae (Antennophorina). May be this is worth to consider and to study in more detail, i.e. considering also the structures behind these plates. At least, it seems likely that the presence of several plates is the early derivative state and the monogynaspid mites are thus further evolved in this respect (strangely, this was not considered in the paper of Camin and Gorirossi 1955, but see Lindquist 1984).

An alternative interpretation could take the Trigynaspida (or part of them) as more closely related to the Parasitina/Dermanyssina than the other Non-Parasitina/Dermanyssina. Arguments to support this could be the presence of modified chelicerae (Celaenopsoidea) and the presternal genital opening in the males (however, this latter character state was regarded as very different from Parasitina/Dermanyssina by Camin and Gorirossi 1955) and the tendency to give up the vacuolated type of sperm. The *Celaenopsis* sperm are really very persuasive. Perhaps it will be possible to give more decisive answers after collecting more results on internal anatomy of these taxa and after investigating Arctacarina that are considered closely related to Parasitina (Krantz 1978; Johnston 1982) or even to represent the sister group of Parasitina/Dermanyssina (Norton *et al.* 1993). Do the Arctacarina still have vacuolated sperm or do they present rows of chambers?

6. Probably the most intriguing problems: why and how did the reproductive system in Gamasida evolved so peculiarly?

This topic tries to give some ideas about how these structures could have evolved. The early derivative Acari all are soil inhabitants. Regarding reproductive systems, it is well known that many arthropods living in soil habitats reproduce by so-called indirect spermatophore transfer (Schaller 1979). This is true for Actinotrichida (e.g., Schuster and Schuster 1966; Witte 1991; Witte and Döring 1999; Alberti and Coons 1999; Alberti 1991 and 2000a) in which this type of sperm transfer is found in the more basal taxa. Instead, several taxa have evolved, convergently, direct sperm transfer (e.g., several Actinedida: Eleutherengona; all Acaridida; see Alberti and Coons 1999 for references). Contrary to this quite varied scenario of reproductive modes, it seems as if in Anactinotrichida there is more uniformity (Alberti 1991 and 2000a). Indirect spermatophore transfer never has been observed. It would be very much important, however, to improve knowledge on this topic. Only few taxa have been studied under this aspect: Ixodida and Gamasida (Uropodina, Parasitina, Dermanyssina). Nothing is known about the early derivative taxa, i.e. Opilioacarida, Holothyrida and the other Gamasida. Nevertheless, concluded from what is known, it seems a quite common male attitude to manipulate a sac-like spermatophore with the mouthparts (e.g., Michael 1892). In some taxa (Ixodida, Uropodina) the spermatophore is attached close to the female primary genital opening and peculiar modifications of the spermatophore, composed of an ecto- and endospermatophore, help to inseminate the female (Faasch 1967; Feldman-Muhsam 1967 and 1986; Compton and Krantz 1978). Only the ectospermatophore containing sperm enters the female. In Parasitina the complete spermatophore is deposited by the male with his modified chelicerae (spermatotreme) in the female vagina via the primary genital orifice (Zukowski 1964; Korn 1982; Alberti *et al.* 1999). In Dermanyssina, the spermatophore is manipulated with the modified chelicerae (spermatodactyl) and is transferred to secondary (see above) insemination pores through which the endospermatophore enters a peculiar sperm access system. The ectospermatophore drops off or is in other ways discarded as in ticks (e.g., Dosse 1959; Lee 1974; Amano and Chant 1978; Krantz and Wernz 1979; Woyke 1994; see also Thomas and Zeh 1984; Alberti and Coons 1999 with further ref.).

If one assumes that the behaviour shown by Ixodida and Uropodina is the plesiotypic mode, which seems most likely, then it makes sense to suggest secondary translocation of the male genital opening towards the anterior. This would make the spermatophore extruded from the genital opening more easy to grasp with the chelicerae. According to this functional argument it seems plausible that this translocation and even the evolution of modified chelicerae in the males could have evolved several times convergently from the plesiomorphic state, e.g. in certain Trigynaspida (Celaenopsoidea), Heterozerconina and in the other Monogynaspida. This scenario seems to be supported by the differences in the relevant structures so far as they are known (e.g., male chelicerae, sperm morphology; see above). Again it would be very important to have a more detailed knowledge on internal structures, e.g. of the female genital system of Heterozerconina (sperm access system, lyrate organ?).

Hence, translocation of the male genital opening into a presternal position is quite easily explained: it makes the plesiotypic way of spermatophore manipulation more efficient or secure. The modification of male chelicerae may be helpful also in this respect. But certain structures may also have further importance during the probing behaviour (see below).

The spermatophores of Ixodida and Uropodina similarly are "actively" involved in sperm transfer. In both, the endospermatophore evaginates from the ectospermatophore due to internal pressures (Faasch 1967; Feldman-Muhsam 1969) and invades the female vagina that previously had been strongly manipulated with the male's mouthparts.

Parasitina that deposit the whole spermatophore within the vaginal cavity of the female probably avoid the period of endospermatophore evagination etc. that may bear some uncertainties. They have "chosen" a more direct way. Males manipulate the vaginal cavity before spermatophore transfer (Alberti *et al.* 1999).

Perhaps the manipulation prior to transfer (probing behaviour) only stimulates the female to be ready for acceptance, but it may be a basis for evolution of a more intricate behaviour, i.e. destroying the spermatophore or sperm of previous mates. At least in *Macrocheles muscaedomesticae* Yasui (1994 and 1997) has shown that this seems not to be the case. However, the male is able to assess the female copulatory state: does she bear already a spermatophore of a previous mate or not? How old is this spermatophore? Such information could allow the male to be more economic in distributing his life-production of spermatophores.

It should be kept in mind in this context that -- according to sexual selection concepts (Parker 1984) -- it is not only in the male's interest to inseminate as many females as possible, but -- more important -- to fertilize as many eggs/oocytes as possible, preferably of high quality with regard to fitness of the resulting offspring. The female on the other hand also wants high quality for her germ cells and hence has an interest to control access of male genital products to her eggs/oocytes.

With the active role of the male's chelicerae in spermatophore transfer as a basic element of reproductive behaviour in Anactinotrichida a further step in evolution seems a realistic assumption. Males could use the chelicerae in searching an even more direct way towards a successful sperm transfer avoiding competition with other males and avoiding control systems of the female that likely are located in the genital ducts that are passed by the sperm when migrating towards the ovary in the plesiomorphic genital systems (Ixodida; e.g., Wagner-Jevseenko 1958; Brinton *et al.* 1974; Alberti 1991 and 2000a) (cryptic choice; see Witaliñski 1999; Alberti *et al.* 1999; Alberti 2000a). As in several other taxa (many Hirudinea, Onychophora, some Insects) the common male ancestor of Dermanyssina may have "decided" to avoid all these problems by penetrating the female's integument and by introducing the spermatophore directly into the body cavity (Alberti and Hänel 1986). Alternatively, he could have deposited the spermatophore in a weak niche in the cuticle or in a gland opening and the spermatophore material may have been able to dissolve the female integument/tissue opening a passage for the sperm cells into the female's haemocoel. This scenario may also have been induced by a strong competition for females, hence the male was forced to be fast during this insemination process. Such a competition is well known: males are fighting against each other and for females (Neumann 1943; Rapp 1959; Costa 1967; Korn 1982; Thomas and Zeh 1984). Of course, such a male would have needed sperm cells adopted to this way of sperm transfer (it is a well founded rule that sperm morphology is dependent of fertilization conditions -- see above; this is certainly also true for the biochemical/molecular characteristics of sperm cells). Such sperm cells are apparently the ribbon sperm and hence it is of much interest that Parasitina with sperm transfer via the primary opening possess ribbon sperm that obviously move through the female's haemocoel (Alberti 1988; Alberti *et al.* 2000). Necessarily the sperm cells have to penetrate the epithelium of the oviduct (unfortunately the site of penetration is still not known). Hence all these data support the scenario outlined above (see also Alberti and Hänel 1986; Alberti 1988, 1995 and 2000a).

What about the females? It seems likely that they reacted in establishing a sperm access system and hence regained some means of control over the insemination process. They obviously had invested much in improving their genital system. They evolved a differentiated gonad with a complex nutritive tissue and reorganised their genital duct establishing a uterus, structures that allowed egg development to commence in the female. These reorganisations made the female genital system likely more effective or economic, since less eggs needed to be produced. However, this in turn very likely forced the female to be even more selective with regard to quality of sperm. Hence, the evolution of a sperm access system that allows control seems to be a "logic" - parsimonious - consequence. We must admit that we only partly understand these access systems that are presently known to occur in two types: a laelapid and a phytoseiid type (see above). In the laelapid type the sperm are transported (how?) to an unpaired spermatheca (receptaculum seminis) where capacitation occurs (e.g., *Varroa jacobsoni*, Alberti and Hänel 1986). The access system is cuticle lined until it reaches the spermatheca. Is the cuticle open towards the spermatheca? Perhaps this is not the case, as some drawings of the taxonomists show a shrunken sac-like structure, i.e., a cuticular sac (e.g., Fain 1963; Athias-Henriot 1968; Lee 1974; Fain *et al.* 1977). If the cuticle is continuous: is the cuticle penetrated by the capacitated sperm that will move through a peculiar tissue towards the oocytes (see Parasitina above)? Even more enigmatic is the situation in the phytoseiid type of sperm access system. Despite much work, we are not able to describe the route of sperm in the female (Alberti and Di Palma, *this volume*; Di Palma and Alberti, *in press*). We think that we observed capacitated sperm in the ovary. But how the sperm reach the oocytes is not known. The so-called minor duct seems quite unsuitable as a passage to the ovary (see also Alberti 1988). Do the spermatozoa leave the so-called vesicles by penetrating their walls?

According to the detailed observations by Amano and Chant (1978), the endospermatophore disappears rather quickly from the vesicle. What are the males doing during the long time of copulation (up to several hours; e.g., Overmeer *et al.* 1982). Do they simply fill the vesicles with endospermatophore material or are they waiting until their sperm has moved to the ovary securing sperm priority by prolonged mating (Parker 1984; Thomas and Zeh 1984)?

As discussed above, it seems most likely that the laelapid type and the phytoseiid type of sperm access systems are homologous. If this is correct: which one is the more plesiomorphic, which the more apomorphic type? Presently, it appears that the phytoseiid type is more derivative. This is concluded from the sperm morphology which is - according to present knowledge - certainly the most modified in Parasitina/Dermanyssina (Alberti 1988).

Again it is much desirable to have more informations on reproductive systems of other Gamasida, in particular of Heterozerconina!

7. A nice problem: much work is left for the future!

Though knowledge on genital systems of mites has considerably increased and newly applied concepts (e.g., comparative spermatology, sexual selection, sperm competition) have helped to produce ideas, many conclusions suggested above or in previous papers are more or less speculative. Much needs to be done to get the necessary information on key taxa (see, e.g., Alberti 2000b; *in press*)! We are confronted with the reality: we still are just at the beginning of describing acarine biodiversity not to speak of understanding it.

Acknowledgments

This study was financially supported by the Deutsche Forschungsgemeinschaft (AL 138/4-1). The improving comments of the referees are gratefully appreciated. Last but not least, the author wishes to express his thanks for the generous hospitality and friendship he experienced during and after the symposium.

References

Alberti G., 1980a - Zur Feinstruktur der Spermien und Spermiocytogenese der Milben (Acari). I. Anactinotrichida. *Zoologische Jahrbücher. Abteilung für Anatomie und Ontogenie der Tiere*, 104: 77-138.

Alberti G., 1980b - Zur Feinstruktur der Spermien und Spermiocytogenese der Milben (Acari). II. Actinotrichida. *Zoolohgische Jahrbücher. Abteilung für Anatomie und Ontogenie der Tiere*, 104: 144-203.

Alberti G., 1984 - The contribution of comparative spermatology to problems of acarine systematics. In: Griffiths D.A., Bowman C.E. (eds.), Acarology VI. Ellis Horwood, Chichester. Vol. 1: 479-489.

Alberti G., 1988 - The genital system of Gamasida and its bearing for phylogenetical considerations. In: Channabasavanna G.P., Viraktamath C.A. (eds.), Progress in Acarology. VII International Congress of Acarology, Bangalore, 1986. Oxford & IBH Publ. Co., New Delhi, vol. 1: 197-204.

Alberti G., 1991 - Spermatology in the Acari: systematic and functional implications. In: Schuster R., Murphy P.W. (eds.), The Acari - Reproduction, Development and Life-history Strategies. Chapman & Hall, London: 77-105.

Alberti G., 1995 - Comparative spermatology of Chelicerata: review and perspective. In: Jamieson B.G.M., Ausio J., Justine J.-L. (eds.), Advances in Spermatozoal Phylogeny and Taxonomy. *Mémoires du Musée natational d'Histoire naturelle*, 166: 203-230.

Alberti G., 2000a - Chelicerata. In: Jamieson B.G.M. (ed.), Progress in Male Gamete Ultrastructure and Phylogeny. In: Adiyodi K.G., Adiyodi R.G. (eds.), Reproductive Biology of the Invertebrates. Oxford & IBH Publishing, New Delhi and Wiley, N.Y., vol. 9, p. B.: 311-388.

Alberti G., 2000b - Spermatologische Untersuchungen an Gamasiden - Stand und Perspectiven. *Abhandlungen und Berichte aus dem Naturkundemuseum Görlitz*, 72: 91-96.

Alberti G., *in press* - Ultrastructural investigations of sperm and genital systems in Gamasida (Acari: Anactinotrichida) - current state and perspectives for future research. *Acarologia.*

Alberti G., Blaszak C., 1995 - Further observations on spermatozoa in gamasid mites. In: Kropczynska D., Boczek J, Tomczyk A. (eds.), The Acari - Physiological and Ecological Aspects of Acari-Host Relationships. (2nd Symposium of EURAAC, Krynica, Poland). Dabor, Warszawa: 15-22.

Alberti G., Coons L.B., 1999 - Acari - Mites. In: Harrison F.W. (ed.), Microscopic Anatomy of Invertebrates. John Wiley & Sons, Inc., New York, Vol. 8c: 515-1265.

Alberti G., Hänel H., 1986 - Fine structure of the genital system in the bee parasite, *Varroa jacobsoni* (Gamasida: Dermanyssina) with remarks on spermiogenesis, spermatozoa and capacitation. *Experimental and Applied Acarology,* 2: 63-104.

Alberti G., Klompen, H., *in prep.* Fine structure of unusual spermatozoa and spermiogenesis of the mite, *Megisthanus floridanus* Banks 1904 (Acari: Gamasida: Antennophorina).

Alberti G., Gegner A., Witaliñski W., 1999 - Fine structure of the genital system in the females of *Pergamasus* mites (Acari: Gamasida: Pergamasidae). *Journal of Morphology,* 240: 195-223.

Alberti G., Kaiser Th., Klauer A.K., 1996 - New ultrastructural observations on coxal glands (nephridia) of Acari. In: Mitchell R., Horn D.J., Needham G.R., Welbourn W.C. (eds.), Acarology IX, Proceedings. The Ohio Biology Survey, Columbus, vol. 1: 309-318.

Alberti G., Gegner A., Witaliñski W., 2000 - Fine structure of the spermatophore and spermatozoa in inseminated females of *Pergamasus* mites (Acari: Gamasida: Pergamasidae). *Journal of Morphology,* 245: 1-18.

Amano H., Chant D.A., 1978 - Mating behaviour and reproductive mechanisms of two species of predacious mites, *Phytoseiulus persimilis* Athias-Henriot and *Amblyseius andersoni* (Chant) (Acarina: Phytoseiidae). *Acarologia,* 20: 196-214.

Athias-Henriot C., 1968 - L'appareil d'insemination laelapoide (Acariens anactinotriches: Laelapoidea, &&) - Premières observations - Possibilité d'emploi à fins taxonomiques. *Bulletin Scientifique de Bourgogne,* 25: 175-228.

Athias-Henriot C., 1975a - Nouvelles notes sur les Amblyseiini, 2. Le relevé organotaxique de la face dorsal adulte (Gamasides, Protoadiniques, Phytoseiidae). *Acarologia,* 17: 20-29.

Athias-Henriot C., 1975b - The idiosomatic euneotaxy and epineotaxy in gamasids (Arachnida, Parasitiformes). *Zeitschrift für Zoologische Systematik und Evolutionsforschung,* 13: 97-109.

Breucker H., Horstmann, E., 1968 - Die Spermatozoen der Zecke *Ornithodorus moubata* (Murr.) *Zeitschrift für Zellforschung und mikroskopische Anatomie,* 88: 1-22.

Breucker H., Horstmann, E. 1972 - Die Spermatogenese der Zecke *Ornithodorus moubata* (Murr.). *Zeitschrift für Zellforschung und mikroskopische Anatomie,* 123: 18-46.

Brinton L.P., Burgdorfer W., Oliver J.H.Jr., 1974 - Histology and fine structure of spermatozoa and egg passage in the female tract of *Dermacentor andersoni* Stiles (Acari-Ixodidae). *Tissue and Cell,* 6: 109-125.

Camin J.H., Gorirossi F., 1955 - A revision of the suborder Mesostigmata (Acarina) based on comparative morphological data. *Special Publication Chicago Academy of Sciences,* 11: 1-70.

Compton L., Krantz G.W., 1978 - Mating behaviour and related morphological specialization in the uropodine mite, *Caminella peraphora. Science (N. Y.),* 200: 1300-1301.

Costa M., 1967 - Notes on macrochelids associated with manure and coprid beetles in Israel. II. Three new species of *Macrocheles pisentii* complex, with notes on their biology. *Acarologia,* 9: 304-329.

Di Palma A., Alberti G., *in press* - Fine structure of female genital system in phytoseiid mites with remarks on egg nutrimentary development, sperm access system, transfer and capacitation (Acari, Gamasida, Phytoseiidae). *Experimental and Applied Acarology.*

Dosse G., 1959 - Über den Kopulationsvorgang bei Raubmilben aus der Gattung *Typhlodromus* (Acari, Phytoseiidae). *Pflanzenschutzberichte,* 22: 125-133.

Evans G.O., 1963 - Observations on the chaetotaxy of the legs in the free-living Gamasina (Acari: Mesostigmata). *Bulletin of the British Museum (Natural History),* 10: 275-303.

Evans G.O., 1992 - Principles of Acarology. C.A.B. International, Wallingford, UK: 563 pp.

Evans G.O., Till W.M., 1979 - Mesostigmatic mites of Britain and Ireland (Chelicerata: Acari: Parasitiformes). An introduction to their external morphology and classification. *Transactions of the Zoological Society of London,* 35: 139-270.

Faasch H., 1967 - Beitrag zur Biologie der einheimischen Uropodinen *Uroobovella Marginata* (C.L. Koch 1839) und *Uropoda Orbicularis* (O.F. Müller 1776) und experimentelle Analyse ihres Phoresieverhaltens. *Zoologische Jahrbücher. Abteilung für Systematik, Ökologie und Geographie der Tiere,* 94: 521-608.

Fain A., 1963 - La spermathèque et ses canaux adducteurs chez les Acariens Mèsostigmatiques parasites des voies respiratoires. *Acarologia,* 5: 463-479.

Fain A., Hyland K.E., Aitken T.H.G., 1977 - Flower mites of the family Ascidae phoretic in nasal cavities of birds (Acarina: Mesostigmata). *Acta Zoologica et Pathologica Antverpiensia,* 69: 99-154.

Feldman-Muhsam B., 1967 - Spermatophore formation and sperm transfer in *Ornithodoros* ticks. *Science,* 156: 1252-1253.

Feldman-Muhsam B., 1969 - The components of the spermatophore, their formation and their biological functions in argasid ticks. In: Evans G.O. (ed.), Proceedings 2nd International. Congress of Acarology 1967. Hungarian Academy of Sciences, Budapest: 357-362.

Feldman-Muhsam B., 1986 - Observations on the mating behaviour of ticks. In: Sauer J.R., Hair J.A. (eds.), Morphology, Physiology, and Behavioral Biology of Ticks. Ellis Horwood, Chichester: 217-232.

Feldman-Muhsam B., Filshie B.K., 1979 - The ultrastructure of the prospermium of *Ornithodorus* ticks and its relation to sperm maturation and capacitation. In: Fawcett D.W., Bedford J.M. (eds.), The spermatozoon. Urban & Schwarzenberg, Baltimore-Munich: 355-369.

Flechtmann C.H.W., Johnston D.E., 1990 - *Zeterohercon*, a new genus of Heterozerconidae (Acari: Mesostigmata) and the description of *Zeterohercon amphisbaenae* n. sp. from Brasil. *International Journal of Acarology,* 16: 143-148.

Franzén A., 1956 - On spermiogenesis, morphology of the spermatozoon, and biology of fertilization among invertebrates. *Zoologiska Bidrag fran Uppsala,* 31: 355-482.

Hirschmann W., 1957 - Gangsystematik der Parasitiformes. Teil 1. Rumpfbehaarung und Rückenflächen. *Acarologie, Schriftenreihe für vergleichende Milbenkunde. Fürth/Bayern,* Folge 1: 1-26.

Johnston D.E., 1982a - Acari. Opilioacariformes. Parasitiformes. In: Parker S.P. (ed.), Synopsis and Classification of Living Organisms. McGraw Hill, N.Y., vol. 2: 111-117.

Karg W., 1965 - Larvalsystematische und phylogenetische Untersuchung sowie Revision des Systems der Gamasina Leach, 1915 (Acarina, Parasitiformes). *Mitteilungen aus dem Zoologischen Museum Berlin,* 41: 193-340.

Karg W., 1989a - Die Bedeutung der Beute- und Wirtsbeziehungen parasitiformer Milben für die bodenbiologische Standortanalysen. *Pedobiologia,* 29: 285-295.

Karg W., 1989b - Acari (Acarina), Milben - Unterordnung Parasitiformes (Anactinochaeta) Uropodina Kramer, Schildkrötenmilben. Die Tierwelt Deutschlands. 67. G. Fischer Verlag, Jena: 203 pp.

Karg W., 1993 - Acari (Acarina), Milben - Parasitiformes (Anactinochaeta) Cohors Gamasina Leach - Raubmilben. 2nd ed. Die Tierwelt Deutschlands. 59. G. Fischer Verlag, Jena: 523 pp.

Karg W., 1998 - Zur Kenntnis der Eugamasides Karg mit neuen Arten aus den Regenwäldern von Ecuador (Acarina, Parasitiformes). *Mitteilungen aus dem Museum für Naturkunde Berlin, Zoologische Reihe,* 74: 185-214.

Korn W., 1982 - Zur Fortpflanzung von *Poecilochirus carabi* G. u. R. Canestrini 1882 (syn. *P. necrophori* Vitzt.) und *P. austroasiaticus* Vitzthum 1930. *Spixiana,* 5: 261-288.

Krantz G.W., 1978 - A Manual of Acarology. 2nd ed. Oregon State University Book Stores, Corvallis: 509 pp.

Krantz G.W., Ainscough, B.D., 1990 - Acarina: Mesostigmata (Gamasida). In: Dindal D.L. (ed.), Soil Biology Guide. John Wiley & Sons, New York: 583-665.

Krantz G.W., Wernz, J.G., 1979 - Sperm transfer in *Glyptholaspis americana.* In: Rodriguez J.G. (ed.), Recent Advances in Acarology. Academic Press, New York, Vol. 2.: 441-446.

Lee D.C., 1974 - Rhodacaridae (Acari: Mesostigmata) from near Adelaide, Australia. 3. Behaviour and development. *Acarologia,* 16: 21-44.

Lindquist E.E., 1984 - Current theories on the evolution of major groups of Acari and on their relationships with other groups of Arachnida, with consequent implications for their classification. In: Griffiths D.A., Bowman C.E. (eds.), Acarology VI. Ellis Horwood, Chichester, Vol. 1: 28-62.

Lindquist E.E., Evans G.O., 1965 - Taxonomic concepts in the Ascidae, with a modified setal nomenclature for the idiosoma of the Gamasina (Acarina: Mesostigmata). *Memoirs of the Entomological Society of Canada,* 47: 1-64.

Lindquist E.E., Moraza M.L., 1998 - Observations on homologies of idiosomal setae in Zerconidae (Acari: Mesostigmata), with modified notation for some posterior body setae. *Acarologia,* 39: 203-226.

Michael A.D., 1889 - Observations on the special internal anatomy of *Uropoda krameri. Journal of the Royal Microscopical Society,* 1889: 1-15.

Michael A.D., 1892 - On the variations in the internal anatomy of the Gamasinae, especially in that of the genital organs and their mode of coition. *Transactions of the Linnean Society of London (ser. 2),* 5: 281-324.

Moraza M.L., Lindquist E.E., 1998 - Coprozerconidae, a new family of zerconoid mites from North America (Acari: Mesostigmata: Zerconoidea). *Acarologia,* 39: 291-313.

Moritz M., 1993 - Unterstamm Arachnata. In: Gruner H.-E. (ed.), Lehrbuch der Speziellen Zoologie (begr. von A. Kaestner). 4. ed. Bd. 1: Wirbellose Tiere. 4. Teil: Arthropoda. G. Fischer Verlag, Jena: 64-442.

Neumann K.W., 1943 - Die Lebensgeschichte der Käfermilbe *Poecilochirus necrophori* Vitzthum nebst Beschreibung aller Entwicklungsstufen. *Zoologischer Anzeiger,* 142: 1-21.

Norton R.A., Kethley J.B., Johnston D.E., OConnor B.M., 1993 - Phylogenetic perspectives on genetic systems and reproductive modes of mites. In: Wrensch D.L., Ebbert M.A. (eds.), Evolution and Diversity of Sex Ratio in Insects and Mites. Chapman and Hall, New York: 8-99.

Overmeer W.P.J., Doodeman M., van Zon A.Q., 1982 - Copulation and egg production in *Amblyseius potentillae* and *Typhlodromus pyri* (Acari, Phytoseiidae). *Zeitschrift für Angewandte Entomologie,* 93: 1-11.

Parker G.A., 1984 - Sperm competition and the evolution of animal mating strategies. In: Smith R.L. (ed.), Sperm Competition and the Evolution of Animal Mating Systems. Academic Press, London: 1-60.

Rapp A., 1959 - Zur Biologie und Ethologie der Käfermilbe *Parasitus coleoptratorum* L. *Zoologische Jahrbücher, Abteilung für Anatomie und Ontogenie der Tiere. Jena,* 86: 303-366.

Schaller F., 1979 - Significance of sperm transfer and formation of spermatophores in arthropod phylogeny. In: Gupta A.P. (ed.), Arthropod Phylogeny. Van Nostrand Reinhold Comp., New York: 587-608.

Schuster R., Schuster I.J., 1966 - Über das Fortpflanzungsverhalten von Anystiden-Männchen (Acari, Trombidiformes). *Naturwissenschaften,* 53: 162.

Sellnick M., 1944 - *Zercon* C.L. Koch. Königsberg: Acari-Blätter für Milbenkunde, 5: 30-41.

Sellnick M., 1958 - Die Familie Zerconidae Berlese. *Acta Zoologica Academia Scientia Hungaricae,* 3: 313-368.

Thomas R.H., Zeh D.W., 1984 - Sperm transfer and utilization strategies in arachnids: Ecological and morphological constraints. In: Smith R.L. (ed.), Sperm Competition and the Evolution of Animal Mating Systems. Academic Press, Orlando: 180-221.

Trägardh I., 1946 - Outlines of a new classification of the Mesostigmata (Acarina) based on comparative morphological data. *Lunds Universitets Årsskrift. N.F. Avd. 2*, 42: 3-36.

Trägardh I., 1950 - Studies on the Celaenopsidae, Diplogyniidae and Schizogyniidae (Acarina). *Arkiv för Zoologi,* 1: 361-451.

Van der Hammen L., 1989 - An Introduction to Comparative Arachnology. SPB Publishing bv, The Hague: 576 pp.

Wagner-Jevseenko O., 1958 - Fortpflanzung bei *Ornithodorus moubata* und genitale Übertragung von *Borrelia duttoni. Acta Tropica,* 15: 119-168.

Weygoldt P., 1999 - Evolution and systematics of the Chelicerata. In Bruin J., van der Geest L.P.S., Sabelis M.W. (eds.), Ecology and Evolution of the Acari. Dordrecht, Kluwer Academ. Publ.: 1-14.

Witaliñski W., 1975 - Spermatogenesis in free living mite, *Pergamasus viator* Halas (Parasitidae, Mesostigmata). I. Fine structure of spermatozoa. *Zeitschrift für mikroskopisch-anatomische Forschung*, Leipzig, 89: 1-17.

Witaliñski W., 1976 - Spermatogenesis in free living mite, *Pergamasus viator* Halas. (Parasitidae, Mesostigmata). II. Spermiogenesis. *Zeitschrift für mikroskopisch-anatomische Forschung*, Leipzig, 90: 657-680.

Witaliñski W., 1979 - Fine structure of the respiratory system in mites from the family Parasitidae. *Acarologia,* 21: 330-339.

Witaliñski W., 1999 - Sperm competition in the Acari. In: Bruin J., van der Geest L.P.S., Sabelis M.W. (eds.), Ecology and Evolution of the Acari. Kluwer Academic Publishers, Dordrecht: 149-156.

Witaliñski W., Dallai R., 1991 - Ultrastructural study of spermatogenesis in gamasid mites (Acari, Gamasida). *Acta Zoologica (Stockholm),* 72: 49-54.

Witaliñski W., Dallai R., 1993 - F-actin and tubulin during spermatogenesis in gamasid mites (Acari: Parasitina). *Acta Zoologica (Stockholm),* 74: 205-214.

Witte H., 1991 - Indirect sperm transfer in prostigmatic mites from a phylogenetic viewpoint. In: Schuster R., Murphy P.W. (eds.), The Acari: Reproduction, Development and Life-history Strategies. Chapman & Hall, London: 137-176.

Witte H., Döring, D., 1999 - Canalized pathways of change and constraints in the evolution of reproductive modes of microarthropods. In: Bruin, J., van der Geest L.P.S., Sabelis M.W. (eds.), Ecology and Evolution of the Acari. Kluwer Academic Publishers, Dordrecht: 15-43.

Woyke J., 1994 - Mating behavior of the parasitic honeybee mite *Tropilaelaps clareae. Experimental and Applied Acarology,* 18: 723-733.

Yasui J.H., 1994 - Adaptive control of copulation duration by males under sperm competition in the mite, *Macrocheles muscaedomesticae. Experimental and Applied Acarology,* 18: 543-554.

Yasui J.H., 1997 - Sperm competition and the significance of female multiple mating in the predatory mite, *Parasitus fimetorum. Experimental and Applied Acarology*, 21: 651-664.

Young J.H., 1968 - The morphology of *Haemogamasus ambulans*. II. Reproductive system (Acarina: Haemogamasidae). *Journal of the Kansas Entomological Society,* 41: 532-543.

Zukowski K., 1964 - Investigations into the embryonic development of *Pergamasus brevicornis* Berl. (Parasitiformes, Mesostigmata). *Zoologica Polonia,* 14: 247-268.

Morphological variability between and within populations of *Tectocepheus* (Acari, Oribatida, Tectocepheidae) from the *velatus*-complex in Central Europe

Gerd Weigmann

Soil Zoology and Ecology Laboratory, Institute of Biology, Free University Berlin, Grunewaldstr. 34, D-12165 Berlin, Germany

Key words: Acari, Oribatida, *Tectocepheus*, systematics, morphology.

Abstract: The morphologically different forms of the *Tectocepheus velatus* complex are regarded by some authors to be a group of closely related species, in Europe, for instance, including *T. velatus* s. str., *T. sarekensis*, *T. tenuis*, *T. knuellei* and *T. alatus*. Other authors summarize all these species under '*T. velatus* (s. l.)'. This opinion seems to be supported by the continuous expression of some morphological characters, such as body length, within a great range of variability.

This paper demonstrates distinct morphological differences between some of 12 Central European populations, divisible into three distinct groups. This supports the idea of three taxonomically different species or subspecific taxa. The patterns of population characters correspond with the above mentioned species s. str.: *T. tenuis*, *T. velatus* and *T. sarekensis*. There are also differences in ecological preferences.

The results are discussed within the background of the thelytoky of *Tectocepheus*, which presents problems for the discussion of speciation or phylogeny between clones. The worldwide character combinations of *Tectocepheus* populations do not indicate stable character combinations of postulated species: as a 'compromize' result from Europe, five somewhat reproducible taxa at a subspecies level should be accepted: *T. velatus velatus* (Michael, 1880); *T. v. sarekensis* Trägardh, 1910; *T. v. tenuis* Knülle, 1954; *T. v. alatus* Berlese, 1913; *T. v. knullei* Vanek, 1960.

Introduction

Since the description of *Tectocepheus velatus* (Michael, 1880), a large number of species and variations have been described within this genus. Nübel-Reidelbach (1994) lists 29 taxa, of which she only accepts two species. This contribution deals predominantly with Central European populations. Therefore, only descriptions from Europe will be discussed. *T. minor* Berlese, 1903, is well separated from *T. velatus* by a different orientation of the lyrifissure *iad* as well as other characters, as demonstrated in detail by Nübel-Reidelbach (1994); the synonymy of *T. cuspidentatus* Knülle, 1954 and *T. bisignatus* Moritz, 1968 to *T. minor* appears to be very probable. But the synonymization of *T. sarekensis* Trägardh, 1910 and *T. tenuis* Knülle, 1954 to *T. velatus* by Nübel-Reidelbach (1994) appears to be problematic. Many other 'micro-species' of the *velatus* group of wordlwide origin will not be discussed in this paper.

F. Bernini, R. Nannelli, G. Nuzzaci, E. de Lillo (eds.), Acarid Phylogeny and Evolution. Adaptations in mites and ticks, 141-152.

Material and methods

Origin of Tectocepheus populations

Twelve populations were studied biometrically, which are labeled 'a' to 'm' in the following list.

Population 'a': collected in 1968 at List on the Isle of Sylt; dune area near the seashore; n=30 specimens, determined as *T. tenuis* (ecology cf. Weigmann 1973: site 34).

Pop. 'b': collected in 1996 by T. Hörnke in Köpenick Forest in Berlin, a pine stand; n=30, det. as *T. velatus*.

Pop. 'c': collected in 1993 near Netra, Central Germany, in a mesophilous oak forest; n=20, det. as *T. velatus*.

Pop. 'd': collected in 1991 in the Bärenklau Forest, north of Berlin, in an acid pine forest; n=20, det. as *T. velatus*.

Pop. 'e': collected in 1988 by L. Miko in Jinin, Southern Bohemia, Czech Republic, in a meadow; n=20, det. as *T. velatus*.

Pop. 'f': collected in 1977 in the Spandau Forest, Berlin. In a sample transect from bog to forest: in the forest border with alder and birch, mainly, surrounding a bog complex; n=30, det. as *T. velatus* (ecology cf. Weigmann 1991: mainly site 1F).

Pop. 'g': collected in 1977 together with 'f'. In a sample transect, near 'f', in the forest border with birch and oak, mainly, surrounding a bog complex; n=30, det. as *T. sarekensis* (ecology cf. Weigmann 1991: site 1G).

Pop. 'h': collected in 1988 by L. Miko in Jinin, Southern Bohemia, Czech Republic, in a meadow; together with 'e'; n=9, det. as *T. sarekensis*.

Pop. 'i': collected in 1968 at List on the Isle of Sylt; upper salt-marsh area near the seashore, adjacent to 'a'; n=20, now det. as *T. sarekensis* (ecology cf. Weigmann 1973: site 13: as '*velatus*').

Pop. 'k': collected between 1993-1996 in flood plains in the Lower Oder Valley, Northeast-Germany; n=20, det. as *T. sarekensis*.

Pop. 'l': collected in 1978 in the central city area of Berlin, at a ruderal site; n=30, det. as *T. sarekensis*.

Pop. 'm': Collected 1984 in the Grunewald Forest, Berlin, in a pine-oak stand with acid soils; n=20, det. as *T. sarekensis*.

Data and statistics

From each specimen the following parameters were taken after preparation in an open lactic-acid microscopic slide: (1) *total body length*; (2) *width of the lamellar complex* (the distance from the outer lamellar border at the bothridium from one side to the other; see fig. 1P); (3) *cuspis tip-form type* (see fig. 1A-K); (4) *direction of the inner cuspis border* in relation to the axis (see fig. 1P: angle α: directed to the outside, strait, or ± to the axial side); (5) presence of a longitudinal *striation in the interlamellar field* (see fig. 2C-D); (6) number of pairs of *depressions on the notogaster* (see fig. 1: N1, N2).

The statistical analysis was restricted to the parameters (1) to (3). Parameter (4) is nearly correlated with parameter (3) and ultimately served only to confirm the cuspis form type. The parameters (5) and (6) are provisional criteria for determination of the 'microspecies' (*T. velatus* s. str., *T. sarekensis*, *T. tenuis*: sensu Knülle 1954). These parameters (5 and 6) were not included in the statistical analysis, but were used to evaluate the results.

The metric parameters (1) and (2) are not normally distributed within the populations; therefore, median values, quartiles, and extremes were used for characterization of the distributions. Differences between the populations were analyzed by a Kruskal-Wallis-H-test; in the case of significance, followed by an unbalanced Nemenyi-test (α=5%).

The cluster analyses were based on the three main parameters (1-3) of all 279 specimens from the 12 populations. The range of the parameters were standardized: the minimum values were set to 0% and the maximum values to 100%, the 11 ordinally scaled cuspis-type classes A-K to 0, 10, ... 100. By this transformation, the influence of each of the parameters is equalized. The *k*-

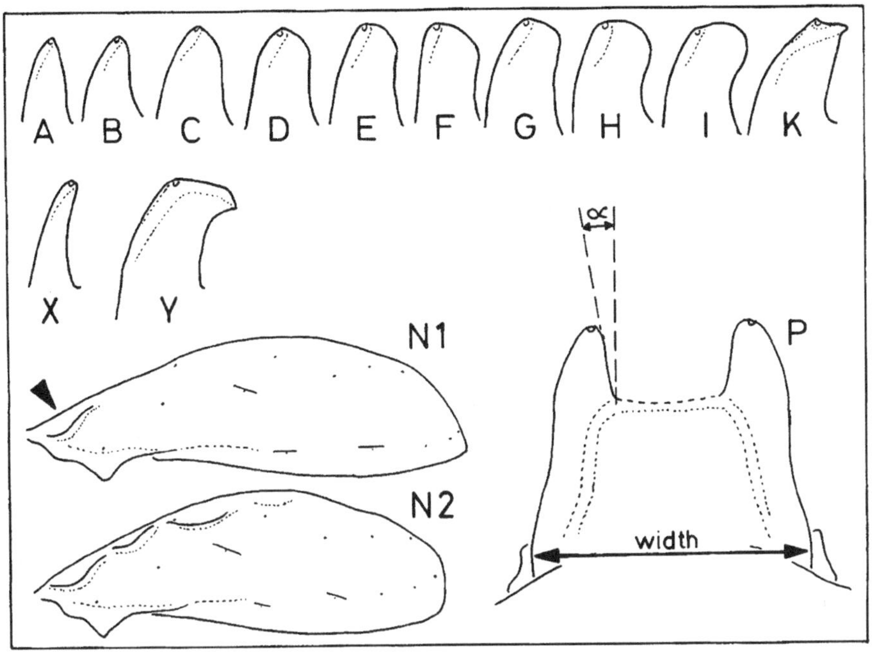

Fig. 1 - Morphological details as used for population differentiation. A - K) Cuspis types within the studied populations; X) Cuspis type of *T. alatus*; Y) Cuspis type of *T. knullei*; N1-N2) lateral view of notogasters: N1) *velatus*-type with one depression (arrow head); N2) *sarekensis*-type with three distinct depressions and the 4th indicated. Schematized lamellar complex with the width parameter (arrow) and the angle of inner cuspis border (A-K, Y, N1-2: modified after Nübel-Reidelbach 1994).

means-clustering and a *tree-clustering* with 'weighted pair-group avarage' (WPGMA) and euclidian distance algorithms were used. The first method puts each specimen into the best fitting group (k = 2, 3, or 4 groups), the number of which was postulated a priori. The results must be compared in regard to the best solution for separating distinct clusters. The tree-clustering runs without a theoretical expectation; the result is proven by whether the specimen-clusters represent the populations or other groupings of presumed taxonomical value.

All statistics and statistical graphs were made with the software package STATISTICA 5.1 (StatSoft).

Results

The classical Central European species

In the revisions of the classical species, *T. velatus* and *T. sarekensis*, the following morphological characters have mainly been used for differentiation: body size, cuspis form, striation on the interlamellar field, and pairs of cuticular depressions on the notogaster (Haarløv 1952; Knülle 1954; Abd-El-Hamid 1965). The same characters have been used and discussed by Nübel-Reidelbach (1994). Some other characters (such as the rostrum-border shape, see tab. 1) in her study seem to be different between the populations studied here and will therefore not

Tab. 1 - Selected characters of *Tectocepheus* species, based on Central European specimens only (after Knülle 1954; in [ ]: original observations).

	tenuis		*velatus*		*sarekensis*	
1. body length (μm)	241-255	[248-263]	281-348	[280-317]	295-362	[302-349]
2. body width (μm)	117-134		161-201		161-214	
3. width of lamellar area (μm)		[66-73]		[81-95]		[93-105]
4a. cuspis tip	tapered tip, as narrow as *le* insertion		axially from *le* slightly convex		axially from *le* strongly convex	
4b. cuspis orientation	inner sides divergent to each other		inner sides strait to divergent		inner sides convergent	
5. rostral border	3-arched		3-arched		rounded	
6. interlamellar field	without ...		without ...		with longitudinal striation	
7. Notogastral depressions	[none]		[1]		[(2-) 3 (-4)]	

be discussed. Within this study, one population was determined to be *T. tenuis* Knülle, 1954, following the classical opinions.

There are two qualitative characters in table 1 that will serve as alternative diagnostic characters, provisionally; but in the following biometrical analysis, they will be omitted: (1) the presence of interlamellar striation; (2) the number of depression pairs on notogaster (see figs 1-2). Based on Knülle's species concept, we find three combinations in the studied populations:

T. tenuis: no depressions on notogaster; no interlamellar striation (fig. 2B),

T. velatus s. str.: one distinct pair of notogastral depressions near the shoulders; no interlamellar striation (fig. 2A),

T. sarekensis: normally, three distinct pairs of notogastal depressions (sometimes 2, sometimes 4 depressions, sometimes the posterior depressions are indistinct); with interlamellar striation (fig. 2C, D).

Body length and width of the lamellar complex

The body-length variation within the 12 studied populations were less than ±5% from the arithmetic means. The total range (minimum to maximum) was about ±16% from the total mean of 310 μm. These facts indicate that the studied material was heterogeneous, possibly with three

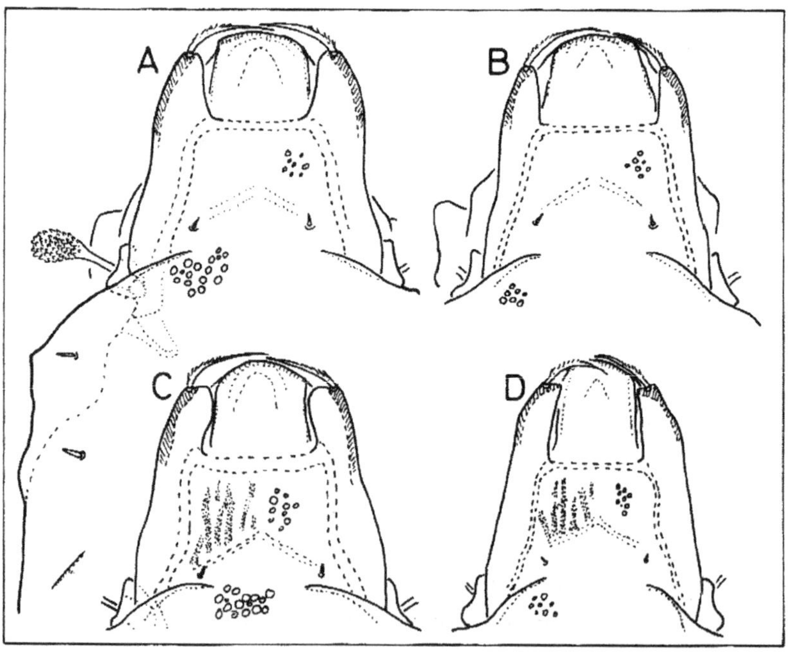

Fig. 2 - Prodorsum morphology of *Tectocepheus* 'species'. A) *T. velatus*; B) *T. tenuis*; C) *T. sarekensis*; D) *T. sarekensis*, variation with dents at the inner cuspis tip (dominant in an Oder River flood plain, population k). The cerotegument dots are only partly drawn.

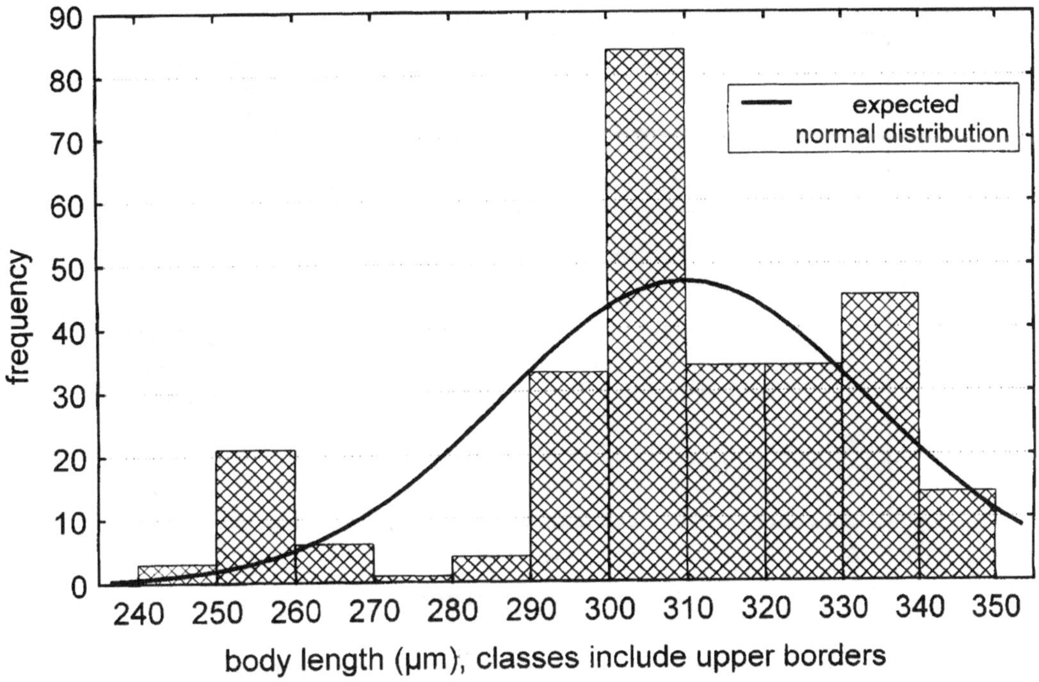

Fig. 3 - Distribution of the body lengths of all studied *Tectocepheus* specimens (N=279). The distribution with three peaks differs from the expected, statistically normal distribution.

subgroup maxima of about 255, 305, and 335 µm body length (fig. 3). The single populations also did not show statistically normal, symmetric distributions. Thus, the following analyses are mostly based on median values with quartiles as characteristics of distribution.

The suspected three subgroups of body-length distribution fit to the three taxa *T. tenuis*, *velatus* and *sarekensis*, as morphologically characterized (n=30, 120, 129). All population length distributions were significantly different (H-test, α=5 %) between subgroups. Each population of the subgroups was significantly different from each population of the other subgroups. Within the three subgroups, some populations were not significantly different from each other (same number at the top of fig. 4).

The 'total notogastral width' seemed to be less reliable than the 'width of the lamellar complex' (see fig. 1P), because the first parameter depends partly on the gravidity status of the specimens, and on the mode of preparation. The distribution pattern of the width of the lamellar complex was similar to that of the body-length distribution. It also showed three maxima of about 68, 88, and 98 µm (results not shown). Three populations subgroups could be established, which coincide with the taxa *T. tenuis*, *velatus*, and *sarekensis*. Within these subgroups, the rate of significant differences was less than that found in the body-length distributions (fig. 5).

Correlations of size parameters and cuspis forms

The parameters 'body length' and 'width of lamellar complex' were highly correlated (Spearman's R=0,905) when calculated for all 279 specimens. But this is an heterogeneity correlation, because the specimens do not belong to a homogenous collection (see last items). In figure 6, the suspected taxa *T. tenuis*, *velatus* and *sarekensis* have been marked differently. The group of *T. tenuis* is absolutely distinct.

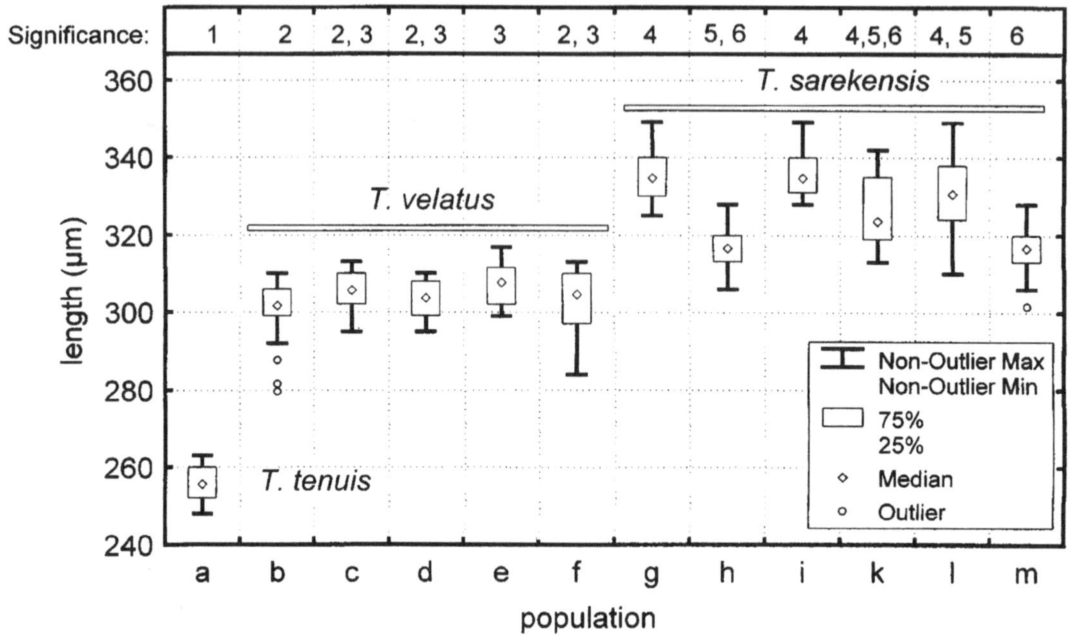

Fig. 4 - Body-length variability within populations. Same numbers above each population indicate no significant differences (α=5%). The labeled subgroups '*T. tenuis*', '*T. velatus*', and '*T. sarekensis*' refer to populations having the same qualitative characters.

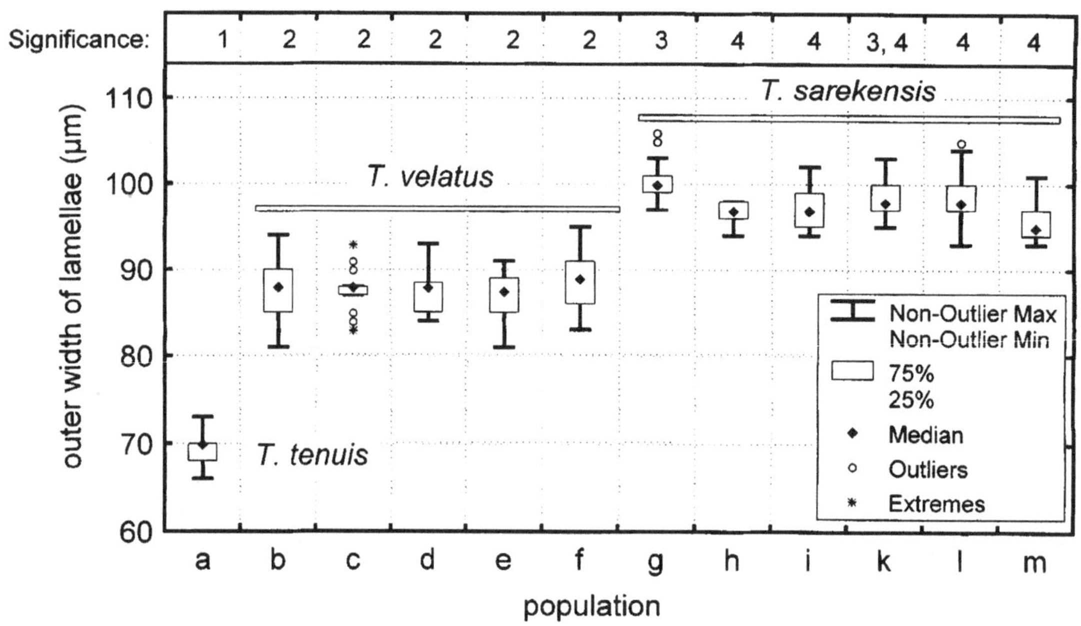

Fig. 5 - Basal width of lamellar region within populations (explanation as in fig. 4).

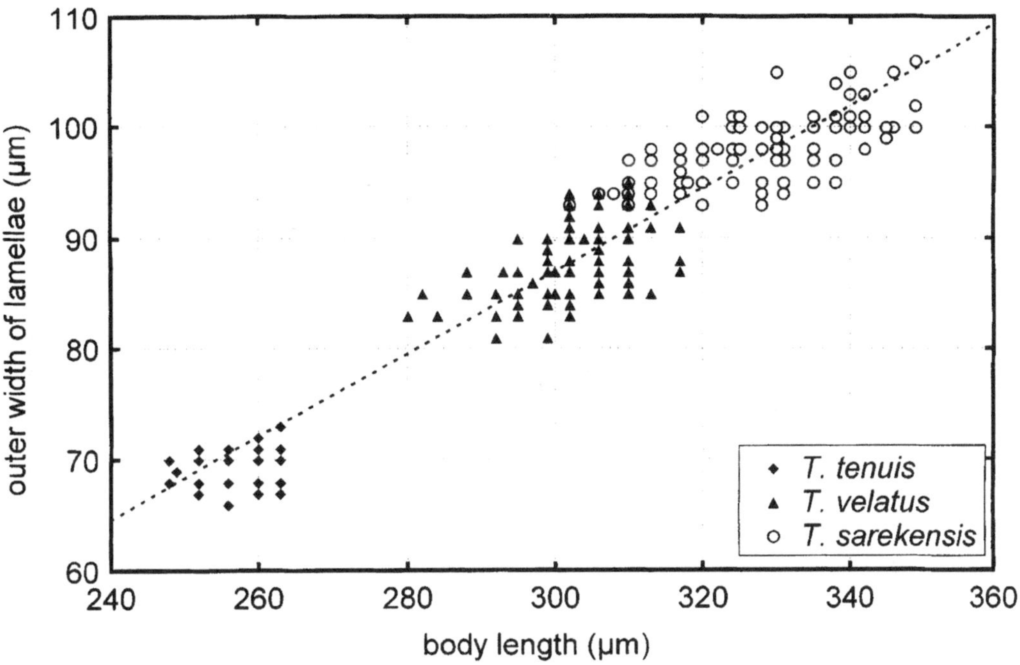

Fig. 6 - Correlation of body length vs. lamellar width. The labeled subgroups '*T. tenuis*'(n=30), '*T. velatus*' (n=120), and '*T. sarekensis*' (n=129) refer to specimens having the same qualitative characters.

The best characterization of the 'cuspis types' seems to be the cuspis-tip shape (see tab. 1: 4a and fig. 1A-K). If the parameters 'body length' vs. 'cuspis types' are plotted, we find three completely distinct groups, which absolutely coincide with the taxa *T. tenuis*, *velatus* and *sarekensis* (fig. 7). The cuspis type ranges overlapped between the *tenuis*- and the *velatus*-group, but not between the *velatus*- and the *sarekensis*-group. With regard to body length, the *tenuis*-group were separated from the others. The same picture could be shown for the correlation of the parameters 'width of lamellar complex' vs. 'cuspis types', with less overlap of width range (results not shown).

Multivariate cluster analysis

The characters used in the last section - body length, width of the lamellar complex, cuspis type - were tested by multivariate cluster analyses after standardization (see material and methods).

The first cluster test was a *k-means-clustering* (see material and methods). The initial expectation was that three clusters would be the best solution to arrange all specimens into clusters. This expectation was supported by the morphological characteristics and by the results listed above. The test was made to create two optimized clusters out of all 279 specimens, first, and then three clusters and four clusters.

The results are visualized in figure 8. The first clustering with k=2 (fig. 8, left) created cluster C1 with all *velatus* and all *tenuis* specimens together. The mean of the standard deviations of the three parameters was approximately s=17, which is symbolized by a circle around the center of the cluster. The euclidian distance to cluster C2 indicates a quite a good separation of the two clusters. C2 includes all *sarekensis*-specimens without exception.

The second clustering with k=3 (fig. 8, middle) created three distinct clusters with minimized standard deviations (compared with k-2-clustering), each of which includes one morphologically defined taxon: C1 - *velatus*, C2 - *tenuis*, C3 - *sarekensis*. The third clustering with k=4 (fig. 8, right) created a cluster C1 with all

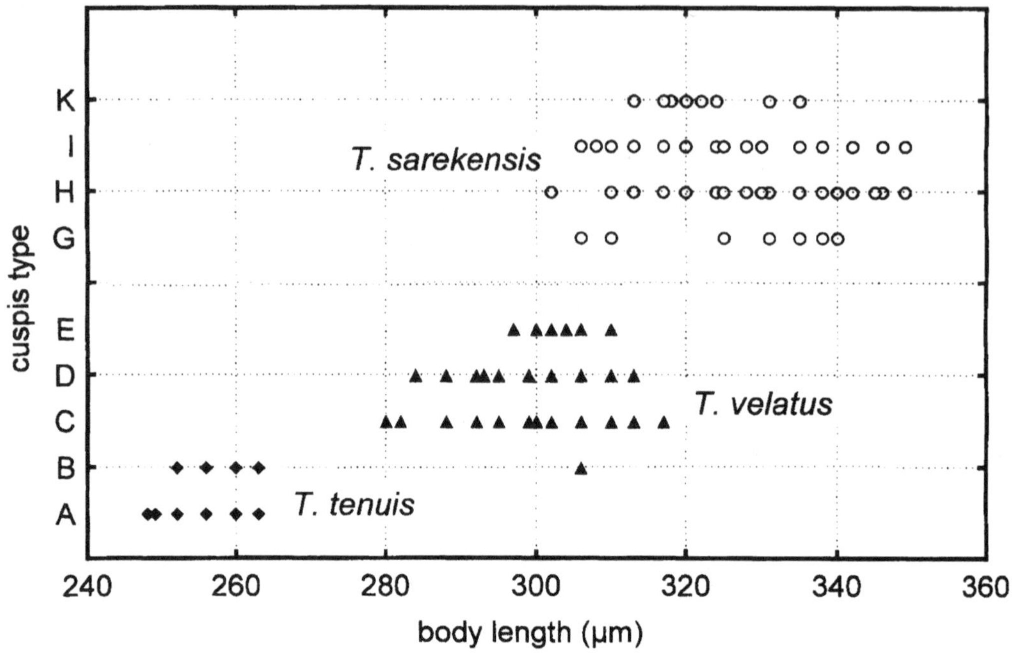

Fig. 7 - Correlation of cuspis type vs. body length (explanation as in fig. 6).

velatus-specimens, a cluster C2 with all *tenuis*-specimens, and two clusters C3 and C4 with parts of the *sarekensis*-specimens. The latter clusters are not well separated from each other. The k-4-clustering appears to be insufficient; the k-3-cluster gives the best solution, statistically, and in the sense of the expectation formulated above. The k-2-clustering is also insufficient, in that the variability in C1 was very high: as we can deduce from the figures 3-7, the large standard deviations

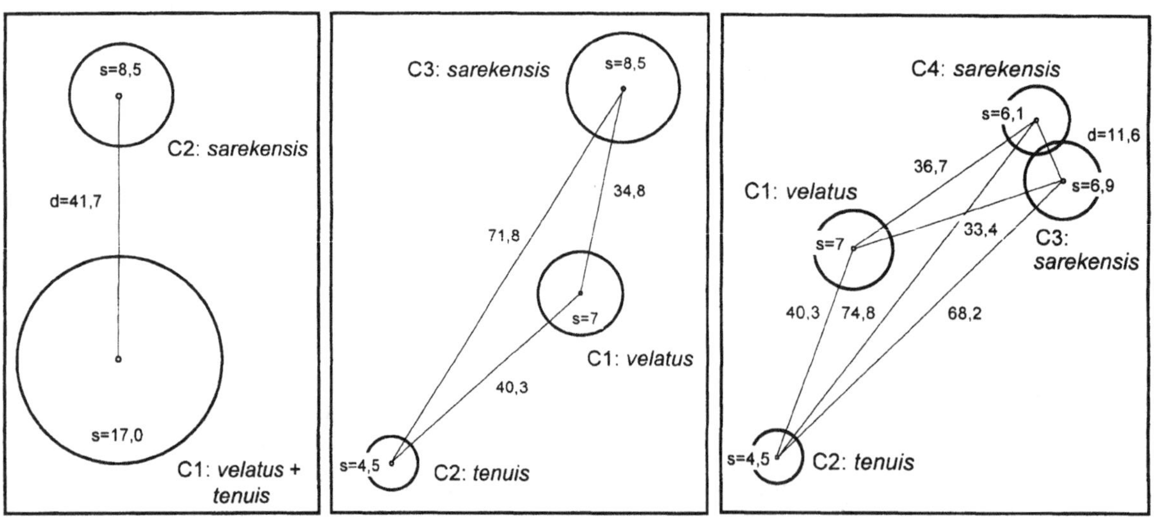

Fig. 8 - Diagram of k-means-clustering results. Left) k=2; middle) k=3; right) k=4. Distances *d* indicate the calculated euclidian distances between clusters, circles around the cluster centers indicate the mean of standard deviations *s* of the three parameters used.

were caused by differently positioned parameter ranges in the *tenuis*- and *velatus*-populations.

All specimens were also clustered by tree-clustering (see material and methods). The large size of the cluster with 279 specimens does not allow a printed figure to be presented (it may be requested from the author). All specimens of the '*tenuis*'-population formed one cluster, which was linked to the rest at a distance of 97%. The maximum subcluster distance within this group was less than 14%, indicating a closely related group of specimens.

The remaining specimens formed two clusters, joined at a distance of 62.5% (all following subclusters had maximum distances of less than 30%). These two clusters included, alternatively, all specimens of the '*velatus*'-populations and all specimens of the '*sarekensis*'-populations, respectively. The latter fact is extraordinarily important: the *velatus*-cluster included specimens from five different localities, the *sarekensis*-cluster from six different localities in Central Europe. But no local population formed a distinct subcluster of it's own. This indicates more similarity between some specimens of each taxon (*velatus, sarekensis*) from different local populations than between all specimens of each population.

Discussion and conclusions

The taxonomical problem of thelytokous species

Tectocepheus species are assumed to be thelytokous, from available information. This form of parthenogenesis is widespread in oribatid mites (Norton *et al.* 1993). Some families are completely thelytokous, thus having numerous, quite different species. But there is no convincing phylogenetic model of species differentiation for such families or genera. We must suggest that all specimens in a population are genetically isolated. The offspring of a somewhat specific female builds a clone group, within which there should be a high similarity; between the clone groups of a population, the similarity may or may not be high (for *Trhypochthoniellus* as an oribatid example, see Weigmann 1999). Thus, the members of a population might partly differ genetically, morphologically and/or ecophysiologically. The total gene pool might differ between populations, depending on the number of different clone mothers colonizing a site by chance, on the mutations within the population of a site, on the competitive capacity of the different specimens, and on other factors (for general theory, see Futuyma 1990; for thelytoky in oribatid mites, see Palmer and Norton 1992).

As a consequence, it is not possible to decide exactly whether two similar 'morpho-species' (differentiated only by morphological characters) are 'species in the common sense' or are of sub-specific status. In bisexual species with syntopic occurrence, we suppose that the speciation took place in geographical separation; when they secondarily came together again, the genetical isolation prevents renewed sexual mixing: the species are 'bio-species'. Yet different forms or races within the species will mix the gene pool by sexual reproduction. In thelytokous populations we can expect a mixture of allopatric and sympatric differentiations with regard to morphological and other characters, and we can expect a high rate of abnormal specimens, as have been observed by some authors with regard to *Tectocepheus* (Knülle 1954; Fujikawa 1995 and 1999; Bayartogtokh 1998). This means that the spread of offspring from populations should result in partly heterogeneous morphological variability. But the mixing of specimens of different local origins (and different genetical and morphological speciality) should decline with geographical distance. This has not yet been substanciated for some lower oribatid mites: 'Among populations in the U.S., there was no relationship between geographic distance and genetic identity' (Palmer and Norton 1992), possibly because the taxa are old and have had time to spread over larger regions with many clones.

Tectocepheus populations of the *velatus*-complex might serve as a good example for studying variability within and between thelytokous populations and for discussing these questions.

The taxonomical structure within the T. velatus group

The 12 Central European populations of *Tectocepheus* studied here can be arranged in three statistical groups. The 'total body length', 'width of the lamellar complex' and 'cuspis type' can discriminate the three groups, between which there is no or minor overlap of character expression, if the characters are plotted each alone or two together. Best results with regard to separation of the groups were created by cluster analysis techniques with all three characters. It seems not to be necessary to use more characters. All groupings coincide with the three described 'species' sensu Knülle (1954): *T. velatus*, *sarekensis* and *tenuis*. These studies on population variability seem to support the idea that we should discriminate three distinct species. On the other hand, we find differences of total length or width parameters that were significant between populations of different origin, but belonged to the same 'species' group.

Haarløv (1952) and Nübel-Reidelbach (1994) came to alternative results based on similar biometric data. Haarløv found all studied populations belonging to one large and polymorphic species, *T. velatus* (including *tenuis*- and *sarekensis*-forms). But he separated *T. alatus*, based on one single Danish population with very distinct special character combinations and without considerable variability: a small and long cuspis which bends to the axis (see fig. 1X), large dots of cerotegument, no notogastral depressions, very broad sensillus, and a body length between 300 and 340 μm. This combination of characters was not present in the populations studied here.

Nübel-Reidelbach (1994) lumped together nearly all described species of *Tectocepheus* with the exception of *T. minor*, whose separation is accepted by all recent authors dealing with it. She examined many characters, some of which seem to be of minor value in discriminating populations, and some of which should discriminate some populations or morphological groups quite well, in my opinion. But some characters (as body length, cuspis forms, and both combined) show gapless continuity when all studied specimens of worldwide origins are plotted together. There is neither an indication of the different taxonomical status in the sense of earlier authors nor is there the chance of discriminating groups from different localities or from the same or different populations in her paper. All studied specimens have been labeled *T. velatus*, posteriorly, including specimens with *tenuis* - or *sarekensis*-characters, as well as with *alatus* - (*sensu* Haarløv 1952) or *knullei*-characters (*sensu* Vanék 1960; cf. fig. 1Y). This means that we cannot judge whether the continuous (=non discrete) expression of characters (e.g. figs 19, 22-24 in Nübel-Reidelbach 1994) is the result of mixed character expression over the origins or whether it is the result of false correlation due to the lumping together of partial heterogeneous groups (cf. fig. 6 in this paper). On the other hand, there are bimodal character expressions in '*T. velatus*' from the same locality (e.g. figs 4, 8 in Nübel-Reidelbach 1994) or different localities (e.g. fig. 10 in Nübel-Reidelbach 1994), which should give sufficient arguments for heterogeneous biometrical data.

Confusion might arise when studies of other authors are compared with those in this paper or with the biometrical analyses of Haarløv (1952) and Nübel-Reidelbach (1994). The Austrian population of '*T. velatus*', as described by Abd-el-Hamid (1965), is in contrast to the differentiation of the species by Knülle (1954) and the one presented in this paper: the cuspis type fits better to '*T. velatus*' according to table 1 in this paper, the striation on the interlamelar field points towards '*T. sarekensis*', any notogastal depressions are not mentioned, and the body length is within the range of both '*velatus*' and '*sarekensis*' (according to tab. 1).

Fujikawa (1995) studied the variability of populations of *T. velatus* from a crop field, grassland, and coniferous forest. There were statistical differences, e.g., in body-length and cuspis-type distributions between the populations. The forest population was smaller in size (mean of body length 269 µm) than the crop-field population, with an average length of 314 µm. But the latter showed narrower cuspis forms than the first. In comparison with the smaller *velatus*-form and the larger *sarekensis*-form in Central Europe, the cuspis-form tendencies are exchanged. This indicates the important fact that the size and cuspis relations in Central Europe are not the same as those found in Japanese populations; the relations might be regional but are not of worldwide constancy.

From Mongolia, Bayartogtokh (1998) discriminated *T. minor*, *T. velatus* and *T. alatus*. His '*T. minor*' does not agree with the redescription of Nübel-Reidelbach (1994) with regard to the position of the lyrifissure *iad*, which is paranal (more or less parallel to the anteriolateral anal plate border) in *minor*, but nearly transversal in the Mongolian population (with also a cuspis form without an outer dent, and the position of the lamellar setae on the outer side are different from those of a typical *minor*). His '*T. minor*' is more likely a special variation of *T. velatus* sensu lato. The Mongolian *T. alatus* is most similar to the *T. alatus* described by Haarløv (1952). This form might be a reproducible morphological type within the range of variations of *T. velatus* sensu lato. The variability of the cuspis types in the Mongolian '*T. velatus*' (in Bayartogtokh 1998: figs 1, 3) includes the European '*velatus*' and '*sarekensis*' ranges from small to broad cuspis, but also includes a special variation with an inner corner on the cuspis tip near the lamellar-seta insertion. The same cuspis type (type K in fig. 1) has been found to be dominant in one German population in this study (population 'k' from the Oder valley: cf. fig. 2D); it seems to be a link to the cuspis of *T. knullei* (cf. fig 1Y).

The contrasting opinions of the taxonomy of the *T. velatus* species group in Europe - some distinct good species (following Knülle 1954; Vanek 1960) vs. one or two species only (following Haarløv 1952; Nübel-Reidelbach 1994) - must be discussed within the background of the reflections presented above on speciation processes in thelytokous genera.

(1) If there are reproducible and disjunct morphological groups with populations in several localities over a greater geographical area, and if these groups show no intermediary morphologies when occurring sympatrically in an area or in the same locality, we should regard the groups as distinct morpho-species. (2) If the characters within the suggested morphological groups are not reproducible in all populations of the same morphological group, and if the characters are not distinct from those in other morphological groups (each important character alone or in combination), we should regard the populations in toto as representatives of one single species.

The situation in the *T. velatus* complex is intermediary: the Central European populations studied here can be separated into three distinct morphological groups: '*tenuis*', '*velatus* s. str.' and '*sarekensis*'. When studying the literature, two other taxa seem to be distinct: *T. knullei* Vanek, 1960, and *T. alatus* Berlese, 1913, sensu Haarløv (1952) and Bayartogtokh (1998). Other described species (cf. Nübel-Reidelbach 1994) may probably be within the variabilities of these five taxa, but shall not be discussed here. However, the incomplete morphological separation and the inconsistent character combinations within the subgroups of '*velatus* s. str.' and '*sarekensis*', when regarded worldwide, contradict the opinion of a well-defined and distinct species.

We know of sympatric occurrence of populations that belong to the three groups '*tenuis*', '*velatus* s. str.', and '*sarekensis*'(cf. figs 4-8): The populations 'f' (*velatus*) and 'g' (*sarekensis*) studied here are from a bog-to-forest sample transect in Berlin-Spandau, but with different dominances in wet or dry plots at a maximal distance of 30 m. Populations 'e' (*velatus*) and 'h' (*sarekensis*) are from the same sample series in Southern Bohemia (collected by L. Miko). Populations 'a' (*tenuis*) and 'i' (*sarekensi*') come from sample plots at a distance of 20 m from each other (a dune area and a salt marsh area on the Isle of Sylt, Northern Germany). Within these syntopic cases, there are no intermediary character expressions at all; the subpopulations are exactly separated in morphological respect.

T. velatus s. l. is one of the most abundant and eurytopic species in Europe; the same seems to be true if *T. velatus* s. str. and *T. sarekensis* are regarded separately (cf. for a review of German references: Weigmann and Kratz 1983). But there are also single occurrences: e.g., in the city area of Berlin in ruderal soils, *T. sarekensis* only occurs (about 15 studied sites); in the more natural surroundings of Berlin, both taxa, *T. velatus* and *T. sarekensis*, occur, partly separated, partly syntopic. Both taxa do not inhabit the same range of habitats and soils, they are partly different in ecological respect. *T. tenuis* seems to be stenotopic in dry dune areas.

The ecological differences as well as the possibility of discriminating some distinct morphological groups of *T. velatus* s. l. in Central Europe (as an example) calls for a taxonomical separation of the groups; the worldwide polymorphism inhibits a convincing separation into well defined species. As a consequence, the groups should be separated at a subspecies level:

Tectocepheus velatus velatus (Michael, 1880)
Tectocepheus velatus sarekensis Trägardh, 1910
Tectocepheus velatus tenuis Knülle, 1954

Based on data from the literature, it seems to be reasonable to separate two further subspecies:

Tectocepheus velatus alatus Berlese, 1913
Tectocepheus velatus knullei Vanék, 1960.

References

Abd-El-Hamid M.E., 1965 - Wiederbeschreibung von *Tectocepheus velatus* (Michael, 1880) (Acari, Oribatei). *Zoologischer Anzeiger*, 175: 426-436.

Bayartogtokh B., 1998 - *Tectocepheus* species in the oribatid mite fauna of Mongolia (Acari: Oribatida, Tectocepheidae). *Scientific Journal, National University of Mongolia, Biology*, 7: 25-44.

Fujikawa T., 1995 - Comparison among populations of *Tectocepheus velatus* (Michael, 1880) from forests, grasslands and crop field. *Edaphologia*, 55: 1-82.

Fujikawa T., 1999 - Individual variations of two reared oribatid species, *Tectocepheus velatus* (Michael, 1880) and *Oppiella nova* (Oudemans, 1902). *Edaphologia*, 62: 11-46.

Futuyma D.J., 1990 - Evolutionsbiologie. Birkhäuser, Basel, Boston, Berlin, 104-497.

Haarløv N., 1952 - Systematics and ecology of the genus: *Tectocepheus* Berlese, 1896 (Acarina). *Entomologisk Meddelelser*, 26(5): 424-437.

Knülle W., 1954 - Die Arten der Gattung *Tectocepheus* Berlese (Acarina: Oribatei). Zoologischer Anzeiger, 152: 280-305.

Norton R.A., Kethley J.B., Johnston D.E., O'Connor B.M., 1993 - Phylogenetic perspectives on genetic systems and reproductive modes of mites. In: Wrensch D., Ebbert M. (eds), Evolution and diversity of sex ratio in insects and mites. Chapman & Hall, New York: 8-99.

Nübel-Reidelbach E., 1994 - Taxonomie und Systematik der Gattung *Tectocepheus* Berlese, 1895 (Acari, Oribatei). *Andrias*, 12: 3-94.

Palmer S.C., Norton R.A., 1992 - Genetic diversity in thelytokous oribatid mites (Acari, Acariformes: Desmonomata). *Biochemical and Systematic Ecology*, 20: 219-231.

Vanék J., 1960 - *Tectocepheus knullei* n. sp. (Acarina, Oribatoidea). *Acta Societas Entomologica Cechoslovakia*, 57: 397-401.

Weigmann G., 1973 - Zur Ökologie der Collembolen und Oribatiden im Grenzbereich Land - Meer (Collembola, Insecta - Oribatei, Acari). *Zeitschrift für wissenschaftliche Zoologie*, 186: 295-391.

Weigmann G., 1991 - Oribatid communities in transects from bogs to forests in Berlin indicating the biotope qualities. In: Dusbábek F., Bukva V. (eds), Modern acarology. SPB Acad. Publ., Hague, vol. 1: 359-364.

Weigmann G., 1999 - Morphological variability in populations of a thelytokous mite, *Trhypochthoniellus longisetus* (Oribatida), with notes on synonymy. In: Bruin J., van der Geest L.P.S., Sabelis M.W. (eds), Ecology and evolution of the Acari. Kluwer Acad Publ, Dordrecht: 581-586.

Weigmann G., Kratz W., 1983 - Die deutschen Hornmilbenarten und ihre ökologische Charakteristik. *Zoologische Beiträge*, 27: 259-489.

Endogynium types in Parasitidae as revealed by SEM (Acari: Gamasida: Parasitina)

Wojciech Witalinski, Katarzyna Borsuk
Institute of Zoology, Jagiellonian University, Ingardena 6, PL-30 060 Kraków, Poland

Key words: receptaculum seminis, insemination, spermatheca evolution, mites.

Abstract: In Parasitidae, insemination occurs during mates' coupling, when spermatophores are deposited with the use of male modified chelicerae, into the endogynium (spermatheca) of the female. The endogynium is the highly sclerotized dorsal region of cuticular lining of the vagina and can be characterized by several cuticular specializations of taxonomic importance such as stipula, spherules, thorns or denticles, lamellae, etc. The details of insemination and sperm storage are, however, still far from being fully explained.

Our aim was to study the distribution of conspicuous perforated cuticle in the genera of Parasitidae, reported to date only in the spherules of endogynia in *Pergamasus* mites and supposedly involved in the signal transmission between the female (ovary) and the spermatophore.

Combined taxonomic literature analysis, light microscopic observations and scanning electron microscopical studies of KOH-macerated material revealed the presence of unpaired areas of perforated cuticle also in endogynia where typical spherules are absent. Such a cuticle was visible either as central, circular perforated plate or as a continuous band of perforated cuticle on endogynium walls. Unpaired areas of perforated cuticle - together with underlying cells - have been considered to be an equivalent of spherules and were named the spherular organ.

Spherules or spherular organ are a constant element in the endogynium of parasitid mites, thus essential for its function.

According to the general structure and distribution of spherules or the spherular organ, we can recognize three types of endogynia:

type 1 - with unpaired spherular organ located on the endogynium wall, usually encircling either totally or partially the endogynial entrance

type 2 - with unpaired, central spherular organ located in the dorsal part of the endogynium

type 3 - with paired, symmetrically located spherules.

The supposed evolutionary routes of endogynium structure, as well as the role of observed cuticular specializations in endogynium function are discussed.

Introduction

Females of gamasid mites can be inseminated with either spermatophore or semen, transmitted by male modified chelicerae (functioning as gonopods) and introduced - depending on the mite group - into a primary or secondary female genital orifice. The mode of insemination is correlated with the structure of female reproductive system, sperm structure and routes of spermatozoa within the female body during migration towards oocytes (see for details Alberti 1984, 1988, 2000 and this volume; Alberti and Hänel 1986).

During insemination in Parasitidae, considered here as the only family belonging to

F. Bernini, R. Nannelli, G. Nuzzaci, E. de Lillo (eds.), Acarid Phylogeny and Evolution. Adaptations in mites and ticks, 153-168.

Parasitina, spermatophores are inserted through a primary genital orifice into the cuticle-lined vagina finally being deposited in its dorsal, well sclerotized convexity. This part of the vagina, termed endogynium, is very important for at least two reasons: physiological and taxonomic. First, it functions as a spermatheca: the deposited sperm is stored there for some time and can allegedly influence female reproductive activity (Alberti *et al.* 1999b). Second, the endogynium is an interspecifically variable structure, very useful for taxonomic identification. Structurally, the endogynium is comprised of the endogynial sac (where spermatophore is deposited) and different cuticular protrusions like thorns, denticles and dentate lamellae. In the endogynium of genus *Pergamasus,* there is an elongated and usually forked protrusion called a stipula, located in endogynium midline and flanked by two spherical-to-elongated spherules (the names of endogynium parts are largely attributable to Athias-Henriot 1967a and b). Recently, Alberti and co-workers (Alberti *et al.* 1999a and b) have found that spherules have a peculiar structure, since their cuticle has a very richly perforated procuticle and the underlying cells show a structure typical for cells engaged in substance transport. Their apical parts form numerous protrusions penetrating the cuticular pores, whereas the basal parts possess many mitochondria-containing infoldings.

In this study we attempt to classify the morphological types of endogynia in Parasitidae according to the structure and distribution of spherule-related perforated cuticle using a scanning electron microscope (SEM) as a basic tool. We also present some further observations on endogynium structure leading to general conclusions concerning the functional aspects of endogynium and its evolutionary diversification.

Material and methods

SEM studies were performed on 10 species (tab. 1), selected on the basis of different endogynium structure as reported in the taxonomic literature (e.g. Athias-Henriot 1967a and b; Micherdzinski 1969; Hyatt 1980; Karg 1993). The mites were collected using a Berlese-Tullgren funnel from forest litter or rotted plant matter, in several localities in Southern Poland. Females were segregated and preliminarily identified under a stereomicroscope, some of them were used for permanent light microscopic slides to confirm species identification and to allow light microscopic observation of endogynium structure; the others were processed for SEM to visualize both the internal and external structure of the cuticular lining of endogynium.

The generic concept of Parasitidae was adopted largely from Evans and Till (1979), Hyatt (1980), and Athias-Henriot (1971) with only some modifications: the introduction of distinct genera like *Rhabdocarpais* Athias-Henriot, 1981, *Leptogamasus* Trägårdh *sensu* Athias-Henriot, 1971, *Aclerogamasus* Athias-Henriot, 1971. The list of genera and species studied and/or referred to in the text is presented in table 1.

Light microscope preparation:

Mites were killed by immersion in 70% ethanol, were then briefly rinsed in distilled water and mounted on microscopic slides in Berlese fluid (Krantz 1978) and finally cleared and dried at 50°C for several days.

SEM procedure:

Females killed as above were dissected with fine needles to remove either the ventral epigynial plate or dorsal cuticle of idiosoma. To observe the details of the cell-faced surface of endogynial cuticle, the material was macerated in 5% KOH at 50°C for one to several days (controlled under stereomicroscope), washed in distilled water (4 x 1 h), and then dehydrated in ethanol and acetone. After a critical point drying and coating with gold, the material was examined in JEOL JSM-35 or JSM-5410 scanning electron microscopes at 25 kV.

Abbreviations

Asterisk, spherular organ; es, endogynial sac; pa, paragynium; sph, spherule; st, stipula; t, muscle attachment (tendon); tr, trachea; vd, cuticular lining of the vaginal duct; vg, vaginal glands; III, IV, openings for coxae III and IV, respectively.

Light microscopic images are indicated as LM; the others are SEM micrographs of KOH-macerated specimens.

Results

In parasitid mites, the female genital orifice is located on the ventral side at the level of III/IV pair of legs. Laterally it is flanked by two paragynial plates and posteriorly by the epigynium, which covers the entrance to the first part of the genital tract, the vagina. The vaginal lumen is lined with a thin cuticle, which is more or less thickened in the dorsal part to form an unpaired convex spermatophore storage organ, the endogynium. The lateral parts of the endogynium make firm connections with the arcuate thickenings of the paragynia. In fact, the endogynium comprises both the cuticular linings and the underlying cells, but in this study the term is largely used in the taxonomic sense and thus concerning the cuticle only.

Posteriorly, the vagina passes into a vaginal duct, which is also lined with a thin cuticle.

General observations on the structure of the endogynium

SEM observations combined with light microscopic examinations and an analysis of the taxonomic literature revealed several characteristic specializations of the endogynial cuticle (figs 1, 3, 19-20, 25-26): endogynial sac, spherules or spherular organ, stipula, and different cuticular protrusions, such as thorns, denticles and lamellae. All those cuticular structures except the spherules and spherular organs are formed by a solid, smooth cuticle, without perforations on either the internal or external side.

The endogynial sac

The endogynial sac is a more (e.g. *Holoparasitus pseudoperforatus*; *H. tirolensis* figs 5, 7; *Pergamasus* spp. figs 20-21, 27, 35-36; *Leptogamasus* spp. figs 31-32) or less (*Vulgarogamasus micherdzinskii* fig. 11; *V. kraepelini* figs 12, 14-15; *Eugamasus* spp.) distinct convex pouch on the dorsal side of the endogynium. The internal surface of the endogynial sac is usually smooth, but in some species denticles are present (e.g. *H. tirolensis* fig. 4; *Pergamasus adinae*; *P. canestrinii*; *P. laminarius*; *P. pinguicrus*).

The stipula

The stipula is either an unpaired protrusion of the posterior endogynial margin, rod-like and usually bifid, or wider, lamellar, sometimes with a dentate margin and/or surface (e.g. *L. belligerens*, *L. orghidani*). When rod-like in shape, its base is usually hollow. Stipulae are typically present in *Pergamasus* and in most *Leptogamasus* species (rod-like stipula: e.g. *P. barbarus* figs 19-20; *P. brevicornis*; *P. mediocris* fig. 25; *P. quisquiliarum* fig. 26; *P. septentrionalis* fig. 34; *L. anoxygenellus* figs 28-29; *L. cristulifer*; *L. doinae*; *L. gimescus*).

Thorns, denticles and lamellae

In most species, the endogynial margin is armed with cuticular thorns and denticles directed largely centripetally (e.g. *H. pseudoperforatus* fig. 2; *P. barbarus* fig. 20; *P. mediocris* fig. 25; *Paragamasus alpestris*). Posteriorly, the entrance into the endogynial sac can be armed with two long spines (e.g. *Neogamasus islandicus*), or covered, at least partially, with a thin, sometimes dentate lamella (e.g. *Parasitus fimetorum* fig. 38; some *Holoparasitus* species: *H. dallaii*) or lamellar protrusions (e.g. *H. ellipticus*; *H. gibber*; *H. hemisphaericus*; *P. alpestris*).

Spherules and spherular organ

Spherules are paired cuticular convexities projecting into the endogynial sac cavity, located posteriorly or posterolaterally. Typically, their

Tab. 1 - Species studied by SEM (**sem**), light microscope (**lm**), or known from literature only (**lit**); basic literature: Athias-Henriot (1967a and b), Micherdzinski (1969), Hyatt (1980), Karg (1993), and other source literature, when necessary.

Pergamasinae		
Paragamasus Hull		
P. alpestris (Berlese, 1904)	**lm**	Bhattacharyya 1963
Aclerogamasus Athias-Henriot		Athias-Henriot 1971
A. similis (Willmann, 1953) *sensu* Athias-Henriot, 1967	**lm**	
Holoparasitus Oudemans		
H. ampullaris Witaliński, 1994	**lm**	Witaliński 1994
H. caesus Micherdziński, 1969	**sem**	
H. crassisetosus Juvara-Bals & Witaliński, 2000	**lm**	Juvara-Bals and Witaliński 2000
H. dallaii Witaliński, 1994	**lm**	Witaliński 1994
H. digitiformis Juvara-Bals & Witaliński, 2000	**lm**	Juvara-Bals and Witaliński 2000
H. ellipticus Juvara-Bals & Witaliński, 2000	**lm**	Juvara-Bals and Witaliński 2000
H. fortunatus Juvara-Bals, 1975	**lm**	Juvara-Bals 1975
H. gibber Juvara-Bals & Witaliński, 2000	**lm**	Juvara-Bals and Witaliński 2000
H. hemisphaericus (Vitzthum, 1923)	**lm**	
H. pseudoperforatus (Berlese, 1906)	**lm**	
H. tirolensis (Sellnick, 1968)	**sem**	
H. tuberculatus Juvara-Bals, 1975	**lm**	Juvara-Bals 1975
H. vasilei Juvara-Bals, 1995	**lit**	Juvara-Bals 1995
Leptogamasus Trägårdh		Trägårdh 1936, Athias-Henriot 1971
L. anoxygenellus (Micherdziński, 1969)	**sem**	
L. belligerens Witaliński, 1973	**lm**	
L. cristulifer (Athias-Henriot, 1967)	**lit**	
L. cuneoliger (Athias-Henriot, 1967)	**lm**	Witaliński 1979
L. doinae Juvara-Bals, 1981	**lit**	Juvara-Bals 1981
L. gimescus (Athias-Henriot, 1970)	**lm**	Witaliński 1979
L. orghidani Juvara-Bals, 1981	**lit**	Juvara-Bals 1981
Pergamasus Berlese		
P. adinae Juvara-Bals, 1970	**lit**	Juvara-Bals 1970
P. barbarus Berlese, 1904	**sem**	
P. brevicornis Berlese, 1903	**sem**	
P. canestrinii (Berlese, 1905)	**lm**	
P. laminarius Witaliński, 1971	**lm**	Witaliński 1971
P. mediocris Berlese, 1904	**sem**	
P. pinguicrus Athias-Henriot, 1967	**lit**	
P. quisquiliarum (G. & R. Canestrini, 1882)	**sem**	
P. septentrionalis (Oudemans, 1902)	**sem**	
Amblygamasus Berlese		Athias-Henriot 1967c
A. mirabilis (Willmann, 1951)	**lit**	
A. mirifactus (Athias-Henriot, 1967)	**lit**	
A. talpicrus (Athias-Henriot, 1967)	**lit**	
A. tescellus (Athias-Henriot, 1967)	**lit**	

Parasitinae		
Parasitus Latreille		
P. fimetorum (Berlese, 1904)	**sem**	
P. loricatus (Wankel, 1861)	**lm**	
Rhabdocarpais Athias-Henriot		Athias-Henriot 1981
R. consanguineus (Oudemans & Voigts, 1904)	**lm**	
Vulgarogamasus Tichomirov		
V. kraepelini (Berlese, 1905)	**sem**	
V. micherdzinskii Karg, 1971	**lm**	
Eugamasus Berlese		
E. berlesei (Willmann, 1935)	**lm**	
Porrhostaspis Müller		
P. lunulata Müller, 1859	**lm**	
Cornigamasus Evans & Till		
C. lunaris (Berlese, 1882)	**lm**	
Parasitellus Willmann	**lit**	
Gamasodes Oudemans		
G. hortivagus (Berlese, 1903)	**lit**	Athias-Henriot 1979
Poecilochirus G. & R. Canestrini	**lit**	
Trachygamasus Berlese		
Trachygamasus sp.	**lm**	
Neogamasus Tichomirov		
N. islandicus (Sellnick, 1940)	**lm**	
N. tikhomirovi Athias-Henriot, 1977	**lm**	Athias-Henriot 1977
Neogamasus sp.	**lm**	

shape is spherical (e.g. *P. barbarus* figs 19-20; *P. brevicornis*; *P. mediocris* fig. 25; *L. anoxygenellus* figs 28-29; most *Paragamasus* species), though sometimes less regular (*P. quisquiliarum* fig. 26) or even evidently elongated, shoe-shaped (e.g. *P. septentrionalis* fig. 34; *P. canestrinii*) or horn-shaped (*P. fimetorum* fig. 40). It may also be irregular in shape or armed with denticles (e.g. in many *Leptogamasus* species: *L. cuneoliger*, *L. gimescus*; *L. orghidani*).

The spherular organ, regarded as the functional equivalent of spherules, is an unpaired structure characterized by a perforated cuticle (in fact, perforations in the cuticle of spherules and spherular organ are present in the procuticle, but not in the epicuticle). It forms either a well-demarcated circular plate located centrally in the dorsal wall of endogynium (e.g. *V. micherdzinskii* fig. 11; *V. kraepelini* figs 12-16; *Eugamasus berlesei*; *Parasitus loricatus*; *Rhabdocarpais consanguineus*) or a less distinct perforated area of the endogynial wall, usually located posteriorly (e.g. *H. pseudoperforatus* fig. 2). In many cases, however, the perforated cuticle forms a band on the endogynial sac wall close to the sac entrance, encircling its diameter completely or partially (e.g. *H. tirolensis* figs 6, 7; *Aclerogamasus similis*; *P. alpestris*; *Amblygamasus mirabilis*; *A. mirifactus*; *A. talpicrus*; *A. tescellus*). Sometimes the perforated cuticle protrudes in the form of sharp projections into the endogynial cavity (e.g. *Cornigamasus lunaris* fig. 1; *Gamasodes hortivagus*; *Neogamasus* sp. fig. 3; *N. tikhomirovi*). In the bubble-shaped endogynia (e.g. *H. caesus* figs 8-10; *H. fortunatus*; *H. tuberculatus*), the perforated cuticle of the spherular organ is located at the entrance to the tube leading to the endogynial sac.

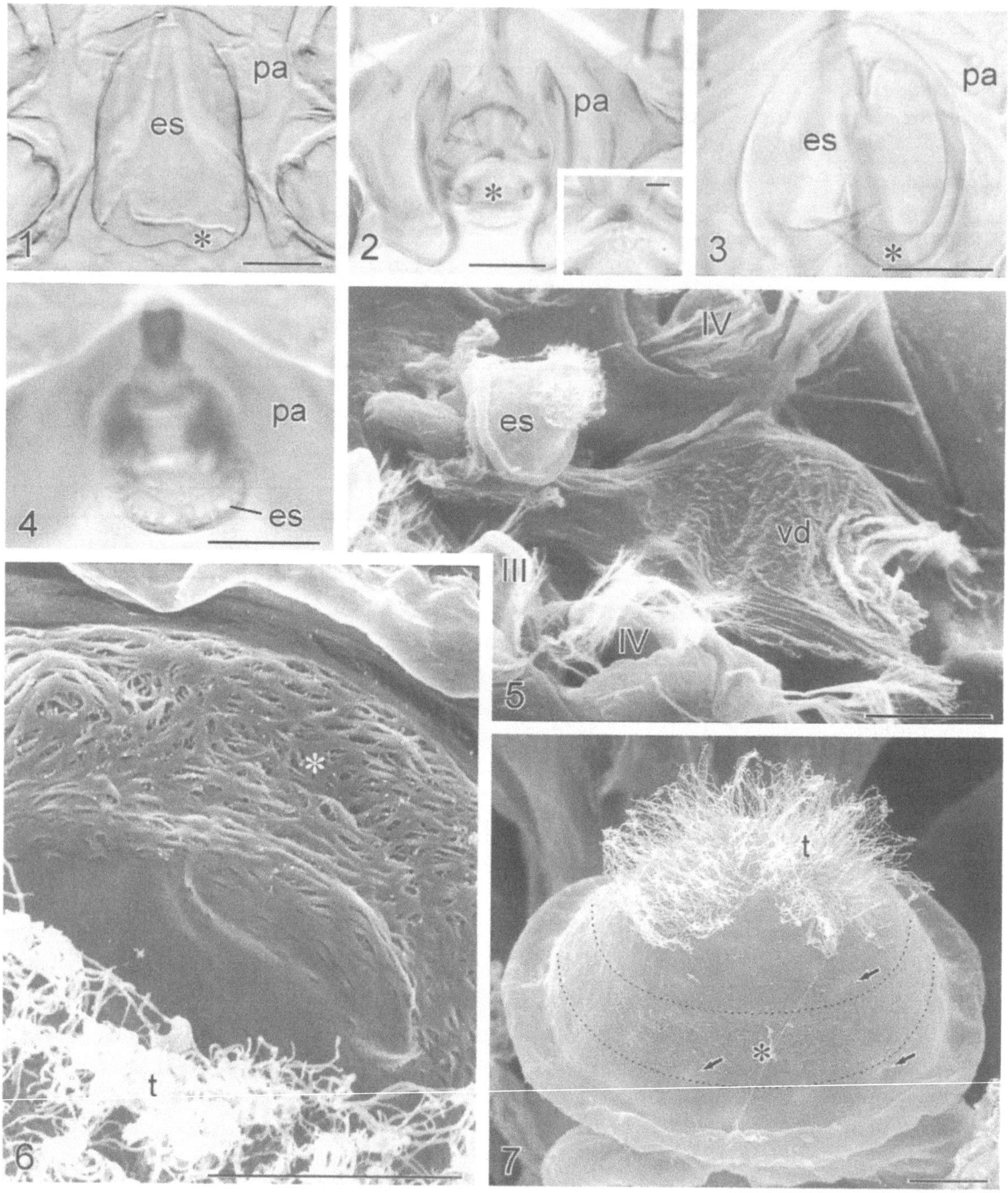

Figs 1-7 - Type 1 endogynium. Fig. 1: *Cornigamasus lunaris*, LM. Asterisk indicates the posteriorly located spherular organ with characteristic asymmetric tooth. Scale bar=50 µm. Fig. 2: *Holoparasitus pseudoperforatus*, LM. Note prominent teeth along the endogynial sac perimeter. Asterisk marks the spherular organ. Scale bar=50 µm. Inset shows the posterior part of the endogynial sac cuticle with its perforated pattern. Scale bar=10 µm. Fig. 3: *Neogamasus* sp., LM. Perforated cuticle of spherular organ (asterisk) surrounds the endogynial sac perimeter and forms prominent teeth posteriorly. Scale bar=50 µm. Figs 4-7: *Holoparasitus tirolensis*. Fig. 4: endogynium in LM. The profiles of teeth on internal walls of endogynial sac are visible. Scale bar=50 µm. Fig. 5: dorsolateral view of the vaginal cuticular lining showing the endogynial sac and a large vaginal duct. Scale bar=50 µm. Fig. 6: higher magnification of endogynial sac with a band of perforated cuticle (spherular organ) as indicated in fig. 7. Scale bar=10 µm. Fig. 7: dome of the endogynial sac with two groups of cuticular protrusions of muscle attachments (tendons). Sac wall shows the band of perforated cuticle (between the dashed lines) and shallow depressions corresponding to internal endogynial teeth (arrows). Scale bar=10 µm.

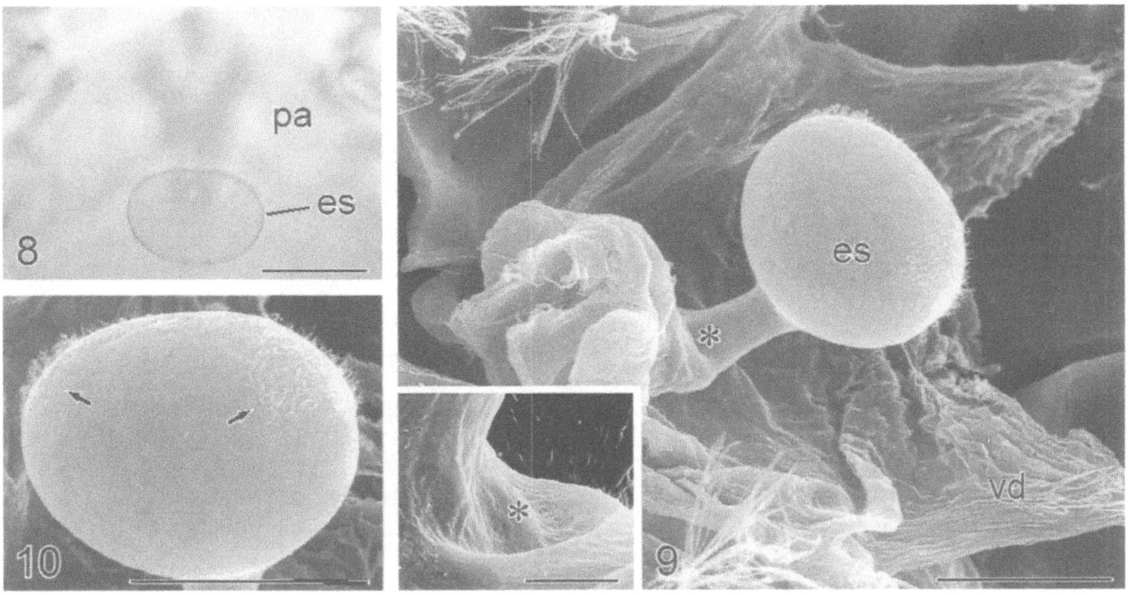

Figs 8-10 - *Holoparasitus caesus*, type 1 endogynium. Scale bars=50 µm. Fig. 8: endogynium in LM, focused on balloon-shaped endogynial sac. Fig. 9: endogynium from dorsal side. An asterisk indicates the region in the basis of tube leading to the sac, where the perforated cuticle of spherular organ is present. Inset: perforated spherular-organ cuticle at higher magnification (scale bar=10 µm). Fig. 10: endogynial sac with two muscle attachments (arrows).

The cuticle of spherules facing to the endogynial lumen frequently shows minute pits (e.g. *P. barbarus* fig. 24; *L. anoxygenellus* fig. 30), sometimes arranged in polygonal areas. The cell-facing procuticular surface of the spherules is richly perforated. The pores vary in shape depending on the species: they are circular and regularly spread within more or less distinct polygonal areas, corresponding to those observed on the adluminal surface of spherule, and are delimited with cuticular ridges. The ridges supposedly represent the borders of the adjacent spherular cells (e.g. *P. barbarus* figs 22-23; *L. anoxygenellus* fig. 33). In some species, the perforation of the spherular cuticle can show a conspicuous, three-dimensional pattern (e.g. *P. septentrionalis* figs 37-39), or the procuticle looks like a mat (e.g. *P. fimetorum* fig. 43). The perforation of the spherular-organ procuticle is similar to that of spherules, but polygonal areas corresponding to spherular cells are usually poorly visible (e.g. *H. tirolensis* fig. 6; *V. kraepelini* figs 15, 17).

Muscle attachments

Maceration of the soft parts of the investigated specimens allowed the observation of very delicate muscle attachments (tendons). They were located mainly in the midline of the posterior part of endogynium on its external (cell-faced) surface (figs 5-7, 9-10, 14, 18, 35, 37, 41-43). In type 3 endogynium, muscle attachments can also be present at the spherule concavity margins (figs 37, 43). In *P. fimetorum*, a midline muscle attachment is absent but there are two strong additional tendons, located close to the base of the dentate lamella covering the endogynium from behind (figs 41-42). In both *Holoparasitus* species studied in SEM, the muscles are attached to the endogynial sac and not to the posterior part of the endogynium (figs 5, 7, 9-10).

Types of endogynia

According to the structure and location of spherules or the spherular organ, three types of endogynia have been distinguished.

Type 1: This endogynium type is characterized

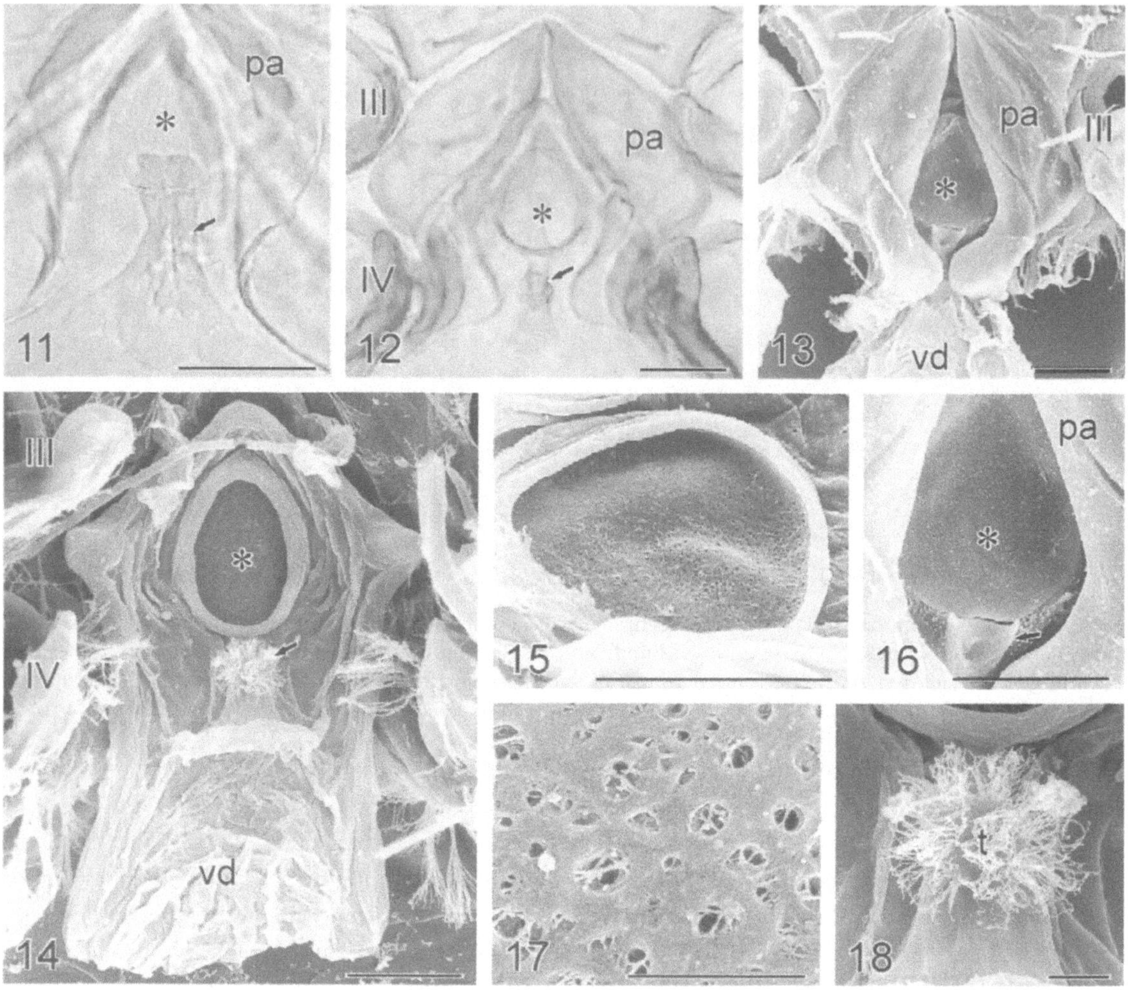

Figs 11-18 - Type 2 endogynium. Fig. 11: *Vulgarogamasus micherdzinskii*, LM. Circular perforated plate of spherular organ (asterisk) and the cuticular thickening forming the "handle" (arrow) are seen. Scale bar=50 µm. Figs 12-18: *Parasitus kraepelini*. Fig. 12: endogynium in LM. The arrow points the muscle attachment on the "handle". Scale bar=50 µm. Fig. 13: endogynium as seen from the ventral side after epigynial plate removal. Scale bar=50 µm. Fig. 14: the same as in fig. 13, but in the dorsal side view. Cuticular lining of the vagina is thicker and more sclerotized to form a frying pan-shaped endogynium, comprising the roundish perforated plate of spherular organ (asterisk) and the "handle" with cuticular protrusions of muscle attachment (arrow). Scale bar=50 µm. Fig. 15: spherular organ in dorsal view. Scale bar=50 µm. Fig. 16: spherular organ as shown in fig. 13 at higher magnification, with the anterior end of the "handle" indicated with arrow. Scale bar=50 µm. Fig. 17: perforations in spherular-organ cuticle from dorsal side. Scale bar=5 µm. Fig. 18: the "handle" with tendon cuticular protrusions from dorsal side. Scale bar=10 µm.

by an unpaired spherular organ located on the endogynium wall, usually encircling totally or partially the entrance to the endogynium cavity. The endogynial sac is moderately to highly sclerotized, sometimes with denticles. To this type also belong the bubble-shaped endogynia of *H. caesus*, *H. fortunatus* and *H. tuberculatus*. This endogynium type is present in *Gamasodes*, *Cornigamasus*, *Neogamasus*, *Aclerogamasus*, *Amblygamasus* and in *Holoparasitus*.

Type 2: This endogynium type looks like a frying pan: it possesses an unpaired, central

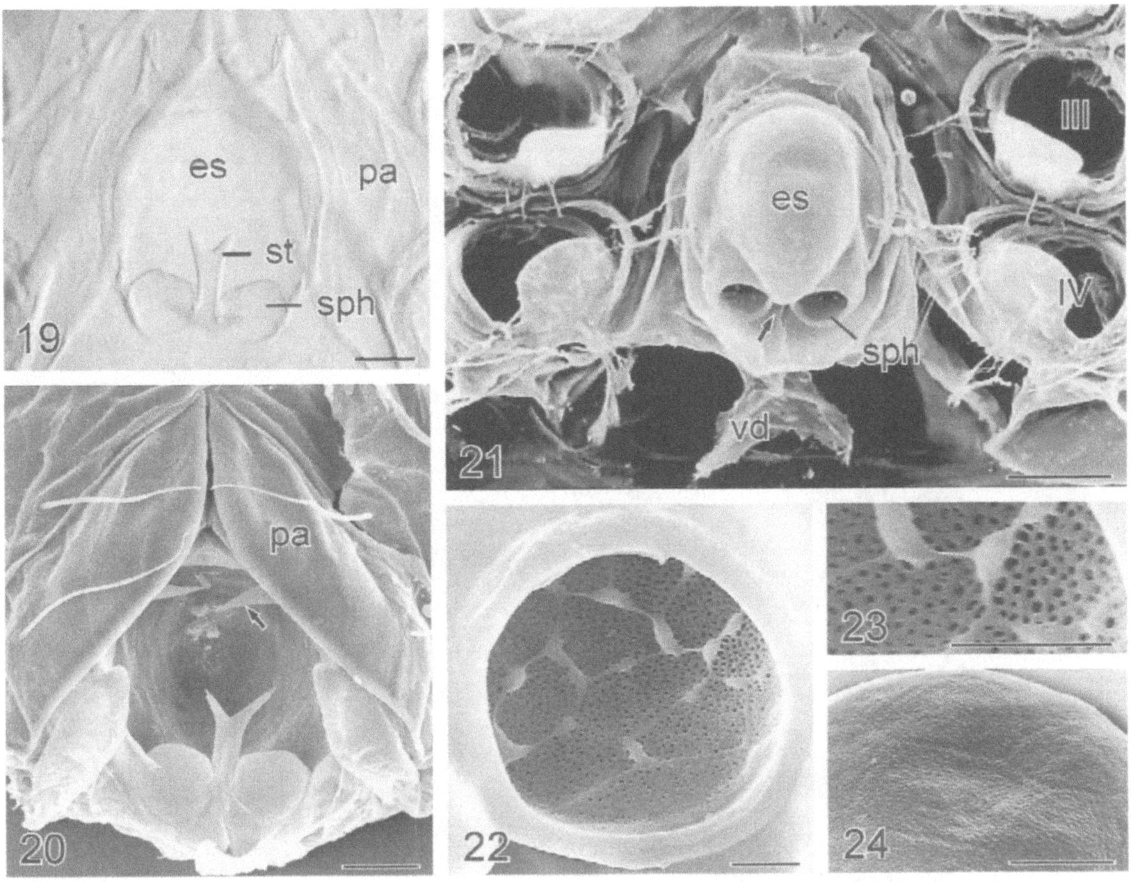

Figs 19-24 - *Pergamasus barbarus,* type 3 endogynium. Fig. 19: endogynium in LM. Scale bar=50 µm. Fig. 20: endogynium in ventral view, after epigynial plate removal; the arrow points to lateral endogynial spine. Scale bar=50 µm. Fig. 21: endogynium as seen from the dorsal side. Note the concavities of spherules (sph) and the entrance to stipula internal channel (arrow). Scale bar=100 µm. Figs 22 and 23: perforated spherular cuticle at two magnifications. Note polygonal areas allegedly corresponding to borders of spherular cells. Scale bars=10 µm. Fig. 24: endogynial cavity-facing surface of spherule with pattern corresponding to that in fig. 22. Scale bar=10 µm.

spherular organ located in the dorsal part of the endogynium, usually attached posteriorly to a cuticular "handle". The endogynial sac is weakly sclerotized, denticles and thorns are largely absent. This type is present in the majority of Parasitinae genera: *Eugamasus, Vulgarogamasus, Trachygamasus* (the "handle" is absent), *Poecilochirus, Parasitellus, Rhabdocarpais,* and in some *Parasitus.*

Type 3: This endogynium type has paired, symmetrically located spherules. The endogynial sac is highly sclerotized, stipula and/or different cuticular protrusions like thorns, denticles and lamellae are usually present. This type is characteristic for *Pergamasus, Paragamasus, Leptogamasus,* and some *Parasitus* species.

Discussion and conclusions

The endogynium of female parasitid mites has undoubtedly undergone evolution driven by sexual selection, resulting in a high, species-related variability of cuticular specializations, very useful for taxonomists. From the functional point of view, the endogynium compartments can be divided into four categories: (1) the endogynial sac, (2) cuticular protrusions like stipula, thorns, denticles and lamellae, (3) the spherular organ and spherules, and (4) endogynium muscle attachments.

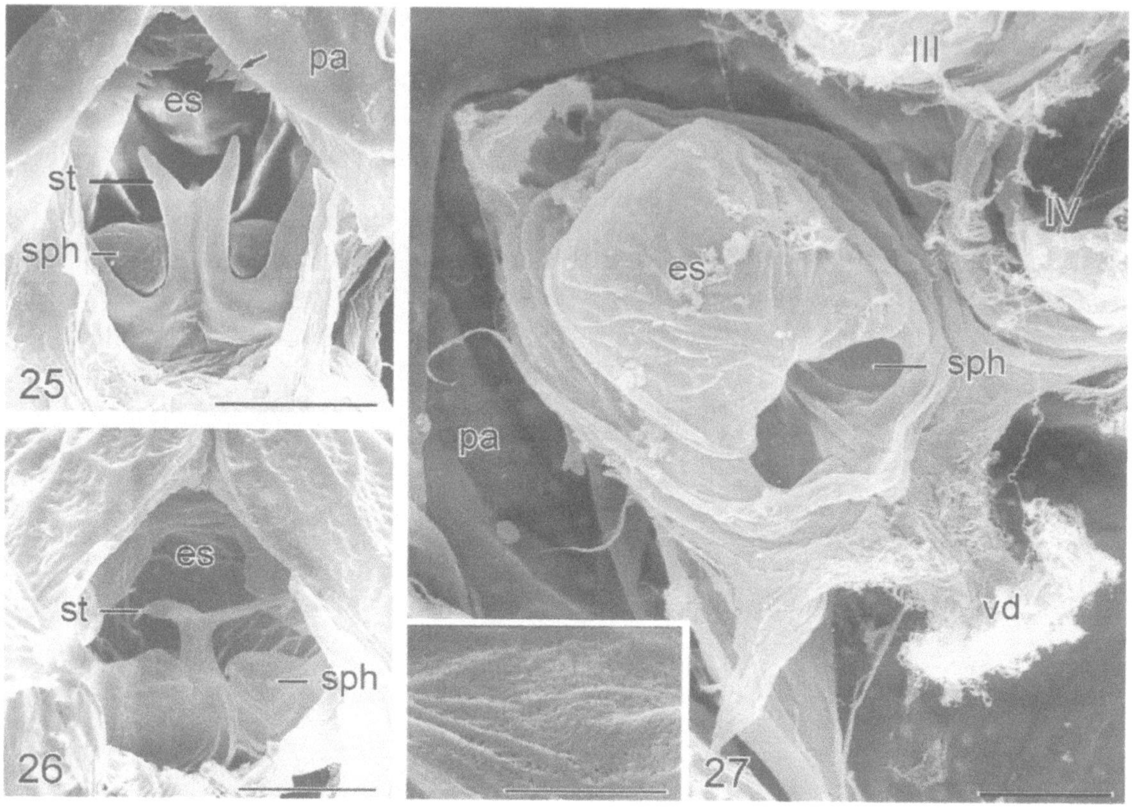

Figs 25-27 - Type 3 endogynium. Fig. 25: *Pergamasus mediocris,* endogynium from ventral side. Arrow points to lateral endogynial spine. Scale bar=50 µm. Figs 26-27: *Pergamasus quisquiliarum.* Fig. 26: endogynium from ventral side. Scale bar=50 µm. Fig. 27: endogynium from dorsal side. There is a joint entrance into both spherular concavities (sph). Scale bar=50 µm. Inset shows the perforation of the spherular cuticle. Scale bar=10 µm.

The endogynial sac

The endogynial sac is a thin-walled concavity (as visible from endogynium/vagina lumen) on the dorsal side of endogynium and provides space for the deposited spermatophore. In mites with type 2 endogynia (*Vulgarogamasus, Eugamasus,* some *Parasitus*), the cuticular lining of the vagina is weakly sclerotized except for the spherular organ and the "handle", resulting in the absence of a well-demarcated endogynial sac. The spermatophore is placed directly on the spherular organ and laterally flanked by a thin and flexible cuticle. A special kind of endogynial sac is present in several *Holoparasitus* species (*H. ampullaris, H. caesus, H. fortunatus, H. tuberculatus*): a balloon-like extension, connected to the vaginal lumen via a relatively narrow tube. The diameter of that tube in *H. caesus* is not larger than the width of movable digit of male chelicera (unpublished observation), which makes its direct participation in the introduction of spermatophore into the endogynial sac somewhat doubtful.

Cuticular protrusions (stipula, thorns and denticles)

Most endogynia are fitted with cuticular protrusions, thorns and/or denticles, distributed more or less close to the endogynial entrance. In particular, stipula in *Pergamasus* spp. or paired protrusions in some *Holoparasitus* (e.g. *H. ellipticus, H. gibber, H. globosus*) grow from the posterior endogynial margin and are directed anterad in the plane of the endogynial entrance. Functionally, all these structures are involved in

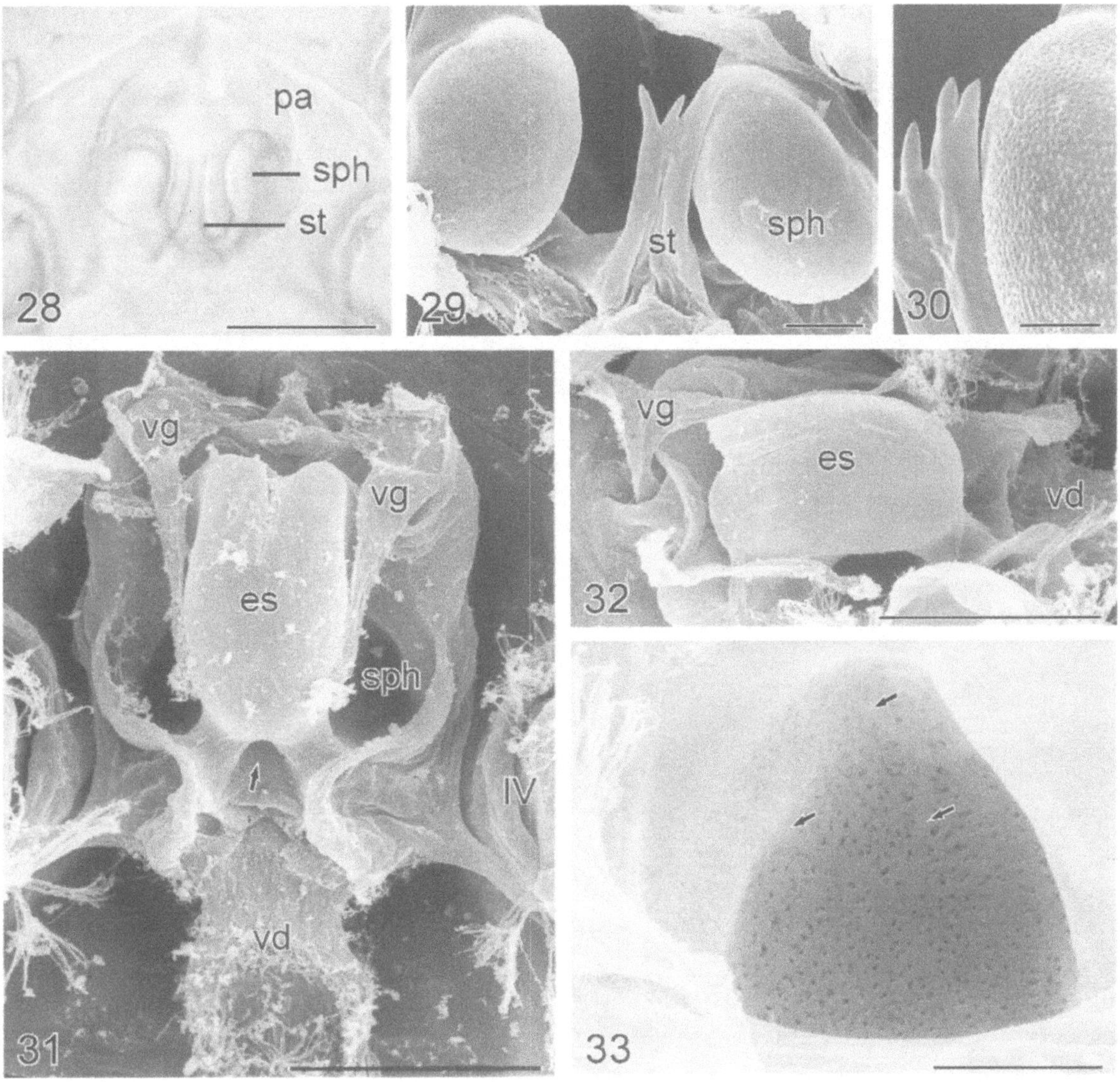

Figs 28-33 - *Leptogamasus anoxygynellus*, type 3 endogynium. Fig. 28: endogynium in LM. Scale bar=50 µm. Fig. 29: endogynium from the ventral side. Scale bar=10 µm. Fig. 30: surface of the spherular cuticle at higher magnification. Scale bar=5 µm. Fig. 31: endogynium from the dorsal side. Note the large spherular concavities (sph) and the entrance into the stipula (arrow), as well as cuticular linings of the vaginal glands (vg). Scale bar=50 µm. Fig. 32: the same as in fig. 31 from the lateral perspective. Scale bar=50 µm. Fig. 33: spherular concavity showing perforation and lines (arrows) demarcating borders of spherular cells. Scale bar=10 µm.

the protection of deposited spermatophore, preventing its possible removal on two occasions. When the egg is oviposited, they shelter the spermatophore from accidental removal by the sculptured chorion of the egg. This function roughly corresponds to the view of Neumann (1941), who suspected these structures to be responsible for managing the egg's passage into the vagina. Denticles, sometimes observed deep in the endogynial sac (e.g. *H. tirolensis*), can also help to keep the spermatophore in position. In the case of insemination, the next mate cannot easily withdraw the resident spermatophore and replace it with his own sperm. The sperm

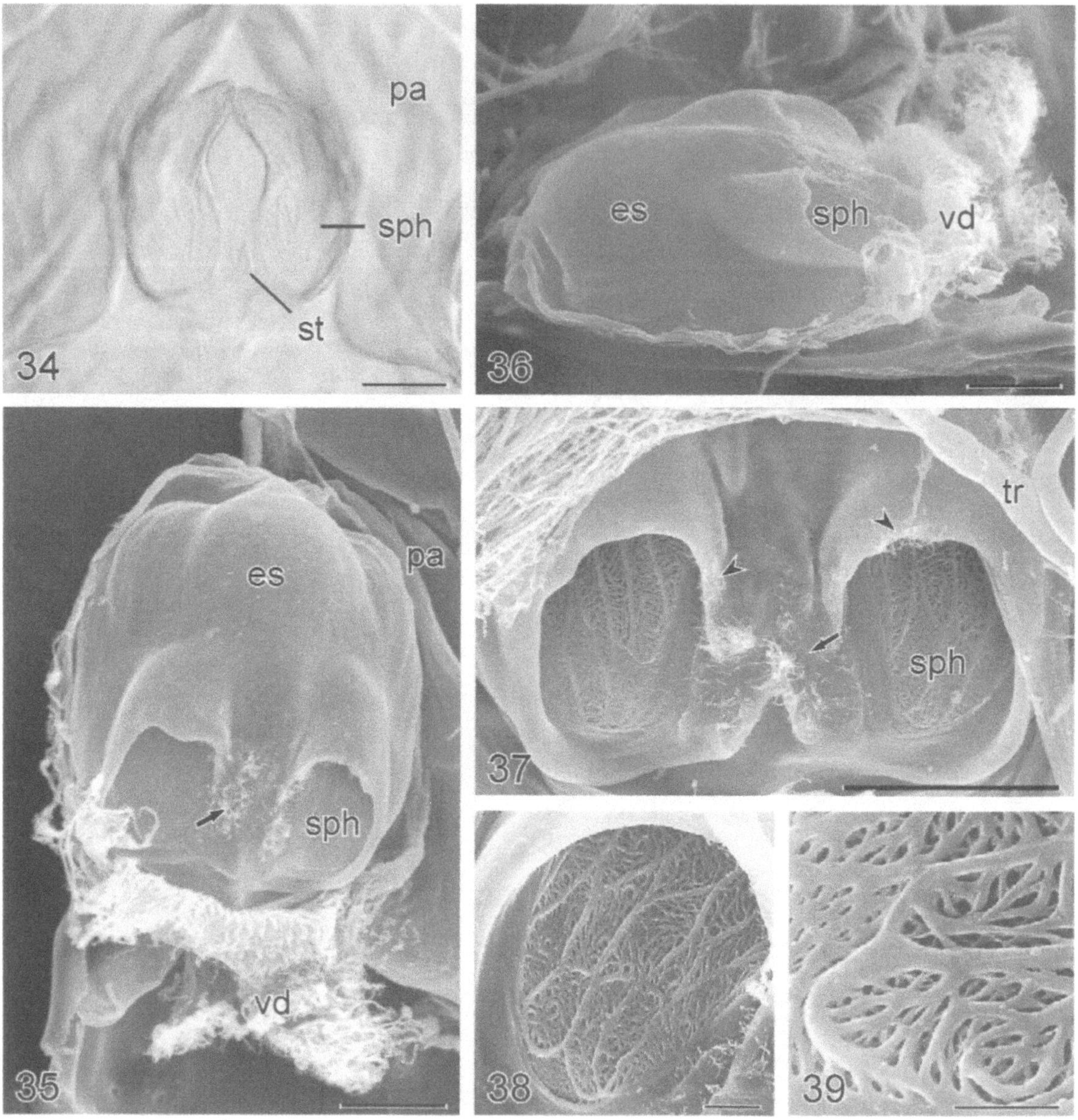

Figs 34-39 - *Pergamasus septentrionalis*, type 3 endogynium. Fig. 34: endogynium with shoe-shaped spherules and a minute stipula. LM. Scale bar=50 µm. Fig. 35: endogynium from the dorsal side. Protrusions of muscle attachments (tendons) are located inbetween the spherular concavities (arrow). Scale bar=50 µm. Fig. 36: the same as in fig. 35 from the lateral perspective. Scale bar=50 µm. Fig. 37: higher magnification of spherule region as in fig. 35. Muscle attachments are visible inbetween the spherular concavities and on their margins (arrow and arrowheads, respectively). Scale bar=50 µm. Figs 38 and 39: show the perforated spherular cuticle at two magnifications; scale bars=10 µm and 5 µm, respectively.

deposited in a bubble-shaped endogynia of some *Holoparasitus* species (see above) is secure due to a very narrow tube connecting the endogynial sac and the vagina, thus no additional cuticular specializations are necessary. However, further studies on the insemination process and spermatophore handling are needed in species possessing type 2 endogynium (with central spherular organ), where additional protrusions, denticles or thorns are generally absent.

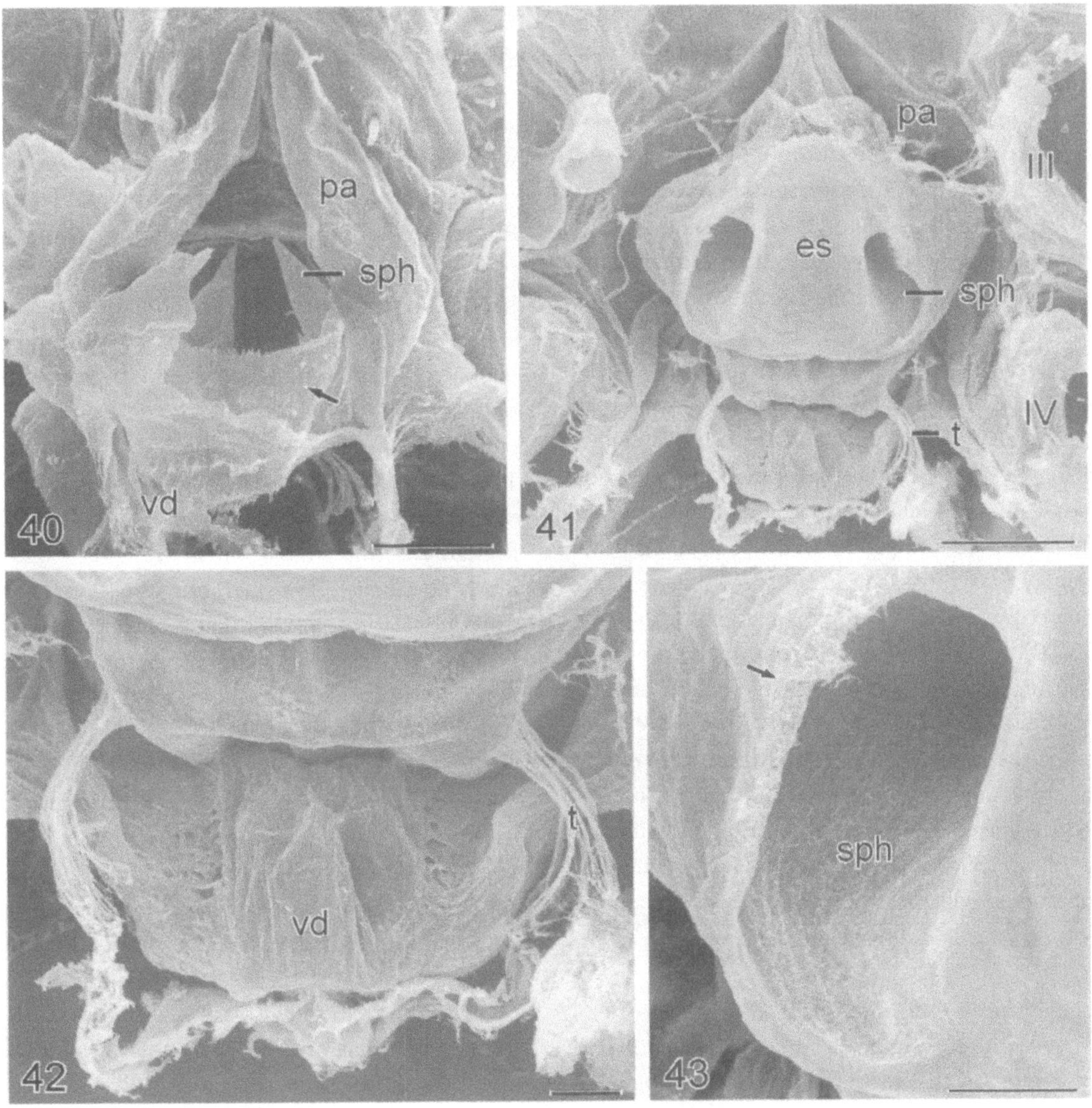

Figs 40-43 - *Parasitus fimetorum*, type 3 endogynium. Fig. 40: endogynium from the ventral side. Two conical, pointed spherules are covered posteriorly by dentate lamella (arrow). Scale bar=50 µm. Fig. 41: endogynium from the dorsal side. Note the two long cuticular protrusions (t) representing the attachments of muscles operating the dentate lamella and, by this means, the whole endogynium. Scale bar=50 µm. Fig. 42: posterior part of endogynium at higher magnification. Scale bar=10 µm. Fig. 43: the entrance into the spherule concavity demonstrating a perforated cuticle and fine protrusions of muscle attachment (arrow). Scale bar=10 µm.

Spherules and the spherular organ

As recently revealed by Alberti and co-workers (Alberti *et al.* 1999a and b), the spherular cuticle in the *Pergamasus* species is conspicuous, since it is richly perforated and the pores contain apical microvilli or protrusions of underlying spherular cells. The authors believe that such a structure is involved in the haemocoel-vagina transport of some substances. The direction of the transport is, however, unknown. The transported substances could influence oogenesis depending on the presence or absence of spermatophore within the endogynium, and/or stimulate sperm motility and in particular the leaking of sperm from the spermatophore (for more discussion see Alberti *et al.* 1999a, b). Our studies have shown that the perforated spherular cuticle, almost certainly combined with specialized spherular cells, is a constant structure in the endogynium of parasitid mites.

The organization of the perforated cuticle (paired lateral spherules, the unpaired spherular organ located centrally or antero/posterolaterally) led to the discrimination of three endogynium types. To collect data concerning as many species as possible, we completed SEM observations with the analysis of most Parasitidae species using light microscopy. When sample material was not available, we analyzed taxonomic drawings. However, in some species with very small endogynia (e.g. *H. vasilei, H. digitiformis, H. crassisetosus*), as well as in species where the endogynium is not discernible (as in many *Trachygamasus* species), the endogynia could not fall into any of the proposed categories. Further detailed SEM studies will probably make possible a more precise typing of endogynia in such species.

Endogynium muscle attachments

The present studies revealed fine cuticular protrusions on the cellular surface of endogynia, representing tendons, which provide an anchor for fine endogynial muscles. These usually paired muscles extend anterodorsally to endosternite and seem to elevate the endogynium toward the dorsal shield during oviposition, as the egg passes by. It can protect the endogynium from evagination out of the mite body and prevent possible damage of the vaginal lining. Similar muscles were described in *Parasitus kempersi* (Neumann, 1941) and *Poecilochirus necrophori* (Belozerov, 1957), whereas their functional aspects in *Poecilochirus* species were discussed by Korn (1982).

Phylogenetic remarks

As far as the evolution of endogynium structure is concerned, the most primitive state is supposedly the endogynium-lacking vagina lined with a thin, uniform cuticle. Progressive sclerotization in the dorsal part of the vaginal cuticle - leading to the formation of the endogynium - was probably favoured, as it made room for the spermatophore and improved conditions of sperm storage, but simultaneously prevented the haemocoel-vagina transport of any substances (in particular physiological signals) across the thin cuticle at the site of spermatophore deposition. Therefore, formation of a distinct "spherular organ" was promoted. Supposedly, it was primarily formed as a small, unpaired area of modified cuticle and underlying hypodermal cells - a condition resembling that in *Porrhostaspis lunulata*. According to light microscopic observations, a distinct endogynium in this species is absent but the vaginal lining dorsally displays a very small, cup-shaped structure formed by a thickened, perforated cuticle; surprisingly, this structure is not reported in the taxonomic literature. In addition, this still unpaired, primary spherular organ has expanded to develop into either a long band encircling the endogynium entrance (type 1), or a single, distinct circular plate located dorsally (type 2). Endogynium of the first type has undergone further modification: the area of the spherular organ has split into paired patches and eventually evaginated into the endogynial cavity (type 3) forming paired spherules. This increased the area of the perforated cuticle suitable for substance exchange.

In conclusion, the results presented support a complex role for the endogynium in insemination, sperm storage and sperm utilization in Parasitina. The common presence of a highly perforated cuticle combined with the underlying specialized spherular cells, despite their different organization in endogynia in distinct genera, strongly supports recent hypothesis on the role of such spherule-related organs in haemocoel-vagina transport and thus participation in spermatophore-ovary signal transmission.

The proposed endogynium types correspond relatively well to the generic structure of the Parasitidae family, with only slight inconsistency. However, in spite of a number of taxonomic papers, the phylogenetic concept of this taxon is still not completed even at the generic level, being at present in the course of revision (I. Juvara-Bals, personal communication). Therefore, conclusive statements on the relation between the evolution of endogynium structure and the phylogeny of genera are not yet justified.

Acknowledgements

The authors wish to thanks J. Faber M. Sc. from the Department of Cytology and Histology, Institute of Zoology, Jagiellonian University, for technical assistance in scanning electron microscopy with JSM-5410 microscope funded by Foundation for Polish Science, grant Subin '94.

References

Alberti G., 1984 - The contribution of comparative spermatology to problems of acarine systematics. In: Griffiths D.A., Bowman C.E. (eds.), Acarology VI. Ellis Horwood, Chichester, vol. 1: 479-490.

Alberti G., 1988 - Genital system of Gamasida and its bearing on phylogeny. In: Channa Basavanna G.P., Viraktamath C.A. (eds.), Progress in Acarology. Oxford & IBH Publishing, New Delhi, vol. 1: 197-204.

Alberti G., 2000 - Chelicerata. In: Jamieson B.G.M. (ed.), Progress in Male Gamete Ultrastructure and Phylogeny. In: Adiyodi K.G., Adiyodi R.G. (eds.), Reproductive Biology of Invertebrates. Oxford & IBH Publishing, New Delhi, Vol. 9 B: 311-388.

Alberti G., Hänel H., 1986 - Fine structure of the genital system in the bee parasite, *Varroa jacobsoni* (Gamasida: Dermanyssina) with remarks on spermiogenesis, spermatozoa and capacitation. *Experimental and Applied Acarology*, 2: 63-104.

Alberti G., Gegner A., Witalinski W., 1999a - Observations on the fine structure of the vagina of pergamasid mites (Pergamasidae, Parasitina, Gamasida). In: Bruin J., van der Geest L.P.S., Sabelis M.W. (eds.), Ecology and Evolution of the Acari. Kluver Academic Publishers, Amsterdam: 593-601.

Alberti G., Gegner A., Witalinski W., 1999b - Fine structure of the genital system in the females of *Pergamasus* mites (Acari: Gamasida: Pergamasidae). *Journal of Morphology*, 240: 195-223.

Athias-Henriot C., 1967a - Observations sur les *Pergamasus*. I. Sous-genre *Paragamasus* Hull, 1918 (Acariens anactinotriches: Parasitidae). *Mémoires du Musée national d'Histoire naturelle*, Série A (Zoologie), 49: 1-198.

Athias-Henriot C., 1967b - Observations sur les *Pergamasus*. II. Sous-genre *Pergamasus* Berlese, 1904 (Acariens anactinotriches, Parasitidae). *Acarologia*, 9: 669-761.

Athias-Henriot C., 1967c - Observations sur les *Pergamasus*. III. Sous-genre *Amblygamasus* Berlese, 1903 (Acariens anactinotriches, Parasitidae). *Acarologia*, 9: 762-800.

Athias-Henriot C., 1971 - *Paragamasus (Tanygamasus) probsti* (Oudemans) (systématique, géographie) avec quelques mises au point synonymiques (Arachnides, Gamasides tocospermiques, Parasitidae). *Zoologische Mededelingen*, 45: 167-179.

Athias-Henriot C., 1977 - Untersuchungen über die Gattung *Neogamasus* Tikhomirov mit zwölf neuen Arten aus Korea (Acari, Parasitiformes, Gamasida). *Annales Historico-naturales Musei Nationalis Hungarici*, 69: 311-341.

Athias-Henriot C., 1979 - Sur le groupe d'especes "*bispinosus*" du genre *Gamasodes* Oudemans, 1939 (Parasitiformes, Parasitidae). *Bulletin de la Société Zoologique de France*, 104: 209-220.

Athias-Henriot C., 1981 - *Rhabdocarpais* n. g. (type *Gamasus mammilatus* Berlese, 1904). Définition. Composition systématique (Parasitiformes, Parasitidae). *Revue Suisse de Zoologie*, 88: 141-156.

Bhattacharyya S.K., 1963 - A revision of the British mites of the genus *Pergamasus* Berlese s. lat. *Bulletin of the British Museum (Natural History)*, Zoology series, 2: 131-242.

Evans G.O., Till W.M., 1979 - Mesostigmatic mites of Britain and Ireland (Chelicerata: Acari - Parasitiformes). *Transactions of the Zoological Society of London*, 35: 139-270.

Hyatt K.H., 1980 - Mites of the subfamily Parasitinae (Mesostigmata: Parasitidae) in the British Isles. *Bulletin of the British Museum (Natural History)*, Zoology series, 38: 237-378.

Juvara-Bals I., 1970 - Contribution à l'étude du sous-genre *Pergamasus* Berlese s. str. section *Crassipes* (Acarina, Parasitidae). *Travaux de l'Institut de Spéologie "Émile Racovitza"*, 9: 159-169.

Juvara-Bals I., 1975 - Sur le genre *Holoparasitus* Oudemans et sur certains caractères morphologiques de la famille Parasitidae Oudem. (Parasitiformes). *Acarologia*, 17: 384-409.

Juvara-Bals I., 1981 - Nouvelle définition du genre *Leptogamasus* Trägårdh, 1936 (Acarina, Gamasida, Parasitidae) et description de six nouvelles espèces. *Revue Suisse de Zoologie*, 88: 77-93.

Juvara-Bals I., 1995 - *Holoparasitus (Holoparasitus) vasilei* n. sp. from Israel (Acari, Gamasida: Parasitidae). In: Nitzu E. (ed.), Soil Fauna of Israel. Editura Academiei Române, Bucuresti: 47-52.

Juvara-Bals I., Witalinski W., 2000 - Description of five new species of *Holoparasitus* s. str. with redescription of *H. apenninorum* (Berlese, 1906) and *H. cultriger* (Berlese, 1906) from Italy and Spain (Acari, Gamasida, Parasitidae). *Revue Suisse de Zoologie*, 107: 3-30.

Karg W., 1993 - Acari (Acarina), Milben - Parasitiformes (Anactinochaeta) Cohort Gamasina Leach - Raubmilben. 2nd ed. Die Tierwelt Deutschlands. 59. G. Fischer Verlag, Jena: 523 pp.

Korn W., 1982 - Zur Fortpflanzung von *Poecilochirus carabi* G. u. R. Canestrini, 1882 (syn. *P. necrophori* Vitzt.) und *P. austroasiaticus* Vitzthum, 1930. *Spixiana*, 5: 261-288.

Krantz G.W., 1978 - A Manual of Acarology. 2nd ed. Oregon State University Bookstores, Corvallis: 509 pp.

Micherdzinski W., 1969 - Die Familie Parasitidae Oudemans, 1901 (Acarina, Mesostigmata). Panstwowe Wydawnictwo Naukowe, Kraków: 690 pp.

Neumann K.W., 1941 - Beiträge zur Anatomie und Histologie von *Parasitus kempersi* Oudms. (Parasitidae). *Zeitschrift für Morphologie und Ökologie der Tiere*, 37: 613-682.

Trägårdh I., 1936 - *Leptogamasus*, a new genus of Acari from Sweden. *Entomologisk Tidskrift*, 57: 227-234.

Witalinski W., 1971 - Mites of the genus *Pergamasus* Berlese, 1903 (Acarina, Parasitidae), group *Crassipes* sensu Micherdzinski, 1969. *Acta Zoologica Cracoviensia*, 16: 669-682.

Witalinski W., 1979 - Systematic remarks on the mites of the genus *Leptogamasus* Trägårdh, 1936, sensu Athias-Henriot, 1971 (Parasitidae, Mesostigmata). *Bulletin de l'Académie Polonaise des Sciences*, Série des sciences biologiques, 27: 749-754.

Witalinski W., 1994 - *Holoparasitus (Holoparasitus) dallaii* sp.n., a new gamasid mite from Sardinia, Italy (Acari: Pergamasidae). *International Journal of Acarology*, 19: 349-354.

Colonization of primary sterile soils by epedaphic gamasina mites

Axel Christian
Staatliches Museum für Naturkunde Görlitz, PF 300 154, D-02806 Görlitz, Germany

Key words: Acari, Gamasina, colonization, dumps.

Abstract: The colonization of dumps in the Berzdorf mining distict near Görlitz in Germany by epedaphic gamasina mites was investigated. One test plot was overgrown with *Melilotus* the other was without vegetation. Almost 99% of the specimens of both test plots belong to the species *Pergamasus crassipes* (Linné, 1758) and *P. quisquiliarum* (G. et R. Canestrini, 1882). The colonization of soil by these species encouraged the opinion that the colonization in the initial phase is probably predominantly abiotically determined.

Introduction

In the eastern parts of Germany brown coal is extracted by opencast mining, which has led to the devastation of 100,000 ha of land. Many of these areas, and those of the dumps, should be recultivated for agriculture and forestry, for which the colonization of primary sterile soils by soil organisms plays a very important part. Succession of Gamasina on dumps has been intensively investigated in recent years (Koehler 1984; Christian 1993; Madej 1990 and 1996). The results show that species or species-groups are indicators for periods of development or for composition of trees in afforestations. The initial phase of colonization, the immigration method and the immigration conditions are little known up to now. Some of the results of the study of immigration of epedaphic gamasina mites are summarized in this paper.

Study area and methods

The studied area is located in Eastern Germany in the Berzdorf mining district near Görlitz. The test plots are characterized by primary sterile material with extensive parts of clay and kaolin. These test plots differ in their distance to the afforested area. The nearby dump was afforested with *Populus* hybrids 6 years ago. One test plot was located adjacent to the afforestation and was overgrown with *Melilotus* (fig. 1). The other test plot was 45 to 50 metres from the afforested area and does not bear any vegetation (fig. 2). Samples were taken in two types of pitfall traps 12 times in 1996 and 1997. Half the traps are protected against immersion (fig. 3), the other traps were unprotected, only with a small round roof against rain (fig. 4).

F. Bernini, R. Nannelli, G. Nuzzaci, E. de Lillo (eds.), Acarid Phylogeny and Evolution. Adaptations in mites and ticks, 169-173.

Fig. 1 - Test plot nearby afforestation overgrown with *Melilotus*.

Fig. 2 - Test plot without any vegetation.

Fig. 3 - Pitfall trap with protection against immission.

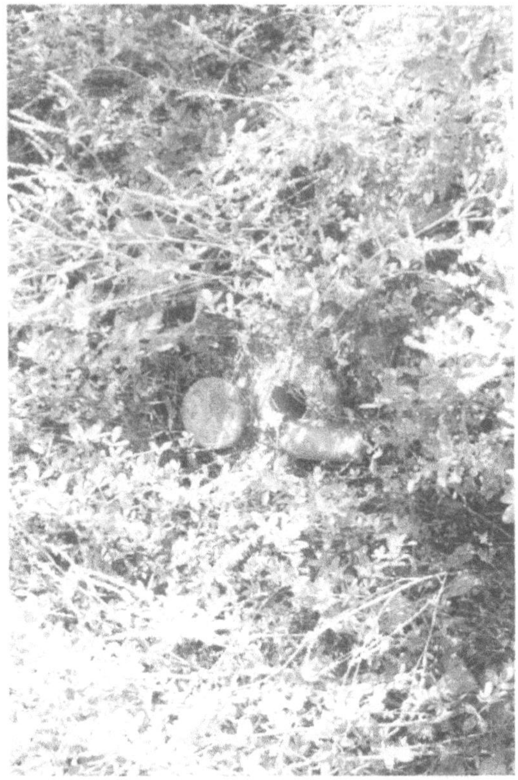

Fig. 4 - Pitfall trap unprotected against immission.

Results

In all traps of both testplots 3,043 individuals from a total of 12 species were found. Most of the gamasina species were to be found in very small numbers. Almost 99% of the specimens of both test plots belong to the species *Pergamasus crassipes* (Linné, 1758) and *P. quisquiliarum* (G. et R. Canestrini, 1882).

Test plot adjacent to the afforestation

In the pitfall traps of the test plot adjacent to the afforestation with overgroth of *Melilotus,* 3,015 individuals of the following 10 species were found:

Hypoaspis neocuneifer Evans et Till, 1966
Laelaps agilis Koch, 1836
Laelaps muris (Ljungh, 1799)
Haemogamasus nidi Michael, 1892
Neojordensia levis (Oudemans et Voigts, 1904)
Pergamasus crassipes (Linné, 1758)
Pergamasus quisquiliarum (G. et R. Canestrini, 1882)
Poecilochirus carabi G. et R. Canestrini, 1882
Gamasodes spiniger (Trägardh, 1910)
Parasitus lunulatus (J. Müller, 1859)

The colonization of the area overgrown with *Melilotus* was predominantly by the mites *P. crassipes* (81%) and *P. quisquiliarum* (18%). Significantly more individuals were in the anti - immission traps as in the unprotected traps (figs 5-6). The colonization of these test plots is most likely affected by active locomotion of the *Pergamasus* mites from the afforested dump. The high activity-density already present 6 months after the beginning of the investigations in winter 1996 and spring 1997 with a well-developed population of *P. crassipes* and *P. quisquiliarum.* It seems that the life conditions, especially moisture and temperature variations under the overgrowth of *Melilotus,* made the development of *Pergamasus* species possible. Succession of gamasina populations could not be observed during the period of investigation. The mites *L. agilis, L. muris, H. nidi* are brought in with small mammals, *H. neocuneifer* and *P. carabi* by phoresy with insects.

Test plot without vegetation

In the pitfall traps of the test plot without vegetation in the middle of the study area only 28 individuals of the following 5 species were found:

Eulaelaps stabularis (Koch, 1840)
Arctoseius semiscissus (Berlese, 1892)
Pergamasus crassipes (Linné, 1758)
Poecilochirus carabi G. et R. Canestrini, 1882
Gamasodes spiniger (Trägardh, 1910)

It is remarkable, that *A. semiscissus* in both years of investigation and *G. spiniger* in 1996 were only to be found in this test plot. In 1997 one specimen of *G. spiniger* was found in the traps near the afforestation. The colonization with gamasina mites of the test plot in the middle of the study area takes place most likely by phoresy or by air, but the life conditions, especially moisture and temperature variations, are unsuitable for the development of gamasina populations. The mite *E. stabularis* is brought in with small mammals.

Discussion and conclusions

The colonization of soil by epedaphic gamasina mites is influenced by many factors. It is very difficult to solve this complex of abiotic and biotic factors, because the specific ecological demands of many gamasina mites are unsufficiently known. Furthermore, the synergistic effects of the separate factors are not much investigated. Some species and species-groups are known as indicators in soil biology (Karg 1968).

The immigration tests near the afforestation and in the middle of the test area show that permanent epedaphic gamasina mites are subject to colonization pressure. If the mites find agreeable life conditions, they build up

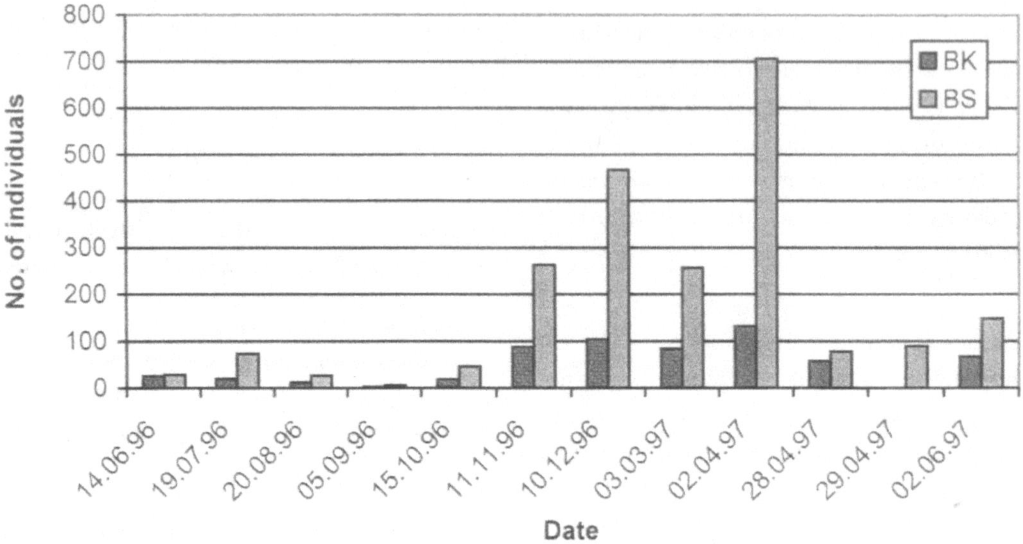

Fig. 5 - Number of individuals of *P. crassipes* in pitfall traps. Abbr.: BK, pitfall traps unprotected against immission; BS, pitfall traps with protection against immission.

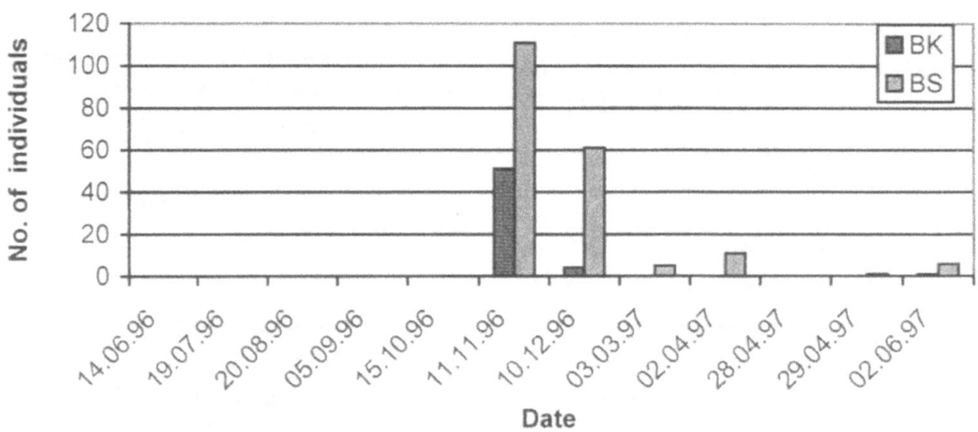

Fig. 6 - Number of individuals of *P. quisquiliarum* in pitfall traps. Abbr.: BK, pitfall traps unprotected against immission; BS, pitfall traps with protection against immission.

populations with many individuals within a few months. According to Koehler (1984), the conditions for the population growth "well-balanced micro-climate" and "relatively high soil moisture" are given if the vegetation canopy is at least 50% closed. These conditions evidently exist under the *Melilotus* overgrowth on the test plot near afforestation, and only in this part of the investigation area did Gamasina develop a population. It is remarkable that on dumps in the Berzdorf mining district, *P. crassipes* is only known from the 10th year on and *P. quisquiliarum* only from the 24th year or later after afforestation (Christian 1993). In the investigations of the succession of Gamasina in the cover stratum of a rubble dump, *P. crassipes* was also not recorded in the first colonization period (Koehler 1984).

The rapid colonization of soil with these species on the test plot near afforestation encouraged the opinion, that the colonization in the initial phase is probably predominantly abiotically determined.

Aknowledgements

Investigations are promoted by BMBF Promotion no. FKZ: 0339668. The author bears sole responsibility for the contents of the contribution.

References

Christian A., 1993 - Untersuchungen zur Entwicklung der Raubmilbenfauna (Gamasina) der Halden des Braunkohlentagebaues Berzdorf/OL. *Abhandlungen und Berichte des Naturkundemuseums Görlitz*, 67(2): 2-64.

Karg W., 1968 - Bodenbiologische Untersuchungen über die Eignung von Milben, insbesondere von parasitiformen Raubmilben, als Indikatoren. *Pedobiologia*, 8: 30-39.

Koehler H., 1984 - Methodische, ökologische und experimentelle Untersuchungen zur Sukzession der Mesofauna der Abdeckschicht einer Bauschuttdeponie unter besonderer Berücksichtigung der Gamasina. Dissertation, Universität Bremen: 1-340.

Madej G., 1990 - Die Besiedlung von Kohlehalden durch Milben der Ordnung Mesostigmata (Arachnida, Acari). II. Ein Wechsel in der Artenstruktur der Mesostigmata je nach Alter und Höhe des Steinkohlenbergwerkes. [Orig. Poln.]. *Acta Biologica Silesiana*, 16(33): 69-85.

Madej G., 1996 - Mesostigmata mite community on the dump of the power plant. *Acta Biologica Silesiana*, 28(45): 130-138.

Colonization of a dolomitic dump by mesostigmatid mites (Acari, Mesostigmata)

Grazyna Madej, Piotr Skubala
Department of Ecology, University of Silesia, Bankowa 9, PL-40-007 Katowice, Poland, gmadej@us.edu.pl, pskubala@us.edu.pl

Key words: Mesostigmata, dolomitic dumps, colonization, community structure.

Abstract: Communities of mesostigmatid mites on the dolomitic dump and in the adjacent forest developed on old galena-calamine wastelands were investigated. The structure of mite communities, trophic and environmental preferences of mites and their reproductive mode were analyzed. Furthermore, the analysis of mite dispersal was described. In the course of the study 2,764 Mesostigmata individuals, representing 64 species of 12 families, were collected. The abundance of mites on the dump was much higher (10,555 individuals per m^2 - bottom; 6,593 individuals per m^2 - top) than in the neighbouring forest (3,377 individuals per m^2). The number of mesostigmatid species recorded on the investigated sites was very similar and varied from 28 to 32. The species composition of the community on old galena-calamine wastelands and on the dump was characteristic of forest and open biotopes, respectively. *Leptogamasus suecicus* and *Hypoaspis aculeifer* were typical species of the bottom of the dump. Whereas *Asca nova*, *Hypoaspis kargi* and *Gamasellodes minor* were dominant species at the second site on the dump. *Rhodacarus coronatus* and *Rhodacarellus silesiacus* belonged to eudominants at both sites on the dump. The following species dominated in the beech forest: *Gamasellodes bicolor, Veigaia nemorensis, Paragamasus vagabundus, Eviphis ostrinus, Leptogamasus cuneoliger* and *Trachytes aegrota*.

Introduction

The exploitation of zinc and lead ores began in the Middle Ages and was continued to present times. Abandoned galena-calamine wastelands are a result of open cast zinc-lead mining activity in the Middle Ages (Madej and Skubala 1996). Dumps of washed material containing mud with large amounts of zinc and lead were pilled no more than 100 years ago. Dolomite dominated over limestone and slate on these dumps. Old galena-calamine wastelands are a good subject for the study of the process of colonization and succession of soil organisms. Studies on the succession of mites and other soil organisms were conducted in natural (Huhta *et al.* 1979; Schulz 1991; Scheu and Schulz 1996) and degraded biotopes (Dunger 1968 and 1991; Christian 1993; Hutson and Luff 1978; Parr 1978; Koehler 1991, 1998 and 1999; Madej and Blaszak 1993). Predacious Gamasina mites were used to estimate changes in soil (Koehler 1992; Karg 1994; Ruf 1997). They are also regarded as a good indicator of successional stages on post-industrial wastelands. Few authors have hitherto studied mites of old galena-calamine wastelands (Skubala and Madej 1995; Alberti *et al.* 1996; Madej and Skubala 1996).

F. Bernini, R. Nannelli, G. Nuzzaci, E. de Lillo (eds.), Acarid Phylogeny and Evolution. Adaptations in mites and ticks, 175-184.

The aim of the study was to describe the colonization of a dolomitic dump by mesostigmatid mites. Mite fauna in the soil of an adjacent beech forest developed on old galena-calamine wastelands has also been the subject of study. Some aspects of the structure of the mesostigmatid and oribatid mite communities in the beech forest have previously been discussed (Skubala and Madej 1995). Furthermore, the process of mite dispersal was analyzed. Skubala (1999) previously presented similar studies concerning oribatid mites on the dolomitic dump and galena-calamine wastelands. The main aims of the investigation were as follows: to analyze changes in the structure and species composition of three mesostigmatid communities, to compare some parameters describing trophic and faunistic characteristics of mesostigmatid communities at the three study sites, to describe the mode of reproduction of dominant species on the dolomitic dump and in the beech forest, to discuss the mechanisms of mites dispersal between the beech forest and the dolomitic dump.

The Study Area

Three study sites were located in Segiet village. This area, between Bytom and Tarnowskie Góry (Upper Silesia, South of Poland), has been the site of easily attainable metal ores - iron, silver, lead and zinc - since the early Middle Ages.

The first site was chosen on abandoned galena - calamine wastelands in the „Segiet" Reserve where the spontaneous plant succession was observed [*Dentario glandulosae - Fagetum* type (Skubala and Madej 1995)].

Two other sites were located on the rinsing dolomitic dump which adjoins the reserve protection zone. The dump belonged to the „Fryderyk" Royal Mine of silver and lead which exploited ores from 1873 to the 1920s. The dump originated from ore - bearing dolomite from which zinc and lead ores (with addition of silver) and iron ore (limonite) were separated in the process of sorting and rinsing. The dump arose over tens of years and coverage of vegetation is uneven. The plant composition is typical of pioneer calciphilous biotopes (spontaneous plant succession). Study sites have been chosen at the bottom and the top of the dump.

At the bottom of the dump, close to the beech forest, covered the ruderal assemblage of plants. Single tree species (*Betula pendula*, *Tilia cordata*) occur in the shrub layer. The herb layer is dense (100%). The dominant species are as follows: *Aegopodium podagraria, Artemisia vulgaris, Centaurea scabiosa, Euphrasia rostkoviana* and *Urtica dioica*.

The third site is situated at the top of the dump about 30 metres from the previous site. The vegetation developed in this part of the dump is classified as xerothermic green from the *Festuco - Brometea* association. There are single self - sown birches (*Betula pendula*) and in the shrub layer single self - sown pine - trees (*Pinus silvestris*). The herb layer is dense (90%) and diverse. Dominant species are *Euphrasia rostkowiana, Festuca ovina, Linum catharticum, Phleum boehmeri, Scabiosa ochroleuca* and *Thymus pulegioides* (Skubala 1999).

Materials and methods

Soil samples were taken from a representative quadrat (10 x 10 m) to a depth of 7.5 cm using a cylindrical steel corer of area 18 cm^2. In total 250 samples were collected. The mites were extracted with a modified Tullgren extractor during 7 days.

The material was analyzed using some zoocenological indices, e.g., abundance, dominance (D), frequency of occurrence (C) and species diversity and equitability (H', e).

Furthermore, the species abundance relationship after Ludwig and Reynolds (1988) was used to compare the structure of the mite communities. Differences between mean abundances of mesostigmatid mites at the three study sites were tested for significance using the t test. Correspondence analysis (Jongmann *et al.* 1987) was used on logarithmically transformed data for the analysis of mesostigmatid responses along the transect. An analysis was carried out with a Multivariate Statistical Package (MVSP) program.

The mesostigmatid mites were categorized as 3 trophic groups (Buryn and Brandl 1992). All species were also put into a group which characterized the typical biotope in which mites usually occurred: forest, "open habitats" and eurytopic species.

Two indices of dispersal direction were used for quantitative and qualitative data (Legendre and Legendre 1983; Borcard *et al.* 1995):

$$DD_4 = [2W/(A+B)][(A-B)/(A+B-W)]$$

W is the sum of the minimum frequency for all species, this minimum being defined for each species as its frequency in the zone where it is the rarest. A and B are the sums of the frequencies of all species in each of two adjacent zones. For presence - absence data, comparison of two adjacent regions is again based upon a combination of 3 values: the number of species present in both regions (a), the number of species found in region 1, but not in 2 (b), and the number of species found in region 2 but not in 1 (c):

$$DD_2 = [2a/(2a+b+c)][(b-c)/(a+b+c)]$$

Results

During the study period 2,764 specimens of Mesostigmata were collected from three sites. They were classified as belonging to 12 families and 64 species (tab. 1). A total of 414 individuals of the Uropodina cohort were counted.

Beech forest

In the course of the study 912 individuals of Mesostigmata were counted. They belonged to 32 species and 8 families. The proportion of Uropodina mites was high (43.8%). Juvenile forms represented 33.8% of the total number. The mean abundance of Mesostigmata was low, being 3,377 individuals per m^2. Mites of the families Veigaiaidae (18.9%) and Parasitidae (18%) dominated in the community. Six species reached similar proportion in the community of more than 5% (*Veigaia nemorensis, Gamasellodes bicolor, Paragamasus vagabundus, Eviphis ostrinus, Leptogamasus cuneoliger, Trachytes aegrota*). They comprised less than 30% of constancy of occurrence. The value of the Shannon index was H'=2.62, whereas the equitability index was e=0.75.

Bottom of the dump

At the site 1,140 specimens representing 31 species and 8 families were counted. Juvenile forms accounted for 26.3%. The community abundance was very high (10,555 individuals per m^2) in comparison with the community in the beech forest. Uropodina mites did not occur at the site. The proportion of the following families was the highest in the community: Rhodacaridae (43.8%), Parasitidae (26.5%), and Laelapidae (14%). Four species reached the highest values of dominance and constancy of occurrence (*Leptogamasus suecicus, Rhodacarellus silesiacus, Rhodacarus coronatus, Hypoaspis aculeifer*). The constancy of occurrence of species was generally low. Only 2 species occurred in more than 50% of soil samples (*Rhodacarellus silesiacus, Leptogamasus suecicus*). The values of the Shannon and equitability index were lower at this site, H'=2.21 and e=0.64.

Top of the dump

Only 712 specimens were collected at the top of the dump during the study period. They belonged to 28 species and 7 families. Uropodina mites did not occur at the site. Juvenile forms represented 22% of the total number. The community abundance was considerably lower (6,593 individuals per m^2) compared to the community at the bottom of the dump. Mites of the families Rhodacaridae (35.1%), Ascidae (32.6%), and Laelapidae (21.8%) reached the highest proportion in the community. Four species were recorded with the dominance index higher than 14% (*Asca nova, Hypoaspis kargi, Rhodacarus coronatus, Rhodacarellus silesiacus*). Only *Asca nova* and *Hypo-*

Tab. 1 - Checklist of the mesostigmatid species in the beech forest and on the dump.

		beech forest	dump bottom	dump top
	Gamasina			
	Zerconidae			
1.	*Prozercon traegardhi* (Halbert, 1923)		+	
	Parasitidae			
2.	*Eugamasus magnus* (Kramer, 1876)	+		
3.	*Vulgarogamasus kraepelini* (Berlese, 1904)	+		
4.	*Amblygamasus mirabilis* Willmann, 1951		+	
5.	*Holoparasitus calcaratus* (C.L.Koch, 1839)		+	
6.	*Leptogamasus cuneoliger* Athias-Henriot, 1967	+		
7.	*Leptogamasus oxygynelloides* (Karg, 1968)	+		
8.	*Leptogamasus suecicus* Trägardh, 1936	+	+	+
9.	*Leptogamasus sp.*		+	+
10.	*Paragamasus cambriensis* Bhattacharyya, 1963	+	+	
11.	*Paragamasus misellus* (Berlese, 1903)		+	
12.	*Paragamasus runcatellus* (Berlese, 1903)		+	
13.	*Paragamasus vagabundus* (Karg, 1968)	+	+	
14.	*Paragamasus wasmanni* (Oudemans, 1902)	+		
15.	*Pergamasus barbarus* (Berlese, 1904)	+		
16.	*Pergamasus quisquiliarum* (Canestrini, 1882)	+		
17.	*Pergamasus sp.*	+		
	Macrochelidae			
18.	*Geholaspis longispinosus* (Kramer, 1876)	+		
19.	*Geholaspis mandibularis* (Berlese, 1904)	+		
20.	*Macrocheles montanus* (Willmann, 1951)	+		
	Eviphididae			
21.	*Eviphis ostrinus* (C.L.Koch, 1836)	+		
	Ascidae			
22.	*Asca aphidioides* (Linnae, 1758)			+
23.	*Asca nova* Willmann, 1939		+	+
24.	*Gamasellodes bicolor* (Berlese, 1918)	+	+	+
25.	*Gamasellodes minor* (Athias – Henriot, 1961)		+	+
26.	*Leioseius elongatus* Evans, 1958	+		
27.	*Leioseius minusculus* Berlese, 1905			+
28.	*Leioseius naglitschi* Karg, 1965			+
29.	*Zercoseius spathuliger* (Leonardi, 1899)			+
30.	*Cheiroseius borealis* (Berlese, 1904)		+	
	Laelapidae			
31.	*Hypoaspis claviger* (Berlese, 1883)		+	+
32.	*Hypoaspis vacua* (Michael, 1891)	+	+	+
33.	*Hypoaspis aculeifer* (Canestrini, 1883)	+	+	+
34.	*Hypoaspis kargi* Costa, 1968		+	+
35.	*Hypoaspis praesternalis* Willmann, 1949		+	+
36.	*Hypoaspis mixta* Shcherbak, 1971			+
37.	*Hypoaspis astronomica* (Koch, 1839)		+	+
38.	*Hypoaspis austriaca* Sellnick, 1935		+	
39.	*Hypoaspis karawaiewi* (Berlese, 1903)			+
40.	*Hypoaspis sp.*		+	+
41.	*Ololaelaps placentula* (Berlese, 1887)	+		
42.	*Ololaelaps sellnicki* Bregetova & Koroleva, 1964		+	
	Veigaiaidae			
43.	*Veigaia cervus* (Kramer, 1876)	+	+	
44.	*Veigaia decurtata* Athias – Henriot, 1961	+	+	+
45.	*Veigaia exigua* (Berlese, 1916)	+		
46.	*Veigaia kochi* (Trägardh, 1901)	+		
47.	*Veigaia nemorensis* (C.L.Koch, 1839)	+	+	+
	Rhodacaridae			
48.	*Rhodacarus coronatus* Berlese, 1912	+	+	
49.	*Rhodacarus reconditus* Athias – Henriot, 1961		+	
50.	*Rhodacarellus silesiacus* Willmann, 1936		+	+

Tab. 1 - Continued.

	Pachylaelapidae			
51.	*Pachylaelaps furcifer* Oudemans, 1903	+		
52.	*Pachylaelaps longisetis* Halbert, 1915		+	
53.	*Pachylaelaps magnus* Halbert, 1915	+		
54.	*Pachylaelaps regularis* Berlese, 1920	+		
55.	*Pachylaelaps sp.*	+		
	Digamasellidae			
56.	*Dendrolaelaps latior* (Leitner, 1949)			+
57.	*Dendrolaelaps sp.*			+
58.	*Dendroseius reticulatus* Sheals, 1956			+
	Phytoseiidae			
59.	*Amblyseius meridionalis* Berlese, 1914		+	+
60.	*Amblyseius obtusus* (C.L.Koch, 1839)		+	+
61.	*Amblyseius sp.*			+
62.	*Typhlodromus sp.*			+
	Uropodina			
	Trachytidae			
63.	*Trachytes aegrota* (C.L.Koch, 1841)	+		
64.	*Trachytes pauperior* (Berlese, 1914)	+		

aspis kargi occurred with the constancy of occurrence higher than 50%. The values of the Shannon index (H'=2.36) and equitability (e=0.72) were slightly higher than at the bottom of the dump.

Comparison between three investigated sites

The abundance of juveniles and adults of Mesostigmata on wastelands in Segiet differed significantly. Only in the case of juveniles of Uropodina in communities at the bottom and the top of the dump were no significant differences observed (tab. 2).

The species composition of the mesostigmatid fauna was examined by means of correspondence analysis (CA). The first axis accounts for over 93% of the total data variability. The first axis segregates the mesostigmatid species, which dominated in the beech forest (right side). The species ordinated close to the first part of the first and second axes: *Hypoaspis aculeifer* (Ha) and *Leptogamasus suecicus* (Ls) are those displaying higher abundance at the bottom of the dump. Three species are ordinated close to each other. They are the following species: *Gamasellodes minor* (Gm), *Hypoaspis kargi* (Hk), *Asca nova* (An). These species appeared in high abundance at the top of the dump.

Tab. 2 - The t value calculated between the mean abundance of Mesostigmata at the investigated sites. * denote significant differences between the mean abundance at the investigated sites at the 0,05*, 0,01** and 0,0001*** probability level.

	Gamasina		Uropodina	
	Juveniles	adults	juveniles	adults
forest/bottom	8.95***	10.51***	3.51**	4.04***
forest/top	5.60***	8.34***	3.57**	
bottom/top (dump)	2.35*	2.12*	0.31	

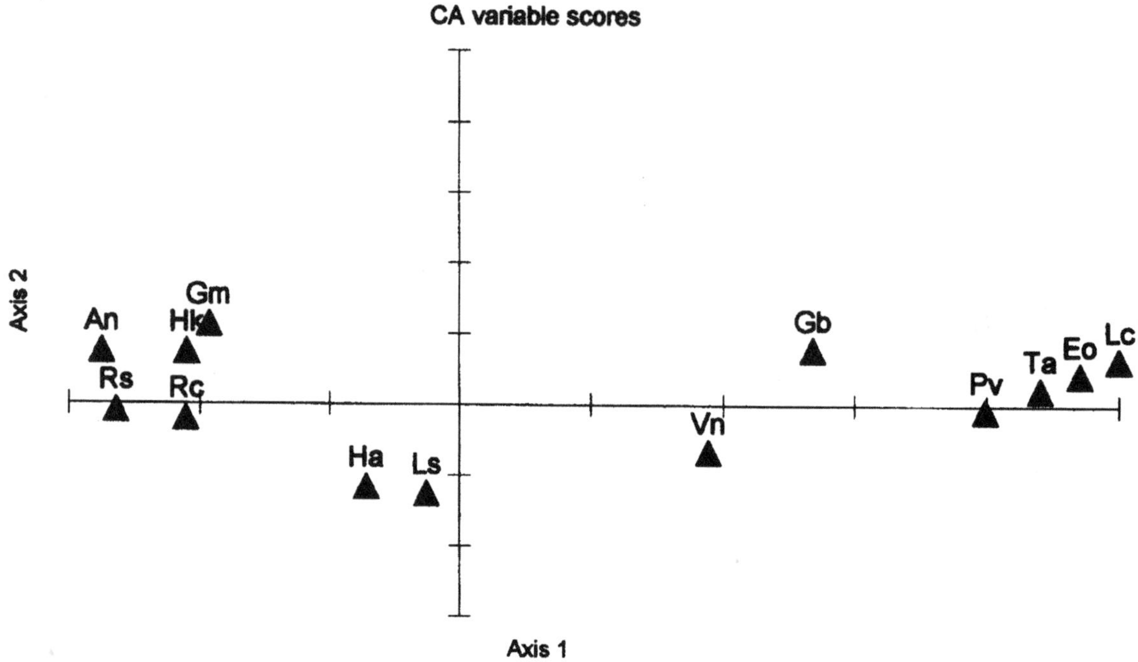

Fig. 1 - Correspondence analysis of numerous species (D>5%) of mesostigmatid mites at the three study sites. An - *Asca nova*, Gm - *Gamasellodes minor*, Gb - *Gamasellodes bicolor*, Hk - *Hypoaspis kargi*, Ha - *Hypoaspis aculeifer*, Eo - *Eviphis ostrinus*, Lc - *Leptogamasus cuneoliger*, Ls - *Leptogamasus suecicus*, Pv - *Paragamasus vagabundus*, Rc - *Rhodacarus coronatus*, Rs - *Rhodacarellus silesiacus*, Vn - *Veigaia nemorensis*, Ta - *Trachytes aegrota*.

Tab. 3 - Faunistic characteristics and food preferences of the mesostigmatid mites on the dump and adjacent the beech forest.

	top % of dominance	bottom % of dominance	beech forest % of dominance
Character groups:			
"open biotope" species	61.4	54.1	2.9
forest species	3.8	7.1	57.2
eurytopic species	30.8	36.8	38.9
unknown	4.0	2.0	1.0
Food preferences:			
nematode feeders	29.9	4.8	32.8
microarthropod feeders	4.2	33.1	51.4
polyphagous species	58.7	53.1	2.1
unknown	7.2	9.0	13.7

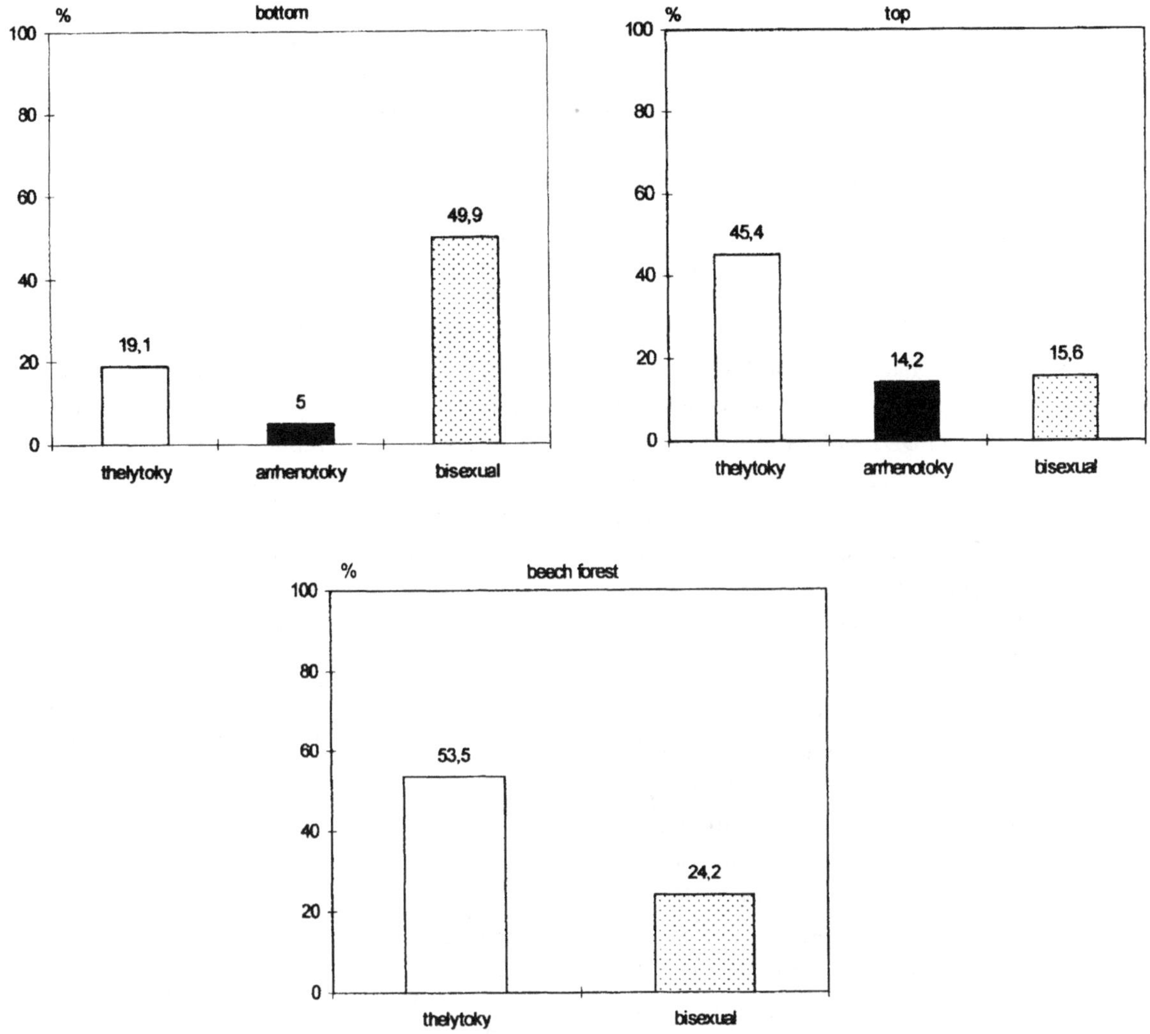

Fig. 2 - Proportion of dominant species (D>5%) with different mode of reproduction.

Two other species were ordinated close to the end of the second axis, i.e. *Rhodacarellus silesiacus* (Rs) and *Rhodacarus coronatus* (Rc). These species appeared in high abundance at the top and the bottom of the dump (fig. 1).

According to habitat preference the highest number of forest species (16) was observed in the beech forest. The mesostigmatid fauna of the dump areas were dominated by "open biotope" species (top - 8, bottom - 12). The proportion of the species exceeded 50% at both sites.

The dominance of eurytopic species was similar at the three investigated sites (tab. 3).

The communities at the top and bottom of the dump differed with regard to the number of mites with a different reproductive mode. The highest number of dominant mites with partenogenetic reproduction occurred at the top of the dump (4 species) and in the beech forest (4 species). The highest proportion of bisexual species was noted at the bottom of the dump (49.9%) (fig. 2).

According to the analysis of dispersal direc-

Tab. 4 - The analysis of mite dispersal.

Pair of sites	DD_2	DD_4
Bottom – beech forest	0.003	-0.060
Top – beech forest	0.009	-0.018
Bottom – top (dump)	0.014	0.059

tion, the migration of mites from the forest to the dump was not observed. Table 4 illustrates the results of the DD_2 and DD_4 indices. Values of DD_2 and DD_4 for the beech and the bottom were very low (0.003, -0.060) (tab. 4). From among seven successional species on the dump, only *Leptogamasus suecicus* and *Hypoaspis aculeifer* also occurred in the beech forest, but in small numbers.

Discussion and conclusions

According to Lamparska Wieland (1997) the natural plant succession on the dolomitic dump is directed towards similar biotopes, as in the neighbouring forest. The formation of the mite community in the beech forest has lasted considerably longer than on the dolomitic dump. It has been 70 years since the dumping was stopped. The development of mite communities on post-industrial wastelands is very slow, similar to plant succession.

The analysis of the dominance structure of the communities showed their similar character. Individuals of mites were most evenly shared between species at the three investigated sites. However the dominance value of some species on the dump was higher than in the beech forest. This resembles the situation in disturbed areas.

None of the dominant species occurring in an adjacent old beech forest dominated on the dump. Though some of them, particularly *Paragamasus vagabundus, Gamasellodes bicolor* have been recorded from a number of dumps in Poland (Madej *in prep.*) and Germany (Christian 1993; Koehler 1999).

The species abundance curves on the dump and in the beech forest were of similar form. They were typical of communities with low stability. The curves were very similar to curves from old galena calamine open cast mining wastelands in Germany (Alberti *et al.* 1996) (fig. 3).

The proportion of "forest" species was only 7.1% of the total number in the community at the bottom of the dump despite the presence of trees and litter. Species typical of open biotopes dominated at the site. Mites of the family Parasitidae, typical mites of the later stages of succession, comprised 26.5% of the total number. The most

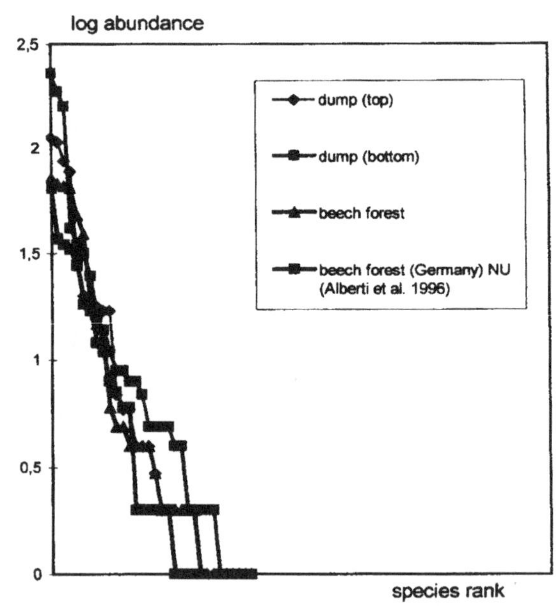

Fig. 3 - The species abundance relation on the dolomitic dump and in the beech forest in Segiet and Nußloch (NU) (Alberti *et al.* 1996).

numerous species of this family was *Leptogamasus suecicus* (26.5%). It was observed in high numbers in anthropogenic soils, mainly on afforested parts of coal-mine or ore dumps (Madej 1990; Cyran and Madej 1990). The species tolerates high concentrations of heavy metals (Alberti *et al.* 1996). The proportion of "forest" species was high in the adjacent beech forest (57.2%). The "forest" stage of succession has been observed on old galena-calamine wastelands. The species richness and the ratio of "forest" and eurytopic species in the investigated beech forest were similar to those observed in other forests. Mites which need litter to colonize new habitats reached the highest proportion in the forest community (Parasitidae, Veigaiaidae, Uropodina). *Leptogamasus cuneoliger, Veigaia nemorensis, Trachytes aegrota* are typical inhabitants of forests.

The analysis of the adjacent sites did not confirm the mite colonization from the surroundings on the dump. From among successional species on the dolomitic dump only *Leptogamasus suecicus* and *Hypoaspis aculeifer* also occurred in the beech forest. However, their abundance was low. The species colonizing the dolomitic dump have probably being transported from other biotopes. Similarly Skubala (1999) did not observe the migration of the oribatid mites from the beech forest on the dolomitic dump.

The importance of potential colonists occurring in the neighbouring natural habitats for the development of the Parasitiformes mites was stressed by Kovác *et al.* (1999). According to some authors, mites are transported by phoresy and wind (Hutson and Luff 1978; Whelan 1978; Hermosilla 1980; Siepel 1995; Walter and Proctor 1999). The species occurring in the secondarily colonized territory of the dump may be accidentally transported by mammals, birds, insects and millipedes (Madej *in prep.*).

The abundance of mites on old galena-calamine wastelands was low compared with the adjacent dump. However, it was similar to abundance of mesostigmatid mites observed in similar wastelands in Germany or Poland (Alberti *et al.* 1996; Madej and Skubala 1996 and 1998; Skubala and Madej 1995). Furthermore, Christian (1993) recorded the low abundance of 3,036 individuals per m^2 on the brown coal mine dump after 33 years of reclamation. The species number on galena-calamine wastelands was similar to figures noted on other wastelands of this kind (Skubala and Madej 1995; Alberti *et al.* 1996; Madej and Skubala 1996 and 1998).

The proportion of parthenogenetic mites was high at the top of the dump and in the forest. The opinions concerning the colonization of different habitats by parthenogenetic organisms are not consistent (Norton *et al.* 1993; Siepel 1995; Walter and Proctor 1999). Parthenogenetic species can be observed in stable habitats as well as in areas devastated by human activity (Hurlbutt 1979).

The proportion of species feeding on other microarthropods was higher on sites with trees. Mites of such trophic preferences usually occurred in later stages of the community development (Madej *in prep.*).

References

Alberti G., Hauk B., Köhler H.-R., Storch V., 1996 - Dekomposition. Qualitative und quantitative Aspekte und deren Beeinflussung durch geogene und anthropogene Belastungsfaktoren. Ecomed, Landsberg: 490 pp.

Borcard D., Geiger W., Matthey W., 1995 - Oribatid mite assemblages in a contact zone between a peat - bog and a meadow in the Swiss Jura (Acari, Oribatida): influence of landscape structures and historical processes. *Pedobiologia*, 39: 318-330.

Buryn R., Brandl R., 1992 - Are the morphometrics of chelicerae correlated with diet in mesostigmatid mites (Acari)? *Experimental and Applied Acarology*, 14: 67-82.

Christian A., 1993 - Untersuchungen zur Entwicklung der Raubmilbenfauna (Gamasina) der Halden des Braunkohlentagebaues Berzdorf O.L. *Abhandlungen und Berichte des Naturkundemuseums*, Görlitz, 67(2): 2-64.

Cyran E., Madej G., 1990 - Colonization of a iron ore dumps by mesostigmatid mites (Acari, Mesostigmata). I. Structure of the mite communities. *Acta Biologica Silesiaca*, 16(33): 139–159 (in Polish).

Dunger W., 1968 - Die Entwicklung der Bodenfauna auf rekultivierten Kippen und Halden des Braunkohlentagebaues. *Abhandlungen und Berichte des Naturkundemuseums*, Görlitz, 43(2): 1-256.

Dunger W., 1991 - Zur Primärsukzession humiphager Tiergruppen auf Bergbauflächen. *Zoologische Jahrbücher Systematik*, 118(3/4): 423-447.

Hermosilla W., 1980 - Die Mesofauna verschiedenalter Rekultivierungsflächen im Braun kohlentagebaugebiet der Ville. *Decheniana*, 133: 79-83.

Huhta V., Ikonen E., Vilkamaa P., 1979 - Succession of invertebrate populations in artifical soil made of sewage sludge and crushed bark. *Annales Zoologici Fennici*, 16: 223-270.

Hurlbutt H.W., 1979 - Distribution of unisexual and bisexual forms in mesostigmatid mites. In: Rodriguez J.G. (ed.). Recent Advances in Acarology. Academic Press, New York, San Francisco, London: 2, 455-460.

Hutson B.R., Luff M.L., 1978 - Invertebrate colonization and succession on industrial reclamation sites. *Scientific Proceedings*, Royal Dublin Society, A, 6: 165-174.

Jongmann R.H., ter Braak C.J.F., van Tongren O.F.R., 1987 - Data analysis in community and landscape ecology. Pudoc. Wageningen: 299 pp.

Karg W., 1994 - Raubmilben, nützliche Regulatoren im Naturhaushalt. Lebensweise, Artenbestimmung und Nutzung. *Die Neue Brehm - Bücherei*, Magdeburg, 624: 1-206.

Koehler H., 1991 - A five year study on the secondary succession of Gamasina on ruderal site: the influence of recultivation. In: Dusbábek F., Bukva V. (eds.), Modern Ecology. Academia Prague and SPB Academic Publishing, The Hague: 373–383.

Koehler H., 1992 - The use of soil mesofauna for the judgement of chemical impact on ecosystems. *Agriculture, Ecosystems and Environment*, 40: 193–205.

Koehler H., 1998 - Secondary succession of soil mesofauna: A thirteen year study. *Applied Soil Ecology*, 9: 81-86.

Koehler H., 1999 - Gamasina in a succession of thirteen years. In: Bruin J., van der Geest L.P.S., Sabelis M.W. (eds), Ecology and Evolution of the Acari. Kluwer Academic Publishers: 531-539.

Kovác L'., Schnitzerova E., Miklisová D., Mati R., 1999 - Gamasina communities (Acari, Parasitiformes) of arable soils with two different soil types. *Pedobiologia*, 43(1): 54-63.

Lamparska-Wieland M., 1997 - Dolomitic dump in Tarnowskie Góry Bobrowniki as a potential preservation object. Geographia. *Studia Dissertations*, 21: 101-115.

Legendre L., Legendre P., 1983 - Numerical ecology. Elsevier, Amsterdam: 853 pp.

Ludwig J.A., Reynolds J.F., 1988 - Statistical Ecology. A primer on methods and computing. Wiley, New York: 337 pp.

Madej G., 1990 - Colonization of a coal dumps by mesostigmatid mites (Arachnida, Acari). I. Structure of the mite communities. *Acta Biological Silesiaca*, 16(33): 37-68 (in Polish).

Madej G. (*in preparation*) - Development of mesostigmatid mite communities on mining wastelands. *Acta Biological Silesiaca* (in Polish)

Madej G., Blaszak C., 1993 - Untersuchungen über die Sukzession der Mesostigmata - Fauna (Acarina) auf verschieden alten Brachfeldern mit Galmei - und Bleiglanzabraum im Bergbau. In: Ehrnsberger R. (ed.), Bodenmesofauna und Naturschutz. Informationen zu Naturschutz und Landschaftspflege in Nordwestdeutschland, 6: 397-400.

Madej G., Skubala P., 1996 - Communities of mites (Acari) on old galena - calamine mining wastelands at Galman, Poland. *Pedobiologia*, 40(4): 311-327.

Madej G., Skubala P. 1998 - Gamasid and Oribatid mites of antropogenically transformed galena - calamine mining wastelands. *Zeszyty Naukowe*, AT-R. Bydgoszcz, 214(2): 229-234.

Norton R.A., Kethley J.B., Johnston D.E.U., O'Connor B.M., 1993 - Phylogenetic Perspectives on Genetic Systems and Reproductive Modes of Mites. In: Wrensch D.L., Ebbert M.A. (eds.), Evolution and Diversity of Sex Ratio in Insects and Mites. Chapman & Hall, New York, London: 8-99.

Parr T.W., 1978 - An analysis of soil microarthropod succession. *Scientific Proceedings*, Royal Dublin Society, A, 6: 185-196.

Ruf A., 1997 - Fortpflanzungsbiologie von Raubmilben und Charakterisierung von Böden. - Ein Konzept zur Indikation von Belastungszuständen von Böden -. *Abhandlungen und Berichte des Naturkundemuseums*, Görlitz 69(2): 209-216.

Scheu S., Schulz E., 1996 - Secondary succession, soil formation and development of a diverse community of oribatids and saprophagous soil macroinvertebrates. *Biodiversity and Conservation*, 5: 235-250.

Schulz E., 1991 - Die Milbenfauna (Acari: Mesostigmata und Cryptostigmata) in Lebensräumen auf Kalkgestein: Populationökologie, Sukzession und Beziehungen zum Lebensraum. *Berichte des Forschungszentrums Waldökosysteme*, Reihe A, 79: 1-245.

Siepel H., 1995 - Applications of microarthropod life - history tactics in nature management and ecotoxicology. *Biology and Fertility of Soils*, 19: 75-83.

Skubala P., 1999 - Colonization of a dolomitic dump by oribatid mites (Acari, Oribatida). *Pedobiologia*, 43(2): 145-159.

Skubala P., Madej G., 1995 - Oribatid and mesostigmatid mites (Acari) of the beech forest on the abandoned galena-calamine wastelands in the "Segiet" reserve. Proceedings of the Symposium on Advances of Acarology in Poland, Siedlce, September 26-27, 1995: 84-87.

Walter D.E., Proctor H.C., 1999 - Mites Ecology, Evolution and Behaviour. CABI Publ., Wallingford & New York: 322 pp.

Whelan J., 1978 - Acarine succession in grassland on cutaway raised bog. *Scientific Proceedings*, Royal Dublin Society, Series A, 6: 175-183.

Zoogeographical survey on the Phthiracaroidea (Acari, Oribatida) of the Ethiopian Region

Wojciech Niedbala
Department of Animal Taxonomy and Ecology, A. Mickiewicz University, Szamarzewskiego 91A, PL-60-569 Poznan, Poland

Key. Words: Acari, Phthiracaroidea, Ethiopian Region, zoogeography.

Abstract: At present 108 species of Phthiracaroidea are known from the Ethiopian region, of which 18 were described by the author previously. The greatest number of species has been found in Tanzania and Madagascar. Among the species which are most abundantly represented in the region there are two pantropical: *Atropacarus (Hoplophorella) stilifer* and *Atropacarus (Hoplophorella) vitrinus* and three Ethiopian: *Hoplophthiracarus ambiguus, Hoplophorella frater* and *Plonaphacarus machadoi.*

The Ethiopian region is the centre of speciation of the genus *Steganacarus*, especially for its subgenus *Rhacaplacarus.*

The unique character of the Ethiopian faunal region is indicated by the large number of endemic species - 73 (67.6%) and a relatively large number of Ethiopian zoogeographic elements - 19 (17.6%), which makes in total 92 (85.2%) of all species found. Analysis of distribution of a few widespread species clearly shows that their dispersion proceeded from the Oriental region.

On both the mainland and on surrounding islands a specialized fauna is present, represented by species of restricted geographical distribution. On the continent the number of species found is 73 (67.6%), on the islands 45 species (42%). There are only 9 (8.3%) species having common continental and island distribution.

The majority of the islands, irrespective of the geographical position, have a specific fauna of Phthiracaroidea, which is well supported by the number of endemic species occurring with most of these on Madagascar. The greatest number of continental endemites was found in the eastern part of Africa, Tanzania, Kenya and Somalia, while a smaller number of them occur in Central Africa. The highest number of endemites in Eastern Africa indicates that a strong speciation took place there. Probably, the fauna colonizing a given area undergoes a strong adaptive radiation under the effect of different conditions in high mountains.

The obtained data suggest that the dispersion of species into the Ethiopian region took place from the Oriental and perhaps also from the Australian region or Southern Africa.

Introduction

The ability of folding the aspidosoma under the opisthosoma to protect their appendages (= ptychoidy) characterizes the moss mites called Ptyctima (*Acari, Oribatida*), which include the Phthiracaroidea. The Phthiracaroidea are macrophytophagous and feed on dead organic matter of plant origin; the majority of them are xylophagous. Living in litter and dead leaves they burrow irregular galleries and cavities in decayed wood (immature stages in particular). They take part in mechanical fragmentation and decomposition of organic matter, its humification and in nutrient cycling processes in soil.

F. Bernini, R. Nannelli, G. Nuzzaci, E. de Lillo (eds.), Acarid Phylogeny and Evolution. Adaptations in mites and ticks, 185-197.

The aim of this paper is to perform a zoogeographical survey on the Phthiracaroidea in the Ethiopian region on the basis of assessment of the knowledge of the Phthiracaroidea fauna, distribution of species and determination of their zoogeographical elements.

The area and material of study

This paper reports the present state of investigation of the fauna of Phthiracaroidea in the Ethiopian region, including the islands located at this geographical latitude. Hence, the work does not deal with species found in palearctic Northern Africa (with the exceptions of Canary Islands) or Southern Africa. According to some sources the fauna of the latter region is more similar to that of Australia and South America than to that of Central Africa (Schnell 1976).

The results have been inferred from many year long studies of Mahunka (1978-1999) reported in 25 papers and Niedbala (1982-1998) reported in 13 papers, mainly concerning descriptions of new species.

Moreover, the material includes the species examined by the author from collections of various museums and institutions and coming from over 200 sites unevenly described in the region. The highest number of sites was in the Democratic Republic of Congo - 37, next to Angola and Madagascar - 23, Rwanda - 16, Kenya, Tanzania, St. Helena Island 12, Malawi - 10. The other countries were represented by single sites.

Principal zoogeographical elements:

Endemic - species occurring at a single site or at most at a few sites in the same subregion (or country);

Ethiopian - species occuring in Ethiopian Region in subregions or countries but not found outside Africa;

Pantropical - species distributed along the tropical and subtropical zones;

Semicosmopolitan - species of wide, worldwide or almost worldwide geographic distribution.

Specificity and abundance of the fauna

The total number of Phthiracaroidea species known from the Ethiopian region is 108, two of which are *nomina dubia*, and 18 which were described as new species by the author (tab. 1).

The highest number of species was found in Tanzania and Madagascar, followed by Uganda, Congo and Angola.

Two pantropical species: *A.(H.) stilifer* and *A.(H.) vitrinus* and three Ethiopian species: *H. ambiguus, H. frater* and *P. machadoi* are most abundantly represented in the region. Two of them A.*(H.) vitrinus* and *P. machadoi* are, equally, the most frequent species, both as far as the number of localities and subregions are concerned. The two endemic species *A. sculptilis* and *S.(S.) carlosi*, are also abundantly represented, but they occur only on the islands of Comoro and Tenerife. The earlier mentioned abundant species occurred exclusively or mainly in the mainland.

The specificity and unique character of the fauna of the Ethiopian region is indicated by a large number of endemic species (73-67.6%) and a relatively large number of species which do not occur outside Africa, i.e. Ethiopian, (19-17.6%), making in total 92 species, which is 85.2% of those found.

A surprising fact is the similarly specific character of the ptyctimous fauna from the adjacent geographical regions: 90% of the species are exclusive for the Palearctic region, 84.9% are exclusive for the Oriental region and 84.6% exclusive for the Australian region (Niedbala 2000).

The widespread species are represented by only 9 pantropical species (8.3%) and 3 semicosmopolitan species (2.7%). The two species *S.(S.) magnus* and *P. nitens* have been introduced from Europe and the Palearctic, respectively, both to St. Helena Island, where each of them was found at 5 sites. No species of oriental origin were found.

Tab. 1 - Phthiracaroidea of Etiopian Region.

	Name of species	Number of specimens	Number of localities	Number of countries	
	Phthiracarus				
1.	*Phthiracarus anonymus* Grandjean, 1933 RÉUNION UGANDA MADAGASCAR	16	6	3	semicosmopol.
2.	*Phthiracarus baloghi* (Mahunka, 1982) ETHIOPIA	1	1	1	endemic
3.	*Phthiracarus benoiti* Balogh, 1958 ANGOLA	1	1	1	nomen dubium
4.	*Phthiracarus cavernosus* (Wallwork, 1977) ST. HELENA	2	1	1	endemic
5.	*Phthiracarus clavifer* Mahunka, 1988 RÉUNION	4	2	1	endemic
6.	*Phthiracarus flagellatus* Wallwork, 1978 ST. HELENA	2	2	1	endemic
7.	*Phthiracarus lentulus* (C.L.Koch, 1841) ANGOLA	7	1	1	semicosmopol.
8.	*Phthiracarus machadoi* Balogh, 1962 ANGOLA	1	1	1	nomen dubium
9.	*Phthiracarus minor* Niedbala, 2001 MADAGASCAR	16	8	1	endemic
10.	*Phthiracarus nitens* (Nicolet, 1855) ST. HELENA	39	5	1	palearctic
11.	*Phthiracarus ochthus* Niedbala, 2001 IVORY COAST TANZANIA MADAGASCAR	33	6	3	ethiopian
12.	*Phthiracarus parabaloghi* Niedbala, 1983 CAMEROON TOGO UGANDA	4	3	3	ethiopian
13.	*Phthiracarus parapocsi* Niedbala, 2001 MADAGASCAR	1	1	1	endemic
14.	*Phthiracarus pocsi* (Mahunka, 1983) TANZANIA	7	2	1	endemic
15.	*Phthiracarus pygmaeus* Balogh, 1958 ANGOLA RÉUNION RWANDA	13	4	3	pantropical
16.	*Phthiracarus schauenbergi* (Mahunka, 1988) MAURITIUS SEYCHELLES	4	2	2	pantropical
17.	*Phthiracarus wallworki* Niedbala, 1984 ST. HELENA	46	4	1	endemic
	Plonaphacarus				
18.	*Plonaphacarus brevisetus* (Mahunka, 1984) TANZANIA	6	1	1	endemic
19.	*Plonaphacarus ecphylus* Niedbala, 2001 KENYA	10	1	1	endemic
20.	*Plonaphacarus kugohi* (Aoki, 1959) COMORO ISLANDS MADAGASCAR SEYCHELLES TANZANIA	38	9	4	pantropical

21.	*Plonaphacarus machadoi* (Balogh, 1958) ANGOLA CAMEROON DEM. REP. CONGO IVORY COAST MALAWI RWANDA TANZANIA UGANDA	152	35	8	ethiopian
22.	*Plonaphacarus ngongi* Niedbala, 2001 KENYA	4	2	1	endemic
23.	*Plonaphacarus optivus* Niedbala, 1994 SOMALIA	3	1	1	endemic
24.	*Plonaphacarus persimilis* Niedbala, 1994 SOMALIA	1	1	1	endemic
25.	*Plonaphacarus styphelos* Niedbala, 2001 TANZANIA	4	1	1	endemic
26.	*Plonaphacarus tanzicus* (Mahunka, 1993) TANZANIA	7	1	1	endemic
	Hoplophthiracarus				
27.	*Hoplophthiracarus ambiguus* Niedbala, 1982 DEM. REP. CONGO GABON KENYA UGANDA	214	16	4	ethiopian
28.	*Hoplophthiracarus discrepus* Niedbala, 1982 DEM. REP. CONGO KENYA MALAWI UGANDA	41	8	4	ethiopian
29.	*Hoplophthiracarus frater* (Balogh, 1958) ANGOLA DEM. REP. CONGO ETHIOPIA	120	15	3	ethiopian
30.	*Hoplophthiracarus hamatus* (Hammer, 1973) MAURITIUS	1	1	1	pantropical
	Steganacarus (Rhacaplacarus)				
31.	*Steganacarus (Rhacaplacarus) amoenus* (Niedbala, 1983) UGANDA	62	3	1	endemic
32.	*Steganacarus (Rhacaplacarus) inflatus* (Niedbala, 1984) DEM. REP. CONGO RWANDA	31	5	2	ethiopian
33.	*Staganacarus (Rhacaplacarus) laetabilis* (Niedbala, 1988) DEM. REP. CONGO	23	2	1	endemic
34.	*Steganacarus (Rhacaplacarus) laterospinosus* (Mah., 1993) RWANDA	15	2	1	endemic
35.	*Steganacarus (Rhacaplacarus) pervigens* (Niedbala, 1988) DEM. REP. CONGO	11	2	1	endemic
36.	*Steganacarus (Rhacaplacarus) spinatus* Niedbala, 2001 DEM. REP. CONGO	4	1	1	endemic
37.	*Steganacarus (Rhacaplacarus) spiniferus* (Mahunka, 1993) TANZANIA	3	1	1	endemic
38.	*Steganacarus (Rhacaplacarus) succinctus* Niedbala, 2001 DEM. REP. CONGO	2	1	1	endemic
39.	*Steganacarus (Rhacaplacarus) zicsii* Mahunka, 1988 TANZANIA	3	1	1	endemic
	Steganacarus (Steganacarus)				
40.	*Steganacarus (Steganacarus) carlosi* Niedbala, 1984 CANARY ISLANDS	108	6	1	endemic
41.	*Steganacarus (Steganacarus) complicatus* Mahunka, 1993 RWANDA	3	1	1	endemic
42.	*Steganacarus (Steganacarus guanarteme* Perez-Inigo and Pena,1996 CANARY ISLANDS	17	3	1	endemic
43.	*Steganacarus (Steganacarus) inurbanus* Niedbala, 1983 UGANDA	3	1	1	endemic
44.	*Steganacarus (Steganacarus) magnus* (Nicolet, 1855) ST. HELENA	28	4	1	european

45.	*Steganacarus (Steganacarus) sol* Balogh, 1958 DEM. REP. CONGO RWANDA TANZANIA	8	4	3	ethiopian
46.	*Steganacarus (Steganacarus) tenerifensis* Perez-Inogo, 1972 CANARY ISLANDS	8	1	1	endemic
47.	*Steganacarus (Steganacarus) vestitus* Niedbala, 1983 DEM. REP. CONGO UGANDA	3	2	2	ethiopian
48.	*Steganacarus (Steganacarus) wallworki* Mahunka, 1984 TANZANIA	2	1	1	endemic
49.	*Steganacarus (Steganacarus) werneri* Mahunka, 1993 RWANDA	10	4	1	endemic
	Arphthicarus				
50.	*Arphthicarus hamidi* (Mahunka, 1987) KENYA	5	1	1	endemic
51.	*Arphthicarus inaequus* (Niedbala, 1984) MALAWI UGANDA	20	4	2	ethiopian
52.	*Arphthicarus indiligens* (Niedbala, 1984) MALAWI	3	2	1	endemic
53.	*Arphthicarus inelegans* (Niedbala, 1988) MAURITIUS RÉUNION	4	2	2	pantropical
54,	*Arphthicarus marginatus* (Mahunka, 1984) TANZANIA	5	1	1	endemic
55.	*Arphthicarus notatus* Niedbala, 2001 GABON	15	1	1	endemic
56.	*Arphthicarus sculptilis* (Niedbala, 1988) COMORO ISLANDS MADAGASCAR RÉUNION	105	11	3	endemic
57.	*Arphthicarus trichosus* (Mahunka, 1988) MAURITIUS	4	2	1	endemic
58.	*Arphthicarus veteratorius* (Niedbala, 1988) COMORO ISLANDS	1	1	1	endemic
	Austrophthiracarus				
59.	*Austrophthiracarus anosculpturatus* (Mahunka, 1987) KENYA	6	1	1	endemic
60.	*Austrophthiracarus aoki* (Mahunka, 1983) MALAWI TANZANIA	10	2	2	ethiopian
	Protophthiracarus				
61.	*Protophthiracarus araios* Niedbala, 2001 MADAGASCAR	7	4	1	endemic
62.	*Protophthiracarus dignus* (Niedbala, 1983) CAMEROON UGANDA	6	2	2	ethiopian
63.	*Protophthiracarus mayottei* Niedbala, 2001 COMORO ISLANDS TANZANIA	13	7	2	ethiopian
64.	*Protophthiracarus prominens* (Balogh, 1958) ANGOLA DEM. REP. CONGO	78	5	2	ethiopian
65.	*Protophthiracarus venustus* (Niedbala, 1983) TANZANIA UGANDA	2	2	2	endemic
	Notophthiracarus				
66.	*Notophthiracarus armatus* (Mahunka, 1986) TANZANIA	4	2	1	endemic
67.	*Notophthiracarus bicarinatus* Niedbala, 2001 MADAGASCAR	2	1	1	endemic
68.	*Notophthiracarus cavernosus* (Mahunka, 1978) TANZANIA	3	1	1	endemic
69.	*Notophthiracarus dactyloscopicus* (Mahunka, 1978) MAURITIUS	3	1	1	endemic

70.	*Notophthiracarus echinus* (Balogh, 1962) MADAGASCAR	1	1	1	endemic
71.	*Notophthiracarus lineolatus* Mahunka, 1993 TANZANIA	11	1	1	endemic
72.	*Notophthiracarus parareductus* (Mahunka, 1992) RÉUNION	1	1	1	endemic
73.	*Notophthiracarus parasomalicus* Niedbala, 2001 MADAGASCAR	12	4	1	endemic
74.	*Notophthiracarus parasummersi* Niedbala, 2001 MADAGASCAR	1	1	1	endemic
75.	*Notophthiracarus (?) pardinus* (Berlese, 1916) SOMALIA	1	1	1	endemic
76.	*Notophthiracarus parilloi* Niedbala, 2001 MADAGASCAR	10	2	1	endemic
77.	*Notophthiracarus (?) paulianus* (Balogh, 1962) MADAGASCAR	1	1	1	endemic
78.	*Notophthiracarus procerus* Niedbala, 2001 MADAGASCAR	8	3	1	endemic
79.	*Notophthiracarus puylaerti* Niedbala, 2001 DEM. REP. CONGO	1	1	1	endemic
80.	*Notophthiracarus rafalskii* Niedbala, 1997 MADAGASCAR	1	1	1	endemic
81.	*Notophthiracarus (?) rapax* (Berlese, 1916) SOMALIA	1	1	1	endemic
82.	*Notophthiracarus sacyae* (Mahunka, 1983) TANZANIA	4	1	1	endemic
83.	*Notophthiracarus similis* Niedbala, 2001 MADAGASCAR	12	7	1	endemic
84.	*Notophthiracarus (?) somalicus* (Berlese, 1923) SOMALIA	2	1	1	endemic
85.	*Notophthiracarus summersi* Niedbala, 2001 MADAGASCAR	15	4	1	endemic
86.	*Notophthiracarus zebrus* (Balogh, 1962) MADAGASCAR	3	3	1	endemic
	Atropacarus (Hoplophorella)				
87.	*Atropacarus (Hoplophorella) angolensis* (Mahunka, 1984) ANGOLA	2	1	1	endemic
88.	*Atropacarus (Hoplophorella) balazsi* (Mahunka, 1983) MADAGASCAR	2	1	1	endemic
89.	*Atropacarus (Hoplophorella) brevipilis* (Balogh, 1958) KENYA MALAWI RWANDA TANZANIA	7	5	4	ethiopian
90.	*Atropacarus (Hoplophorella) collaris* (Balogh, 1958) ANGOLA	3	3	1	endemic
91.	*Atropacarus (Hoplophorella) cucullatus* (Ewing, 1909) TOGO UGANDA	6	2	2	semicosmopol.
92.	*Atropacarus (Hoplophorella) ensiferus* (Mahunka, 1984) ANGOLA	1	1	1	endemic
93.	*Atropacarus (Hoplophorella) floridus* (Jacot, 1933) CAPE VERDE Isl COMORO ISLANDS MAURITIUS SOMALIA	87	6	4	pantropical
94.	*Atropacarus (Hoplophorella) insolens* Niedbala, 1988 DEM. REP. CONGO	4	3	1	endemic
95.	*Atropacarus (Hoplophorella) lemurius* (Mahunka, 1993) MADAGASCAR	4	1	1	endemic
96.	*Atropacarus (Hoplophorella) minisetosus* (Mahunka, 1978) MALAWI RÉUNION	4	2	2	ethiopian
97.	*Atropacarus (Hoplophorella) multirugosus* (Mahunka, 1978) DEM. REP. CONGO MAURITIUS RÉUNION	21	4	3	ethiopian
98.	*Atropacarus (Hoplophorella) mwali* (Mahunka, 1994) COMORO ISLANDS	5	1	1	endemic

99.	*Atropacarus (Hoplophorella) oblongus* (Niedbala, 1983) KENYA	1	1	1	endemic
100.	*Atropacarus (Hoplophorella) pallens* (Niedbala, 1983) KENYA MALI TOGO	17	3	3	ethiopian
101.	*Atropacarus (Hoplophorella) pustulatus* (Mahunka, 1983) RWANDA	7	1	1	endemic
102.	*Atropacarus (Hoplophorella) remigerus* (Berlese, 1923) SOMALIA	5	1	1	endemic
103	*Atropacarus (Hoplophorella) singularis* (Sellnick, 1959) ALDABRA COMORO ISLANDS MAURITIUS	7	3	3	pantropical
104.	*Atropacarus (Hoplophorella) stilifer* (Hammer, 1961) ANGOLA CAMEROON	122	6	2	pantropical
105.	*Atropacarus (Hoplophorella) subciliatus* (Mahunka, 1983) TANZANIA	1	1	1	endemic
106.	*Atropacarus (Hoplophorella) subitus* (Niedbala, 1983) KENYA	1	1	1	endemic
107.	*Atropacarus (Hoplophorella) tuberculosissimus* (Mah.,1978) ANGOLA KENYA MAURITIUS TANZANIA UGANDA	33	7	5	ethiopian
108.	*Atropacarus (Hoplophorella) vitrinus* (Berlese, 1913) ANGOLA CAMEROON CAPE VERDE Isl CHAD COMORO ISLANDS DEM. REP. CONGO RWANDA TOGO UGANDA	138	39	9	pantropical
	Σ	2001	213	24	

The fauna of the African mainland and the islands

Seventy three species (67.6% of all species found) occur in continental Africa, on the islands near Africa - 45 species; 36 of them (one third), occur exclusively on the islands. There are only 9 species in common (8.3%): *P. anonymus, P. ochthus, P. pygmaeus, P. kugohi, A.(H.) floridus, A.(H.) minisetosus, A.(H.) multirugosus, A.(H.) tuberculosissimus, A.(H.) vitrinus.* Apart from the two Ethiopian species: *A.(H.) minisetosus* and *A.(H.) multirugosus*, they are widespread, pantropical and semicosmopolitan. Thus, the fauna of the continent and that of the islands are rather specialized and represented by species of restricted geographical distribution.

There are 44 endemic species (60.3%) in the continent and 29 (64%) on the islands. The fauna of the islands includes many more pantropical elements (8 species - 17.8%) than that of the continent. These species have been probably passively brought in, found their place on the islands and have not penetrated into the continent. At the same time, there is a low number of Ethiopian species on islands (5 species - 11.1%). Except Cape Verde Island and the Seychelles, each of the islands, irrespective of the geographical position, has a specific fauna of the Phthiracaroidea, manifested by the number of endemic species. The endemism of the fauna of Madagascar is the strongest, expressed by as many as 17 endemic species. The neighbouring Reunion and Mauritius, Canary Islands, as well as St. Helena Island (the latter two on the opposite side of the continent) have specific and distinctive faunas expressed by high degrees of endemism (tab. 1).

Subregions and their zoogeographical elements

The faunas of particular subregions of the Ethiopian region reveal specific and distinctive features as far as the presence of different zoogeographical elements is concerned, and 5 groups of the subregions can be distinguished (tab. 2). The first non-specific group includes only the Cape Verde Islands, on the west, with two pantropical species occurring there. The second group includes Angola, Cameroon, Togo and Uganda, subregions of the western part of the Central Africa whose fauna is a mixture of different zoogeographical elements. The third group comprises Reunion, Mauritius, Comoro, Seychelles, i.e, the islands in the south-east part of Africa, and St. Helena Island in the Atlantic Ocean. These islands make a basis for widespread pantropical species (Seychelles), but many of them developed endemic elements, a few species on Comoro, Mauritius and Reunion and St. Helena Island.

To the fourth group belong subregions with mainly Ethiopian and endemic faunas, without widespread species, or with single widespread species (given below in parenthesis next to the name of the country). This group includes the countries of Central and Eastern Africa: Ivory Coast, Kenya, Malawi, Democratic Republic of Congo (1), Rwanda (2), Ethiopia, Somalia (1).

The fifth group is made of specific regions, whose fauna includes a majority of endemic species, and to this group belong Tanzania and Madagascar.

Zoogeographic elements in subregions (tabs 1-2)

Endemites

The greatest number of endemites occurs in the Eastern Africa: Tanzania, Kenya, Madagascar and Somalia. Lower numbers can be found in countries of Central Africa - Malawi, Uganda and Rwanda. A relatively large number of endemites has been reported from the Democratic Republic of Congo, but it should be noted that the largest number of samples was collected in this region.

The highest number of endemites found in the Eastern Africa points to a strong speciation taking place there. Probably, the fauna which came into this area underwent strong adaptive radiation under the specific conditions of high mountains.

Ethiopian elements

The 19 Ethiopian species occurring in Africa over a limited area of two or three countries (only *P. machadoi* is a widespread species found in 8 countries of Central Africa) can be divided into 3 groups:

1) 7 species [*P. ochthus, P. parabaloghi, P. machadoi, H. frater, A. inaequus, P. dignus, A.(H.) pallens*], occurring in Central Africa, characterized by relatively disjunctive distribution.

2) 4 species (*S.(R.) inflatus, S.(S.) vestitus, A. aoki, P. prominens*) also from Central Africa, but characterized by compact range of distribution. These species may be endemites.

3) 8 species including the Ethiopian elements whose distribution indicates perhaps their origin from the Oriental or Australian regions. These species are *H. ambiguus, H. discrepus, S.(S.) sol* found in the eastern countries of Central Africa, *A.(H.) multirugosus, A.(H.) tuberculosissimus* occurring on the eastern islands and in central countries of Central Africa and [*P. mayoeti, A.(H.) brevipilis, A.(H.) minisetosus*] having apparently an eastern type of distribution.

Tab. 2 - Phthiracaroidea fauna in selected countries investigated of the Ethiopian Region.

COUNTRY	NAME OF SPECIES		Number of specimens	Number of species	Number of localities	NUMBER OF SPECIES					
						Endemic	Ethiopian	Pantropical	Semicosmopol.	Palearctic or European	Nomina dubia
ALDABRA	*A. (H.) singularis*	pantropical	1	1	1			1			
ANGOLA	*A. (H.) angolensis*	endemic	2								
	A. (H.) collaris	endemic	3								
	A. (H.) ensiferus	endemic	1								
	A. (H.) stilifer	pantropical	20								
	A. (H.) tubercullosissimus	ethiopian	12								
	A. (H.) vitrinus	pantropical	48	13	23	3	4	3	1		2
	H. frater	ethiopian	14								
	P. benoiti	nomen dubium	1								
	P. lentulus	semicosmopol.	7								
	Ph. Machadoi	nomen dubium	1								
	P. pygmaeus	pantropical	5								
	P. machadoi	ethiopian	28								
	P. prominens	ethiopian	77								
CAMEROON	*A. (H.) stilifer*	pantropical	102								
	A. (H.) vitrinus	pantropical	12								
	P. parabaloghi	ethiopian	1	5	6		3	2			
	P. machadoi	ethiopian	12								
	P. dignus	ethiopian	4								
CANARY ISLANDS	*S. (S.) tenerifensis*	endemic	8								
	S. (S.) carlosi	endemic	108	3	9	3					
	S. (S.) guanarteme	endemic	17								
CAPE VERDE ISLAND	*A. (H.) vitrinus*	pantropical	8								
	A. (H.) floridus	pantropical	16	2	3			2			
CHAD	A. (H.) vitrinus	pantropical	13	1	1			1			
COMORO ISLANDS	*A. sculptilis*	endemic	88								
	A. veteratorius	endemic	1								
	A. (H.) floridus	pantropical	4								
	A. (H.) mwali	endemic	5	8	12	3	1	4			
	A. (H.) singularis	pantropical	1								
	A. (H.) vitrinus	pantropical	2								
	P. kugohi	pantropical	32								
	P. mayottei	ethiopian	9								
DEM. REP. CONGO	*A. (H.) multirugosus*	ethiopian	2								
	A. (H.) insolens	endemic	4								
	A. (H.) vitrinus	pantropical	39								
	H. ambiguus	ethiopian	11								
	H. discrepus	ethiopian	28								
	H. frater	ethiopian	104								
	N. puylaerti	endemic	1	16	37	6	9	1			
	P. machadoi	ethiopian	50								
	P.prominens	ethiopian	1								
	S. (R.) spinatus	endemic	4								
	S. (R.) inflatus	ethiopian	2								
	S. (R.) succinctus	endemic	2								
	S. (R.) laetabilis	endemic	23								
	S. (R.) pervigens	endemic	11								
	S. (S.) vestitus	ethiopian	2								
	S. (S.) sol	ethiopian	2								
ETHIOPIA	*H. frater*	ethiopian	2								
	P. baloghi	endemic	1	2	2	1	1				
GABON	*A. notatus*	endemic	15								
	H. ambiguus	ethiopian	3	2	1	1	1				
IVORY COAST	*P. ochthus*	ethiopian	4								
	P. machadoi	ethiopian	2	2	3		2				

KENYA	*A. anosculpturatus*	endemic	6								
	A. hamidi	endemic	5								
	A. (H.) brevipilis	ethiopian	4								
	A. (H.) oblongus	endemic	1								
	A. (H.) pallens	ethiopian	5	11	12	6	5				
	A. (H.) subitus	endemic	1								
	A .(H.) tuberculosissimus	ethiopian	2								
	H. ambiguus	ethiopian	25								
	H. discrepus	ethiopian	8								
	P. ecphylus	endemic	10								
	P. ngongi	endemic	4								
MADAGASCAR	*A. sculptilis*	endemic	4								
	A. (H.) balazsi	endemic	2								
	A. (H.) lemurius	endemic	4	20	23	17	1	1	1		
	N. bicarinatus	endemic	2								
	N. echinus	endemic	1								
	N. parasomalicus	endemic	12								
	N. parasummersi	endemic	1								
	N. parilloi	endemic	10								
	N. (?) paulianus	endemic	1								
	N. procerus	endemic	8								
	N. rafalskii	endemic	1								
	N. similis	endemic	12								
	N. summersi	endemic	15								
	N. zebrus	endemic	3								
	P. anonymus	semicosmopol.	3								
	P. minor	endemic	16								
	P. ochthus	ethiopian	24								
	P. parapocsi	endemic	1								
	P. kugohi	pantropical	4								
	P. araios	endemic	7								
MALAWI	*A. aoki*	ethiopian	1								
	A. inaequus	ethiopian	10								
	A. indiligens	endemic	3								
	A. (H). Brevipilis	ethiopian	1	7	10	1	6				
	A. (H). minisetosus	ethiopian	1								
	H. discrepus	ethiopian	4								
	P. machadoi	ethiopian	9								
MALI	*A. (H.) pallens*	ethiopian	10	1	1		1				
MAURITIUS	*A. inelegans*	pantropical	2								
	A. trichosus	endemic	4								
	A. (H.) floridus	pantropical	65								
	A. (H.) multirugosus	ethiopian	4								
	A. (H.) singularis	pantropical	5	9	8	2	2	5			
	A. (H.) tuberculosissimus	ethiopian	16								
	H. hamatus	pantropical	1								
	N. dactyloscopicus	endemic	3								
	P. schauenbergi	pantropical	2								
RÉUNION	*A. inelegans*	pantropical	2								
	A. sculptilis	endemic	13								
	A. (H.) minisetosus	ethiopian	3								
	A. (H.) multirugosus	ethiopian	15	8	5	3	2	2	1		
	N. parareductus	endemic	1								
	P. anonymus	semicosmopol.	11								
	P. clavifer	endemic	4								
	P. pygmaeus	pantropical	7								
RWANDA	*A. (H.) brevipilis*	ethiopian	1								
	A. (H.) pustulatus	endemic	7								
	A. (H.) vitrinus	pantropical	10								
	P. pygmaeus	pantropical	1								
	P. machadoi	ethiopian	12	10	16	4	4	2			
	S. (R.) laterospinosus	endemic	15								
	S. (R.) inflatus	ethiopian	29								
	S. (S.) complicatus	endemic	3								
	S. (S.) sol	ethiopian	1								
	S. (S.) werneri	endemic	10								
SEYCHELLES	*P. kugohi*	pantropical	1								
	P. schauenbergi	pantropical	2	2	2			2			

SOMALIA (Berlese 1923, Niedbala 1994)	*A. (H.) floridus*	pantropical	2	7	2	6		1			
	A. (H.) remigerus	endemic	5								
	N. (?) pardinus	endemic	1								
	N. (?) rapax	endemic	1								
	N. (?) somalicus	endemic	2								
	P. optivus	endemic	3								
	P. persimilis	endemic	1								
ST. HELENA	*P. cavernosus*	endemic	2	5	12	3				2	
	P. flagellatus	endemic	2								
	P. nitens	palearctic	39								
	P. wallworki	endemic	46								
	S. (S.) magnus	european	28								
TANZANIA	*A. aoki*	ethiopian	9	22	12	14	7	1			
	A. marginatus	endemic	5								
	A. (H.) brevipilis	ethiopian	1								
	A. (H.) subciliatus	endemic	1								
	A. (H.) tuberculosissimus	ethiopian	2								
	N. armatus	endemic	4								
	N. cavernosus	endemic	3								
	N. lineolatus	endemic	11								
	N. sacyae	endemic	4								
	P. ochthus	ethiopian	5								
	P. pocsi	endemic	7								
	P. kugohi	pantropical	1								
	P. machadoi	ethiopian	11								
	P. styphelos	endemic	4								
	P. tanzicus	endemic	7								
	P. mayottei	ethiopian	4								
	P. venustus	endemic	1								
	P. brevisetus	endemic	6								
	S. (R.) zicsii	endemic	3								
	S. (R.) spiniferus	endemic	3								
	S. (S.) sol	ethiopian	5								
	S. (S.) wallworki	endemic	2								
TOGO	*A. (H.) cucullatus*	semicosmopol.	2	4	3		2	1	1		
	A. (H.) pallens	ethiopian	2								
	A. (H.) vitrinus	pantropical	3								
	P. parabaloghi	ethiopian	2								
UGANDA	*A. inaequus*	ethiopian	10	14	9	3	8	1	2		
	A. (H.) cucullatus	semicosmopol.	4								
	A. (H.) tuberculosissimus	ethiopian	1								
	A. (H.) vitrinus	pantropical	3								
	H. ambiguus	ethiopian	175								
	H. discrepus	ethiopian	1								
	P. anonymus	semicosmopol.	2								
	P. machadoi	ethiopian	28								
	P. parabaloghi	ethiopian	1								
	P. dignus	ethiopian	2								
	P. venustus	endemic	1								
	S. (R.) amoenus	endemic	62								
	S. (S.) inurbanus	endemic	3								
	S. (S.) vestitus	ethiopian	1								

Widespread species

As follows from analysis of distribution of widespread species (3 semicosmopolitan and 9 pantropical) five of them may have dispersed from the east, most probably from the Oriental region. This group includes 3 species [*P. schauenbergi, A. inelegans, A.(H.) singularis*] occurring on the south-eastern islands, but not reaching the continent, and two species (*P. anonymus* and *P. kugohi*) occurring on the same islands but also in the eastern part of the continent.

The second group comprises species which are not very widely distributed in Africa, and occurring in the limited area of two Central African countries, these are *P. pygmaeus, A.(H.) stilifer, A.(H.) cucullatus* and *P. lentulus*, although the latter might have been introduced to Africa.

The third group includes pantropical species widespread in Africa, i.e. occurring from the Cape Verde Island on the west, to the Comoro on the east [*A.(H). floridus, A.(H.) vitrinus*].

Genera and zoogeographical subregions

The Ethiopian fauna comprises three groups of genera (tab. 3).

The first includes widespread genera: *Phthiracarus, Plonaphacarus* and *Atropacarus (Hoplophorella)* - the two most primitive and one evolutionary most distant. The widespread distribution of the species which belong to phylogenetically distant genera is one of the arguments supporting the view that Africa was colonized by the phthiracaroid species before separation by continental drift. The second group of genera (*Hoplophthiracarus, Steganacarus* and *Protophthiracarus*) occurs mainly in Central Africa.

The third group is composed of genera showing distinct Eastern African distribution: *Arphthicarus, Austrophthiracarus* and *Notophthiracarus*. A clear path of migration for *Arphthicarus* and especially *Notophthiracarus* can be identified from the Australian region through the islands of East Africa, Tanzania ending in Congo (where e.g. only one species of *Notophthiracarus* occurs).

Species of *Notophthiracarus* which are abundantly present in the Ethiopian region (tabs 1, 3) are probably of southern Australian or Southern African origin. In both these regions this genus belongs to the genera which is most abundant in species. Over 45% of the Phthiracaroidea species found in the Australian region (Niedbala 2000) and about 35% of those found in South Africa (*unpubl. data*) belong to *Notophthiracarus.*

Africa is the centre of speciation for the genus *Steganacarus* (mainly in the Democratic Republic of Congo, Rwanda, Uganda and Tanzania) and in particular *Steganacarus (Rhacaplacarus)* (in Canary Islands). It has also been established that the subgenus *Hoplophorella*, which is profoundly represented by endemic species, undergoes strong speciation, in particular in Angola. St. Helena Island is the area of adaptive radiation of *Phthiracarus*. Other subregions such as Comoro, Madagascar, Mauritius, Kenya and Somalia, are non-specific in this respect.

Discussion

The core of the African fauna developed before the break-up of Pangea, which is evidenced by the presence of both the phylogenetically oldest and youngest genera (Niedbala 1986). On the other hand, Africa seems

Tab. 3 - Number of zoogeographic elements of Phthiracaroidea at generic level.

	Semicosm	Pantropic	Palaearct.	Europ.	Etiop.	Endem.	Nomina dubia	Σ	%
Phthiracarus	2	2	1		2	8	2	17	15,7
Plonaphacarus		1			1	7		9	8,3
Hoplophthiracarus		1			3			4	3,7
Steganacarus				1	3	15		19	17,6
Steganacarus (Racaplacarus)					1	8		9	8,3
Steganacarus (Steganacarus)				1	2	7		10	9,3
Arphthicarus		1			1	7		9	8,3
Austrophthiracarus					1	1		2	1,8
Protophthiracarus					3	2		5	4,6
Notophthiracarus						21		21	19,4
Atropacarus (Hoplophorella)	1	4			5	12		22	20,4
	3	9	1	1	19	73	2	108	

to be colonized relatively late by some Phthiracaroidea. This supposition is supported by the presence of a large number of species representing the evolutionarily most recently derived genera in two phylogenetic lineages: *Steganacarus* in one line and *Notophthiracarus* and *Hoplophorella* in the other. A total of 48.1% (almost half of species represented) belong to these genera (tab. 3).

For the two subgenera taking extreme positions in the cladogram by Niedbala (1986), Africa is the centre of speciation. The range of the *Steganacarus (Rhacaplacarus)* typical of the Ethiopian region, reaches the Australian region, but appeared only in the northern part of Queensland and was found in a single separate site in Japan (Niedbala 2000). Also the subgenus *Atropacarus (Hoplophorella)*, which is particularly abundantly represented by endemic species, underwent strong speciation on the continent (tab. 1).

Furthermore the *Phthiracarus* underwent adaptive radiation on St. Helena Island (tab. 1). From among the Euphthiracaroidea the Ethiopian region is the centre of speciation for the subgenera *Euphthiracarus (Pocsia)* and *Indotritia (Afrotritia)* (Niedbala 1998).

Morphological differences among some species e.g. *S.(S.) complicatus, S.(S.) werneri, S.(S.) vestitus* are very small, which means that they are at the stage of species differentiation. An analogous example for the *Atropacarus (Hoplophorella)* are *A.(H.) pustulatus* and *A.(H.) subcillatus*.

No representatives of the subgenus *Atropacarus (Atropacarus)* have been reported there.

Contemporary distribution of Phthiracaroidea of the Ethiopian region seems a result of a complex mixture: origin before the break-up of Pangea (Niedbala 1991) and dispersion of certain species or even genera from the east, probably from the Oriental region or from south i.e. the Australian region or southern Africa.

This suggestion is based on the following data: 1) strong speciation of endemic species in the eastern part of the Ethiopian region, 2) the occurrence of a significant group of Ethiopian species usually found in the eastern part of Africa, 3) the presence of five semicosmopolitan species of distinctly eastern distribution, 4) the presence of representatives of three genera of definitely East-African distribution, in particular *Notophthiracarus* - a genus abundant in the Australian region and in South Africa.

A comparison of the Phthiracaroidea fauna from the Oriental, Australian and Ethiopian regions was made in an earlier study (Niedbala 2000) and has shown a clear faunistic distinctness of these two regions. Apart from a few widespread species, there are no species occurring exclusively in the two former regions.

Similarly, a comparison of the fauna from the Palearctic (Niedbala 1992) and Ethiopian regions has disclosed that, apart from two species introduced from the Palearctic to St. Helena Island (*P. nitens* and *S.(S.) magnus*) and a few widespread species, there are no common ones. Hence, it can be concluded that the Phthiracaroidea faunas of the neighbouring zoogeographical regions are specific and definitely distinct.

Analysis of the descriptions of the sites studied has not implied any ecological correlations.

References

Niedbala W., 1986 - Systeme des Phthiracaroidea (Aoribatida, Euptyctima). *Acarologia*, 27(1): 61-84.

Niedbala W., 1991 - Preliminary hypothesis on the biogeography of the Phthiracaroidea (Acari, Oribatida). In: Dusbabek F., Bukva V. (eds.), Modern Acarology. Academia, Prague and SPB Academic Publishing bv: 219-231.

Niedbala W., 1992 - Phthiracaroidea (Acari, Oribatida). Systematic Studies. PWN, Elsevier: 612 pp.

Niedbala W., 1998 - Ptyctimous mites of the Ethiopian region. 1. Euphthiracaroidea (Acari, Oribatida). *Journal African Zoology*, 112(1): 15-75.

Niedbala W., 2000 - The ptyctimous mites fauna of the Oriental and Australian regions and their centres of origin (Acari, Oribatida). *Genus*, (supplement): 493 pp.

Schnell R., 1976 - La flore et la vegetation de l'Afrique tropicale. J. Gauthier-Villars, Paris: 234 pp.

The occurrence of *Macrocheles* mites (Acari: Macrochelidae) in relation to the activity of dung beetles: a field study in Southern France

Habiba Glida, Michel Bertrand
Laboratoire de Zoogéographie, Université Montpellier III, F-34199 Montpellier cedex 5, France

Key words: *Macrocheles*, beetle, phoresy, dung, seasonal activity.

Abstract: A field survey of *Macrocheles* mites and dung beetle communities was carried out at three sites in Southern France. The number of mites collected in the dung pads increased simultaneously with dung beetle activity. A wide range of beetles species were involved in mite transport. Larger beetles transported more mites than smaller beetles, but some medium size species (*Onthophagus* genus) also transported numerous mites. Some of the smaller beetle species [*Aphodius* (4-8 mm)] were observed to carry some mites. *Aphodius* species, which are active throughout the whole year, may have an important role in the establishment of *Macrocheles* population during the colder season.

Introduction

Animal droppings constitute an ephemeral habitat where specialized invertebrate communities (beetles, flies, mites and nematodes) live together. The dung pad is a temporary breeding site for consumers, prey and predators (Krantz 1983). The size of the pad limits trophic resources and flies and beetles are often in competition: beetle feeding and dung-burying activities may also reduce fly breeding (Wallace *et al.* 1979). Macrochelidae are active consumers of eggs and larvae of synanthropic flies and might be used as natural agents for biological control of flies breeding in dung (Axtell 1963 and 1969; Filipponi 1964; Wallace *et al.* 1979; Krantz 1983; Wallace and Holm 1984). Macrochelid females are phoretic on dung beetles and/or coprophagous flies (phoresy being the obligate transport strategy to carry the mites from pad to pad). They have been collected from the major taxa of coprophagous Scarabaeidae. *Macrocheles glaber* (Müller) is involved in the transport of nematodes (Cicolani 1992).

The *Macrocheles* fauna of Southern France are poorly understood, except for one study on the relationships between the *pisentii* group and *Scarabaeus* (Krantz 1991). A field survey of both mites and beetle activity has never been carried out. This study is the first attempt to explain the distribution and dynamics of macrochelid mites under natural conditions.

Material and methods

The field study was carried out at three experimental sites near Montpellier (France). These sites were chosen as representatives of available

F. Bernini, R. Nannelli, G. Nuzzaci, E. de Lillo (eds.), Acarid Phylogeny and Evolution. Adaptations in mites and ticks, 199-207.

dung resources (sheep, horse or cow pads) and scarabaeid communities from littoral (alluvial plain) to mattoral zones on rocky hills [*Quercus ilex* L. and *Q. pubescens* (Willd)].

Study area and climatic conditions

The three experimental sites near Montpellier are under a cool sub-humid Mediterranean climate (wet autumn and spring, and dry summer). L'Estagnol is a Nature Reserve in an ancient salt marsh located about 8 km south-west of Montpellier and 5 km from the coast. It is composed of lagoons and situated at sea level. The site is grazed by horses which are never treated antihelmintic. The two sites of La Cardonille and Saint-Martin-de-Londres are located about 35 and 40 km north of Montpellier, and are situated at 250 and 330 m above sea level, respectively. These farms support cattle herds (Saint-Martin-de-Londres), and sheep flocks (La Cardonille).

Survey of dung beetle and mite activity

Dung beetle activity and phoretic females

Samples were taken at each of the three sites using modified pitfall traps (3 permanent traps per site) of standard design (type CSR) (Lobo *et al.* 1988). Traps were composed of plastic buckets, embedded in the soil, containing fresh dung and sand to keep beetles alive. Fresh cattle dung (500 g) (sheep, cow or horse) was supported on a grill. Previous studies have shown that five baited traps could collect 75-80% of total local species and 95% of the abundance and biomass of the dung beetle community in the Mediterranean area (Lumaret and Kirk 1987; Lobo *et al.* 1998). Three permanent traps collecting about 50% of total fauna may, therefore, give a good representation of dung beetle activity.

The beetles were identified to species level and counted, mites were collected with a fine moistened brush. The phoretic rate (**P**) was expressed as the number of mites attached per beetle.

Macrocheles *mite activity*

The survey of *Macrocheles* mite activity was carried out at l'Estagnol: six experimental cattle pads (500 g of fresh dung collected from local horses) were deposited every 3 weeks, on a plastic sheet to avoid direct colonization by soil inhabiting mites. Pads were spaced at 5 m intervals. One and two weeks old dung samples were collected and *Macrocheles* mites extracted by means of Berlese-Tullgren funnels.

Results

At the 3 experimental sites 3,342 *Macrocheles* mites were found. *M. glaber* (Müller), *M. insignitus* (Berlese), *M. merdarius* (Berlese), *M. perglaber* Filipponi & Pegazzano, *M. robustulus* (Berlese) and *M. scutatus* (Berlese) were the dominant species identified. *Glyptholaspis* sp. and *Holostaspella* sp. were also present. In this paper, we consider the global *Macrocheles* fauna.

Seasonal variation in density of Macrocheles *mites at l'Estagnol (fig. 1)*

The number of trapped *Macrocheles* varied from February to June 2000. The efficiency of each trap varied but results showed the same tendencies. Macrochelids appeared at the end of winter, with the maximal abundance being observed in spring. The maximum number of *Macrocheles* caught was 187 in the fourth trap (P4). By the end of spring the number of mites decreased considerably (beginning of the dry season). In the first trap, the largest number of mites was observed at the beginning of spring with the arrival of one scarab *Bubas bubalus* (Ol.) which was carrying 65 mites.

Contribution of dung beetles to the establishment of Macrocheles *mites (fig. 2)*

Until March, neither dung beetles nor *Macrocheles* mites were collected. The first appearance of mites coincided with the first appearance of the dung beetles in early spring. We observed a peak

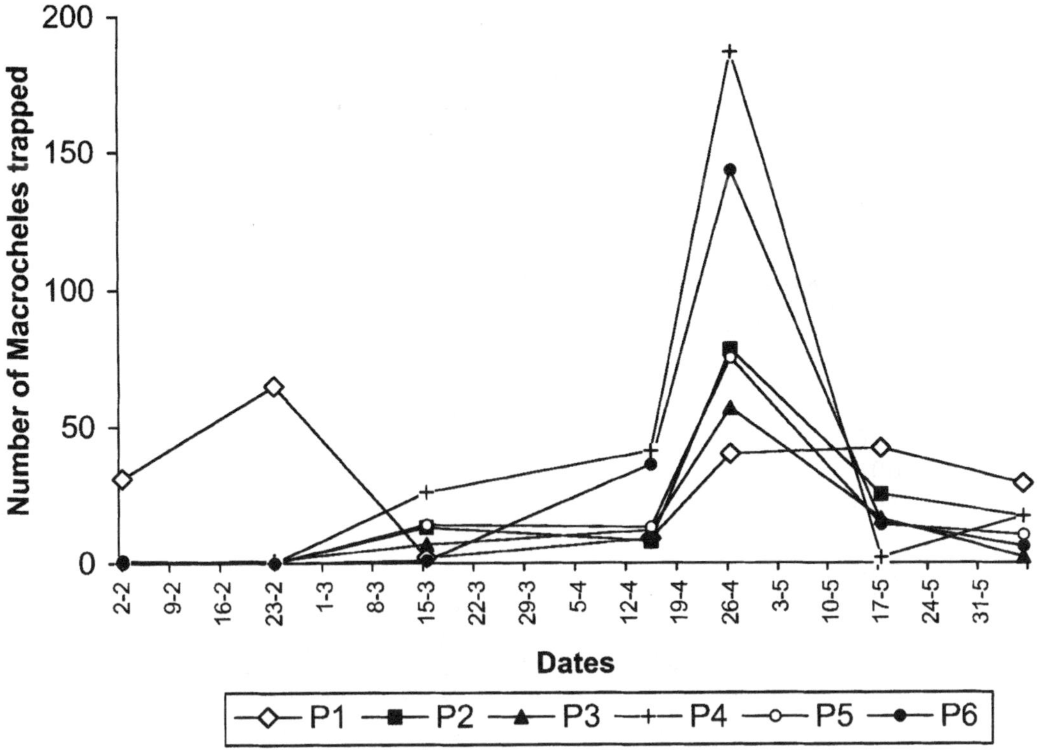

Fig. 1 - Seasonal variation: number of *Macrocheles* trapped at l'Estagnol (P1 to P6).

of beetle numbers at the end of spring, at which point the density of mites reached its maximum. The number of *Macrocheles* decreased considerably as the number of dung beetles decreased.

These results showed the parallel tendencies of both dung beetle and mite activities: two unfavourable periods (too cold or too dry) and a favourable season (late winter and spring).

Mite - beetle association

More than 2,000 beetles were collected but *Macrocheles* mites were found on 28 of the 34 species of beetles collected from February to June near Montpellier (604, 955, and 960 beetles at L'Estagnol, La Cardonille and St Martin, respectively). Figure 3 shows the mean, maximum and minimum number of *Macrocheles* mites found on each beetle species: *Aphodius* species usually carried few mites, but *A. merdarius* (Fabr.), *A. haemorrhoidalis* (L.) and *A. erraticus* (L.) were observed with some *Macrocheles* (3 or 4). On *Onthophagous* genus (4-11 mm), phoresy rate was about 5 mites per beetle with a maximum of 20 mites observed on *O. coenobita* (Herbst.). The number carried by the biggest scarabs [*Scarabaeus laticollis* L., *Anoplotrupes stercorosus* (Scriba) and *Geotrupes mutator* Marsh] was large, with the highest number of phoretic mites being observed on *B. bubalus*.

Aphodius are small size beetles (4-8 mm) when compared with *Bubas*, *Geotrupes*, and *Scarabaeus* (15-20 mm). Carrying ability is firstly determined by the surface of the beetle (is there a competition amongst mites for attachment sites?) and it is possible that the fitness of smallest species may be affected by mite load. Surface area and weight of the host are correlated with its dimensions. The dry weight was used to represent these parameters. The mean number of *Macrocheles* per beetle was found to be positively correlated with the mean dry weight of beetle species (R^2=0.67) (fig. 4). The points above the

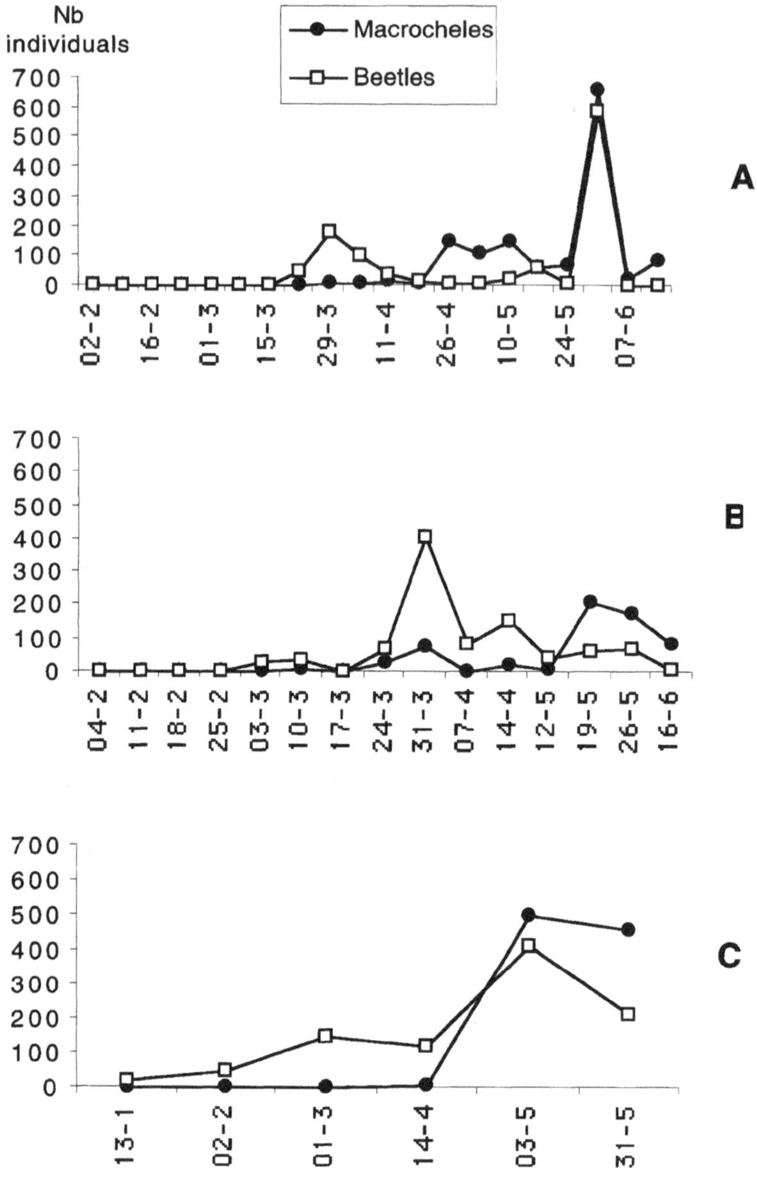

Fig. 2 - Seasonal activities of dung beetles and macrochelid mites: number of dung beetles and *Macrocheles* collected. (A: L'Estagnol; B: La Cardonille; C: St Martin de Londres).

curve correspond to the scarab species that carry the greatest number of *Macrocheles* mites.

Basis for a model of scarab efficiency in the transport of Macrocheles *mites*

The contribution of scarab hosts to the transport of *Macrocheles* was estimated by counting the number of *Macrocheles* per scarab at the three sites: l'Estagnol, La Cardonille and Saint Martin-de-Londres. We calculated for each beetle species the frequency at each time (**F**) and the mean number of phoretic *Macrocheles* mites ($\mathbf{N_m}$). Each beetle species was characterized by the frequency of individuals observed with mites ($\mathbf{M_m}$). We obtained the phoresy rate per beetle species per

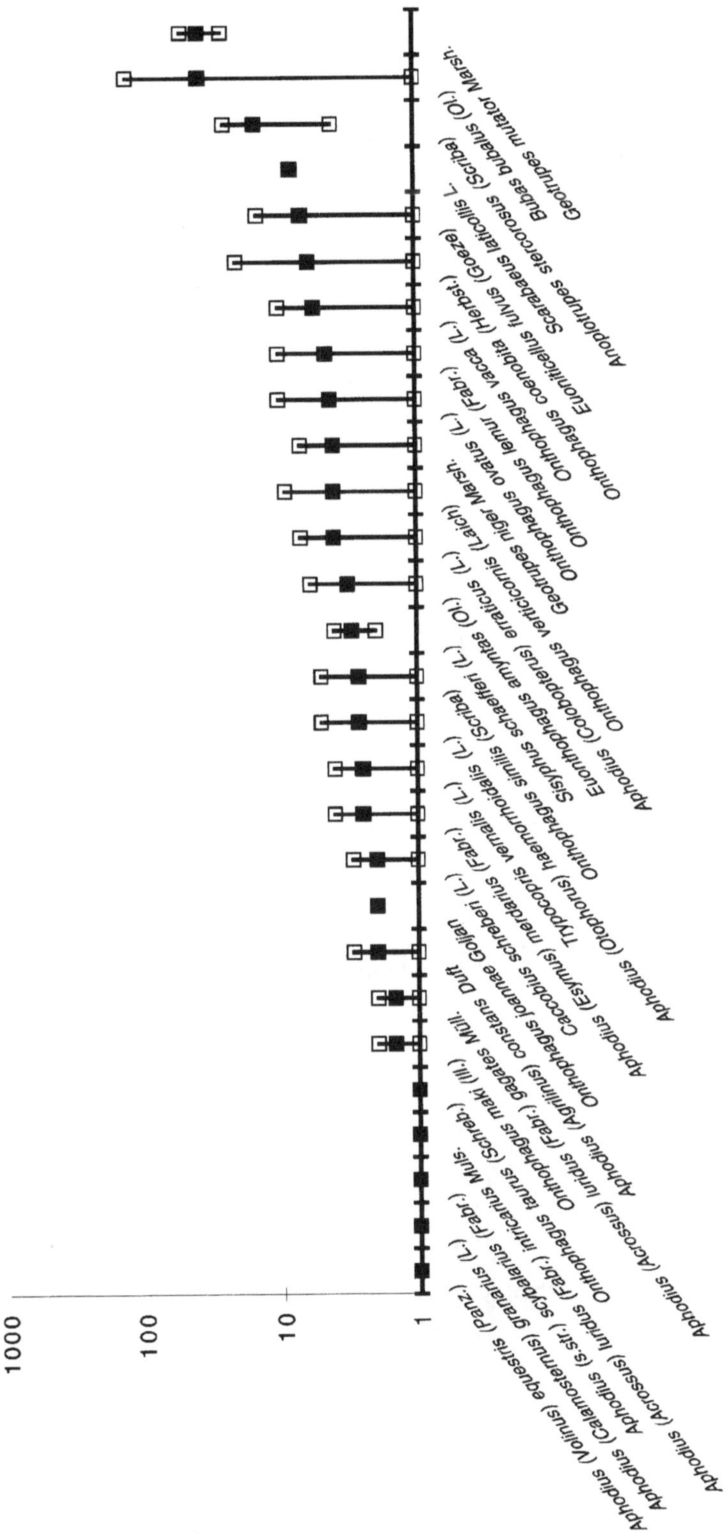

Fig. 3 - Phoretic *Macrocheles* on dung beetles: number of *Macrocheles* per beetle (mean, minimum and maximum).

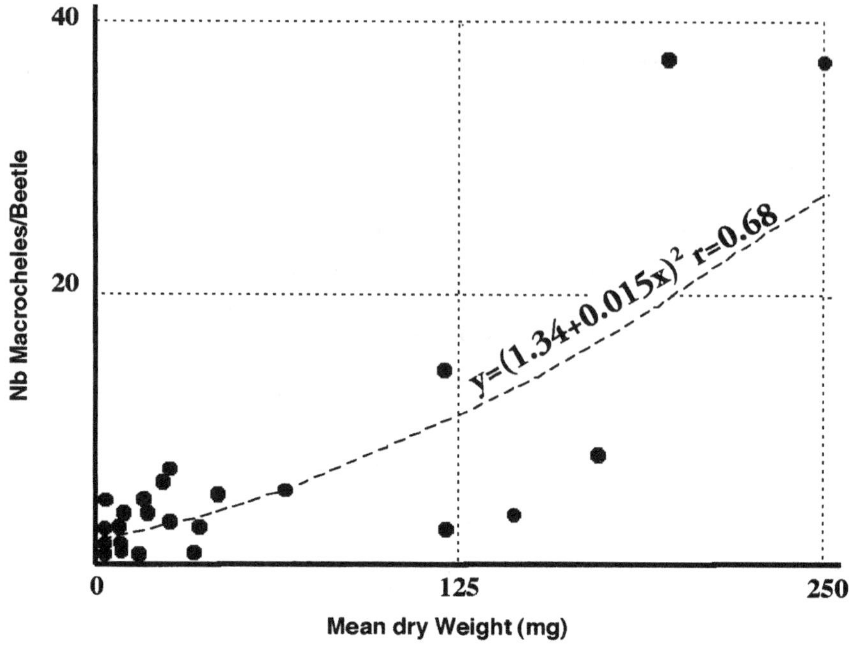

Fig. 4 - Relation between the dry weight of dung beetle species and the mean number of phoretic *Macrocheles.*

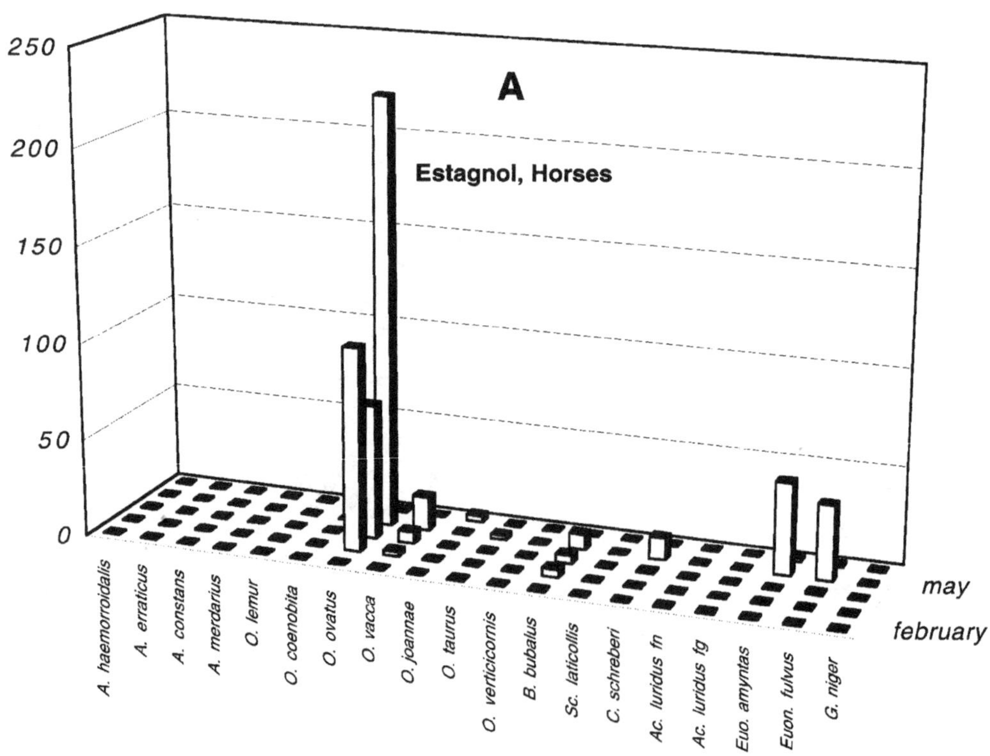

Fig. 5 - Theoretic contribution of different dung beetle species in the transport of *Macrocheles* mites (A: L'Estagnol; B: La Cardonille; C: St Martin de Londres).

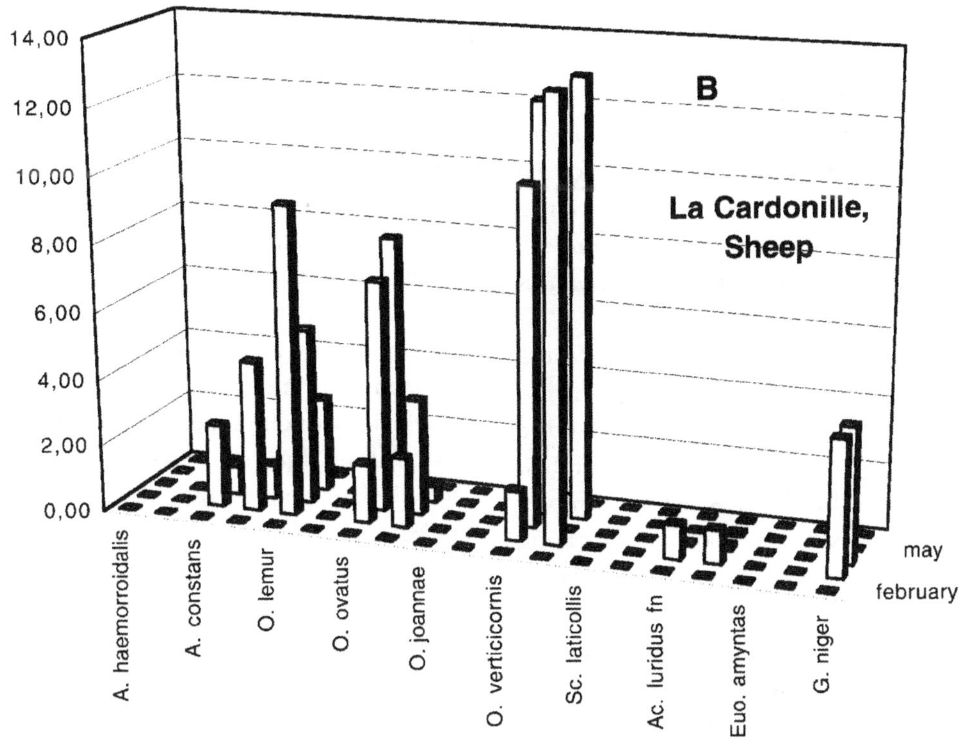

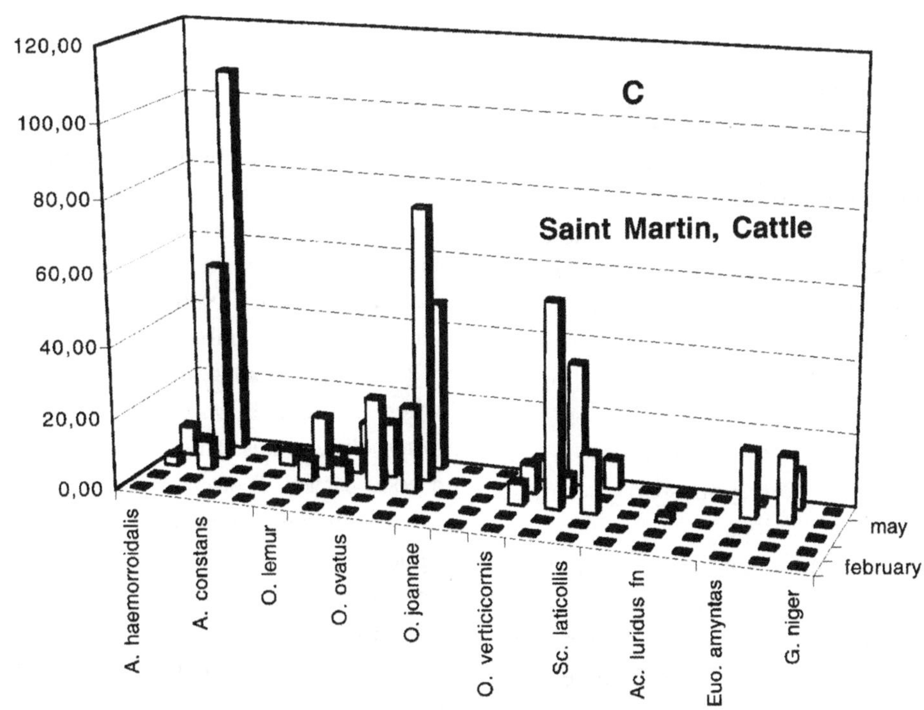

Fig. 5 - continued.

site $=\mathbf{P_m}$ with $\mathbf{P_m} = \mathbf{N_m} \cdot \mathbf{M_m}$. The theoretic contribution of beetle species (fig. 5) can be estimated in each site by $\mathbf{P_m} \cdot \mathbf{F}$.

At L'Estagnol (fig. 5A), *Onthophagus ovatus* (L.) and *B. bubalus* are mainly responsible for the colonization of pads by *Macrocheles* with a smaller contribution of being made by *Euoniticellus fulvus* (Goeze) and *Euonthophagus amyntas* (Ol.) from May to June.

At La Cardonille (fig. 5B), *Macrocheles* were carried by 11 beetle species. A remarkable contribution of *O. coenobita* was observed at the end of the season, *O. verticicornis* (Laich) and *O. ovatus* may be also involved. Available samples indicated that *B. bubalus* was observed carrying many mites but few specimens were collected.

At Saint Martin-de-Londres (fig. 5C), *Macrocheles* mites were transported by a large number of scarabs (13 species). Amongst them, the most important were *A. erraticus*, *O. vacca*, and *E. fulvus*. The rare and large scarabs *S. laticollis* and *B. bubalus* were also involved in mite transportation.

Discussion and conclusions

Biological parameters of *Macrocheles* species are known to be affected by temperature and relative humidity (Cicolani 1983; Filipponi and Cicolani 1974; Singh *et al.* 1967). Ho (1990) demonstrated that these factors may modify indirectly the phoresy behaviour of *M. muscadomesticae*. Our first results on the seasonal dynamics of *Macrocheles* fauna in Southern France showed that their relative abundance in dung, at each of the different sites, varies according to season. The beginning of spring corresponds to the most favourable climatic conditions.

In the dry season (summer), the desiccation of dung proceeds faster and the number of *Macrocheles* decreases. Dryness and high temperature limit the use of dung by beetles (Lumaret and Kirk 1987). We showed that dung beetle community parameters (abundance and specific composition) can affect the density of *Macrocheles* mites and the low abundance in the cold seasons could be attributed to the low level of beetle activity. Lumaret and Kirk (1987) have demonstrated that very few dung beetles are active during winter in the French "*garrigues*". For *Macrocheles*, little is known about their survival strategies during periods of reduced host availability and prey activity (Krantz 1983).

At Saint Martin, *Macrocheles* mites were transported by a large number of scarabs in comparison with the two other sites (fig. 5C). This is due mainly to differences in dung beetle communities at the three sites. Dung beetle communities are also determined by changes in the presence of specific domestic animals (cattle, sheep or horse) (Lumaret *et al.* 1992).

Specific associations between *Macrocheles* mites and scarabs are well known. Macrochelids have been recovered from all 16 major tribes and subtribes of dung-associated Scarabaeidae throughout the world (Krantz 1983). These mites have phoretic associations with Aphodiine, Geotrupine and Scarabaeine dung beetles that share their habitat (Krantz 1991 and 1998). A wide range of phoretic associations between *Macrocheles* mites and scarab beetles has been observed in the Southern France. The mean number of mites per beetle was greater on the larger beetles. This preference of *Macrocheles* mites for the larger insects agrees with a similar study carried out in Australia (Wallace 1986). Colonization of a dung pad by a single scarab may modify considerably the *Macrocheles* population, as observed at l'Estagnol where a single *B. bubalus* was found to carry 65 *Macrocheles* mites. The genus *Onthophagus* is also a good carrier of *Macrocheles* mites.

Macrocheles were often found on the genus *Aphodius*, such as *A. haemorrhoidalis*, *A. merdarius*, *A. constans* and *A. erraticus*. When the beetles are present, these small species are abundant. Work carried out in Central Italy, indicated mites were never found on *Aphodius* beetles (Cicolani 1992). Our study showed that this genus could play the role of "secondary host" when preferred scarab carriers are in low abundance. *Aphodius* species appear earlier in the season and may provide for the establishment of *Macrocheles* populations.

The maintenance of integrity in the relationship between mites and beetles depends on chemical, behavioural and ecological factors (Krantz 1998). Members of the *pisentii* group show a true phoretic specificity on their scarabaeine beetle carriers (Costa 1967; Krantz 1991), whereas species of the *glaber* group may be associated with a range of potential scarab hosts (Krantz 1983). During our survey, species of the *glaber* group (*M. glaber* and *M. perglaber*) were the most abundant. They were associated with several Scarabaeidae. Flies might be involved in the transport of *Macrocheles* mites (i.e. *M. muscadomesticae* and the *glaber* group). We did not take into consideration their contribution. This will be the subject of our future works.

Agricultural practices and the extension of pastoral areas induce changes in dung beetle and coprophilous fly communities (Lumaret *et al.* 1992). The control of fly pests: *Musca domestica* L., and *Stomoxys calcitrans* (L.), constitutes a major problem in pastoral settings. Spring treatments using antihelminthic products may induce dung beetle mortality, a decrease in *Macrocheles* populations and a proliferation of flies in the absence of their natural enemies.

References

Cicolani B., 1983 - Action des différentes températures constantes sur la durée de développement, de l'oeuf à l'adulte, des macrochélides fimicoles *Macrocheles subaddius* et *Macrocheles vernalis* (Acarina: Mesostigmata). *Acarologia,* 24: 145-158.

Cicolani B., 1992 - Macrochelid mites (Acari: Mesostigmata) occurring in animal droppings in the pasture ecosystem in central Italy. *Agriculture, Ecosystem and Environment*, 40: 47-60.

Costa M., 1967 - Notes on macrochelids associated with manure and coprids in Israel. II. Three new species of the *Macrocheles pisentii* complex with notes on their biology. *Acarologia*, 9: 304-329.

Filipponi A., 1964 - The feasibility of mass producing macrochelid mites for field trials against house flies. *Bulletin World Health Organization (Geneva)*, 31: 499-501.

Filipponi A., Cicolani B., 1974 - Influenza della temperatura sulla fecondità, longevità e capacità moltiplicativa nell'intervallo ottimale di *Macrocheles matrius* (Acarina, Mesostigmata). *Rivista di Parassitologia*, 35: 291-306.

Ho T.M., 1990 - Phoretic association between *Macrocheles muscadomesticae* (Acari: Macrochelidae) and flies inhabiting poultry manure in Peninsular Malaysia. *Experimental and Applied Acarology*, 10: 61-68.

Krantz G.W., 1983 - Mites as biological control agents of dung-breeding flies, with special reference to the Macrochelidae. In: Hoy M.A., Cunningham G.L., Knutson L. (eds), Biological Control of Pests by Mites. University of California, Berkeley: 91-98.

Krantz G.W., 1991 - Nature of the association between pisentii-group mites (Acari: Macrochelidae: *Macrocheles*) and dung beetles of the genus *Scarabaeus* (Coleoptera: Scarabaeidae) in Southern France. *Acarologia*, 32: 3-11.

Krantz G.W., 1998 - Reflections on the biology, morphology and ecology of the Macrochelidae. *Experimental and Applied Acarology*, 22: 125-137.

Lobo J.M., Lumaret J.P., Jay Robert P., 1998 - Sampling dung beetles in the French Mediterranean area: effects of abiotic factors and farm practices. *Pedobiologia*, 42: 252-266.

Lobo J.M., Martin Pierra F., Veiga C.M., 1988 - Las trampas pitfall con cebo, sus posibilidades en el estudio de las comunidades coprofagas de Scarabaeoidea (Col.). I. Caracteristicas determinantes de su capacidad de captura. *Revue d'Écologie et de Biologie de Sols*, 25: 77-100.

Lumaret J.P., Kirk A., 1987 - Ecology of dung beetles in the French Mediterranean region (Coleoptera: Scarabaeidae). *Acta Zoologica Mexicana*. (n.s.), 24: 1-55.

Lumaret J.P., Kadiri N., Bertrand M., 1992 - Changes in resources: consequences for the dynamics of dung beetle communities. *Journal of Applied Ecology*, 29: 349-356.

Singh P., McEllistrem M.T., Rodriguez J.G., 1967 - The response of some macrochelids to temperature and humidity (Acarina: Macrochelidae). *Acarologia*, 9: 1-16.

Wallace M.M.H., 1986 - Some macrochelid mites (Acari: Macrochelidae) associated with Australian dung beetles (Coleoptera: Scarabaeidae). *Acarologia,* 17: 3-15.

Wallace M.M.H., Holm E., 1984 - Competition between species of predatory macrochelid mites in the cow-dung pad. In: Griffiths D.A., Bowman C.E. (eds), Acarology VI. Ellis Horwood. Publishers Chichester, vol. 1: 581-584.

Wallace M.M.H., Tyndale-Biscoe M., Holm E., 1979 - The influence of *Macrocheles glaber* on the breeding of the Australian bushfly, *Musca vetustissima* in cow dung. In: Rodriguez J.C. (ed.), Recent Advances in Acarology. Academic Press, N.Y., vol. 2: 217-222.

Development of oribatid mite communities (Acari, Oribatida) on a mine dump

Piotr Skubala
Department of Ecology, University of Silesia, Katowice, Poland

Key words: Oribatida, dumps, succession, pioneer species.

Abstract: The oribatid mite fauna on the mine dump and in the adjacent mixed forest has been the subject of the study. Three study plots, differing in age and plant cover, have been chosen on the dump. A continuos increase in abundance of oribatids, species richness with successional stages was observed on the dump. After 20 years of the development the abundance and species richness on the mine dump was still 3-4 times lower than in the adjacent forest. The dissimilarity of species composition in the three communities on the dump was high. Many of the successional oribatid species have probably been transported from other biotopes than the neighbouring forest. The progressive importance of *Sellnickochthonius immaculatus*, *S. cricoides*, *Liochthonius piluliferus* in communities from the forest to the community at the youngest site indicates their special features that enable the mites to colonize unfavourable habitats. As indicated by Canonical Correspondence Analysis (CCA), the most important factor responsible for the development of the oribatid mite community on the mine dump was soil bulk density and available Mg content accompanied by water capacity. *Graptoppia (Stenoppia) italica* Bernini, 1973, *Suctobelbella duplex* (Strenzke, 1951), and *Suctobelbella latirostris* (Forsslund, 1941) were new species for the Polish fauna.

Introduction

Dumps created as a result of the industrial activity are an inseparable and characteristic element of the landscape in the Upper Silesian Region. Over 44% of wastes produced in Poland are deposited in the region. The proportion of waste tips created by mining is highest in this region and estimated as 72% (Jarzebski 1997). Fundamental to the understanding and management of natural and disturbed or threatened ecosystems is a description of the biodiversity and understanding of the processes that promote diversity patterns and control the functioning of communities (Giller 1996). Much of the research on land reclamation has been devoted to the reestablishment of vegetation and wildlife (vertebrates), only a few studies have addressed the recolonization of arthropods on mine lands (Parmenter and MacMahon 1987). And especially little is known about oribatids on mine dumps (Dunger 1968; Bielska 1982; Luxton 1982; Skubala 1997).

Oribatid mites, one of the major group in most soils which plays an important "ecological service" in decomposition processes, has been the subject of the study. This project focused on the development of the oribatid mite fauna on a coal mine dump. They were contrasted to the oribatid community found in the nearby undisturbed forest.

F. Bernini, R. Nannelli, G. Nuzzaci, E. de Lillo (eds.), Acarid Phylogeny and Evolution. Adaptations in mites and ticks, 209-215.

Materials and methods

Study site

Pedozoological investigations were undertaken on the mine dump in Katowice (Upper Silesia, South Poland). The "Maria" dump belongs to the "Murcki" coal mine and was formed as a huge hill up to 50 metres high and its area is about 14 ha. The dump was pilled from mining wastes of differentiated granulation, e.g. clay stones, carboniferous sandstone, coal-mud, coal crumbs and middling. The exploitation of the dump was stopped in 1985. Three sites differing in age and plant cover have been chosen on the dump.

The vegetation cover was poorest at site I. Six plant species were recorded in the C layer with *Tussilago farfara* L. predominant. There was almost no organic matter in the surface layer. Better-developed vegetation with 13 herbaceous plant species (*Calamagrostis epigeios* (L.) Roth and *T. farfara* dominated) and *Ceratodon purpureus* Brid in the D layer covered site II. The thin layer of humus and a more advanced succession of ruderal vegetation were noted at site III. Four tree species occurred in the shrub layer with *Betula pendula* Roth most numerous. The cover of the undergrowth and moss layer was quite high (40%). The dump is surrounded by a forest (part of it is the beech forest reserve). The fourth study plot was chosen in adjacent mixed forest of the *Querco roboris-Pinetum* type.

Samples, Statistics

The fauna was sampled at four seasons in 1998-99. Samples, each of 18 cm^2 in surface area, were taken to a depth of 7.5 cm using a cylindrical core sampler. In total 150 samples were collected. The mites were extracted with a modified Tullgren extractor, the procedure lasting 7 days. Mites were preserved in 75% ethyl alcohol. Adult oribatids were determined to species level. Sixteen environmental factors were measured on the dump.

The communities were characterized by abundance, species richness, dominance (D) and species diversity (Shannon index, H'). Correlation between the occurrence of species and the measured set of environmental variables was analyzed with Canonical Correspondence Analysis (CCA) (Ter Braak 1988), using the MVSP program. The content of heavy metals (Cd, Ni, Pb, and Zn) was low at all sites and therefore their influence may be eliminated. Fifteen species that were presented just in one sample were arbitrarily discarded from the analysis. The index of dispersal direction was used for qualitative (presence-absence) data (Legendre and Legendre 1983). Comparison of two sites was based upon a combination of 3 values: the number of species present in both sites (a), the number of species found in site 1 but not in 2 (b), and the number of species found in site 2 but not in 1 (c) (Borcard *et al.* 1995).

$$DD_2 = [2a / (2a + b + c)] [(b - c) / (a + b + c)]$$

To test the null hypothesis that there is no asymmetry between two sites, the log-linear form of the McNemar test was used, which is preferable when the values of b and c can be small (Borcard *et al.* 1995).

$$\chi^2 = 2 [b \ln b + c \ln c - (b + c) \ln (b + c) / 2]$$

Results

In the course of the one-year investigation 5,601 individuals of Oribatida, representing 86 species, were collected. The mean abundance of Oribatida was the lowest at site I (2,972 individuals/m^2 ± S.E. 1,398) and the highest at site III (21,305±7,031) (tab. 1). In the neighbouring forest it was almost 3 times higher (61,814±10,444) than at the oldest site. The proportion of juveniles decreased from the youngest site (68.2%) to the oldest one (19.1%). The share of juveniles in the forest was similar to site III (21.1%). As for the number of species, there were only 3 species

Tab. 1 - Characteristics of the oribadid communities on the dump and in the adjacent forest.

	Dump			Forest
	Site I	Site II	Site III	
Abundance of juveniles	2,028 ± 901	4,347 ± 2,277	4,069 ± 1,557	13,055 ± 3,907
Abundance of adults	944 ± 503	3,028 ± 1,215	17,236 ± 6,705	48,759 ± 6,537
Abundance in total	2,972 ± 1,398	7,375 ± 3,467	21,305 ± 7,031	61,814 ± 10,444
		Dominants		
Sellnickochthonius cricoides	7.3	1.8	-	0.15
Sellnickochthonius immaculatus	20.6	15.1	-	0,04
Liochthonius piluliferus	16.1	15.6	-	-
Tectocepheus velatus	4.4	13.3	2.2	0.3
Medioppia obsoleta	-	-	9.8	0.4
Lauroppia falcata	-	-	-	11.2
Oppiella nova	32.3	43.1	64.2	14.5
Conchogneta dalecarlica	-	-	-	26.8
Ceratozetes mediocris	-	1.8	7.4	-
Shannon index (H')	2.053	1.769	1.425	2.893

more at site III compared with the to site I (15 species). The number of species in the forest was about 4 times higher (70) than on the dump (35). It was characteristic that the value of the Shannon index was considerably highest at the youngest site of the dump (H'=2.058). Its value in the forest was much higher than on the dump (tab. 1).

The dominant species at the younger sites (I, II) were generally similar, e.g. *Sellnickochthonius immaculatus* Forsslund, 1942, *S. cricoides* (Weis-Fogh, 1948), *Liochthonius piluliferus* (Forsslund, 1942), *Oppiella nova* (Oudemans, 1902), and *Tectocepheus velatus* (Michael, 1880). Only *S. cricoides* occurred with smaller proportion at site II and the dominance of *T. velatus* was 3 times lower than at site I. Besides *O. nova*, two other dominants were different at the III [*Ceratozetes mediocris* Berlese, 1908 and *Medioppia obsoleta* (Paoli, 1908)]. Different oribatids dominated in the forest [*Conchogneta dalecarlica* (Forsslund, 1947) and *Lauroppia falcata* (Paoli, 1908)] with one exception - a high proportion of *O. nova* (tab. 1).

The dissimilarity of species composition in the three communities on the dump was high. There were only four (in two cases) and six species common for pairs of communities (fig. 1). Similarly only 18 of 86 species recorded on the dump and in the forest were common for both

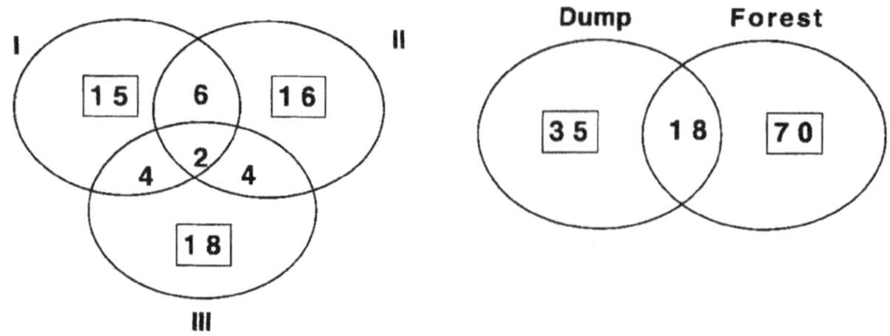

Fig. 1 - Number of species on the mine dump and in the forest and common species.

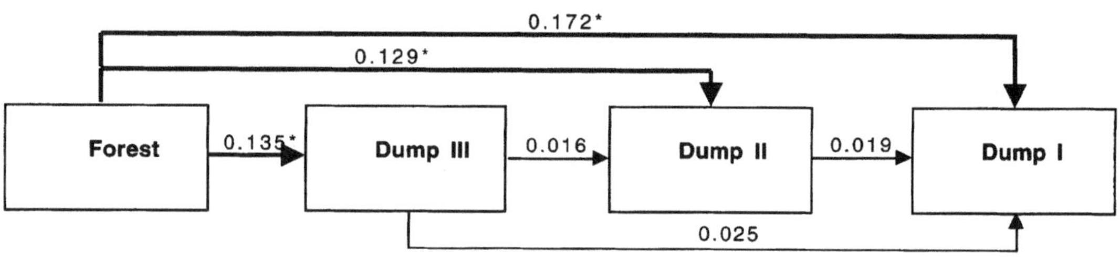

Fig. 2 - Analysis of dispersal direction and values of DD_2 index of dispersal direction for the 4 pairs of investigated sites. Explanations: the asterisk denotes a statistically significant value at the 0.05 probability level (McNemar's test).

habitats. As regards the value of the index of dispersal direction, its value was significant between forest and particular sites on the dump. The value of DD_2 between pairs of investigated sites on the dump was very small (fig. 2).

The Canonical Correspondence Analysis (CCA) is used to disentangle the relative importance of environmental factors as regulating forces during succession (Verschoor and Krebs 1995). CCA makes it possible to "explain" a fraction of the variation of the species in terms of the ordination of the environmental variables. The first and second axes were significant and "explain" respectively 55% and 36.4% of the total variance in species occurrence. The distribution of species in the factorial plane of the axis 1 and 2 of the analysis shows three well-differentiated groups (fig. 3). One group is composed of 5 numerous species (3 brachychthoniids, *O. nova* and *T. velatus*) that were mainly affected by soil compactness and available Mg content. *Mediopia obsoleta*, *C. mediocris*, and several other species were influenced by water capacity and C/N ratio. The third group with two dominants - *C. dalecarlica* and *L. falcata* - was mainly affected by pore volume and available K and P content and to a lesser extent by temperature and moisture of soil.

Discussion and conclusions

Oribatid mites are interesting to study on dumps both for the sake of scientific knowledge and for practical reasons. Sensitivity of oribatid communities to many kinds of disturbances makes them an appropriate tool for monitoring purposes (Borcard 1997).

A continuos increase in abundance of oribatids and species richness with successional stages was observed on the mine dump. However, only differences in abundance were significant and species diversity was even higher on the younger sites. After 20 years of the devel-

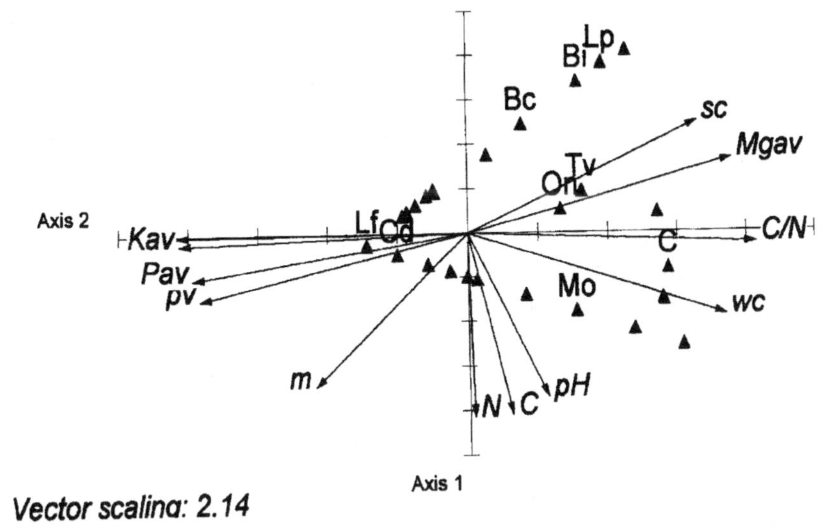

Fig. 3 - Ordination diagram based on Canonical Correspondence Analysis (quantitative environmental variables are represented by arrows, eigenvalue of axis 1=0.674, axis 2=0.446). Species - see table 1. Explanations:

pH: pH in KCl	wc: water capacity
C: organic content (%)	K_{av}: K available
N: total nitrogen (%)	Mg_{av}: Mg available
C/N: carbon/nitrogen ratio	P_{av}: P available
sc: soil compactness (g/cm^3)	t: mean soil temperature at 3 cm depth
pv: pore volume (%)	m: moisture content

opment, the abundance and species richness on the dump was still 3-4 times lower than in the adjacent forest. It is noteworthy, that the value of the Shannon index was considerable higher for obviously less developed oribatid communities. It is in opposition to previously observed relationship between species diversity and the development of communities on dumps by the author (Skubala 1995 and 1997; Skubala and Ciosk 1999). The association of diversity indices with stability of the community has been questioned (Hulbert 1971; Pimm 1984; Schulze and Mooney 1993) and these observations confirmed that we must be very careful in applying diversity indices to compare oribatid communities, especially those stressed by unfavourable conditions.

The dissimilarity of species composition in the three communities on the dump was high. Only two ubiquitous parthenogens, *Oppiella nova* and *Tectocepheus velatus,* were common for three sites. From among 70 species from the forest only 18 also occurred on the dump. Analysis of dispersal direction indicated that the most probable direction of migration of a majority of species was from the forest to particular sites on the dump. The flow of mite species between sites on the dump was not observed. Furthermore, it may be concluded that many of the successional oribatid species have probably been transported from other biotopes than the adjacent forest.

The major role of oribatids of the family Brachychthonidae (*Sellnickochthonius immaculatus*, *S. cricoides*, and *Liochthonius piluliferus*) in the colonising of the dump was very symptomatic. *Oppiella nova* and *T. velatus* were also found in high numbers on the younger sites (I, II). However, they occurred numerously on the oldest site and in the forest, as well. The progressive importance of brachychthoniids in communities from the forest to the community at the youngest site indicates their special features which enable the mites to colonize unfavourable habitats. In the Canonical Correspondence Analysis all early colonizers (brachychthoniids, *O. nova*, and *T. velatus*) of the dump appeared to be associated with soil bulk density and available Mg content. Brachychthoniids have been previously observed in higher numbers on dumps with low pH (Luxton 1982), whereas on dumps with high pH they occurred in significantly lower numbers (Bielska 1995; Skubala 1995, 1997 and 1999; Skubala and Ciosk 1999). Their negative reaction to higher pH (Luxton 1982) was also confirmed in this study (the value of pH varied between 3.3 to 3.5 on the dump).

It is worth mentioning, that all early colonizers and some other oribatids recorded on the younger sites of the dump are known to be parthenogenetic (Norton and Palmer 1991). Parthenogenetic reproduction may be an important advantage for oribatids in colonizing post-industrial wastelands (Norton and Palmer 1991), which has been previously observed by the author (Skubala 1997; Skubala *et al.* 1998; Skubala and Ciosk 1999) and also confirmed again in this study.

Late colonizers of the dump (dominants at the oldest site), *Medioppia obsoleta* and *Ceratozetes mediocris*, were mainly associated with water capacity and C/N ratio. The species have been hitherto observed on some dumps usually occurring at older sites in the highest numbers (Bielska 1995; Skubala 1995, 1997 and 1998).

Conchogneta dalecarlica and *Lauroppia falcata*, two most numerous species from the forest, appeared to be unable to colonizeing the dump. However, the species were sporadically noted on dumps, but always in small numbers (Bielska 1982; Skubala 1997 and 1998). The species were mainly affected by pore volume and available K and P content on the investigated mine dump. *Lauroppia falcata* is known as "forest" (Schatz 1983), whereas the typical biotope occurrence of *C. dalecarlica* is still uncertain.

Despite the fact that mites were investigated in poor and disturbed biotopes, some interesting oribatid species were found. *Graptoppia (Stenoppia) italica* Bernini, 1973 collected on the dump is a new species for the Polish fauna, whereas 7 others are new oribatids for the Upper Silesian Region. Moreover, *Suctobelbella duplex* (Strenzke, 1951) and *Suctobelbella latirostris* (Forsslund, 1941) from the forest are also new species for the Polish fauna and 5 other species are new for this region.

References

Bielska I., 1982 - Communities of moss mites (Acari, Oribatei) of degraded and recultivated areas in Silesia. I. Communities of moss mites of mine dumps. *Polish Ecological Studies*, 8(4): 499-510.

Bielska I., 1995 - Mining dump and electrical power plant dump Oribatida. In: Kropczynska D., Boczek J., Tomczyk A. (eds.), The Acari - Physiological and ecological aspects of Acari - Host relationships. Dabor, Warszawa: 173-182.

Borcard D., 1997 - Oribatid mites in peat-bog remnants of the Swiss Jura mountains. *Abhandlungen und Berichte des Naturkundemuseums Görlitz*, 69(6): 19-23.

Borcard D., Geiger W., Matthey W., 1995 - Oribatid mite assemblages in a contact zone between a peat-bog and a meadow in the Swiss Jura (Acari, Oribatida): influence of landscape structures and historical processes. *Pedobiologia*, 39: 318-330.

Dunger W., 1968 - Die Entwicklung der Bodenfauna auf rekutivierten Kippen und Halden des Braunkohletagebaues. *Abhandlungen und Berichte des Naturkundemuseums*, Leipzig, 43(2): 1-256.

Giller P.S., 1996 - The diversity of soil communities, the "poor man's tropical rainforest". *Biodiversity and Conservation*, 5: 135-168.

Hurlbert S.H., 1971 - The nonconcept of species diversity: a critique and alternative approaches. *Ecology*, 52: 577-586.

Jarzebski L. (ed.), 1997 - Environmental quality report in the Katowice province. Library of Environmental Monitoring, Katowice: 371 pp.

Legendre L., Legendre P., 1983 - Numerical ecology. Elsevier, Amsterdam: 419 pp.

Luxton M., 1982 - The ecology of some soil mites from coal shale tips. *Journal of Applied Ecology*, 19: 427-442.

Norton R.A., Palmer S.C., 1991 - The distribution, mechanisms and evolutionary significance of parthenogenesis in oribatid mites. In: Schuster R., Murphy P.W. (eds.), The Acari: Reproduction, Development and Life-History Strategies. Chapman & Hall, London: 107-136.

Permenter R.R., MacMahon J.A., 1987 - Early successional patterns of arthropod recolonization on reclaimed strip mines in southwestern Wyoming: the ground-dwelling beetle fauna (Coleoptera). *Environmental Entomology*, 16: 168-177.

Pimm S.L., 1984 - The complexity and stability of ecosystems. *Nature*, 307: 321-326.

Schatz H., 1983 - Catalogus Fauna Austriae, Teil IXi U. - Ordn.: Oribatei, Hornmilben. Vienna: Österreichischen, Akademie der Wissenschaften: 118 pp.

Schulze E.D., Mooney H.A., 1993 - Ecosystem Function of Biodiversity: A Summary. *Ecological Studies*, 99: 497-510.

Skubala P., 1995 - Moss mites (Acarina: Oribatida) on industrial dumps of different ages. *Pedobiologia*, 39: 170-184.

Skubala P., 1997 - Oribatid mite communities (Acari, Oribatida) on post-industrial dumps of different kinds. I. Abundance and species richness. *Abhandlungen und Berichte des Naturkundemuseums Görlitz*, 69(6): 63-68.

Skubala P., 1998 - Oribatid mite communities (Acari, Oribatida) on post-industrial dumps of different kinds. II. Zoocenological analysis. *Fragmenta Faunistica*, 41(14): 193-207.

Skubala P., 1999 - Colonization of a dolomitic dump by oribatid mites (Acari, Oribatida). *Pedobiologia*, 43: 145-159.

Skubala P., Ciosk M., 1999 - Oribatid mites (Acari, Oribatida) colonizing the zinc metallurgic dump. *Fragmenta Faunistica*, 42(6): 41-45.

Skubala P., Dziuba S., Stodólka A., 1998 - Acari and Collembola in the „Zabie Doly" protected area or how nature struggle with industry. *Zeszyty Naukowe ATR, Ochrona Srodowiska*, Bydgoszcz, 2: 251-257.

Ter Braak C.J.F., 1988 - CANOCO - a FORTRAN program for canonical community ordination by (partial) (detrended) (canonical) correspondence analysis, principle component analysis and redundancy analysis. Groep Landbouwwiskunde, Wageningen: 95 pp.

Verschoor B.C., Krebs B.P.M., 1995 - Successional changes in a saltmarsh carabid beetle (Coleoptera, Carabidae) community after embankment of the Markiezaat area. *Pedobiologia*, 39: 385-404.

Three-dimensional structure of the gut system of the mites *Ameronothrus lineatus* (Oribatida: Ameronothridae) and *Hyadesia fusca* (Astigmata: Hyadesiidae)

Jens Bücking
University of Bremen, Department of Biology (FB2), Leobener Str. - NW2, D-28359 Bremen, Germany
jbueck@uni-bremen.de

Key words: Acari, gut system, malpighian tubules, topography, reconstruction.

Abstract: Despite the fact that the general structure of the gut system of Oribatida and Astigmata is well known from several microanatomical investigations, there is considerable confusion in terminology and definition of gut subunits. Some of these contradictory views may be explained by dynamic changes in gut morphology during the digestive process. The three-dimensional structure of the gut system of the oribatid mite *Ameronothrus lineatus* (Thorell) and the astigmatid mite *Hyadesia fusca* (Lohmann) was analyzed by means of serial sections. Computer-assisted reconstructions show that in certain physiological states the colon of *Hyadesia* 'disappears', which could lead to a misinterpretation of the postventricular parts of the gut. This may have happened in some investigations of house-dust mites and itch mites. In *Ameronothrus*, an additional gut subunit was found between colon and postcolon and it might be homologous to the gut region bearing the Malpighian tubules in the Astigmata.

Introduction

The general structure of the digestive system of oribatid and astigmatid mites is well known from several studies (see reviews in Evans 1992; Alberti and Coons 1999). Following the nomenclature proposed by Evans (1992), it is composed of (I) a foregut with pharynx and esophagus, (II) a midgut divided into ventriculus, one or two pairs of ventricular caeca, colon and postcolon, and (III) a hindgut with anal atrium and anus. Foregut and hindgut are lined by a cuticular intima and are of ectodermal origin (stomodeum and proctodeum), the midgut is of entodermal origin (Hughes 1950a for *Acarus siro* L.). A similar generalized gut structure, characterized by the division of the postventricular midgut in colon and postcolon, can also be found in early derivative Prostigmata like Rhagidiidae (Ehrnsberger 1984). Therefore the so-called 'sarcoptiform gut type' (sensu Reuter 1909 and Vitzhum 1940-43) probably represents the plesiomorphic state within Actinotrichida (see discussion in Ehrnsberger 1984; Alberti and Coons 1999).

Although this gut organization is regarded as uniform within Sarcoptiformes, there is considerable confusion in nomenclature of the postventricular gut subunits and their classification as parts of midgut or hindgut. As an example, the term 'rectum' has been variously used for the postcolon (e.g. Woodring and Cook 1962; Smrz 1989), for the anal atrium (e.g. Prasse 1967; Wurst 1993) or for both parts together (e.g. Michael 1901; Bäumler 1970; Kuo and Nesbitt 1970). In many of the latter studies, the cuticular intima of

F. Bernini, R. Nannelli, G. Nuzzaci, E. de Lillo (eds.), Acarid Phylogeny and Evolution. Adaptations in mites and ticks, 217-225.

the anal atrium was overlooked. Bäumler (1970) probably misinterpreted the dense brushborder of the postcolon as a thick, striated intima. For *Ceratozetes cisalpinus* Berlese, Woodring and Cook (1962) described a microvillous border but also an intima of colon and postcolon, visible during the premoult period. In studies of the astigmatid mites *Dermatophagoides farinae* Hughes (Brody *et al.* 1972) and *Sarcoptes scabiei* (L.) (Desch *et al.* 1991), the postventricular midgut was found to be undivided. Since the cuticle-lined hindgut of *D. farinae* shows a division into a globular anterior and a narrow posterior part, Brody *et al.* (1972) proposed a homology with 'postcolon' and 'anal atrium'. In *S. scabiei*, however, the hindgut showed no division. Wurst (1993) argued, that the true postcolon of these stout species might be very short and could have been overlooked.

The presence or absence of Malpighian tubules in Sarcoptiformes is of special interest. Entodermal Malpighian tubules are a character of most arachnid taxa and can be found even in late derivative groups like Solifugae, Ricinulei and Acari. Within the Acari they are well developed in most anactinotrichid taxa. As is typical of arachnid Malpighian tubules, they originate on the postcolon and contain whitish, birefringent excretory concretions (Alberti and Coons 1999). In the Actinotrichida short pouches or tubes at the connection of colon and postcolon have been reported from some families of Astigmata (Acaridae, Carpoglyphidae, Algophagidae, Dermoglyphidae, Analgesidae), but were not found in others (for references see Alberti and Coons 1999). Since these tubes appear to be either empty or only partly filled with a dark, non-crystalline material, they are thought to be non-functional or to play only a minor role in excretion (Hughes 1950b; Prasse 1967). In all investigated Oribatida as well as in most Actinedida (e.g. Bdellidae, Rhagidiidae, Eriophyoidea, Tetranychidae), Malpighian tubules are absent. Reports on Malpighian tubules in some Eupodoidea, Hydrachnidia and Trombiculidae refer to a pair of anteriorly directed and branched extensions of the postventricular midgut (references in Alberti and Coons 1999). It is not clear, however, if these tubular projections differ histologically and functionally from the other multiple branches of this gut region. Thus, the homology of Malpighian tubules within Actinotrichida, and within Acari as a whole, is rather uncertain (Alberti and Coons 1999).

In live observations of gut activity (e.g. Prasse 1967), feasible in mites with a low body opacity, dynamic changes of the three-dimensional structure become obvious. The inflow and mixture of food material, as well as the formation and passage of food balls and faecal pellets due to peristalsis and other types of gut movement, lead to tremendous changes of shape and volume of all gut subunits. These changes might be responsible for some contradictory results, e.g. on the presence or absence of constrictions and valves between subsequent units of the gut. The aim of the present work is to make a first attempt in visualization of such dynamic changes by means of computer-assisted 3D-reconstructions. Such reconstructions, based on serial histological sections, can give exact and reliable snapshots of the gut topography at the moment of fixation and may help to understand contradictory results reported in the literature.

The algophagous species *Ameronothrus lineatus* (Thorell) (Brachypylida, Oribatida) and *Hyadesia fusca* (Lohmann) (Hemisarcoptoidea, Astigmata), both inhabitants of rocky shores with a similar feeding habit, are used as examples. This paper focuses mainly on the general gut structure; the histological details will be more fully discussed elsewhere.

Materials and methods

Live individuals were briefly immersed in an aqueous solution of 1% Triton-100, washed in double-distilled water, air dried up to the disappearance of surface water, killed in cooled ethanol (3 min) and transferred to a modified Karnovsky fixative (2.5% glutaraldehyde - 1.5% paraformaldehyde in 0.05 M sodium cacodylate buffer pH 7.0, without calcium chloride). For better penetration, legs and, in some specimens, gnathosoma were removed. After 6-8 h of primary fixation at room temperature, the samples were left in buffer overnight at 4°C.

Specimens assigned for transmission electron microscopy (TEM) studies were postfixed in 2% osmium tetroxide for 2 h, washed in buffer, dehydrated in a graded series of ethanol up to 70%, stained *en bloc* in 3% uranyl acetate (in 70% ethanol), dehydrated in ethanol and gradually infiltrated with SPURR's epoxy resin using propylene oxide as transitional solvent. Ultrathin sections were obtained using the ultramicrotome Leica Ultracut R and stained for 2 h in alcoholic uranyl acetate (in 30% ethanol) and 10-12 min in Reynold's lead citrate. Examination took place on a Zeiss EM 10. Specimens assigned for light microscopy were transferred through the dehydration series and gradually embedded via propylene oxide in SPURR's or LR-White. Semithin sections (1 μm) were stained at 70°C in an aqueous 0.2% toluidin blue - 0.2% methylene blue - 1.2% sodium tetraborate solution (pH 9.1).

Several series of light-microscopical (LM) sections from males, females and tritonymphs of both species were analyzed and compared for the three-dimensional structure of the gut, but only three of these series were finally used for the computer-assisted 3D-reconstruction. These reconstructions are based on aligned and digitized contour drawings of body and gut from: an *A. lineatus* male (169 transverse slices spaced at 3 μm); a *H. fusca* tritonymph (92 horizontal slices spaced at 2.3 μm); and a *H. fusca* male (226 transverse slices spaced at 1.6 μm). The image stacks were processed using the software SURFDRIVER 3.5.4 (http://www.surfdriver.com, Authors: Scott Lozanoff and David Moody) for Windows®. Details of the reconstruction method are described in Bücking (2000).

Results

The gut system of the *A. lineatus* male (fig. 1) makes up a high proportion of the body volume. A short esophagus anteroventrally enters the pear-shaped ventriculus, which posterolaterally bears a pair of long caeca with a wavy surface structure. The colon is divided into a narrow anterior and a globular posterior part. The anterior part is pressed against the ventriculus, thus producing a shallow impression of the posterior wall of the ventriculus. The colon is followed by a short, hour-glass-shaped intercolon which is provided with a pair of short, dorsolateral projections. Anteriorly the intercolon is pressed against the, thereby deeply invaginated, posterior wall of the colon. The intercolon is followed by a voluminous postcolon which extends, behind a posterior narrowing, into a laterally flattened, rhombic anal atrium.

The analyses of a number of further individuals shows that this reconstruction gives only one example of many possible three-dimensional states. In the reconstructed specimen the ventriculus was filled with solid food particles, i.e. fragments of green algae thalli, and the postcolon carried a faecal pellet. The colon and anal atrium were empty. In specimens with an empty ventriculus or postcolon, these parts are considerably smaller; the colon and anterior part of the anal atrium can be widely expanded if filled with a faecal pellet. The shape and volume of the gut regions are also defined by the state of contraction of the intrinsic musculature. In some specimens the colon has no middle constriction and appears to be pear-shaped, with the narrow end anterior. The junction of caeca and ventriculus may be constricted, as in the reconstructed specimen, or may be widely open. In some specimens the anal atrium has a broad open connection to the postcolon.

The intercolon is often dislocated to a paramedian position or it is enveloped by invaginations of colon and postcolon if these are pressed against each other. Histologically it differs clearly from

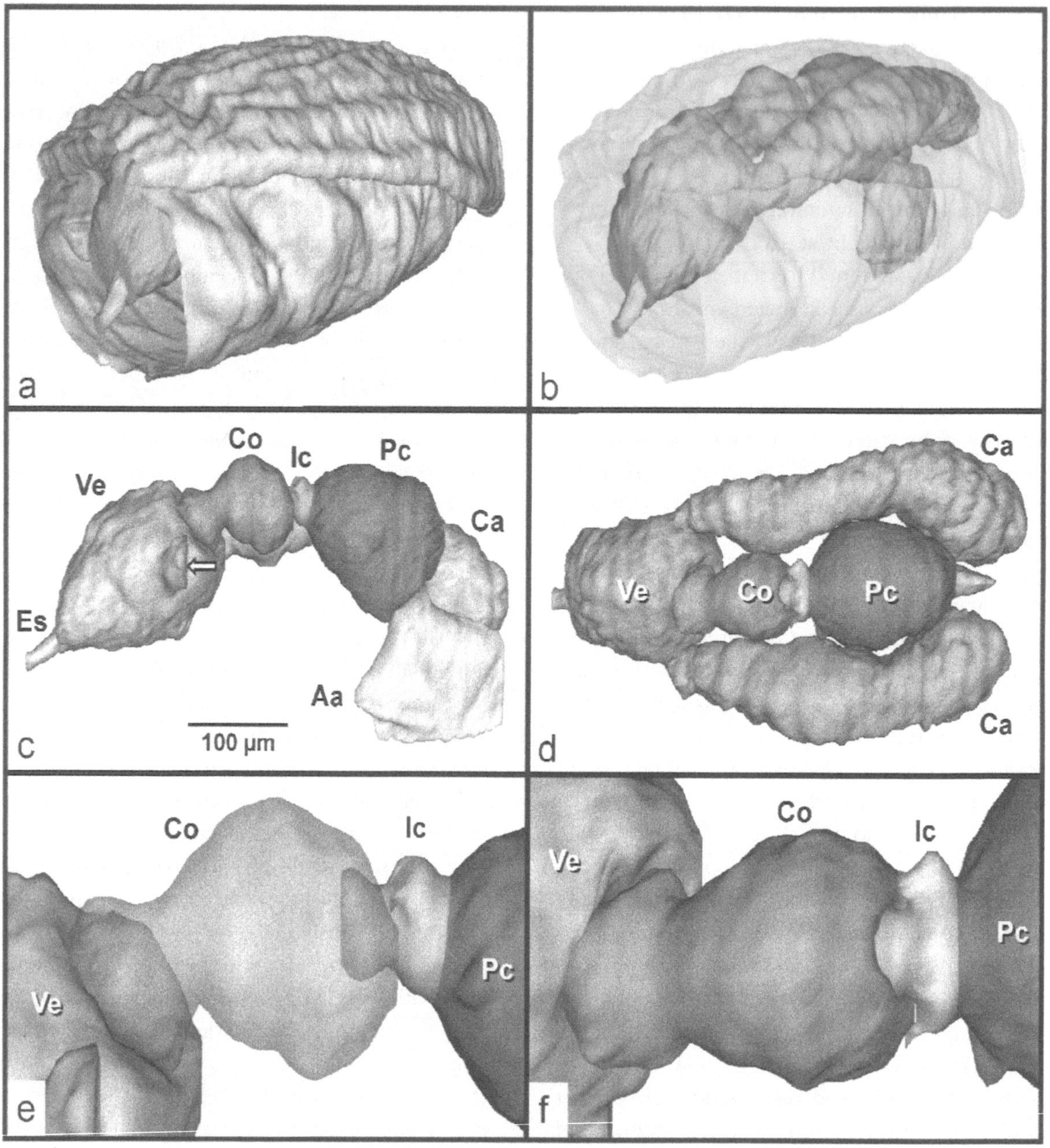

Fig. 1 - Three D-reconstruction of the gut system of *Ameronothrus lineatus*: a) anterolateral view of body contour. The anterior part is cut off; b) anterolateral view of body contour (set to 40% opacity) and gut system; c) lateral view of the gut system. Anterior part of esophagus and left caecum omitted. The arrow points to the origin of the left caecum; d) dorsal view of the gut system; e) lateral view of colon (50% opacity) and intercolon; f) dorsal view of colon and intercolon. Abbr.: Aa, anal atrium; Ca, caecum; Co, colon; Es, esophagus; Ic, intercolon; Pc, postcolon; Ve, ventriculus.

colon and postcolon (fig. 2). The centre of the intercolon is built of large, vacuolated cells which might be glandular, whereas the dorsolateral projections are characterized by a thin epithelium of slender cells provided with a microvillous border. The cytological details of these cells, with few rough

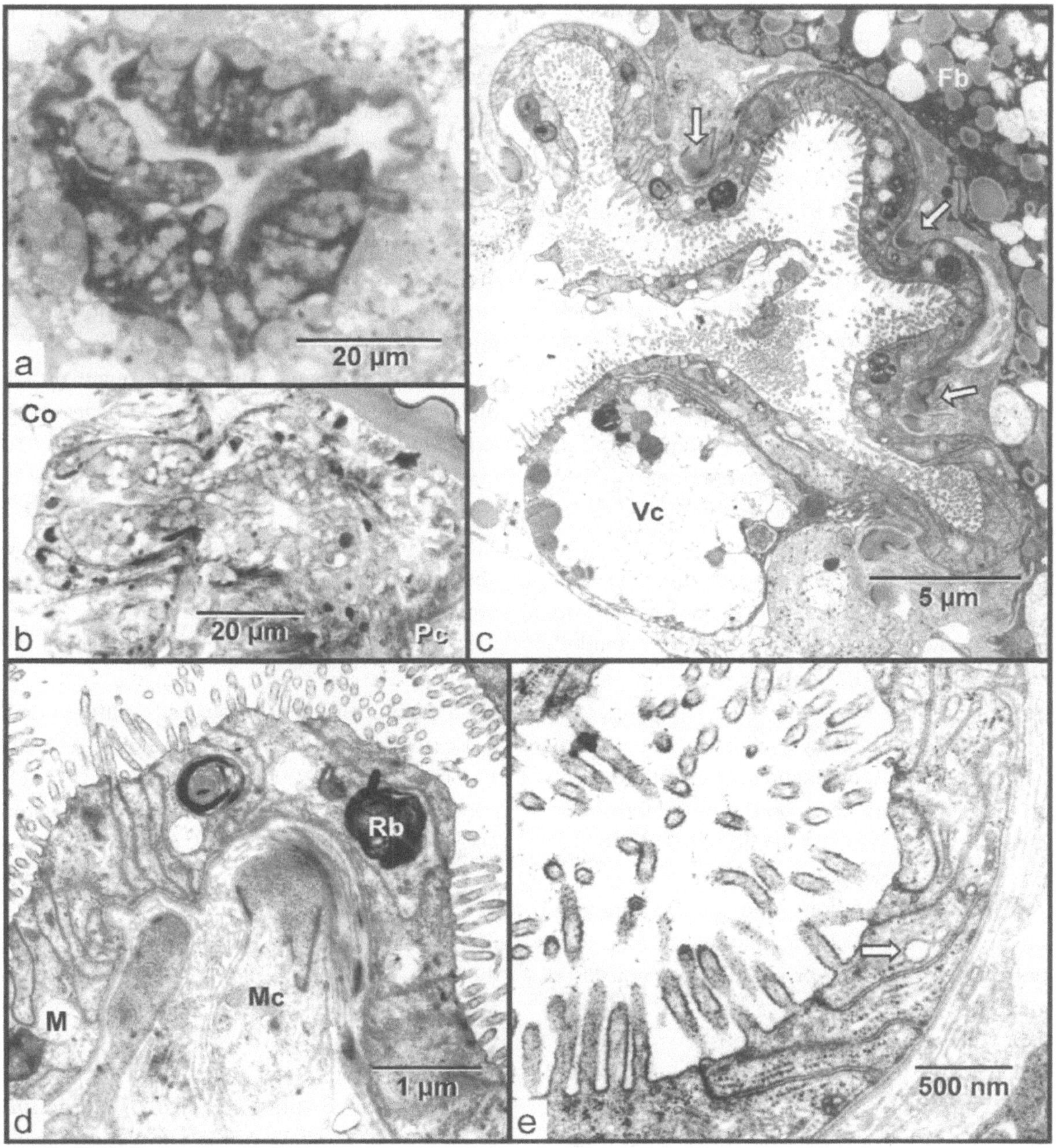

Fig. 2 - Histology and fine structure of the intercolon of *Ameronothrus lineatus*: a) cross section at the level of the dorsolateral projections (LM); b) sagittal section (LM). Note hiding of the intercolon by the invaginated tissue of the colon; c) cross -section of the right projection (TEM). The arrows mark indentations of the surface produced by visceral muscles; d) detail of epithelium and muscularis of projections (TEM); e) cells of projections with microvillous border, RER and vesicular inclusions (arrow) (TEM). Abbr.: Fb, fat body; Co, colon; M, mitochondrion; Mc, muscle cell; Pc, postcolon; Rb, residual body; Vc, vacuolated cell.

endoplasmatic reticulum, mitochondria, residual bodies and vesicular inclusions, give no clear hints about the function of the projections.

The gut system of the *Hyadesia* tritonymph was reconstructed using a series of horizontal sections. The long esophagus has a variable diameter and joins the midgut anteroventrally (fig. 3a-b). The outline of the ventriculus is difficult to define. Ventrally it is elongated into a huge left caecum, on the right side a small caecum is attached via a narrow connection. Dorsally the ventriculus has an open connection to a globular colon. The connection to the postcolon is tightly constricted by circular muscles; an intercolon does not exist. A pair of long, anteroventrally directed Malpighian tubules branch from the anterior wall of the postcolon. The small postcolon is pressed against the bowl-shaped anterodorsal wall of the huge anal atrium. The lumen of the anal atrium is open anteriorly but laterally flattened at its posteroventral end.

The gut system of the reconstructed male (fig. 3d-e) differs considerably from that of the tritonymph. The esophagus is very long and has a similar diameter over its entire length. The midgut is divided in only two parts; an extended anterior midgut and a globular posterior midgut. A pair of small caeca is ventrolaterally connected to the anterior midgut via small narrow ducts. The Malpighian tubules originating at the anterior wall and the high microvillous border indicate that the posterior midgut can be homologized with the postcolon. At the expected position of a colon a simple, narrowing elongation of the anterior midgut is present. Light microscopically, the epithelium of the anterior midgut does not differ along the longitudinal axis. The anal atrium is shaped like a rudder with the posteroventral two thirds laterally flattened.

Comparing serial sections of a number of males, several physiological states ranging from a seemingly absence of a colon to a well separated colon were found (fig. 3c and f). The formation of a colon is strictly correlated with the presence of a colonic faecal pellet.

Discussion and conclusions

With regard to the dynamic changes of the gut structure in *H. fusca*, some contradictory results reported in the literature can be reevaluated. The undivided anterior midgut of *Dermatophagoides farinae* (Brody *et al.* 1972; Tongu *et al.* 1986) and of *Sarcoptes scabiei* (Desch *et al.* 1991) probably represents both ventriculus and colon. The posterior midgut probably is homologous to the postcolon, characterized by the occurrence of a high microvillous border. The division of the hindgut of *D. farinae* into a globular anterior and a narrow posterior part, mistakenly homologized by Brody *et al.* (1972) as 'postcolon' and 'anal atrium', is putatively the result of the presence of a faecal pellet in the expanded anterior part of the anal atrium. The question, if the colon in these species is permanently reduced or if it was, in the sense of a certain physiological state, only temporary absent in the specimens investigated so far, remains to be answered.

In most studies of Astigmata showing a well developed colon separated from the ventriculus by a 'sphincter', the colon cells differed from those of ventriculus and postcolon (Hughes 1950b; Kuo and Nesbitt 1970; Baker 1975). By contrast, Wurst (1993) found for *Listrophorus leucarti* Pagenstecher that the colon cells do not differ ultrastructurally from those of the posterior part of the ventriculus. Further studies must show if, dependent on the actual physiological state, the plasticity of the colon cells let them appear temporarily different from or similar to the ventriculus cells.

Hints on a similar dynamic of the gut can also be found in the Oribatida. In *Hermannia gibba* (Koch) the cells of the narrowed posterior end of the ventriculus are similar to those of the colon, which is separated from the ventriculus by a tight constriction (Bäumler 1970). A division of the colon into a narrow anterior tube and a wide elliptical or globular posterior part, as shown here for *A. lineatus*, was previously described by Michael (1884), who referred to the anterior part as 'small intestine', for a number of oribatid species dissected by hand. It is astonishing that

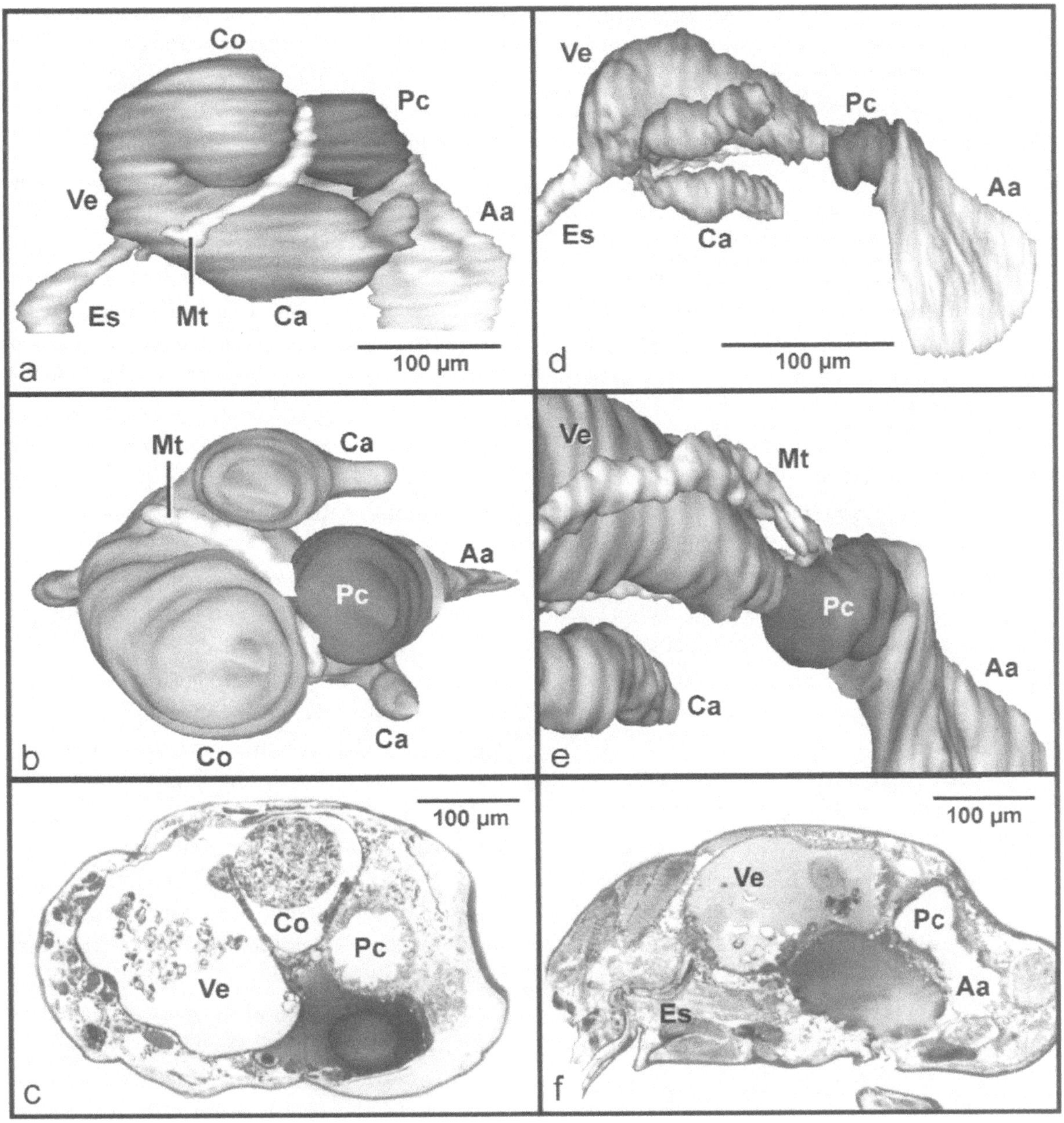

Fig. 3 - Three D-reconstructions and semithin sections of the gut system of *Hyadesia fusca*: a) gut system of a tritonymph in lateral view, ventrally cut off; b) gut system of the same tritonymph in dorsal view; c) horizontal section of a male at the level of the broadest connection of ventriculus and colon (LM); d) gut system of a male in lateral view. Anterior part of the long esophagus and left Malpighian tubule omitted; e) detail of middle and posterior midgut in anterolateral view. Left caecum omitted; f) sagittal section of male (LM). Abbr.: Mt, malpighian tubule (for other abbreviations see Fig. 1).

such a division of the colon was not mentioned in any of the subsequent histological studies on higher oribatid mites. Only Smrz (1989) found a division of the colon in the enarthronotid mite *Hypochthonius rufulus* C.L. Koch, referred to as colon 1 and colon 2.

In conclusion, the term 'postventricular midgut', proposed by Evans (1992) to take into account the homology of the excretory organ of certain Prostigmata and colon/postcolon of Sarcoptiformes, must be used with caution for certain Astigmata. The statement of Brody *et al.* (1972) and others, that the origin of Malpighian tubules mark the border between midgut and hindgut in all Acari and that in all Acari the midgut has only one division, must be rejected. The entodermal origin of postcolon and Malpighian tubules was shown for the tick *Ornithodorus moubata* (Murray) (Aeschlimann 1958) and for the astigmatid mite *Acarus siro* (Hughes 1950a).

The false homologization of the gut subunits of *D. farinae* has led to some confusion in the literature. Wharton and Brody (1972) described the wrapping of the faecal pellet with a peritrophic membrane during its passage through the valve between 'ventriculus' and 'colon'. As discussed above, this valve is actually the connection of colon and postcolon. In other studies of sarcoptiform species, the formation of a peritrophic membrane was proposed to start in the ventriculus (Hoebel-Mävers 1967; Dindsdale 1974; Wurst 1993) or in the colon (Hughes 1950b; Prasse 1967). Further studies are needed to determine, if this contradiction might be explained by a two- or three-step formation mode resulting in two or more membrane layers.

Unlike the condition in *H. fusca*, Prasse (1967) has found for the astigmatid species *Caloglyphus berlesei* Michael, *C. michaeli* Oudemans, *Tyrophagus infestans* Berlese and *Acarus siro* that the Malpighian tubules branch dorsolaterally from a short intercalary subunit between colon and postcolon. The presence of a similar intercolon with thinwalled, dorsolateral pouches in *A. lineatus* and three other *Ameronothrus* species (unpublished results) raise the possibility that this gut region might be homologous to that bearing the Malpighian tubules in the Astigmata. Warren (1947) also found a small, ovoid intercolon in *Cepheus tegeocranus* (Herm.) and discussed a possible relationship to the Malpighian tubules in other Acari. Michael (1884) found a tubular anterior elongation of the postcolon in a number of hand-dissected oribatid mites. In *Hypochthonius rufulus*, Smrz (1989) described a division of the postcolon, referred to as a dorsoventrally flattened 'rectum 1', which might be homologous to the intercolon, and a pear-shaped 'rectum 2' (see also fig. 1 in Tarman 1968). In several other detailed, histological studies of oribatid mites (e.g. Woodring and Cook 1962; Hoebel-Mävers 1967; Dindsdale 1974) no hints of an intercolon were mentioned, but this small part can easily be overlooked if colon and postcolon are pressed against each other. This often results in a hiding of the intercolon into invaginations of the surrounding gut regions.

The hypothesis of Norton (1998) that the Astigmata phylogenetically should be regarded as a subgroup of the Oribatida and probably are closely related to the Malaconothridae, is now widely accepted. If the Malpighian tubules found in Anactinotrichida and Astigmata are homologous, this character must have been reduced several times convergently within the Actinotrichida. As an alternative hypothesis, an intercalary section between colon and postcolon might have evolved within the Oribatida and the development of dorsolateral projections, ranging from short pouches (*Ameronothrus* spp., *Acarus siro*) to long tubes (e.g. the putative Malpighian tubules of *Caloglyphus* spp. and *H. fusca*), might be a phylogenetic trend within late derivative taxa. Further studies of phylogeny, microanatomy and physiology are needed to elucidate the putative homology or non-homology of Malpighian tubules within the Acari. The ultrastructure and function of the so-called Malpighian tubules and the nature of their content (excretory or secretory?) in the Astigmata is unknown and remains to be investigated. Special emphasis should be given to early derivative Oribatida (e.g. Palaeosomata and Enarthronota) and the paraphyletic Endeostigmata, the latter being a 'missing link' in our knowledge of the microanatomy of mites.

References

Aeschlimann A., 1958 - Développement embryonaire d' *Ornithodorus moubata* (Murray) et transmission trans-ovarienne de *Borrelia duttoni*. *Acta Tropica*, 15: 15-64.

Alberti G., Coons L.B., 1999 - Acari: Mites. In: Harrison F.W. (ed.), Microscopic anatomy of invertebrates. Wiley-Liss, New York, Vol. 8C: 512-1215.

Baker R.A., 1975 - The structure and function of the alimentary canal in *Histiogaster carpio* (Kramer, 1881). *Acarologia*, 17: 126-137.

Bäumler W., 1970 - Zur Morphologie, Biologie und Ökologie von *Hermannia gibba* (C.L. Koch) (Acarina: Oribatei) unter Berücksichtigung einiger Begleitarten. *Zeitschrift für Angewandte Entomologie*, 66: 257-277.

Brody A.R., McGrath J.C., Wharton G.W., 1972 - *Dermatophagoides farinae*: the digestive system. *Journal of the New York Entomological Society*, 80: 152-177.

Bücking J., 2000 - Methoden der computergestützten 3D-Rekonstruktion histologischer Schnittserien mit Anwendungsbeispielen aus der Acarologie. *Abhandlungen und Berichte des Naturkundemuseums Görlitz*, 72(1): 97-104.

Desch C.E. Jr., Andrews H., Arlian L.G., 1991 - The digestive system of *Sarcoptes scabiei* (L.): Light and electron microscope study. In: Dusbábek F., Bukva V. (eds), Modern Acarology. Academia, Prague. Vol. 1: 271-279.

Dindsdale D., 1974 - The digestive activity of a phthiracarid mite mesenteron. *Journal of Insect Physiology*, 20: 2247-2260.

Ehrnberger R., 1984 - Anatomie des Verdauungstraktes der Rhagidiidae (Acari, Trombidiformes). *Osnabrücker Naturwissenschaftliche Mitteilungen*, 11: 67-90.

Evans G.O., 1992 - Principles of Acarology. C.A.B. International, Wallingford: 563 pp.

Hoebel-Mävers M., 1967 - Funktionsanatomische Untersuchungen am Verdauungstrakt der Hornmilben (Oribatei). Doctoral Thesis, Technische Hochschule Carolo-Wilhelmina zu Braunschweig, Germany: 45 pp.

Hughes T.E. 1950a - The embryonic development of the mite *Tyroglyphus farinae* L., 1758. *Proceedings of the Zoological Society of London*, 119: 873-886.

Hughes T.E., 1950b - The physiology of the alimentary canal of *Tyroglyphus farinae* L. *Quarterly Journal of Microscopical Science*, 91: 45-61.

Kuo J.S., Nesbitt H.H.J., 1970 - The internal morphology and histology of adult *Caloglyphus mycophagus* (Mégnin) (Acarina: Acaridae). *Canadian Journal of Zoology*, 48: 505-518.

Michael A.D., 1884 - British Oribatidae. Ray Society, London, Vol. 1: 333 pp.

Michael A.D., 1901 - British Tyroglyphidae. Ray Society, Vol 1: 291 pp.

Norton R.A., 1998 - Morphological evidence for the evolutionary origin of Astigmata (Acari, Acariformes. *Experimental and Applied Acarology*, 22: 559-594.

Prasse J., 1967 - Zur Anatomie und Histologie der Acarina mit besonderer Berücksichtigung von *Caloglyphus berlesei* (Michael, 1903) und *C. michaeli* (Oudemans, 1924). I. Das Darmsystem. *Wissenschaftliche Zeitschrift der Martin-Luther-Universität Halle*, 16: 789-812.

Reuter E., 1909 - Zur Morphologie und Ontogenie der Acariden mit besonderer Berücksichtigung von *Pediculopsis graminum* (E. Reut.). *Acta Societatis Scientiarum Fennicae*, 36: 1-288.

Smrz J., 1989 - Internal anatomy of *Hypochthonius rufulus* (Acari, Oribatida). *Journal of Morphology*, 200: 215-230.

Tarman K., 1968 - Anatomy, histology of oribatid gut and their digestion. *Bioloski Vestnik* (Ljubljana), 16: 67-76.

Tongu Y., Ishii A., Oh H., 1986 - Ultrastructure of the house-dust mites *Dermatophagoides farinae* and *D. pteronyssinus*. *Japanese Journal of Sanitary Zoology*, 37: 237-244.

Vitzhum H. Graf, 1940-43 - Acarina. Bronns Klassen und Ordnungen des Tierreichs. Bd. 5., Abtl. IV., Buch 5., Lief. 1-7. Akadem. Verl. Ges. Becker & Erler, Leipzig: 1011 pp.

Warren E., 1947 - On the genital system and gut of the oribatid mite, *Cepheus tegeocranus* (Herm.) and the reaction of these organs to a ray-fungus parasite. *Annals of the Natural Museum*, 11: 1-36.

Wharton G.W., Brody A.R., 1972 - The peritrophic membrane of the mite *Dermatophagoides farinae*: Acariformes. *Journal of Parasitology*, 58: 801-804.

Woodring, J.P., Cook E.F., 1962 - The internal anatomy, reproductive physiology, and molting process of *Ceratozetes cisalpinus* (Acarina: Oribatei). *Annals of the Entomological Society of America*, 55: 164-181.

Wurst E., 1993 - Investigations on the anatomy and the behaviour of the fur mite *Listrophorus leuckarti* (Acari: Listrophoridae). *Stuttgarter Beiträge zur Naturkunde*, Serie A (Biologie), 503: 1-68.

Ultrastructural investigations of Haller's organ in *Dermacentor reticulatus* (Fabr.) (Acari: Ixodida: Ixodidae)

A. Buczek[1], L. Buczek[1], A. Kusmierz[1], K. Olszewski[1], K. Jasik[2]
[1] *Department of Biology and Parasitology, Medical University, Lublin, Radziwillowska 11, PL-20-080 Lublin, Poland, parasit@panaceum.am.lublin.pl*
[2] *Department of Electron Microscopy, Silesian Medical Academy, Medyków 18, PL-40-752 Katowice, Poland*

Key words: *Dermacentor reticulatus,* Haller's organ, sensilla.

Abstract: The study presents the external structure of Haller's organ in all developmental stages of *Dermacentor reticulatus* (Fabr.). Our investigations, made by light and scanning electron microscopy, showed that the number of sensilla in the anterior pit varies in postembryonic stages of *D. reticulatus.* The larvae have five sensilla, whereas nymphs and adult ticks have six sensilla. The anterior pit sensilla show differences in topography, size and surface structure. The modifications of Haller's organ are connected with life cycle, feeding, host specificity and habitat of *D. reticulatus.*

Introduction

Ticks are widespread vectors for many diseases that affect animals. In order to locate hosts, they use different sense organs, mainly Haller's organ situated dorso-distally on tarsus I (Elizarov 1965; Elizarov and Vasewta 1977). The morphology and structural organization of Haller's organs have been examined in various ixodid ticks (Axtell *et al.* 1971; Foelix and Axtell 1971 and 1972; Sixl *et al.* 1971 and 1973; Chu-Wang and Axtell 1974; Balashov and Leonovich 1976, 1978 and 1981; Waladde 1971; Waladde and Rice 1982; others). A considerable number of studies has hitherto concentrated on the anterior pit setae and capsule roof aperture of adult ticks, and sometimes nymphs.

This article reports observations on the external structure of Haller's organ of three life stages of *Dermacentor reticulatus* (Fabricius, 1794) which is an important tick species in temperate zones. It is associated with grassy areas and adjoining wooded areas. It is distributed in a mosaic-like pattern. All developmental stages of *D. reticulatus* must find new hosts for feeding. Adult ticks occupy habitats with access to large domestic and wild herbivorous and carnivorous animals. Immature stages infest burrow-dwelling, ground-living small mammals and sometimes birds.

Materials and methods

Ticks were collected in localities around Lublin (South-Eastern Poland) using the flagging method. In the laboratory they were allowed to feed on naive rabbits. Larvae, nymphs and adults reared at 25°C and 90-95 R.H. were killed in ethanol and then prepared for studies using light and scanning electron microscope (SEM). For light

F. Bernini, R. Nannelli, G. Nuzzaci, E. de Lillo (eds.), Acarid Phylogeny and Evolution. Adaptations in mites and ticks, 227-231.

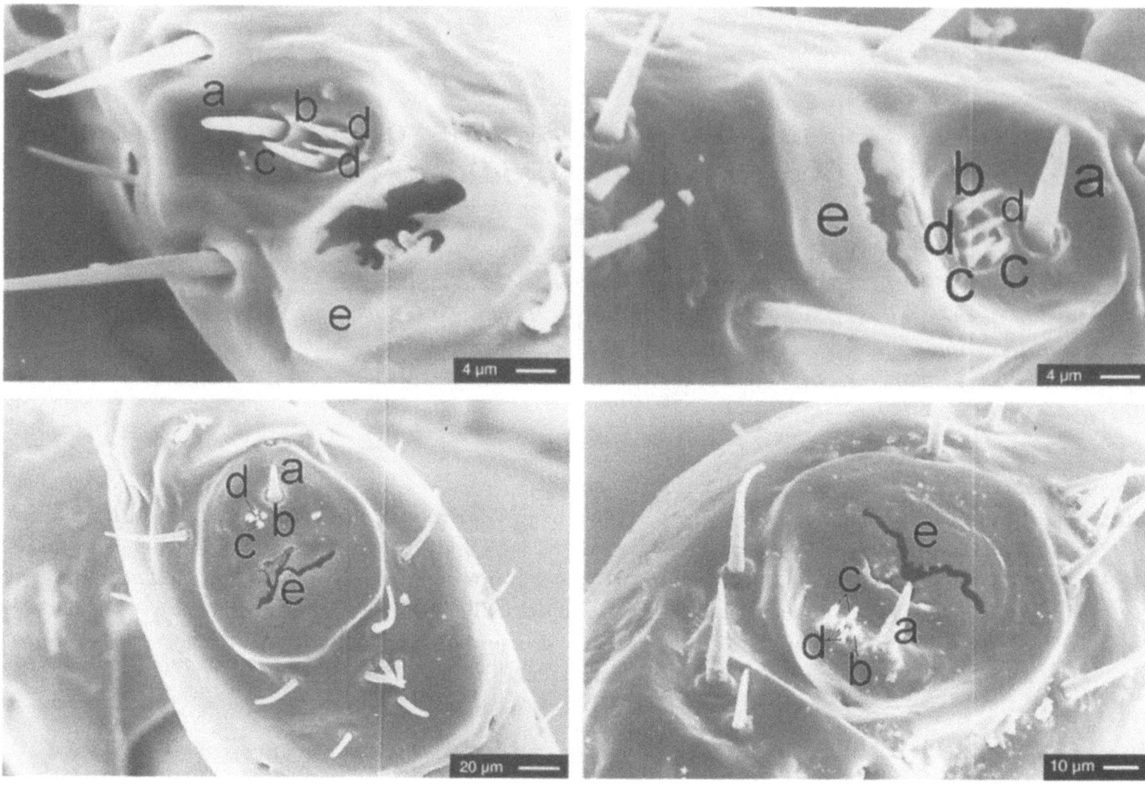

Fig. 1 - Scanning electron micrograph: Haller's organ of *Dermacentor reticulatus*. A, larva; B, nymph; C, female; D: male. Abbrev.: a, porose sensilla; b, conical sensilla; c, grooved sensilla; d, fine sensilla; e, capsule.

microscopic observations the specimens were mounted in Hoyer's medium. The ticks studied by SEM were dehydrated, cleaned and coated with gold (Corwin *et al.* 1979). Ten adult ticks (females and males) and larvae, and eight nymphs of *D. reticulatus* were examined with light microscope and Nanolab 20 scanning electron microscope at 15 kV. Two morphological features, i. e. capsule length, and tarsus I width in all developmental stages of this tick species were measured. On the basis of these measurements, the index of capsule length/tarsus I width was calculated.

Results and discussion

Haller's organ *sensu stricto* consists of an arterior pit and a capsule. The variability of the structure (both anterior pit and capsule) among tick species is considerable (Klompen and Oliver 1993). The anterior pit contains from five to eleven (usually 6-10) sensilla. The capsule consists of a strongly depressed area containing the sensilla that are chemoreceptors.

The development of Haller's organ through tick ontogeny was investigated in some Argasidae (Gothe *et al.* 1991) and Ixodidae (Hess and Vlimant 1986; Buczek *et al.* 1998). As in other examined ticks, the number of anterior pit sensilla and structure of the capsule of *D. reticulatus* differs among the life stages (fig. 1).

There are four types of sensilla in the anterior pit of larvae, nymphs and adult ticks. The larvae have five sensilla, whereas nymphs and adult ticks have six sensilla. They are classified as being porose, grooved, conical and fine. One porose, one conical, one grooved sensilla and two fine

Tab. 1 - The Haller's organ capsule length - C (in µm); tarsus I width - TW (in µm); and index of capsule length / tarsus I width - C/TW in larvae, nymphs and adult ticks of *Dermacentor reticulatus*.

	Larvae (n=10)			Nymphs (n=8)			Females (n=10)			Males (n=10)		
	C	TW	C/TW	C	TW	C/TW	C	TW	C/TW	C	TW	C/TW
Range	21.96-24.40	65.88-78.08	0.30-0.36	34.16-43.92	85.40-100.04	0.39-0.49	85.40-122.00	163.48-212.28	0.41-0.75	97.60-126.88	170.8-214.72	0.46-0.75
Mean	23.18	71.25	0.33	39.68	91.50	0.44	107.60	201.06	0.54	117.85	195.44	0.61
SD	1.29	3.41	0.02	2.84	5.53	0.04	11.70	14.47	0.09	9.13	14.84	0.08

sensilla occur in larval stage of *D. reticulatus*. The types of a particular sensilla of the anterior pit are the same in the nymphs and adults. They have one porose, one conical, two fine and two grooved sensilla. The topography of the sensilla in the anterior pit changes during the post-embryonic development. As in other ticks, sexual dimorphism in the structure of Haller's organ of *D. reticulatus* is absent (fig. 1).

The sensillar composition of the anterior pit of Haller's organ also shows low variability in the postlarval stages of other tick species of the sub-family Amblyomminae (Balashov and Leonovich 1981; Backer and Jackson 1996). The porose sensilla are classified as wall-pore single-walled (WP-SW), the fine sensilla as no-pore single-walled (NP-SW), the grooved sensilla as wall-pore double-walled (WP-DW), the conical sensilla as no-pore double-walled (NP-DW) (Hess and Vlimant 1982 and 1983). These sensilla show functional differentiation. They are olfactory, mechano-, hygro-, and thermo-receptors (Leonovich 1977; Axtell 1979; Waladde and Rice 1982; Hess and Vlimant 1986).

The capsule consists of a roof and cavity, the dimensions of which increase in the course of the development from larvae to adult ticks of *D. reticulatus* (tab. 1). The capsular cavity of larvae, nymphs, females and males is 28.18, 39.68, 107.60, 117.85 µm in length, respectivly. The index of capsule length/tarsus I width also increases during the postembryonic development of this tick species. This structural feature was investigated only in adult ticks of Ixodidae (Schulze 1941) and Argasidae (Buczek 1992/1993). Thus, comparative evaluation of the capsule dimensions of various instars in other tick species is impossible. The capsule roof contains characteristic slits differing in shape in the various stages of *D. reticulatus* (fig. 1). The capsule roof in the larvae is much smaller than in the postlarval stages.

The changes in shape and size of capsule roof were also found in other tick species of metastriate ticks of the genera *Dermacentor*, *Rhipicephalus*, and *Hyalomma* (Leonovich 1978; Balashov and Leonovich 1981; Buczek *et al.* 1998).

The structure of Haller's organ is characteristic of the tick species and their instars. Therefore, the morphology of Haller's organ is used in studies of the phylogeny and taxonomy of ixodid (Axtell *et al.* 1971; Balashov and Leonovich 1978; 1981; Homsher and Sonenshine 1979; Homsher *et al.* 1988 and 1991) and argasid ticks (Klompen and Oliver 1993; Roshdy *et al.* 1972 and 1984; Clifford *et al.* 1983; De la Cruz and Dusbabek 1989; Leonovich 1979 and 1980).

The morphological variation of Haller's organ is connected with biology, feeding and mate behaviour, host selection, and habitats of ticks. Highly modified Haller's organs occur in 2- and 3-host ticks with wide range of hosts, and in ticks that actively seek various hosts like *D. reticulatus*.

References

Axtell R.C., 1979 - Tarsal sensory receptors of ticks. In: Piffl E. (ed.), Proceedings 4th International Congress of Acarology, Akadémiai Kiadó, Budapest, Hungary: 669-672.

Axtell R.C., Foelix R.F., Coons L.B., Roshdy M.A., 1971 - Sensory receptors in ticks and mites. In: Daniel M., Rosicky B. (eds.), Proceedings 3rd International Congress of Acarology, Prague: 35-39.

Baker G.T., Jackson L.K., 1996 - Sensilla in the anterior pit, palpal apex, dm1, and dm2 setal groups of several metastriate ticks (Ixodidae). Piffl E. (ed.), Proceedings 4th International Congress of Acarology, Saalfelden, 1974, Akadémiai Kiadó, Budapest, Hungary: 377-390.

Balashov Yu.S., Leonovich S.A., 1976 - Morphological peculiarities of Haller's organ in ixodid ticks of tribe Amblyommatini (Acarina, Ixodidae). *Entomologischeskoe Obozrenie*, 4: 946-952 (in Russian).

Balashov Yu.S., Leonovich S.A., 1978 - External structure of Haller's organ in ixodid ticks of the subfamily Ixodinae (Acarina, Ixodidae) in connection with the systematics of this group. *Trudy Zoologicheskogo Instituta Akademii Nauk SSSR*, 77: 29-36 (in Russian).

Balashov Yu.S., Leonovich S.A., 1981 - Structure and evolutionary trends of Haller's organ in ixodid ticks of the subfamily Amblyomminae. *Parazitologicheski Sbornik*, 30: 5-21 (in Russian).

Buczek A., 1992/1993 - Morphological variability of *Argas (Argas) reflexus* (Fabricius, 1794) (Acari: Ixodida: Ixodidae) females and males from Silesian population. *Zeitschrift fur Angewandte Zoologie*, 4: 409-419.

Buczek A., Jasik K., Buczek L., 1998 - Sense organs in post-embryonic stages of *Hyalomma marginatum* Koch, 1844 (Acari: Ixodida: Ixodidae). I Tarsal sensory system. *Parassitologia*, 40: 279-282.

Chu-Wang I.W., Axtel R.C., 1974 - Comparative fine structure of the claw sensilla of a soft tick *Argas (Persicargas) arboreus* Kaiser, Hoogstral, and Kohls, and a hard tick *Amblyomma americanum* (L.). *Journal of Parasitology*, 59: 545-555.

Clifford C.M., Hoogstraal H., Roshdy M.A., Wassef H.Y., Keirans J.E., Corwin M.D., 1983 - Observations on the subgenus *Argas* (Ixodoidea: Argasidae: Argas). 18. Haller's organ roof and anterior pit setae. *Journal of Parasitology*, 69: 967-976.

Corwin D., Clifford C.M., Keirans J.E., 1979 - An improved method for cleaning and preparing ticks for examination with the scanning electron microscope. *Journal of Medical Entomology*, 16: 352-353.

De La Cruz J., Dusbabek F., 1989 - Haller's organ and anterior pit setae in the genera *Antricola* and *Parantricola* (Ixodoidea: Argasidae). *Folia Parasitologica*, Prague 36: 275-279.

Elizarov Y.A., 1965 - Investigation of chemoreception of insects and ticks. Physiology of contact chemoreceptory sensilla in ixodid ticks. *Zoologichesky Zhurnal*, 44: 1461-1473 (in Russian).

Elizarov Y.A., Vasewta A.A., 1977 - A distant orientation of blood-sucking ixodid ticks to the host's attractant factors. *Parazitologiya*, 10: 136-141 (in Russian).

Foelix R.F., Axtell R.C., 1971 - Fine structure of tarsal sensilla in the tick *Amblyomma americanum* L. *Zeitschrift für Zellforschung*, 124: 275-292.

Foelix R.F., Axtell R.C., 1972 - Ultrastructure of Haller's organ in the tick *Amblyomma americanum* L. *Zeitschrift für Zellforschung*, 125: 213-222.

Gothe R., Beelitz P., Schol H., 1991 - Morphology and structural organisation of Haller's organ during postembryonic development of *Argas (Persicargas) walkerae* (Ixodoidea: Argasidae). *Experimental and Applied Acarology*, 11: 99-109.

Hess E., Vlimant M., 1982 - The tarsal sensory system of *Amblyomma variegatum* Fabricius (Ixodidae, Metastriata). I. Wall pore and terminal pore sensilla. *Revue Suisse de Zoologie*, 89: 713-729.

Hess E., Vlimant M., 1983 - The tarsal sensory system of *Amblyomma variegatum* Fabricius (Ixodidae, Metastriata). II. Wall pore and terminal pore sensilla. *Revue Suisse de Zoologie*, 90: 157-167.

Hess E., Vlimant M., 1986 - Leg sense organs of ticks. In: Sauer J.R., Hair J.A. (eds.), Morphology, physiology and behavioral biology of ticks. Ellis Horwood, Chichester: 361-390.

Homsher P.J., Sonenshine D.E., 1979 - Scanning electron microscopy of ticks for systematic studies: 3. Structure of Haller's organ in five species of the subgenus *Multidentatus* of the genus *Ixodes*. In: Rodriguez J.G. (ed.), Recent advances in Acarology. Academic Press, New York, vol. 2: 485-490.

Homsher P.J., Robbins R.G., Keirans J.E., 1991 - Scanning electron microscopy of Haller's organ for subgeneric systematic studies in the genus *Ixodes*. In: Dusbabek F., Bukva V. (eds.), Modern Acarology, Academia, Prague, vol. 2: 335-342.

Homsher P.J., Keirans J.E., Robbins R.G., Irwin-Pinkley L.I., Sonenshine D.E., 1988 - Scanning electron microscopy of ticks for systematic studies: structure of Haller's organ in eight species of the subgenus *Sternalixodes* of the genus *Ixodes* (Acari: Ixodidae). *Journal of Medical Entomology*, 25: 348-353.

Klompen J.S.H., Oliver J.H.Jr., 1993 - Haller's organ in the tick family Argasidae (Acari: Parasitiformes: Ixodida). *Journal of Parasitology*, 79: 591-603.

Leonovich S.A., 1977 - An electron-microscope investigation of Haller's of *Ixodes persulcatus* (Ixodidae). *Parazitologiya*, 11: 340-347.

Leonovich S.A., 1978 - Fine structure of Haller's organ in the tick *Hyalomma asiaticum* P. Sch. et E. Schl. (Parasitiformes, Ixodidae, Amblyomminae). *Entomologischeskoe Obozrenie*, 1: 221-224

Leonovich S.A., 1979 - Ultrastructural ivestigations of Haller's organ in argasid ticks, *Argas tridentatus* (Argasidae). *Parazitologiya*, 13: 483-487.

Leonovich S.A., 1980 - Fine structural investigation of Haller's organ in argasid ticks, *Ornithodoros moubata* and *Alveonasus lahorensis*. *Parazitologiya*, 19: 456-463.

Roshdy M.A., Foelix R.F., Axtell R.C., 1972 - The subgenus *Persicargas* (Ixodoidea: Argasidae: *Argas*). 16. Fine structure of Haller's organ and associated tarsal setae of aduld *A. (P.) arboreus* Kaiser, Hoogstraal and Kohls. *Journal of Parasitology*, 58: 805-816.

Roshdy M.A., Hoogstraal H., Clifford C.M., Keirans J.E., Corwin M.D. 1984 - Bat ticks of the genus *Argas* (Ixodoidea: Argasidae). 13. Haller's organ roof and anterior pit setae in *Carios* and *Chiropterargas*. *Journal of Parasitology*, 70: 407-411.

Schulze P., 1941 - Das Geruchsorgan der Zecken. Untersuchungen über die abwandlungen eines Sinnesorgans und seine stammesgeschichtliche Bedeutung. *Zeitschrift für Morphologie und Ökologie Tiere*, 37: 491-564.

Sixl W., Denng E., Waltinger H., 1971 - Das Hallersche organ von *Ixodes ricinus*. *Angewandte Parasitologie*, 12: 225-228.

Sixl W., Denng E., Waltinger H., 1973 - Scanning electron microscope investigations on ticks, with special regard to the sensillae, Haller's organ and the apical palpal segment. In: Daniel M., Rosicky B. (eds.), Proceedings 3rd International Congress of Acarology, Prague: 53-54.

Waladde S.M., 1971 - The sensory nervous system of the adult cattle tick *Boophilus microplus* (Canestrini) Ixodidae. Part II. Scanning electron microscopy. *Journal of Australian Entomology Society*, 16: 73-79.

Waladde S.M., Rice M.J., 1982 - The sensory basis of tick feeding behaviour. In: Obenchain F.D., Galun R. (eds.), Physiology of ticks. Pergamon Press, Oxford: 71-118.

Morphologic adaptations to parasitism on reptiles: Pterygosomatidae (Prostigmata: Raphignathina)

Michel Bertrand
Laboratoire de Zoogéographie, Université Montpellier III, Route de Mende, F-34199 Montpellier cedex 5, France, Michel.Bertrand@univ-montp3.fr

Key words: adaptation, parasitism, scale mites, morphology, reptiles, convergence.

Abstract: The Pterygosomatidae are ectoparasitic on reptiles. The species belonging to *Geckobia* and *Pterygosoma* genera are strongly modified: the body shape is rounded or wider than long, the posterior legs are strong and larger than legs II and legs I, the ventral setae may be scale like (*Geckobia* genus). This study reviews the state of our understanding of the adaptive values of these apomorphies. The characteristics can be related to the development of physical forces, avoiding stripping from the host and improvements in fixing. These hypotheses have to be related to the remarkable convergence of body shape and peripheral setae observed on some mesostigmatid mites parasitic on snakes and a tick described from giant tortoises.

Introduction

Two kinds of phenomena have to be distinguished in adaptive processes: constructive and regressive, with the latter well known amongst parasitic animals, with different degrees of regression of organs. Fain (1979 and 1981) considered that constructive processes are important amongst ectoparasites. Many examples might illustrate this tendency amongst feather mites or mites which are parasitic on higher vertebrates. Studies on scale-mites (Pterygosomatidae) which are blood feeding on lizards allow us to outline these characters on parasitic mites of reptiles. Their life cycle is spent on the host and parthenogenesis is the rule. The larvae are mobile and look for an host after hatching. The nymphs and adults are attached by mouthparts to the hosts; protonymph and tritonymph are calyptostasic on the host. Five genera are strongly specialized on a particular type of host: *Pterygosoma* on Agamidae, *Geckobia*, *Zonurobia* and *Scaphothrix* on Gekkonidae and lizards, *Cyclurobia* on Iguanidae. Except in *Cyclurobia*, the body is often rounded (i.e.: *Geckobia*) or wider than long, with many large peripheral setae (i.e.: *Pterygosoma*). Hirst (1926) speculated that these features might be the consequences of adaptive processes: the mites with the body longer than wide (*Cyclurobia*, *Geckobiella*, *Hirstiella*) have been considered as plesiomorphic. Hirst (1926) thought that the body shape was related to the site attachment on the host. Species belonging to *Pterygosoma* and *Geckobia* genera have the pairs of legs III and IV often enlarged (Bertrand and Ineich 1987; Bertrand *et al.* 1995). Other mites found on reptiles show convergent morphological features: amongst Prostigmata *Ophioptes* (Prostigmata) exhibits large

F. Bernini, R. Nannelli, G. Nuzzaci, E. de Lillo (eds.), Acarid Phylogeny and Evolution. Adaptations in mites and ticks, 233-240.

coxae and strong legs; Oementolaelaptidae (Mesostigmata) mites are parasitic on snakes and have a rounded body with large peripheral setae (Fain 1961); amongst ticks, *Argas (Microargas) transversus* Banks, 1902 (Hoogstraal and Kohls 1966; Hoogstraal *et al.* 1973; Klompen and Oliver 1993), the only tick which spends its entire life on reptiles (giant tortoise from Galapagos Islands), exhibits the same rounded shape, and peripheral setae.

The first hypothesis of this study was that these features constitutes convergent homoplasies. The Pterygosomatidae mites were a good model because many species were described (especially in *Pterygosoma* and *Geckobia* genera), because their stenoxeny yields close adaptations to some few hosts and because characters vary sufficiently to allow us to speculate on their adaptive significance.

The characters related to scale mite way of life

The five most evident apomorphic characters are:

1) body shape, more or less wider than long (figs 1A, 2);
2) legs of unequal length (legs III and IV thicker and longer than legs II and legs I), directed forward, the claws being clutched in lizard's skin around the gnathosoma;
3) specialization of the palp as organ of fixation and sometimes reduction of the palp tarsus (*Pterygosoma)*;
4) shape of ventral setae (*Geckobia*: figs 1B-C);
5) large and numerous posterior setae (*Pterygosoma* and *Geckobia*).

Body shape

The scale mites dig deeply with their gnathosoma in the host's derma between scales. They are attached by means of hooked chelicerae. In *Pterygosoma* and *Geckobia*, the flattened body is jammed between the upper and the lower scales: mites are rarely found on the ventral surface of the host and on the sites reached by itching activities of the host. They are attached on the head (ears, neck), between toes, or at the baseis of the tail protected by skin folds (fig. 3).

What is the adaptive value of the rounded or wide body shape?

It permits the mite to be like a new scale on the reptile's skin. The body is soft and the engorged mites are inflated: the internal pressure develops a passive strength which keeps the mite in position between the scales (fig. 4). The passive protection enhanced by the body shape is important for the survival: protonymphs and tritonymphs are calyptostasic on the host and could not resist actively to tearing up. But, when the mites are engorged, the soft cuticle is submitted to friction with the lizard's scales. The wide body shape moves away the posterior coxae to the bottom: the last pairs have to be longer than the first pairs to clutch the claws in the lizard skin near the mouthparts of the mite.

Role of the legs in active fixing

Pterygosoma and *Geckobia* genera have legs of unequal length, directed forward and the four pairs reach the level of the gnathosoma (fig. 4). The first pair is reduced, third and fourth are longer. Long and stout legs III and legs IV facilitate a strong attachment on the host. Jack (1964) compared the chaetotaxy of the legs: primitive Pterygosomatidae mites (body longer than wide and with long sub-equal legs: i.e. *Hirstiella*) have numerous setae (5 by article between trochanter and tarsus) while specialized species have reduced chaetotaxy (except tarsi and tibiae, only 3, 2, 1 or 0 setae). The legs have lost their primitive role (walking) to play an active role in fixing the mite on its host. The enlarged pair of legs IV increases the leverage and the developed strength. Legs I and legs II, are short often less the half that of legs III and IV respectively.

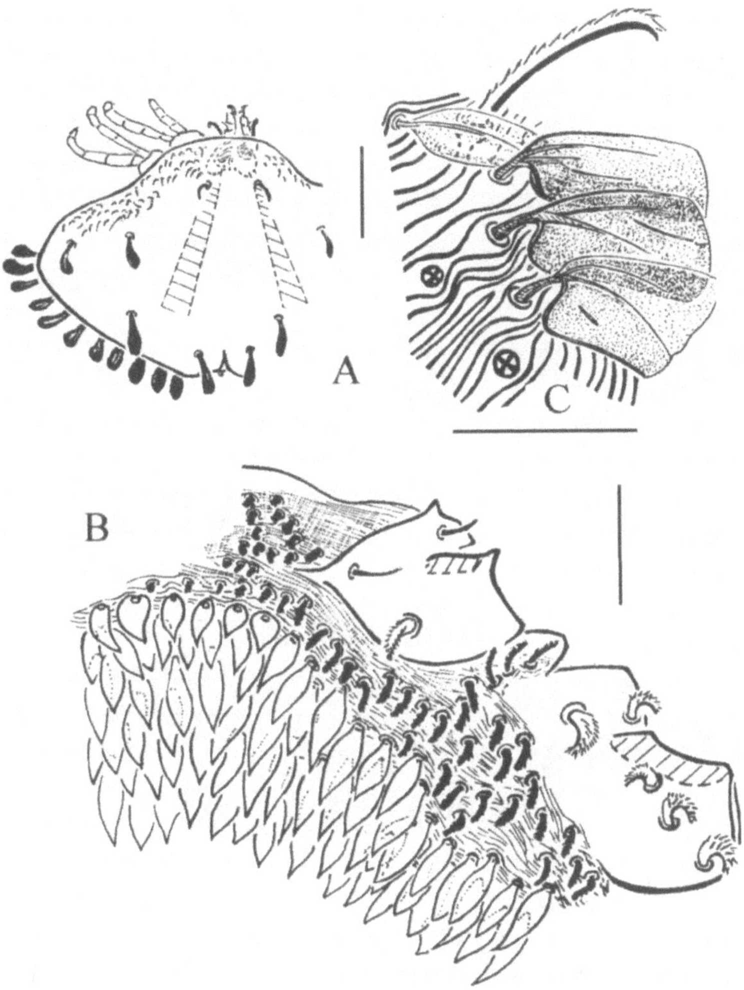

Fig. 1 - A) *Pterygosoma gladiator* Bertrand, Paperna & Finkelmann, 1999: female dorsal view; B) *Geckobia glebosum* Bertrand, Paperna & Finkelmann, 1999: female ventral view of coxae and ventral setae; C) *Geckobia squameum* Bertrand, Paperna & Finkelmann, 1999: female ventral setae, detail. Scale bars=200 µm for A, and 50 µm for B and C. From Bertrand *et al.* (1999) modified.

Palp

The palp presents a thumb like process with a tibial claw opposite to tarsus. This article is always well distinct in *Geckobia*, but it varies among *Pterygosoma*: it is often reduced to a small zone where remaining tarsal setae are present (Bertrand *et al.* 1999). The sensorial function of this article tends to be reduced.

Shape of ventral setae

In the *Geckobia* genus, we observed three types of ventral setae: "normal", "feather-like" and "scale-like setae". They are directed backward. Scale like setae cover the entire surface of the body (except coxae) like tiles on a roof (fig. 1B). They shelter the ventral surface of the mite (limiting friction on the reptile's scales). The ventral setae are directed backward, and,

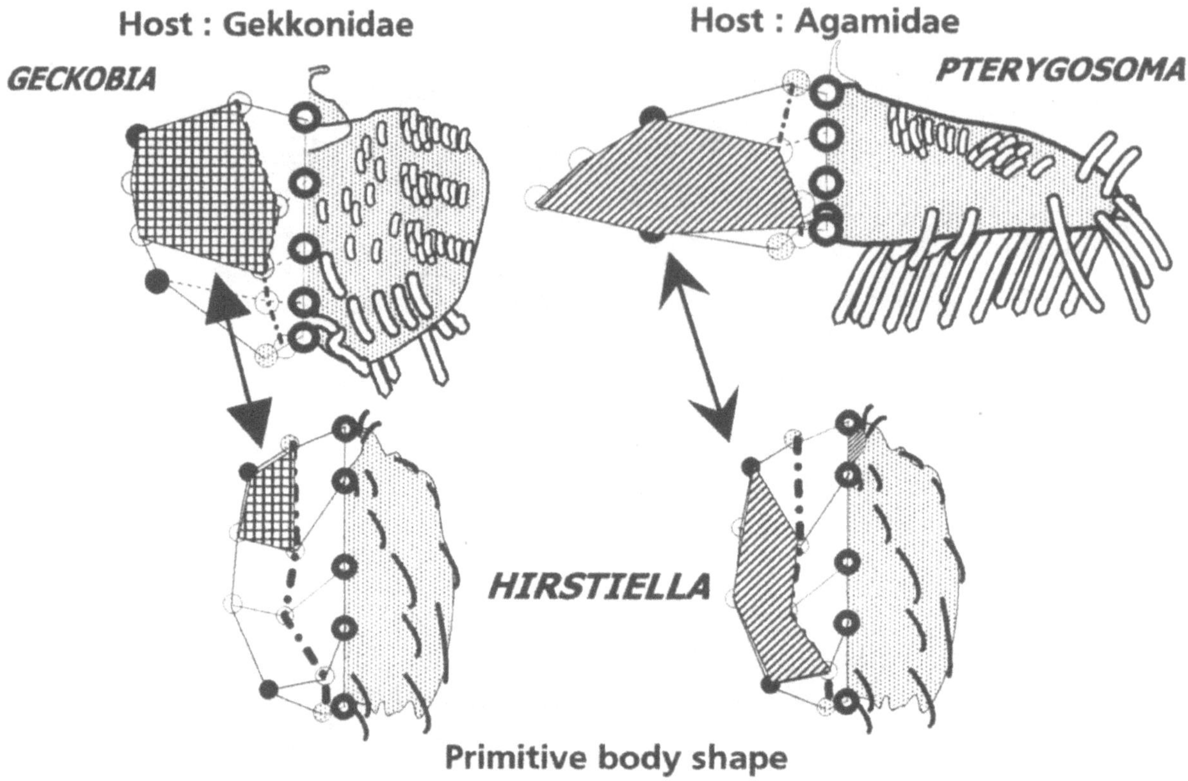

Fig. 2 - Evidence of deformation in Pterygosomatidae: *Hirstiella* is a long legged mite found on lizards, *Geckobia* is found on Gekkonidae, *Pterygosoma* is found on Agamidae.

mechanically, they avoid stripping. Their shape ensures a close contact with the host's scales, and evenly distributes pressure on the entire ventral surface. The surface is closely pressed against the scale, while the basal part of the setae (the stalk) is rigid enough to avoid direct rubbing on scales of the surface of the body (fig. 1C). By their shape, these flattened hairs might give to the mites some assistance in fixation by a simple physical phenomenon. They present tiny teeth or slight ripples on the surface (Bertrand *et al.* 1999). Recently, Gee (2000) underlined the role in adherence of the tiny hairs at the surface of scales of the feet of geckos. The setae on the ventral surface of the scale mites might play the same role.

Peripheral long setae

The large posterior and peripheral setae of *Geckobia, Zonurobia* and *Pterygosoma* are remarkable. They are very similar to the peripheral setae of Oementolaelaptidae and *Microargas*.

Four roles may be assigned:

1) they are directed backward. As ventral setae, they play a passive role in fixing: they avoid stripping;
2) these setae are very long (fig. 1A) (often about 100 µm): they constitute a barrier to other mites looking for an attachment site. They may play an important role in competition with other mites on the same host: the best attachment sites are restricted to skin folds and

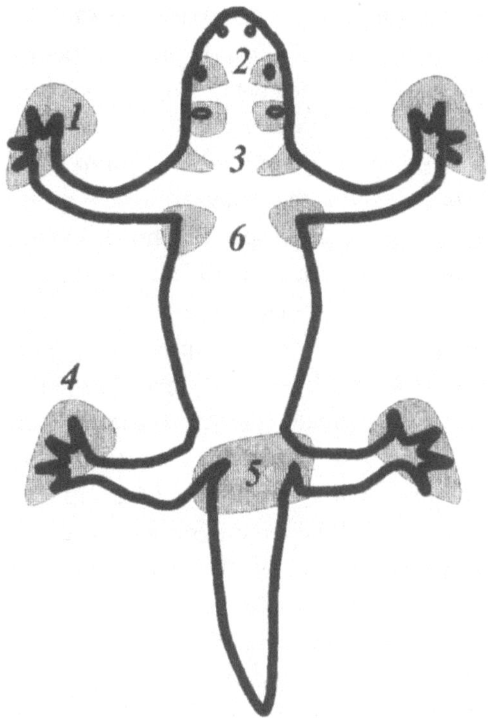

Fig. 3 - The best sites on geckos: 1, toes of anterior legs; 2, ocular zone; 3, neck, ears; 4, toes of posterior legs; 5, skin fold near the tail base; 6, skin fold near the arm's base.

between toes scales of the geckos. Nymphs and adults of *Geckobia* are mainly found between toes whereas the larvae are often found on the head and body parts. They will be eliminated by itching if they cannot find a favourable site. These setae are probably sensory protective in function as hypothesized about *Micoargas* (Hoogstraal and Kohls 1966);

3) by their shape, these flattened hairs might give to the mites assistance in fixing by a simple physical phenomenon (fig. 1A);

4) these setae emerge from the scales. They may have a role as mechanical receptors.

The adaptations to parasitism on reptiles

The Pterygosomatidae mites (*Hirstiella*, *Geckobiella*, *Cyclurobia*, *Scaphothrix*, *Ixodiderma*, *Pterygosoma*, *Zonurobia* and *Geckobia*) present either a primitive morphological pattern (body longer than wide, long legs of equal length: *Hirstiella*, *Geckobiella*, *Cyclurobia*) or combined adaptive characters to parasitic life on reptiles [*Pterygosoma*, *Zonurobia* and *Geckobia* exhibit different grades of the characteristics of the body shape, the length of peripheral body setae and

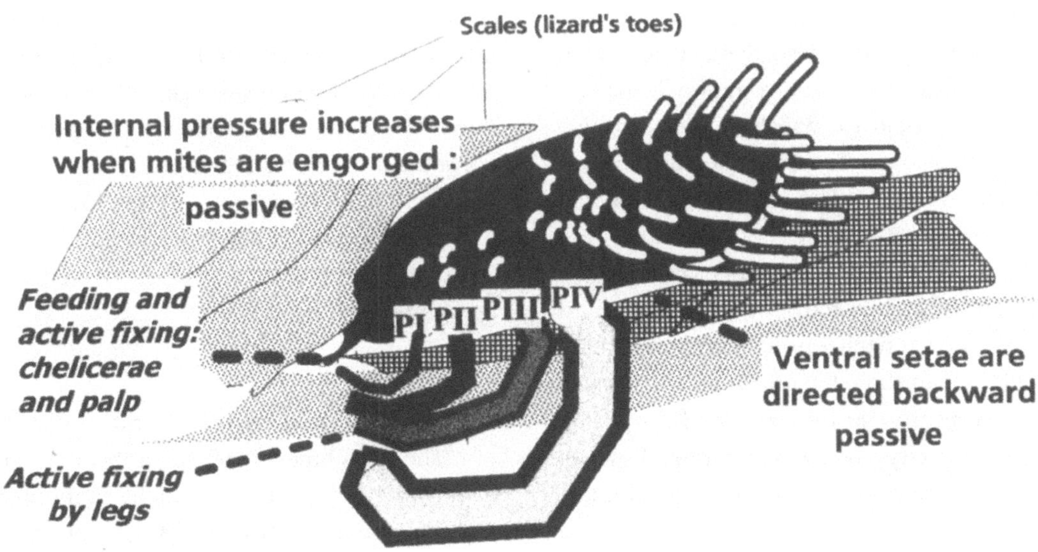

Fig. 4 - Fixing of Pterygosomatid mites: scheme of passive and active strengths.

the shape of ventral setae; *Scaphothrix* and *Ixodiderma* are represented by few species (Lawrence 1935 and 1936), the genus *Pimeliaphilus* is still uncertain]. The variability reaches a climax in *Pterygosoma* which presents an unusual body shape wider than long. In the *Geckobia* genus ventral setae are either normal or scale like: in the latter case, they cover the ventral surface like tiles.

These constructive characters constitute synapomorphies. They participate to the fitness of the ectoparasites and favours the survival of the fittest mite on each species of reptile. However, it is remarkable that the most primitive mites, according to the above characters (plesiomorphic), present a primitive chaetotaxy on legs, whereas the most modified towards parasitic life (apomorphic) have a regressive leg chaetotaxy (Jack 1964). If constructive characters are nothing else than specialization toward a special function (fixation on the host), and regressive characters are related to evolution (Fain 1979), *Geckobia*, *Pterygosoma* and *Zonurobia* genera attest that sometimes the genera mostly affected by regressive evolution (leg chaetotaxy) may be affected by the most constructive adaptive processes.

Any parasitic mites on reptiles are submitted to the same constraints. Dermanyssidae, Laelapidae, Paramegistidae (Mesostigmata) are often found on snakes. The most common is *Ophionyssus* (Dermanyssidae) which is not strongly adapted to parasitism. These species are known as blood feeding mites. The Oementolaelaptidae, collected on ventral scales of a snake, are submitted to important constraints (Fain 1961). They must resist to important stripping forces and presents the most characteristic shape. The body is oval, wider than long, flattened and the body is ventrally surrounded by strong files setae analogous to pterygosomid's peripheral setae. Amongst Prostigmata, Trombiculidae, Myobiidae, Erythraeidae and Pterygosomatidae large and strong setae are common. Genus *Ophioptes* is strongly deformed with very large coxae and strong legs. The pterygosomatid mites are the true scale mites with characteristic shapes. Ticks are commonly found on tortoises, snakes and varanids. Amongst them *Argas (Microargas) transversus* lives on the giant tortoises of the Galapagos Islands. This species shows evident convergence with scale mites and Oementolaelaptidae: the body shape, the ventral surface and the peripheral setae. Apparently, convergences are shown by Mesostigmata (Oementolaelaptidae), Prostigmata (Pterygosomatidae) and Ixodida (*Microargas*) which are submitted to the strongest constraints by their life cycle, and both the body shape and the large peripheral setae are the characters shared by the most modified species.

The well known genera of Pterygosomatidae are *Geckobia* and *Pterygosoma*. They exhibit the characters described above. We considered them as synapomorphic (parasitomorphisms) (Bertrand and Ineich 1987; Bertrand *et al.* 1999). These characters may be arranged in three lists (tab. 1).

The first list is the list of characters depending directly on the parasitic life on reptiles. The second list shows the improvements that the adaptive processes might determine: some are constructive (development of peripheral setae) but others are regressive (fusion of palp tarsus with tibia), and indicates the main evolutionary tendencies. The characters of the third list might be considered independent and constitutes specific adaptation.

We might envisage a fourth list established on physiological and behavioural adaptations (parthenogenesis, presence of neotenic males, recognition of the host).

Discussion and conclusions

The attachment site on a host is the first dimension of the ecological niche of the parasites, and the characters of these sites are the results of heterogeneity of living organisms. To be ex-

Tab. 1 - The constraints and the morphological responses of the Pterygosomid. The characters from the first list are the adaptations to the constraints; in the second list, characters may be considered as consequences of those of the first list; the third list summarises some specific adaptations. Characters of the second and third list may constitutes convergence.

MAIN CONSTRAINTS	ANCESTRAL CHARACTERS		FIRST LIST OF ADAPTIVE PROCESSES	SECOND LIST OF ADAPTIVE PROCESSES	THIRD LIST OF MORPHOLOGICAL CHARACTERS	CONSEQUENCES FOR THE FITNESS
Scale mites are ectoparasites They are blood feeding mites They spend their life on animals covered by scales The hosts present grooming activities	Gnathosoma, anterior part	Blood sucking and fixing	Sharp chelicera hocked distally, palp with a strong tibial claw, long tube-like lips	Reduction of palp tarsus: priority to fixing rather than sensorial role	Reduction of sensorial role of anterior part of the body, loss of eyes	Active fixing by mouthparts, palp and the fourth legs Loss of mobility
	Legs	Long legged animals, mobile Fixing by claws, Trichobothria on PI, PII, PIII and PIV	PIII and PIV long, PI and PII short	Loss of genual and femoral setae	Trichobothria restricted to posterior pairs of legs	
	Body	Soft body, longer than wide, inflated when engorged, anogenital zone terminal	Flat body, rounded or wider than large, hidden below host's scales	Scutum reduced	Migration of anogenital region in sub dorsal position	Resistance to pulling out by itching
	Ventral surface	Coxae in two groups	Neotaxy on ventral surface, large coxae III and IV	Long peripheral setae, setae directed backward	Ventral setae flattened or tile like	Avoid rubbing against host's scales Setae flattened distally enlarge cohesiveness with the host
	Dorsal surface	Setae arranged in symmetric files	Anterior and peripheral fields of dense setae	Long peripheral setae		Avoid rubbing and competition for space with others mites

ploited, a site must be free, but inadequate morphological adaptations represent an handicap. If morphological adaptations enhance the fitness on some few hosts, they limit the expanding possibilities. The "specificity" varies in scale mites: amongst the most primitive forms (mobile with long legs), *Cyclurobia* was only found on Iguanidae (De la Cruz 1984); are they really generalists? Very different are the "highly adapted" ones (i.e. *Geckobia*): are they so specialized? They are parasites on geckos on the five continents. Recently, we described *Geckobia enigmaticus* from individuals collected on tortoise (Bertrand and Pedroño 1999): even "specialized" species may be euryxenic (alternative host). The highly specialized species spend their entire life on reptiles, the mobile larvae show less exigencies than the others stases in site attachment: the mobility is essential in the search of the host. The hypothesis that the morphological features are mainly determined by constraints seemed to be proved by the examples of Oementolaelaptidae and *Microargas transversus*. The "scale mite's way of life" is so exacting that convergent morphological adaptations are shown in the highly specialized Mesostigmata and Ixodida. Convergent adaptation in different systematic groups collected on reptiles was established. However, many reptiles present mite pockets, which reduce the mite's damage (Arnold 1986). These facts suggest a rich field for evolutionary and behavioural researches and further investigations.

References

Arnold E.N., 1986 - Mite pockets of lizards, a possible means of reducing mite damage. *Biological Journal of the Linnean Society*, 29: 1-21.

Banks N., 1902 - Entomological results. (Papers from the Hopkins Galápagos expedition, 1898-1899). *Proceedings of Washington Academy of Sciences*. 4: 49-86.

Bertrand M., Ineich I., 1987 - Contribution à la connaissance des Pterygosomatidae du Pacifique Sud. *Acarologia*, 28: 241-249.

Bertrand M., Pedroño M., 1999 - Euryxénie et sténoxénie du genre *Geckobia* Mégnin 1878 (Actinedida: Pterygosomatidae): récolte de *G. enigmatica* n. sp. sur une tortue terrestre malgache (*Geochelone yniphora*). *Acarologia*, 40: 147-153.

Bertrand M., Paperna I., Finkelman S., 1995 - Etude préliminaire du genre *Pterygosoma* Peters, 1849 (Actinedida: Pterygosomidae): compléments à la description de *Pterygosoma bibronii* Jack, 1962 et notes sur les caractères évolutifs des Pterygosomidae. *Acarologia*, 36: 133-143.

Bertrand M., Paperna I., Finkelman S., 1999 - Pterygosomatidae: descriptions et observations sur les genres *Pterygosoma*, *Geckobia*, *Zonurobia* et *Hirstiella*. *Acarologia*, 40: 275-304.

De la Cruz J., 1984 - Sistematica de la familia Pterygosomidae (Acari: Prostigmata), con la descripcion de un nuevo genero y especie. *Poeyana*, 278: 1-22.

Fain A., 1961 - Une nouvelle famille d'Acariens parasites de serpents du genre *Mehelya* au Congo: Omentolaelaptidae fam. nov. *Revue de Zoologie et Botanique Africaines*, 61: 283-296.

Fain A., 1979 - Specificity, adaptation, and parallel Host-Parasite evolution in Acarines, especially Myobiidae with a tentative explanation for the regressive evolution caused by the immunological reactions of the host. In Rodriguez J.G. (ed.), Recents Advances in Acarology. Academic Press, N.Y., vol. 2: 321-329.

Fain A., 1981 - Adaptations morphologiques et biologiques au parasitisme chez les Acariens. *Phegea*, 10: 23-34.

Gee H., 2000 - Biomechanics: Gripping feat. *Nature*, 405(6787): 631.

Hirst A.S., 1926 - On the parasitic mites of the suborder Prostigmata (Trombidoidea) found on Lizards. *Journal of the Linnean Society*, 36: 173-200.

Hoogstraal H., Kohls G.M., 1966 - *Argas (Microargas) transversus* Banks (New Subgenus) (Ixodoidea, Argasidae), a diminutive parasite of the giant Tortoise: redescription of the Holotype Male and Description of the larva. *Annals of the Entomological Society of America*, 59: 247-252.

Hoogstraal H., Clifford C.M., Keirans J.M., 1973 - *Argas (Microargas) transversus* (Ixodoidea: Argasidae) of Galapagos Giant tortoises: description of the female and nymph. *Annals of the Entomological Society of America*, 66: 727-732.

Jack K.M., 1964 - Leg-chetotaxy with special reference to the Pterygosomidae (Acarina). *Annals of the Natal Museum*, 16: 152-171.

Klompen J.S.H., Oliver J.H.Jr., 1993 - Systematic relationships in the soft ticks (Acari: Ixodida: Argasidae). *Systematic Entomology*, 18: 313-331.

Lawrence R.F., 1935 - The prostigmatic mites of South African Lizards. *Parasitology*, 37: 1-45.

Lawrence R.F., 1936 - The prostigmatic mites of South African Lizards. *Parasitology*, 38: 1-39.

Fine structure of the phytoseiid-type sperm access system (Acari, Gamasida, Phytoseiidae)

Gerd Alberti[1], Antonella Di Palma[2]
1 Zoological Institute and Museum, J.S. Bach Str. 11/12, University of Greifswald, D-17489 Greifswald, Germany
2 Agricultural Faculty, University of Foggia, via Napoli 25, I-71100 Foggia, Italy

Key words: ultrastructure, female genital system, functional morphology, podospermy.

Abstract: Females of Dermanyssina possess, according to present knowledge, two types of sperm access systems: the laelapid and the phytoseiid type. Contrary to the laelapid type, the phytoseiid type is hardly understood regarding structural and functional aspects.

In the present study we report on the sperm access system of two phytoseiid species: *Phytoseiulus persimilis* Athias-Henriot mating females, and *Typhlodromus rhenanoides* Athias-Henriot overwintering females.

In both species the general anatomy of the sperm access system is the same, even though some differences in the gross morphology are present. The system is composed of paired elements: solenostomes, major and minor ducts, emboli, calyces and vesicles. The fine structure of all these parts is described. Attention is drawn especially to the minor duct that has been traced here for the first time in full length. Moreover, a peculiar electronlucent tissue, connecting as a transversal belt the two minor duct ends, is detected. This tissue makes the phytoseiid-type sperm access system a continuous structure as the laelapid-type. From the results presented here, it seems doubtful that the minor duct serves as a simple path for sperm to the ovary. Apparently the phytoseiid-type sperm access system plays a role in sperm reception but is not or at least less involved in capacitation and storage as the laelapid-type system.

Introduction

In phytoseiid mites (fig. 1A), the reproductive system is composed of a region involved in egg development and oviposition and another for sperm reception, and, perhaps, storage and capacitation (sperm access system). This is due to a peculiar mode of insemination called podospermy which means that the male introduces, by means of modified chelicerae, the spermatophore into paired sperm induction pores. These are connected to a sperm access system (phytoseiid-type).

Available papers on this system (Dosse 1958 and 1959; Schuster and Smith 1960; Fain 1963; Athias-Henriot 1971; Wainstein 1973; Amano and Chant 1978; Karg 1982) focus mainly on its gross morphology, widely used in taxonomy, while its histology, fine structure and functional anatomy are almost unknown (Alberti 1988; Nuzzaci *et al.* 2001). Thus, the aim of this study is to give a detailed description of this sperm access system as a contribution to clarify its role in sperm reception, capacitation and storage.

F. Bernini, R. Nannelli, G. Nuzzaci, E. de Lillo (eds.), Acarid Phylogeny and Evolution. Adaptations in mites and ticks, 241-252.

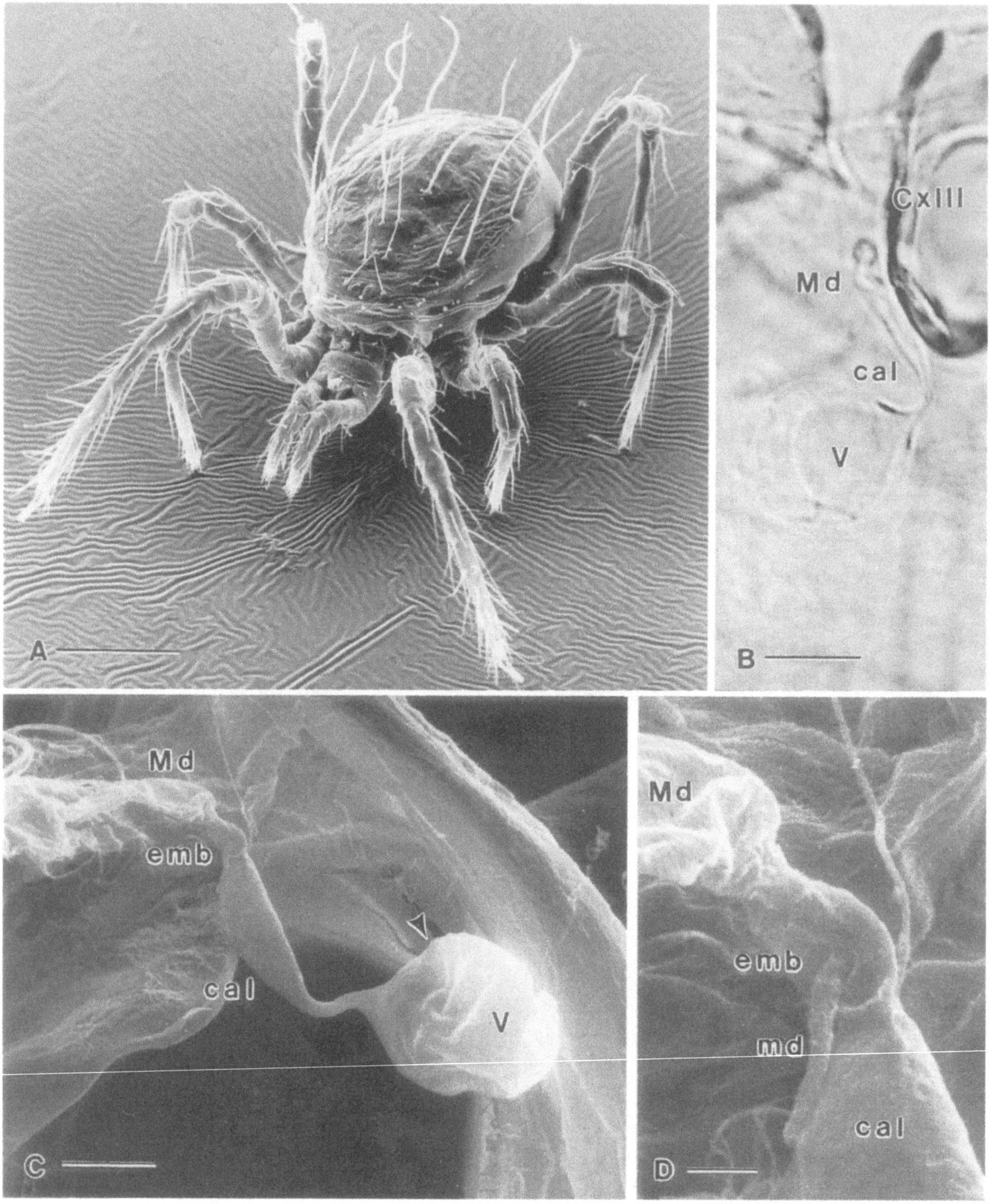

Fig. 1 - *Phytoseiulus persimilis*. A: a female in frontal view. SEM: Scale bar=100 μm. B: mated female showing the solenostome (insemination pore), major duct, calyx and vesicle. The latter is spherical because it is still filled with remnants of the endospermatophore. LM: Scale bar=20 μm. C: major duct, embolus, calyx and vesicle shown in a lactic acid macerated specimen. Minor duct is broken. Note the border between sclerotized part of calyx and the weak cuticle of the vesicle (arrowhead). SEM: Scale bar=10 μm. D: same specimen as in fig. 1C showing embolus and a part of minor duct in higher magnification. SEM: Scale bar=2 μm. Abbr.: cal, calyx; CxIII, coxa III; emb, embolus; Md, major duct; md, minor duct (broken); V, vesicle.

Finally, this clarification will be important for comparisons between the phytoseiid-type sperm access system and the laelapid-type already described (Alberti and Hänel 1986) in another taxon, *Varroa jacobsoni* Oudemans.

Materials and methods

Overwintering females of *Typhlodromus rhenanoides* Athias-Henriot and active females of *Phytoseiulus persimilis* Athias-Henriot were used for this study. Some females of *P. persimilis* were fixed at different times after mating.

Transmission electron microscopy (TEM): *T. rhenanoides* females were dissected and prefixed in Karnovsky's (1965) solution while *P. persimilis* females were dissected and prefixed in glutaraldehyde (3.5% in phosphate buffer at pH 7.4). After postfixation in OsO_4 (2%), the tissues were dehydrated with graded ethanols and embedded in Araldite. Ultrathin sections were made using a Leica ultracut UCT and stained with uranyl acetate and lead citrate (Reynolds 1963). Observations were performed with a Zeiss EM 10 CR.

Scanning electron microscopy (SEM): *Phytoseiulus* females macerated in lactic acid were dehydrated in graded alcohol, dried with a BALTEC CPD 030 critical point apparatus, and glued on a stub. Then the dorsal shield was removed and the specimens were coated with gold using a Polaron SC 7640 sputter coater apparatus. A Zeiss DSM 940A scanning electron microscope was used during the study.

Finally, slides for light microscopy (LM) were prepared mounting the specimens in Hoyer's medium. Observations were performed with an Olympus BX50 light microscope.

Results

The general anatomy of the sperm access system is largely the same in both species; however, some differences in the gross morphology are due to species-specific features.

The system is composed of paired elements: solenostomes (insemination or sperm induction pores), major and minor ducts, emboli, calyces and vesicles (terminology according to Fain 1963).

The sperm access system opens by two solenostomes. They are simple invaginations of the cuticle, surrounded by a circular cuticular fold and located on each ventral side of the body, close but in front of coxae IV (fig. 1B). These pores lead into the major duct. This is a short cuticular duct with thick walls. In some specimens the duct (as well as the calyx) may contain dense material, which likely represents remnants of the transferred spermatophore (figs. 2A-B). In *Typhlodromus*, the cuticular wall of the major duct shows, shortly behind the solenostome, peculiar cuticular folds (fig. 2A), while in *Phytoseiulus* it remains smooth.

The major duct cuticle consists of several layers and its epithelium is flat showing endoplasmic reticulum, roundish mitochondria and Golgi bodies. Nuclei are irregular in outline with scattered chromatin.

The so-called minor duct starts in a modified region in the major duct defined as embolus. Here the lumen of the major duct is very much reduced by a strong cuticular protrusion or plug. The minor duct starts from the center of this protrusion (figs 1D, 2B).

Scanning electron microscopic observations of the macerated system exhibit a kind of depression showing several rings in the cuticle and the thin minor duct arising from the center of this depression (fig. 1D).

At its beginning, the minor duct lumen is hardly visible, rather it appears as a thin dense strand (diameter about 40 nm) surrounded by thick and uniform cuticle (fig. 2B). Further along its course, a very narrow lumen (diameter around 100 nm) becomes more evident (fig. 2C). In *Typhlodromus* it looks even smaller (diameter 30 nm) than in *Phytoseiulus*; moreover, in *Typhlodromus* it is always empty while in *Phytoseiulus*

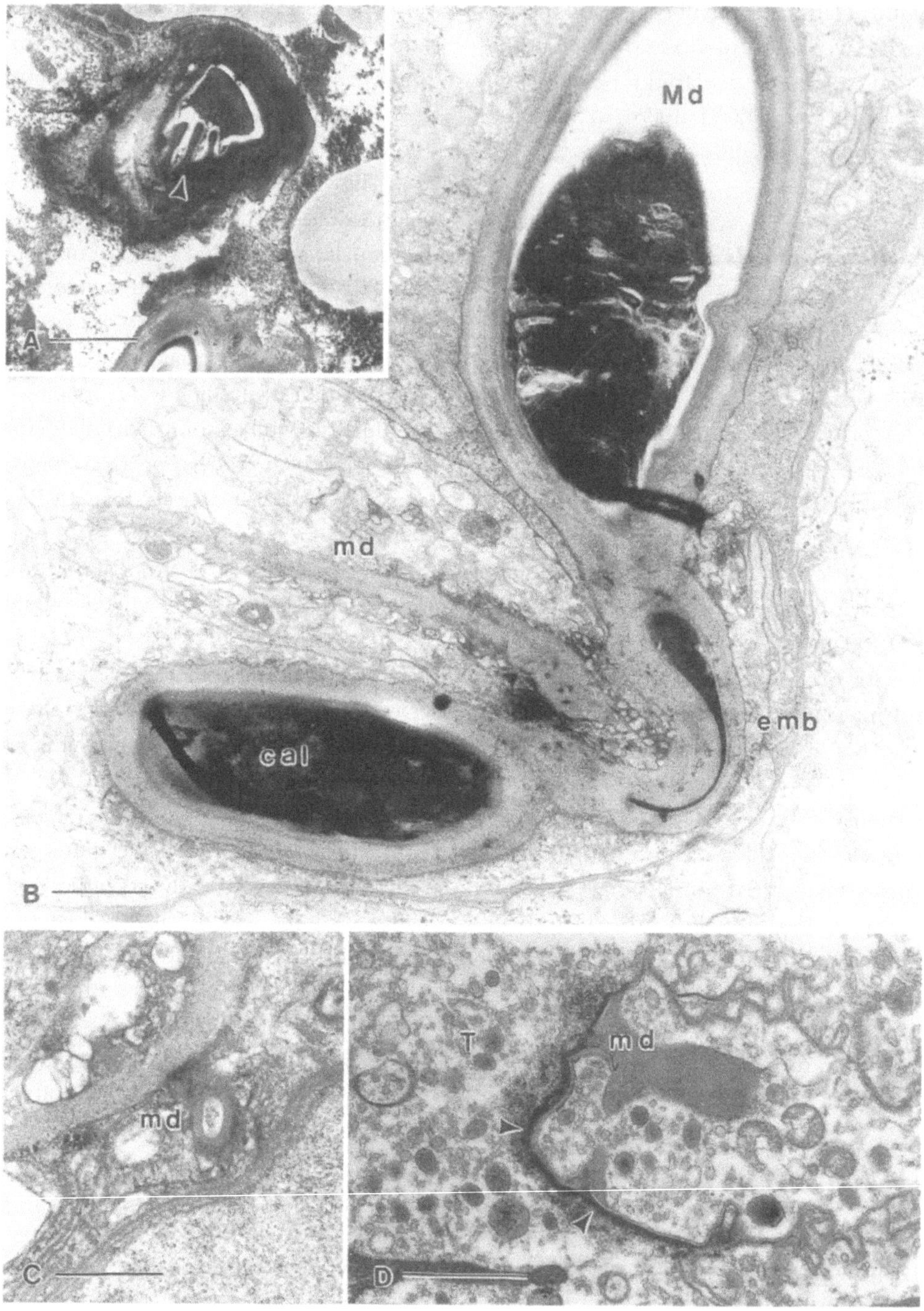

Fig. 2 - TEM. A: *Typhlodromus rhenanoides*. The major duct with cuticular folds (arrowhead) is characteristic for this species. B: *Phytoseiulus persimilis*. Embolus with plug and minor duct. Note the very thin "lumen" of the latter. If sperm cells would really move through the minor duct to the ovary they would have to move back from the vesicle via the calyx to find an opening in the center of the plug. The dense material within major duct and calyx likely represents remnants of a transferred spermatophore. C: *Phytoseiulus persimilis*. Cross section through the minor duct close to the vesicle. Note the very narrow lumen of the duct. D: *Phytoseiulus persimilis*. Section of the end of the cuticular minor duct. Note extense electronlucent tissue containing numerous vesicles and surrounding the irregular folds of the cuticular intima (arrowheads) of the minor duct. All scale bars=1 μm. Abbr.: cal, calyx; emb, embolus; Md, major duct; md, minor duct; T, electronlucent tissue.

some diffuse, not identified material, can be present in the duct. The cytoplasm surrounding the cuticular duct is continuous i.e. there are no cell borders. Some roundish mitochondria, endoplasmic reticulum and an irregularly shaped nucleus with diffuse chromatin are seen.

The minor duct runs towards the ovary: at the beginning it is mainly directed dorsad, then it runs posteriad close to the walls of calyx and vesicle. Finally, it moves towards the dorsally located lyrate organ (Alberti 1988; Nuzzaci *et al.* 2001). The cuticular minor duct ends in the region where the lyrate organ meets the ovary. Close to its end, the minor duct lumen slightly enlarges and its cuticular wall becomes thinner, more irregular and finally ends with many folds (fig. 2D). But there is no opening towards the haemocoel nor a direct connection to the ovary or the lyrate organ. The surrounding cytoplasm widens into a more extensive electronlucent tissue adjacent to the basal lamina of the lyrate organ. This electronlucent tissue extends dorsally along the inner side of the ovary and reaches the corresponding tissue coming from the other side in order to build up a kind of transverse belt (fig. 5). Some electronlucent nuclei, many lysosome like structures and heterogeneous inclusions are present in this tissue while no cell borders have been observed (fig. 3A).

After the beginning of the minor duct, the major duct widens to produce the calyx that is also cuticle lined. In both species the calyx is quite long, in contrast to the rather short major duct, and directed dorsally and a bit posteriorly (fig. 5). At least in *Phytoseiulus*, its diameter gets smaller for a short distance, then it widens again to continue into the vesicle (fig. 1C).

In cross sections through the calyx, it has usually a circular shape that may also be deformed by pressures from adjacent tissues. The cuticle is composed of three layers: a thin compact and uniform external and a similar internal layer separated by a thick less compact layer. More interiorly, the cuticle becomes less thick and is represented by only a single homogeneous ring in cross sections. In *Phytoseiulus,* the cuticle is more uniform. The lumen of the calyx of some specimens contains some heterogeneous material (figs 2B, 3B). The calyx continues into the vesicle that is also cuticle lined but less sclerotized hence it frequently is much folded.

The vesicle wall shows different appearance, presumably depending on the mating state. In *Typhlodromus* specimens, it was always folded and apparently composed of an electrondense cuticle of varying thickness; the underlying flat epithelium shows a very irregular apical region with many microvilli-like protrusions (fig. 4A). In *Phytoseiulus*, some females, especially if just mated, show at least one vesicle with roundish shape containing presumably remnants of the endospermatophore. The cuticle of the vesicle seems to be thinner than in *Typhlodromus* and the epithelium varies. Sometimes it is very flat and inconspicuous while in other specimens it is, at least in some regions, thick with extensive dark inclusions or vacuoles together with some endoplasmic reticulum, free ribosomes and mitochondria (figs 3B, 4B).

The contents of the vesicles can also vary. In *Typhlodromus* it is an unstructured material of different electron density (fig. 4A). In *Phytoseiulus* the content is very much more complex, apparently due to the presence or absence of an endospermatophore that may also be of different age. So it was possible to observe a vesicle that contained an electronlucent material surrounded by a thick, multilayered envelope. Another one included dense material surrounded by a thick rather homogeneous wall (fig. 4B). In another female there was a very much heterogeneous material surrounded by a very thin layer. Since the mating state of the females was not exactly determined, e.g, the investigation was not yet based on virgin females, it is not possible to correlate alterations in the vesicles with times after mating. Sperm cells have never been distinguished with certainty inside the vesicle.

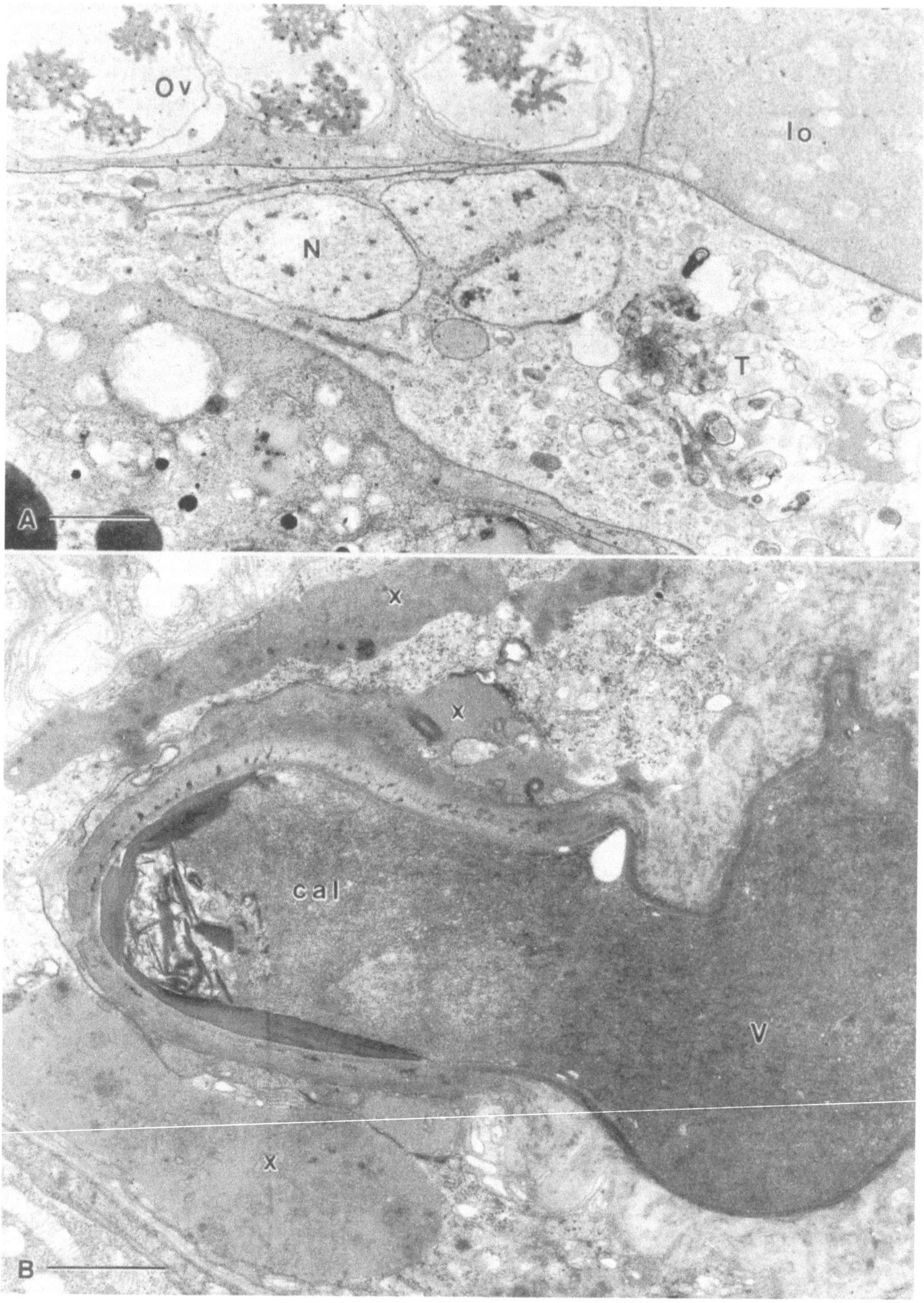

Fig. 3 - TEM. *Phytoseiulus persimilis*. A: detail of electronlucent tissue. Note three nuclei not separated by cell membranes from each other, hence indicating syncytial character of the tissue. Furthermore, there are many heterogeneous inclusions, presumably lysosomes. Scale bar=1 µm. B: section of the calyx opening into the thin-walled vesicle. Note extensive vacuoles (x) in the epithelium underlying the cuticle. Scale bar=2 µm. Abbr.: cal, calyx; lo, lyrate organ; Ov, peculiar cells in the ovary; N, nucleus; T, electronlucent tissue; V, vesicle; x, vacuoles in vesicle/calyx epithelium.

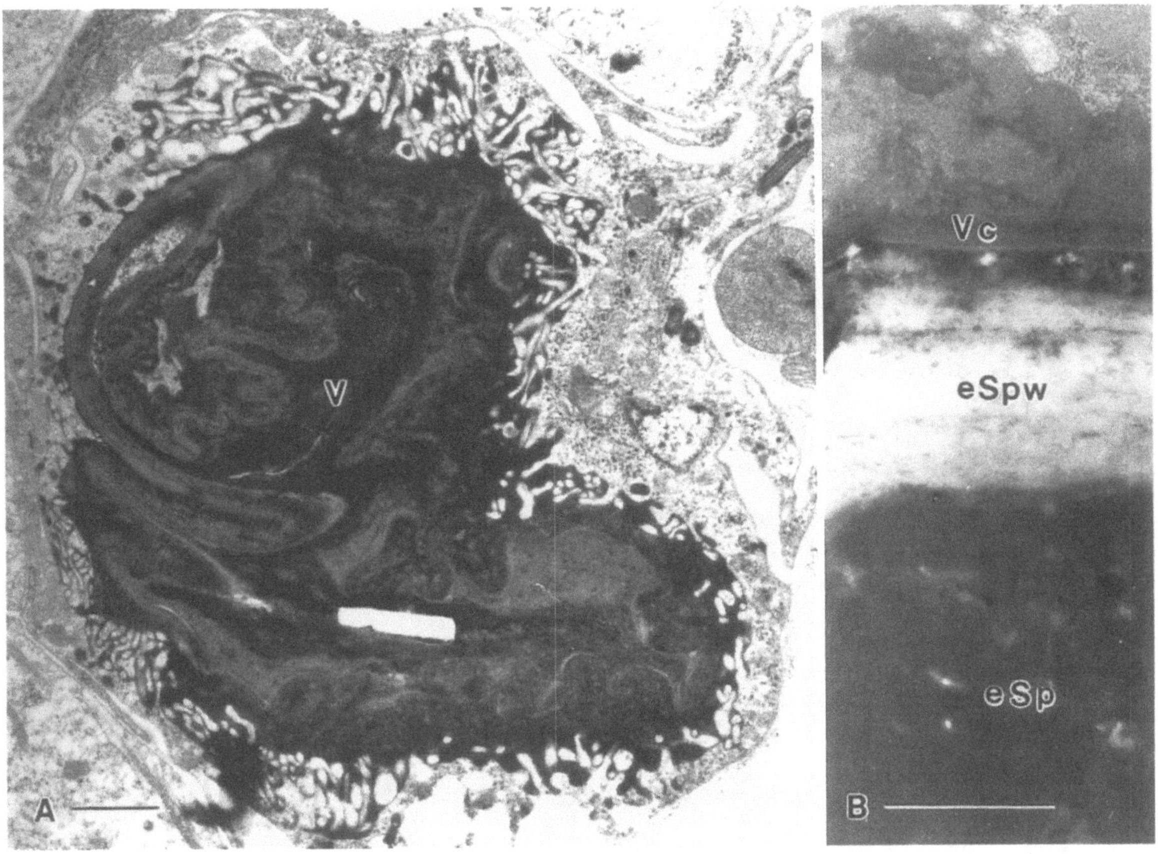

Fig. 4 - TEM. A: *Typhlodromus rhenanoides*. In this overwintering female the vesicle is highly folded with an epithelium showing numerous microvilli-like processes. The layered and folded material presumably represents the cuticular intima. The light bar is an artifact. B: *Phytoseiulus persimilis*. Detail of a vesicle inflated by the endospermatophore. Note lucent wall of endospermatophore, its dense heterogeneous contents and the thin cuticular intima of the vesicle and the flat underlying epithelium. Both scale bars=1 μm. Abbr.: eSp, endospermatophore; eSpw, wall of endospermatophore; V, vesicle; Vc, cuticular intima of vesicle.

Discussion and conclusions

Phytoseiid mites, as all Dermanyssina, exibit an insemination mode called "podospermy" in that the male transfers the spermatophores into the female via copulatory pores (solenostomes, insemination pores or sperm induction pores). This contrasts with "tocospermy" where the female genital opening is used both for insemination and oviposition. In Dermanyssina, the insemination pores are connected to a complex sperm access system. In this group two types of sperm access systems are known: a laelapid and a phytoseiid type (Evans 1992; Alberti and Coons 1999).

From light microscopic investigations, it is known that the endospermatophores are placed in the vesicle, via the major duct, by the male provided with modified chelicerae (Michael 1892; Dosse 1959; Amano and Chant 1978; Krantz and Wernz 1979). According to Amano and Chant (1978), in *Phytoseiulus* females that were fixed during copulation or soon after, the spermatophore is still present inside the vesicle. On the other hand, in overwintering females of *Typhlodromus*, vesicles always show a folded shape of their walls and the vesicles appear to be largely empty.

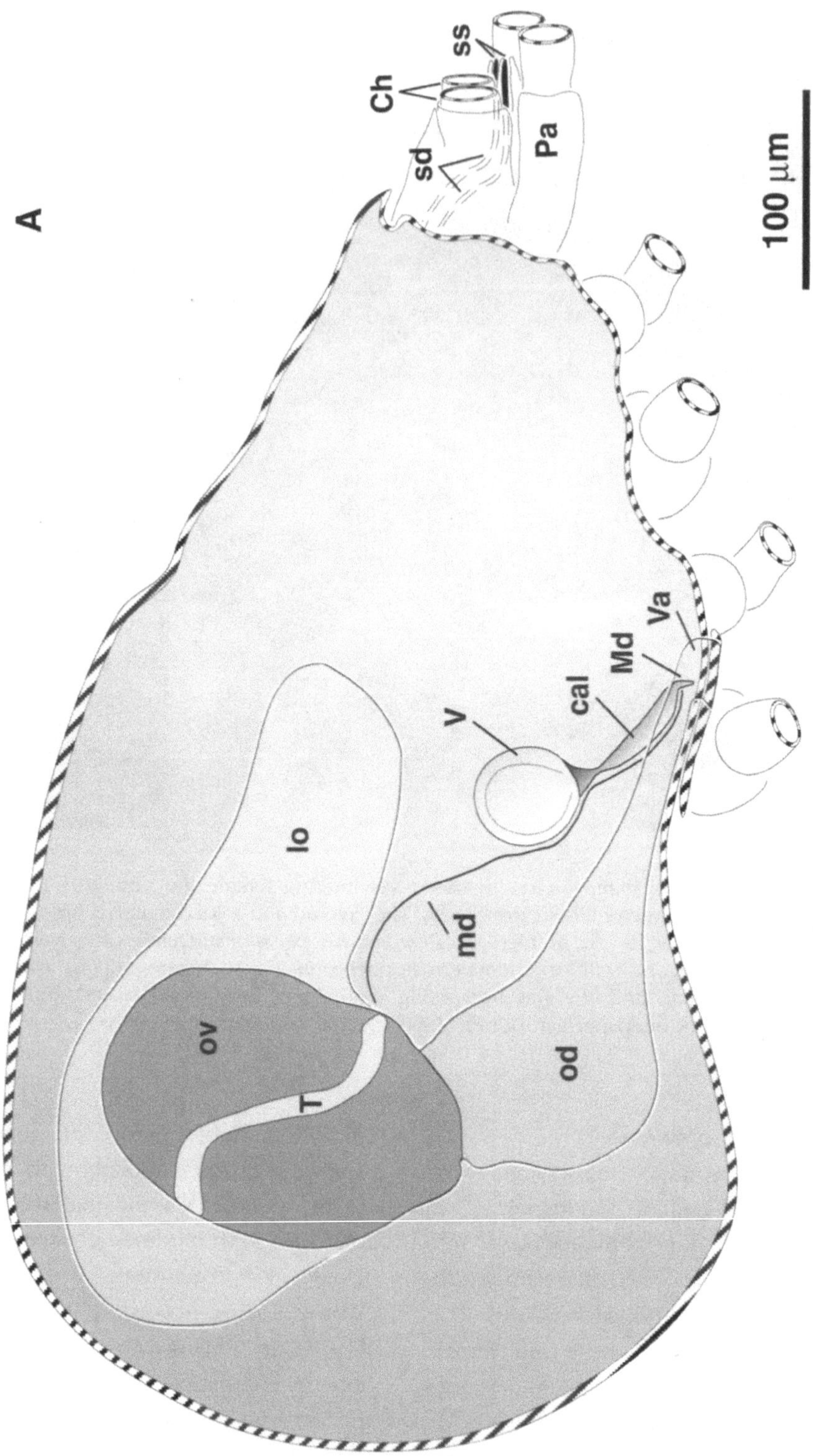

Fig. 5 - Reconstruction of the female genital system of Phytoseiidae. A: lateral view. B: dorsal view. The vesicle of the left body side is filled, that of the right side empty. Abbr.: cal, calyx; Ch, chelicerae; lo, lyrate organ; Md, major duct; md, minor duct; od, oviduct; ov, germinative part of ovary; Pa, pedipalps; sd, salivary ducts; ss, salivary styli; T, electron lucent tissue; V, vesicle; Va, vagina.

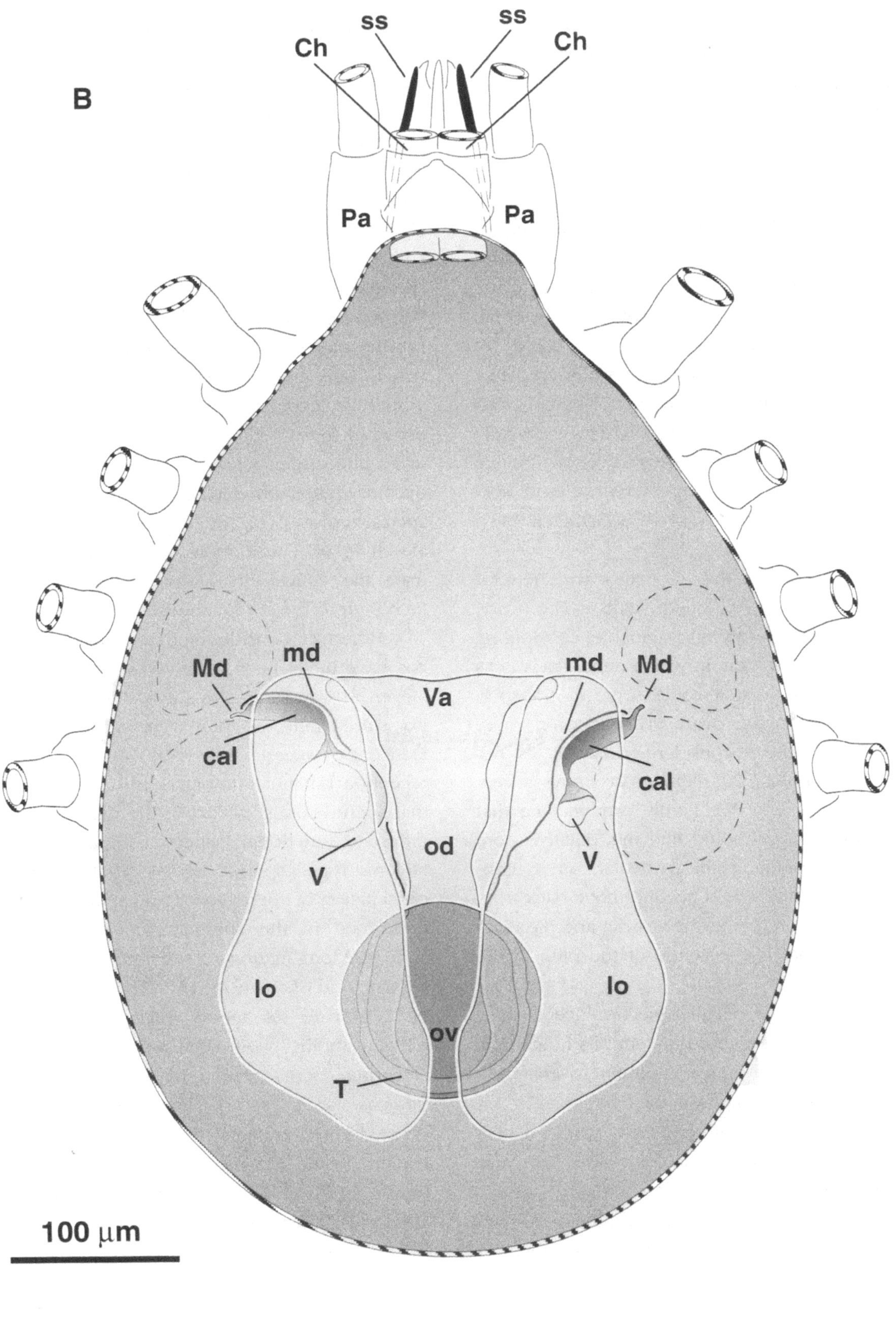
B
ss
ss
Ch
Ch
Pa
Pa
Md
md
md
Md
Va
cal
cal
V
od
V
lo
lo
ov
T
100 μm

Anyway, the mechanism of sperm transfer and the fate of the spermatophore and sperm within the female is still not clear in phytoseiids. In fact, considering the small lumen of both, solenostome and major duct, and the cuticular plug at the level of minor duct insertion (embolus) that narrows the major duct lumen, a dilatation must be possible to allow endospermatophore introduction into the vesicle. On the other hand in all mating or just mated females no distinct spermatozoa are observed inside the vesicle but only a very electrondense material, sometimes differently arranged, that could represent the endospermatophore walls and may camouflage sperm cells. In fact, Dosse (1959) already pointed out that this structure is strongly "chitinized" and not destroyed even during maceration of the whole mite with hot lactic acid (see also Amano and Chant 1978; Schulten 1985; Alberti and Coons 1999).

The minor duct, that has been traced here for the first time in its full length, leads to the ovary. Anyway, it seems very unimaginable, considering its very narrow lumen, in which never sperm has been observed, that it could serve to transport spermatozoa to the ovary. This is even more difficult to believe considering that at its beginning (embolus) the minor duct shows a very fine lumen, if any. We hardly can imagine that intact spermatozoa could find this putative pore and pass through it. In particular, when considering that they would need to move back from the putative storage site, the vesicle, and pass into a putative pore in the centre of the plug in the embolus.

Regarding the peculiar electronlucent tissue that surrounds the minor duct end, its function is still unclear. Moreover, it is important to stress that it joins the one coming from the other side. In this respect, also the phytoseiid-type sperm access system would represent a continuous structure between both body sides as the laelapid-type.

In *Varroa* and Parasitina it is already known (Alberti and Hänel 1986; Alberti 1988; Alberti *et al.* 2000) that spermatozoa reach the ovary penetrating through a cell bridge which connects the spermatheca with the ovary, in the first, and by a haemocoelic route in the second. Anyway it is necessary to point out that in Parasitina spermatozoa are introduced into the (primary) genital opening and that, in *Varroa*, the spermatheca apparently lacks a cuticular intima. In contrast, in *Phytoseiulus* and *Typhlodromus*, vesicle walls are cuticle-lined. It is quite difficult to imagine how spermatozoa could pass through these walls. Furthermore, no sperm cells have been observed in the haemocoel at the region surrounding vesicle walls nor in other areas of the body what would be expected if vesicle wall penetration occurred. However, one should keep in mind that in a quite distant taxon, the spider mites (Tetranychidae, Actinotrichida), sperm penetrate the specialized wall of the receptacle (Alberti and Storch 1976). Furthermore, the passage of sperm from the cuticle lined sperm duct of *Varroa jacobsoni* to the spermatheca is also not known.

Moreover, considering that distinct spermatozoa have never been observed inside the vesicle - even in mated females - it seems that they are not stored here, as they were supposed to be. Hence this organ could play a role in sperm reception but not in storage. Instead, it may be that sperm storage occurs in the ovary. This is a different functional aspect compared to the laelapid-type sperm access system, where capacitation of spermatozoa and sperm storage is performed in the spermatheca. On the other hand, the long mating act reported for the species (Amano and Chant 1978: 131.07±20.83 min) could be a way to let sperm reach the ovary and prevent another male from reproductive success (securing sperm priority by prolonged mating; Thomas and Zeh 1984). The same function could be performed by the "heterogeneous material" present inside the calyx and the vesicle, after transfer of the endospermatophore, and that takes almost 24 hours to be dissolved (Amano and Chant 1978).

Finally, there are some important aspects regarding the functional morphology of the phytoseiid sperm access system that are still not clear: the role played by the complex embolus and minor duct needs further investigation, the function of the peculiar electronlucent tissue is still under speculation, and also the way spermatozoa can move out of the vesicle and reach the ovary needs to be clarified (see Alberti this volume). The possibility that highly modified sperm cells may traverse the epithelium of the vesicle reaching the ovary through a haemocoelic route rather than being moved through the minor duct should be taken into consideration. Moreover an ultrastructural study of the female genital system, especially focusing on sperm capacitation and nutrimentary development of eggs, as well as of the male modified chelicerae (spermatodactyl) and of the spermatophore are very much needed. They certainly will improve a better understanding of the mating, sperm transfer and fertilization processes.

Acknowledgments

This study was financially supported by a grant from the Deutsche Forschungsgemeinschaft (Al 138/4-1) to G.A. Furthermore, both authors received support from the Vigoni-program (CRUI and DAAD). We wish also to express our thanks to Mrs. Ch. Putzar, G. Schittek and Mr. P. Michalik for their skilful technical assistance. Finally, the constructive comments of the referees are gratefully acknowledged.

References

Alberti G., 1988 - Genital system of Gamasida and its bearing on phylogeny. In: Channabasavanna, G.P., Viraktamath C.A. (eds.), Progress in Acarology. Oxford & IBH Publishing, New Delhi, vol 1: 197-204.

Alberti G., Coons L.B., 1999 - Acari-Mites. In: Harrison F.W. (ed.), Microscopic Anatomy of Invertebrates. Wiley-Liss, N.Y., vol 8c.: 515-1265.

Alberti G., Hänel H., 1986 - Fine structure of the genital system in the bee parasite *Varroa jacobsoni* (Gamasida: Dermanyssina) with remarks on spermiogenesis, spermatozoa and capacitation. *Experimental and Applied Acarology,* 2: 63-104.

Alberti G., Storch V., 1976 - Ultrastruktur-Untersuchungen am männlichen Genitaltrakt und an Spermien von *Tetranychus urticae* (Tetranychidae, Acari). *Zoomorphologie,* 83: 283-296.

Alberti G., Gegner A., Witaliñski W., 2000 - Fine structure of the spermatophore and spermatozoa in inseminated females of *Pergamasus* mites (Acari: Gamasida: Pergamasidae). *Journal of Morphology,* 245: 1-18.

Amano H., Chant D.A., 1978 - Mating behaviour and reproductive mechanisms of two species of predaceous mites, *Phytoseiulus persimilis* Athias-Henriot and *Amblyseius andersoni* (Chant) (Acarina: Phytoseiidae). *Acarologia,* 28: 201-220.

Athias-Henriot C., 1971 - Nouvelles notes sur les Amblyseiini (Gamasides Podospermiques, Phytoseiidae). 1. La dépilation des genuaux et tibias des pattes. *Acarologia,* 13: 4-15.

Dosse G., 1958 - Die Spermathecae, ein zusätzliches Bestimmungsmerkmal bei Raubmilben. *Pflanzenschutzberichte,* 20: 1-11.

Dosse G., 1959 - Über den Kopulationsvorgang bei Raubmilben aus der Gattung *Typhlodromus* (Acari, Phytoseiidae). *Pflanzenschutzberichte,* 22: 125-133.

Evans G.O., 1992 - Principles of Acarology. CAB International, Wallingford, UK: 563 pp.

Fain A., 1963 - La spermathéque et ses canaux adducteurs chez les acariens mésostigmatiques parasites des voies respiratoires. *Acarologia,* 5: 463-479.

Karg W., 1982 - Diagnostik und Systematik der Raubmilben aus der Familie Phytoseiidae Berlese in Obstanlagen. *Zoologische Jahrbücher, Abteilung für Systematik, Ökologie und Geographie der Tiere,* 109: 188-210.

Karnovsky M.J., 1965 - A Formaldehyde-Glutaraldehyde fixative of high osmolarity for use in electron microscopy. *Journal of Cell Biology,* 27: 137-138.

Krantz G.W., Wernz J.G., 1979 - Sperm transfer in *Glyptholaspis americana*. In: Rogriguez J.G. (ed.), Recent Advances in Acarology. Academic Press, N.Y., vol. 2: 441-446.

Michael A.D., 1892 - On the variations in the internal anatomy of the Gamasinae, especially in that of the genital organs and their mode of coition. *Transactions of the Linnean Society*, London (ser. 2*),* 5: 281-324.

Nuzzaci G., Di Palma A., Aldini P., 2001 - Functional morphology and fine structure of the female genital system in *Typhlodromus* spp. (Acari: Phytoseiidae). In: Halliday R.B., Walter D.E., Proctor H.C., Norton R.A., Colloff M.J. (eds.), Proceedings of the 10th International Congress of Acarology. CSIRO Publishing, Melbourne: 196-202

Reynolds E.S., 1963 - The use of lead citrate at high pH as electron opaque stain in electron microscopy. *Journal of Cell Biology,* 17: 208-212.

Schulten G.G.M., 1985 - Mating. In: Helle W., Sabelis M.W. (eds.), Spider Mites - Their Biology, Natural Enemies and Control. World Crop Pests. Elsevier, Amsterdam, vol 1B: 55-65.

Schuster R.O., Smith L.M., 1960 - The spermatheca as taxonomic features in Phytoseiid mites of Western North America. *Proceedings of Entomological Society of Washington,* 62: 181-188.

Thomas R.H., Zeh D.W., 1984 - Sperm Transfer and Utilization Strategies in Arachnids: Ecological and Morphological Constraints. In: Smith R.L. (ed.), Sperm Competition and the Evolution of Animal Mating Systems. Academic Press, New York: 179-221.

Wainstein B.A., 1973 - On the structure of some organs of Phytoseiidae (Parasitiformes) important for taxonomy. *Zoologichevsky Zhurnal,* 52: 1871-1872 (in Russian).

Fine morphology of the bothridial receptor organs of *Pediculaster mesembrinae* (Acari: Siteroptidae) adult female

Enrico de Lillo[1], Pasquale Aldini[2]
[1] *Dipartimento di Biologia e Chimica Agroforestale e Ambientale, Facoltà di Agraria, Università degli Studi di Bari, via Amendola, 165/A, I-70126 Bari, Italy, delillo@agr.uniba.it*
[2] *Dipartimento di Oftalmologia e Otorinolaringoiatria, Facoltà di Medicina e Chirurgia, Policlinico, P.zza G. Cesare, I-70124 Bari, Italy*

Key words: mycophagy, Heterostigmata, sensilla, ultrastructure, mechanoreceptors, vibroreceptors.

Abstract: Using electron microscopy, the setiform sensilla on the prodorsal shield in a phoretomorphic female of *Pediculaster mesembrinae* (Canestrini) (Pygmephoroidea, Siteroptidae) are described and illustrated. The simple setae (first and second pair of vertical setae - v_1, v_2, second pair of scapular setae - sc_2) and the trichobothrium (first pair of scapular setae - sc_1) have a solid shaft and smooth surface; their dendrites have tubular bodies connected to their shaft bases. According to their ultrastructural details, they can be classified as aporous sensilla: the simple setae are tactile receptors while the trichobothrium is assumed to be involved in vibration perception. Morphological differences with caeculid and oribatid trichobothria are pointed out.

Introduction

Our knowledge on the fine structure and function of sensilla in mites, as opposed to ticks, remains limited. Function is largely inferred from structure and there has been little experimental work of the type undertaken, for example, on Haller's organ in ticks. The receptor organs of heterostigmatic mites have been little studied except for the solenidia on the leg tibio-tarsus of some siteroptids (Camerik and Coetzee *in press*; de Lillo and Aldini 2001).

The present paper deals with the fine structure of the vertical and scapular setae on the prodorsal shield of a phoretomorphic female of *Pediculaster mesembrinae* (Canestrini) (Pygmephoroidea, Siteroptidae). An attempt is made to interpret the function of these sensilla from their fine structure.

Materials and methods

Living phoretomorphic females of *P. mesembrinae*, collected from mushroom substrates of *Pleurotus eryngii* (D.C. ex Fr.) Quél., were studied by transmission electron microscopy. The specimens were dissected using a microscalpel and the anterior part of the body was fixed in Karnowsky's solution, rinsed in cacodylate buffer, postfixed in 1% buffered osmium tetroxide, rinsed as above, dehydrated in an ethanol series and embedded in Araldite M. Ultrathin transverse and longitudinal sections were obtained using a diamond knife on an ultrotome III LKB 8800, stained with 5% uranyl acetate and lead citrate, observed and micrographed using a Zeiss EM109T transmission electron microscope.

Freshly killed specimens of *P. mesembrinae* were observed using a stereoscan Cambridge S100.

F. Bernini, R. Nannelli, G. Nuzzaci, E. de Lillo (eds.), Acarid Phylogeny and Evolution. Adaptations in mites and ticks, 253-261.

Sensilla were described and interpreted mainly according to Grandjean (1935), McIver (1985), and Lindquist (1986).

Explanations of abbreviations used in the figures: bs, bothridial seta; cd, ciliary constriction; ch, chamber; cu, cuticle; ds, dendritic sheath; hs, setal shaft; l, first leg; la, lamella; m, mitochondrion; mt, microtubules; n, nucleus; pr, peritreme; r, rim; sc_1, scapular external seta (trichobothrium); sc_2, scapular internal seta; sf, suspension fibres; tb, tubular body; v_1, vertical external seta, v_2, vertical internal seta.

Results

The common set of sense organs on the prodorsal shield of a phoretomorphic female in *P. mesembrinae* consists of three pairs of simple setae (first and second pair of vertical setae - v_1, v_2, second pair of scapular setae - sc_2) and one pair of trichobothria (first pair of scapular setae - sc_1) (figs 1-2). Both vertical setae are just posterior to the peritremes: v_2 is antiaxially and v_1 is mesally located. Both scapular setae are antiaxial: sc_2 is well ahead of the rear shield margin and sc_1 lies between the vertical setae and sc_2.

Vertical (v_1, v_2) and scapular (sc_2) setae

Both kinds of setae are provided with a solid and hair-like shaft (fig. 1) and its lumen is not innervated (fig. 3). The hair shafts do not display any pores on their walls and tips while they are minutely spiculated over most of their length (figs 1-3). The shaft base is set into a flexible socket resembling a sort of tubercle on the prodorsal shield (figs 1, 4). Each of the two dendrites ends with a conspicuous tubular body which is laterally connected to the shaft base (figs 5-6). The dendritic sheath of the vertical setae is granular, irregular in shape and thick (fig. 5), while the dendritic sheath of the scapular setae is very thin (fig. 6). Two ciliary constrictions have been detected, and those of the v_1 display a microtubule arrangement composed of nine and eleven sets of peripherally located doublets (fig. 7). One sheath cell has been observed.

Scapular seta sc_1

Its cuticular component consists of a rigid hair (bothridial seta) whose distal part is expanded and globose ('capitate' type) such as in other heterostigmatic mites (figs 1-2). The hair shaft stands straight and erect in a deep cup-like socket (bothrium or bothridium). The bothridial setae are anteriorly and antiaxially oriented, divergent and pointed more or less obliquely into the air, describing an angle of about 100° between their axis on the median plane (figs 1-2). The part extending beyond the socket is about 15 μm long, it lacks pores and its lumen is not innervated. The distal part of the shaft is mostly elliptical in transverse section and is composed of three concentric layers. Its surface is provided with few ridges and furrows (fig. 8). Close to the bothridium and inside the bothridial lumen, the shaft is thin and almost circular in cross section; at most four concentric layers have been detected (figs 9-10).

The bothridial opening has a barely protruding cuticular rim (figs 1, 11), and it is at least three times larger than the shaft base diameter. The bothridial walls are provided with projections, which resemble thin lamellae (figs 10-12). They delimit up to four chambers which are superimposed and directed inwards from the prodorsal shield. The shaft passes straight through the chambers. The distance between the hair shaft and the lamellar borders becomes narrower towards the setal base. Similarly, the chambers become gradually smaller towards the bothridial base. The chamber surface is smooth and provided with very few ribs. Its wall contains dense areas (figs 13-14) which seem to be relatively irregular aggregations of 'tubules'. Their lumen is electron lucent and their walls are strongly electron dense. These structures are surrounded by an amorphous substance and some of them display micropores

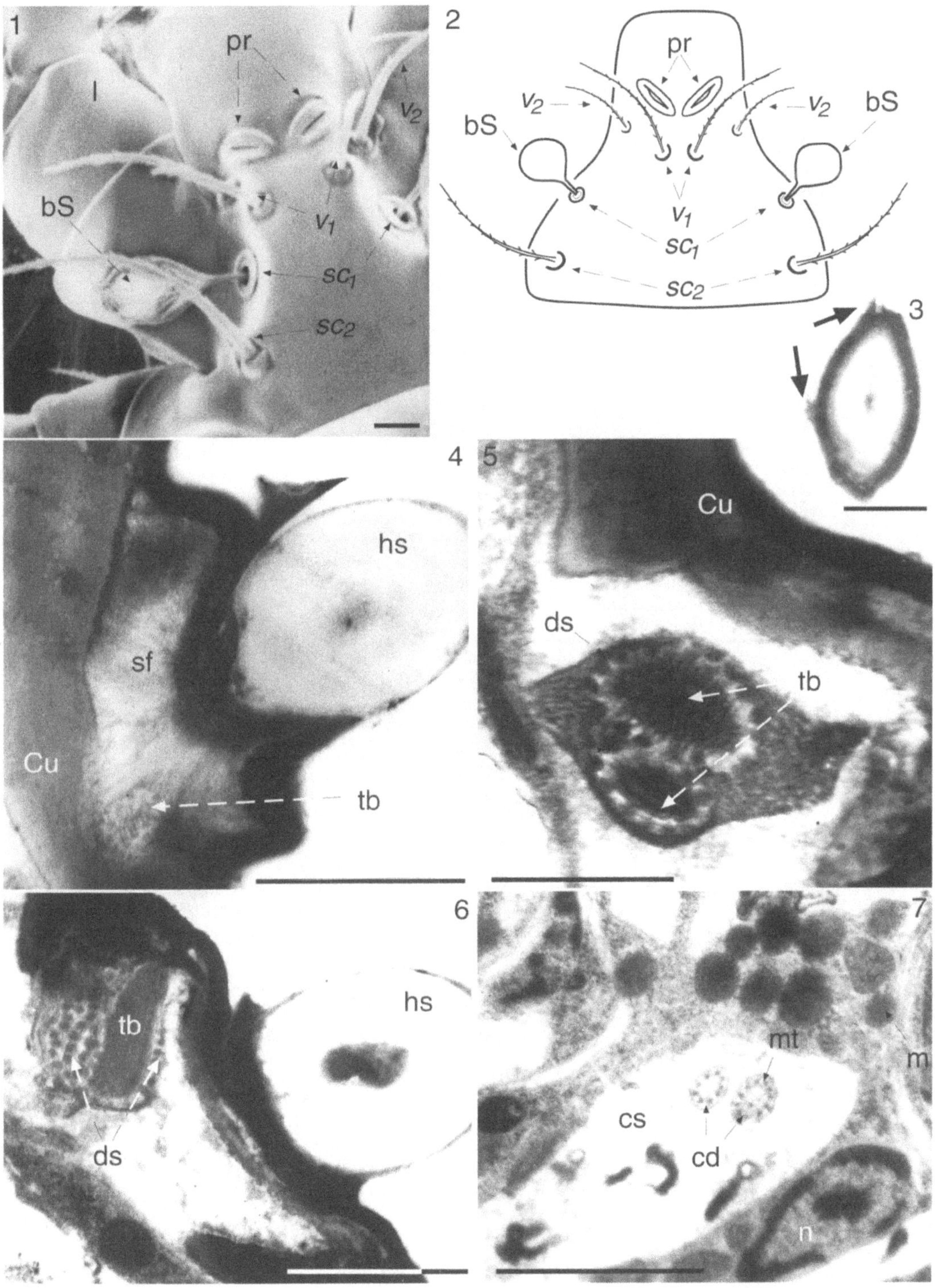

Figs 1-7 - *Pediculaster mesembrinae* (Canestrini): 1) scanning electron micrograph of the propodosoma in a dorsal view; 2) semischematic drawing of the prodorsal shield and its sense organs; 3-8) transmission electron micrographs showing transverse sections of the simple setae (3) through the v_2 setal shaft, (4) at the level of the setal base v_2 and its socket, (5) details of v_2 tubular bodies, (6) at the level of sc_2 socket, (7) at the level of v_1 ciliary region. Scale bar=5 μm for fig 1, and 1 μm for figs 3-7. Arrows indicate spicules.

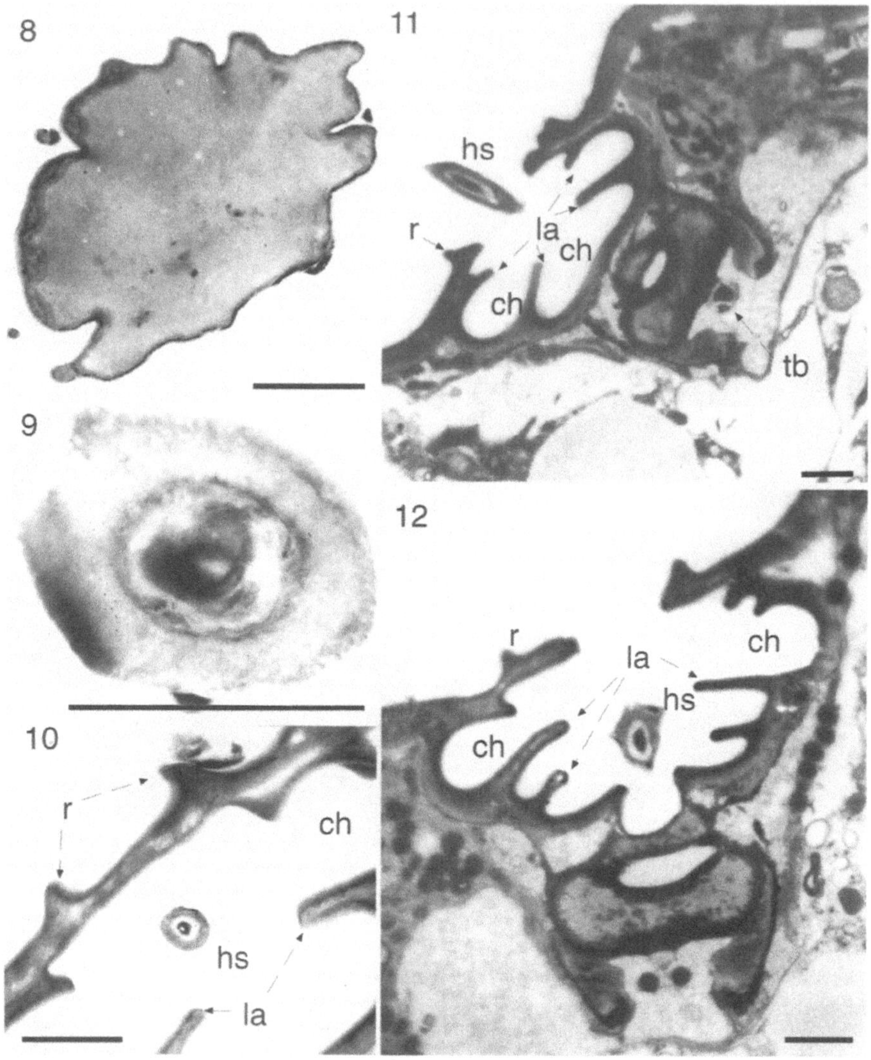

Figs 8-12 - Transmission electron micrographs of *Pediculaster mesembrinae* (Canestrini) trichobothrium: 8-9) transverse sections (8) of the bothridial seta at level of the globose part and (9) at a more proximal level than the previous figure; 10-12) tangential sections showing the relationships between the bothridial shaft, the socket chambers and their lamellae. Scale bar=1 μm.

on the chamber surface (fig. 14). It was not possible to correlate these structures with glands and no material was found inside the socket or adhering to the socket surface. The proximal part of the hair shaft is connected to an articulating membrane provided with radiating fibres (figs 15-18). Two receptor cells are located under the setal base, each terminating with a tubular body (fig. 19). The microtubules of the tubular bodies appear immersed in a moderately electron dense matrix and the bundle appears to be surrounded by an electron lucent area (figs 19-22). These structures are ensheathed in a very thin electrondense tube. Only one tubular body seems to reach the setal base (fig. 19) and both tubular bodies are eccentrically encased in a peculiar and large dense structure (figs 19-22) which derives from the bothridial cuticle and forms a sort of inverted capsule (figs 11-12, 15-23). This capsule is elliptical in cross section and its walls are not uniform in

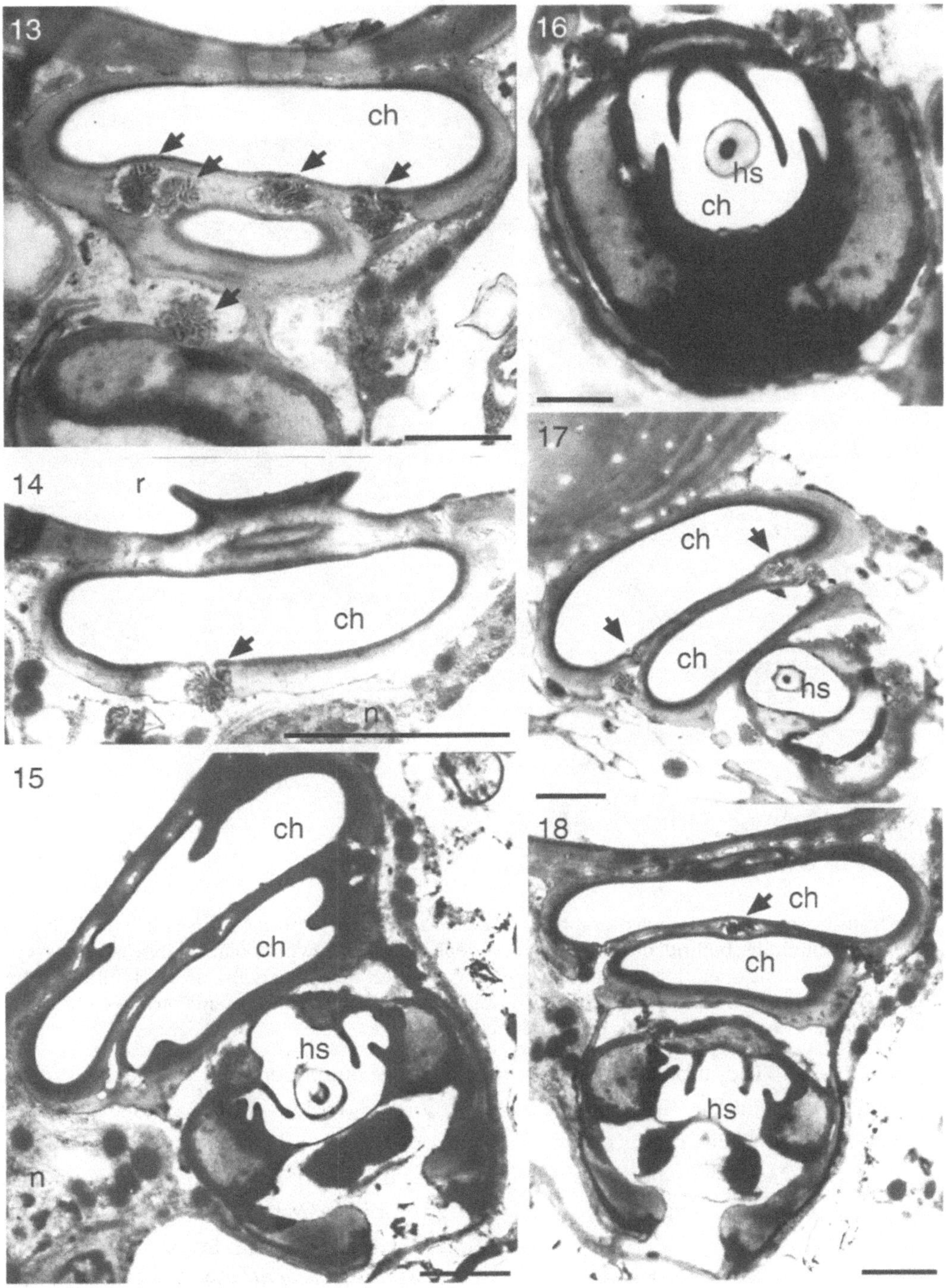

Figs 13-18 - Transmission electron micrographs of *Pediculaster mesembrinae* (Canestrini): various aspects of the bothridium. Tangential sections in figs 13-15 and 17-18, oblique transverse section in fig. 16. 13-14) Detail of the dense areas of lamellar 'tubules'. 15-18) sections through the innermost bothridial chamber, showing the hair shaft base. Arrows indicate dense areas of 'tubules' through the bothridial lamellae. Scale bar=1 μm.

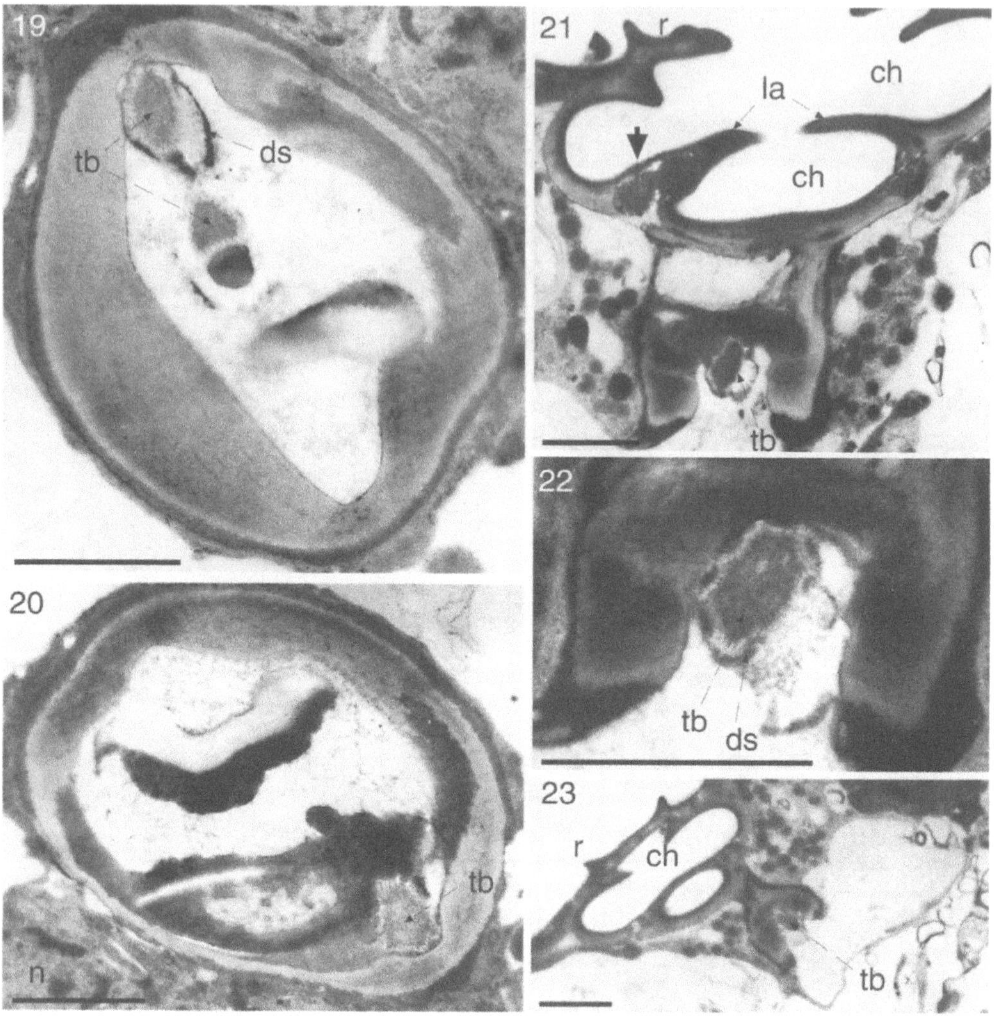

Figs 19-23 - Transmission electron micrographs of *Pediculaster mesembrinae* (Canestrini) trichobothrium: 19) transverse section through the bothridial base showing a dense 'cup' and two tubular bodies; 20) transverse section more distal than the previous one; 21) tangential section of the socket revealing dense areas of 'tubules', the inverted cup and the related tubular body; 22) detail of the tubular body connection; 23) tangential section of the socket. Arrows indicate dense areas of 'tubules'. Scale bar=1 μm.

thickness (figs 19-20). The tubular bodies are embedded in a large sensillar sinus containing flocculent material most likely produced by the enveloping cells (fig. 23). It was not possible to discriminate enveloping cells completely.

Discussion and conclusions

The ultrastructural features of the vertical setae and scapular setae *sc2* (i.e. the presence of an aporous hair shaft innervated by receptor cells whose outer dendritic segments terminate with tubular bodies connected to the shaft base) correspond to those of mechanoreceptive sensilla which have been extensively studied in other arthropods (Mills 1973; Foelix and Chu-Wang 1973; McIver 1985; Evans 1992; Nuzzaci *et al.* 1992; de Lillo *et al.* 1996; Alberti 1998; Alberti and Coons 1999).

The trichobothrium exhibits a similar innervation (fig. 24). It is characterized by a large socket composed of regularly superimposed

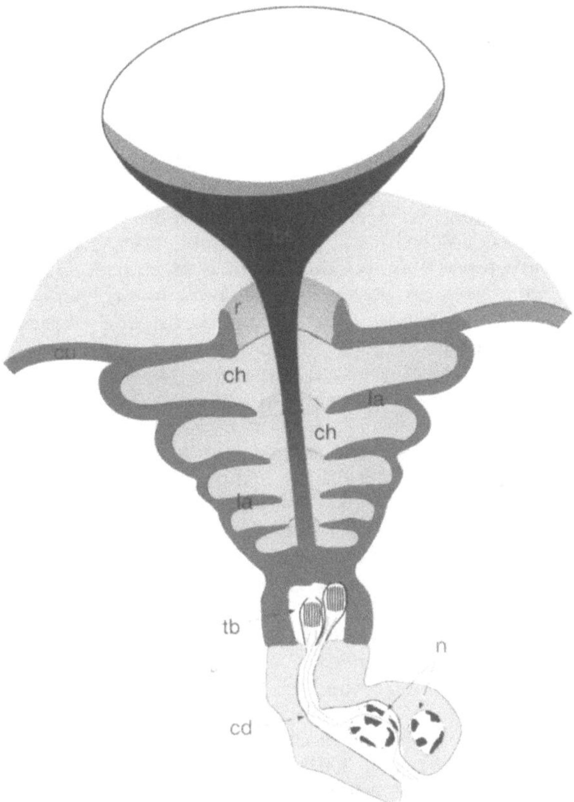

Fig. 24 - Semischematic drawing showing the bothridial arrangement.

chambers (partly reported by Smiley 1978, in fig. 48 at page 146, drawn using a light microscope), and a capitate bothridial seta with an aporous shaft. Trichobothria occur widely in insects and other arachnids (McIver 1985). They have been found on the body and legs of Actinotrichida, but are rare in the Anactinotrichida (more references in Alberti and Coons 1999). Despite their widespread occurrence in a variety of actinotrichid mites, they have been studied mostly for their external morphology. Haupt and Coineau (1975), Alberti *et al.* (1994 and 1995), and Alberti (1998), using scanning and transmission electron microscopes, described and illustrated trichobothria occurring on an actinedid mite (Caeculidae) and several oribatids.

The fine structure of the trichobothrium in *P. mesembrinae* (fig. 24) displays differences when compared to those in the caeculids (Haupt and Coineau 1975) and oribatids (Alberti 1998), especially in its cuticular components. In comparison with *P. mesembrinae*, the socket of some oribatids has a more elaborate and tortuous arrangement of the chambers, which are also provided with cerotegumental material and evident cuticular ribs. The caeculid bothridium is delimited by an elevated rim but is not chambered. The siteroptid setal shaft stands straight in its socket and is not curved as in most oribatids. It is possible that the bothridial seta in *P. mesembrinae* can be deflected in all directions on the basis of the shape of the setal base and its relationship with the bothridial chambers. On the other hand, the tubular body of the hair shaft is arranged in an electron dense structure (probably of cuticular origin) apparently elongated in one direction. This suggests a directional sensitivity of the receptor, as mentioned for caeculid and oribatid sensilla. Therefore, the *P. mesembrinae* tubular bodies appear to be almost engaged in a sort of cup, which is continuous with the socket walls and peculiar for this mite. Moreover, the bothridium forms a sort of bell in which the bothridial seta works as a clapper. In our opinion, the shaft might also press on the chamber lamellae during its movement and in this way induce a stress not only directly to the tubular bodies through the setal base but also indirectly by means of the whole socket. This action might cause a socket deformation that would be perceived by the encased tubular bodies. This effect should be non uniform but of variable intensity and direction depending on the point of contact of the hair shaft on the lamellae.

The meaning and function of the dense areas of 'tubules' at the level of the lamellar cuticle remain to be ascertained. One might speculate that they are extremely complex pore canals or deep depressions of the epicuticle. Otherwise they might be related to dermal glands involved in the secretion of substances that allow the chambers to keep 'clean' by preventing the accumulation of debris which would form an obstacle to shaft movement (Alberti *et al.* 1994; Alberti 1998; Alberti *in litt.*).

Finally, the ultrastructural peculiarities of *P. mesembrinae* trichobothrium suggest a vibro- and anemoreceptive function, and this interpretation is in agreement with the conclusions of previous studies (Haupt and Coineau 1975; McIver 1985; Alberti *et al.* 1994 and 1995; Alberti 1998). Vice versa, there are not enough data to compare the current results with the Wergin *et al.* (2000) hypothesis, which suggested that this sensillum may work as a capacitor that could be deflected by the electromagnetic field produced into the column of a low temperature scanning microscope. As in other arthropods, the trichobothrium is able to detect slight air currents. Moreover, spider prey localization (Barth 1982), suggest to us that this siteroptid, tipically phoretic, might utilize the trichobothrium to detect the specific frequency of the wing vibrations produced by flies. In fact, the socket chambers look like an acoustic resonator and their lamellae might vibrate for resonance to a particular sound frequency. This device might increase the perception of the vibrations and the cuticular frame of the socket might transmit a stimulus to the tubular bodies connected to the bothridium. Finally, the mite could combine the vibroreception with the infochemical perception to identify its vector.

Acknowledgements

This research was partly supported by MURST and MiRAF grants. We are grateful to Dr. H. Dastych (Germany) for species identification, Prof. G. Alberti (Germany), G.O. Evans (United Kingdom), E.E. Lindquist (Canada), G. Nuzzaci, and M. Solinas (Italy) for their helpful critical comments. The senior author has planned the research, analyzed and interpreted the data; the junior author has mainly provided for the preparation of the iconographic material.

References

Alberti G., 1998 - Fine structure of receptor organs in oribatid mites (Acari). In: Ebermann E. (ed.), Arthropod Biology: contributions to morphology, ecology and systematics. Biosystematics and Ecology Series, *Österreichische Akademie Wissenschaften*, Wien, 14: 27-77.

Alberti G., Coons L.B., 1999 - Acari - Mites. In: Harrison F.W. (ed.), Microscopic anatomy of Invertebrates. New York, John Wiley and Sons, Inc., vol. 8C: 515-1265.

Alberti G., Moreno A.I., Kratzmann M., 1994 - The fine structure of trichobothria in moss mites with special emphasis on *Acrogalumna longipluma* (Berlese, 1904) (Oribatida, Acari, Arachnida). *Acta Zoologica*, Stockholm, 75(1): 57-74.

Alberti G., Moreno A.I., Kratzmann M., 1995 - Fine structure of trichobothria in moss mites (Oribatida). In: Kropczynska D., Boczek J., Tomczyk A. (eds.), The Acari - Physiological and ecological aspects of Acari - Host relationships. Dabor, Warszawa: 23-30.

Barth F.G., 1972 - Spiders and vibratory signals: sensory reception and behavioural significance. In: Witt P.N., Rovner J.S. (eds.), Spider communication. Mechanisms and ecological significance. Princeton University Press: 66-122.

Camerik A.M., Coetzee S.H., *in press* - Skeletal structure of solenidion omega-1 in *Pediculaster australis*, seen through the confocal laser scanning microscope (CLSM). Proceedings I African Acarological Symposium, Pretoria, November 1998.

de Lillo E., Aldini P., 2001 - Functional morphology of some leg sense organs in *Pediculaster mesembrinae* (Acari: Siteroptidae) and *Phytoptus avellanae* (Acari: Phytoptidae). In: Halliday R.B., Walter D.E., Proctor H.C., Norton R.A., Colloff M.J. (eds.), Proceedings of the 10th International Congress of Acarology. CSIRO Publishing, Melbourne: 217-225.

de Lillo E., Nuzzaci G., Aldini P., 1996 - Fine morphology of the mouthpart sensilla in females of *Typhlodromus exhilaratus* Ragusa (Phytoseiidae). In: Mitchell R., Horn D.J., Needham G.R., Welbourn W.C. (eds.), Acarology IX, Ohio Biological Survey, Columbus, Ohio, vol. 1: 287-295.

Evans G.O., 1992 - Principles of Acarology. CAB International, Wallingford, UK, 563 pp.

Foelix R.F., Chu-Wang I.-W., 1973 - The morphology of spider sensilla. I. Mechanoreceptors. *Tissue and Cells*, 5(3): 451-460.

Grandjean F., 1935 - Les poils et les organes sensitif potrés par les pattes et le palpe chez les Oribates. Première partie. *Bulletin de la Société Zoologique de France*, 60: 6-39.

Haupt J., Coineau Y., 1975 - Trichobothrien und Tastborsten der Milbe *Microcaceculus* (Acari, Prostigmata, Caeculidae). *Zeitschrift für Morphologie der Tiere*, 81: 305-322.

Lindquist E.E., 1986 - The world genera of Tarsonemidae (Acari: Heterostigmata): a morphological, phylogenetic, and systematic revision, with a reclassification of family-group taxa in the Heterostigmata. *Memoirs of Entomological Society of Canada*, 136: 517 pp.

McIver S.B., 1985 - Mechanoreception. In: Kerkut G.A., Gilbert L.I. (eds.), Comprehensive Insect Physiology Biochemistry and Pharmacology. VI. Nervous system: sensory. Pergamon Press: 71-132.

Mills L.R., 1973 - Structure of dorsal setae in the two-spotted spider mite *Tetranychus urticae* Koch 1936. *Acarologia*, 15: 649-658.

Nuzzaci G., de Lillo E., Porcelli F., 1992 - Functional morphology of the mouthpart sensilla in females of *Varroa jacobsoni* Oudemans (Acari: Varroidae). *Entomologica*, Bari, 27: 41-67.

Smiley R.L., 1978 - Taxonomic studies of *Pygmephorus* species from the western hemisphere, with a key to females and an overview of the current problems for classification (Acari: Pyemotidae and Pygmephoridae). *International Journal of Acarology*, 4(2): 125-160.

Wergin W.P., Ochoa R., Erbe E.F., Joy D.J., 2000 - Oscillating trichobothria in the low temperature SEM: biologica capacitors or charging artifacts. *Scanning*, 22(2): 140-141.

Fine structure of a putative pedipalpal thermo/hygrosensillum in *Tarsonemus nodosus* Schaarschmidt, 1959 (Acari: Tarsonemidae)

G. Nuzzaci[1], A. Di Palma[2], F. Porcelli[1], W.L. Magowski[3], P. Aldini[1]

[1] *Department of Agro-Forestry and Environmental Biology and Chemistry, University of Bari, via Amendola 165/A, I-70126 Bari, Italy*

[2] *Agricultural Faculty, University of Foggia, via Napoli 25, I-71100 Foggia, Italy*

[3] *Department of Animal Taxonomy and Ecology, A. Mickiewicz University, Szamarzewskiego 91 A, PL-60569 Poznan, Poland*

Key words: Heterostigmata, sensilla, functional morphology.

Abstract: A putative thermo/hygroreceptor has been detected in *Tarsonemus nodosus* Schaarschmidt, 1959 by means of scanning and transmission electron microscopy. It is dome-shaped and located on the distal-lateral part of each pedipalp. The receptor shows a thick and electrondense cuticular wall, moreover no articulation or specialized socket has been observed. As no wall pores have been found, it is inferred that it belongs to the type NP-sensilla with inflexible socket. The dendritic sheath is thick and no outer receptor lymph sinus has been detected. According to their morphology, it was possible to distinguish at least two receptor cells. One presents extensive lamellations of the dendrite whose outer segment fills the dome completely with stacks of parallel lamellae. Moreover in the dendritic outer segment, microtubules are present. The other dendrite terminates at the base of the dome. According to the morphological and structural organization of the described structure and the homology with those described in other arthropods, and particularly in ticks, its thermo and hygroreceptor function is hypothesized.

Introduction

Most land arthropods have the ability to differentiate between environmental changes in temperature and/or humidity. Even though the presence of hygro- and thermorceptors in ticks and mites was suggested a long time ago (Wilkinson 1965), there is still little known about the physiology and ultrastructure of these receptors.

A few studies on the ultrastructure and electrophysiology of these sensilla (Waladde *et al.* 1981; Hess and Loftus 1984; Jorgensen 1985; Hess and Vlimant 1986) and some behavioural observations (Lees 1946) to deduce the presence of hygro- and thermoreceptors, have been carried out on ticks as they are of medical and veterinary importance and due to the role these sensilla play in host finding.

Regarding tarsonemid mites, nothing about this is known, perhaps abetted by of their minute size (150-300 μm) which limits the electrophysiological observations so that the role of the sensilla can be inferred only from their structure and location. In order to improve the knowledge on these scarcely studied sensorial structures, in the current paper an ultrastructural description of a presumed thermo/hygroreceptor on pedipalps of *Tarsonemus nodosus* Schaarschmidt, 1959 is reported.

F. Bernini, R. Nannelli, G. Nuzzaci, E. de Lillo (eds.), Acarid Phylogeny and Evolution. Adaptations in mites and ticks, 263-267.

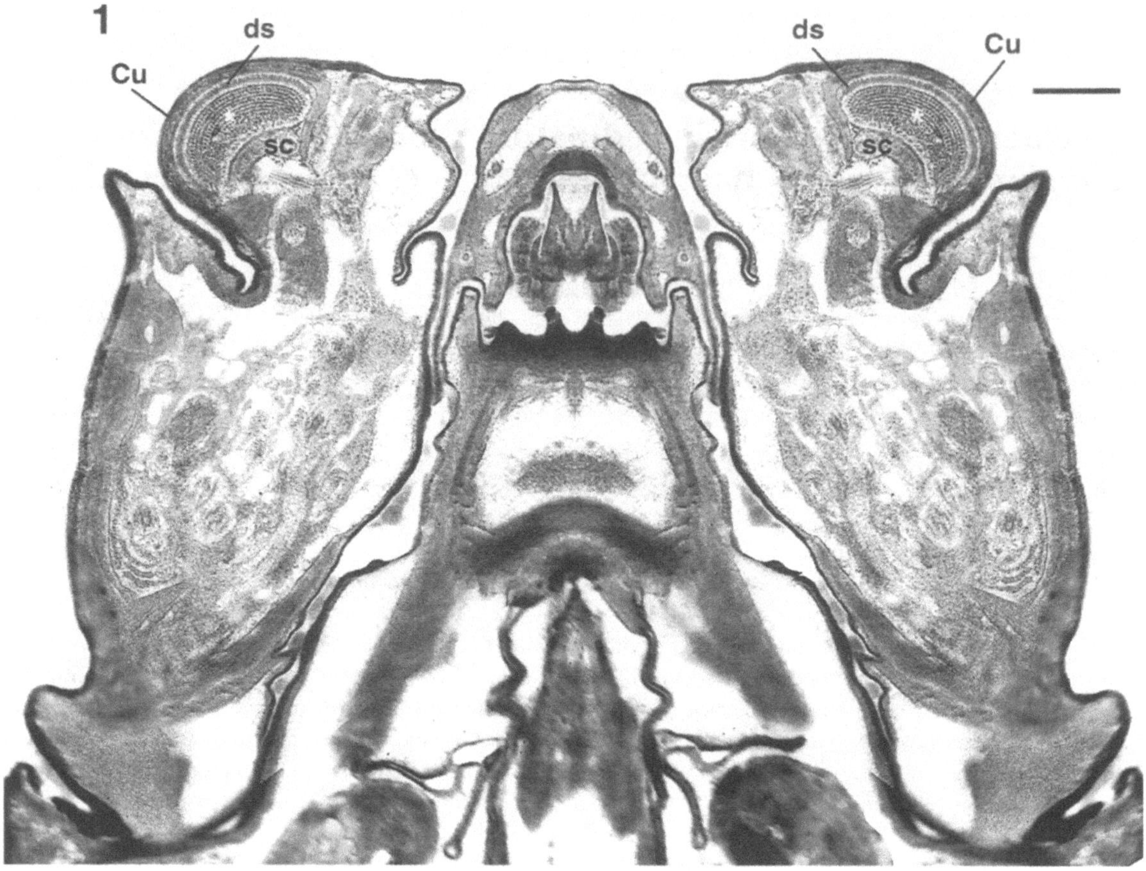

Fig. 1 - *Tarsonemus nodosus*, electron micrograph: reconstruction of a frontal section at the level of the dome-shaped structures. Scale bar=7 µm. Abbr: Cu, cuticle; ds, dendritic sheath; sc, sensory cell.

Materials and methods

Females of *Tarsonemus nodosus* Schaarschmidt, 1959 collected from apple calices in Poznan (Poland), were used for the present study. The specimens were pre-fixed in Karnovsky's (1965) solution, post-fixed in Osmium tetroxide, dehydrated with graded ethanols and finally embedded in Araldite M medium.

Ultrathin sections were obtained by means of a LKB ultrotome III and observations were performed with a Zeiss EM 109T transmission electron microscope (TEM).

Moreover, some specimens treated as already described up to absolute ethanol, were dried with a critical point apparatus and then observed using a stereoscan Cambridge S100 (SEM).

Results

A dome-shaped structure is observed by scanning electron microscopy, latero-distally on each pedipalp. The structure appears rounded and smooth (figs 1-2).

In TEM sections, the cuticular wall of the dome is fairly massive (0.56 nm), multilayered and moderately electron-dense. No wall pores have been observed. Moreover there are no suspension fibres or specialized socket cuticle at the base of the dome (figs 1, 3).

Beneath the cuticle of each dome, a conspicuous arrangement of folded lamellae is observed (figs 1, 3). These lamellae are formed by the outer dendritic segment (fig. 4); they are folded and intertwined so that sometimes more

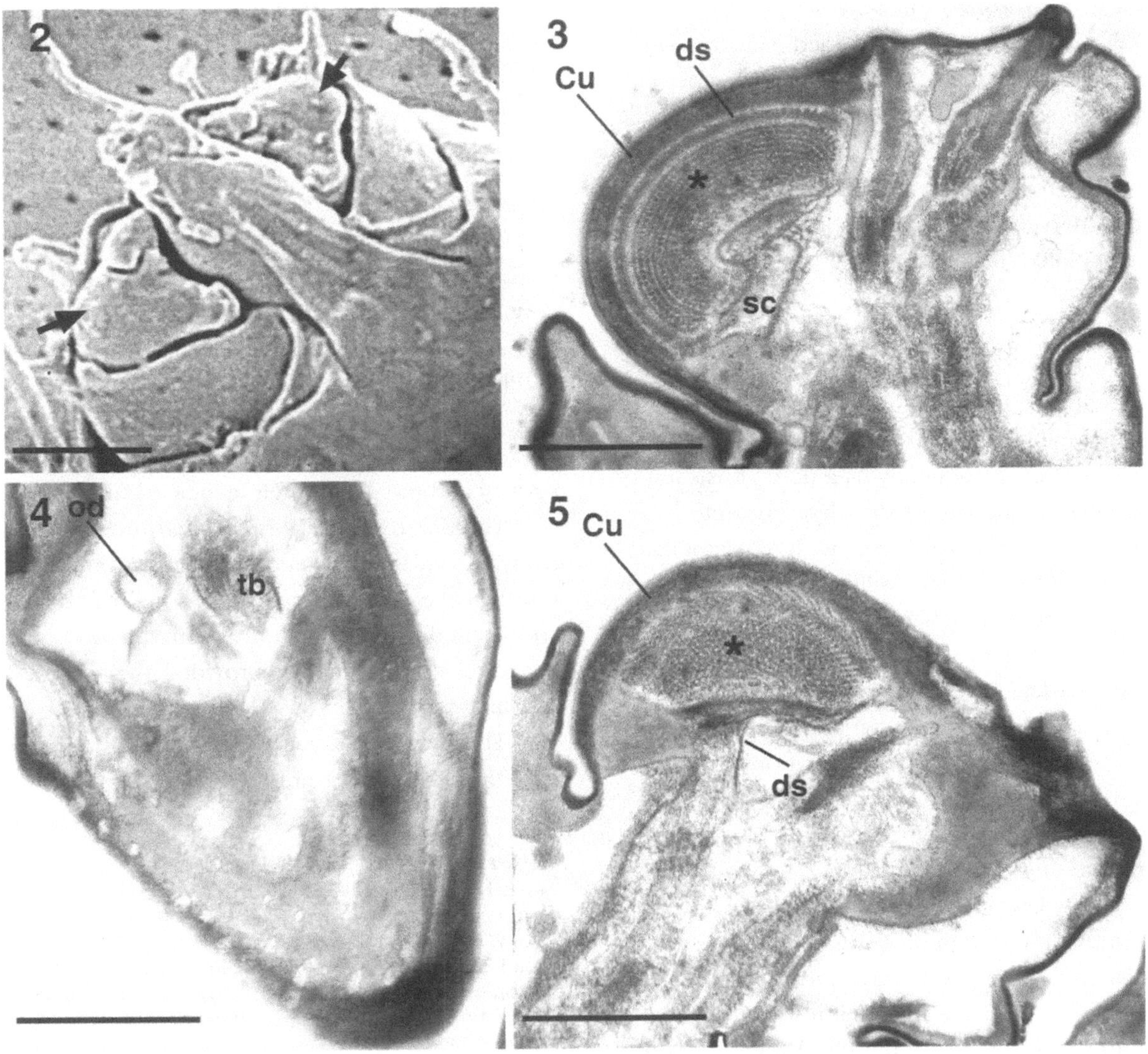

Figs 2-5 - *Tarsonemus nodosus*, electron micrographs: 2) scanning electron micrograph showing a ventral view of the pedipalps and the dome-shaped structures (arrows). Scale bar=2.5 µm; 3) tangential section showing the arrangement of folded lamellae (asterisk), the tick dendritic sheath and the second sensory cell. Scale bar=3.5 µm; 4) tangential section at the level of the outer dendritic segment before its lamellation. Scale bar=3.5 µm; 5) tangential section of the dome-shaped sensillum. Note the dendritic sheath contiguous with the cuticular wall. Scale bar=6.25 µm. Abbr: Cu, cuticle; ds, dendritic sheath; od, outer dendritic segment; sc, sensory cell; tb, tubular body.

than ten lamellar profiles appear in a given section to fill the dome completely. Within the intercellular thin spaces, an electrondense material, which seems to interconnect the membranes, is visible. In the dendritic outer segment, microtubules are present (figs 1, 3).

Beneath the stacks of lamellae, no outer receptor lymph sinus is observed while a thick dendritic sheath, continuous with the cuticle of the dome, is present (figs 3, 5).

It seems that another sensory cell reaches the dome and terminates somewhat below its base. Unfortunately it has been not possible to ascertain this aspect (fig. 3).

Discussion and conclusions

The observed dome-shaped sensory structure belongs to the no pore sensillum type with inflexible socket (Altner and Loftus 1985) in that its cuticular wall lacks pores and a socket. Moreover it is noteworthy that there is a sensory cell that terminates with a highly lamellated dendrite to completely fill the lumen of the dome (figs 1, 3, 5).

Poreless sensilla have often been shown to be thermo- or hygrosensitive (Hallberg 1979; Altner and Prillinger 1980; Hess and Vlimant 1983, 1986; Steinbrecht 1984). Moreover thermo/hygroreceptors occur frequently as a triad consisting of two antagonistic (moist/dry) hygroreceptors and one, the thermosensitive, with a lamellated dendrite (Altner and Prillinger 1980). In our observations it has been possible to detect the lamellated dendrite and an unbranched cell but it is still unclear if another sensory cell completes the triad (figs 1, 3). Hess and Vlimant (1986) reported that in ticks the innervation of no pore hygro-thermosensitive sensilla can be variable and suggested that either the warm or the cold sensitive unit could be lacking.

Regarding the shape of sensilla housing thermo- and hygroreceptors, there is great variability: trichoidea, basiconica, coeloconica, styloconica, sensillum capitulum and coelocapitulum, have been reported (Altner and Loftus 1985). Altner *et al.* (1978) and Hallberg (1979) described two dome-shaped thermo-hygroreceptors even though in both cases a short hair or a peg is present in the middle of the dome. Our dome-shaped sensillum does not have any peg-like structure but it could resemble the rearrangement of a very short and broad peg.

Unfortunately there is still little known about thermo/hygroreceptors in arthropods and especially in mites because these sensilla are usually inconspicuous and few in number.

In conclusion, the observed sensilla, even though variable in their gross morphology and still lacking of some ultrastructural details, could fit well in the category of thermo/hygroreceptors. In fact the poreless cuticular wall with the inflexible socket and the evident lamellated outer dendritic segment are typical features.

Acknowledgements

The authors are very much grateful to: Dr. E. de Lillo (University of Bari, Italy) for his critical contributions; to Prof. Hanna Mazur, Prof. Barbara Tomaszewska, Ms. Liliana Ciszewska, Department of Plant Metabolism, Institute of Molecular Biology and Biotechnology (A. Mickiewicz University, Poznan, Poland) for the hospitality in their laboratory and the use of their technical equipments.

The first three authors planned the research, analyzed and interpreted the observations, the fourth collected the specimens, identified the species and provided a critical contribution to the text, the last has mainly provided the technical assistance.

This study was partly supported by MURST ex 40% and a grant from the University of Bari.

References

Altner H., Loftus R., 1985 - Ultrastructure and function of insect thermo- and hygroreceptors. *Annual Review of Entomology*, 30: 273-295.

Altner H., Prillinger L., 1980 - Ultrastructure of invertebrate chemo-, thermo-, and hygroreceptors and its functional significance. *International review of Cytology*, 67: 69-139.

Altner H., Tichy H., Altner I., 1978 - Lamellated outer dendritic segments of a sensory cell within a poreless thermo- and hygroreceptive sensillum of the insect *Carausius morosus*. *Cell and Tissue Research*, 191: 287-304.

Hallberg E., 1979 - The fine structure of the antennal sensilla of the pine saw fly *Neodiprion sertifer* (Insecta: Hymenoptera). *Protoplasma*, 101: 111-126.

Hess E., Loftus R., 1984 - Warm and cold receptors of two sensilla on the foreleg tarsi of the tropical bont tick *Amblyomma variegatum*. *Journal of Comparative Physiology*, 155: 187-195.

Hess E., Vlimant M., 1983 - The tarsal sensory system of *Amblyomma variegatum* Fabricius (Ixodidae, Metrastriata). II. No pore sensilla. *Revue Suisse de Zoologie*, 90: 157-167.

Hess E., Vlimant M., 1986 - Leg sense organ of ticks. In: Sauer J.R., Hair J.A. (eds.), Morphology, physiology and behavioral biology of ticks. Ellis Horwood, Chichester: 361-390.

Jorgensen W.K., 1985 - The ultrastructure of the sense organs of tarsus I, the palps and chelicerae of the larval cattle tick, *Boophilus microplus* (Canestrini) Ixodidae including a study of the ultrastructure and function of the tick podium. Ph.D. Thesis, University of Queensland, Australia.

Karnovsky M.J., 1965 - A Formaldehyde-Glutaraldehyde fixative of high osmolarity for use in electron microscpy. *Journal of Cell Biology*, 27: 137-138.

Lees A.D., 1946 - Chloride regulation and the function of the coxal glands in ticks. *Parasitology*, 37: 172-184.

Steinbrecht R.A., 1984 - Chemo-, hygro-, and thermoreceptors. In: Bereiter-Hahn J., Matoltsy A.G., Richards K.S. (eds.), Biology of the integument. Springer, Berlin, vol. 1: 523-553.

Waladde S.M., Kokwaro E.D., Chimtawi M., 1981 - A cold receptor on the tick, *Rhipicephalus appendiculatus*: electrophysiological and ultrastructural observations. *Insect Science and its Application*, 1: 191-196.

Wilkinson P.R., 1965 - Permeability of the sagittiform sensilla in ixodid ticks, and their possible relation to hygroreception. *The Canadian Entomologist*, 97: 1116-1118.

Mouthparts of *Tarsonemus nodosus* Schaarschmidt, 1959 (Acari: Tarsonemidae): fine structure and functional morphology

G. Nuzzaci[1], A. Di Palma[2], W.L. Magowski[3], P. Aldini[1]

[1] *Department of Agro-Forestry and Environmental Biology and Chemistry, University of Bari, via Amendola 165/A, I-70126 Bari, Italy*

[2] *Agricultural Faculty, University of Foggia, via Napoli 25, I-71100 Foggia, Italy*

[3] *Department of Animal Taxonomy and Ecology, A. Mickiewicz University, Szamarzewskiego 91 A, PL-60569 Poznan, Poland*

Key words: Heterostigmata, gnathosoma, labrum, fixed digits, cheliceral stylets, lever, pharynx, muscles, sensilla.

Abstract: The mouthparts of *Tarsonemus nodosus* Schaarschmidt, 1959 were studied by means of transmission and scanning electron microscopy. The gnathosoma, elongated and conically shaped, has a cheliceral stylophore fully integrated with the subcapitulum to build up a gnathosomal capsule. A dorsomedian apodeme, the vestige of coalesced axial walls of the cheliceral bases, is evident. Modified and fused fixed digits form a dorsal cover of the distal part of the subcapitulum. The movable digits are tiny and stylet-like, each proximally connected to a cheliceral lever. Stylets slide, anteriorly, into two cheliceral sleeves located in the ventral region of the fixed digits; the subcapitulum is provided distally with folds to accommodate the cheliceral stylets. Stylet movement is due to the cheliceral lever where retractor and protractor muscles are inserted. In cheliceral stylet cross sections up to five sensorial structures are detected. The subcapitulum presents a pre-oral groove where variously shaped ridges and furrows are precisely adapted to the corresponding ones of the labrum, such that the food channel is hermetically sealed. Moreover, the dorsal rims of the subcapitulum and the ventral processes of the fixed digits are shaped to assure a tight coalescence. The walls of the pharyngeal chamber appear to be very much electrondense; the pharynx shows only proximally an elastic dorsal surface where dilator muscles are inserted to effect a pump mechanism. The gnathosoma is provided distally with several sensorial structures. According to the gnathosomal general anatomy, a functional interpretation is proposed and compared with that of other phytophagous mites.

Introduction

Tarsonemid mites show a great diversity of feeding habits: fungivores, plant feeders, predators, parasitoids and parasites are present (Lindquist 1986; Kaliszewski *et al.* 1995).

Their economic importance is primarily linked to agriculture, both crops and ornamental plants, where they cause serious injuries much as do the well known Tetranychidae, Tenuipalpidae and Eriophyoidea. Some tarsonemid mites cause distortion of buds, flowers and leaves such as bronzing and wrinkling of leaves mainly on succulent and growing tissues. On the other hand, some other tarsonemid species may be pests of cultivated mushrooms and may disseminate fungi that live in stored food products.

The specialized feeding structures of other phytophagous mites have already been described

F. Bernini, R. Nannelli, G. Nuzzaci, E. de Lillo (eds.), Acarid Phylogeny and Evolution. Adaptations in mites and ticks, 269-281.

in detail (Blauvelt 1945; Summers *et al.* 1973; Alberti and Storch 1974; Hislop and Jeppson 1976; Nuzzaci 1979a and b; Mothes and Seitz 1981; André and Remacle 1984; Alberti and Crooker 1985; Nuzzaci and de Lillo 1989, 1991a, b and c; Nuzzaci 1994; Nuzzaci and Alberti 1996; etc.).

Until now, however, little attention has been directed toward the anatomy and functional morphology of gnathosomal structures in tarsonemid mites. Few investigations, performed by means of light microscopy (Sachs 1951) and scanning electron microscope (Hislop and Jepson 1976; Chow *et al.* 1980) were made, but a complete morphological and functional description is still lacking.

Therefore, the aim of the present study is to consider, in one representative of this mite group, the anatomy, fine structure and functional relationship of the gnathosomal structures in order to clarify the feeding mechanism (penetration of the host surface, production and injection of saliva and food intake).

Materials and methods

Females of *Tarsonemus nodosus* Schaarschmidt were collected from apple calices in Poznan (Poland) prefixed in Karnovsky's (1965) solution for approximately two hours at 4°C and then dissected and left in the fixative over night. They were then rinsed in cacodylate buffer solution, post-fixed in osmium tetroxide, rinsed again in buffer solution, dehydrated with graded ethanols, and finally embedded in Araldite M medium.

Ultrathin sections were obtained by means of a LKB ultrotome III and observations were performed with a Zeiss EM 109T transmission electron microscope.

Some specimens were treated as described above to the absolute ethanol stage, these were then dried with a critical point apparatus. Observations were made using a stereoscan Cambridge S100.

Results

Gnathosoma

The gnathosoma of *T. nodosus* consists of the stylophore and the subcapitulum fully integrated to form a gnathosomal capsule (Lindquist 1986) (fig. 1).

Distally the gnathosoma is conical, comprised of the subcapitular lateral lips and covered dorsally by the fused fixed digits of the chelicerae. The lateral lips are completely fused ventrally while dorsally show folds and variously shaped ridges opposed to the ventral ones of the fixed digits (fig. 2). More proximally, the dorsal surface of the subcapitulum presents two deep lateral furrows, with elevated margins, where the ventrolateral processes of the fixed digits are located and fitted in (figs 3-6). Their dorsal surface is enlarged laterally to partly cover the pedipalps that are shaped so as to be appressed to the fixed digits-subcapitulum complex (figs 4-5). Proximally the subcapitulum is completely fused with the stylophore (figs 6-8). On the external surface it is still possible to distinguish a border line between the stylophore and the subcapitulum (arrow in fig. 1). The stylophore consists of the coalesced cheliceral bases and forms the dorsal and proximal part of the gnathosoma (figs 1, 7-8). The dorsomedian apodeme (Lindquist 1986) includes the median salivary duct and represents a vestige of the vertical septum between the cheliceral segments to indicate the originally paired nature of the stylophore (figs 8-11, 13).

Pre-oral groove, labrum and mouth

The pre-oral groove and the labral morphology change in form at different levels. They do not reach the gnathosomal apex.

Distally the pre-oral groove is wide, represented by a depression with strong electron-dense walls and not covered by the labrum (fig. 2). More proximally, the pre-oral groove is covered by the labrum which is located between the cheliceral

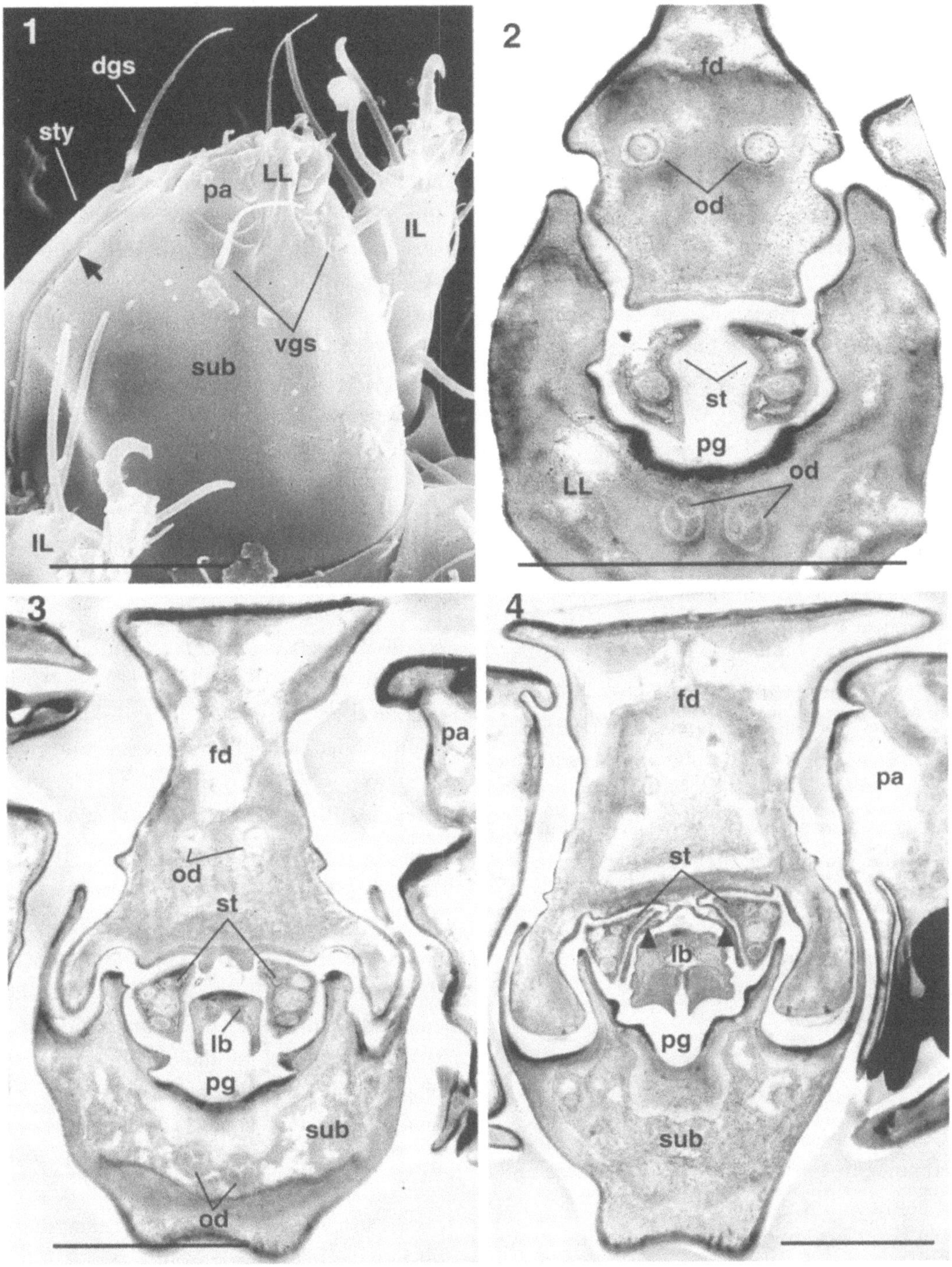

Figs 1-4 - *Tarsonemus nodosus*, electron micrographs: 1) scanning electron micrograph showing a lateroventral view of the gnathosoma (arrow points to the border line between the stylophore and the subcapitulum), scale bar=10 µm; 2) cross section of the gnathosoma at the level of the pre-oral groove, scale bar=1 µm; 3) cross section showing the ventral surface of the fixed digits and the dorsal surface of the subcapitulum reciprocally adapted in shape, scale bar=1 µm; 4) cross section slightly more proximal then the previous, scale bar=1 µm. Abbr: dgs, dorsal gnathosoma seta; fd, fixed digits; IL, first leg; lb, labrum; LL, lateral lips; od, outer dendritic segment; pa, pedipalp; pg, pre-oral groove; st, stylet; sty, stylophore; sub, subcapitulum; vgs, ventral gnathosomal seta.

stylet. The groove is subquadrangular in cross section while the ventral surface is concave (fig. 3). At this level the pre-oral groove is still wide, however it then narrows and deepens. Its lateral walls, which are electrondense, present ridges and processes opposed to the ones of the labrum, which, at this level, is thicker, larger, irregularly shaped and provided with a ventral medial gutter (figs 4-6). A food channel composed of the labrum interlocked with the lateral margins of the pre-oral groove is realized (figs 4-6). Behind the mouth the labrum ends with the cervix that bears a capitular saddle while the pre-oral groove leads into the pharynx (figs 7-8).

Pharynx and pharyngeal chamber

The pharynx is located in the subcapitulum at the level of the cervix. Anteriorly, in cross section, the pharyngeal chamber appears elongated dorso-ventrally; its walls are thickened and electrondense (fig. 7). More proximally the chamber becomes wider but less elongated, and beyond the cervix it is pear shaped (fig. 8). Closely apposed to the pharyngeal lateral walls, there are two electron-dense regions (arrowheads in fig. 8). The posterior region of the chamber is hemicircular in shape, with highly electrondense ventral walls and with a tendineous area on the dorsal surface. Here there is the insertion of a strong tendon connected to the pharyngeal dilator muscles (fig. 9).

Chelicerae

The cheliceral basal segments are fused to form the stylophore (figs 1, 6-7). The fixed digits, also completely fused, make up the distal and dorsal part of the gnathosoma, whereas the movable digits are modified into stylet-like structures (figs 2-6).

In cross section the cheliceral stylets appear sickle-like distally (fig. 2). More proximally they are almost subtriangular in shaped with the paraxial surface slightly concave (figs 3-5) and proximally they are elliptical in cross section (fig. 6).

Distally the stylets are opposed and located in a common groove in the subcapitulum which is sclerified ventrally and bordered by elevated rims (fig. 2). Posteriorly the subcapitulum is provided dorsally with two lateral folds shaped to be appressed to the antiaxial surface of the stylets (fig. 3). The two stylets are separated and held in position by the centrally located labrum (fig. 3). The fixed digits present two folds on their ventral surface which are shaped to fit with the stylets (fig. 3). Moreover from these folds, two ridges arise. They are located between the stylets and produce two laminar processes (figs 3-4). Each process is closely apposed to the paraxial surface of a stylet and more posteriorly joins the lateral ventral wall of the fixed digit (fig. 5). Therefore, at this level, each stylet is completely enclosed in a cheliceral sleeve (figs 5-6).

Proximally each stylet comprises a spatulate sclerite, the cheliceral lever (Lindquist 1986), one in each side of the stylophore (fig. 7). In cross section the levers are located dorsally in the gnathosoma, longer than wide, and with their long axis oblique with that of the stylophore. They bear the attachment of the stylet muscles.

Median salivary duct

A median salivary duct is present posterior to the cervix and located dorsally in the gnathosoma (figs 8-10). The duct is lined with cuticle and embedded in the dorsomedian apodeme (figs 8-10). At the level of the dilator muscles of the pharynx, the duct is enlarged (fig. 10) and then divided into two smaller ducts that extend into the body where their lumens become very much reduced and they diverge laterally (fig. 11). No observations on the posterior tract of the two ducts have been carried out.

Sensorial structures

Two outer dendritic segments are observed at the level of the fixed digits. They are symmetrically located and each contains one micro-

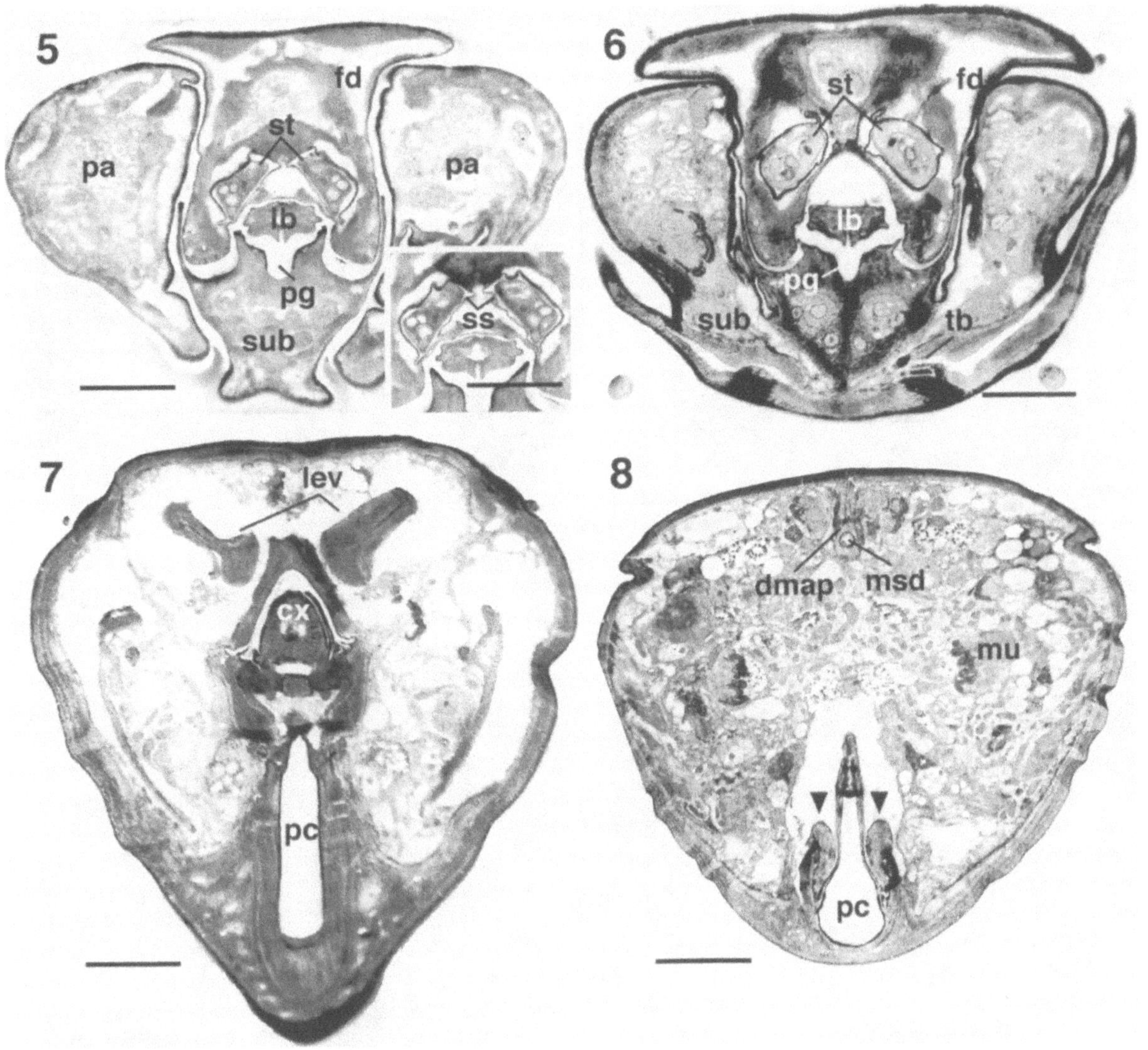

Figs 5-8 - *Tarsonemus nodosus*, electron micrographs of cross sections: 5) showing the gnathosomal complex composed of the fixed digits, the subcapitulum and the pedipalps, scale bar=1 μm. Right bottom: detail of the stylets surrounded by the cheliceral sleeve, scale bar: 1.4 μm; 6) at the level of the subcapitulum partly fused with the stylophore, scale bar=1.4 μm; 7) at the level of the cheliceral levers and the pahryngeal chamber, scale bar=2 μm; 8) showing the pharyngeal chamber and the median salivary duct embedded in the dorso median apodeme, scale bar=3 μm. Abbr: cx, cervix; dmap: dorso median apodeme; fd, fixed digits; lb, labrum; lev, stylet lever; msd, median salivary duct; mu, muscle; pc, pharyngeal chamber; pg, pre-oral groove; ss, stylet sleeve; st, stylet; sub, subcapitulum; tb, tubular body.

tubule (fig. 2). It has not been possible to detect any other cellular component or their connection with any sensilla. In addition, at least four ciliary constrictions are present ventrally to the pre-oral groove in the subcapitulum. They show three groups each one composed of three microtubules (3+3+3) (figs 2-3). No connected sensilla have been recognized.

Posteriorly, two tubular bodies are connected to the two ventral gnathosomal setae which are probably mechanoreceptors (figs 6).

Moreover, in the stylet cross sections, some

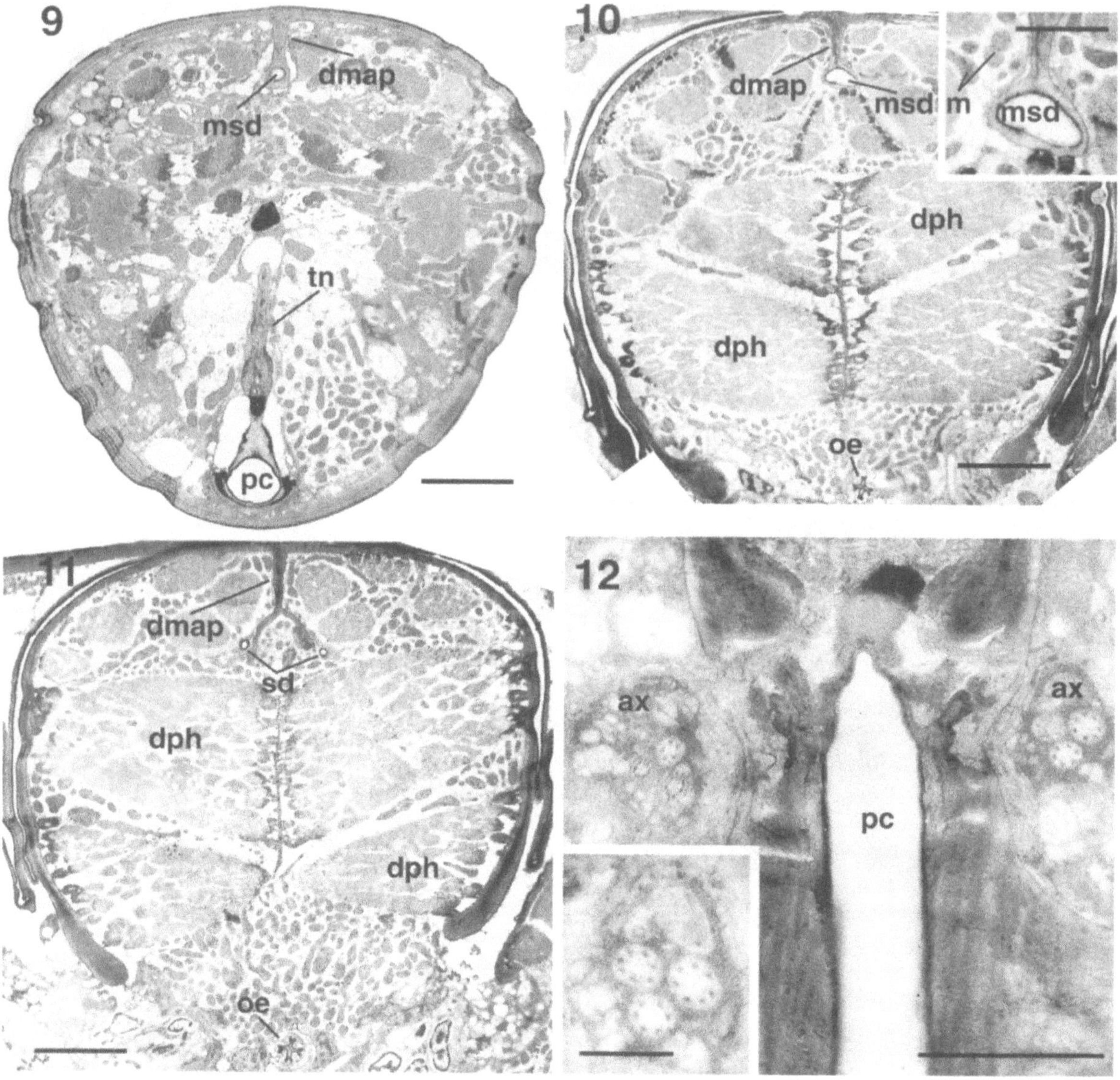

Figs 9-12 - *Tarsonemus nodosus*, electron micrographs of cross sections: 9) showing the insertion of the tendon on the pharyngeal chamber, scale bar: 3 µm; 10) at the level of the median salivary duct and the pharingeal dilator muscles, scale bar=4 µm. Upper right: detail of the median salivary duct, scale bar=1 µm; 11) section at the level of the oesophagus, the pharyngeal dilator muscles and the median salivary duct divided into two smaller ducts, scale bar=4 µm; 12) micrograph of the dendritic segments on both side of the pharyngeal chamber, scale bar=1 µm; left bottom: detail of the microtubules arranged in circle, scale bar=0.4 µm. Abbr: ax, axons; dmap, dorso median apodeme; dph, pharyngeal dilator muscles; m, mitochondrion; msd, median salivary duct; oe, oesophagus; pc, pharyngeal chamber; sd, salivary duct; tn, tendon.

dendritic structures are observed running lengthwise. Apically they are four in number (figs 2-4), while posteriorly they become five (fig. 5). Microporous areas, observed in the stylet cross sections, could allow the connection of these dendritic segments with the external surface.

Six dendritic segments are present on each side of the pharyngeal chamber (fig. 12). The observed microtubules are arranged in a circle and can be in number of five or seven (fig. 12). Moreover, nerve fibres can be observed reaching the pharyngeal lumen.

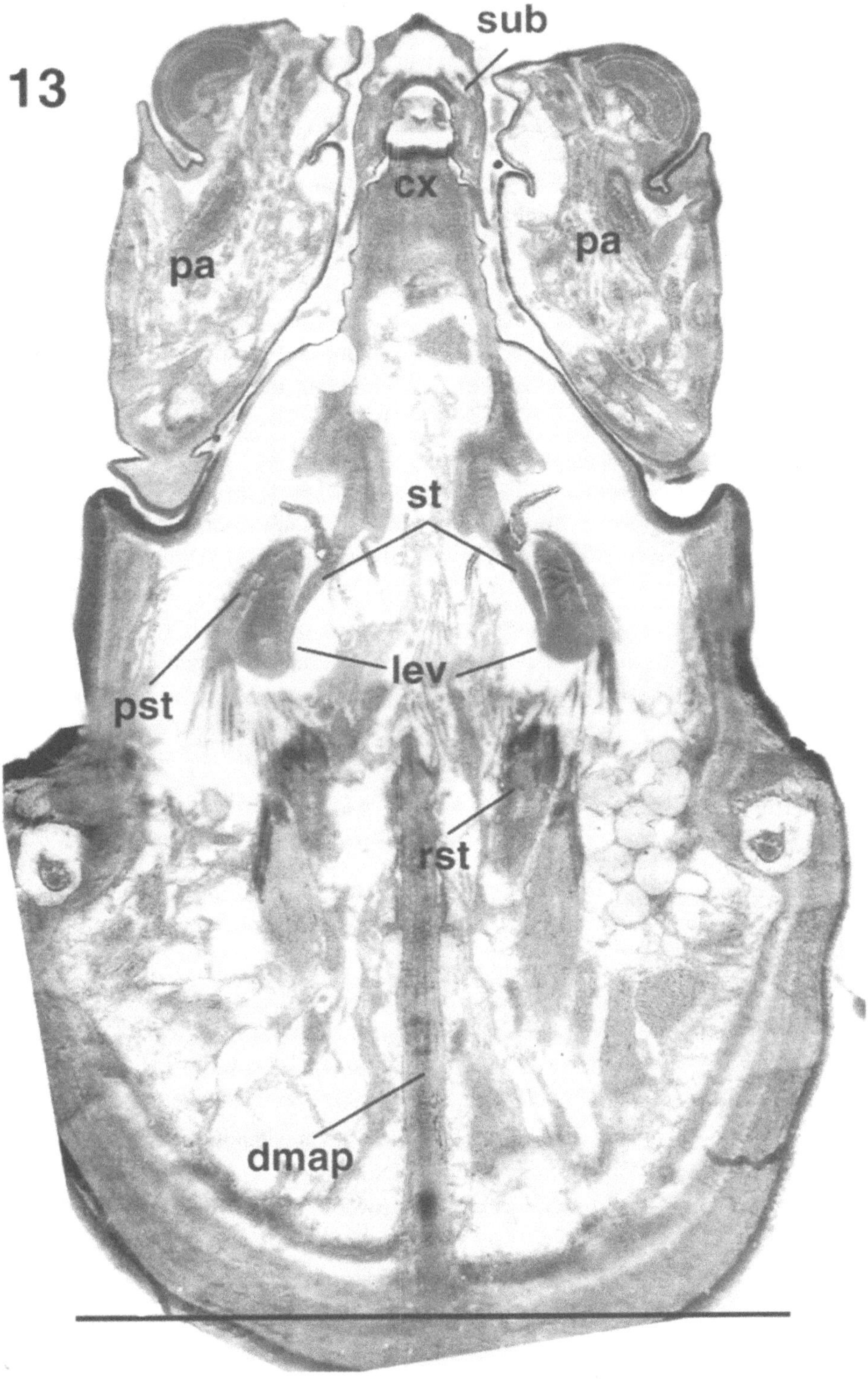

Fig. 13 - Gnathosoma of *Tarsonemus nodosus*: electron micrograph of a frontal section at the level of the cheliceral levers, scale bar=5 µm. Abbr: cx, cervix; dmap, dorso median apodeme; lev, stylet lever; pa, pedipalp; pst, protractors of stylets; rst, retractors of stylets; st, stylet; sub, sucapitulum.

Muscles

Protractor and retractor muscles of the cheliceral stylets can be observed. Protractor muscles insert on the superior antiaxial surface of the cheliceral lever while retractors insert on the inferior postero paraxial surface (fig. 13).

Protractors arise from the posterolateral walls of the stylophore; they are composed of at least two units on each side and reach the lever by means of a common tendon. Retractor muscles arise from the posterior margin of the dorso-median apodeme (septum) and comprise at least one unit on each side.

Pharyngeal dilator muscles are also present and consist of two pairs of symmetrical, well developed, muscles (figs 10-11). They arise from the lateral walls of the gnathosoma (figs 10-11), one pair slightly anterior to the other, and then converge towards a common and mesal tendon that inserts on the dorsal surface of the pharyngeal chamber (fig. 9). The strong tendon shows a fenestrated structure (fig 9) that is connected to a tubular body. No antagonist of the dilators has been identified.

Discussion and conclusions

In *T. nodosus* the gnathosoma is formed by the completely fused stylophore and subcapitulum (figs 1, 7-9, 14-15). Other taxa possessing a stylophore capsule (Cross 1965; Summers and Witt 1971; Lindquist 1986) may also show a median dorsoventrally oriented apodeme, representing the mesial fusion of the cheliceral bases. In *T. nodosus* the apodeme is present only as a vestige of the coalesced paraxial walls of the chelicerae (figs 8-11, 13-15).

The gnathosomal external morphology agrees with that described for *Phytonemus pallidus* (Banks) (Hislop and Jeppson 1976) by means of scanning electron microscopy.

The mouthparts are well adapted for piercing a variety of tissue. Since the cheliceral stylets are tiny and short, they serve merely for superficial penetration. Stylets are withdrawn into the gnathosoma and can be protruded by means of the rotation of the cheliceral lever due to the action of the stylet muscles (figs 13-15). Considering the insertion points of the stylet protractor and retractor muscles on the antiaxial and paraxial surface of the lever respectively, and the oblique arrangement of the lever's long axis, the forward and backward rotation of the lever should be realized not in a vertical plane but actually in an oblique one (figs 7, 13-15). During protraction and retraction, the stylets are held in position by the cheliceral sleeves and, distally, by the labrum and the lateral lip folds (figs 2-6, 14-15). The stylet paraxial surfaces are concave only distally and do not show any complementary ridges or furrows to allow the stylets to fit together and form a tube (fig. 2). That is why, according also to the oblique rotation of the lever, it seems that each stylet may be extended independently from the other and parted slightly. This is confirmed by observations of Chow *et al.* (1980) on *Steneotarsonemus spinki* Smiley, even though the authors thought that a single stylet was subdivided apically. Independent stylet motion is also reported by Bruce *et al.* (1993) in *Pyemotes tritici* Lagrèze-Fossat & Montané (Pyemotidae). Of course, the two stylets may also slide in close proximity but not joined.

Regarding the cheliceral lever, an analogous structure has been described in other mite groups: Tetranychidae (Summers *et al.* 1973), Cheyletidae (Summers and Witt 1971; Akimov and Gorgol 1990), Demodicidae (Desch and Nutting 1977), Cheyletiellidae (Hallas 1982), Penthaleidae (Nuzzaci and de Lillo 1991a; Di Palma 1995). However, in all these groups, the rotation of the lever is realized in a vertical plane and not obliquely as in *T. nodosus*. André (1977) described in Stigmaeidae a lever moving in a horizontal plane, while in other tarsonemid mites the lever may function in a vertical plane (*Acarapis*, Sachs 1951) or horizontal plane (*Steneotarsonemus*, Lindquist 1986). In pyemotid mites (Krczal 1959) the protractor-retractor

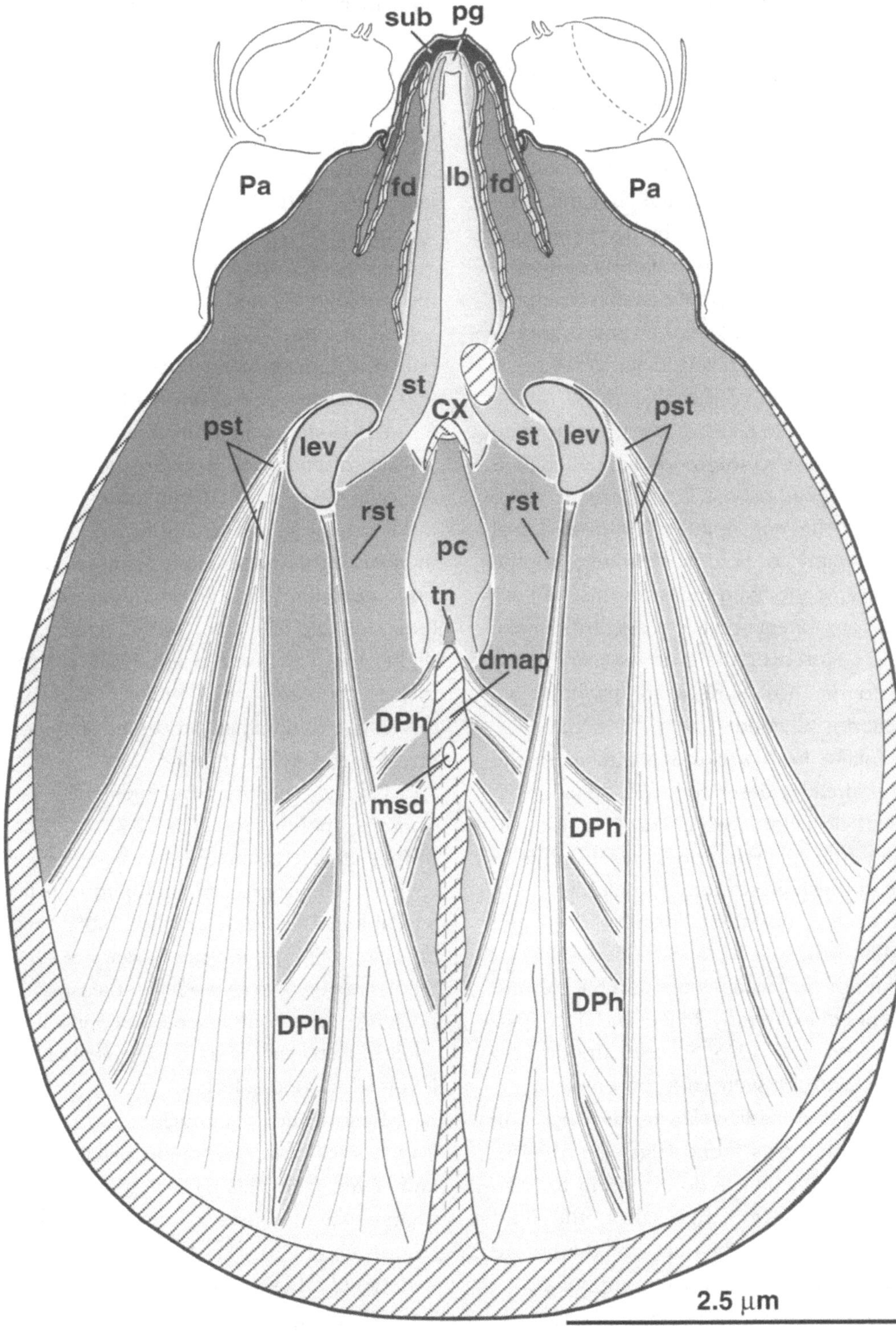

Fig. 14 - *Tarsonemus nodosus*, semischematic drawing: dorsal view of the gnathosoma cut according to an oblique plane. Abbr: cx, cervix; dmap, dorso median apodeme; dph, pharyngeal dilator muscles; fd, fixed digits; lb, labrum; lev, stylet lever; msd, median salivary duct; pa, pedipalp; pc, pharyngeal chamber; pg, pre-oral groove; pst, protractors of stylets; rst, retractors of stylets; st, stylet; sub, subcapitulum; tn, tendon.

mechanism of the stylets is realized by a lever acting in an oblique plane. According to Alberti and Coons (1999), the lever of spider mites represents the basal articulating part of the movable digits.

In *T. nodosus*, the stylets, while piercing the host surface, are supported by the subcapitulum-fixed digits complex as well as by the pedipalpal segments that are closely apposed and configured to it (figs 3-5). Cellular components that reach the surface through the wound are then sucked up by means of the vacuum realized in the pharyngeal chamber by the action of the pharyngeal dilator muscles (figs 10-11). In fact this low pressure state is transferred distally to the food channel (figs 4-6). Distally the gnathosoma is composed of the lateral lips of the subcapitulum whose dorsal walls are adapted to hold and adhere to the ventral surface of the fixed digits. In this way the interlocked components realize a single hermetic unit (figs 2-5). Moreover, the gnathosomal apex is appressed to the host surface so that effective sucking action is allowed.

Food intake in *T. nodosus* is similar to the mechanism already described for other mites (Nuzzaci and de Lillo 1989, 1991a, b and c), although here the pharynx does not include a plunger as in these groups and the elastic region is limited to the posterior dorsal part of the chamber (figs 7-9).

The inferior oral commissure and the salivary pump described in Tetranychidae and Tenuipalpidae (Summers *et al.* 1973; Nuzzaci and de Lillo 1989, 1991b and c) were not observed in *T. nodosus* although a median salivary duct opening behind the labrum was found (figs 8-11, 14-15). Considering that the stylets do not seem to form a tube-like structure to carry saliva, the fluid substances, coming from the duct, could lubricate the movements of the mouthpart components.

In *Pyemotes tritici*, Bruce *et al.* (1993) described an unpaired and elongated labrum that is located distally between the stylets, but there is an evident difference concerning the labral movement. In *P. tritici*, the labrum is apparently withdrawn into the gnathosoma when the cheliceral stylets are completely retracted and everted when the cheliceral stylets are protracted. In *T. nodosus*, as in other mites (Nuzzaci 1979a; Nuzzaci and de Lillo 1989, 1991a and b), the labrum is apparently immobile.

In *T. nodosus*, the fixed digits are different, morphologically and functionally, from those observed in tetranychid mites. They form the dorsal part of the gnathosomal apex; they are modified and completely fused, but not laminar-like (figs 2-6, 14-15). Moreover, they seem not involved in salivary discharge as in other mites (Nuzzaci and de Lillo 1989, 1991b), but rather in sealing the distal region of the gnathosoma. In this respect their function and structure is more similar to the one described in *Penthaleus major* Dugès (Nuzzaci and de Lillo 1991a) where the fixed digits are not fused even though firmly interlocked. Moreover, in *T. nodosus*, the fixed digits develop two laminar processes that completely envelop the stylets (figs 4-6, 15).

Finally, sensory structures detected in *T. nodosus* mouthparts (figs 2-3, 6, 12) require further detailed studies. However, according to several observations on mouthpart sensilla already available (Alberti and Crooker 1985; Nuzzaci and de Lillo 1989 and 1991a; Nuzzaci *et al.* 1992; de Lillo *et al.* 1996; ect.) we may at least consider the presence of chemoreceptors connected to the cheliceral stylets and the pharyngeal chamber and involved in food selection.

Tarsonemid mites include taxa with diverse forms, sizes, and orientation of cheliceral stylets and lever. They also possess diverse forms of pharyngeal structures as already pointed out by Lindquist (1973). Further studies targeting a variety of these forms would contribute to understanding the habits of species in diverse genera and in predicting the feeding habits of taxa still unstudied.

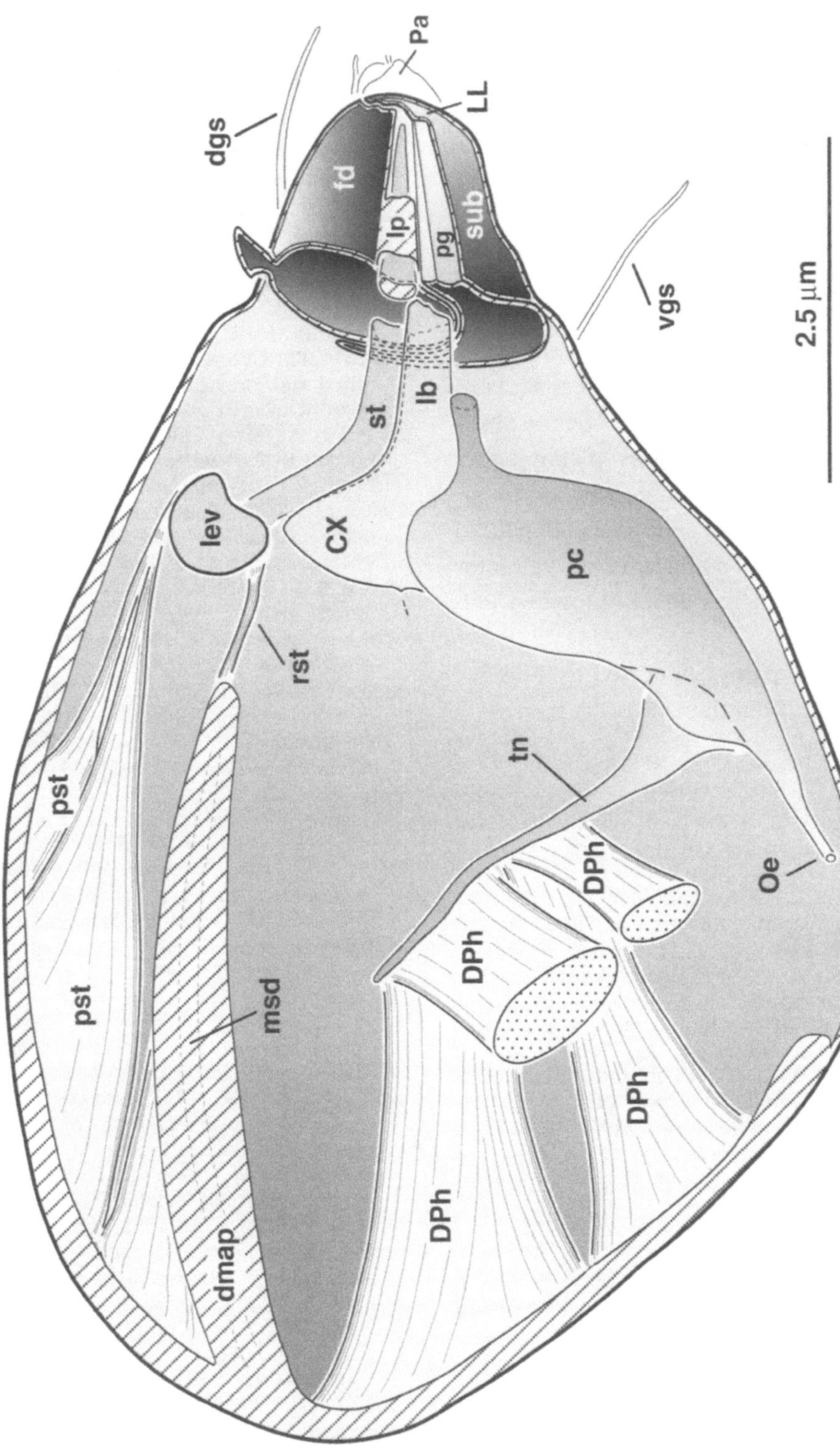

Fig. 15 - *Tarsonemus nodosus*, semischematic drawing: lateral view of the gnathosoma cut according to an oblique plane and slightly turned towards the observer. Abbr: cx, cervix; dmap, dorso median apodeme; dgs, dorsal gnathosoma seta; dph, pharyngeal dilator muscles; fd, fixed digits; lb, labrum; lev, stylet lever; LL, lateral lips; lp, laminar process; msd: median salivary duct; oe, oesophagus; pa, pedipalp; pg, pre-oral groove; pc, pharyngeal chamber; pst, protractors of stylets; rst, retractors of stylets; st, stylet; sub, subcapitulum; tn, tendon; vgs, ventral gnathosomal seta.

Acknowledgements

The authors are very much grateful to: Prof. Dr. G. Alberti (University of Greifswald, Germany), Dr. E.E. Lindquist (Eastern Cereal and Oilseed Research Centre, Ontario Canada) and Dr E. de Lillo (University of Bari, Italy) for their critical contributions. To Prof. Hanna Mazur, Prof. Barbara Tomaszewska, Ms. Liliana Ciszewska, (Department of Plant Metabolism, Institute of Molecular Biology and Biotechnology A. Mickiewicz University, Poznan, Poland) for the hospitality in their laboratory and the use of their technical equipments.

The first and second authors planned the research, analyzed and interpreted the observations, the third collected the specimens, identified the species and provided a critical contribution to the text and the fourth has mainly provided the technical assistance.

This study was partly supported by MURST ex 40% and a grant from the University of Bari.

References

Akimov I.A., Gorgol V.T., 1990 - Predatory and Parasitic Mites - Cheyletidae. Kiev: Naukova Dumka, 120 pp. (in Russian).

Alberti G., Coons L.B., 1999 - Acari-Mites. In: Harrison F.W. (ed.), Microscopic Anatomy of Invertebrates Wiley-Liss, New York. Vol 8c: 515-1265.

Alberti G., Crooker A.R., 1985 - Internal anatomy. In: Helle W, Sabelis M.W. (eds.), Spider Mites, Their Biology, Natural Enemies and Control. World Crop Pests. Elsevier, Amsterdam, vol. 1A: 29-62.

Alberti G., Storch V., 1974 - Über Bau und Funktion der Prosoma-Drüsen von Spinnmilben (Tetranychidae, Trombidiformes). *Zeitschrift für Morphologie und Ökologie der Tiere*, 79: 133-153.

André H.M., 1977 - Note sur le genere *Mediolata* (Actinedida: Stigmaeidae) et description d'une nouvelle espèce corticole. *Acarologia*, 18: 462-474.

André H.M., Remacle C., 1984 - Comparative and functional morphology of the gnathosoma of *Tetranychus urticae* (Acari: Tetranychidae). *Acarologia*, 25: 179-190.

Blauvelt W.E., 1945 - The internal anatomy of the common red spider mite (*Tetranychus telarius* Linn.). *Memoirs Cornell University Agricultural Experiment Station*, 270: 1-35.

Bruce W.A., Kethley J.B., Kaliszewski M., 1993 - Morphology of the gnathosoma of *Pyemotes tritici*: cheliceral stylets and an associated cheliceral structure (Acari: Pyemotidae). *International Journal of Acarology*, 19(2): 127-136.

Chow Y.S., Tzean S.S., Chang C.S., Wang C.H., 1980 - A morphological approach of the tarsonemid mite *Steneotarsonemus spinki* Smiley (Tarsonemidae) as a rice plant pest. *Acta Arachnologica*, 29: 25-41.

Cross E.A., 1965 - The generic relationships of the family Pyemotidae (Acarina: Trombidiformes). *The University of Kansas Science Bulletin*, 45(2): 29-275.

de Lillo E., Nuzzaci G., Aldini P., 1996 - Fine morphology of the mouthpart sensilla in females of *Typhlodromus exhilaratus* Ragusa (Phytoseiidae). In: Mitchell R.D., Horn D.J., Needham G.R., Welbourn W.C. (eds.), Acarology IX. Ohio Biology Survey, Columbus, vol. 1: 287-295.

Desch C.E., Nutting W.B., 1977 - Morphology and functional anatomy of *Demodex folliculorum* (Simon) of man. *Acarologia*, 19: 422-462.

Di Palma A., 1995 - Morfologia funzionale delle parti boccali di *Penthaleus major* (Dugès) (Eupodoidea Penthaleidae). *Entomologica*, Bari, 29: 69-86.

Hallas T.E., 1982 - Functional mouthparts in the dog fur mite *Cheyletiella yasguri* Smiley, 1965 (Acari, Prostigmata). *Acta Zoologica (Stockh.)*, 63(3): 147-152.

Hislop R.G., Jeppson L.R., 1976 - Morphology of the mouthparts of several species of phytophagous mites. *Annals of the Entomological Society of America*, 69: 1125-1135.

Kaliszewski M., Athias-Binche F., Lindquist E.E., 1995 - Parasitism and parasitoidism in Tarsonemina (Acari: Heterostigmata) and evolutionary considerations. *Advances in Parasitology*, 35: 335-367.

Karnovsky M.J., 1965 - A Formaldehyde-Glutaraldehyde fixative of high osmolarity for use in electron microscpy. *Journal of Cell Biology*, 27: 137-138.

Krczal H., 1959 - Systematik und Ökologie der Pyemotiden. In: H.-J. Stammer (ed.), Beiträge zur Systematik und Ökologie Mitteleuropäischer Acarina. I. Tyroglyphidae und Tarsonemini. II. Leipzig: Akadem. Verlagsgesellsch. Geest & Portig: 385-625.

Lindquist E.E., 1973 - Observations on the generic classification of tarsonemid mites (Prostigmata). Proceedings of the 3rd International Congress of Acarology, Prague 1971: 293-295.

Lindquist E.E., 1986 - The world genera of Tarsonemidae (Acari: Heterostigmata): A morphological, phylogenetic, and systematic revision, with a reclassification of family-group taxa in the Heterostigmata. *Memoirs of the Entomological Society of Canada*, 136: 1-517.

Mothes U., Seitz K.-A., 1981 - Fine structure and function of the prosomal glands of the two-spotted spider mite *Tetranychus urticae* (Acari, Tetranychidae). *Cell Tissue Research*, 221: 339-349.

Nuzzaci G., 1979a - Contributo alla conoscenza dello gnatosoma degli Eriofidi (Acarina: Eriophyoidea). *Entomologica*, Bari, 15: 73-101.

Nuzzaci G., 1979b - A study of the internal anatomy of *Eriophyes canestrini* Nal. In: Piffl E. (ed.), Proceedings of the 4th International Congress of Acarology. Budapest: Akadémiai Kiadó: 725-727.

Nuzzaci G., 1994 - Recenti acquisizioni di morfologia funzionale negli acari. Atti XVII Congresso nazionale italiano di Entomologia, Udine: 273-286.

Nuzzaci G., Alberti G., 1996 - Internal anatomy and physiology. In: Lindquist E.E., Sabelis M.W., Bruin J. (eds.), Eriophyoid Mites - Their Biology, Natural Enemies and Control. World Crop Pests. Elsevier, Amsterdam, vol. 6: 101-150.

Nuzzaci G., de Lillo E., 1989 - Contributo alla conoscenza dello gnatosoma degli Acari Tenuipalpidi (Tetranychoidea: Tenuipalpidae). *Entomologica*, Bari, 24: 5-32.

Nuzzaci G., de Lillo E., 1991a - Contributo alla conoscenza delle parti boccali di *Penthaleus major* (Dugès) (Eupodoidea: Penthaleidae). Atti XVI Congresso Nazionale italiano di Entomologia. Bari-Martina Franca (TA): 265-277.

Nuzzaci G., de Lillo E., 1991b - Linee evolutive dello gnatosoma in alcuni Acari Prostigmata. Atti XVI Congresso Nazionale italiano di Entomologia. Bari-Martina Franca (TA): 279-290.

Nuzzaci G., de Lillo E., 1991c - Fine structure and functions of the mouthparts involved in the feeding mechanisms in *Tetranychus urticae* Koch (Tetranychoidea: Tetranychidae). In: Dusbábek F., Bukva V. (eds.), Modern Acarology. Academia, Prague, vol. 2: 301-306.

Nuzzaci G., de Lillo E., Porcelli F., 1992 - Functional morphology of mouthpart sensilla in females of *Varroa jacobsoni* Oudemans (Acari: Varroidae). *Entomologica*, Bari, 27: 41-67.

Sachs H., 1951 - Zur Morphologie von *Acarapis*. 1. Bau und Funktion der Mundwerkzeuge der Tracheenmilbe Acarapis woodi Rennie, 1921. *Zeitschrift für Bienenforschung*, 1: 103-112.

Summers F.M., Gonzales-R R.H., Witt R.L., 1973 - The mouthparts of *Bryobia rubrioculus* (Sch.) (Acarina: Tetranychidae). *Proceedings of the Entomological Society of Washington*, 75: 96-111.

Summers F.M., Witt R.L., 1971 - The gnathosoma of *Cheyletus cacahuamilpensis* Baker (Acarina: Cheyletidae). *Proceedings of the Entomological Society of Washington*, 73(2): 158-168.

Some future prospects in Agricultural Acarology, with focus on phytoseiid mites - host plant relationships

Serge Kreiter, Marie-Stéphane Tixier
ENSAM/INRA, UFR d'Ecologie animale et de Zoologie agricole, Laboratoire d'Acarologie, 2 Place Pierre Viala, F-34060 Montpellier cedex 01, France, kreiter@ensam.inra.fr and garcin@ensam.inra.fr

Key words: future prospects, Agricultural Acarology, Phytoseiidae, plant-phytoseiid relationships, crop colonization, taxonomy.

Abstract: This paper gives a quick overview of the recent advances in basic and applied researches in Agricultural Acarology and examines the principal research lines that have been involved in various teams during the last few years. The best progresses was made in population biology and genetics, behaviour of mites within three trophic levels, molecular markers. These studies mainly involved tetranychid and phytoseiid mites, however, other plant inhabiting mites remain presently less known. The main objective of this paper is not to give precise directions of future prospects in Agricultural Acarology that may concern, for example; impact of the climatic global change on mite populations or communities, the impact of transgenic plants on plant inhabiting mites and the increasing problem of resistance to new acaricide (which is becoming more and more rapid among tetranychid mites). This may also concern transgenic mites, mainly transgenic phytoseiid mites, or influence of urban and agricultural pollution on mite communities.

The present paper aims to tell how authors see the future prospects of their own researches with possible (but necessary!) collaboration with other European acarologists. Some topics are taken as examples for illustrating the main subjects: plant-phytoseiid mites relationships, biology of *Kampimodromus aberrans*, colonization of crops by phytoseiid mites, side effects of intentional introduction of predators and phytoseiid mites taxonomy. The availability of new tools and better collaboration between acarologists may improve our knowledge in these areas.

New position may be devoted to Acarology and not only on new tools and not only on species that facilitate quick results for rapid and numerous publications. We must also work with species that have an interest in Agricultural Acarology and not necessarily easily manipulated in the laboratory. There is a big demand from developing countries for education in Agricultural Acarology and this must be a big challenge and an important point for European acarologists who want to collaborate with those countries, mainly in Africa, South America and Asia, and improve transmission of knowledge and ideas to them, but also discovering new species, new habitats, new crop protection methods, and new life-styles of mites.

Introduction

As in other branches of Biology, fundamental research in modern Agricultural Acarology has benefited from technical progress with the development of computing, biochemistry and molecular tools. The first consists of the mathematical models developed due to progress in data processing. They have led either to the establishment of models applied to crop protection or models for research on population

F. Bernini, R. Nannelli, G. Nuzzaci, E. de Lillo (eds.), Acarid Phylogeny and Evolution. Adaptations in mites and ticks, 283-302.

dynamics. The second concerns the different aspects of the behaviour of pests and predators. This has been developed due to progress in data recording methods. The third concerns population genetics which has expanded considerably with the recent progress in biochemistry and molecular biology techniques. This paper gives rapidly an overview on some recent advances in Agricultural Acarology. After this, the authors choose to develop the future prospects of some of their own research and not necessarily the future prospects of Agricultural Acarology, which would be a too broad, general, wide and speculative matter. We have limited the scope of our paper purpose to tetranychid and phytoseiid mites. We will just mention some possible ways of investigations and the philosophy of our research in the future.

Recent advances in some field of research

Behaviour of plant-inhabiting mites and chemical and physical interactions within food webs

A great number of studies deals with these aspects. They concern the mating behaviour of pests and predators (Aponte and McMurtry 1992; Enders 1993), biparental defence in a sub-social spider mite (Yamamura 1987; Saito 1990) and dispersal of tetranychid (Berry *et al.* 1990; Zhang *et al.* 1992; Margolies 1995).

Phytoseiid mites, as in all Mesostigmata, are blind and lack the distinctive chemosensory solenidia characteristics of acariform mites, but other recognition signals exist.

Behavioural observations show that females of *Iphiseius degenerans* (Berlese) can discriminate not only between conspecific and heterospecific eggs but also between eggs that vary in degree of relatedness. This enables females to oviposit in clusters containing related eggs and thus avoid cannibalism by non-kin and/or produce adaptative sex ratios (Faraji *et al.* 2000).

It has now been convincingly demonstrated that not only phytoseiid mites such as *Phytoseiulus persimilis* Athias-Henriot sense their prey from long distances but they also respond to plant volatiles that can inattract the predators from far away (Dicke *et al.* 1999; Janssen 1999) or alert their neighbours (Bruin *et al.* 1995). Spider mites may also avoid plants with heterospecific competitors such as thrips (Pallini *et al.* 1997) and with predators (Pallini *et al.* 1999) but show a slight attraction for plants with conspecifics (Pallini *et al.* 1997).

According to these relations, many authors infer existence of mutualism relationships between plants and phytoseiid mites. Volatile compounds emitted by plants are considered by some researchers as induced defences (Dicke and Sabelis 1988; Sabelis *et al.* 1999). Plants harbour specific structures as leaf domatia and hairs that also favour the occurrence of phytoseiid mites (Walter 1996). Leaf hairs and domatia may constitute protection for some phytoseiid mites and could be considered as constitutive defence of the plants against herbivores (Sabelis *et al.* 1999). However, sometimes these structures may also negatively influence searching efficiency and predation rate of *P. persimilis* (Krips *et al.* 1999).

Intra- and interspecific predation were investigated recently, as never before, by various teams in the world (Schausberger 1999; Walzer and Schausberger 1999a and b).

Using feeding patterns and other biological traits, McMurtry and Croft (1997) have proposed a very pertinent classification separating phytoseiids into two types of generalists and two types of specialist predators life styles or predation types and are now trying to quantify some life traits.

Aerial dispersal of phytoseiid mites was demonstrated in field studies for some predatory mites for which this means of movement had not been reported before (Tixier *et al.* 1998 and 2000; Tixier 2000).

Population genetics and molecular markers

Much research on isozyme variability has been carried out on plant-inhabiting mites. Investigations concerned inter- and intraspecific polymorphism in tetranychid and tenuipalpid mites (Osakabe and Komazaki 1996; Tsagkarakou *et al.* 1997 and 1999;

Gotoh *et al.* 1998), relationships between eriophyoid mite families (Kuang *et al.* 1992), diapause polymorphism (Gotoh *et al.* 1991) and acaricidal resistance (Tsagkarakou *et al.* 1996).

Molecular tools should enable accurate identification of species, the establishment of phylogenies, the studies of population genetics and similarities, the characterisation and identification of *Wolbachia* (Gotoh *et al.* 1999a and b) and lead to genetic transformation of predators (Li and Hoy 1996). For the tetranychid mites, the contribution of Kaliszewski *et al.* (1992), Navajas *et al.* (1996) in phylogeny and Hance *et al.* (1998) for species characterization were determinant, and eriophyid identifications has been investigated by Fenton *et al.* (1995). For phytoseiid mites, RAPD-PCR techniques allowed Edwards *et al.* (1998) to discriminate three species of morphologically close *Typhlodromalus* and Tixier and Kreiter (this volume) to discriminate several strains of *Kampimodromus aberrans* (Oudemans).

The mite resistance to acaricide

The populations behaviour of acaricide-resistant pests has been investigated by Kolmes *et al.* (1994). The resistance of mites to acaricide could be considered as an old problem, solved or under control. The increasing problem of resistance to new acaricides such as tebufenpyrad (Herron and Rophail 1998) in *Tetranychus urticae* Koch strains and the cross-resistance conferred to pyridaben, fenpyroximate and other METI (mitochondrial electron transport inhibitors) acaricides show that chemical companies have some objective reasons to be worried about this process. This resistance seems to appear more and more rapidly among tetranychid mites, especially in *Panonychus ulmi* (Koch) strains (Auger, Bonafos and Kreiter, *unpubl. data*).

Some possible future prospects in the field of Agricultural Acarology

As stated above, the main objective of this paper is not to give precise directions of future prospects in Agricultural Acarology. However, we can quickly describe about some original themes.

These future themes may concern for example *the impact of the climatic global change on mite communities*. This field of research is at the moment poorly studied. What has been done mainly concerns the population dynamics of *T. urticae* (Heagle *et al.* 1994; Boutaleb Joutei 1999) under increasing atmospheric CO_2 concentration. This increase affects the fecundity of *T. urticae* (reduced by 40% during several generations) and adults and immatures survival (respectively reduced by 50% and 10%). Finally, the intrinsic rate of increase and thus the finite rate of increase are considerably reduced (Boutaleb Joutei 1999). The increase in global temperature and the effect of increasing concentration in ozone are presently poorly documented.

The impact of transgenic plants on plant inhabiting mite population, meta-population or community dynamics may constitute an interesting and actual area of investigations. There is presently no plant with transgene for resistance to phytophagous mites and only few transgenic plant resistant to some insects were evaluated for mites. Transgenic plants of *Chrysanthemum morifolium* Ramat with *Bacillus thuringiensis* delta-endotoxin gene appeared resistant to *T. urticae* in Russia (Dolgov *et al.* 1995) while transgenic (Bt) cotton varieties in China were more susceptible to *T. urticae* and aphids (Jie and Yuan 1999) suggesting that where the transgenic variety is planted, monitoring and control measures should be carried out against cotton pests other than *Helicoverpa armigera* (Hübner). This constitutes only few data on the positive or negative effects of those plants for phytophagous or predatory mites and a considerable field of research is open, as a part (but not only) of risk assessment.

This may also concerns *transgenic mites and mainly transgenic phytoseiid mites*. Mites may be the first arthropod biocontrol agents engineered using recombinant DNA techniques to be released into the field. Genetic manipulation of

Galendromus occidentalis (Nesbitt) has yielded transgenic strains (pesticide resistant) but none have been released yet (Li and Hoy 1996; McDermott and Hoy 1997; Presnail *et al.* 1998).

Induced resistance. Spider mites are major pests of many vineyards but outbreaks are often prevented or controlled by phytoseiid mites (Duso and Pasqualetto 1993). In California however, an unexpected interaction between two species of spider mite was found. The best method for controlling Pacific mite populations, which are more injurious to the vine late in the season, was to avoid miticide applications and to release Willamette mite in vineyards early in the season. When chemicals are not used, Willamette mites caused little damage to grapes but these vines showed increased resistance to Pacific mite resulting in little or no Pacific mite damage. Although high densities of predatory mites may explain some of this effect, resistance to spider mite feeding induced by early season herbivory appears to be more important (Karban *et al.* 1997). These unexpected results show that we need to think differently in the future and that we presently know very little about some plant inhabiting mite systems. Mechanisms of this induced resistance have to be explored. Predators may also induce interspecific competition for food and intraguild predation among herbivores within refuge space (Pallini *et al.* 1998).

What we will try to do in the future? Research on Plant-Phytoseiid Mites relationships

After this brief overview above, we are going to present one of the main research areas conducted in our laboratory focusing on *K. aberrans.*

Kampimodromus aberrans is a common phytoseiid mite on grape and many non-agricultural plants that surround vineyards (Duso *et al.* 1993; Kreiter *et al.* 2000). This important biological control agent is a generalist (vs. specialist; McMurtry and Croft 1997) predator of spider mites, including the common pest in Southern France, *Eotetranychus carpini* (Oudemans). *Kampimodromus aberrans* feeds on many non-tetranychid mites, pollen and insects as well (Kropczynska *et al.* 1988; Schausberger 1992, 1997 and 1998). This predaceous mite seems to be closely associated with specific host plants and it varies regionally in this regard (de Moraes *et al.* 1986). In Europe it occurs on economic plants such as apple (*Malus domestica* Barkh.), grape (*Vitis vinifera* L.) and hazelnut (*Corylus avellana* L.) and many other non-economic plants, whereas it is specific to hazelnut in North America (McMurtry and Croft 1997). Relative to preference, *K. aberrans* is most common on plants that have hairy leaves, especially those with domatia (de Moraes *et al.* 1986; Coiutti 1993; Barret and Kreiter 1995). Early studies with *K. aberrans* were field-based and pest control-related, but the need to understand relationships between plant types, leaf structures, pollen foods and their effects on this mite were recognized (Barret and Kreiter 1995; Camporese and Duso 1996).

Phytoseiid mites live mostly on leaf undersurfaces that have raised veins, dense hairs, tunnelled margins and cave-like structures in the vein axils (domatia). These structures affect searching, feeding, mating, oviposition, hiding, etc. of these small arthropods. Leaves with domatia often harbour more phytoseiid species or greater densities of a species than leaves without domatia or leaves from which domatia have been removed (O'Dowd and Willson 1991; Walter and O'Dowd 1992a and b, Walter 1996). Such shelters and leaf pubescence, may be as important as food is for some phytoseiids (Duso 1992; Karban *et al.* 1995) and especially for generalist feeders that have evolved more in response to plants than specialist feeders (McMurtry and Croft 1997). Yet many questions remain about adaptations of phytoseiids to plant attributes. For example, how does leaf architecture allow some phytoseiids to colonize a plant and others not? Can two or more species cohabit a leaf by habitat partitioning? If so, do some refuges minimize interspecific competition more than others? Answers to these questions have implications to biological control. Few studies have examined how host plant traits influence control (Duso 1992; Walter 1996; Lester *et al.* 2000). Answers are also important for understanding colonization of crops by phytoseiid mites from neighbouring areas and for determining which plants may constitute reservoirs and why (Tixier *et al.* 1998 and 2000; Tixier 2000).

Results of surveys of predaceous mites that occur on vegetation surrounding vineyards are reported and correlated to factors of plant type, leaf pilosity, leaf domatia and pollen retention. Also, reported and discussed, are features of the biology of *K. aberrans*, including aspects of pollen feeding and water uptake from plants. Remaining questions and future prospects, are discussed.

A survey of phytoseiids on plants near vineyards

Two kinds of vegetation were sampled: woody margins, and hedges 1, 2 and 3. Samples of 30 or more leaves/plant/date were taken twice in 1996 and 1998 and 3 times in 1997. Mite abundance was analysed using an index of Woody Richness (WR, Tixier *et al.* 1998)=Σ (abundance-dominance index of a plant species x the richness of this plant in phytoseiid mites/leaf) where index are: 1=plants are rare, <5% of the canopy; 2=abundant, 5-25%; 3=abundant, 25-50%; 4=very abundant, 50-75%; and 5=extremely abundant, 75-100%.

The woody margin and hedge 1 had high WR indexes, while hedge 2 was low (fig. 1). Phytoseiid abundance in these habitats was linked to *Quercus pubescens* Willd occurrence. Temporal trends of phytoseiid abundance were similar in the 4 uncultivated areas with decreases in May-July, except for the woody margin in 1996 and hedge 3 in 1997 (fig. 1). The extent of decreases depended on the dominant phytoseiid present (see below). On a plant basis, the species that were sampled did not harbour the same phytoseiid densities per cm^2 (P=0.013) (tab. 1).

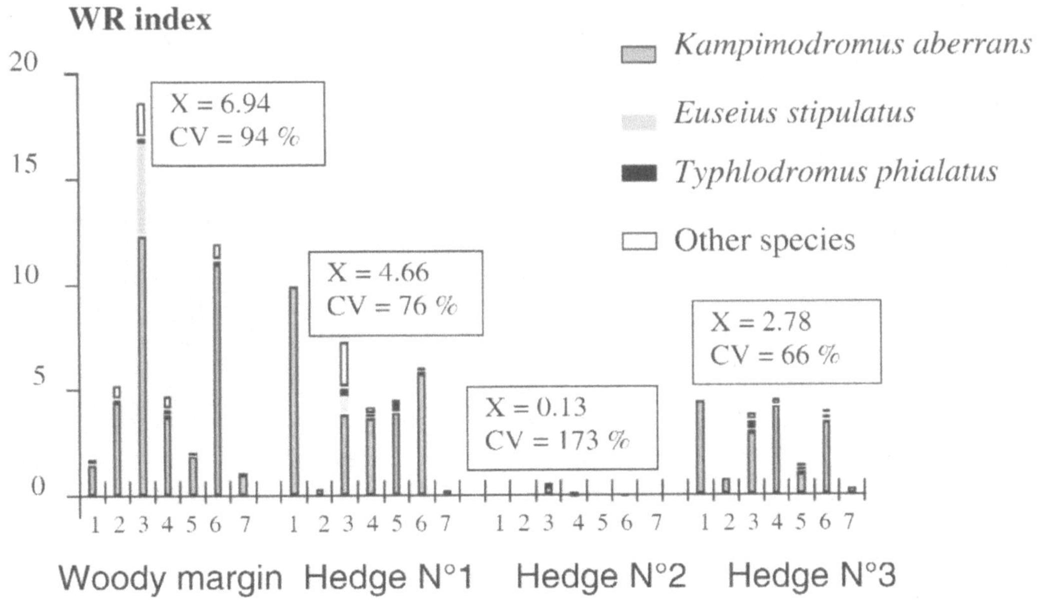

Fig. 1 - Variations of the WR index in the 4 uncultivated areas surrounding the plot during 1996 (1: May. 2: July). 1997 (3: May. 4: July. 5: August) and 1998 (6: May. 7: July) and the main species collected (included inside figures. means. variation coefficient. ANOVA Duncan's groups).

Tab. 1 - Number of phytoseiid mites, % of *Kampimodromus aberrans*, % of immatures and predator frequency per plant.

	Phytoseiid mites per cm^2	Duncan groups	% of *K. aberrans* *	Predator frequency **	% of immatures
Euphorbia characias	0.05	**A**	95.5	1	10
Celtis australis	0.039	**AB**	100	1	35
Rubus sp.	0.036	**AB**	50	0.86	37
Lonicera sp.	0.032	**AB**	77.8	1	18
Bryonia dioica	0.03	**AB**	39.4	1	23
Osyris alba	0.03	**AB**	0	0.14	0
Quercus pubescens	0.0266	**AB**	94.4	1	21
Viburnum tinus	0.025	**AB**	42.1	0.80	36
Cornus sp.	0.021	**AB**	85.7	0.75	8
Ulmus sp.	0.018	**AB**	87.4	0.71	18
Rubia peregrina	0.017	**AB**	50.0	0.29	18
Galactites tomentosa	0.014	**B**	100	0.14	0
Picris hieracioides	0.007	**B**	0	0.14	0
Calamintha nepela	0.007	**B**	0	0.43	0
Torilis arvensis	0.0068	**B**	46.2	0.75	20
Lychnis sp.	0.0051	**B**	0	0.14	20
Malva sylvestris	0.0051	**B**	90	0.20	0
Prunus spinosa	0.0044	**B**	54.1	0.86	18
Clematis flammula	0.003	**B**	16.7	0.3	0
Geranium sp.	0.003	**B**	0	0.1	0
Hedera helix	0.0027	**B**	100	0.1	0
Crataegus monogyna	0.0026	**B**	50	0.14	30
Ruscus aculeatus	0.0022	**B**	60	0.43	50
Clematis vitalba	0.0013	**B**	5.3	0.29	30
Quercus ilex	0.0004	**B**	0	0.29	0
Vincetoxinum sp.	0.00018	**B**	0	0.14	0
Fraxinus sp.	0.0001	**B**	100	0.1	0
Sambucus nigra	0.0001	**B**	98	0.2	0
Lactuca serriola	<0.00000	**B**	0	0.14	0
Galium aparine	<0.00000	**B**	80	0.14	0
Conyza sumatrensis	<0.00000	**B**	0	0.14	0
Agrimonia eupatoria	0.92		54	1	22
Echium vulgare	0.063		50	0.3	0
Borago officinalis	4.05		58	1	44

(*) Percentage established for adults.
(**) This rate is 1 when phytoseiid mites are observed for all the sampling dates.

Tab. 2 - Phytoseiid mite species sampled on plants in the four uncultivated surroundings (WM: Woody Margins, H1, H2, H3: respectively Hedges 1, 2 and 3).

	Kampimodromus aberrans	*Typhlodromus pyri*	*Euseius stipulatus*	*Typhlodromus phialatus*	*Typhlodromus recki*	*Typhlodromus intercalaris*	*Neoseiulus californicus*	*Typhlodromus rhenanoides*	*Typhlodromus exhilaratus*	*Euseius finlandicus*	*Phytoseius juvenis*	*Neoseiulus aurescens*	*Typhlodromips luthezicus*	Localization
Agrimonia eupatoria	x	x												WM
Arum sp.	NO PHYTOSEIID MITE													WM
Arundo donax	NO PHYTOSEIID MITE													WM, H2
Asparagus sp.	NO PHYTOSEIID MITE													WM, H3
Borago officinalis	x	x	x	x	x									WM
Bryonia dioica	x	x	x	x	x				x					WM
Calamintha nepela				x	x									H3
Celtis australis	x	x												WM
Centaurea aspera	NO PHYTOSEIID MITE													WM, H2
Clematis flammula	x	x	x	x										WM, H2
Clematis vitalba	x	x	x											WM, H2
Conyza sumatrensis	x	x		x	x									WM, H1, 2, 3
Cornus sp.	x	x	x			x								WM
Crataegus monogyna	x	x	x		x		x	x						WM, H1, 3
Echium vulgare	x				x							x	x	H2
Euphorbia characias	x	x		x	x									WM, H1, 2, 3
Fraxinus sp.	x													WM, H1
Galactites tomentosa				x	x									WM, H2
Galium aparine	x	x												WM, H2
Geranium sp.	x				x									WM, H1
Hedera helix	x													WM
Jasminium sp.	NO PHYTOSEIID MITE													WM
Lactuca serriola			x											WM, H2
Lonicera sp.	X	x	x			x								WM
Lychnis dioica		x	x											WM
Malva sylvestris	X		x	x	x									H2
Nigella damascena	NO PHYTOSEIID MITE													H2
Olea sp.	NO PHYTOSEIID MITE													WM, H3
Osyris alba	X						x							WM, H1
Phillyrea angustifolia	NO PHYTOSEIID MITE													WM, H3
Picris hieracioides			x	x										WM, H2
Prunus dulcis	NO PHYTOSEIID MITE													H3
Prunus spinosa	X		x											WM, H1, 2
Psoralea bituminosa	NO PHYTOSEIID MITE													H2
Quercus ilex		x	x	x										WM
Quercus pubescens	X	x	x	x	x	x								WM, H1, 3
Rubia peregrina	X	x												WM, H1, 3
Rubus sp.	X	x	x	x		x			x	x	x			WM, H1,2,3
Ruscus aculeatus	X			x										WM
Saponaria officinalis	NO PHYTOSEIID MITE													WM
Spartium junceum	NO PHYTOSEIID MITE													WM
Sambucus nigra	X		x				x							H1, 2
Tamarix sp.				x										H1
Torilis arvensis	X		x	x										WM, H1, 2, 3
Ulmus sp.	X	X	x			x								WM, H1
Viburnum tinus	X	X	x	x										WM
Vincetoxinum sp.			x											WM

Highest densities of phytoseiids were observed on *Euphorbia characias* L., *Celtis australis* L., *Rubus* sp., *Lonicera* sp., *Q. pubescens*, *Bryonia dioica* Jacq., *Cornus* sp., *Viburnum tinus* L., and *Ulmus* sp. (tab. 1). Thirteen phytoseiid species were collected from all plants and *K. aberrans* was the species observed most often (tab. 2). The proportion of this species was lower in 1997 but always dominant in samples after May (fig. 1). All plants sheltered at least 1 other phytoseiid species besides *K. aberrans*. *Euseius stipulatus* (Athias-Henriot), *Typhlodromus pyri* Scheuten, *T. phialatus* Athias-Henriot and *T. recki* (Wainstein) were the most frequent (tab. 2). Differences were observed among more than 20 sampled plants for % of *K. aberrans* among all phytoseiids on a plant (P=0.000). Among plants with high densities of phytoseiids, proportions of *K. aberrans* were quite high on *C. australis, Q. pubescens, E. characias, Ulmus* sp. and *Cornus* sp. (tab. 1).

Correlation between plant types, leaf structures and pollen capture

Counts of trichomes, domatia and pollen grains densities were taken at randomly selected locations on 3 leaves per plant on 3 plants per species in order to explain large differences in phytoseiid densities among plants. For measuring domatia types, we used a rating scale based on O'Dowd and Willson (1989): 0=no hair at intersection of leaf veins; 1=some hairs, but no shelter; 2=somewhat closed domatia, intersection of leaf veins lightly covered by trichomes; 3=leaf vein intersection completely obstructed by hairs. Analysis of variance and mean tests (Duncan) were used to compare phytoseiid densities/cm^2, leaf hairs and pollen densities/cm^2 and domatia densities/leaf. Domatia ratings among plants were compared by χ^2 tests. Multiple regressions were run between leaf data types and densities of phytoseiid mites/cm^2, predator frequency per plant, percent of *K. aberrans* (vs. all phytoseiids) per plant and number of phytoseiid species per plant. Multifactorial regressions PCA and decision tree analysis were used to assess importance of leaf parameters to predict phytoseiid densities.

Trichome ($P\leq$0.0001), domatia ($P\leq$0.0001), and pollen ($P\leq$0.0001) densities per leaf differed among 20 plant species (tab. 3). *Celtis australis, Q. pubescens* and *Ulmus* sp. had high levels for each plant variable. *Rubus* sp. and *E. characias* had high trichome densities and *Lonicera* sp. and *Ulmus* sp. had many domatia. Domatia were observed on 9 plant species; χ^2 tests showed that the domatia ratings differed among these plants (P=0.0001; tab. 3). *Celtis australis, Ulmus* sp., *Prunus spinosa* L. and *Q. pubescens* had the highest domatia rating. The number of phytoseiids (all life stages) per cm^2 (r^2=0.74, P=0.012) and their frequency on plants (r^2=0.70, P=0.016) were correlated with trichome density (trichome density vs. phytoseiids per cm^2: r^2=0.43, P=0.0007; trichome density vs. frequency on plants: r^2=0.38, P=0.026). No significant correlation was found between any leaf parameter and either percent of *K. aberrans* per plant (r^2=0.39, P=0.65, F=0.52) or total phytoseiid species per plant (r^2=0.28, P=0.84).

Pollen density was significantly correlated with domatia rating (r^2=0.65, P=0.0042). Excluding *Rubus* sp. and *E. characias*, there was significant correlation between pollen and trichome densities (r^2=0.78, P=0.033). These two plants were excluded from further analyses because, while they had high trichome densities, little if any pollen was present on their surfaces probably because these two plants are small bushes.

Among 20 plant species, 5 groupings with similar characteristics were identified by multifactorial analyses (fig. 2). One group with 8 plants had low values for all significant plant variables and few phytoseiids. *Torilis arvensis* (Hudson), *Rubus* sp. and *E. characias* formed another group that had high phytoseiid densities, only for *Rubus* sp. and *E. charachias*, and high trichome densities but only a few small closed domatia and low pollen densities. *Celtis australis, P. spinosa*

Tab. 3 - Leaf characteristics of 20 plants and mean comparison groups (Duncan). Abbr.: *SV, leaf secondary vein; MV, leaf main vein.

	Mean density of trichomes/cm^2		Mean density of pollen/cm^2		Number of domatia (SV*)	Number of domatia (MV*)	Total number of domatia		Mean rating of domatia	Leaf area (cm^2)
Rubus sp.	7000	**A**	10	**CDE**	0	0	0	**G**	0	8.73
Euphorbia characias	6292	**AB**	7.4	**DE**	0	0	0	**G**	0	5.47
Quercus pubescens	5388	**B**	43	**A**	12	0	12	**CD**	2.88	19.63
Ulmus sp.	2997	**CD**	26	**ABCD**	26	11	37	**A**	2.49	9.87
Prunus spinosa	2530	**CD**	23	**BCDE**	10	0	10	**D**	2.51	13.3
Celtis australis	1637	**DE**	46	**A**	7	7	14	**C**	2.63	15.03
Torilis arvensis	1463	**E**	31	**ABCD**	0	0	0	**G**	0	4.68
Cornus sp.	340	**F**	18	**CDE**	6	0	6	**F**	1.38	14
Lonicera sp.	337	**F**	13	**CDE**	12	9	21	**B**	0.48	22.1
Crataegus monogyna	98	**F**	9	**CDE**	6	0	6	**F**	1.36	6.3
Clematis vitalba	34	**F**	20	**CDE**	5	2	7	**EF**	0.82	11.4
Clematis flammula	17	**F**	11	**CDE**	10	0	10	**DE**	0.65	3.35
Osyris alba	0	**F**	2.19	**E**	0	0	0	**G**	0	0.66
Rubia peregrina	0	**F**	1.67	**E**	0	0	0	**G**	0	1.56
Phillyrea angustifolia	0	**F**	1.12	**E**	0	0	0	**G**	0	4.3
Fraxinus sp.	0	**F**	2.46	**E**	0	0	0	**G**	0	4.52
Saponaria officinalis	0	**F**	1.83	**E**	0	0	0	**G**	0	6.44
Arum sp.	0	**F**	0	**E**	0	0	0	**G**	0	107
Lactuca serriola	0	**F**	0	**E**	0	0	0	**G**	0	62
Arundo donax	0	**F**	0	**E**	0	0	0	**G**	0	50

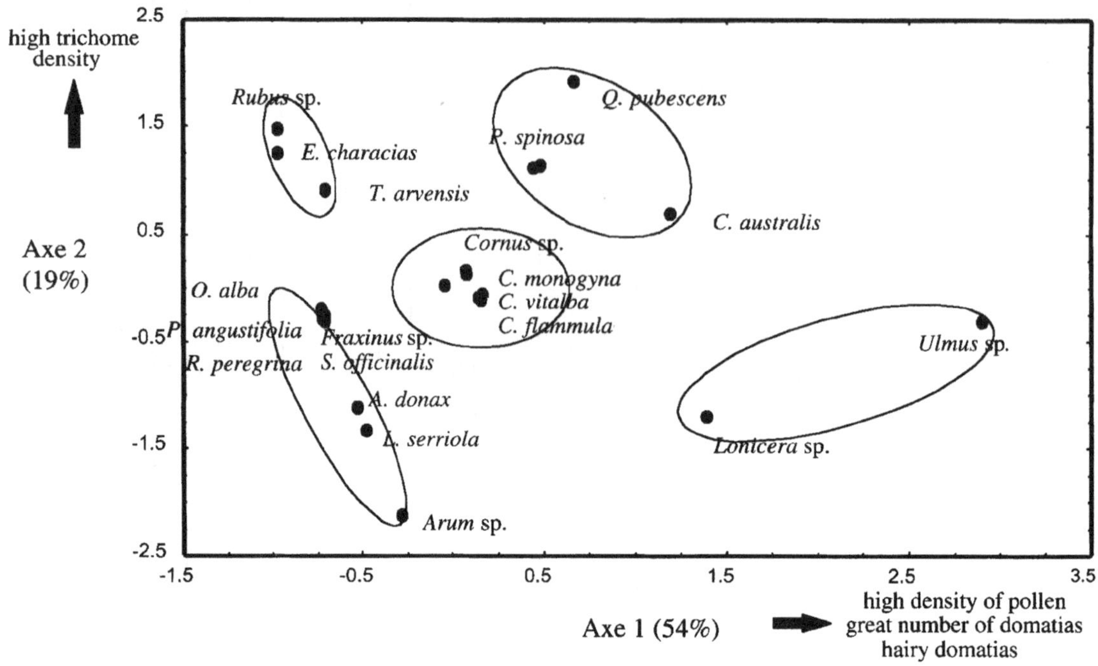

Fig. 2 - Projection of the various studied plants in the plan 1-2 of the ACP.

and *Q. pubescens* were another group that had high pollen densities and average numbers and structures of domatia. *Ulmus* sp. and *Lonicera* sp. had high levels of phytoseiids, many obstructed domatia and dense pollen counts but average to low levels of trichomes. Lastly, *Cornus* sp., *Crataegus monogyna* Jacq., *Clematis flammula* L., and *Clematis vitalba* L. had intermediate numbers of domatia and complexity of domatia (rating). Of this latter group, only *Cornus* sp. showed high phytoseiid mite densities. Results of decision tree analysis (fig. 3) were similar to those of the multifactorial analysis. Abundances of phytoseiid mites on 20 plants were most linked to the trichome densities, then to domatia rating. Neither phytoseiid diversity nor percent *K.aberrans* were linked to the parameters studied.

Pollen on selected plants relative to leaf pilosity

Many phytoseiids including *K. aberrans* are pollen feeders (McMurtry 1992; Schausberger 1992; McMurtry and Croft 1997) and it is thought that pollen capture and retention by leaf hairs affect these mites greatly. Duso *et al.* (1997) reported different seasonal trends in pollens on a grape variety and found a significant effect on *K. aberrans* in Northern Italy vineyards. To evaluate such relationships among plants near French vineyards, plant species, leaf pilosity and pollen retention measurements were made. Identification of pollen in air or leaf samples (Faegri and Iversen 1989) is difficult because many pollens occur (Andersen 1974). Abietaceae pollens that have two distinctive air bags can be identified with a binocular microscope (Faegri and Iversen 1989). Many phytoseiids feed upon these pollens (Engel 1989), including *K. aberrans* (Barret 1994). Pollen counts were made during April 1992 on both sides of leaves of European hairy nettle (22 leaves), *Ulmus procera* Salisb. (13 leaves) and glabrous winter lime trees (20 leaves). All sample trees were located in a local area near Montpellier and received similar conditions of wind and distances to pollen sources. To convert to a common unit of

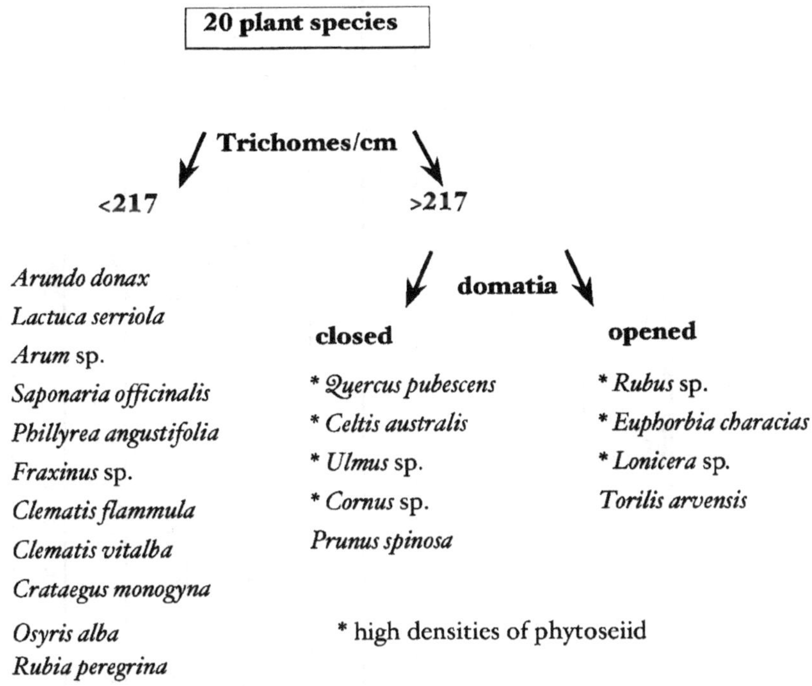

Fig. 3 - Decision tree concerning the fig. explanation of the variable number of phytoseiid mites per leaf.

pollen density, surface area per leaf was measured (±0.2 cm^2). A regression between pollen density and leaf surface (underside and upperside) was calculated for each plant type.

These tree species had different pollen densities per unit area (fig. 4). Leaves of *C. australis* trapped and retained more pollen than *U. procera* and *T. vulgaris*. Leaves of *T. vulgaris* were large but had a low density of pollen that increased significantly with leaf surface (r^2=0.38, P<0.003). The regression slope (3.73) represents an increase of 4 grains/cm^2. Most all pollen grains appeared to be trapped by domatia. On *U. procera*, the correlation was similar to *T. vulgaris* (r^2=0.34, P<0.004) but the increase in pollen density was 45.9 grains/cm^2. For *C. australis*, pollen grain density was 83.5 grains/cm^2, but there was no correlation between pollen density and leaf surface ($r^2=1.7.10^{-3}$, P=0.89). Over all plants, pollen density was higher on under- than upper-leaf surfaces and the upper-side was always less hairy.

A second comparison of pollen density and pilosity was made between European hairy and the American glabrous nettle trees. Two types of leaves were used, either unwashed or washed in acetone for three months to remove sticky materials that might affect pollen retention. Two kinds of unwashed leaves were used, live and dried. All tests were conducted at a site near Montpellier where both European hairy and American glabrous nettle trees grew in mixed stands. Also, a light wind of 3 km/h was simulated in an environmental chamber to nettle trees that were held under similar climatic conditions. Both unwashed and washed leaves (100 of each tree species) were placed on both sides (50 of each) randomly on 50 x 40 cm plastic panels. Also, 100 leaves from currently growing plants were collected at random from the two tree species. Both hairy and glabrous nettle (both *Celtis*) tree leaves were compared (washed and unwashed) after being held under two experimental conditions (natural and laboratory-

Tab. 4 - Abietaceae pollen grains per cm^2 on leaves of *Celtis australis* or *C. occidentalis* in field or laboratory tests.

Number and origin of leaves used for experiment	Type of experiment	Number of Pollen *Celtis australis*	Grains (SE) *Celtis occidentalis*	Probability & significance
50 alive leaves from field. Collected on trees. Side down	-	43.92	9.72	-
50 alive leaves from field. Collected on trees. Side up	-	15.42	2.78	-
Probability and significance	-	1.84×10^{-8} S	8.80×10^{-7} S	-
50 dead leaves. Collected on trees. Put in acetone 3 months. Side down	Lab simulation	16.39	5.52	-
50 dead leaves. Collected on trees. Put in acetone 3 months. Side up	Lab simulation	6.52	3.48	-
Probability and significance	-	6.45×10^{-5} S	0.1586 NS	-
152 alive leaves from field. Collected on trees	-	8.31 (6.52)	4.15 (3.76)	7.62×10^{-11} S
50 leaves from field. Collected on trees & dried	Field. On trees	0.80 (1.01)	0.18 (0.17)	10^{-4} S
50 dead leaves. Collected on trees. put in acetone 3 months	Field. On trees	0.93 (0.77)	0.26 (0.26)	10^{-4} S
50 alive leaves from field. Collected on trees	Lab simulation	4.47 (6.98)	1.80 (2.26)	0.0126 S
50 dead leaves. Collected on trees. Put in acetone 3 months	Lab simulation	1.06 (0.70)	0.69 (0.81)	0.014 S

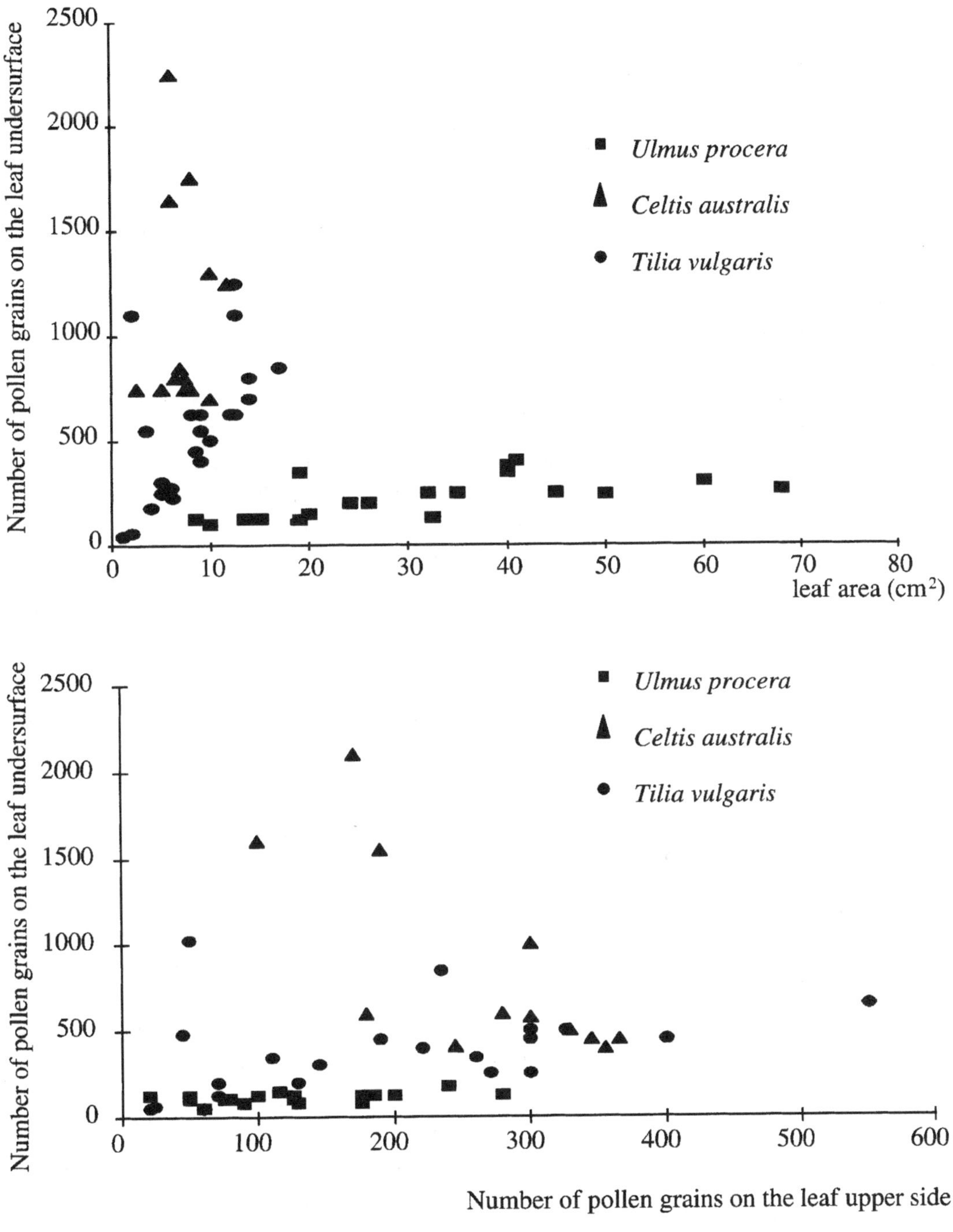

Fig. 4 - Relations between a) the number of pollen grains of Abietaceae on the leaf undersurface and the leaf area (upper) and b) the number of pollen grains of Abietaceae on the leaf under- and upper-surface for three plants: *Celtis australis*, *Tilia vulgaris*, *Ulmus procera* (bottom).

reared). Data from both plant species and pilosity-nettle tree tests were submitted to ANOVA and a mean comparison tests (Newman and Keuls'; a=0.05).

Celtis australis retained significantly more pollen by 100-150% over smooth leafed *C. occidentalis,* either when washed or unwashed (tab. 4). Leaf sizes, shapes and veination were similar for these *Celtis* sp. and only pilosity was different. Nevertheless, for both conditions for washed leaves, pollen density still was higher on hairy than glabrous leaves (respectively, 258 and 53% more), unwashed leaves, especially live ones from trees, retained pollen better. *Kampimodromus aberrans*, on taxonomically related or morphologically similar trees, has a tendency to inhabit those with high pilosity whereas some other species of phytoseiid dominate on those with less hairy leaves.

Uptake of plant fluids

Use of concentrated pollens as foods may require free water for assimilation. On European hairy nettle tree, *K. aberrans* often has its mouthparts inserted in the leaf. This behaviour also occurs on other plants but on nettle tree it is common to observe, perhaps because the leaf is easier to penetrate or because pollen is dense and need for water is great. Rubidium (rb^+), has been widely used to test water uptake by organisms (Beery *et al.* 1972; Graham *et al.* 1978; Fernandes *et al.* 1978). Water can be taken up by plants, and then by predaceous mites either from the plant or from phytophagous mites that have fed (Frazer and Raworth 1974; Jackson 1991). Thirty-two year-old hairy nettle trees having 5-6 leaves were cultivated in planter boxes and caged with muslin. All visible foods on leaves were removed with a brush and leaves were exposed for 1 min every 12 h and 5 d to ultraviolet light (wavelength 254 nm) to eliminate micro-organisms. The soil of 20 trees was watered daily with a rubidium chloride solution at 2 g/l (1250 ml/tree total). Female *K. aberrans* were collected on other European hairy nettle and after 24 h starvation, female adult mites (1 per plant) were placed both on labelled and unlabelled trees. A glue ring was placed on the trunk to avoid mites to escape. After 4 days treatment, analysis was performed on recovered mites with an atomic absorption spectrophotometer. A single labelled female predator was taken with a sterile plastic shaft and dissolved in 2 ml of pure nitric acid for 48 h at 20°C. Solutions were homogenized and diluted with 0.5 ml distilled water before analysis.

Kampimodromus aberrans from labelled trees had rubidium labelled individuals (0.167 ± 0.104 ng/female) while mites from unlabelled trees did not (0 ng/female). UV light use excluded any labelled micro-arthropods or any feeding micro-organisms as sources of rubidium. While we observed *K. aberrans* on many occasions with its mouthparts inserted into a leaf, these data confirm that actual fluids are taken up by this behaviour.

Discussion

The role of plants in providing substrates for colonization, liquid and pollen foods, and pilosity and domatia either for pollen trapping or protection are important for phytoseiid mites, especially for generalist feeding species (Karban *et al.* 1995; Walter 1996; McMurtry and Croft 1997). For *K. aberrans* these effects may act in direct, indirect and complex ways. We have tried to identify some of these effects. While field studies are useful to explore initial aspects of phytoseiid biology, they are limited in explaining mechanisms of behaviour because of the poor control of variables. Despite these constraints, we have shown that pilosity affects *K. aberrans*; but exactly why is not yet clear. However, our results show how strong plant relationships are for *K. aberrans*:

1. It was known that this mite was common in Southern Europe and often associated with economic crops and surrounding plants having pilose leaves (Duso 1992; Duso *et al.* 1993; Barret and Kreiter 1995; Camporese and Duso 1996). Our

surveys extend populations studies for this mite to many non-economic plants and most have pilose leaves. Others have observed this trend on grapes (Camporese and Duso 1996; Duso and Vettorazzo 1999), in other regions of France (Burgundy, S. Kreiter, unpublished data) and in other countries (Austria, Schausberger 1992). On apple *K. aberrans* is displaced by *E. finlandicus* on less hairy varieties (Schausberger 1992), which suggests that pilosity provides protection for *K. aberrans* or that larger phytoseiids do not like pilose leaves. However, other factors may contribute to these patterns of occurrence or dominance (e.g., thigmotaxis, need for oviposition site, low humidity tolerance, use of pollen foods, etc.).

2. While the pilosity-*K. aberrans* relationship was known, our study reports more about effects of trichome and domatia densities and domatia structure. Higher trichome and domatia densities were correlated with higher predator densities. Trichome spacing of 170-180 mm may allow deep movement of *K. aberrans* while excluding larger predators. Domatia ratings that reflect domatia complexity also seem to affect densities of *K. aberrans*. Such leaf structures may allow habitat partitioning and more phytoseiid species to provide pest regulation at lower densities than single species (Croft and Slone 1998).

3. On closely related trees, we demonstrated that pilosity was the most important factor affecting pollen capture/retention, while leaf positioning, leaf surface type, stickiness and domatia types also could be important. Differences in pollen retention could dramatically influence densities of phytoseiids, especially those that primarily feed on pollens early before prey mites increase to appreciable densities (*K. aberrans*, Duso *et al.* 1997; *T. pyri*, Addison *et al.* 2000). Energy costs of walking on hairy leaves for a large species may be excessively high. Behavioural studies with competing species and different pilosity and pollen levels would be helpful in testing this hypothesis directly (see below).

4. We demonstrated that *K. aberrans* does take up plant fluids. We have not observed this species to ingest water from open sources like many other phytoseiids do. Just what benefit is obtained from plant fluids is unclear, but if pollens require water for food processing, then this uptake could meet this need. Such uptake could confer considerable plant specificity to *K. aberrans* and it seems to be a limiting factor in rearing this mite on substrates other than leaves (S. Kreiter, *unpubl. data*, Schausberger and Croft 1999).

Finally, while we have shed light on several aspects of plant-phytoseiid interactions and on some effects that leaf pilosity may have on behavior and development, we have not been able to test many of these aspects directly. Most of our conclusions are based on correlation or inferences from field-derived data. Such types of data are useful in early stages of research to indicate general relationships and to identify hypotheses. Laboratory tests are required now in simplified systems where variables can be controlled and single factor relationships evaluated. For example, by observing behaviour of *K. aberrans* and *E. finlandicus* when put together on overlapping smooth and a hairy leafed *Celtis* and by manipulating pollen levels, one could test innate leaf preferences, attractions to pollens and interspecific interactions. One challenge would be to maintain a plant physiology that is favorable to both phytoseiids. *Kampimodromus aberrans* seems to require a plant with its turgor intact and a detached leaf soon looses turgor to the extent that *K. aberrans* stops feeding and laying eggs. Presumably, this is because it lacks plant fluids. While these constraints pose problems, they are not insurmountable. Such studies will help elucidate plant relationships for generalist-feeding phytoseiids that seem more complex than for specialist-feeding phytoseiids (Grafton-Cardwell and Ouyang 1996; O'Dowd and Willson 1991; McMurtry and Croft 1997).

We may consider that *K. aberrans* is a well-known species and that we have made great progresses in the knowledge of *K. aberrans* / host plant relationships. However we must consider that these are only the first steps as we still have not addressed how to rear this species. This phytoseiid mites seems to need turgor leaves for rearing as it seems to take very often plant fluids from the leaf. It is hard, even impossible, to rear on *T. urticae* and pollens are not regularly suitable.

As we cannot rear this species, it is impossible to conduct applied research on prey preference, food preference, dispersal behaviour, leaf structure effects, side effects of pesticides, etc. So there is a need for basic research on the biology of this species, involving may be a plant physiologist and acarologists having experience of this species.

Studying this species in the field, we even thought that several species of *Kampimodromus* may exist all around the experimental plot. Indeed, some species described previously from the Mediterranean area are morphologically very close (Ragusa and Tsolakis 1994). You will see in a paper by Tixier and Kreiter (this volume) that molecular RAPD markers, and it is the case also from results of a morphometry study and of some cross breeding experiments, allow us to tell that there is only one species. However, we observed a great range of variations of both molecular and morphometric data, and other species within the genus *Kampimodromus* are in the range of variation of this morphometric data. Morphological differences between these species are so tiny that it is natural to consider that some synonymy may exist. How many species finally within the genus *Kampimodromus*? Molecular markers such as RAPD will allow us to determine if other species close to *K. aberrans* can really have the status of species. These data crossed with morphometric analysis and cross breeding results could lead to a revision of the genus *Kampimodromus*. This work on taxonomy and distinction between species is really basic. But it is a big challenge if we want to understand and manage populations of *K. aberrans* for controlling spider mites population. This also requires the effort of all specialists involved on this species, in Italy, Austria, Poland, USA but also other teams excellent in lab experiments or in genetics.

One could be surprised by the fact that knowledge on one of the most important species for protection of vineyards against tetranychid mites remains so obscure. It is not so surprising looking at the literature of the few past decades. In a period of 28 years (1973-2000) 804 papers concerned *P. persimilis*, 420 *Typhlodromus pyri* Scheuten, 152 *Neoseiulus cucumeris* (Oudemans) and *Amblyseius andersoni*, 112 *Neoseiulus californicus* (McGregor) but only 104 for *K. aberrans* and 30 for *Phytoseius plumifer* Canestrini & Fanzago (Source: CAB database). Generalist predators are less known because they are less lucrative in terms of money and in terms of scientific papers. We need to change this.

Conclusions

That is to say that an International acarologist's working group could start a project funded by the Europe Union on *K. aberrans*! And on many other species or subjects involving mites of agricultural importance! What is sure is that our future research will always start in the field, combine field and lab experiments with all tools available, old and recent. Indeed, according to us, questions must be determined thanks to field naturalist observations on phytoseiid mites distribution (on plants, regions, climatic conditions, with other organisms: phytoseiids, preys ...) and lab experiments must be carried out to understand the relationships between phytoseiid mite species and their environment. If this kind of research groups, concern bywith the biology of a closely related species within a genus, involving several fields of biology, several recent and old tools and techniques, the Wilson's (1989) scenario, developed in the opening lecture of the 9th International Congress of Acarology by Krantz (1998), is a reality in Europe. This marriage allies descriptive and molecular biology for a study of a taxonomic group at all levels of organisation. This is a confirmation that Acarology is a modern science following Conroy's introduction (1990) in Ceské Budéjovice (former Czechoslovakia).

New position may be devoted to Acarology and not only on new tools and not only on species that facilitate quick results for numerous publications, such as *P. persimilis*. We must also work with species that have an interest in Agricultural Acarology and not necessarily easily to managed in the laboratory.

There is a big need of education in Agricultural Acarology in all countries, in basic topics such as taxonomy as well as in new tools available for research (but tools must not constitute the end and the aim). This also concerns developing countries and this must be a big challenge and an important point for European acarologists who want to collaborate with those countries, mainly in Africa, South America and Asia, and improve transmission of knowledge and ideas, and also discovering new species, new habitats and new life-style of mites.

References

Addison J.A., Hardman J.M., Walde S.J., 2000 - Pollen availability for predaceous mites on apple: spatial and temporal heterogeneity. *Experimental and Applied Acarology*, 24: 1-18.

Andersen S.T., 1974 - Wind conditions and pollen deposition in a mixed deciduous forest. *Grana*, 4: 57-77.

Aponte O., McMurtry J.A., 1992 - Mating behaviour and reproductive mechanisms of *Amblyseius colimensis*. *Boletin de Entomologia Venezolana*, 7: 1-12.

Barret D. 1994 - Influence de l'architecture du phylloplan dans l'organisation des peuplements de phytoséiides et dans leurs associations avec les plantes. PhD thesis, ENSA Montpellier, France: 174 pp.

Barret D., Kreiter S., 1995 - Morphometrics of some phytoseiid mites and characteristics of their habitat: consequences for biological control. In: Kropczynska D., Boczek J., Tomczik A. (eds.), The Acari - Physiological and ecological aspects of Acari-host relationships. Dabor, Warszawa: 461-473.

Beery W.L., Stimmann M.W., Wolf W.W., 1972 - Marking native phytophagous insects with rubidium: a proposed technique. *Annals of the Entomological Society of America*, 65: 236-238.

Berry J.S., Holtzer T.O., Ball H.J., 1990 - Ambulatory dispersal of spider mites on whole, untreated maize plants after exposure to fenvalerate and permethrin. *Journal of Econonomic Entomology*, 83: 217-220.

Boutaleb Joutei A., 1999 - Etude de l'effet de l'augmentation du CO_2 atmosphérique sur l'interaction plante-phytophage. PhD thesis, University of Louvain-La-Neuve, Belgium: 193 pp.

Bruin J., Sabelis M.W., Dicke M., 1995 - Do plants tap SOS signals from their infested neighbours. *Trends in Ecology and Evolution*, 10: 167-170.

Camporese P., Duso C., 1996 - Different colonisation patterns of phytophagous mites on three grape varieties: a case study. *Experimental and Applied Acarology*, 20: 1-22.

Coiutti C., 1993 - Acari fitoseidi su piante arboree spontanee e coltivate in fruili-venezia giulia. *Frustula Entomologica*, 16(24): 65-77.

Conroy J.C., 1991 - Acarology - A modern Science. In: Dusbabek F., Bukva V. (eds.), Modern Acarology. Academie Prague and SPB Academic Publishing, The Hague, The Netherlands, vol. 1: 5-15.

Croft B.A., Slone D.H., 1998 - Perturbation of regulated apple mites: immigration and pesticide effects on outbreaks of *Panonychus ulmi* and associated mites. *Environmental Entomology*, 27: 1548-1556.

De Moraes G.J., McMurtry J.A., Denmark H.A., 1986 - A catalog of the mite family Phytoseiidae : References to Taxonomy, Synonymy, Distribution and Habitat. Embrapa, Brasilia: 353 pp.

Dicke M., Sabelis M.W., 1988 - How plants obtain predatory mites as bodyguards. *Netherland Journal of Zoology*, 38: 148-165.

Dicke M., Takabayashi J., Posthumus M.A., Schütte C., Krips O., 1999 - Behavioural ecology of plant-phytoseiid interactions mediated by herbivore-induced plant volatiles. In: Bruin J., van der Geest L., Sabelis M.W. (eds.), Ecology and Evolution of the Acari, Kluwer, Dordrecht: 251-268.

Dolgov S.V., Mityshkina T.U., Bukavtsova E.B., Buryanov Y.I., 1995 - Production of transgenic plants of *Chrysanthemum morifolium* Ramat with the gene of *Bacillus thuringiensis* delta-endotoxin. *Acta Horticulturae*, 420: 46-47.

Duso C., 1992 - Role of *Amblyseius aberrans*, *Typhlodromus pyri* and *Amblyseius andersoni* in vineyards. III. Influence of variety characteristics on the success of *A. aberrans* and *T. pyri* releases. *Journal of Applied Entomology*, 114: 455-462.

Duso C., Pasqualetto C., 1993 - Factors affecting the potential of phytoseiid mites as biological control agents in North-Italian vineyards. *Experimental and Applied Acarology*, 17: 241-258.

Duso C., Malagnini V., Paganelli A., 1997 - Indagini preliminari sul rapporto tra polline e *K. aberrans* su Vite. *Allionia*, 35: 229-239.

Duso C., Torresan L., Vettorazzo E., 1993 - La vegetazione spontanea come riserva di ausiliari: considerazioni sulla diffusione degli Acari Fitoseiidi in un vigneto e sulle piante spontanee contigue. *Bollettino di Zoologia Agraria e Bachicoltura*, 25: 183-203.

Duso C., Vettorazzo E., 1999 - Mite population dynamics on different grape varieties with or without phytoseiids released. *Experimental and Applied Acarology*, 23: 741-763.

Edwards O.R., Melo E.L., Smith L., Hoy M.A., 1998 - Discrimination of three *Typhlodromalus* species using random amplified polymorphic DNA markers. *Experimental and Applied Acarology*, 22(2): 101-109.

Enders M.M., 1993 - The effect of male size and operational sex ratio on male mating success in the common spider mite, *Tetranychus urticae*. *Animal Behaviour*, 46: 835-846.

Engel R., 1989 - Alternative prey and other food resources of the phytoseiid mites *Typhlodromus pyri*. *Zusammen Tagungs. Sitten*, 28: 124-127.

Faegri K., Iversen J., 1989 - Textbook of pollen analysis. 4th edition. John Wiley & Sons, New York: 430 pp.

Faraji F., Janssen A., Van Rijn P.C., Sabelis M.W., 2000 - Kin recognition by the predatory mite *Iphiseius degenerans*: discrimination among own, conspecific, and hetero-specific eggs. *Ecological Entomology*, 27: 147-155.

Fenton B., Mallocgh G., Jones A.T., Amrine J.W.Jr, Gordon S.C., A'Hara S., McGavin W.J., Birch A.N.E., 1995 - Species identification of *Cecidophyopsis* mites from different *Ribes* species and countries using molecular genetics. *Molecular Ecology*, 4: 383-387.

Fernandes O.A., Wright R.J., Baumgarten R.J., Mayo Z.B., 1978 - Use of rubidium to label *Lysiphhlebus testaceipes*, a parasitoid of green bugs, for dispersal studies. *Environmental Entomology*, 26: 1167-1172.

Frazer B.D., Raworth D., 1974 - Marking aphids with rubidium. *Canadian Journal of Zoology*, 52: 1135-1136.

Gotoh T., Ishikawa Y., Takayama K., Suzuki M., 1991 - Differences in esterase zymograms between diapausing and non-diapausing individuals of the hawthorn spider mite, *Tetranychus viennensis*. *Applied Entomology and Zoology*, 26: 153-155.

Gotoh T., Gomi K., Nagata T., 1999a - Incompatibility and host plant differences among populations of *Tetranychus kanzawai* Kishida. *Applied Entomology Zoology*, 34: 551-561.

Gotoh T., Gutierrez J., Navajas M., 1998 - Molecular comparison of the sibling species *Tetranychus pueraricola* and *T. urticae*. *Entomological Science*, 1: 55-57.

Gotoh T., Sugasawa J., Nagata T., 1999b - Reproductive compatibility of the two-spotted spider mite (*Tetranychus urticae*) infected with *Wolbachia*. *Entomological Science*, 2: 289-295.

Grafton-Cardwell E.E., Ouyang Y., 1996 - Influence of citrus leaf nutrition on survivorship, sex ratio, and reproduction of *Euseius tularensis*. *Environmental Entomology*, 25: 1020-1025.

Graham H.M., Wolfenbarger D.A., Nosky J.B., 1978 - Labelling plants and their insect fauna with rubidium. *Environmental Entomology*, 7: 379-383.

Hance T., Neuberg P., Noël-Lastelle C., 1998 - The use of fecundity, lobe biometry and the RAPD-PCR technique in order to compare strains of *Tetranychus* sp. *Experimental and Applied Acarology*, 22: 649-666.

Heagle A.S., Brandenburg R.L., Burns J.C., Miller J.E., 1994 - Ozone and carbon dioxide effects on spider mites in white clover and peanut. *Journal of Environmental Quality*, 23: 1168-1176.

Herron G.A., Rophail J., 1998 - Tebufenpyrad (Pyranica") resistance detected in two-spotted spider mite *Tetranychus urticae* from apples in Western Australia. *Experimental and Applied Acarology*, 22: 633-641.

Jackson C.G., 1991 - Elemental markers for entomophagous insects. *Southwestern Entomologist*, 14: 65-69.

Janssen A., 1999 - Plants with spider-mite prey attract more predatory mites than clean plants under greenhouse conditions. *Entomologia Experimentalis et Applicata*, 90: 191-198.

Jie C.J., Yuan X.J., 1999 - Influence of transgenic (Bt) cotton variety R93-4 on the insect community (in Chinese). *Journal of Southwest Agricultural University*, 21: 358-363.

Kaliszewski M.J., Tobolewski J., Seyoum S., Chojnacki I., Kaliszewska M.M., Stanton D.J., Colwell R.K., 1992 - The polymerase chain reaction and sequencing of mite DNA. *International Journal of Acarology*, 18: 231-239.

Karban R., English-Loeb G., Walker M.A., Thaler J., 1995 - Abundance of phytoseiid mites on *Vitis* sp.: effects of leaf hairs, domatia, prey abundance and plant phylogeny. *Experimental and Applied Acarology*, 19: 189-197.

Karban R., English-Loeb G., Hougen-Eitzman D., 1997 - Mite vaccinations for sustainable management of spider mites in vineyards. *Ecological applications*, 7: 183-193.

Kolmes S.A., Dennehy T.J., Sam Y., 1994 - Contrasting behaviour of two-spotted spider mites on discontinuous residues of a pyrethroid and a chlorinated hydrocarbon acaricide. *Journal of Economic Entomology*, 87: 559-565.

Krantz G.W., 1998 - Specialisation and systematics in Acarology: reflections and predictions. In: Needham G.R., Mitchell R., Horn D.J., Welbourn W.C. (eds.), Acarology IX. Ohio Biological Survey, Columbus, Ohio, vol. 2: 1-4.

Kreiter S., Tixier M.-S., Auger P., Muckensturm N., Sentenac G., Doublet B., Weber M., 2000 - Phytoseiid mites of vineyards in France. *Acarologia*, 41(2): *in press*.

Krips O.E., Kleijn P.W., Willems P.E.L., Gols G.J.Z., Dicke M., 1999 - Leaf hairs influence searching efficiency and predation rate of the predatory mite *Phytoseiulus persimilis*. *Experimental and Applied Acarology*, 23: 119-131.

Kropczynska D., Van de Vrie M., Tomczyk A., 1988 - Bionomics of *Eotetranychus tiliarum* and its phytoseiid predators. *Experimental and Applied Acarology*, 5: 65-81.

Kuang H.Y., Xie M.C., Gong G.J., 1992 - A preliminary study on the relationship between the families of Eriophyoidea. *Acta Zootaxonomica Sinica*, 17: 294-298.

Lester P.J., Thistlewood H.M.A., Harmsen R., 2000 - Some effects of pre-release host-plant on the biological control of *Panonychus ulmi* by the predatory mite *Amblyseius fallacis*. *Experimental and Applied Acarology*, 24: 19-33.

Li J., Hoy M.A., 1996 - Adaptability and efficacy of transgenic and wild-type *Metaseiulus occidentalis* compared as part of risk assessment. *Experimental and Applied Acarology*, 20: 563-573.

Margolies D.C., 1995 - Evidence of selection on spider mites dispersal rates in relation to habitat persistence in agroecosystems. *Entomologia Experimentalis et Applicata*, 76: 105-108.

McDermott G.J., Hoy M.A., 1997 - Persistence and containment of *Metaseiulus occidentalis* in Florida: risk assessment for possible releases of transgenic strains. *Florida Entomologist*, 80: 42-53.

McMurtry J.A., 1992 - Dynamics and potential impact of "generalist" phytoseiids in agroecosystems and possibilities for establishment of exotic species. *Experimental and Applied Acarology*, 14: 371-382.

McMurtry J.A., Croft B.A., 1997 - Life-styles of Phytoseiid mites and their roles in biological control. *Annual Review of Entomology*, 42: 291-321.

Navajas M., Gutierrez J., Lagnel J., 1996 - Mitochondrial COI in tetranychid mites: a comparison between molecular phylogeny and changes of morphological and life history traits. *Bulletin of Entomological Research*, 86: 407-417.

O'Dowd D.J., Willson M.F., 1989 - Leaf domatia and mites on Australasian plants: ecological and evolutionary implications. *Biological Journal of the Linnean Society*, 37: 191-236.

O'Dowd D.J., Willson M.F., 1991 - Associations between mites and leaf domatia. *Trends in Ecology and Evolution*, 6: 179-182.

Osakabe M., Komazaki S., 1996 - Differences in esterase isozymes between *Panonychus citri* populations infesting citrus and *Osmanthus*. *Applied Entomology and Acarology*, 20: 113-119.

Pallini A., Janssen A., Sabelis M.W., 1997 - Odour-mediated responses of phytophagous mites to conspecific and heterospecific competitors. *Oecologia*, 110: 179-185.

Pallini A., Janssen A., Sabelis M.W., 1998 - Predators induce interspecific herbivore competition for food in refuge space. *Ecological Letters*, 1: 171-177.

Pallini A., Janssen A., Sabelis M.W., 1999 - Do western flower thrips avoid plants infested with spider mites? Interactions between potential competitors. In: Bruin J., van der Geest L., Sabelis M.W. (eds.), Ecology and Evolution of the Acari, Kluwer, Dordrecht: 375-380.

Presnail J.K., Hoy M.A., Jeyaprakash A., 1998 - Developing transgenic phytoseiids for biological control programs. In: Needham G.R., Mitchell R., Horn D.J., Welbourn W.C. (eds.). Acarology IX. Ohio Biological Survey, Columbus, Ohio, vol. 2: 15-19.

Ragusa S., Tsolakis H., 1994 - Revision of the genus *Kampimodromus* with a description of a new species. *Acarologia*, 35: 305-322.

Sabelis M.W., Janssen A., Bruin J., Bakker F.M., Drukker B., Scutareanu P., Van Rijn P.C.J., 1999 - Interactions between arthropod predators and plants: a conspiracy against herbivorous arthropods. In: Bruin J., van der Geest L., Sabelis M.W. (eds.), Ecology and Evolution of the Acari, Kluwer, Dordrecht: 207-229.

Saito Y., 1990 - "Harem" and "non-harem" type mating system in two species of sub-social spider mite. *Research Population Ecology*, 32: 263-278.

Schausberger P., 1992 - Vergleichende untersuchungen zum lebensverlauf die erstellung von lebenstafeln und die vermehrungskapazitaet von *Amblyseius aberrans* und *Amblyseius finlandicus*. *Pflanzenschutzberichte*, 52: 53-71.

Schausberger P., 1997 - Inter- and intraspecific predation on immatures by adult females in *Euseius finlandicus*, *Typhlodromus pyri* and *Kampimodromus aberrans*. *Experimental and Applied Acarology*, 21: 131-150.

Schausberger P., 1998 - Survival, development and fecundity in *Euseius finlandicus*, *Typhlodromus pyri* and *Kampimodromus aberrans* feeding on the San José scale *Quadraspidiotus persniciosus*. *Journal of Applied Entomology*, 122: 53-26.

Schausberger P., 1999 - Predation preference of *Typhlodromus pyri* and *Kampimodromus aberrans* when offered con- and heterospecific immature life stages. *Experimental and Applied Acarology*, 23: 389-398.

Schausberger P., Croft B.A., 1999 - Predation and discrimination between con- and heterospecific eggs among specialist and generalist phytoseiid mites. *Biological Control*, 28(3): 523-528.

Tixier M.-S., 2000 - La présence de phytoséiides dans les abords non cultivés des parcelles a-t-elle un intérêt appliqué en viticulture? PhD. Thesis, ENSA Montpellier, France: 189 pp.

Tixier M.-S., Kreiter S., Auger P., 2000 - Colonisation of vineyards by phytoseiid mites: their dispersal patterns in the plot and their fate. *Experimental and Applied Acarology*, 24(3): 191-211.

Tixier M.-S., Kreiter S., Auger P., Weber M., 1998 - Colonisation of Languedoc vineyards by phytoseiid mites: influence of wind and crop environment. *Experimental and Applied Acarology*, 22: 523-542.

Tsagkarakou A., Navajas M., Lagnel J., Gutierrez J., Pasteur N., 1996 - Genetic variability in *Tetranychus urticae* from Greece: insecticide resistance and isozymes. *Journal of Economic Entomology*, 89: 1354-1358.

Tsagkarakou A., Navajas M., Lagnel J., Pasteur N., 1997 - Population structure in the spider mite *Tetranychus urticae* from Crete based on the variations of multiple allozymes. *Heredity*, 78: 84-92.

Tsagkarakou A., Navajas M., Rousset F., Pasteur N., 1999 - Genetic differenciation in *Tetranychus urticae* from greenhouses in France. *Experimental and Applied Acarology*, 23: 365-378.

Walzer D.E., Schausberger P., 1999a - Cannibalism and interspecific predation in the phytoseiid mites *Phytoseiulus persimilis* and *Neoseiulus californicus*: predation rates and effects on reproduction and juvenile development. *BioControl*, 43: 457-468.

Walzer D.E., Schausberger P., 1999b - Predation preferences and discrimination between con- and heterospecific prey by the phytoseiid mites *Phytoseiulus persimilis* and *Neoseiulus californicus*. *BioControl*, 43: 449-478.

Walter D.E., 1996 - Living on leaves: mites, tomenta and leaf domatia. *Annual Review of Entomology*, 41: 101-114.

Walter D.E., O'Dowd D.J., 1992a - Leaves with domatia have more mites. *Ecology*, 73: 1514-1518.

Walter D.E., O'Dowd D.J., 1992b - Leaf morphology and predators: effect of leaf domatia on the abundance of predatory mites. *Environmental Entomology*, 21: 478-484.

Wilson E.O., 1989 - The coming pluralization of biology and the stewardship of systematics. *Bioscience*, 39: 242-245.

Yamamura N., 1987 - Bi-parental defence in a sub-social spider mite. Trends in *Ecology and Evolution*, 2: 261-162.

Zhang Z.-Q., Sanderson J.P., Nyrop J.P., 1992 - Foraging time and spatial pattern of predation in experimental populations. A comparative study of three mite predator-prey systems (Acari: Phytoseiidae, Tetranychidae). *Oecologia*, 90: 185-196.

Mite communities on linden trees (*Tilia* sp.) in an urban environment

D. Kropczynska, B. Czajkowska, A. Tomczyk, M. Kielkiewicz
Department of Applied Entomology, Warsaw Agricultural University, PL-02-787 Warsaw, Nowoursynowska 166, Poland

Key words: *Eotetranychus tiliarium* (Hermann), Phytoseiidae, predators, population dynamics, urban conditions.

Abstract: The research on the factors influencing plant inhabiting mite population under urban conditions has been conducted in our department over the past several years. In the paper are summarized experimental evidence on the impact of town climate, traffic pollutants, host plant quality on the occurrence and density on the linden mite and its phytoseiid predators.

Introduction

As shown by our previous research, mite communities associated with urban trees are characterized by considerable prevalence in mite quantity of singular herbivorous species of spider mite (Tetranychidae), sometimes accompanied by representatives of Eriophyoidea (Kropczynska *et al.* 1985). Their frequent outbreaks usually occur on trees growing along urban roadways. It was shown that other groups of plant inhabiting mites (Tydeidae, Tarsonemidae and Acaridae) are numerous. It was also shown that anthropogenic factors did not have adverse effect on predatory mites (Phytoseiidae), which were found in large diversity of species and abundance both on street and park trees. A mechanism affecting the complex anthropogenic factors involved is not yet fully explained. We found that an increase in spider mite population can be caused by higher soil salinity which modifies the linden leaf biochemical composition and prevalence among phytoseiids of species of low efficiency (Kropczynska *et al.* 1988). The effect of many other factors on linden spider mite population is unknown.

Mite communities inhabiting linden trees were the subject of this study. Among them linden spider mite [*Eotetranychus tiliarium* (Hermann)] appeared to be super-dominant as it showed a multiple prevalence over other species especially on street trees. It highly outnumbered its own population developing in other locations. Relations between linden spider mite and its predators on linden trees were studied with regard to a potential influence of the following factors of an urban environment: urban microclimate, linden species, plant pollution with exhaust fumes from vehicles and biochemical composition of linden leaves.

F. Bernini, R. Nannelli, G. Nuzzaci, E. de Lillo (eds.), Acarid Phylogeny and Evolution. Adaptations in mites and ticks, 303-313.

Materials and methods

Basic observations were carried out in 1995-1999 on *Tilia platyphyllos* Scop. and *Tilia euchlora* K. growing in two stands located by the main roads in Warsaw (stand 1 and 2) and a park in Ursynow (stand 3). Information on the study sites is summarized in table 1. During the consecutive seasons, samples of 100 leaves were collected from each stand every two weeks. Mites found in each sample were counted and identified.

Population densities of Eotetranychus tiliarium *on six linden species*

In order to obtain information on the suitability as host for *E. tiliarium*, leaves from six linden species (*T. platyphyllos* Scop. , *T. euchlora* K., *T. cordata, T. varsoviensis, T. americana* and *T. tomentosa*) growing along roads with high traffic were sampled. In 1997, 100 leaf samples were collected three times in the middle of June, July and August and in 1998 every two weeks starting from the end of May. Mite density and damaged leaf area were the criterion of plant suitability as host for the spider mite. Variance analysis followed by a Duncan mean comparison test were used to compare the number of spider mites on different linden species.

Influence of urban traffic pollutants on mites

In 1995-98 leaves were collected from *T. platyphyllos* and *T. euchlora* growing in all experimental stands three times in the season. The leaves were analyzed for the content of heavy metals (Zn, Cu, Pb, Ni, Cd, Fe), polycyclic aromatic hydrocarbons benzo/a/pyrene (BaP) and benzo/e/pyrene (BeP). Heavy metals were determined by an adsorption spectrophotometric analysis (ASA) in a Perkin Elmer 300 type apparatus. For polycyclic aromatic hydrocarbons a liquid chromatography method was employed. Comparison between experimental stands and between linden species was performed by means of LSD (least significant difference).

Data acquired from laboratory observation allowed the construction of a life table for *E. tiliarium* feeding on leaves contaminated with pollutants to various degrees. In those experiments *T. platyphyllos* and *T. euchlora* leaves were sampled twice (June and August) from trees growing along the street (stand 1) and from the control (stand 3). Tests were carried out according to detached leaf culture method described by Helle and Overmeer (1985). For the calculation of developmental time and mortality a hundred eggs were observed in each combination. Fecundity and longevity were calculated for 25 females. Life table parameters

Tab. 1 - Characteristics of linden stands under study.

Stand no.	Location	Stand characteristics
1	W. Sikorski Street	Row of lime trees growing very close to a roadway along a cross-town aerial road; heavy traffic of cars, buses, vans and lorries. Poor greenary: lawns, single trees and bushes; good ventilation.
2	Ujazdowskie Avenue	Linden trees growing on a lawn, separated from a roadway by a pavement and a row of trees, close neighbourhood of the Lazienki Park and Botanical Garden. Heavy traffic cars and buses, high concentration of exhaust fumes, poor ventilation.
3	Ursynow, Park	Linden growing in the park centre; large diversity of vegetation.

were calculated according to the equations given by Andrewartha and Birch (1954).

Chemical composition of linden leaves

Leaf samples of *T. platyphyllos* and *T. euchlora* from experimental stands were taken for chemical analyses during two growing seasons 1995 (dry) and 1996 (wet). Leaves were collected three times in 1995 (May 25, July 15, September 12) and twice in 1996 (June 6 and July 31). Spider mite densities were monitored at the same time. Concentration of organic substances were estimated in fresh leaves (samples of 0.5-1 g) according to the methods described by Bradford (1976) for proteins, Nelson (1944) for reducing sugars, Yemm and Willis (1954) for total soluble sugars, Johnson and Schaal (1957) for total phenols. The leaf area was measured and content of dry matter was calculated. On the basis of measured leaf area and its dry weight, the specific leaf area (SLA) was estimated. For detailed description of the above methods see Tomczyk *et al.* (1997).

Results and discussion

Population densities of Eotetranychus tiliarium *on six linden species*

Eotetranychus tiliarium was recorded from all sampled trees in both seasons. Its population densities appeared to be significantly different on tested species of lindens (fig. 1). On *T. platyphyllos* and *T. americana* spider mite population reached high levels. Mite feeding caused severe defoliation of trees. Low densities of *E. tiliarium* were found on *T. cordata, T. varsoviensis, T. euchlora* and *T. tomentosa*. Damage symptoms on those linden species were primarily concentrated along the main veins of leaf and in the later stage of population development on restricted area of the leaf. No symptoms of mite damage were observed on *T. tomentosa* leaves. Possible reasons for the different suitability of *T. platy-*

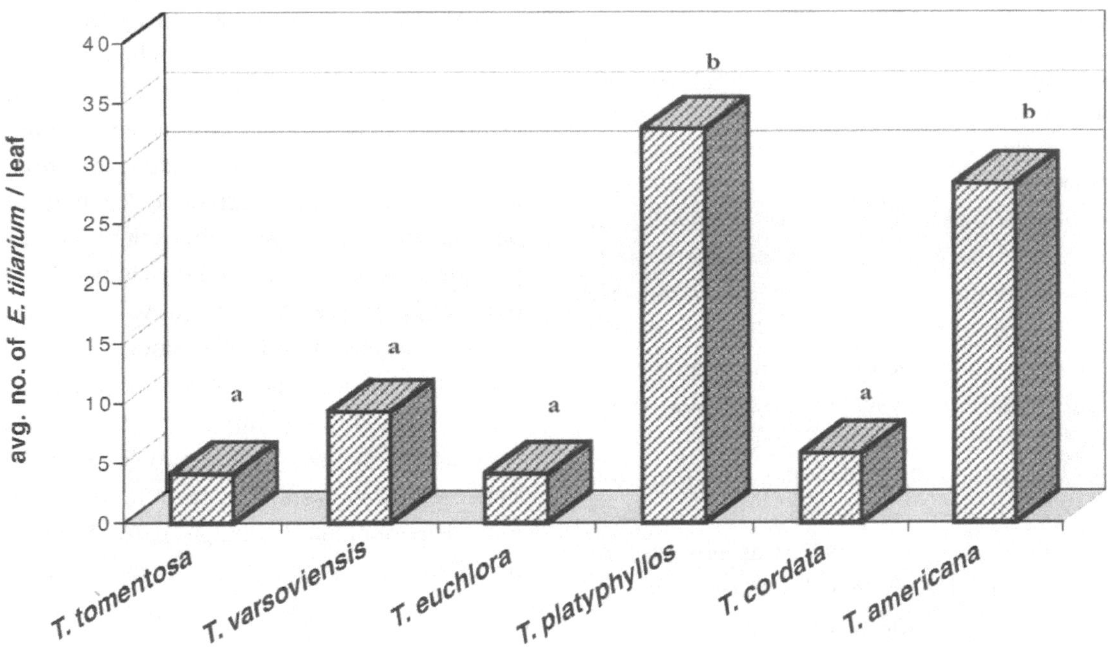

Fig. 1 - Average densities of *E. tiliarium* on surveyed linden species.

phyllos, T. cordata and *T. tomentosa* are related to leaf morphology and anatomy and discussed in the paper by Czajkowska and Kielkiewicz in this volume. Based on the density of *E. tiliarium* population, the tested linden species can be classified into three groups:

- most susceptible: *T. platyphyllos* and *T. americana*;
- susceptible to a low degree: *T. euchlora, T. cordata* and *T. varsoviensis*;
- least susceptible: *T. tomentosa*.

Seasonal densities of Eotetranychus tiliarium *and predatory mites*

The development of the first *E. tiliarium* generation began when the sum of effective daily temperatures (above 10°C) exceeded 400 degrees. Both the intensity and frequency of rainfall determined the rate of mite population growth (fig. 2, 3 years 1996 and 1999). During dry periods there was a multiple increase of *E. tiliarium* population in street stand (fig. 2A, 1995). Such a rise was not observed on the control trees nor on those growing near large park complexes (fig. 2B, year 1995, fig. 4, year 1995).

Another possible factor influencing linden mite population development seemed to be correlated to the number of females entering the overwintering sites. During the spring and early summer of 1997 the population of *E. tiliarium* was very low due to unfavourable weather conditions. A rapid growth of the linden mite was observed in August. Mites feeding on undamaged leaves reached a very high level. This resulted in the presence of numerous linden mites the following spring (fig. 2, year 1997, fig. 3, year 1998). On street trees, during 1995 and 1996, linden mites reached very high densities, 493 mites/leaf and 860 mites/leaf, respectively. In both years such high population of mites was observed on the leaves, not longer than for two weeks and was followed by a rapid decrease in mite number (fig. 2A, years 1995, 1996). This was related to the decrease in the nutritional value of leaves and activity of predators.

The quantity of predatory mites was strictly connected with linden species, tree stand and prey supply. In each stand the linden trees were inhabited by a complex of predatory mites consisting of 2-5 species. Such rich specific complex of predatory mites occurred on *T. platyphyllos* with the prevalence of *Paraseiulus soleiger* (Ribaga), whereas on other observed linden species a distinct dominance of *Euseius finlandicus* (Oud.) was recorded (fig. 5).

Euseius finlandicus and *P. soleiger* are apparently tolerant to stress conditions as their high numbers were found on trees growing in a polluted and dry environment.

The influence of predatory species on *E. tiliarium* population inhabiting street trees depended on the initial density of the pest at the beginning of the growing season. If abundant, the first generation of linden spider mite became a rich source of nourishment for predators, which subsequently could expand in numbers to the level which efficiently limited prey (fig. 3A, year 1998). *Eotetranychus tiliarium* outbreaks due to favourable weather conditions resulted in the delayed effect of phytoseiid activity (fig. 2A, year 1995). Sabelis (1985) stated that in high temperature and low humidity the numerical response of most phytoseiid predators are inadequate or lacking.

The outbreak of linden mite population was not observed in the street stand 2 where old *T. platyphyllos* trees grew in the neighbourhood of parks (fig. 4, year 1995). In the vicinity of big park trees experimental lindens grew partly in the shade. This resulted in the linden spider mite population developing in lower temperature and higher humidity as compared to the population developing on trees exposed to high insolation. The conditions were less favourable for spider mite development (Rabbinge 1976; Tanigoshi *et al.* 1975).

Influence of urban traffic pollutants on mites

Leaves of street trees absorbed significantly more heavy metals than those of the control,

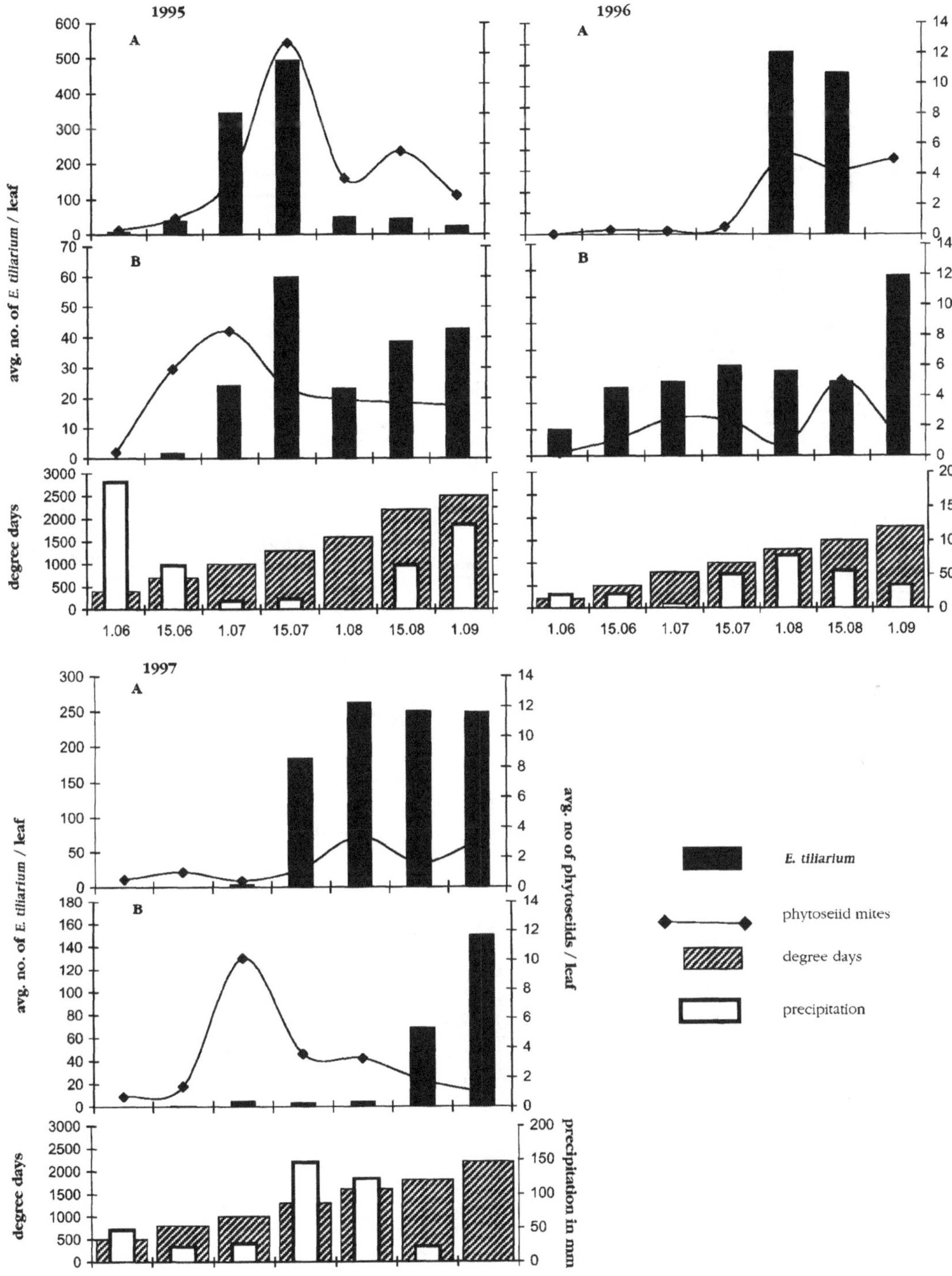

Fig. 2 - Seasonal densities of *E. tiliarium* and phytoseiid mites on *T. platyphyllos* during 1995-1997. A - stand 1, B stand 3.

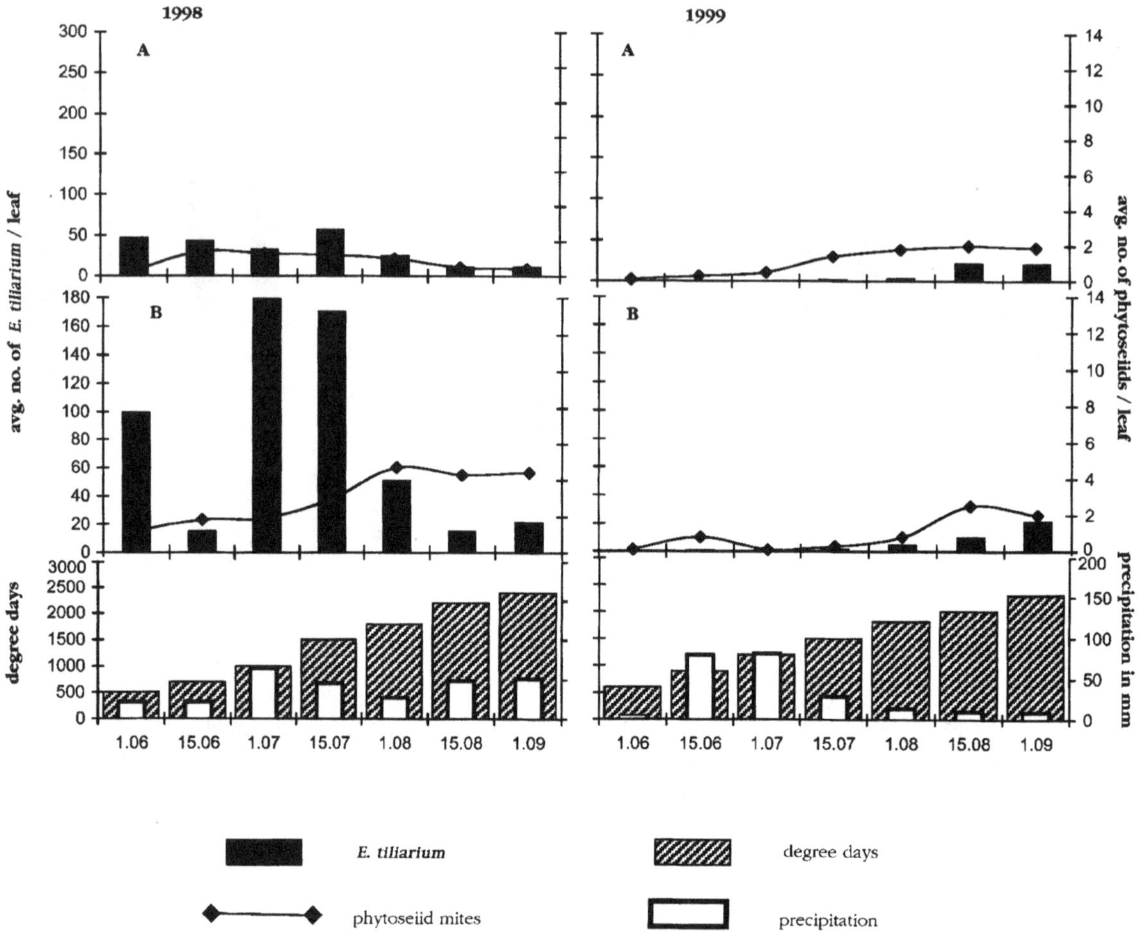

Fig. 3 - Seasonal densities of *E. tiliarium* and phytoseiid mites on *T. platyphyllos* during 1998 and 1999. A - stand 1, B - stand 3.

however, there were no significant differences in this respect between the foliage from both examined street stands (tab. 2). It was shown that the content of examined pollutants was higher in *T. platyphyllos* leaves as compared to *T. euchlora* (tab. 3).

Net reproduction rate R_o and intrinsic rate of increase r_m of linden mites were higher on leaves of *T. platyphyllos* growing in street stand. The higher rate of increase was caused by higher fecundity of female feeding on polluted leaves. The effect of pollution was not so evident in the

Tab. 2 - Value of LSD for heavy metals and polycyclic aromatic hydrocarbons in the leaves of *T. platyphyllos* and *T. euchlora*. * - significantly different at P=0.05.

Stand	Zn	Cu	Pb	Ni	Fe	BaP	BeP
2	5,51*	2,58	2,3*	0,37*	58,8*	5,75	9,28*
1	7,04*	4,62*	5,12	0,52	154*	11,30	17,2
3	2,47	2,32	1,38	0,45	60,1*	1,70*	3,20

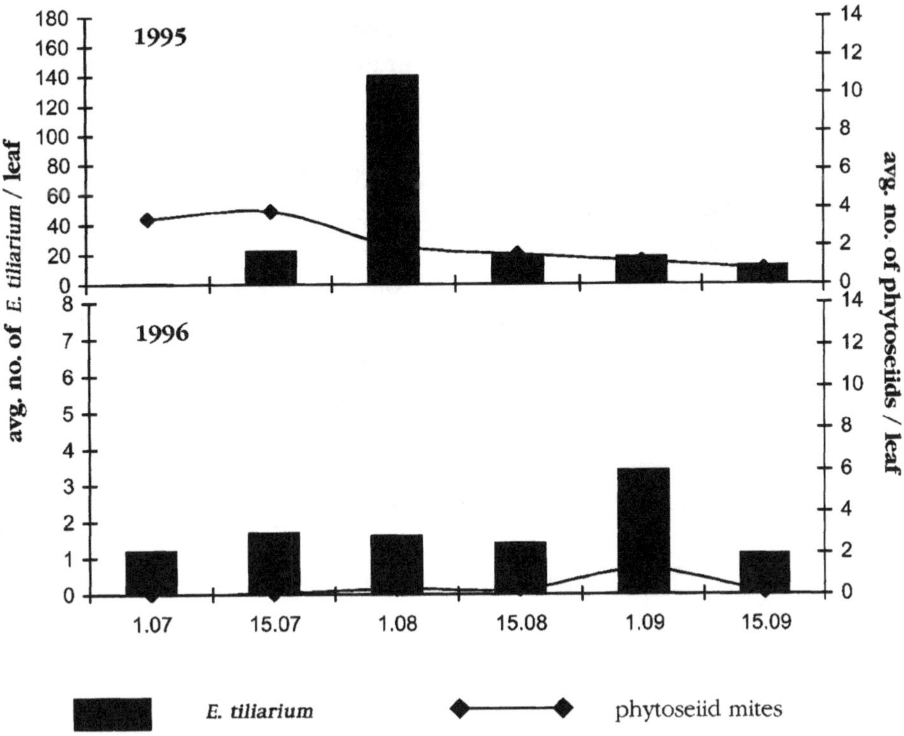

Fig. 4 - Seasonal densities of *E. tiliarium* and phytoseiid mites on *T. platyphyllos* in stand 2 during 1995 and 1996.

Tab. 3 - Value of LSD for metals and polycyclic aromatic hydrocarbons in linden leaves growing in experimental stands. * differences significant at P=0.05.

A. *Tilia platyphyllos*

Stand	Zn	Cu	Cr	Pb	Ni	Fe	BaP	BeP
2/1	6,5	6,6	1,4	3,5	0,5	153*	4,9	8,1
2/3	10,4*	5,35	1,14	4,5*	0, 6*	99	17,4*	27,2*
1/3	9,5*	5,1 *	2,1	5,9*	0,5	17*	17,7*	25,5*

B *Tilia euchlora*

Stand	Zn	Cu	Cr	Pb	Ni	Fe	BaP	BeP
2/1	6,8	2,5	0,9	1,4	0,3	88*	5,0	8,9
2/3	3,2*	2,9*	0,9	2,1*	0,3	59*	9,2*	12,6*
1/3	5,3*	2,5*	1,01	2,8*	0,4	112*	11,4*	17,0*

case of mites developing on polluted and non polluted leaves of *T. euchlora* (tab. 4.). This can be related to lower content of pollutants on the leaves of this linden species (tab. 3).

Chemical composition of linden leaves

Differences in the content of organic substances depended upon the year, linden species and experimental stands (tab. 5). In both years in the leaves of *T. euchlora* originating from polluted trees, minor changes in the level of soluble proteins were found as compared to the control. In the leaves of *T. platyphyllos* collected from street trees the content of soluble proteins strongly decreased during warm and dry season of 1995. In more wet conditions (1996) protein concentration in the leaves from the polluted stands was much lower than in those from the control.

Total content of soluble sugars in the leaves of street trees decreased as compared to the control, mainly because of the decrease in concentration of non-reducing sugars. In the warm season the concentration of phenols in the polluted leaves of both linden species strongly declined. In the following year (wet conditions) this decrease was only observed in *T. euchlora* leaves.

During a dry summer, in contrast to a wet summer, leaves from street trees were bigger and thicker (the lower value SLA). Their dry matter

Tab. 4 - Life history parametres of *E. tiliarium* feeding on the leaves of linden trees exposed to traffic pollution.

Linden species		r_m/week	R_0	T	λ
Tilia platyphyllos					
	Street trees	1.05	40.00	23.17	3.00
	Park trees	1.01	28.2	23.24	2.78
Tilia euchlora					
	Street trees	1.09	34.65	22.54	3.03
	Park trees	1.13	32.27	22.40	3.06

Tab. 5 - Content of some organic substances (mg/g fresh weight, means ±SE) in the leaves of *Tilia platyphyllos* and *Tilia euchlora* growing in urban condition.

Tilia platyphyllos

Year/Stand		Reducing sugars	Total sugars	Soluble proteins	Total phenols
1995	2	7,4 ± 1,02	16,7 ± 1,04	2,3 ± 0,25	8,1 ± 1,19
	1	8,3 ± 0,79	19,2 ± 0,99	1,3 ± 0,23	7,3 ± 0,87
	3	7,2 ± 0,73	21,7 ±1,31	0,9 ± 0,08	12,9 ±3,29
1996	2	11,0 ± 1,50	24,0 ± 0,37	1,5 ± 0,27	7,8 ± 1,73
	1	11,8 ± 1,36	21,4 ± 0,97	1,5 ± 0,23	5,9 ± 1,28
	3	9,9 ± 1,06	25,4 ± 2,76	2,9 ± 0,33	4,0 ± 0,84

Tilia euchlora

Year/Stand		Reducing sugars	Total sugars	Soluble proteins	Total phenols
1995	2	9,5 ± 1,22	23,5 ±1,00	1,5 ± 0,19	7,9 ± 0,87
	1	14,0 ± 1,62	30,3 ±1,89	1,3 ± 0,23	9,8 ± 0,83
	3	14,6 ± 2,32	37,0 ±3,98	1,1 ± 0,13	11,2 ± 0,58
1996	2	6,6 ± 0,87	20,8 ±1,58	1,9 ± 0,19	2,7 ± 0,51
	1	6,7 ± 1,32	22,6 ±1,10	2,1 ± 0,25	2,9 ± 0,12
	3	6,9 ± 0,40	25,8 ±1,70	1,8 ± 0,26	4,0 ± 0,67

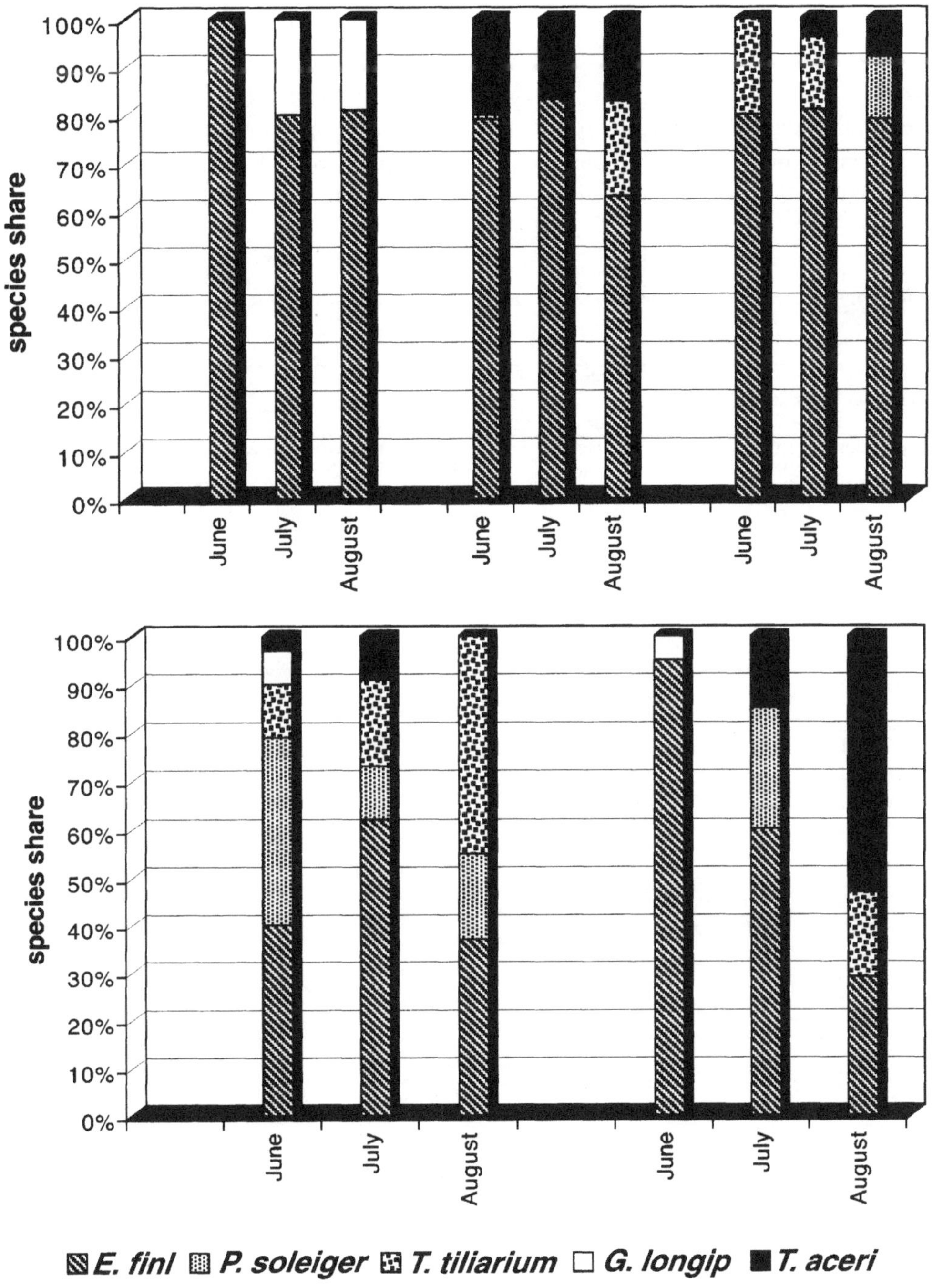

Fig. 5 - Diversity of phytoseiid mites on five linden species growing along streets.

content was significantly higher (tab. 6). High ratio of soluble sugars to phenols (fig. 6) and higher dry matter content in polluted leaves can stimulate mite population development.

Conclusions

Collected data indicate that an urban microclimate has a decisive influence on the density of *E. tiliarium* population. Urban microclimate is characterized by: evident differences in the daily temperature which in summer can exceed 10-12°C between the center of town and the suburbs, low R.H. especially during May and June varying between 58-71% and an advanced (by two weeks) initiation of the growing season (Kossowska 1968 *after* Zimny 1973). All these factors can positively influence spider mite population development. Within crowns of linden trees growing along busy streets mites have optimal conditions for develop-

Tab. 6 - Content of dry matter (%±SE) in leaves and SLA (m^2/g of dry weight) for leaves of linden trees growing in urban condition.

Linden species/year		Stand 1		Stand 2		Stand 3	
		Dry matter	SLA	Dry matter	SLA	Dry matter	SLA
T. platyphyllos	1995	38,9 ± 0,86	166	32,4 ± 1,16	274	35,0 ± 0,69	259
	1996	32,8 ± 0,48	-	33,4 ± 0,57	-	34,2 ± 1,33	-
T. euchlora	1995	35,8 ± 0,53	141	28,3 ± 0,46	240	31,0 ± 0,66	192
	1996	33,1 ± 0,24	-	28,7 ± 0,90	-	31,1 ± 0,46	-

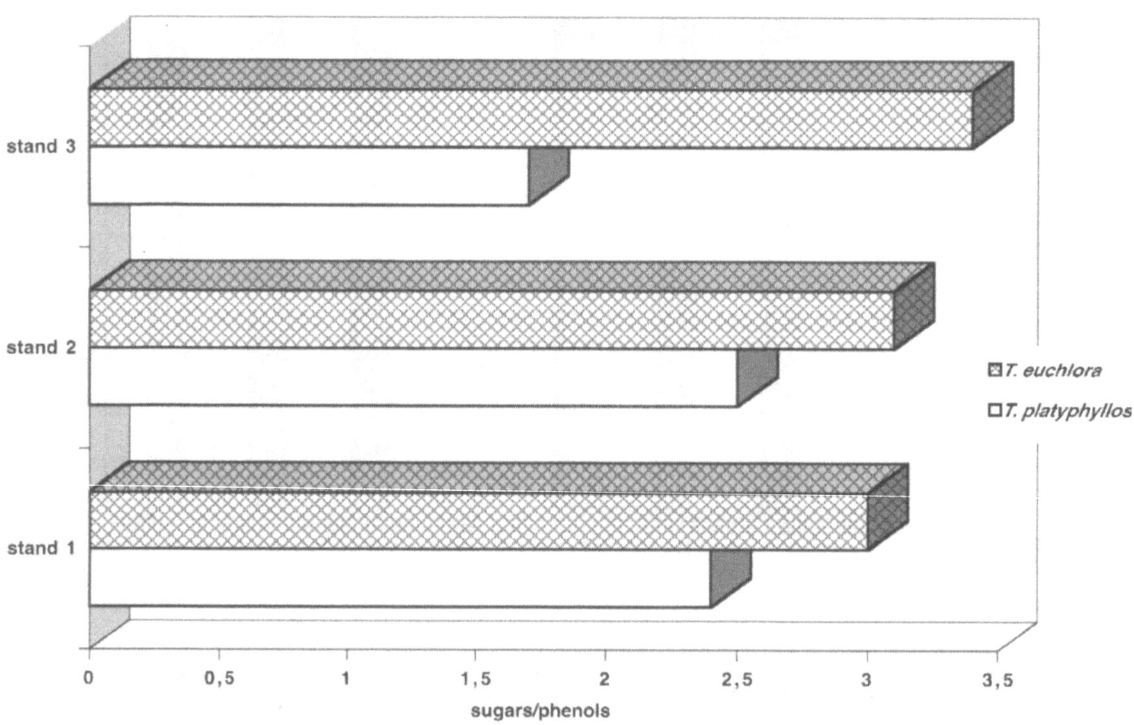

Fig. 6 - Ratio of sugars to phenols in the linden leaves in 1995.

ment: favourable temperatures and low humidity. Such a vital influence of thermal conditions was proved by low densities of spider mite populations during cool seasons and their rapid increase when the weather turned optimal.

Weather conditions also show an indirect effect on *E. tiliarium* by modifying the course of host plant metabolism. During the hot and dry season the content of phenolic compounds in the foliage of street trees was reduced, thus one of the factors affecting the expansion of spider mites was limited. Leaves of street trees enhanced their dry weight which could stimulate the development of pests. Such an effect could also be caused by an increase in the concentration of non-protein nitrogen and a modified ratio of soluble sugars to phenols.

According to the data presented in the work by Kabata-Pendias and Pendias (1993) the extent of leaf pollution with heavy metals and polycyclic aromatic hydrocarbons in experimental trees was considered not detrimental for plants, but in some cases Pb and Cr concentration were about to reach a harmful level. We could prove that this degree of pollution apparently had a positive effect on developmental parameters of *E. tiliarium*.

Phytoseiidae were able to keep *E. tiliarium* populations under control only on trees exposed to limited insolation and enhanced air humidity. Such conditions in tree crowns are favourable for predatory mites but less favourable for pests.

References

Andrewartha H.G., Birch L.C. 1954 - The distribution and abundance of animals. The University of Chicago Press, Chicago: 782.

Bradford M.M., 1976 - A rapid and sensitive method for the quantitation of microgram quantities of protein utilizing the principle of protein dye binding. *Annals of Biochemistry*, 72: 248-254.

Helle W., Overmeer W.P.J., 1985 - Rearing Techniques. In: Helle W., Sabelis M.W. (eds.), Spider Mites. Their Biology, Natural Enemies and Control. Elsevier. Amsterdam, 1A: 331-335.

Johnson G., Schaal L.A., 1957 - Accumulation of phenolic substances and ascorbic acid in potato tuber tissue upon injury and their possible role in disease resistance. *American Potato Journal*, 34: 200-209.

Kabata-Pendias A., Pendias H., 1993 - Biogeochemistry of trace elements. Polskie Wydawnictwo Naukowe, Warszawa: 364 pp (in polish).

Kropczynska D., van de Vrie M., Tomczyk A. 1985 - Woody ornamentals. In: Helle W., Sabelis M.W. (eds.), World Crop Pests. Spider Mites their Biology, Natural Enemies and Control. Elsevier, Amsterdam, 1B: 385-393.

Kropczynska D., van de Vrie M., Tomczyk A. 1988 - Bionomics of *Eotetranychus tiliarium* and its phytoseiid predators. *Experimental and Applied Acarology*, 5(1/2): 65-81.

Nelson N., 1944 - A photometric adaptation of the Somogyi for determination of glucose. *Journal of Biological Chemistry*, 153: 3755-3780.

Rabbinge R., 1976 - Biological control of fruit-tree red spider mite. Simulation. Monograph, Pudoc, Wageningen, The Netherlands: 228 pp.

Sabelis M.W., 1985 - Life history. Capacity for population increase. In: Helle W., Sabelis M.W. (eds.), World Crop Pests. Spider Mites their Biology, Natural Enemies and Control. Elsevier, vol. 1B: 35-41.

Tanigoshi K.K., Hoyt S.C., Browne R.W., Logan J.A., 1975 - Influence of temperature on population increase of *Tetranychus mcdanieli*. *Annals of Entomological Society of America*, 68: 972-978.

Tomczyk A., Kielkiewicz M., Czajkowska B., 1997 - Water content and thickness of linden leaf blades in relation to spider mite population in urban conditions. *Progress in Plant Protection, Postepy w Ochronie Roslin*, 37(2): 65-67 (in polish).

Yemm E.N., Willis A.J., 1954 - The estimation of carbohydrates in plant extracts by antrone. *Biochemical Journal*, 57: 508-514.

Zimny H., 1973 - Urban environment ecology and trees. Proceedings of Symposium "Protection and cultivation of trees in towns and historical parks". Lancut 2-3 April 1973: 1-9 (in polish).

Effect of prey density on functional and numerical responses of the predatory mite *Typhlodromus negevi* (Acari: Phytoseiidae)

Aly H. Rasmy, G.M. Abou-Elella
Plant Protection Department, National Research Centre, Dokki, Cairo, Egypt

Key words: *Typhlodromus negevi*, prey density, *Tetranychus urticae*.

Abstract: Experimental studies on acarine predator-prey interactions were performed to determine the functional and numerical responses of the predacious phytoseiid mite *Typhlodromus negevi* Swirski and Amitai to density of the two-spotted spider mite *Tetranychus urticae* Koch. The experiments showed that prey density can have important effects on the development, reproduction and predation of the predator. Predator immatures developed faster in presence of excess prey. The results indicated that the predation rates of immatures and adults of *T. negevi* increased as prey density increased. This coincided with increasing numbers of deposited eggs although the fecundity of females tended to decrease slightly at high prey density. Also, at high prey densities, the predator produced female biased sex ratios.

Introduction

Although many studies have been conducted on phytoseiid mites and certain species have been recognized as valuable natural enemies, some researchers have doubted the abilities of these predatory mites to suppress high densities of their prey (Mori and Chant, 1966; Mori, 1969; El-Banhawy *et al.*, 1999). On the other hand, Laing and Osborn (1974) and Momen (1996) reported that increasing prey density did not have adverse effect on the functional and numerical responses of some phytoseiid mites.

This study represents an attempt to shed light on the effect of increasing the density of the tetranychid mite *Tetranychus urticae* Koch on the functional and numerical responses of the predatory mite *Typhlodromus negevi* Swirski and Amitai.

Materials and methods

Six groups of 20 predator larvae each were individually reared on leaf discs, 15 mm diameter, of mulberry leaves *Morus alba* L. provided with one of the six levels of prey availability 1, 3, 5, 15, 30 and 50 nymphs of the tetranychid mite *T. urticae* per predator larva. Prey nymphs at each predator group were renewed daily until predator larvae developed into adults under laboratory condition of 27±1°C and 75±5% R.H. Virgin predator females were then coupled with newly emerged males and the aforementioned prey levels were maintained until oviposition ceased. The number of prey consumed per predator and number of eggs deposited per female were recorded daily. Progeny were allowed to develop into adults at which time the sex could be easily determined.

F. Bernini, R. Nannelli, G. Nuzzaci, E. de Lillo (eds.), Acarid Phylogeny and Evolution. Adaptations in mites and ticks, 315-318.

Larvae of the phytoseiid mite *T. negevi* were obtained from a laboratory culture maintained on *T. urticae* as prey.

Throughout this study the term density is defined as the number of prey per experimental arena. Statistical analyses, using F-test, were conducted and values of standard error (SE) are presented in figures 1, 2 and 3.

Results

Development

Results showed that the protonymphs and deutonymphs of *T. negevi* developed significantly faster as prey density increased from 3 to 15 prey nymphs, then at higher prey densities the developmental periods of predator immatures were almost similar to that of their counter parts reared at prey density of 15 nymphs per leaf disc (fig. 1). The lowest prey density of 1 prey led to 24% juvenile mortality for *T. negevi* immatures, while no mortality was noted with higher prey densities. Therefore, the protonymph and deutonymph of *T. negevi* required a prey density not less than 2-3 prey for 100% survival.

Reproduction

It was noted that *T. negevi* females reared at a density of 1 prey nymph per leaf disc did not lay eggs, while rearing at densities of 30 and 50 prey increased the daily rate of oviposition and the fecundity of females (fig. 2). Also, it was revealed that at least 7-8 prey nymphs must be consumed to allow *T. negevi* females to reach adulthood and lay eggs. Results presented in figure 1 showed that preoviposition and oviposition periods correlated negatively with increasing prey density.

Predation rate of immatures and adult females

Figure 3 showed that the predation rate of *T. negevi* immatures, protonymphs and deutonymphs, was markedly increased when prey density increased to the levels of 30 and 50 as predator immatures killed a total number of prey nymphs averaged 12.6 and 15.5, respectively. On the other hand, prey densities from 1 to 15 led to a significantly less effect on predation rate increase (fig. 3). This correlation was more pronounced with predator adults as the predation rate of females increased sharply as prey density increased.

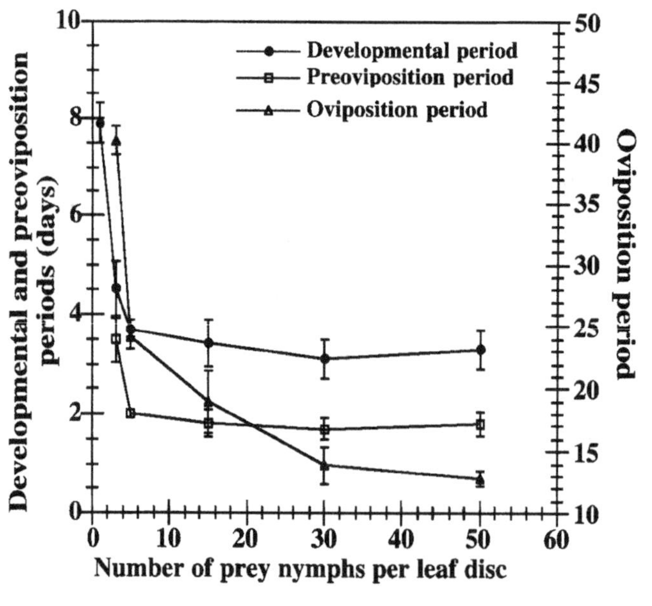

Fig. 1 - Developmental, preoviposition and oviposition periods of the predatory mite *Typhlodromus negevi* at different prey densities.

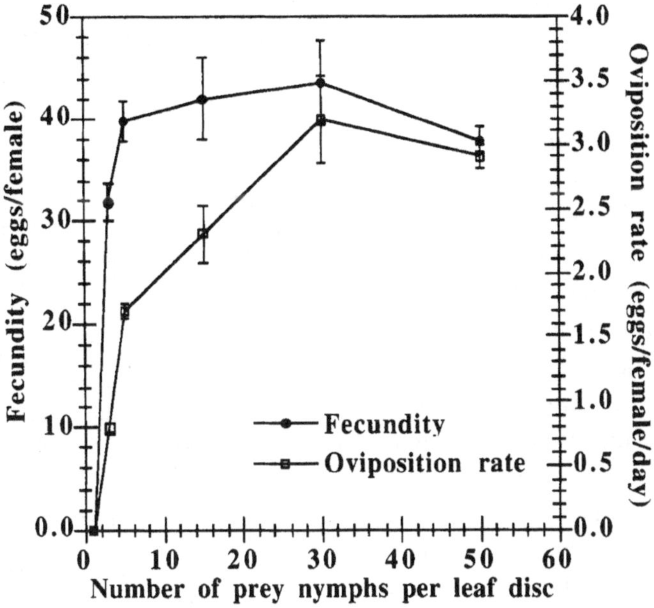

Fig. 2 - Fecundity and oviposition rate of the predatory mite *Typhlodromus negevi* at different prey densities.

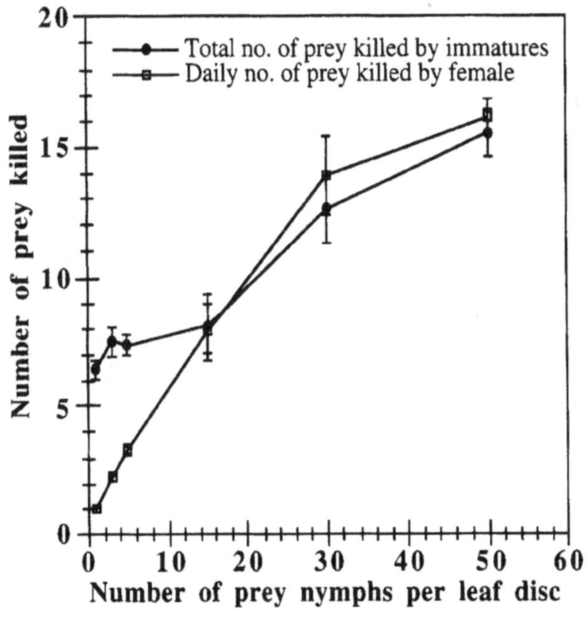

Fig. 3 - Daily rate of predation of adult female and total number of prey killed by immature stages of the predatory mite *Typhlodromus negevi* at different prey densities.

Sex ratio

At low prey densities unbiased sex ratio was noted, while at high prey densities *T. negevi* produced female biased sex ratios. When the predator was reared at prey densities of 3, 5, 15, 30 and 50 prey nymphs per leaf disc, sex ratios of

1:0.9, 1:1, 2:1, 3:1 and 3:1 females: male were produced, respectively.

Discussion and conclusions

The present study reveals that the predatory mite *T. negevi* requires few prey for development and reproduction. Also, it was noted that oviposition rate increased as the prey density increased and this coincided with a shortening in preoviposition and oviposition periods. It seemed that increasing the density of the tetranychid mite *T. urticae* enhanced the fecundity of the phytoseiid mite *T. negevi*. In addition, female biased sex ratios were produced at high densities of prey.

Concerning predation, it was noted that the rate of predation increased when *T. negevi* was offered excess of prey.

Similar results were reported by Laing and Osborn (1974), Fries and Gilstrap (1982) and Momen (1994 and 1996). They noted that oviposition and consumption rates of other phytoseiid species depend on the number of prey available. It is worth noting that Sabelis and van de Vrie (1979) observed the inability of *Amblyseius potentillae* (Garman) to suppress increasing population of *T. urticae* on roses. They attributed their findings to the hindrance of the predator by the web of the spider mite.

However, a dome-shaped functional response curve was not noted in these predator-prey interactions and did not prevent the suppressions of *T. urticae* nymphs at high densities. Additional work may be needed to shed light on the ability of *T. negevi* to suppress high population levels of the spider mite in the field.

References

El-Banhawy E.M., Hafez S.M., Saber S.A., 1999 - Effect of the nymph prey density of the two-spotted spider mite *Tetranychus urticae* on the consumption and reproduction rates of the predacious mite *Cydnoseius negevi* in absence and presence of nymphs of the whitefly *Bemisia tabaci*. *Journal of Pest Science*, 72: 55-56.

Fries D.D., Gilstrap E.E., 1982 - Influence of prey availability on reproduction and prey consumption of *Phytoseiulus persimilis*, *Amblyseius californicus* and *Metaseiulus occidentalis* (Acarina: Phytoseiidae). *International Journal of Acarology*, 8: 85-89.

Laing J.E., Osborn J.A.L., 1974 - The effect of prey density on the functional and numerical responses of three species of predatory mites. *Entomophaga*, 19: 267-277.

Mori H., Chant D.A., 1966 - The influence of prey density, relative humidity and starvation on the predacious behaviour of *Phytoseiulus persimilis* Athias-Henriot (Acarina: Phytoseiidae). *Canadian Journal of Zoology*, 44: 483-491.

Mori H., 1969 - The influence of prey density on the predation of *Amblyseius longispinosus* (Evans) (Acarina: Phytoseiidae). Proceeding of 2nd International Congress of Acarology, 1967: 149-153.

Momen F.M., 1994 - Fertilization and starvation affecting production in *Amblyseius barkeri* (Acari: Phytoseiidae). *Anzeiger für Schädlingskunde, Pflanzenschutz Umweltschutz*, 67: 130-132.

Momen F.M., 1996 - Effect of prey density on reproduction, prey consumption and sex ratio of *Amblyseius barkeri* (Acari: Phytoseiidae). *Acarologia*, 48: 3-6.

Sabelis M.W., van de Vrie M., 1979 - Evaluation of phytoseiid predators for control of spider mites. In: Rodriguez J.G. (ed.), Recent Advances in Acarology. Academic Press, New York, vol. 1: 492-498.

Dietary influence on life history and predation of the phytoseiid mite, *Amblyseius deleoni* (Acari: Phytoseiidae)

Aly H. Rasmy[1], F.M. Momen[1], M.A. Zaher[2], M.S. Nawar[2], G.M. Abou-Elella[1]
[1] *National Research Centre, Dokki, Cairo, Egypt*
[2] Faculty of Agriculture, Cairo University, Cairo, Egypt

Key words: *Amblyseius deleoni*, Phytoseiidae, spider mites, pollen.

Abstract: Laboratory experiments were conducted to study the biology and predation capacity of the phytoseiid mite *Amblyseius deleoni* (Muma & Denmark) as affected by feeding on different diets. The predatory mite was able to develop and reproduce when fed on nymphs of *Tetranychus urticae* Koch, nymphs of *Eutetranychus orientalis* (Klein), motile stages of *Eriophyes dioscoridis* Soliman & Abou-Awad, pollen grains of date palm and castor bean, eggs of scale insect, *Parlatoria ziziphus* Lucas and nymphs of the whitefly, *Bemisia tabaci* (Genn.). The present study revealed that the type of food significantly affected development, female longevity, sex ratio, fecundity and predation capacity of *A. deleoni*.

Introduction

The interest in phytoseiid mites has expanded greatly because they are predators of different pests and can survive and reproduce on different varieties of food sources.

In the present work, the predatory mite *Amblyseius deleoni* (Muma & Denmark) has been firstly recorded in Egypt on fruits and citrus trees associated with acarine pests. This phytoseiid mite has been recorded also on citrus trees in Florida (Muma, 1964). Therefore, laboratory studies were performed to assess the effect of different types of diets on various biological aspects of *A. deleoni*, e.g., development, sex ratio, reproduction and predation capacity.

Materials and methods

The predatory mite *A. deleoni* was obtained from a laboratory culture maintained on mulberry leaves, *Morus alba* L., infested with the two-spotted spider mite as prey. The leaves were placed upside down on cotton wool soaked in water in Petri-dishes.

Thereafter, eight groups, consisting of thirty predator eggs each were singly transferred to mulberry leaf discs, 2 cm in diameter. Each predatory larva was then supplied with a sufficient known number of prey to feed the predators, i.e., 20 nymphs of either of the three acarine prey, 20 nymphs of the whitefly, 10 aphids and 15 eggs of the scale insects. Each group of the predatory

F. Bernini, R. Nannelli, G. Nuzzaci, E. de Lillo (eds.), Acarid Phylogeny and Evolution. Adaptations in mites and ticks, 319-323.

larvae was fed on one of the following diets, i.e., nymphs of *Tetranychus urticae* Koch, nymphs of *Eutetranychus orientalis* (Klein), nymphs of *Aphis duranta* Theo, motile stages of *Eriophyes dioscoridis* Soliman & Abou-Awad, 1st and 2nd nymph instars of the whitefly, *Bemisia tabaci* (Genn.), eggs of the scale insect, *Parlatoria ziziphus* Lucas as well as pollen grains obtained from palm *Phoenix dactylifera* L. and castor bean *Ricinus communis* L. Another group of predatory larvae was kept on plant leaf discs did not harbour food for the predators.

Prey were replaced daily by fresh ones and numbers of prey consumed were recorded. Emerged females were coupled with emerged males for mating and the number of eggs deposited per female was recorded. Offspring were reared to adulthood to determine the sexes. Observations were recorded twice daily. Experiments were carried out in the laboratory at a temperature of 27±1°C and 70±5% R.H.

Results and Discussion

Development

Data presented in tables (1 and 2) showed that development and reproduction of *A. deleoni* were noted on all tested diets except aphid nymphs. Additionally, predator larvae kept on leaf discs, which did not harbour food for the predators, were not able to develop to adulthood except one male.

Feeding on the motile stages of the eriophyid mite, nymphs of the red spider mite and the citrus brown mite, eggs of the scale insect and pollen of the date palm led to a significantly shorter developmental period of predator immatures compared to feeding on the nymphs of the whitefly and pollen of castor bean. Predator females fed on castor bean pollen and nymphs of the red spider mite lived longest followed by their counterparts fed on the citrus brown mite, the eriophyid mite and the scale insect eggs, respectively. Feeding on whitefly nymphs led to the shortest life span (tab. 1). Development of predator males showed almost the same trends when fed on the aforementioned diets (tab. 2).

Reproduction

Results presented in table 3 revealed that predator females fed on the three species of mite prey and pollen of date palm exhibited the highest fecundity, while feeding on eggs of scale insects, whitefly nymphs or pollen of castor bean led to a significantly lowest oviposition rate. Moreover, postoviposition period exhibited by predator females reared on either of the eriophyid mite, the tetranychid mite and scale insect eggs was longer than that of their counterparts fed on the other diets.

Sex ratio

It is worth noting that the sex ratio of progeny was 2 females: 1 male when the predator females fed on either of the three species of mite prey, and 1.5:1 when they were reared on date palm pollen. On the other hand, rearing on pollen of castor bean and eggs of the scale insect led to an unbiased sex ratio (tab. 3).

Predation capacity of immature stages and adults

Predator larvae developed to protonymphal stage without feeding on any of the different tested prey species (tab. 4). Adult female consumed a remarkable daily average of *T. urticae* nymphs, *E. orientalis* nymphs and eggs of *P. ziziphus*.

However, the present work revealed that the type of food significantly affects the development, female longevity, fecundity, sex ratio and predation capacity of the phytoseiid mite *A. deleoni*. The aforementioned prey species were suitable diets for such biological aspects and the red-spider mite ranked as the most suitable prey followed by the other two mite prey species and pollen of date palm.

Other workers have reported also that type of food affected the development and fecundity of

Tab. 1 - Average duration in days of different stages of *Amblyseius deleoni* female fed on different diets.

Predatory stages	*T. urticae* nymphs	*E. orientalis* nymphs	*E. dioscoridis* motile stages	*P. ziziphus* eggs	*B. tabaci* nymphs	Date palm pollen	Castor bean pollen	*A. duranta* nymphs
Egg	2.0±0.0	2.0±0.0	2.0±0.0	2.0±0.0	2.0±0.0	2.0±0.0	2.0±0.0	2.0±0.0
Larva	1.0±0.0	1.0±0.0	1.0±0.0	1.0±0.0	1.0±0.0	1.0±0.0	1.0±0.0	1.0±0.0
Protonymph	1.38±0.14	1.86±0.10	1.31±0.13	1.07±0.07	2.38±0.18	2.0±0.0	2.15±0.27	8.78±2.34
Deutonymph	1.54±0.14	1.36±0.13	1.08±0.08	2.00±0.18	1.77±0.12	1.23±0.12	2.31±0.17	10.83±1.60
Total immatures	3.92±0.21 ab	4.22±0.11 b	3.39±0.14 a	4.07±0.15 b	5.15±0.22 c	4.23±0.12 b	5.46±0.42 c	---
% Reaching maturity	100	100	100	100	100	100	100	0
Adult longevity	32.31±0.83 d	28.23±1.50 c	28.08±2.64 c	23.38±1.26 b	13.31±0.49 a	29.23±1.09 c	34.54±1.12 d	---
Life span	38.23±0.84 d	34.45±1.53 cd	33.47±2.57 c	29.45±1.21 b	20.46±0.50 a	35.46±1.02 cd	42.00±1.19 d	---

Different letters in the same row denote significant differences.
L.S.D. at 5%: Total immatures: 0.61 Adult longevity: 4.01 Life span: 3.95

Tab. 2 - Average duration in days of different stages of *Amblyseius deleoni* male fed on different diets.

Predatory stages	*T. urticae* Nymphs	*E. orientalis* nymphs	*E. dioscoridis* motile stages	*P. ziziphus* eggs	*B. tabaci* nymphs	Date palm pollen	Castor bean pollen	*A. duranta* nymphs
Egg	2.0±0.0	2.0±0.0	2.0±0.0	2.0±0.0	2.0±0.0	2.0±0.0	2.0±0.0	2.0±0.0
Larva	1.0±0.0	1.0±0.0	1.0±0.0	1.0±0.0	1.0±0.0	1.0±0.0	1.0±0.0	1.0±0.0
Protonymph	1.43±0.20	1.43±0.20	1.0±0.0	1.29±0.18	2.14±0.14	1.43±0.20	1.71±0.18	8.78±2.34
Deutonymph	1.00±0.00	1.00±0.00	1.14±0.14	1.43±0.20	1.57±0.20	1.00±0.00	2.00±0.00	10.83±1.60
Total immatures	3.43±0.18 ab	3.43±0.20 ab	3.14±0.14 a	3.72±0.18 b	4.71±0.29 c	3.43±0.20 ab	4.71±0.18 c	---
% Reaching maturity	100	100	100	100	100	100	100	0
Adult longevity	28.43±1.34 bc	27.57±1.31 bc	32.71±3.51 c	24.43±1.72 b	18.43±1.29 a	28.57±1.41 bc	30.71±1.02 c	---
Life span	33.86±1.32 bc	33.00±1.35 bc	37.85±3.49 c	30.15±1.64 ab	25.14±1.24 a	34.00±1.35 bc	37.42±1.07 c	---

Different letters in the same row denote significant differences
L.S.D. at 5%: Total immatures: 0.58 Adult longevity: 5.23 Life span: 5.17

Tab. 3 - Oviposition period, fecundity and sex ratio of *Amblyseius deleoni* fed on different types of food.

Types of food	Average duration in days			Average number of eggs/female		Proportion females %
	Preoviposition	Oviposition	Postoviposition	Total	Daily rate	
T. urticae nymphs	2.15±0.15 a	24.46±0.97 c	5.69±1.03 b	45.31±1.37 b	1.87±0.05 c	66
E. orientalis nymphs	2.33±0.12 ab	22.23±1.34 bc	3.77±0.68 a	40.54±2.25 b	1.85±0.07 c	66
E. dioscoridis moving stage	2.00±0.11 a	20.00±1.72 b	6.08±1.26 b	41.15±2.97 b	2.10±0.07 d	66
P. ziziphus eggs	3.07±0.29 b	9.46±1.02 a	10.77±1.42 c	8.23±0.74 a	0.91±0.06 b	50
B. tabaci nymphs	2.54±0.14 ab	6.31±0.31 a	4.46±0.49 a	9.23±0.61 a	1.47±0.08 b	56
P. dactylifera pollen	2.77±0.23 b	21.92±1.08 bc	4.54±0.60 a	41.00±1.79 b	1.87±0.03 c	60
R. communis pollen	5.85±0.48 c	24.62±1.58 bc	4.08±0.73 a	12.00±0.86 a	0.50±0.03 a	52
L.S.D. at 5%	0.71	3.44	0.81	4.83	0.16	

Different letters in a column denote significant differences.

Tab. 4 - Consumption rate of *Amblyseius deleoni* on different food substances.

Predatory stages		*T. urticae* nymphs	*E. orientalis* nymphs	*P. ziziphus* eggs	*B. tabaci* nymphs
Larva	♀	0	0	0	0
	♂	0	0	0	0
Protonymph	♀	4.69±0.64	11.64±1.40	4.54±0.62	3.77±0.32
	♂	4.00±0.58	6.00±0.72	5.00±1.09	2.86±0.34
Deutonymph	♀	10.77±1.19	10.71±1.00	16.23±1.54	4.62±0.35
	♂	5.00±0.44	9.71±0.68	9.43±1.54	2.86±0.55
Total of immatures	♀	15.46±1.40 b	22.35±1.56 c	20.77±1.33 c	8.39±0.45 a
	♂	9.00±0.79 b	15.71±1.25 c	14.43±1.49 c	5.72±0.75 a
Adult/day	♀	11.19±0.48 b	18.81±0.70 c	11.76±0.48 b	4.55±0.30 a
	♂	2.74±0.17 b	5.49±0.20 b	3.83±0.22 c	1.83±0.08 a
L.S.D. at 5%	Total immatures	♀: 3.63	Adult	♀: 1.45	
		♂: 3.25		♂: 0.51	

Different letters in the same row denote significant differences.

phytoseiid mites, e.g., Swirski *et al.* (1967) on *A. rubini* Swirski & Amitai and *A. swirskii* Athias, Afifi *et al.* (1988) on *Phytoseiulus persimilis* Athias-Henriot, *Neoseiulus fallacis* (German) and *Typhlodromus occidentalis* Nesbitt, Abou-Awad *et al.* (1989) on *Typhlodromus negevi* Swirski & Amitai and Momen (1995) on *A. barkeri* (Hugh.).

Additional studies have to be carried out to evaluate the predatory efficiency of *A. deleoni* under field conditions.

References

Abou-Awad B.A., Nasr A.K., Gomaa E.A., Abou-Elela M.M., 1989 - Life history of the predatory mite *Cydnodromella negevi* and the effect of nutrition on its biology (Acari: Phytoseiidae). *Insect Science and its Application*, 10(5): 617-623.

Afifi A.M., Patterson C.G., Potts M.F., Rodriguez J.C., 1988 - Comparative attractancy of 3 phytoseiid predator species to the two-spotted spider mite, *Tetranychus urticae* Koch. *Transaction Kentucky Academy of Science*, 49: 96-100.

Momen F.M., 1995 - Feeding, development and reproduction of *Amblyseius barkeri* (Acarina: Phytoseiidae) on various kinds of substances. *Acarologia*, 36(2): 101-105.

Muma M.M., 1964 - The population of Phytoseiidae on Florida citrus. *The Florida Entomologist*, 47(1): 5-11.

Swirski E., Amatai S., Dorzia N., 1967 - Laboratory studies on the feeding, development and reproduction of the predacious mites *Amblyseius rubini* and *Amblyseius swirski* on various kinds of food substances. *Israel Journal of Agricultural Research*, 17: 101-119.

Dispersion of *Amblyseius andersoni* (Chant) (Parasitiformes, Phytoseiidae) on grapevines and hazelnut near apples and peaches in the province of Caserta (Southern Italy)

Mariano Nicòtina, Ernesto Cioffi*
Dipartimento di Entomologia e Zoologia agraria, Università degli Studi di Napoli "Federico II", Portici (Na), Italy, nicotina@unina.it

Key words: *Amblyseius andersoni*, dispersion, grapevines, hazelnut, Caserta, Italy.

Abstract: Surveys were carried out on phytoseiid mites present on peach and apple orchards in the Campania region from 1994 to April 2000.

From the analyzed samples, 7 different species have been identified and are listed hereafter in decreasing levels of prevalence: *Amblyseius andersoni* (Chant), *Amblyseius stipulatus* Athias-Henriot, *Kampimodromus aberrans* (Oudemans), *Cydnodromus californicus* (McGregor), *Typhlodromus cryptus* Athias-Henriot, *Typhlodromus exhilaratus* Ragusa and *Typhlodromus talbii* (Athias-Henriot).

Among the species found there was a constant presence of *A. andersoni* on all the apple and peach trees in the entire Campania region.

Such a presence suggests that the widespread cultivations of apple and peach trees in the province of Caserta strongly influence and increase the distribution and colonization of this phytoseiid to hazelnut trees and vines.

Introduction

The various species of phytoseiid mites living on plants in agricultural ecosystems have led many researchers to deepen faunistic studies in order to collect data related to their distribution and dispersion patterns. Over the last 20 years studies in the Campania region have concentrated their attention to vine and hazelnut groves (Nicòtina and Viggiani 1985; Nicòtina *et al.* 1990; Nicòtina 1996 and 1999; Nicòtina and Cioffi 1999). These researches revealed that *Amblyseius andersoni* (Chant) was present in the province of Caserta on both vines and hazelnut trees whereas it was infrequently found in the other 4 provinces of the region. This phytoseiid is quoted by many Authors as common on fruit trees such as apple and peach (Croft and Ali Niazee 1983; Castagnoli *et al.* 1984; Strapazzon and Rensi 1989; Duso and Sbrissa 1990; Ivancich Gambaro 1991; Thistlewood 1991; Costa-Comelles *et al.* 1994; Nicòtina and Caprio 1996). In the province of Caserta the above mentioned two crops prevail. As a matter of fact from a 1999 census, 2,300 ha were cultivated with apple trees (51% of the whole Campania region destined to the cultivation of apple trees) and 12,119 ha with peach trees, 71% of the entire cultivated area in the Campania region (Cuozzo 2000).

It is for this reason that from 1994 to April 2000 surveys have been carried out in the fruit groves of Campania with particular attention to those located in the province of Caserta. The primary objective was to know the species of phytoseiid mites present on apple and peach trees.

* Authors have equally contributed to this research.

F. Bernini, R. Nannelli, G. Nuzzaci, E. de Lillo (eds.), Acarid Phylogeny and Evolution. Adaptations in mites and ticks, 325-329.

Previous studies throughout Campania on 40 hazelnut groves (1996-97) and 75 vineyards (1989-99) (Nicòtina 1999; Nicòtina and Cioffi 1999) showed that *A. andersoni* on vines and hazelnut trees was present only in the province of Caserta, where there was a massive presence of apple and peach trees. Consequently, the objective of this study is to investigate whether there is an effective link between these latter cultivations and the dispersion of *A. andersoni*.

Materials and methods

To check the evolution of the phytoseiid populations, an apple grove was chosen in Giugliano, province of Naples, on the borders of the province of Caserta, specialized in the cultivation of apples (cv Annurca). Samples were taken periodically every 15 days from June 1999 to June 2000.

Moreover phytoseiid mites were also collected in 52 orchards of the Campania region: 30 apple groves, 22 peach-nectarine groves where samples were taken only once for identifying the species of phytoseiids. All collections were carried out by the shaking method (Tsolakis and Ragusa 1999). Winter collection of samples, was done by vigorously shaking 20 branches 2 or 3 times every 30 cm by means of a wooden stick. This caused the mites to fall on a dark tray placed under the branches and where they were collected by using a tiny brush and preserved in Oudeman's solution.

Collections of spring-summer samples were carried out in all farms. In the spring-summer period a total of 50 leaves was taken from the central part of the foliage from different trees for each single farm.

Results and discussion

Results regarding phytoseiid collected on peach, apple, vine and hazel-nut are given in figures 1-5. In figures 1-3 *A. andersoni* was the dominant species associated with different crops in the province of Caserta. Figure 4 shows *A. andersoni* compared with other phytoseiid species collected

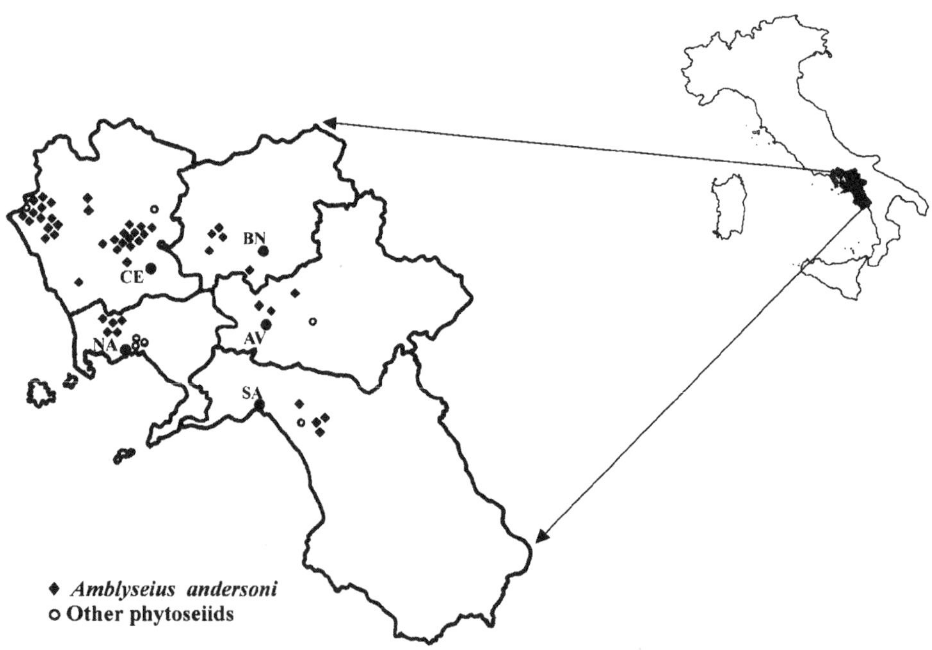

Fig. 1 - Distribution of *Amblyseius andersoni* and other species of phytoseiids associated with apple and peach trees in the Campania region.

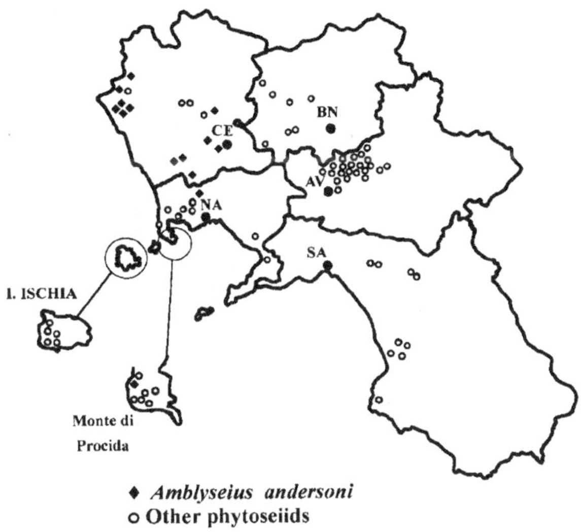

Fig. 2 - Distribution of *Amblyseius andersoni* and other species of phytoseiids associated with vines in the Campania region.

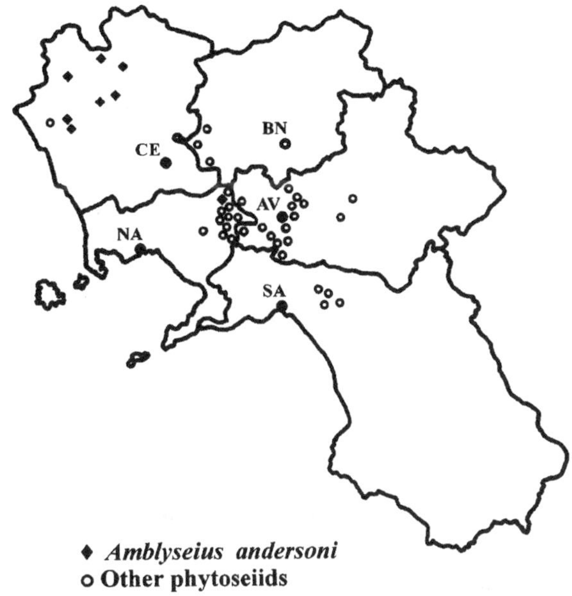

Fig. 3 - Distribution of *Amblyseius andersoni* and other species of phytoseiids associated with hazel-nut-groves in the Campania region.

on apple and peach in the Campania region. Surveys showed that the phytoseiids associated with apple and peach were the following 7 species: *A. andersoni, Amblyseius stipulatus* Athias-Henriot, *Kampimodromus aberrans* (Oudemans), *Cydnodromus californicus* (McGregor), *Typhlodromus cryptus* Athias-Henriot, *Typhlodromus exhilaratus* Ragusa and *Typhlodromus talbii*

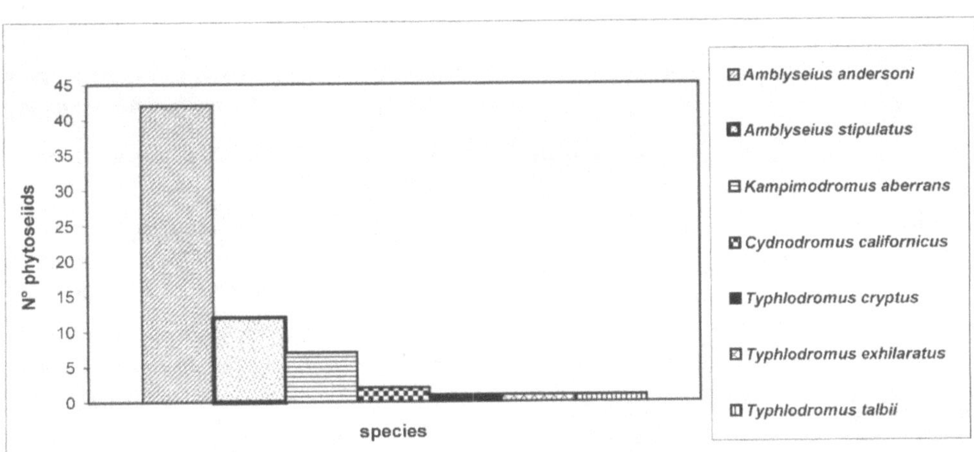

Fig. 4 - Phytoseiid mites collected in Campania on branchlets and leaves of apple and peach trees from April 1994 to April 2000.

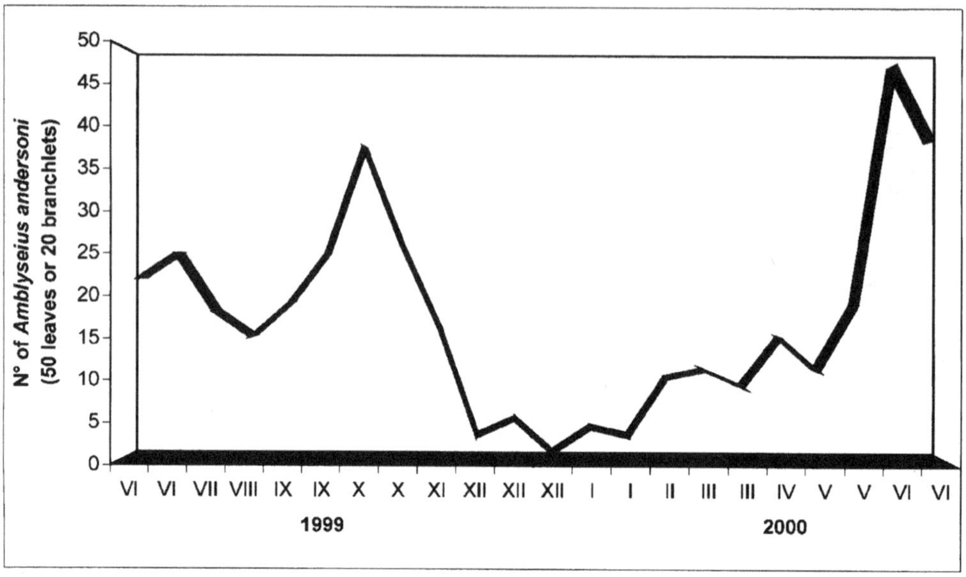

Fig. 5 - Total number of *Amblyseius andersoni* in Giugliano (Napoli) on one apple orchard from June 1999 to June 2000.

(Athias-Henriot). In figure 5 the fluctuation of *A. andersoni* in the Giugliano farm (Naples) is reported. It is evident from figure 5 that the population of *A. andersoni* showed two peaks, in October 1999 and in June 2000 and that it was at a very low level from December 1999 to April 2000.

The large amount of *A. andersoni* associated with apple and peach trees in the province of Caserta and the presence of this phytoseiid species in the same area during the summer on hazelnut and vines, compared to the other provinces of the region, where it is interesting to note that this mite was not found (Nicòtina 1999; Nicòtina and Cioffi 1999), leads us to believe and hypothesize that the widespread presence of peach and apple cultivations could influence the dispersion of the species on hazelnut and vines.

Our work shows that such abundance of *A. andersoni* on crops other than apple and peach, where the phytoseiid has unchallengeably settled, tend to move to cultivations which are different to those they normally inhabit.

But further studies are needed to make final conclusions.

References

Castagnoli M., Liguori M., Nannelli R., 1984 - Contributo alla conoscenza degli acari del pesco in Toscana e osservazioni sull'andamento delle loro popolazioni. *Redia*, 64: 493-504.

Costa-Comelles J., Santamaría A., Ferragut F., García Marí F., 1994 - Poblaciones de ácaros en la cubierta vegetal de huertos de manzanos. Boletin de Sanidad Vegetal, *Plagas*, 20(2): 339-355.

Croft B.A., Ali Niazee M.T., 1983 - Differential resistence to insecticides in *Typhlodromus arboreus* Chant and associate phytoseiid mites of apple in the Willemette Valley, Oregon. *Enviromental Entomology*, 12(5): 1420-1423.

Cuozzo N., 2000 - Statistica agraria. *Agricoltura Campania*, 2/3: 48.

Duso C., Sbrissa F., 1990 - Gli acari Fitoseidi (Acari, Phytoseiidae) del melo nell'Italia settentrionale, distribuzione, biologia, ecologia ed importanza economica. *Bollettino di Zoologia Agraria e Bachicoltura*, 22(1): 53-89.

Ivancich Gambaro P., 1991 - Confronto fra la dinamica delle popolazioni di *Amblyseius andersoni* Chant in meleti condotti con programmi di difesa a 'lotta guidata'. *L'Informatore Agrario*, 47: 73-75.

Nicòtina M., 1996 - Phytoseiid mites associated with vines in vineyards of the Campania region. Procedings IX International Congress of Acarology, Columbus (Ohio), 17-22 July, 1994, vol. 1: 237-240.

Nicòtina M., 1999 - Acari fitoseidi (Acarina, Phytoseiidae) della vite nel centro sud Italia: dieci anni di studi faunistici. *Informatore Fitopatologico*, 11: 30-35.

Nicòtina M., Caprio E., 1996 - Prove di selettività su *Amblyseius andersoni* (Chant) e lotta chimica contro *Panonychus ulmi* (Koch) su melo in provincia di Caserta. Informatore Fitopatologico, 4: 48-53.

Nicòtina M., Cioffi E., 1999 - Distribution of the Phytoseiid mites (Acarina, Phytoseiidae) in hazel-nut-growing areas in Campania. Redia, 81: 115-124.

Nicòtina M., Ragusa S., Viggiani G., 1990 - Acari Fitoseidi in alcuni vigneti della Campania e del Lazio. *L'Informatore Agrario*, 46(19): 75-76.

Nicòtina M., Viggiani G., 1985 - Distribuzione e densità stagionali delle popolazioni degli acari fitoseidi in noccioleti della Campania. Atti XIV Congresso Nazionale Italiano di Entomologia, Palermo - Erice-Bagheria: 729-732.

Strapazzon A., Rensi F., 1989 - Prospettive di controllo biologico di *Panonychus ulmi* (Koch) su melo nel Veneto. *L'Informatore Agrario*, 55(12): 117-126.

Thistlewood H. M. A., 1991 - A survey of predator mites in Ontario apple orchards with diverse pesticide programs. *Canadian Entomologist*, 123(6): 1163-1174.

Tsolakis H., Ragusa S., 1999 - Overwintering of phytoseiid mites (Parasitiformes, Phytoseiidae) on hazelnut (*Corylus avellana*) in Sicily (Italy). In: Bruin J., van der Geest L.P.S., Sabelis M.W. (eds.), Ecology and Evolution of the Acari. Kluwer Acad. Publ.: 625-635.

How can molecular data contribute to the analysis of the colonization of vineyards by *Kampimodromus aberrans* (Acari: Phytoseiidae)?

Marie-Stéphane Tixier, Serge Kreiter, Philippe Auger
Ecole Nationale Supérieure Agronomique/Institut National de la Recherche Agronomique, UFR d'Ecologie animale et de Zoologie agricole, Laboratoire d'Acarologie, 2 Place Pierre Viala, F-34060 Montpellier cedex 01, France.

Key words: Phytoseiidae, *Kampimodromus aberrans*, colonization, vineyards, Random Amplified Polymorphism DNA.

Abstract: *Kampimodromus aberrans* (Oudemans) is the most important predatory mite found in vineyards in the South of France. This species is also present in uncultivated areas surrounding vineyards and its colonization of vineyards has been studied for 3 years. Several factors affect the exchange between uncultivated and cultivated areas including the floristic composition of the woody margins and wind direction. However, it is very difficult to determine the origin of migrants and to observe their settlement in the plot. Therefore a molecular study using RAPD markers (Random Amplified Polymorphism DNA) has been performed. Samples of females were obtained from an experimental vineyard, and from two independent surrounding areas: two cultivated vine plots and two naturally occurring plants (*Quercus pubescens* Willd.) and *Celtis australis* L., both with a high density of *K. aberrans*. The samples were taken in July 1998. Genetic distances (simple matching) and dendrograms were established. Genetic distances between individuals from the same sample were lower than those between individuals from different samples. Females from two plots in the experimental field were similar in genetic distances. Females from two different plants sampled in a woody margin were also very close. And at least, females sampled from two other vineyards, planted with the same varieties of grape (cv. Carignan) shared high similarities. Populations of *K. aberrans* seem to be highly structured. Even if migration from the surrounding environment mostly occurs from plants near the experimental plot, it is not possible to determine exactly the source areas. Once in the plot, pesticide application or leaf structure (vine variety) may select specific genotypes and phenotypes. The high differentiation of populations supports the fact that settlement of the migrants has not been observed for three years. It seems that limits for settlement and gene flow occur although high numbers of mites disperse into the plot.

Introduction

Phytoseiid mites are of great agricultural interest as they can limit the populations of phytophagous mites in numerous crops, especially in vineyards (Ivancich Gambaro 1973; Sabelis 1985; McMurtry and Croft 1997). *Kampimodromus aberrans* (Oudemans) is the most important phytoseiid mite found in vineyards in the South of France (Kreiter *et al.* 2000). This species is not only present in vineyards but also, as many other phytoseiids, in uncultivated areas surrounding the plots (Krantz 1973; Moraes *et al.* 1986; Coiutti 1993; Duso *et al.* 1993; Ragusa *et al.* 1995; Duso and Fontana 1996; Ragusa and Tsolakis 1996; Tixier *et al.* 1998 and 2000). Some studies dealing with other natural enemies show that surrounding uncultivated area

F. Bernini, R. Nannelli, G. Nuzzaci, E. de Lillo (eds.), Acarid Phylogeny and Evolution. Adaptations in mites and ticks, 331-340.

management can improve crop protection and yield (Altiéri and Letourneau 1982; Dennis and Fry 1992; Dennis *et al.* 1994). Several studies suggested that phytoseiid mites present in surrounding areas could colonize the neighbouring crops especially when pesticides used have low toxicity against them (Boller *et al.* 1988; Corino 1989; Tuovinen and Rokx 1991; Tuovinen 1993a and b; Tuovinen 1994; Kreiter and Sentenac 1995). However, there is little data concerning the exchange of phytoseiid mites between uncultivated and cultivated areas. Some studies show that these exchanges exist and great numbers of phytoseiid mites could disperse into the neighbouring cultivated fields from woody areas (Fauvel and Cotton 1981; Tixier *et al.* 1998 and 2000). This "natural colonization" deals with very important agronomic and environmental matters as plant management takes into account all the agrosystem within a framework of sustainable farming.

A three-year study conducted in an experimental vineyard in the South of France showed that great numbers of *K. aberrans* moved from uncultivated areas to vineyards mainly by aerial dispersal (Tixier *et al.* 1998 and 2000). We identified three parameters that could influence the densities of migrants and the periodicity of dispersal: 1) The floristic composition of the surrounding areas; 2) The height of the trees; 3) The direction and the speed of the wind (Tixier *et al.* 1998 and 2000).

However, it was very difficult to determine the exact origin of these migrants. Further studies using other experimental tools were conducted to determine the origin of the migrants. The main questions were: Does *K. aberrans* come from a woody margin or from the surrounding vineyards? Does this origin differ according to the location of traps in the plot? Our main hypothesis is that plots located near surrounding vineyards were mainly colonized by mites coming from these areas, and that plots located near the woody margin were colonized by mites coming from these surrounding uncultivated areas. Another question is: does the settlement of *K. aberrans* into the vine field vary according to the origin of migrants? Over three years we observed a low number of mites and a small increase of populations in plots close to the woody margin, although a large number of *K. aberrans* were trapped. On the other-hand, large numbers of mites were observed in the zones of the field closest to vine crops even when low numbers of *K. aberrans* were caught in these areas.

To address these questions, we used a molecular typing method (Random Amplified Polymorphism DNA) on *K. aberrans* individuals living in the experimental vine field and neighbouring uncultivated and cultivated areas.

Materials and methods

Samplings

The DNA of migrants is altered in water filled aerial traps (Tixier *et al.* 1998 and 2000), therefore other kinds of traps were tested, however none of these caught living migrants either (Tixier 2000). Thus, we studied only females of *K. aberrans* that were already settled in the plot. Males could not be studied as they were too small to produce enough DNA to obtain reliable results.

Samples were taken in an experimental site located in the South of France (Pouzolles, Hérault) and already selected for the colonization study over three years (Tixier *et al.* 1998 and 2000). Females of *K. aberrans* were sampled in 2 locations of the experimental vineyard, in 2 surrounding vineyards and on 2 trees located in the bordering woody margin (fig. 1).

Living females were taken from leaves using a stereoscopic microscope. For each leaf, the number of females collected is very different, due to the aggregated distribution of this mite species (Tixier *et al.* 2000). They were then placed on a plastic disk bordered by moistened tissue paper for at least one day without food to prevent potential amplification of prey DNA. Then, each

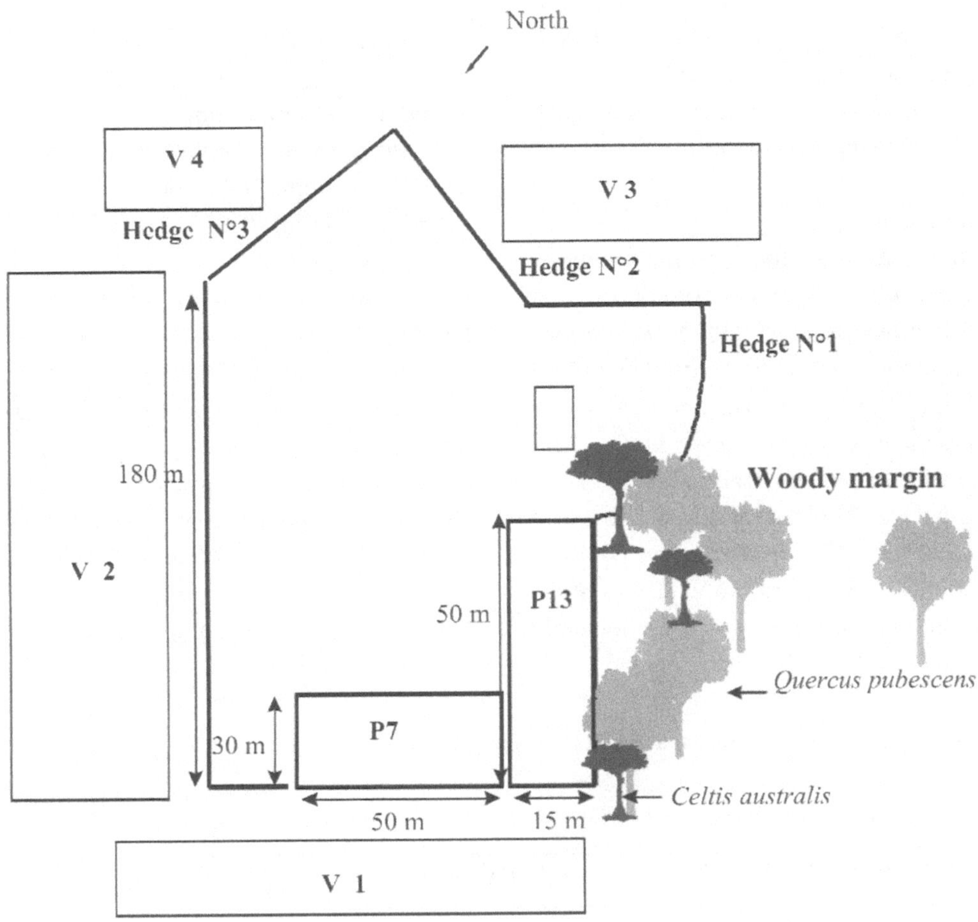

Numbers of females collected

In the woody margin:

Quercus pubescens :	19 females
Celtis australis :	19 females

In the neighbouring vineyards:

V1:	16 females
V2:	16 females

In the experimental vine field

P7:	20 females
P13:	20 females

Fig. 1 - Location of the different populations sampled and number of females collected per population. *Quercus pubescens* = oak; *Celtis australis* = hairy nettle tree.

mite was frozen at -30°C in a 200 µl sterile tube. As the highest number of migrants was observed in late June - early July (Tixier *et al.* 1998 and 2000), samples were collected in late July in order to observe the impact of migrants arrival on RAPD patterns of mites living into the plot.

Molecular protocols

Extraction. Among the several methods tested (Kazmer *et al.* 1995; Perrot-Minnot and Navajas 1995; Edwards *et al.* 1997), we choose extraction protocols used by Edwards *et al.* (1997) to study, with RAPD markers, differences between several species of *Typhlodromalus*. The DNA of one female was used individually for the RAPD-PCR analysis. The mite was crushed in 40 µl of a 5% Chelex® 100 resin solution (Biorad) (Walsh *et al.* 1991) in a sterile tube. Tubes were vortexed at high speed for 10 s, and immediately placed in ice. The tubes were then incubated at 56°C for 15 min, then heated to 100°C for 4 min to extract the DNA. These DNA solutions were then centrifuged at 12,000 rpm for 5 min and stored at -30°C until required.

Amplification. Amplification was performed in a 25 µl volume containing 200 mM Tris-HCl (pH=8.4), 500 mM KCl, 0.5 U of Taq DNA polymerase (Gibco-BRL), 15 ng of one primer, 0.1% Triton X-100, 0.1 mM of each DNTP, 25 mM of $MgCl_2$ and 6 µl of the total DNA extract of one mite. The amplification was performed in a Hot Bonnet PTC-100 (Mj Research, Inc.) thermal reactor programmed as follows: 5 min at 94°C, 45 s at 36°C and 2 min at 72 °C for an initial cycle; 50 s at 94°C, 45 s at 36°C, and 2 min at 72°C for 43 cycles; and 50 s at 94°C, 45 s at 36°C and 10 min at 72°C for a terminal cycle. A control was conducted where no DNA was used at each amplification.

Tab. 1 - Primers and fragments selected for RAPD analysis, size and frequency, for 4 populations used for polymorphism study.

Primers name and sequences (5'-3')	Fragments	Size (pb)	Fragment frequency for 4 populations (V1, V2, oak and hairy nettle tree)
P03	1	380	0.70
CTGATACGCC	2	410	0.75
	3	760	0.45
	4	1450	0.36
	5	600	0.36
A05	1	240	0.73
AGGGGTCTTG	2	385	0.75
	3	460	0.65
	4	770	0.64
A02	1	620	0.65
	2	700	0.46
	3	840	0.56
H01	1	890	0.55
GGTCGGAGAA	2	2300	0.06
	3	590	0.34
A04	1	450	0.75
AATCGGGCTC	2	605	0.66
	3	690	0.48
	4	890	0.66
	5	990	0.56
	6	570	0.19
K11	1	1900	0.24
AATGCCCCAG	2	550	0.54
	3	695	0.69
	4	960	0.18
	5	510	0.29

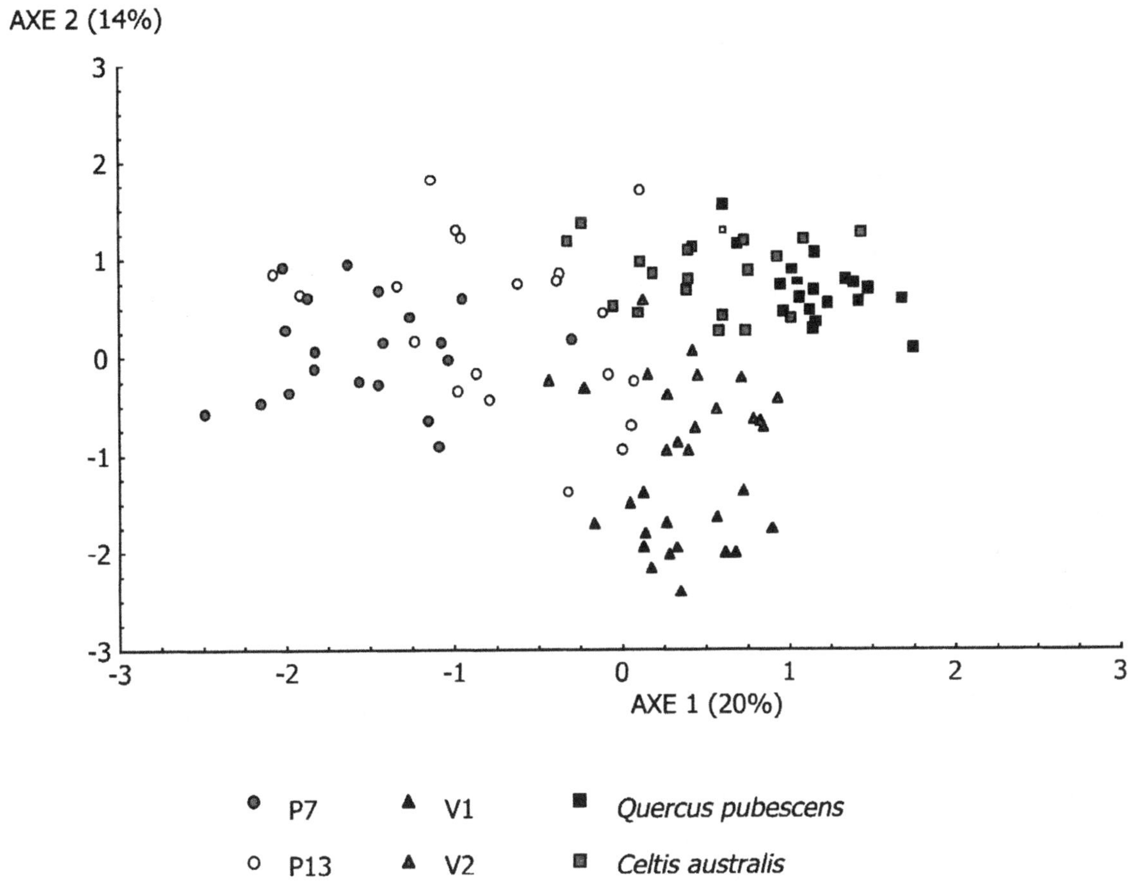

Fig. 2 - Females of *K. aberrans* sampled in the experimental vine plot, the neighbouring vineyards and on 2 trees in the woody margin in the plan 1-2 of the multifactorial analysis. *Quercus pubescens* = oak; *Celtis australis* = hairy nettle tree.

Migration

Amplification products were separated electrophoretically on Tris-Borate EDTA agarose 1.5% gels for 4 hours at 85 V. The DNA was stained with ethidium bromide (0.1 µg/µl) and the bands visualized under UV fluorescence (312 nm). A Molecular weight marker (36-2,645 pb) was run on each gel to make it easier to compare patterns among gels. Precautions were needed to prevent contamination of PCR experiments with previously amplified DNA. Therefore, pre- and post-amplification procedures were separated, and extraction occurred in sterile conditions.

Primers screening

Among the 20 10-mer primers of arbitrary sequences tested, only seven were retained and 26 markers were selected (tab. 1). Scored markers were identified as bands that were polymorphic (frequency occurrence comprised between 5 and 95%) (Edwards and Hoy 1993). The polymorphism was tested using 4 populations, 10 females each, collected in V1, V2, on one oak and one hairy nettle tree (fig. 1). The scored markers were brightly stained, between 240 and 2,300 bp in size, and reproducible. The repeatability of these fragments was tested by amplifying the DNA of one individual twice, for each primer where three individuals were used per primer.

Data analysis

We assumed that markers of the same size and appearance from different mite populations are homologous. Each band in a mite was scored as 1 if present or 0 if absent. The fraction of bands matching in two mites (M) was calculated using the formula: M=2Nab/Nt, where Nab is the total number of matches (i.e. both bands absent or both bands present) in individuals a and b, and Nt is the total number of fragments scored. A value of 1 indicates that two individuals have identical patterns; a value of 0 indicates that two individuals have completely different patterns. The simple matching distance (1-M) was calculated with RAPDPLOT program (Kambhampati *et al.* 1992; Black 1995). We bootstrapped the RAPD presence/absence data for every individual by performing 200 re-samples consisting of 13 of the 26 RAPD loci. We bootstrapped the RAPD presence/absence data for each population by performing 2,000 re-samples of 13 of the 26 RAPD loci (Njbs® software, Cornuet *pers. comm.*). Neighbour-Joining cluster analysis was performed (Njbs®, Cornuet *pers. comm.*). Multifactorial analysis were also performed with the presence/absence data for each mite (statistica5® 1998).

Results

Intra-population distances

For each population, intra-population distances were lower than distances to other populations (tab. 2). They ranged between 0.15 for *Q. pubescens* to 0.33 for P13 (tab. 2). Several reasons could explain this high homogeneity between females collected at the same location: 1) exchanges between individuals from the same sample unit (sexual reproduction); 2) high dispersal between leaves of the same tree, or vinestocks of the same vine plot; 3) high densities of mites; 4) local selective pressures that select adapted phenotypes in each sample unit.

Some females within each sample unit are very different from the other females of the same population. Sub-units within populations seem to exist possibly caused by little exchange, inbreeding and/or genetic drift due to low dispersal and small population size. The aggregated distribution of *K. aberrans* (Malison *et al.* 1995; Tixier *et al.* 2000) could also explain the presence of sub-units (leaf, groups of leaves, vinestocks, groups of vinestocks) in the different samples. Indeed, some females may be very close because they were sampled on the same leaf. On the opposite, other females may have very different RAPD patterns because they were collected on remote leaves of the same tree. Further studies should be done to compare the similarity of females living on the same leaf and those living on remote leaves.

This could also be due to the presence of new migrants in the different units, and especially in the experimental plot. However, we will see that this last hypotheses is probably inaccurate.

Distances between populations

We observe 3 different RAPD patterns according to the type of populations and locations.

1) The females collected in the neighbouring vine crops (V1, V2) of the vine Carignan cultivar have very close RAPD patterns (1-M=0.27) (tab. 2; figs 2-3). The females collected in these two plots are different from the females of the other populations, except from individuals collected in V2 and on *Q. pubescens*. However this low average seems to be due to the similarities between few females.

2) The females from the 2 plots in the experimental vineyard (Cabernet-Sauvignon cv.) shared similar RAPD patterns (1-M=0.34) (tab. 2; figs 2-3). These females also differed from the four other populations.

3) the females collected on 2 tree habitats (*Q. pubescens*, *C. australis*) in the woody margin also have close RAPD patterns (1-M=0.24) (tab. 2; figs 2-3) and are different from the other populations.

Tab. 2 - Mean genetic distance matrix (simple matching) for populations of *Kampimodromus aberrans* collected in July 1998 from the woody margin, in the experimental vine field and in two neighbouring vine crops (V1 & V2). * mean genetic distance within populations (minimum - maximum).

	V1	V2	Oak	Hairy nettle tree	P13	P7
Number of females	16	16	19	19	20	20
V1	0.21* (0-0.46)	0.27	0.36	0.40	0.41	0.42
V2		0.19* (0-0.50)	0.24	0.32	0.36	0.41
Oak			0.15* (0.03-0.27)	0.24	0.38	0.46
Hairy nettle tree				0.21* (0.03-0.46)	0.36	0.12
P13					0.33* (0.07-0.61)	0.34
P7						0.22* (0-0.57)

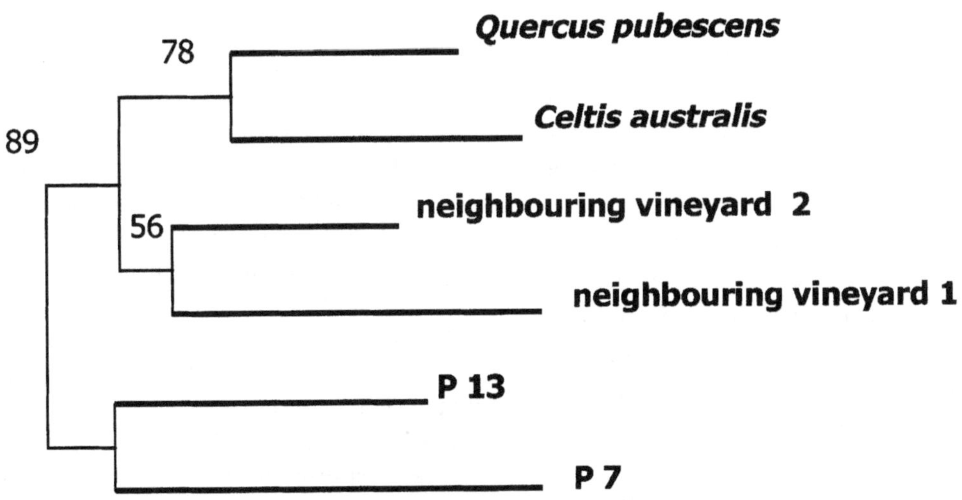

Fig. 3 - Dendrogram and bootstraps values established with the mean genetic distance for populations collected in the experimental vine plot, the neighbouring vineyards and on 2 trees in the woody margin. *Quercus pubescens* = oak; *Celtis australis* = hairy nettle tree.

Discussion and conclusions

Within a small area (5 ha maximum), we observed high polymorphism between populations according to their location. Several factors such as genetic drift, low dispersal, low gene flow and specific selective pressures, could explain these differences. The genetic similarities are not correlated with geographic distances. Dispersal between the different samples could not fully explain the similarities and differences between populations.

It is difficult to explain the high similarities between V1 and V2, and the differences between P7 and V1 despite the fact that V1 and P7 are geographically closer than V1 and V2. We also need to explain the similarities between P13 and P7 and the differences between P13 and *Q. pubescens,* although the geographic distance between P13 and *Q. pubescens* is lower than the geographic distance separating P13 from P7 (fig. 1).

RAPD patterns of females collected in uncultivated areas differ from those of females located in cultivated areas. However, the sole factor of cultural practice is not a sufficient selection pressure to explain the differentiation of populations within the experimental site. Indeed, high differences were observed between females collected in the experimental plot and in the two neighbouring cultivated vine crops. Other parameters such as microclimatic conditions and plant leaf could probably affect local populations and explain the differences observed between females collected in nearby places.

The vine variety and the pilosity of leaves could affect polymorphism and the genetic patterns. Ecological conditions, such as host plant, can indeed affect the morphology and morphometry of populations for some phytoseiid mite species (Hoying and Croft 1977; Abou-Setta *et al.* 1991). These factors could also affect the RAPD patterns too, and determine the similarities between populations.

What about these data and colonization of the experimental vineyard? Based on the data obtained, it is impossible to determine the origin of the females collected in the experimental field. It seems that these females are very different from those located in the putative source areas surrounding the crop. If females that originate from these latter areas have colonized the experimental vineyard, then only some individuals that possess specific adapted phenotypes (and genotypes) can succeed and settle in the plot. These phenotypes are certainly different from the wild type found on oak, hairy nettle tree and in neighbouring vineyards. It seems that each area has its own selection pressure and modification is necessary to ensure settlement. Such conclusions are exactly those expressed by Roderick (1992 and 1996) about migrant evolution during post-colonization processes.

There are still a lot of differences between females located in bordering areas of the experimental vineyard and those located in nearby surroundings, even after the arrival of high numbers of *K. aberrans* in the plot. This is especially true for P13. High numbers of mites dispersed into this area from nearby oaks that overlap the plot. However, no similarities were observed between *K. aberrans* females located in P13 and an adjacent oak. These results confirm those already obtained in previous studies (Tixier *et al.* 1998 and 2000). Over the 3 years, the number of mites in P13 has not increased as we expected from the high numbers of mites dispersing in this zone of the experimental vineyard. It seems that not all the migrants settle into the plot.

This study suggests that there is little genetic exchange between females located in the vineyard and its surroundings. High selection pressures certainly explain the differentiation of the populations, and affect the settlement of migrants in new areas. This study confirms that the settlement of migrants may be a limiting factor of the colonization process. These results are similar to those obtained when releases of *K. aberrans* were conducted (Kreiter *et al.* 1993).

Further research is required to allow the use of phytoseiid mites, especially of *K. aberrans*, in a framework for conservation biological control and sustainable agriculture. Even though there are high densities of *K. aberrans* in the surroundings of the vine plots and significant dispersal into cultivated areas occurs, this may not be sufficient to ensure colonization.

Further studies are needed to determine if this situation and differentiation will be the same in the following years. It seems important too to determine the influence of sampling on the results obtained, especially due to the aggregated distribution of *K. aberrans*. At least, other studies are required to identify factors that may reduce the settlement of migrants. Do the pesticides applied into the plot kill the migrants coming from untreated areas? Could reproduction incompatibilities between migrants and individuals already present into the plot explain the low genetic flow observed? Do plant changes during dispersal affect the settlement in the vineyards? Some studies show indeed for more specific predators [for *Neoseiulus fallacis* (Garman) and *Neoseiulus californicus* (McGregor)] that host plant can influence and delay their settlement (Castagnoli *et al.* 1999; Lester *et al.* 2000). Finally, what about the survival of mites during dispersal? Answers to these questions are needed to enhance the use of phytoseiid mites naturally occurring in vineyard surroundings.

ACKNOWLEDGEMENTS

We would like to thank a few people who have been of great help towards the study. Firstly Claude Rouquette, the farmer of the experimental plot near Béziers. We also thank J.-M. Cornuet and S. Santoni for their help using and interpreting RAPD on phytoseiid mites. Lastly, we give our sincere thanks to Ai-Linh Tran for help in translation.

References

Abou-Setta M.M., Childers C.C., Denmark H.A., Browning H.W., 1991 - Comparative morphology reproductive compatibility between populations of *Euseius mesembrinus* (Acari: Phytoseiidae) from Florida and Texas. *Experimental and Applied Acarology*, 10: 213-220.

Altiéri M.A., Letourneau D.K., 1982 - Vegetation management and biological control in agroecosystems. *Crop Protection*, 1(4): 405-430.

Black W.C., 1995 - Fortran programs for the analysis of RAPD-PCR markers in populations. Colorado State University, Ft. Collins, CO 805523.

Boller E.F., Remund U., Candolfi M.P., 1988 - Hedges as potential sources of *Typhlodromus pyri*, the most important predatory mite in vineyards of northern Switzerland. *Entomophaga*, 33(2): 249-255.

Castagnoli M., Liguori M., Simoni S., 1999 - Effect of two different host plants on biological features of *Neoseiulus californicus* (McGregor). *International Journal of Acarology*, 25(2): 145-149.

Coiutti C., 1993 - Acari Fitoseidi su piante arboree spontanee e coltivate in Friuli-Venezia Giulia. *Frustula Entomologica*, 16(24): 65-77.

Corino L., 1989 - Les acariens phytophages sur vigne en Italie: évolution des populations et équilibres naturels avec les acariens prédateurs. Compte rendu du "Colloque sur les acariens des cultures", Montpellier, 24-26 octobre 1989. *Ann. ANPP*, 2(1): 116-123.

de Moraes G.J., McMurtry J.A., Denmark H.A., 1986 - A catalog of the mite family Phytoseiidae: references to taxonomy, synonymy, distribution and habitat. Embrapa-ddt-pub.: 353 pp.

Dennis P., Fry G.L.A., 1992 - Field margins: can they enhance natural enemy population densities and general arthropod diversity on farmland? *Agriculture, Ecosystems and Environment*, 40: 95-115.

Dennis P., Thomas M.B., Sotherton N.W., 1994 - Structural features of field boundaries which influence the overwintering densities of beneficial arthropod predators. *Journal of Applied Ecology*, 31: 361-370.

Duso C., Fontana P., 1996 - Mite communities on wild plants surrounding vineyards in North-Eastern Italy with special emphasis on Phytoseiids (Phytoseiidae). In: Mitchell R., Horn D.J., Needham G.R., Welbourn W.C. (eds.), Acarology IX. Ohio Biological Survey, Columbus, Ohio, vol. 1: 261-264.

Duso C., Torresan L., Vettorazzo E., 1993 - La vegetazione spontanea come riserva di ausiliari: considerazioni sulla diffusione degli Acari Fitoseidi (Acari: Phytoseiidae) in un vigneto e sulle piante spontanee contigue. *Bollettino di Zoologia Agraria e Bachicoltura*, 25: 183-203.

Edwards O.R., Hoy M.A., 1993 - Polymorphism in two parasitoids detected using random amplified DNA polymerase chain reaction. *Biological Control*, 3: 243-257.

Edwards O.R., Melo E.L., Smith L., Hoy M.A., 1997 - Discrimination of three *Typhlodromalus* species (Acari: Phytoseiidae) using random amplified polymorphic DNA markers. *Experimental and Applied Acarology*, 21: 101-109.

Fauvel G., Cotton D., 1981 - Evolution des populations de typhlodromes, *Amblyseius aberrans* essentiellement, dans une haie d'ormes et un verger de pommiers et observations de leur transport par le vent. In: Coste J. (ed.), Compte rendus de la 6e journée phytiatrie. Phytopharmacie. Circum mediterranéennes. Technic Offset pub.: 471-479.

Hoying S.A., Croft B.A., 1977 - Comparisons between populations of *Typhlodromus longipilus* Nesbitt and *T. occidentalis* Nesbitt: taxonomy, distribution and hybridization. *Annals of Entomological Society of America*, 70(1): 150-159.

Ivancich Gambaro P., 1973 - Il ruolo del *Typhlodromus aberrans* Oudm. (Acarina: Phytoseiidae) nel controllo biologico degli acari fitofagi dei vigneti del Veronese. *Bollettino di Zoologia Agraria e Bachicoltura*, 2: 151-165.

Kambhampati S., Black W.C., Rai K. S., 1992 - Random amplified polymorphic DNA of mosquito species and populations (Diptera: Culicidae): techniques, statistical, analysis and applications. *Journal of Medical Entomology*, 29(6): 939-945.

Kazmer D.J., Hopper K.R., Coutinot D.M., Heckel D., 1995 - Suitability of random amplified polymorphic DNA for genetic markers in the aphid parasitoid, *Aphelinus asychis* Walker. *Biological Control*, 5: 503-512.

Krantz G.W., 1973 - Dissemination of *Kampimodromus aberrans* by the filbert aphid. *Journal of Economic Entomology*, 66(2): 575-576.

Kreiter S., Sentenac G., 1995 - Gestion des populations d'auxiliares: recolonisation naturelle ou introduction de phytoséiides en vignobles. In: Leclant F., Reboulet J.-N. (eds.), Compte rendu de la "Journée d'informations sur les auxiliaires entomophages". Valence 15/XI/1995, *ANPP Publication*: 49-63.

Kreiter S., Sentenac G., Weber M., Valentin G., 1993 - Les phytoséiides des vignobles français. Synthèse de huit années de recensement. In: Leclant F. (ed.), Compte rendu de la "Troisième conférence internationale sur les ravageurs en agriculture". Montpellier 7-8-9/XII/1993, *ANPP Publication*: 597-609.

Kreiter S., Tixier M.-S., Auger P., Muckensturm N., Sentenac G., Doublet B., Weber M., 2000 - Phytoseiid mites of vineyards in France (Acari: Phytoeiidae). *Acarologia*, 41(4): 75-94.

Lester P.J., Thistlewood H.M.A., Harmsen R., 2000 - Some effects of pre-release host-plant on the biological control of *Panonychus ulmi* by the predatory mite *Amblyseius fallacis*. *Experimental and Applied Acarology*, 24: 19-33.

Malison M., Baldacci R., Posenato G., Girolami V., 1995 - Frequency distribution of *Kampimodromus aberrans* (Oud.) (Acari: Phytoseiidae) in vineyards and optimization of the samplings methods. *Frustula Entomologica*, 18: 153-168.

McMurtry J.A., Croft B.A., 1997 - Life-styles of phytoseiid mites and their roles in biological control. *Annual Review of Entomology*, 42 : 291-321.

Perrot-Minnot M.-J., Navajas M., 1995 - Biparental inheritance of RAPD markers in males of the pseudo-arrhenotokous mite *Typhlodromus pyri*. *Genome* 38: 838-844.

Ragusa S., Papaioannou-Souliotis P., Tsolakis H., Tsagkarakou N., 1995 - Acari fitoseidi (Parasitiformes, Phytoseiidae) della Grecia associati a piante forestali a diverse altitudini. *Bollettino di Zoologia Agraria e Bachicoltura*, 27(1): 85-91.

Ragusa S., Tsolakis H., 1996 - A survey of phytoseiid mites (Phytoseiidae) associated with various plants in Sicily. In: Mitchell R., Horn D.J., Needham G.R., Welbourn W.C. (eds.), Acarology IX., Ohio Biologal Survey, Columbus, Ohio, vol. 1: 253-256.

Roderick G.K., 1992 - Post-colonization evolution of natural enemies. In: Kauffman W.C., Nechols J.R. (eds.), Selection criteria and ecological consequences of importing natural enemies. T. Say Publication: 71-86.

Roderick G.K., 1996 - Geographic structure of insect populations: gene flow, phylogeography and their uses. *Annual Review of Entomology*, 41: 325-352.

Sabelis M.W., 1985 - Predator-prey interaction. In: Helle W., Sabelis M.W. (eds.), Spider Mites: their biology, natural enemies and control. Elsevier, vol. 1B: 103-129.

Statistica5®, 1998 - Version 6.0. Statsoft, Inc., USA.

Tixier M.-S., 2000 - La présence de phytoséiides dans l'environnement non cultivé des parcelles a-t-elle un intérêt appliqué en viticulture? PHD. ENSA-Montpellier, France: 189 pp.

Tixier M.-S., Kreiter S., Auger P., Weber M., 1998 - Colonization of Languedoc vineyards by phytoseiid mites (Acari: Phytoseiidae): influence of wind and crop environment. *Experimental and Applied Acarology*, 22: 523-542.

Tixier M.-S., Kreiter S., Auger P., 2000 - Colonization of vineyards by phytoseiid mites: their dispersal patterns in the plot and their fate. *Experimental and Applied Acarology*, 24(3): 191-211.

Tuovinen T., 1993a. - Identification and occurrence of phytoseiid mites (Gamasida: Phytoseiidae) in Finnish apple plantations and their surroundings. *Entomologica Fennica*, 31(4): 95-113.

Tuovinen T., 1993b - Phytoseiid mites (Acari: Gamasina) in Finnish apple plantations with reference to integrated control of phytophagous. *Agricultural Science in Finland*, 2(1): 7-31.

Tuovinen T., 1994 - Influence of surrounding trees and bushes on the phytoseiid mites fauna on apple orchard trees in Finland agriculture. *Agriculture, Ecosystems and Environment*, 50(1): 39-47.

Tuovinen T., Rokx J.A.H., 1991 - Phytoseiid mites (Acari: Phytoseiidae) on apple trees and in surrounding vegetation in Southern Finland. Densities and species composition. *Experimental and Applied Acarology*, 12: 35-46.

Walsh P.S., Metzger D.A., Higuchi R., 1991 - Chelex® 100 as a medium for simple extraction of DNA for PCR-Based typing from forensic material. *Biotechniques*, 10: 506-513.

Interspecific predation and cannibalism in four phytoseiid species of the Mediterranean area (Acari: Phytoseiidae)

Marisa Castagnoli, Marialivia Liguori, Sauro Simoni
Istituto Sperimentale per la Zoologia Agraria, via Lanciola 12/A, 50125 Firenze, Italy

Key words: *Amblyseius andersoni, Neoseiulus californicus, Typhlodromus exhilaratus, T. kerkirae,* inter- intra-specific predation, oviposition.

Abstract: Interspecific predation, cannibalism and associated oviposition rates were evaluated in *Amblyseius andersoni* Chant, *Neoseiulus californicus* (McGregor), *Typhlodromus exhilaratus* Ragusa and *T. kerkirae* Swirski & Ragusa, which are widespread on agricultural crops in Central Italy. Young ovipositing females of each species coming from long-term mass-rearing on pollen were starved and then isolated on artificial arenas with a large number of con- or heterospecific protonymphs. The females were checked daily over a ten day period to assess predation, oviposition and survival.

Depending on protonymph species supplied, the number of prey eaten per day ranged from 0.12 (*T. kerkirae* preying on itself) to 7.44 (*A. andersoni* on *N. californicus*), while daily oviposition rates from 0 (*T. exhilaratus* on itself) to 1.21 (*A. andersoni* on *T. kerkirae*). Predation and oviposition rates on conspecifics were always very low, while on heterospecifics *A. andersoni* exceeded all other species. The mortality in the period was limited. With the exception of *A. andersoni* previously reared on heterospecifics, when the surviving females were fed on pollen for a further ten days, a considerable increase in oviposition rates was observed.

Introduction

Many interacting factors affect the composition and structure of predator-prey communities in agroecosystems. The agricultural practices, including pesticide management, combined with climatic conditions, the cultivar morphology and the characteristics of phytophagous species play well known roles in determining the prevalence of one phytoseiid species over others (Helle and Sabelis 1985). However, interspecific predation and cannibalism can also be involved in these mechanisms, stabilizing phytoseiid population when other preferred prey is unavailable (MacRae and Croft 1997).

In Central and Southern Italy, the phytoseiids *Typhlodromus exhilaratus* Ragusa, *T. kerkirae* Swirski & Ragusa, *Amblyseius andersoni* Chant and *Neoseiulus californicus* (McGregor) are widespread in several agroecosystems such as vineyards, orchards, soybean and strawberry fields (Castagnoli *et al.* 1984; Ragusa 1986; Castagnoli 1989; Calvitti and Tsolakis 1992; Castagnoli *et al.* 1993; Liguori 1995; Duso 1997). They often share the same habitat and exploit common foods and thus direct interactions may occur in time of food scarcity. Reports on interspecific predation and cannibalism are available for *A. andersoni,* but only in comparison with *Metaseiulus occidentalis* (Nesbitt), *N. fallacis* (Garman) and *T. pyri* Scheuten (Zhang and Croft 1995a and b; Croft and Croft

F. Bernini, R. Nannelli, G. Nuzzaci, E. de Lillo (eds.), Acarid Phylogeny and Evolution. Adaptations in mites and ticks, 341-349.

1996; Croft *et al.* 1996) as well as for *N. californicus* in comparison with *N. fallacis, T. athiasae* Porath & Swirski, and *Phytoseiulus persimilis* Athias-Henriot (Monetti and Croft 1997; Croft *et al.* 1998; Palevski *et al.* 1999; Walzer and Schausberger 1999a and b). Investigation on interspecific predation is still lacking for *T. exhilaratus* and *T. kerkirae*. Given the recognized importance of these four predators in biological control, studies to increase the knowledge of intra- and interspecific predation were carried out.

Materials and methods

Experimental procedures

The phytoseiid species considered originated from agricultural crops in Central Italy (Tuscany). The cultures were reared in the laboratory at 25±1°C, 75±10% RH and 16 hours photoperiod on artificial arenas and fed with *Quercus* sp. pollen supplied twice a week. All specimens used for this experiment come from 3-4 year old cultures.

The top of a small plastic cylinder glued onto the base of a Petri dish filled with water was used as an experimental unit. The artificial arena obtained (approximately 7 cm^2) was also bordered by strips of wet cotton wool to supply the mites with the necessary water and to hinder their escape.

Fifty young couples of each species were isolated on artificial arenas and checked for the first egg laid. After the start of oviposition, the males were removed and the females were starved for 24 hours. The eggs laid during starvation were discarded. Then single females of each species (10 replicate individuals) were assigned to five food treatments: i.e. the same pollen of mass-rearing, 8 conspecific protonymphs, and 8 protonymphs of each of the other three species. This immature stage was chosen because its development times, lasting more than 1 day, assured a great food uniformity. All protonymphs were well fed and just moulted and their number reasonably adequate for all predator females. For each species the body, including the palps, and the first legs of protonymphs were measured on 10 specimens mounted on slides. Simultaneously, 8 well-fed protonymphs without female were also confined in similar artificial arenas (12 replicates) to assess their intraspecific predation and natural mortality. All prey was renewed once a day. The prey loss, the eggs laid, female survival and escape attempts (i.e. when the female was found entrapped in wet cotton wool or floating in water) were controlled daily for 10 days. After this first 10-day period, the surviving females were fed on pollen for another 10 days to assess the oviposition rate of the period. The eggs laid during the two periods were transferred daily to new arenas and checked until hatched to evaluate their survival. The entire experiment was performed in a climatic cabinet at 25±1°C, 75±10% RH and 16 hours photoperiod on the above-mentioned artificial arenas. The tests were simultaneously run for the four predators.

Data analysis

Predation rates within conspecific protonymphs and their natural mortality were analyzed by one-way ANOVA; the mean values obtained for each prey species were discarded from prey loss data to obtain net predation rates. Using the same analysis, the trend to escape (number of escapes per female in the period considered) was also evaluated.

The effects of predator and prey on the net predation and oviposition rates were analyzed by means of two-way ANOVAs. Then single-factor ANOVAs and Bonferroni modified LSD were run to test differences among means in two ways: (a) for each predator species versus all prey supplied and (b) for each food (the four prey for predation and also the pollen for oviposition) versus all predators. To minimize the variance (Yamamura 1999) the data was $\ln(x + 0.5)$ transformed prior to various ANOVAs. When a suitable oviposition occurred, the egg hatching data was compared by χ^2 test.

For all statistical analysis SPSS (1994) was used.

Results

Some differences were registered in the size of protonymphs. The body, including the palps, ranged from 334 to 342 μm in *A. andersoni, N. californicus* and *T. exhilaratus*, while in *T. kerkirae* it was significantly smaller (281 μm). *Amblyseius andersoni* had the longest first legs (279 μm), followed by *N. californicus* (225 μm), whereas the legs of *Typhlodromus* species were the shortest. The *N. californicus* protonymphs appeared the most active. When groups of protonymphs were confined without food for a day, the natural mortality was nil, while different cannibalism was observed and computed for evaluating the net predation rate. Among the species examined, *T. kerkirae* showed a significantly higher intraspecific predation of protonymphs (0.5 specimens per day), followed by *T. exhilaratus* (0.17), *A. andersoni* (0.08), and *N. californicus* (0.04). These last three values were not significantly different ($P < 0.05$).

When starved ovipositing females were confined with protonymphs, the predator species, the prey species, and their interaction showed a significant main effect on predation and oviposition of the four phytoseiid species considered. When the same females were all fed on pollen after a diet of con- or heterospecific protonymphs, the influence of previously supplied food was still reflected on oviposition (tab. 1).

All females, with the exception of two of *T. kerkirae* on conspecifics, began to prey. At the end of the first ten days of the experiment all *T. kerkirae* and *A. andersoni* females were still alive, while in the period between the sixth and ninth day, two *T. exhilaratus* females, one preying on *T. kerkirae* and one on *A. andersoni,* died as well as one of *N. californicus* preying on itself.

The data on female predation, oviposition and escape attempts and their differences are summarized in table 2. A significantly low net predation rate on conspecifics was registered for *T. kerkirae* and *A. andersoni*. When predation occurred versus heterospecific protonymphs the most effective species was *A. andersoni* which consumed, independently of the kind of prey supplied, a very considerable number of protonymphs per female. Furthermore, this result could be underestimated because the number of supplied protonymphs was slightly higher than the mean of consumed prey. *Typhlodromus exhilaratus* and *N. californicus* when fed on *A. andersoni* protonymphs showed the lowest predation rates.

Few escape attempts, none of which were influenced by food, were observed for *T. kerkirae* and *N. californicus*. Significant differences with values of more than one per female in the period were registered for *T. exhilaratus* preying on *A. andersoni* and in the latter preying on conspecifics (tab. 2).

For all the species the significantly highest number of eggs was laid when pollen was supplied. When reared with conspecific protonymphs the oviposition rate was very low: not one *T. exhilaratus* female oviposited, and only 20% of *N. californicus* and 80% of the other two species began to oviposit. With heterospecific protonymphs as prey, a small increase in the oviposition rate was observed for *T. exhilaratus* and *N. californicus* feeding on *T. kerkirae* protonymphs, but only *A. andersoni* laid more than 1 egg per day per female, a figure significantly higher than that obtained when it preyed on itself. All females of this last species oviposited and the same was observed for *T. exhilaratus* on *T. kerkirae* and on *N. californicus* and for *T. kerkirae* on *A. andersoni*. On the other hand, only 60% of *T. exhilaratus* females oviposited on *A. andersoni* and 80% and 40% of *T. kerkirae* on *N. californicus* and on *T. exhilaratus*, respectively; while in *N. californicus* the ovipositing females ranged from 60% to 70% on *T. exhilaratus* and on *T. kerkirae* and *A. andersoni*, respectively.

In the second period of the experiment, when all phytoseiids were fed with pollen after a

Tab. 1 - Effects of predator (*T. kerkirae*, *T. exhilaratus*, *N. californicus* and *A. andersoni* females) and food (protonymphs of *T. kerkirae*, *T. exhilaratus*, *N. californicus*, *A. andersoni* and *Quercus* sp. pollen) on intra- and interspecific predation and oviposition rate (Two-way ANOVAs).

Dependent variable / effects	df	MS	F	Sig of F
Net predation rate				
Predator	3	14.09	294.76	.000
Food (prey)	3	7.82	163.57	.000
Predator by food	9	3.48	72.91	.000
Within + residual	144	0.05		
1st period oviposition rate				
Predator	3	3.84	143.60	.000
Food (prey and pollen)	4	9.44	352.53	.000
Predator by food	12	0.95	35.66	.000
Within + residual	180	0.03		
2nd period oviposition rate				
Predator	3	0.76	5.37	.001
1st period food supplied	4	1.66	11.73	.000
Predator by food	12	0.33	2.35	.008
Within + residual	173	0.14		

diet of con- or heterospecific protonymphs, mortality was not observed in *T. kerkirae*. On the contrary, one female died when *T. exhilaratus* had been previously fed on itself and on *A. andersoni*. Similarly, one female died when *N. californicus* was fed on itself. The same happened when *A. andersoni* was fed on *T. kerkirae* and on *N. californicus*. Furthermore, two females died when *T. exhilaratus* and *N. californicus* had been previously nourished on *T. kerkirae* and pollen, respectively.

The females which did not oviposit in the first period also began to lay eggs. On the whole, the change of protonymphs prey with pollen induced a considerable increase in oviposition rates, with the exception of *A. andersoni* preying on heterospecifics. In all other species when previously fed with *A. andersoni* an oviposition less than 1 egg/day was registered. In Amblyseiinae no differences emerged for females continuously fed on pollen and the others.

The eggs shrivelling in the two considered periods are shown in table 3. With regard to the latter data, statistical analysis was not always possible and when performed it was sometimes hardly affected by the large variability in oviposition rates, especially in the first period under consideration. In all situations, most of eggs hatched, with the exceptions of *T. kerkirae* preying on *N. californicus* (68.75%) and *N. californicus* preying on *T. exhilaratus* (78.58%). In the second period, the hatching rate was almost always lower on females continuously fed on pollen.

Tab. 2 - Intra- and interspecific predation, oviposition and escape rate of females of *T. kerkirae* (*Tk*), *T. exhilaratus* (*Te*), *N. californicus* (*Nc*) and *A. andersoni* (*Aa*) on different food. The means followed by the same letter are not significantly different for that species of predator among prey species/food (first column) and for each prey/food among the predator species (second column).

Predator sp. food	net predation rate	eggs/female/day (1st per)	eggs/female/day (2nd per)	escapes/female/day (1st per)
T. kerkirae				
Tk	0.12 a a	0.09 ab a	1.49 a a	0.75 a
Te	0.93 b a	0.04 a ab	1.43 ab a	0.83 a
Nc	0.96 b a	0.16 b b	1.64 a a	0.83 a
Aa	0.69 b b	0.16 b b	0.64 c a	0.13 a
pollen	-- -- --	1.94 c a	0.78 bc a	0.29 a
T. exhilaratus				
Tk	1.94 a b	0.45 c ab	1.04 abc a	0.17 a
Te	0.97 b a	0.00 a a	1.15 bc a	0.08 a
Nc	1.39 ab b	0.22 b bc	1.25 c a	0.08 a
Aa	0.35 c a	0.06 a a	0.44 ab a	1.13 b
pollen	-- -- --	1.52 d a	0.33 a a	0.10 a
N. californicus				
Tk	2.25 a b	0.23 b b	1.50 ab a	0.33 a
Te	1.59 a b	0.14 a bc	1.20 ab a	0.33 a
Nc	0.71 b a	0.02 a a	1.66 b a	0.42 a
Aa	0.56 b ab	0.09 a ab	0.71 a ab	0.83 a
pollen	-- -- --	1.49 c a	0.79 ab a	0.00 a
A. andersoni				
Tk	6.83 a c	1.21 b c	1.21 a a	0.59 ab
Te	7.05 a c	1.18 b c	1.35 a a	0.33 a
Nc	7.44 a c	1.03 b c	0.82 a b	0.88 b
Aa	0.33 b a	0.07 a a	1.17 a b	1.25 b
pollen	-- -- --	1.52 c a	0.96 a a	0.25 a

If we consider the overall daily oviposition (fig. 1), two common trends emerged. In all species, the fecundity of females continuously reared on pollen decreased in the second period considered, most of eggs being laid in the first ten days. Furthermore, when the initial prey was *A. andersoni* protonymphs, the oviposition was limited to the first or second day of the experiment and more time was needed for its increase when the prey was changed with pollen.

As prey, the *T. kerkirae* protonymphs are more suitable for the females of other species, while *T. exhilaratus* protonymphs were more preyed on by the females of Amblyseiinae, but only *A. andersoni* was able to convert both species of prey into a satisfactory egg production. *N. californicus* was more suitable prey in the order for *A. andersoni* followed by *T. exhilaratus*, while *A. andersoni* was absolutely unsuitable for itself and among the least consumed by heterospecifics.

Discussion and conclusions

Amblyseius andersoni confirms its traits of generalist type III predator according to the McMurtry and Croft (1997) classification. In fact, it killed less its own nymphs than the other and successfully converted the heterospecific phytoseiid prey into egg production with a rate similar

Tab. 3 - Influence of food supplied in the first period of experiment on shrivelling eggs laid in the same period and in the subsequent ten days. Shrivelling egg percentages of the 1st period of experiment and their significance (χ^2 test; P=0.05) are over the diagonal, while those of the 2nd period under. The food supplied in the 1st period was protonymphs of *T. kerkirae (Tk)*, *T. exhilaratus (Te)*, *N. californicus (Nc)*, *A. andersoni (Aa)* and pollen (po), in the 2nd only pollen.

		T. kerkirae					*T. exhilaratus*				
		Tk	*Te*	*Nc*	*Aa*	po	*Tk*	*Te*	*Nc*	*Aa*	po
1st per	%	0	0	31.3	0	2.1	2.3	0	9.1	0	2
	n	9	4	16	16	194	43	0	22	6	152
	Tk		---	ns	---	ns		---	ns	ns	ns
	Te	ns		ns	---	ns	ns		---	---	---
	Nc	*	*		*	*	ns	---		ns	ns
	Aa	ns	ns	ns		ns	ns	ns	ns		ns
	po	ns	ns	ns	ns		*	*	*	ns	
		Tk	*Te*	*Nc*	*Aa*	po	*Tk*	*Te*	*Nc*	*Aa*	po
2nd per	%	4.7	4.9	0.6	0	3.2	1.1	0	0	2.5	6.8
	n	149	143	161	56	62	89	110	125	40	44

		N. californicus					*A. andersoni*				
		Tk	*Te*	*Nc*	*Aa*	po	*Tk*	*Te*	*Nc*	*Aa*	po
1st per	%	8.7	21.4	0	0	3.4	5	3.4	1	0	0
	n	21	10	2	9	149	121	118	103	7	152
	Tk		ns	ns	ns	ns		ns	ns	ns	*
	Te	*		ns	ns	*	ns		ns	ns	*
	Nc	*	ns		---	ns	---	ns		ns	
	Aa	---	ns	ns		ns	---	ns	---		---
	po	*	ns	*	*		ns	ns	ns	ns	
		Tk	*Te*	*Nc*	*Aa*	po	*Tk*	*Te*	*Nc*	*Aa*	po
2nd per	%	0	1.1	2.7	0	8.4	0	2.9	0	0	1.1
	n	150	108	146	71	107	107	135	76	105	93

to that obtained with pollen as food. Furthermore, its protonymphs were able to avoid predation probably because of the length of their legs which made them faster runners than others. These factors, combined with its large size and polyphagy, place the phytoseiid, in the range of considered species, among the most competitive at low phytophagous level. In the absence of, or at low tetranychid density, *A. andersoni* outcompeted all others also when intraspecific competition between the unfed larvae of the phytoseiid and those of *A. fallacis, T. occidentalis* and *T. pyri* were compared (Zhang and Croft 1995a); it was also able to discriminate heterospecific eggs from its own (Schausberger and Croft 1999). However, our starved females seemed to feed on a considerable lower number of their own protonymphs than that observed in well fed females in similar laboratory conditions (Croft *et al.* 1996). In our country this species is widespread, sometimes in large populations, on apple, peach, and vines (Duso 1997). However, on the latter (Duso and Vettorazzo 1999) it seems to have a scarce persistence in the absence of food, probably because of its large size and activity (Croft and Croft 1996).

Less clear features were evidenced for the other species. *Neoseiulus californicus* is generally considered as a specialist or selective type II predator (McMurtry and Croft 1997), but for some traits recently evidenced, an intermediate rating between specialist and generalist type seems more appropriate (Croft *et al.* 1998). From our data, it preyed on more Typhlodrominae protonymphs than its own or other Amblyseiinae protonymphs, even if the fecundity was in all

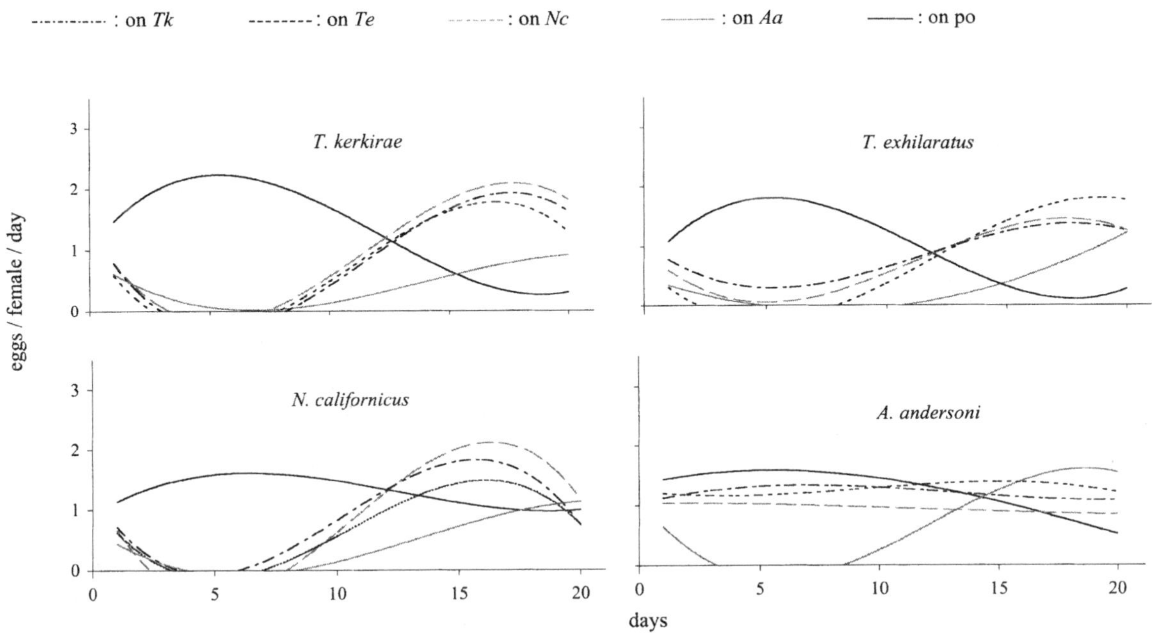

Fig. 1 - Oviposition trend obtained from experimental data smoothed with cubic regression in the four phytoseiid species fed on con- and heterospecific protonymphs and pollen in the first ten days and on pollen in the second ten day.

cases very low. Other authors reported that both females and juveniles are able to discriminate between con- and heterospecifics, preferably preying on heterospecifics (Palevsky *et al.* 1999; Schausberger and Croft 1999; Walzer and Schausberger 1999b). This species is characterized by the highest rate of increase on *T. urticae,* on which it is able to efficiently prey and oviposit also after being reared for a long time on alternative food (Castagnoli and Simoni 1999). On dust mites, pollen and tarsonemids, it reaches a good rate of increase only after many generations which have experienced each diet (Castagnoli and Falchini 1993; Castagnoli and Liguori 1994; Castagnoli *et al.* 1999). In Central Italy *N. californicus* is dominant on low-growing plants where it is associated with high populations of *T. urticae.* According to our data the phytoseiid could be outcompeted by *A. andersoni*, but this does not occur in our country because the two species usually do not share the same habitat.

Both the *Typhlodromus* species are probably to be included among the generalist type III species. Differently from the other two species, on a small scale they also share the same habitat, namely grape leaf, where *T. exhilaratus* is the dominant species. The fact that it preys on almost two *T. kerkirae* protonymphs per day and oviposits about 1 egg every two days could be two of the reasons for the *T. exhilaratus* dominance. With regard to intra- and interspecific predation, *T. exhilaratus* seems more similar to *N. californicus* than to *T. kerkirae.* The high oviposition rate on different prey species (Castagnoli and Liguori 1986), the best intrinsic rate on *Eotetranychus carpini* (Oudemans) and the association with this tetranychid on grape leaves (Castagnoli *et al.* 1989 and 1991) could configure *T. exhilaratus* as a generalist shifting versus specialist phytoseiid.

Typhlodromus kerkirae was the only species to share with *A. andersoni* a significantly lower predation of females on conspecifics. However, it showed extremely low predation and oviposition rates on protonymphs, while on pollen the daily oviposition in the first ten days was the highest observed in this experiment. Despite the lack of studies available on this species, it is the only one, among those considered, in which the intrinsic rate of increase is higher on pollen than on prey (Liguori *et al.* 1996) and thus could be considered specialized on pollen.

The high survival of females of all four species when fed on immature phytoseiids and the high oviposition when newly supplied with a more suitable food, can be interpreted as an index of their ability to overcome periods of food scarcity. However, only *A. andersoni* was able to exploit other phytoseiids to increase its population.

References

Calvitti M., Tsolakis H., 1992 - Acari predatori fitoseidi presenti in alcune culture erbacee del Lazio. *Redia*, LXXV: 529-535.

Castagnoli M., 1989 - Recent advances in knowledge of the mites fauna in the biocenoses of the grapevine in Italy. In: Cavalloro R. (ed.), Influence of environmental factors on the control of grape pests, diseases and weeds. Balkema, Rotterdam, Brookfield: 169-180.

Castagnoli M., Falchini L., 1993 - Suitability of *Polyphagotarsonemus latus* (Banks) (Acari Tarsonemidae) as prey for *Amblyseius californicus* (McGregor). *Redia,* LXXVI: 273-279.

Castagnoli M., Liguori M., 1986 - Tempi di sviluppo e ovideposizione di *Typhlodromus exhilaratus* Ragusa (Acarina: Phytoseiidae) allevato con vari tipi di cibo. *Redia*, LXIX: 361-368.

Castagnoli M., Liguori M., 1994 - Utilizzazione di polline nell'allevamento massale di *Typhlodromus exhilaratus* Ragusa e *Amblyseius californicus* (McGregor) (Acari: Phytoseiidae). In: Viggiani G. (ed.), MAF, Convegno "Lotta Biologica", Acireale 1991, Istituto di Patologia Vegetale, Roma: 139-144.

Castagnoli M., Simoni S., 1999 - Effect of long-term nutritional history on functional and numerical response of *Neoseiulus californicus*. *Experimental and Applied Acarology*, 23: 217-234.

Castagnoli M., Amato F., Monagheddu M., 1989 - Osservazioni biologiche e parametri demografici di *Eotetranychus carpini* (Oudemans) (Acarina: Tetranychidae) e del suo predatore *Typhlodromus exhilaratus* Ragusa (Acarina: Phytoseiidae) in condizioni di laboratorio. *Redia*, LXXII: 545-557.

Castagnoli M., Liguori M., Nannelli R., 1984 - Contributo alla conoscenza degli acari del pesco in Toscana e osservazioni sull'andamento delle loro popolazioni. *Redia*, LXVII: 493-504.

Castagnoli M., Liguori M., Simoni S., 1993 - Distribuzione, campionamento e associazione degli acari della soia in Italia. *Redia*, LXXVI: 111-120.

Castagnoli M., Simoni S., Biliotti N., 1999 - Mass-rearing of *Amblyseius californicus* on two alternative food source. In: Bruin J., van der Geest L.P.S., Sabelis M.W. (eds), Ecology and Evolution of the Acari, Kluwer Academic Publishers, Dordrecht, The Netherlands: 425-431.

Castagnoli M., Liguori M., Amato F., Guidi S., 1991 - Dinamica spaziale e temporale di *Eotetranychus carpini* (Oudemans) (Acarina: Tetranychidae) e dei fitoseidi suoi predatori sulla vite. Atti XVI Congresso Nazionale Italiano di Entomologia, Bari-Martina-Franca (Ta), 23-28 settembre 1991: 339-345.

Croft B.A., Croft M.B., 1996 - Intra- and interspecific predation among adult female Phytoseiid mites (Acari: Phytoseiidae): effects on survival and reproduction. *Environmental Entomology*, 25: 853-858.

Croft B.A., Kim S.S., Kim D.I., 1996 - Intra- and interspecific predation on four life stage groups by the adult females of *Metaseiulus occidentalis, Typhlodromus pyri, Neoseiulus fallacis* and *Amblyseius andersoni. Experimental and Applied Acarology*, 20: 435-444.

Croft B.A., Monetti L.L., Pratt P.D., 1998 - Comparative life histories and predation types: are *Neoseiulus californicus* and *N. fallacis* (Acari: Phytoseiidae) similar type II selective predator of spider mites? *Environmental Entomology*, 27: 531-538.

Duso C., 1997 - Distribution, biology and management of important phytophagous mites on temperate fruits in Italy. *Redia*, LXXX, Appendice: 1-24.

Duso C., Vettorazzo E., 1999 - Mite population dynamics on different grape varieties with or without phytoseiid release (Acari: Phytoseiidae). *Experimental and Applied Acarology*, 23: 741-763.

Helle W., Sabelis M.W., 1985 - World crops pests. Spider mites, their biology, natural enemies and control. Elsevier, Amsterdam, vol. 1B: 458 pp.

Liguori M., 1995 - Phytoseiid mites (Acari: Phytoseiidae) on *Actinidia deliciosa* (A. Chev.) Liang Ferguson and on associated spontaneous vegetation in Tuscany. In: Kropczynska D., Boczek J., Tomczyk A. (eds), The Acari. Physiological and ecological aspects of Acari - host relationships. Dabor, Warszawa: 231-234.

Liguori M., Guidi S., Simoni S., 1996 - Life history and potential for increase of a laboratory strain of the predaceous mite *Typhlodromus kerkirae* Swirski and Ragusa (Acari: Phytoseiidae) reared on two different kinds of food. *Redia*, LXXIX. 247-256.

McMurtry J.A., Croft B.A., 1997 - Life-styles of phytoseiid mites and their roles in biological control. *Annual Review of Entomology*, 42: 291-321.

McRae I.V., Croft B.A., 1997 - Intra- and interspecific predation by adult females of *Metaseiulus occidentalis* and *Typhlodromus pyri,* (Acari: Phytoseiidae) when provisioned with varying densities and ratios of *Tetranychus urticae* (Acari: Tetranychidae) and phytoseiid larvae. *Experimental and Applied Acarology*, 21: 235-245.

Monetti L.N., Croft B.A., 1997 - *Neoseiulus californicus* (McGregor) and *N. fallacis* (Garman): larval response to prey and humidity, nymphal feeding drive and nymphal predation on phytoseiid eggs. *Experimental and Applied Acarology*, 21: 225-234.

Palevsky E., Reuveny H., Okonis O., Gerson U., 1999 - Comparative behavioural studies of larval and adult stages of the phytoseiids (Acari: Mesostigmata) *Typhlodromus athiasae* and *Neoseiulus californicus*. *Experimental and Applied Acarology*, 23: 467-485.

Ragusa S., 1986 - A five year study on population fluctuations of Phytoseiid mites in a citrus orchard in Sicily. *Acarologia*, XXVII: 193-201.

Schausberger P., Croft B.A., 1999 - Predation on and discrimination between con- and heterospecific eggs among specialist and generalist phytoseiid mites (Acari: Phytoseiidae). *Environmental Entomology*, 28: 523-528.

SPSS Inc., 1994 - SPSS for Windows 6.1. Advanced Statistics 6.1. SPSS Inc., Chicago, Il.

Yamamura K., 1999 - Transformation using (x + 0.5) to stabilize the variance of population. *Researches on Population Ecology*, 41: 229-234.

Walzer A., Schausberger P., 1999a - Cannibalism and interspecific predation in the phytoseiid mites *Phytoseiulus persimilis* and *Neoseiulus californicus*: predation rates and effects on reprodution and juvenile development. *Biocontrol*, 43: 457-468.

Walzer A., Schausberger P., 1999b - Predation preference and discrimination between con and heterospecific prey by the phytoseiid mites *Phytoseiulus persimilis* and *Neoseiulus californicus*. *Biocontrol*, 43: 469-478.

Zhang Z.-Q., Croft B.A., 1995a - Intraspecific competition in immature *Amblyseius fallacis, Amblyseius andersoni, Typhlodromus occidentalis* and *Typhlodromus pyri* (Acari: Phytoseiidae). *Experimental and Applied Acarology*, 19: 65-77.

Zhang Z.-Q., Croft B.A., 1995b - Interspecific competition and predation between immature *Amblyseius fallacis, Amblyseius andersoni, Typhlodromus occidentalis* and *Typhlodromus pyri* (Acari: Phytoseiidae). *Experimental and Applied Acarology,* 19: 247-257.

Effects of Neem oil (*Azadirachta indica* A. Juss) on *Tetranychus urticae* Koch (Acariformes, Tetranychidae) in laboratory tests*

Haralabos Tsolakis[1], Ernesto Ragusa[2], Salvatore Ragusa[1**]
[1] *Istituto di Entomologia agraria, Viale delle Scienze, 90128 Palermo, Italy, tsolakis@unipa.it and ragusa@unipa.it*
[2] *Via Lincoln 37, 90133 Palermo, Italy.*

Key words: neem oil, toxicity, repellence, *Tetranychus urticae*, laboratory trials.

Abstract: The effects of a commercial neem oil titolated in azadirachtin at 10,000 ppm on the phytophagous mite *Tetranychus urticae* Koch were investigated in laboratory trials. The material provoked a high mortality on the post embryonic developmental stages of the mite when citrus leaves were used as substrate, while mortality was lower (40%) on bean leaves. Neem oil caused a high mortality of young females (100%) and also a remarkable reduction of the oviposition rate in contaminated females (from 10.9 to 93.5%). The two choice-tests, carried out in order to evaluate the repellent effect of the material, showed that neem oil has long term repellent properties.

Introduction

Tetranychus urticae Koch is one of the most injurious pests on various agricultural crops. To control the infestations of this species, biological control techniques have been adopted, mainly in greenhouses, using the phytoseiid mite *Phytoseiulus persimilis* Athias-Henriot (Oatman *et al.* 1968; Mori and Moriyama 1970; Pralavorio 1973; Markkula and Tiittanen 1976; Stenseth 1980; Pickett *et al.* 1987; Waite 1988; Ragusa and Ciulla 1989; Decou 1994). However, sometimes the predator is not able to control the infestations of the two-spotted spider mite in a short time and chemical control is usually necessary to protect the crop. It is clear that, in this case, selective products have to be applied in order to reduce the population of *T. urticae* and preserve the phytoseiid mites. Among selective products, some botanical pesticides such as neem oil, quassin, absinthium and urtica, did not show strong negative effects on various phytoseiid species both in laboratory and field tests (Mansour *et al.* 1987, 1993; Dimetry *et al.* 1994; Tsolakis *et al.* 1997). On the other hand some formulations of the above products showed a strong negative effect toward the two-spotted spider mite (Mansour and Ascher 1983; Mansour *et al.* 1987; Dimetry *et al.* 1993; Tsolakis *et al.* 1997). If we consider botanical pesticides, the most used is neem oil or the various extracts of neem seed kernel (Schmutterer and Ascher 1984 and 1987; Mansour and Ascher 1983; Mansour *et al.* 1987 and 1997; Dimetry *et al.* 1993; Momen *et*

* Work carried out with ex 60% funds.

** Authors have equally contributed to the present paper.

F. Bernini, R. Nannelli, G. Nuzzaci, E. de Lillo (eds.), Acarid Phylogeny and Evolution. Adaptations in mites and ticks, 351-362.

al. 1997; Tsolakis *et al.* 1997). As far as the latter are concerned, their miticidal effects show great differences among the different formulations and the different types of solvent used to extract them. As a matter of fact Mansour and Ascher (1983 and 1995) showed that extracts obtained from neem oil using lipophilic solvents were more active against tetranychid mites when compared to those obtained with hydrophilic solvents. It should be said that is not clear which limonoids present in the neem seeds are active against mites, but certainly the tetranotriterpenoid azadirachtin was harmless to spider mites (Sanguanpong and Schmutterer 1992), while it was effective against different species of insects (Ladd 1983; Rembold *et al.* 1987; Schmutterer 1987). The commercially available products actually are titolated in azadirachtin and contain other limonoids at different concentrations. Most probably the different ratios of the various limonoids have an effect on different species even if the concentration in azadirachtin is the same.

The aim of the present work was to check the effect of a commercially available product on various biological parameters of *T. urticae* in laboratory tests.

Materials and Methods

The strain of *T. urticae* Koch was collected on weeds in a field near Palermo (Italy) and kept in the laboratory on lemon fruits at 24±1°C, 70±5% of R.H. and a photoperiod of 16:8 LD.

A commercial neem oil titolated in azadirachtin at 10,000 ppm was used in the tests at the dose of 500 ml/hl (0.5%) corresponding to 0.005% of azadirachtin.

Experimental units

Each experimental unit (EU) consisted of the upper surface of a bean leaf (3 x 4 cm ca.) or a citrus leaf (1.5 x 3.5 cm). Each leaf, including the petiole, was put on wet cotton wool in a Petri dish. Each test consisted of 5 repetitions.

Test procedure

To verify the effects of the neem oil on the pre-imaginal stages, new laid eggs (max 24 hours old) were used. To obtain them, 10 females were put on each EU for 24 hours to oviposit, and then removed. Ten eggs were left on each leaf. Eggs were directly sprayed with 4 ml of solution per leaf using a Potter tower. Observations were made daily until the mites reached adulthood. These tests were carried out on both bean and citrus leaves (tests 1, 2 and 3, 4 respectively). Females obtained from the tests on the pre-imaginal stages (tests 1 and 2) were put on untreated bean leaves to verify the side effects of neem oil on their oviposition rate (tests 5 and 6).

The effects on the adults (survival, oviposition rate, repellent effect) were investigated only on bean leaves. Young females and males (max 1 day old) were tested. To ensure the homogeneous age of the mites, 50 ovipositing females were kept on each of five fresh lemon fruits to oviposit for 24 hours; afterwards they were removed. When adults were obtained, 5 young females and 5 young males were put on each EU and observed daily for 10 days. As far as sprayings are concerned two types of tests were carried out: *no-choice* and *two-choice* tests.

As far as *no-choice* tests are concerned spraying was done a) directly on the adults previously located on the leaves (post-infestation treatment) (test 7) or b) on the leaf putting adults on it 5' after the spraying (dried leaf) (pre-infestation treatment) (test 8).

In the *two-choice* tests, only the right half of the leaf was sprayed, to check if the material showed repellent effect. Adults were put on the mid-rib of the leaf 5' after the spraying (test 9).

Distilled water was used for the control and also in these tests, adults were put on the mid-rib of the leaf 5' after the spraying (test 10).

Sprayings were done with 4 ml of solution per leaf, using a Potter tower, as described above, also in tests 7-10.

In all adult tests the laid eggs were removed

Tab. 1 - Influence of neem oil on the post embryonic development of *T. urticae* on bean leaves (1 and 2) and citrus leaves (3 and 4).

Tests	Initial number of eggs	Dried eggs	Mortality of the post embryonic stages: Larvae	Protonymphs	Deutonymphs	Attaining adulthood ♀	♂	Mortality % *
1 Neem oil	50	9	6	3	2	24	6	40 a
2 Control	50	1	0	0	0	34	15	2 b
3 Neem oil	50	24	26	0	0	0	0	100 c
4 Control	50	4	4	0	0	35	7	16 ab

* Different letters denote significant differences for $P< 0.05$ and $P< 0.01$ (Student t - test).

Tab. 2 - Oviposition rate of the females obtained from the tests on the post embryonic development (see tab. 1).

Tests	Number of Females	Number of eggs laid in 48 hours	Eggs/female/day Mean ± S.E.	Reduction of eggs/♀ %
5 Females obtained from test n.1	24	249	3.8 ± 0.52 a	29.7
6 Females obtained from test n. 2 (control)	34	447	7.0 ± 0.73 b	-

* Different letters denote significant differences for $P< 0.05$ and $P< 001$ (Student t - test).

daily. The eggs obtained from the tests on adults (tests 8 and 10), were put on untreated leaves to check the effects on egg hatching (tests 11 and 12).

All tests were carried out in a conditioned room at 24±1°C, 70±5% of R.H. and a photoperiod of 16:8 LD.

The reduction in the total number of eggs/female was calculated with the following formula (Dimetry *et al.* 1993):

Repellent effect was calculated as the percentage of females staying off the treated part of the leaf in the two-choice tests.

Results were statistically analyzed by Student t test and Analysis of Variance (ANOVA) followed by Tukey's test when the ANOVA was significant ($P < 0.05$). The software "Statistica" from StatSoft Inc. (1995) was used.

$$\text{\% of reduction} = \frac{\text{Total n. of eggs laid on the control} - \text{Total n. of eggs laid on the treated leaves}}{\text{Total n. of eggs laid on the control} + \text{Total n. of eggs laid on the treated leaves}} \times 100$$

Results and discussion

Neem oil had a strong negative effect on eggs and larvae when citrus leaves were used as substrate, causing a 100% mortality; in contrast the mortality on bean leaves was only 40% and decreased from eggs to deutonymphs (tab. 1). It should be said, however, that the mortality of about 50% eggs on glabrous citrus leaves was most probably caused by the asphyctic effect of oil more than by the effect of its compounds. As a matter of fact on the tomentose bean leaves the mortality of eggs was only 18%. This fact suggests that a) leaf structure has a great influence on the efficiency of sprayings and b) the material is not really very persistent as far as toxicity is concerned. In a previous experiment (Tsolakis *et al.* 1997) where the same material was used on citrus leaves, a greater mortality of eggs (76%) was obtained, but it should be noted that the spraying procedure used was different as more than 4 ml of solution per leaf was applied.

Neem oil showed also a remarkable influence on the oviposition rate of females that were in contact with sublethal doses of the product during their postembryonic development (test 5). As a matter of fact, as shown in table 2, the oviposition rate of the females was reduced (about 30%) in comparison to the oviposition obtained in the control tests (t=7.3; df=4; P=0.001). Similar results were obtained by Mansour and Ascher (1983) who found a reduction of fecundity in females of *Tetranychus cinnabarinus* (Boisd.) ranging from 20 to 70% at different concentrations of extracts of fresh neem seed kernels.

Neem oil caused a high mortality (100%) of newly emerged females in both pre-infestation and post-infestation treatment tests (tests 7 and 8, tab. 3). However, while mortality was already 80% after 24 hours in the latter and the surviving females died within 7 days, in the pre-infestation tests the same level of mortality was reached after 3-4 days and the surviving females died in 10 days (fig. 1). Mansour *et al.* (1987) found a mortality of *T. cinnabarinus* ranging from 80 to 100% after 48 hours using 0.5% concentration of acetone and pentane extracts of neem, while it was lower (about 40%) using methanol and ethanol extracts. A mortality of 66% was obtained after 2 days

Tab. 3 - Influence of neem oil on the oviposition rate and survival of adults of *T. urticae*.

Tests *	Initial number of adults		Number of adults at the end of the test		Mortality ***	Total number of eggs/♀ ** ***	Reduction of eggs/♀	Eggs/♀/day ***
	♀	♂	♀	♂	%	Mean ± S.E.	%	Mean ± S.E.
7 *Neem oil*	25	25	0	0	100 a	2.4 ± 1.44 a	93.5	0.6 ± 0.16 a
8 *Neem oil*	25	25	0	0	100 a	15.9 ± 7.41 a	63.7	4.2 ± 0.40 b
9 *Neem oil*	25	25	13	14	46 b	57.6 ± 4.39 b	10.9	5.8 ± 0.44 c
10 *Control*	25	25	24	24	4 c	71.7 ± 4.45 c	-	7.2 ± 0.44 d

* 7. Spraying was done on adults (post-infestation treatment).
8. Introduction of adults 5' after the spraying (pre-infestation treatment).
9. Spraying was done on the right half of the leaf (introduction of adults on mid rib after the leaves had dried) (two choice-tests).
10. Introduction of adults on mid rib of the leaves.

** Considering the life span of the females for the tests 7 and 8, and for 10 days for the tests 9 and 10.

*** Different letters denote significant differences for P< 0.05 and P< 0.01 (Student t - test).

using 0.5% concentration of RD9-Repelin (a neem seed kernel extract formulation containing 0.15% of azadirachtin) on the same mite species, while mortality reached levels of 86% at higher concentration (3%) (Mansour *et al.* 1993). As regards the reduction of eggs/♀ in the post-infestation treatment tests, it was of 93.5% and the number of eggs/♀/day was very low (0.6) while in the pre-infestation treatment tests the reduction of eggs/♀ was of 63.5% and the number of eggs/♀/day was statistically higher compared to that obtained in the previous test (4.2 eggs/♀/day) (t=5.2; df=18; P=0.00) (tab. 3). Mansour *et al.* (1987) using a 0.5% concentration of pentane and acetone extracts of neem on females of *T. cinnabarinus* found a reduction in fecundity of 97% and 89% respectively while a lower reduction was obtained when methanol and ethanol extracts were used (35 and 50% respectively). On the other hand Dimetry *et al.* (1993) found that a 0.4% concentration of Neem azal-S (3,500 ppm of azadirachtin) caused a 100% mortality of females of *T. urticae* and a 92.7% of reduction in egg laying within 7 days, while mortality was lower (85%)

Tab. 4 - Effects of neem oil on the hatching of eggs laid by females deriving from tests n. 8 and n. 10 (cf. tab. 3).

Tests	Initial number of eggs	Number of dried eggs	Mortality of postembryonic stages: Larvae	Protonymphs	Deutonymphs	Reaching adulthood: ♀	♂	Mortality %*
11 Eggs laid in the tests n. 8	74	23	1	0	0	35	15	32.4 a
12 Eggs laid in the tests n. 10	332	14	0	0	0	227	91	4.2 b

* Different letters denote significant differences for $P< 0.05$ and $P< 0.01$ (Student t - test).

Tab. 5 - Repellence activity of neem oil on females of *T. urticae*.

	Distribution of eggs on the leaf					
	Left **			Right**		
Tests*	Number of eggs laid in 10 days	%	Eggs/♀/day Mean ± S.E.	Number of eggs laid in 10 days	%	Eggs/♀/day Mean ± S.E.
9 *Neem oil*	760	84.1 β	6.3 ±2.0a	144	15.9 γ	2.4 ±2.1b
10 Control	789	44.6 δ	7.0 ±1.9a	982	55.4 δ	7.4 ±2.4a

* For explanation see tab. 3.
** Different letters denote significant differences between the left and the right part of the leaf and among the tests on the same part of the leaf; different symbols denote significant differences between the left and the right part of the leaf for each test (P<0.01 Student t test).

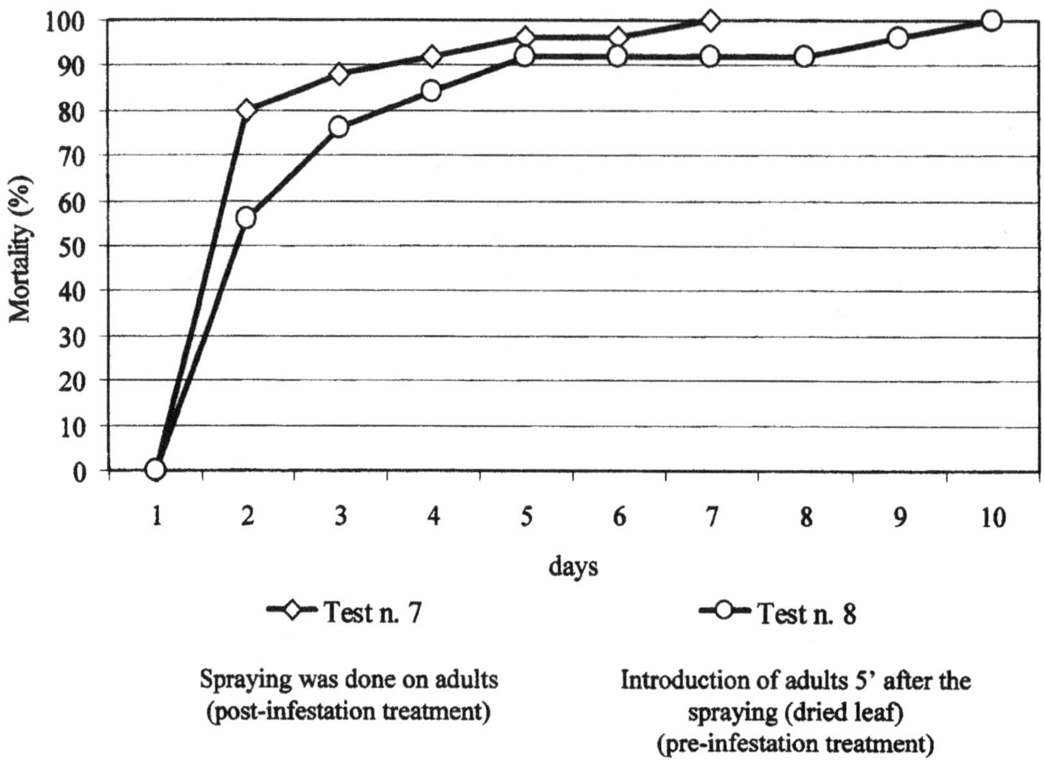

Fig. 1 - Trend of mortality of adult females of *T. urticae*.

using Margosan-O (3,000 ppm of azadirachtin) and there was an egg laying reduction (83.4%) in the same period.

Mortality was 46% after 10 days in the two choice test and the reduction of eggs/female was low (10.9%, 5.8 eggs/♀/day), but the results were statistically different if compared to those obtained in the control (t=5.2; df=49; P=0.00) (test 9, tab. 3).

In table 4 we report the results obtained on hatching of eggs laid by contamined females. We see that a remarkable percentage of eggs (ca 32%) did not hatch and this difference was statistically significant if compared to that obtained in the control (t=3.8; df=4; P=0.01). Similar results were obtained on *T. urticae* by Dimetry *et al.* (1993) who observed that 20.7% of eggs did not hatch when Neem azal-S was used and only 13.64% of eggs did not hatch when Margosan-O was used.

In the two-choice tests a low mortality of the young females was found after 24 hours, while a majority of females preferred staying on the untreated part of the leaf for the entire duration of the tests (fig. 2) (F=28.2; df=3, 196; P=0.00). On the control tests the distribution of females was almost homogeneous on the entire leaf surface (fig. 3). In similar tests Dimetry *et al.* (1993) obtained a very low mortality on females of *T. urticae* after 24 hours using a 0.4% concentration of Margosan-O and Neem azal-S (4 and 6% respectively).

As a consequence such a preference the distribution of eggs laid in the two parts of the leaves had a similar trend: only few eggs (2.4 eggs/♀/day) were laid on the treated part of the leaf while 6.3 eggs/♀/day were laid on the un-

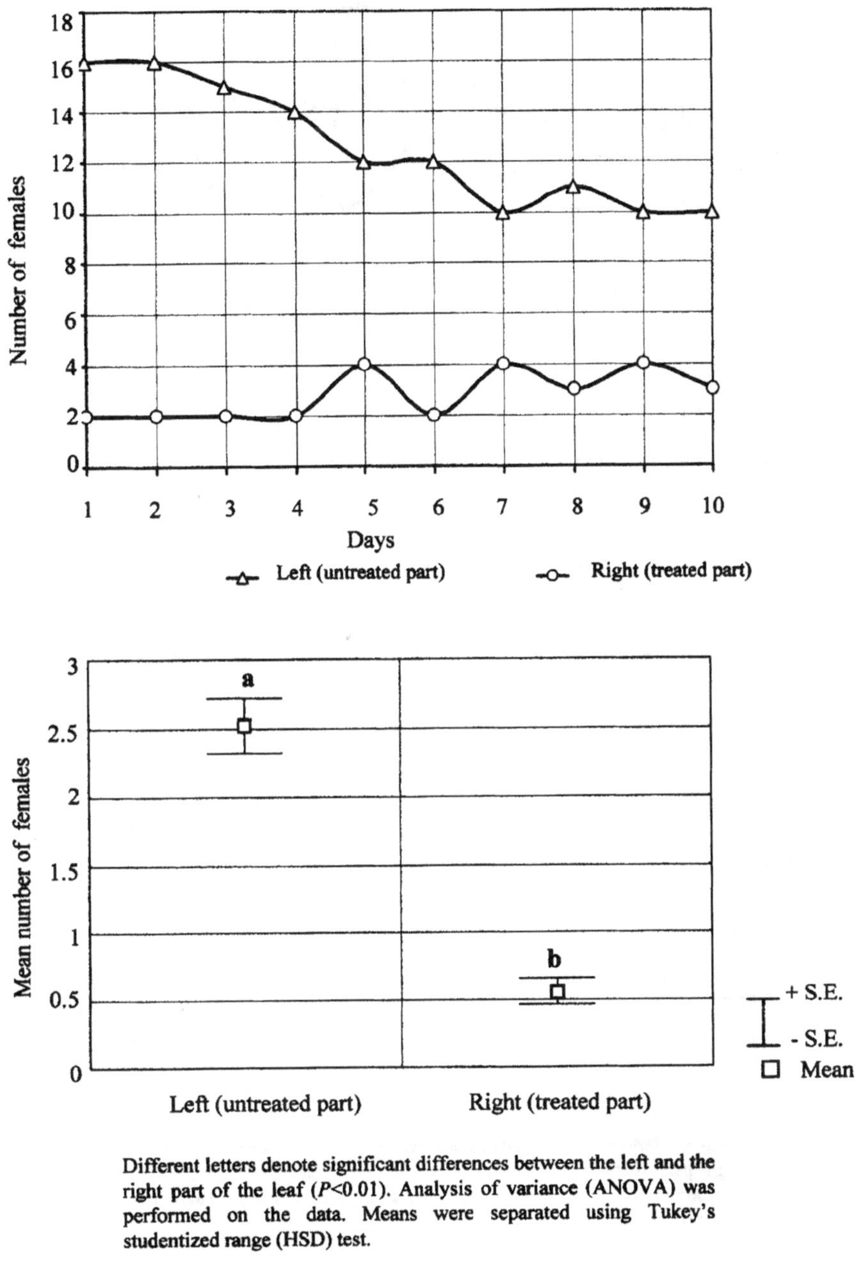

Fig. 2 - Distribution of females of *T. urticae* on bean leaf treated with neem oil on the right part.

treated part (fig. 4, tab. 5). Differences among the two parts were highly significant (t=6.05; df=49; P=0.00). On the other hand, the distribution of eggs was almost homogeneous in the control tests. No statistical differences were found (t=0.5; df=49; P=0.6) as an average of 7.5 eggs/"/day

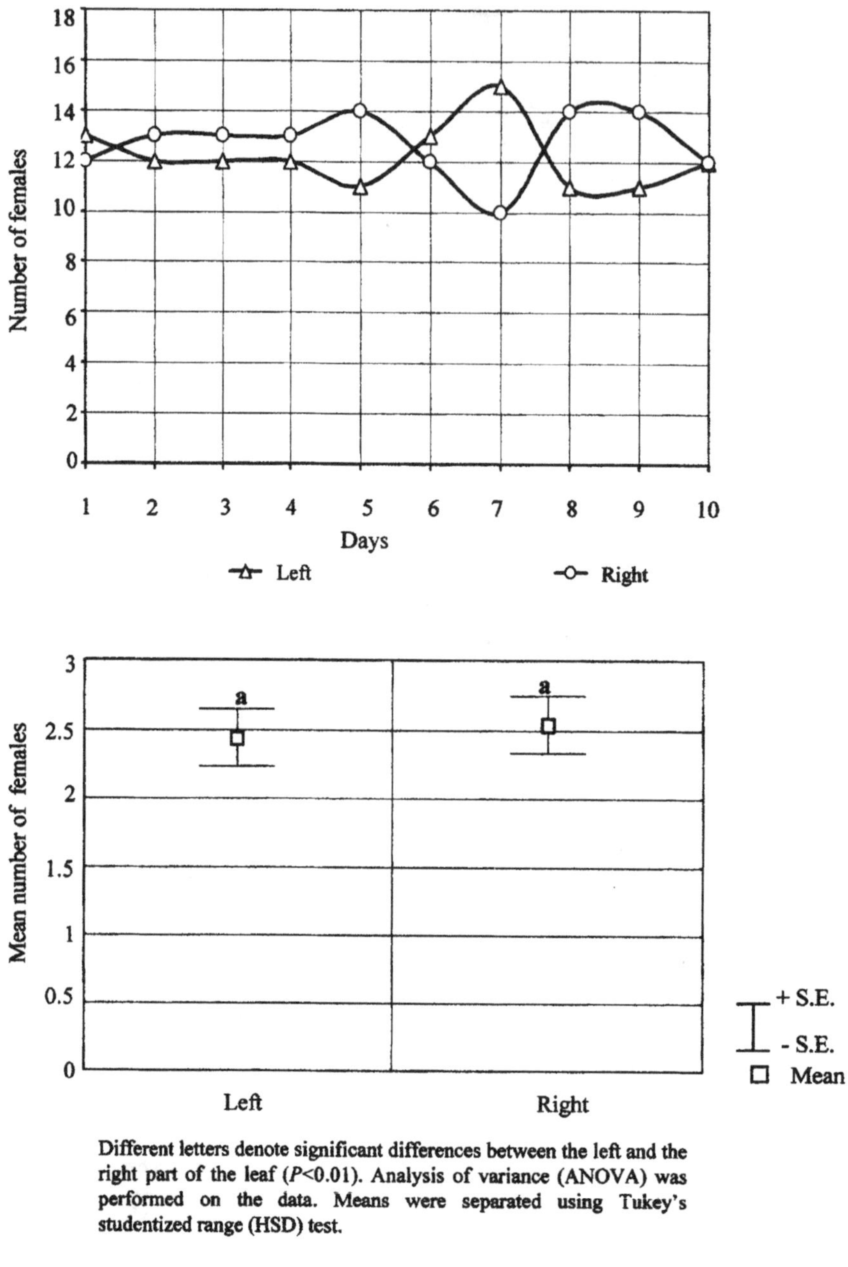

Fig. 3 - Distribution of females of *T. urticae* on untreated bean leaf.

was laid on each part of the leaf (fig. 5, tab. 5)

The repellent effect was very high during the first 4 days (80 to 100%) and was mantained over 70% for the entire duration of the test (fig. 6). Dimetry *et al.* (1993) using Margosan-O at 0.4% conc. on *T. urticae* found a strong repellence

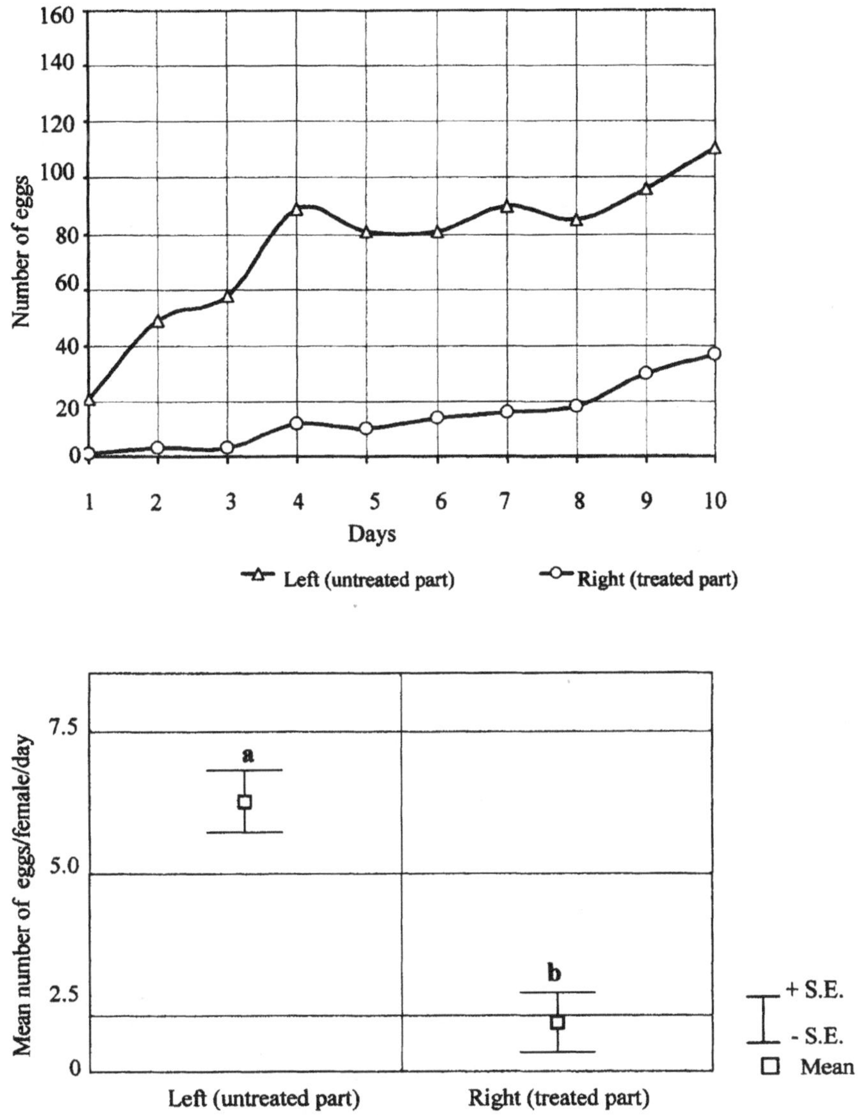

Different letters denote significant differences between the left and the right part of the leaf (*P*<0.01). Analysis of variance (ANOVA) was performed on the data. Means were separated using Tukey's studentized range (HSD) test.

Fig. 4 - Distribution of eggs of *T. urticae* on bean leaf treated with neem oil on the right part.

(from 88% to 80% after 6 and 24 h respectively) while using Neem azal-S at the same concentration the repellent effect was higher after 6 h (98%), but lower after 24h (64%). However, Mansour *et al.* (1993) using Margosan-O at a concentration of 0.5% on *T. cinnabarinus* found a 32% of repellence after

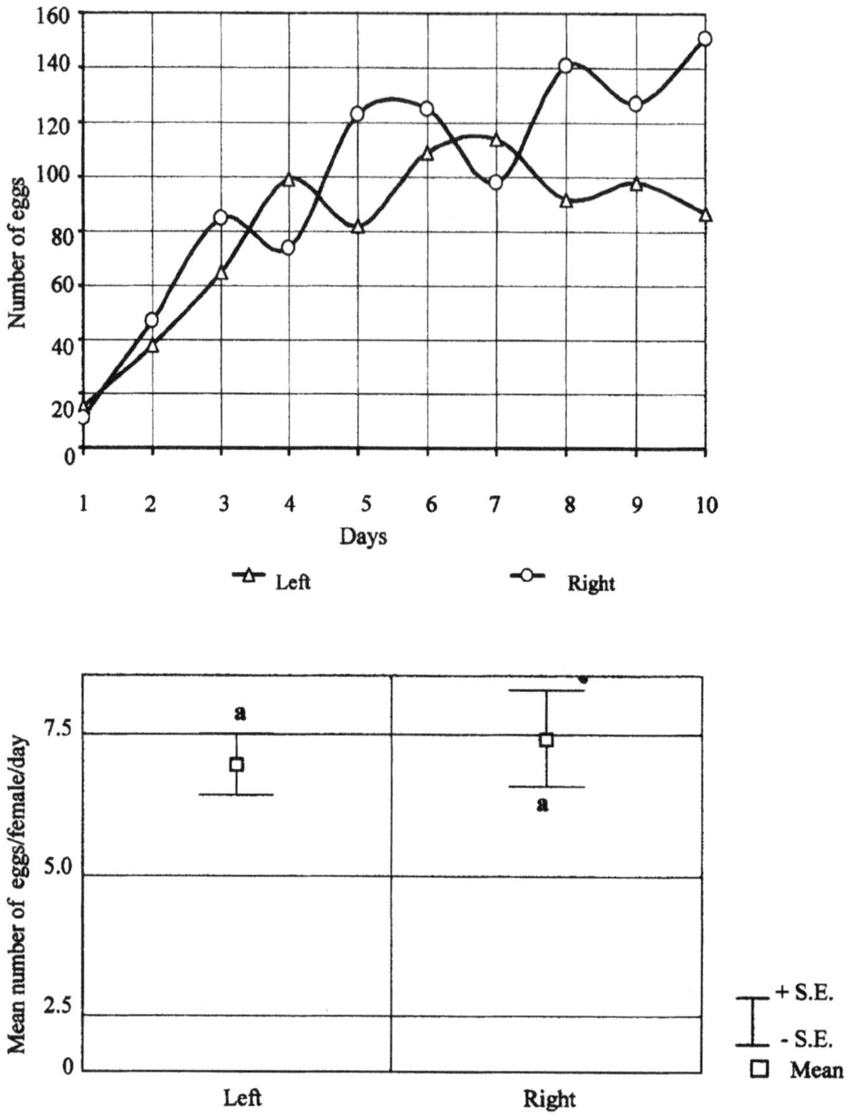

Different letters denote significant differences between the left and the right part of the leaf ($P<0.01$). Analysis of variance (ANOVA) was performed on the data. Means were separated using Tukey's studentized range (HSD) test.

Fig. 5 - Distribution of eggs of *T. urticae* on untreated bean leaf.

2 days and 42% after 4 days while using RD9-Repelin at 0.5% and Azatin at 0.4% the repellence after 2 days was lower (25% and 24% for RD9-Repelin and Azatin respectively).

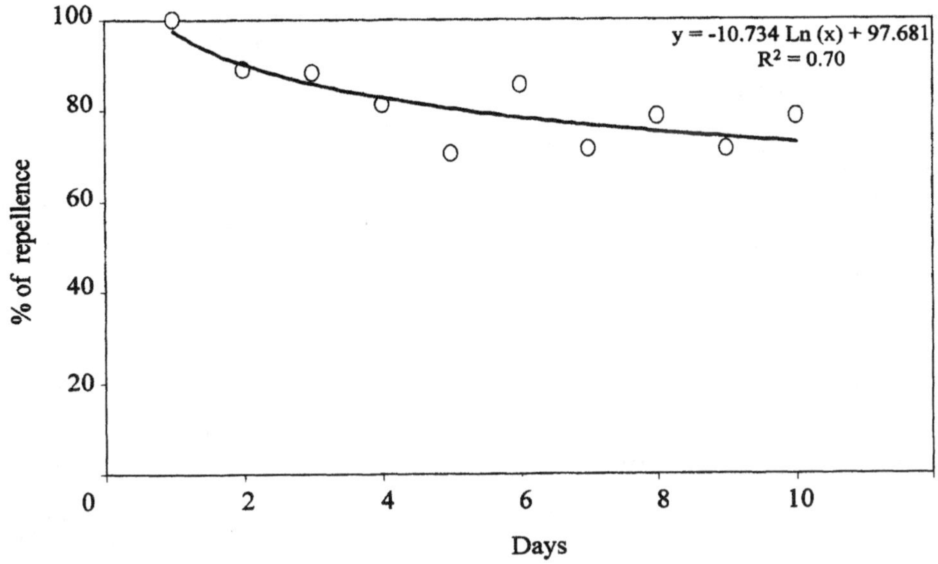

Fig. 6 - Repellence effect of neem oil towards adult females of *T. urticae.*

Conclusions

From what is report above we can assume that neem oil:

- was harmful to eggs and larvae on glabrous leaves but less toxic to the postembryonic stages on tomentose leaves;
- showed a negative side effects on the oviposition rate of females that were in contact with sublethal doses of neem oil during their postembyonic development;
- was harmful to adults also when they could freely walk on treated parts of the leaves and reduced the oviposition significantly;
- reduced significantly the hatching of eggs laid by females living on treated leaves;
- showed a lasting repellent effect towards females that preferred to live and oviposit on the untreated part of the leaves.

References

Decou G.C., 1994 - Biological control of the two-spotted spider mite (Acarina: Tetranychidae) on commercial strawberries in Florida with *Phytoseiulus persimilis* (Acarina: Phytoseiidae). *Florida Entomologist*, 77(1): 33-41.

Dimetry N.Z., Amer S.A.A., Momen F.M., 1994 - Laboratory trials of two neem seed extracts on the predatory mites *Amblyseius barkeri* (Hughes) and *Typhlodromus richteri* Karg. *Bollettino di Zoologia Agraria e di Bachicoltura*, 26(1): 127-137.

Dimetry N.Z., Amer S.A.A., Reda A.S., 1993 - Biological activity of two neem seed kernel extracts against the two-spotted spider mite *Tetranychus urticae* Koch. *Journal of Applied Entomology*, 116: 308-312.

Ladd T.L., 1983 - The influence of azadirachtin on the growth and development of immature forms of the japanese beetle, *Popilia japonica*. Proceedings of the 2nd International Neem Conference, Rauisch-holzhausen, Federal Republic of Germany 25-28 May, 1983. Deutsche Gesellschaft fur Technische Zusammenarbeit (GTZ) GmbH, Dag-Hammarskjold-Weg 1+2, D 6236 Eschborn 1: 425-434.

Mansour F., Ascher K.R.S. 1983 - Effects of neem (*Azadirachta indica*) seed kernel extracts from different solvents on the carmine spider mite, *Tetranychus cinnabarinus*. *Phytoparasitica*, 11(3-4): 177-185.

Mansour F., Ascher K.R.S., 1995 - Acarina, mites. In: Schmutterer H. (ed.), The Neem Tree - Source of Unique Natural Products for Integrated Pest Management, Medicine, Industry and Other Purposes. VCH, Weinheim, Germany: 161-164.

Mansour F., Ascher K.R.S., Abo-Moch F., 1993 - Effects of Margosan-O™, Azatin™ and RD9-Repelin® on spiders, and on predacious and phytophagous mites. *Phytoparasitica*, 21(3): 205-211.

Mansour F., Ascher K.R.S., Abo-Moch F., 1997 - Effects of neemgard on phytophagous and predacious mites and on spiders. *Phytoparasitica*, 25(4): 333-336.

Mansour F., Ascher K.R.S., Omari N., 1987 - Effect of neem (*Azadirachta indica*) seed kernel extracts from different solvents on the predacious mite *Phytoseiulus persimilis* and the phytophagous mite *Tetranychus cinnabarinus*. *Phytoparasitica*, 15(2): 125-130.

Markkula M., Tiittanen K., 1976 - "Pest in first" and "Natural infestation" methods in the control of *Tetranychus urticae* Koch with *Phytoseiulus persimilis* Athias-Henriot on Glasshouse cucumbers. *Annales Agriculturae Fenniae*, 15: 81-85.

Momen F.M., Reda A.S., Amer S.A.A., 1997 - Effect of Neem Azal-F on *Tetranychus urticae* and three predacious mites of the family Phytoseiidae. *Acta Phytopathologica et Entomologica Hungarica*, 32(3-4): 355-362.

Mori H., Moriyama S., 1970 - Biological control of *Tetranychus urticae* Koch on white clover by *Phytoseiulus persimilis* Athias-Henriot (Acarina: Tetranychidae, Phytoseiidae). *Mushi*, 43(2): 29-38.

Oatman E.R., McMurtry J.A., Voth V., 1968 - Suppression of the two-spotted spider mite on strawberry with mass releases of *Phytoseiulus persimilis*. *Journal of Economic Entomology*, 61(6): 1517-1521.

Pickett C.H., Gilstrap F.E., Morrison R.K., Bouse L.F., 1987 - Release of predatory mites (Acari: Phytoseiidae) by Aircraft for the biological control of spider mites (Acari: Tetranychidae) infesting corn. *Journal of Economic Entomology*, 80: 906-910.

Pralavorio M., 1973 - Essai d'utilisation de *Phytoseiulus persimilis* dans differentes cultures sous serres de la regione mediterraneenne. *Bullettin SROP*, 4: 19-25.

Ragusa S., Ciulla A.M., 1989 - Controllo di *Tetranychus urticae* Koch (Acariformes: Tetranychidae) su fragola in ambiente protetto, a mezzo di *Phytoseiulus persimilis* Athias-Henriot (Parasitiformes: Phytoseiidae). *Informatore Fitopatologico*, 39(7-8): 46-48.

Rembold H., Uhl M., Muller Th., 1987 - Effect of azadirachtin A on hormone titers during the gonadotrophic cycle of *Locusta migratoria*. In: Schmutterer H., Ascher K.R.S. (eds), Natural pesticides from the neem tree (*Azadirachta indica* A. Juss) and other tropical plants, Proceedings of the 3rd International Neem Conference, Nairobi, Kenya, 10-15 July, 1986. Deutsche Gesellschaft fur Technische Zusammenarbeit (GTZ) GmbH, Dag-Hammarskjold-Weg 1+2, D 6236 Eschborn 1: 289-298.

Sanguanpong U., Schmutterer H., 1992 - Laborversuche uber die Wirkung von Niemoel und Niemsamenextracten bei der gemeinen Spinnmilbe *Tetranychus urticae* Koch (Acari: Tetranychidae).-*Zeitschrift für Pflanzenkrankheiten und Pflanzenschutz*, 99: 637-646.

Schmutterer H., 1987 - Fecundity-reducing and sterilizing effects of neem seed kernel extracts in the Colorado potato beetle, *Leptinotarsa decemlineata*.- In: Schmutterer H. and Ascher K.R.S. (eds.), Natural pesticides from the neem tree (*Azadirachta indica* A. Juss) and other tropical plants, Proceedings of the 3rd International Neem Conference, Nairobi, Kenya, 10-15 July, 1986. Deutsche Gesellschaft fur Technische Zusammenarbeit (GTZ) GmbH, Dag-Hammarskjold-Weg 1+2, D 6236 Eschborn 1: 351-360.

Schmutterer H., Ascher K.R.S., 1984 - Natural pesticides from the neem tree (*Azadirachta indica* A. Juss) and other tropical plants. Proceedings of the 2nd International Neem Conference, Rauischholzhausen, Federal Republic of Germany 25-28 May, 1983. Deutsche Gesellschaft fur Technische Zusammenarbeit (GTZ) GmbH, Dag-Hammarskjold-Weg 1+2, D 6236 Eschborn 1: 587 pp.

Schmutterer H., Ascher K.R.S., 1987 - Natural pesticides from the neem tree (Azadirachta indica A. Juss) and other tropical plants. Proceedings of the 3rd International Neem Conference, Nairobi, Kenya, 10-15 July, 1986. Deutsche Gesellschaft fur Technische Zusammenarbeit (GTZ) GmbH, Dag-Hammarskjold-Weg 1+2, D 6236 Eschborn 1, 703 pp.

StatSoft Inc., 1995 - Statistica, users manual 1997. Ed. Stat Soft Inc., Tusla, OK.

Stenseth C., 1980 - Methods for using *Phytoseiulus persimilis* for control of *Tetranychus urticae* on cucumber. *Bulletin S.R.O.P./W.P.R.S.* 3: 221-224.

Tsolakis H., Leto G., Ragusa S., 1997 - Effects of some plant materials on *Tetranychus urticae* Koch (Acariformes, Tetranychidae) and *Typhlodromus exhilaratus* Ragusa (Parasitiformes, Phytoseiidae). ANPP 4th International Conference of the Pests in Agriculture, Montpellier 6-7-8 Jan. 1997: 239-245.

Waite G.K., 1988 - Control of *Tetranychus urticae* Koch by *Phytoseiulus persimilis* Athias-Henriot in low-chill stonefruit. *Queensland Journal of Agricultural and Animal Sciences*, 45(2): 185-188.

Influence of different types of grass mulching on the communities of Oribatid mites in the vineyards

Roberto Nannelli, Sauro Simoni
Istituto Sperimentale per la Zoologia Agraria, Via Lanciola 12/A, 50125 Firenze, Italy

Key words: Oribatida, soil management, ecological indices, vineyards.

Abstract: The effects of different soil management techniques (spring-summer soil tillage and three different types of grass mulching - with spontaneous herbs, with *Festuca arundinacea*, and with *Trifolium subterraneum*) on the abundance and distribution of Oribatid mites in the vineyard agroecosystem were studied. Samples of turf were taken from the ground cover along the rows six times between June and November 1996. The epigeal and ipogeal parts of the herbage core samples were divided and processed differently.

On a total surface sampled of nearly 0.2 m^2, for every treatment, the density ranged from 3,300/m^2 adult individuals in grass mulching with *Festuca* to 6,900/m^2 adult individuals with spontaneous herbs. On the herbage cover of the vineyards, 4,316 specimens belonging to 51 species of Oribatids were recorded. Forty-nine species were found in the organic layer associated with the surface of the soil (*Tectocepheus velatus, Eupelops curtipilus* and *Scheloribates* cf. *minifimbriatus* were the dominant species) and 29 species were recorded on the epigeal and ipogeal parts of the herbage. The comparison of the Shannon-Wiener diversity index and Equitability calculated for every treatment has highlighted the uniformity of the mite fauna in the habitats studied. The soil management techniques considered make significant differences on the number of specimens of Oribatid mites.

Introduction

The Cryptostigmata or oribatid mites together with Collembola are the most common representatives of the mesofaune of forest soils. They are usually most abundant near the surface of the litter layer where accumulations of organic debris are plentiful, whereas they are relatively sparse in cultivated soils, where the continual mixing of mineral and topsoil produces an unstable and poorly structured profile (Wallwork 1976). Like most soil animals, the soil mites are important components in the food web (Lagerlöf and Andrén 1988) playing a fundamental role in decomposing organic matter, recycling and increasing nutrient availability (Petersen and Luxton 1982).

Agronomists underline the need to minimise tillage, which reduces expense, time and labour, and soil erosion, so as to improve the level of organic matter in the soil. In the region of Tuscany, experimental vineyards have been constructed for some years to evaluate the influence of different types of grass mulching on the productivity of grapevines and quality of vines.

For alternative management practices to become successful, a better knowledge of the functioning of the soil-crop system is essential.

In this context the effect of different mulching practices was investigated with special emphasis on oribatid mites involved in decomposition processes through catalyzing microbial activity.

F. Bernini, R. Nannelli, G. Nuzzaci, E. de Lillo (eds.), Acarid Phylogeny and Evolution. Adaptations in mites and ticks, 363-371.

Materials and methods

The study site is located near Montalcino (Siena, Italy) in a fine area producing Brunello vine (Azienda Col d'Orcia, S.Angelo in Colle, Siena).

The climate of the region is typically Mediterranean with mild winters and warm, dry summers. In 1996, the year of the study, the minimum temperature was recorded in February (-6°C) and the warmest temperature in August (+40°C); the annual total precipitation was 470 mm (max in September = 140 mm). The temperatures and precipitation levels during the investigation period are shown in figure 1.

The experiment was conducted in two adjacent vineyards cultivated with the same cultivar of grapevine (Sangiovese) but using four different soil management techniques (s.m.t. 1÷4). In an experimental field the soil cultivation of the inter-row was alternatively tillage (spring-summer soil tillage, s.m.t. 1) and natural grassing (with spontaneous herbs, s.m.t. 2). In the other vineyard, half was mulched with *Trifolium subterraneum* (cv. clare) (s.m.t. 3) and the other half with *Festuca arundinacea* (cv. penna) (s.m.t. 4). Both vineyards selected had received the same chemical treatments and no herbicidal treatments were used (tab. 1).

In each of the four different types of cultivation practices three rows with four plots were selected; on each sampling day 9 samples were randomly selected from the 12 plots of each test site. Each field was sampled six times between June and November 1996 (6-VI, 3-VII, 5-VIII, 3-IX, 2-X, 7-XI). Samples of turf (core sample 7x5 cm x 5-7 cm deep) were taken out of the ground cover along the rows in the middle of a plot to minimize possible edge effects. Table 2 presents the most representative species of herbs growing along the rows. The herbage samples (35 cm^2) were numbered; the epigeal part of plants was cut as close as possible to the soil surface and collected in a plastic bag, while the organic layer and the ipogeal part of plants of a similar area were taken below the sampled herbage with a stainless steel core sampler. This sample was collected in another plastic bag labelled with the same number as the herbage sample from the same area.

The herbs were washed in a detergent solution to remove mites (Vial and Monterrat

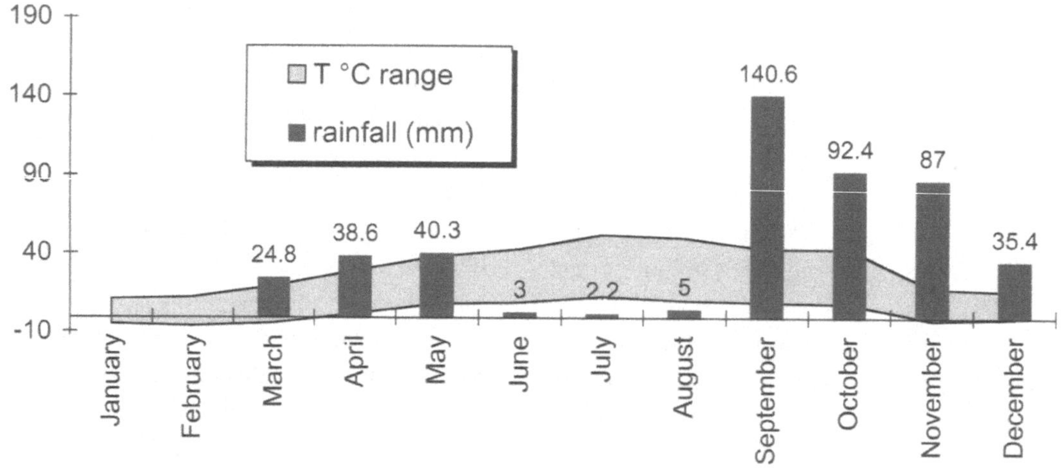

Fig. 1 - Climatic data of 1996 registered in the experimental vineyards.

Tab. 1 - Chemicals used in pest management of the vineyards considered in 1996.

May - Cymoxanil + Copper - Cymoxanil + Mancozeb + Phosethyl + Dinocap + Myclobutanil - Cymoxanil + Mancozeb + Phosethyl + Dinocap	**July** - Copper oxychloride + Sulfur WP - Sulfur WP
June - Copper oxychloride + Dinocap - Cymoxanil + Copper + Myclobutanil	**August** - Bordeaux mixture + Sulfur WP

1971), which were subsequently sieved, counted and mounted for study. Oribatid mites were extracted from the core samples by using conventional Berlese-Tullgren funnels. Only adult specimens were counted and determined to specific level.

For data analyses the number of specimens collected in the epigeal and ipogeal parts of plants were pooled from all 9 samples from each site. The significance of treatment and sampling date effects on the number of mites per population was tested by means of a two-way

Tab. 2 - Presence and diffusion (%) of herbs in the vineyards with different management.

Spring-summer soil tillage between the rows	Grass mulching with spontaneous herbs	Grass mulching with *Festuca arundinacea*	Grass mulching with *Trifolium subterraneum*
Along the rows sporadic presence of - *Avena fatua* - *Bromus* spp. - *Cynodon dactylon* - *Hordeum* sp. - *Poa* spp. - *Cirsium arvensis* - *Inula* sp. - *Taraxacum officinale* - *Convolvulus arvensis* - *Capsella bursa pastoris*	80-90 % - *Avena fatua* - *Bromus* spp. 10-20 % - *Cynodon dactylon* - *Hordeum* sp. - *Poa* spp. - *Cirsium arvensis* - *Inula* sp. - *Taraxacum officinale* - *Convolvulus arvensis* - *Capsella bursa pastoris*	80-90% - *Festuca arundinacea* 10-20% - *Avena fatua* - *Bromus* spp. - *Inula* sp. - *Plantago* sp. - *Rumex* sp. - *Anagallis* spp.	<u>In midsummer</u> 40-50% - *Trifolium subterraneum* 60-50% - *Avena fatua* - *Bromus* spp. - *Inula* sp. - *Plantago* sp. - *Rumex* sp. - *Anagallis* spp. <u>In spring and late summer</u> 80-90% - *Trifolium subterraneum* 20-10% - *Plantago* sp. - *Rumex* sp. - *Inula* sp.

ANOVA. Prior to ANOVA the data was $\ln(x + 0.5)$ transformed to minimize the variance (Yamamura 1999). Community composition and relative species abundance were investigated adopting the most common ecological indices: diversity was measured by means of Shannon-Wiener's index (H), Equitability (E), Simpson's concentration (C) and Simpson's diversity indices (D). All indices were calculated as indicated in Castagnoli *et al.* 1999. For the classification of dominance of species the scheme described by Engelmann (1978) and newly applied to mites by Juvara-Bals (1992) was adopted.

Results

On herbs living along the rows of the two vineyards considered 1,336 specimens of adult oribatids belonging to 31 species were collected; on the ipogeal part of herbaceous plants together with the organic layer associated with the surface of the soil 2,980 specimens belonging to 49 species were recorded. In total, 51 species of oribatids were collected on the ground cover and hemiedaphic zone of soil layer of the two vineards (tab. 3).

The density of the oribatid mites ranged from 3,349 adult specimens/m^2 in grass mulching with *Festuca* to 6,931 adult specimens/m^2 on herb in the plots with tilled inter-row, while 6,873 specimens/m^2 were registered in the plots with grass mulching with spontaneous herbs and 5,703 specimens/m^2 with *Trifolium*. These values are close to the density of the oribatid fauna of the meadows of the South and Central Europe reported by Petersen and Luxton (1982), Petrov (1997), Hülsmann and Wolters (1998) and Khalil *et al.* (1999).

The adult specimens of oribatids collected in the four different management techniques are reported in figure 2. Predictably, the highest number of specimens was recorded in the two autumnal samples when the local conditions, temperature and humidity of the soil were most favourable to mesofaune. In contrast, because of dryness of the soil, the August samples were low in number, only a few spontaneous herbs survived and the mite fauna was poorly represented by xerophic species.

The ANOVA (tab. 4) shows that both the sampling date and management technique of the soil significantly affect the number of oribatids registered.

Table 3 presents the specific composition of oribatids and their percentage of dominance according to Engelmann's classification (Engelmann 1978) recorded in the experimental fields with the different treatments. Out of the total of 51 species

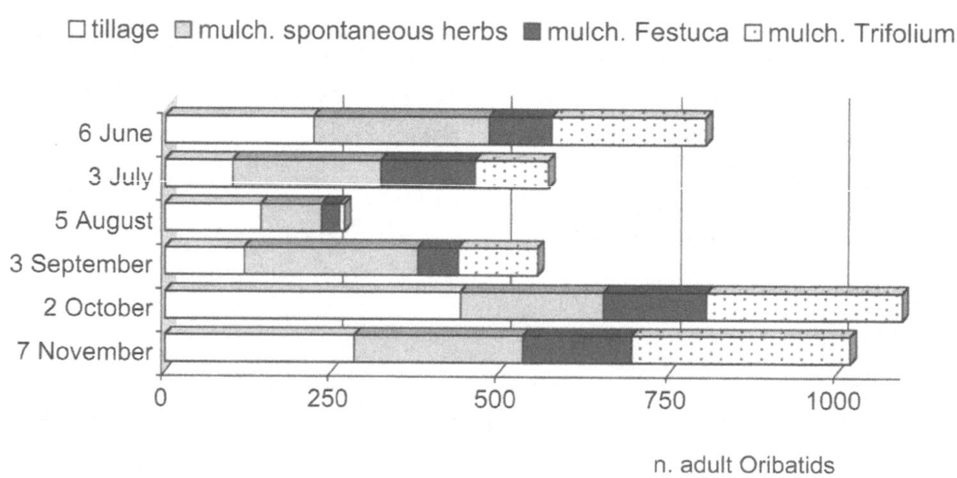

Fig. 2 - Adult oribatid mites over 1996 for each soil management system.

Tab. 3 - List of oribatid mite species collected on the herbage cover and organic layer in the two experimental vineyards and percentage of dominance throughout the sampling period.

Species	Soil management technique			
	1	2	3	4
Gilarovella sp.	1.15 D	3.31 C	0.47 E	--
Hypochthonius luteus Oud.	--	0.54 E	--	--
Cosmochthonius sp.	--	--	--	0.09 F
Phyllozetes emmae (Berl.)	0.08 F	0.23 F	0.32 E	0.09 F
Haplochthonius simplex Willmann	--	--	--	0.09 F
Sphaerochthonius splendidus (Berl.)	1.83 D	1.39 D	2.37 D	0.65 E
Brachychthonius impressus Moritz	--	--	--	0.46 E
Poecilochthonius italicus (Berl.)	--	--	--	0.19 F
Rhysotrizia ardua (Koch)	0.38 E	1.23 D	0.16 F	0.93 E
Lohmannia hungarorum Mahunka	0.15 F	--	--	--
Papillacarus aciculatus (Berl.)	2.52 D	0.23 F	2.84 D	0.56 E
Epilohmannia cylindrica (Berl.)	--	--	0.63 E	0.74 E
Nothrus biciliatus Koch	4.5 C	0.77 E	13.27 B	4.17 C
Arthrodamaeus cf. *mediterraneus* Subias, Arillo & Subias	1.53 D	8.01 C	4.27 C	6.03 C
Epidamaeus sp.	--	0.08 F	--	0.19 F
Metabelba sp.	0.08 F	0.15 F	--	0.09 F
Berlesezetes sp.	0.31 E	--	--	--
Amerobelba decedens Berl.	1.22 D	--	0.63 E	0.09 F
Eremulus flagellifer Berl.	--	--	2.37 D	2.41 D
Fosseremus laciniatus (Berl.)	1.15 D	0.62 E	2.37 D	1.76 D
Liacarus coracinus (Koch)	0.76 E	0.46 E	--	--
Xenillus tegeocranus (Herm.)	0.15 F	0.08 F	--	--
Tectocepheus velatus (Mich.)	12.6 B	11.39 B	22.27 B	31.17 B
Corynoppia kosarovi (Jeleva)	--	--	2.21 D	--
Graptoppia paraanalis Subias & Rodriguez	0.31 E	2.85 D	2.05 D	1.76 D
Hypogeoppia hypogea (Paoli)	--	0.08 F	--	--
Medioppia obsoleta (Paoli)	0.61 E	0.31 E	1.11 D	--
Microppia minus (Paoli)	--	--	0.16 F	--
Neotricoppia confinis (Paoli)	2.9 D	1.15 D	1.9 D	0.65 E
Oppia arcidiaconoae Bernini	1.3 D	2.16 D	2.05 D	2.41 D
Oppiella nova (Oud.)	--	--	--	0.65 E
Oxyoppioides decipiens (Paoli)	0.08 F	0.08 F	--	--
Ramusella (Insculptoppia) insculpta (Paoli)	--	--	--	0.09 F
Suctobelbella perpendiculata (Forsslund)	0.99 E	0.15 F	0.79 E	1.95 D
Suctobelbella subcornigera (Forsslund)	0.23 F	0.23 F	0.47 E	0.56 E
Passalozetes africanus (Grandjean)	2.37 D	5.31 C	0.32 E	0.37 E
Protoribates capucinus (Berl.)	1.6 D	0.69 E	0.47 E	1.48 D
Lucoppia burrowsi (Mich.)	2.75 D	7.39 C	1.26 D	1.11 D
Zygoribatula arcuatissima (Berl.)	--	0.08 F	--	--
Zygoribatula connexa (Berl.)	3.51 C	4.08 C	11.37 B	7.88 C
Zygoribatula exarata (Berl.)	0.15 F	--	--	--
Zygoribatula socia (Berl.)	7.48 C	5.23 C	10.9 B	7.33 C
Hemileius sp.	0.08 F	0.15 F	--	--
Scheloribates cf. *minifimbriatus* Minguez, Subias & Ruiz	4.12 C	8.55 C	8.69 C	19.57 B
Scheloribates laevigatus (Koch)	0.84 E	--	--	--
Ceratozetes cf. *laticuspidatus* Menke	0.46 E	1.92 D	1.26 D	0.83 E
Latilamellobates incisellus (Kramer)	18.17 B	19.17 B	--	--
Punctoribates punctum (Koch)	8.4 C	4.85 C	2.53 D	3.62 C
Eupelops curtipilus (Berl.)	15.19 B	7.08 C	--	--
Peloptulus phaenotus (Koch)	--	--	0.47 E	--
Pilogalumna crassiclavata (Berl.)	0.08 F	--	--	--

Soil management technique 1: spring- summer soil tillage
" " 2: grass mulching with spontaneous herbs
" " 3: grass mulching with *Festuca arundinacea*
" " 4: grass mulching with *Trifolium subterraneum*

A: eudominant, 32 – 100 %
B: dominant, 10 – 31.9 %
C: subdominant, 9.9 – 3.2 %
D: rare, 3.19 – 1 %
E: accidental, 0.99 – 0.30 %
F: sporadic, < 0.30 %

sampled, 20 species were always present. This result indicates that a large number of species are probably insensitive to cultivation methods.

In the vineyard with tilled inter-row (s.m.t. 1) 36 species of mites were found and 34 species were found in the field with grass mulching with spontaneous herbs (s.m.t. 2); 17 species had the same class of dominance, 6 species had dominance in s.m.t. 1> s.m.t. 2 and, on the contrary, 7 species in s.m.t. 2> s.m.t. 1. Therefore, the grassy habitat along the rows constitutes a shelter for oribatids. In the vineyard with grass mulching a lower number of species, 29 in s.m.t. 3 and 32 in s.m.t. 4, were always collected and only 10 species, about a third of the total, had the same class of dominance; 9 species had a class of dominance in s.m.t. 3> s.m.t. 4 and only 5 species in s.m.t. 4> s.m.t. 3.

In all four soil treatments, the most important species collected was *Tectocepheus velatus*: the highest number of its specimens were collected in the vineyard with permanent mulching (*Festuca* + *Trifolium*) rather than in vineyard with spontaneous herbs; its populations were more numerous in the hemiedaphic zone of the soil layer than on the cover vegetation. In contrast, *Latilamellobates incisellus* was the most important species in the plots with spontaneous herbs and the highest number was recorded on the epigeal part of plants; this species was always associated with abundant populations of *Eupelops curtipilus* and *Scheloribates* cf. *minifimbriatus*. The Oppiidae family was well represented and specimens of *Oppia arcidiaconoae*, *Neotricoppia confinis* and *Graptoppia paraanalis* were always present in all plots. The genus *Zygoribatula* was well represented with four species: in particular considerable populations of *Z. connexa* and *Z. socia* were registered in plots with *Festuca*, whereas in the other treatments the densities recorded were always similar and slightly lower; *Nothrus biciliatus* showed similar behaviour.

Figure 3 shows the splitting of the percentages of the oribatid mites on the basis of the dominance classes. An important presence of subdominant species in the plots with mulching with spontaneous herbs is worthy of note. Differently, a similar splitting of dominance was observed in the other treatments where a great number of oribatids was ascribed to dominant classes.

In tab. 5, the species diversity indices are reported. They provide another way of detecting and evaluating the effects of different management of the soil on common soil mite fauna.

From the present data it can be observed that the highest values for the Shannon-Wiener index (H) and for the Simpson diversity index (D) are obtained when they are calculated to evaluate the species diversity concerning grass mulching with natural herbs. Similarly, as has been said before, these values confirm the greater biodiversity registered in this treatment. In contrast, the

Tab. 4 - Results of analysis of variance on the effect of sampling date and treatment on the number of oribatids.

	Sum of Squares	DF	F	Sig. of F
Sampling date (S)	25.68	5	4.89	0.000
Treatment (T)	15.22	3	4.83	0.002
S x T	13.24	15	0.84	0.633
Within + residual	1261.05	1200		

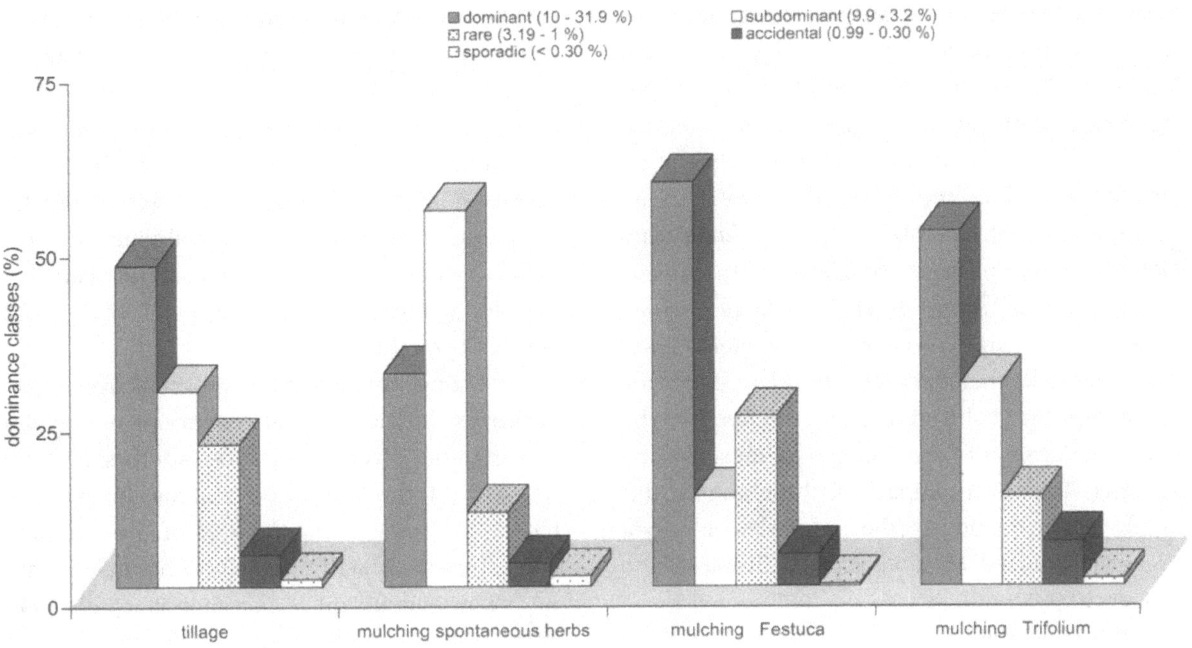

Fig. 3 - Splitting of dominance classes of the specimens registered in each soil management system.

complex of the indices adopted indicates the grass mulching with *Trifolium subterraneum* as the more homogeneous habitat.

Conclusions

The composition of the oribatid fauna in mulching vineyards of the Brunello area seems to be similar to agricultural soils and some of the taxa here listed have been reported in other investigations (Franchini and Rockett 1996; Tolstikov *et al.* 1997; Petrov 1997). *T. velatus* is one of the most common species encountered in fields cultivated in Ohio (Franchini and Rockett 1996) and it is more frequent in conventional and reduced-till than in no-till fields. Rocket (1986) reported that *T. velatus* was collected exclusively in cultivated plots as compared to adjacent grassy

Tab. 5 - Species diversity of oribatid mite community on herbs and organic layer of the vineyards (H = Shannon-Wiener's diversity index; E = Equitability; C = Simpson's concentration index; D = Simpson's diversity index).

Soil management technique	N. species	N. specimens	H	E	C	D
Spring-summer tillage	36	1310	3.972	0.769	0.094	10.629
Grass mulching with spontaneous herbs	34	1299	3.976	0.782	0.087	11.502
Grass mulching with *Festuca arundinacea*	29	633	3.806	0.783	0.107	9.351
Grass mulching with *Trifolium subterraneum*	32	1078	3.450	0.690	0.157	6.382

and wooded areas. Tolstikov *et al.* (1997) found *T. velatus* to be the most abundant species of the soil in an apple orchard of a nature reserve near Moscow. Petrov (1997) found *Tectocepheus sarekensis* dominant in ley meadow and natural meadows with rich multi-species vegetation covers. Lagerlöf and Andrén (1988) found reasonably high number of *T. velatus* and *Oppiella nova* in fields cultivated with Spring-barley, *Festuca pratensis* and *Medicago sativa*. Rockett (1986) and Hülsmann and Wolters (1998) also reported members of Oppidae occurring frequently in cultivated plots; generally the seasonal density of these mites is greatest in early summer. The severe impact of ploughing on this family might be due to the reduction of micro-organisms caused by cultivation (Hülsmann and Wolters 1998).

It is generally accepted that the effect of cultivation on the soil fauna depends on the type and extent of tillage. Franchini and Rockett (1996) found a significant reduction in abundance and species diversity in conventional-till fields in comparison to no-till and reduced-till. Simultaneously, they also observed that some oribatid species appeared to positively react to conventional ploughing. This evidence indicates that numerous oribatid species are sensitive to agricultural impact, particularly to cultivation practices.

Our results concerning the oribatids of the turf of the vineyard have shown that the species diversity of this mite community is not greatly affected by the kind of soil mulching. On the contrary, the fluctuation of the oribatid population densities is great, probably due to different micro-climatic conditions and to a different amount of organic matter as a consequence of agronomic techniques. Furthermore, plant cover and its litter fall are beneficial for soil animals both as a moderator of the microclimate and as a nutrient and energy source (Khalil *et al.* 1999). Luxton (1981) also suggests food and environmental conditions as the two principle factors which influence the migration and hence the vertical distribution of mites. The ability to migrate into the herbage and to utilize a rapidly increasing food supply would be in part an advantage to such species. The collecting of over 50% of subdominant species in the vineyard with spontaneous herbs leads us to consider that a species-rich grassy vegetation usable on the mulching of plots in experimental vineyards, can ensure different microhabitats and can undoubtedly favour the development of a multi-species community.

Planting or conserving ground cover plants improves the ecological environment by improving the diversity of bio-resources and the diversity of the arthropod community. Furthermore, the grassy ground cover of the vineyards constitutes an effective habitat for natural enemies of phytophagous mites (Castagnoli *et al.* 1999) and oribatid mites, especially the young specimens, and may be an appreciable source of food for the predator community.

References

Castagnoli M., Liguori M., Nannelli R., 1999 - Influence of soil management on mite populations in a vineyard agroecosystem. In: Bruin J., van der Geest L.P.S., Sabelis M.W. (eds.), Ecology and Evolution of the Acari, Kluwer Academic Publishers: 617-623.

Engelmann H.-D., 1978 - Zur Dominanz Klassifizierung von Bodenarthropoden. *Pedobiologia*, 18: 378-380.

Franchini P., Rockett C.L., 1996 - Oribatid mites as "indicator" species for estimating the environmental impact of conventional and conservation tillage practices. *Pedobiologia*, 40: 217-225.

Hülsmann A., Wolters V., 1998 - The effects of different tillage practices on soil mites, with particular reference to Oribatida. *Applied Soil Ecology*, 9: 327-332.

Juvara-Bals I., 1992 - Ecosystème verger de pommier: analyse de l'interaction Gamasida - *Tetranychus urticae* Koch (Acariens) au niveau du sol et de sa couverture végétale. Thése n. 9641 École Polytechnique Fédérale de Zürich: 110 pp.

Khalil M.A., Hala M.A.-L., Al-Assiuty A.-N.I., 1999 - Changes in oribatid faunal structure associated with land conversion from annual crop into orchard. *Pedobiologia*, 43: 85-96.

Lagerlöf J., Andrén O., 1988 - Abundance and activity of soil mites (Acari) in four cropping systems. *Pedobiologia*, 32: 129-145.

Luxton M., 1981 - Studies on the oribatid mites of a Danish beech wood soil. V. Vertical distribution. *Pedobiologia*, 21: 359-380.

Petersen H., Luxton M., 1982 - A comparative analysis of the soil fauna population and their role in decomposition processes. *Oikos*, 39: 288-388.
Petrov P., 1997 - The reactions of communities of oribatid mites to plant succession on meadows. *Ekologia Polska*, 45(3-4): 781-793.
Rockett C.L., 1986 - Agricultural impact on the horizontal distribution of oribatid mites (Acari: Oribatida). *International Journal of Acarology*, 12: 175-180.
Tolstikov A.V., Petrova-Nikitina A.D., Makaeva J.A., 1997 - On the knowledge of mite biodiversity in a large city: Oribatids of the apple orchards of Moscow. *Abhandlungen und Berichte des Naturkundemuseums Görlitz*, 69(6): 31-40.
Vial J., Monterrat G., 1971 - Méthode de dénombrement des Acariens par lavage: *Phytiatrie-Phytopharmacie*, 20: 183-189.
Wallwork J.A., 1976 - The distribution and diversity of soil fauna. Academic Press, London: 355 pp.
Yamamura K., 1999 - Transformation using (x + 0.5) to stabilize the variance of populations. *Research of Population Ecology*, 41: 229-234.

Phytoseiid mites - pollen relationships: observations in a vineyard and the surrounding vegetation

C. Duso[1], V. Malagnini[1], A. Paganelli[2], L. Aldegheri[2], M. Bottini[2]
[1] *Department of Environmental Agronomy and Crop Science, University of Padova, Italy*
[2] *Department of Biology, University of Padova, Italy*

Key words: Phytoseiidae, pollen, natural vegetation, grapevine.

Abstract: The present work is part of a long-term research project devoted to studying the influence of natural vegetation on phytoseiid colonization in vineyards. Observations were carried out in an experimental site including a vineyard and a contiguous hedgerow made up of *Acer campestre, Sambucus nigra, Cornus sanguinea, Carpinus betulus* and *Corylus avellana* and structured in pure or mixed modules. The aims of this study were:

- to evaluate pollen diversity and abundance on the foliage of grapevines and of plants included in the hedgerow, by using acetolysis;
- to study mite colonization in the vineyard and on different plant species with special emphasis on phytoseiid-pollen relationships.

On grapevine foliage the pollen produced by Poaceae was dominant while that of plants included in the contiguous hedgerow was scarcely represented. *Vitis vinifera* pollen was detected at low to moderate densities. Pollen peaks were recorded in May and June. In some situations, the dynamics of phytoseiids occurring on the grapevine (in particular *Amblyseius andersoni*) appeared to be affected by pollen abundance, as potential prey was scarce.

As far as pollen diversity and abundance on plants included in the hedgerow are concerned, *Sambucus nigra* and Poaceae were the dominant taxa. *Salix* sp. pl., *Pinus* sp. pl. and *Quercus* sp. pl. were frequently collected among AP (arboreal pollen) while Chenopodiaceae and Plantaginaceae were found among NAP (non arboreal pollen). *Cornus sanguinea* pollen was detected on the respective plants. Generally, pollen densities reached higher levels from late April to late June and decreased in summer. Pollen abundance was different among plant species, within the same module, while there were slight differences in the comparison between pure and mixed modules. In most cases, the occurrence of phytophagous mites was low while tydeids were commonly recorded. Among phytoseiids, *Euseius finlandicus* and *Neoseiulella tiliarum* were dominant species.

Introduction

In North-Eastern Italy, arboreal crops, such as vineyards, are dominated by generalist phytoseiids, e.g. *Kampimodromus aberrans* (Oudemans), *Amblyseius andersoni* (Chant) and *Typhlodromus pyri* Scheuten (Duso 1989). Laboratory studies showed the importance of alternative foods, mainly eriophyids and pollen, for the development and reproduction of these species (Boller and Frey 1990; Duso and Camporese 1991; Schausberger 1992). The role of pollen for generalist phytoseiids in vineyards has been a matter of study especially during the last decade. A relationship between pollen abundance and phytoseiid phenology in vineyards was suggested by some authors (Eichhorn and Hoos 1990; Wiedmer and Boller 1990; Engel and Ohnesorge 1994; Duso *et al.* 1997). The experimental

F. Bernini, R. Nannelli, G. Nuzzaci, E. de Lillo (eds.), Acarid Phylogeny and Evolution. Adaptations in mites and ticks, 373-387.

approach and methods employed were somewhat different, for instance pollen abundance was estimated using different techniques and factors affecting phytoseiid population dynamics were only partially under control. The quality of pollen for phytoseiids is also a matter for discussion (Boller and Frey 1990; Maixner 1990; Engel and Ohnesorge 1994). However, the results obtained in these studies are promising. McMurtry (1992) and Grout and Richards (1992) emphasized the possibility of improving IPM strategies with the use of plants producing pollen for phytoseiids or their alternative prey. Since Poaceae is the main component of non arboreal pollen (NAP) found on grapevine foliage in Europe (Eichhorn and Hoos 1990; Wiedmer and Boller 1990; Engel and Ohnesorge 1994; Duso *et al.* 1997), one of these implications would be grass management (Remund and Boller 1992; Girolami *et al.* 2000). At the same time, pollen produced by arboreal plants (AP) is relatively abundant on grapevine foliage during spring (Eichhorn and Hoos 1990; Wiedmer and Boller 1990; Boller and Frey 1990; Engel and Ohnesorge 1994; Duso *et al.* 1997). The role of hedgerows and stands for biological pest control in vineyards constitutes an interesting topic. The increase in plant complexity within an ecosystem can improve arthropod diversity and stability as these strategies promote the survival and reproduction of pest natural enemies (Altieri and Letourneau 1982). During the last decade the positive effects of hedgerows for sustainable agriculture have been stressed by EC rules. There is a need to plan new agro ecosystem models allowing for a reduction of pest outbreaks. However, knowledge of the arthropod fauna occurring on plants surrounding crops, including natural hedgerows or stands, is somewhat scarce. Probably, mites have been studied more closely than other groups (Boller *et al.* 1988; Duso *et al.* 1993; Duso and Fontana 1996; Tixier *et al.* 1998).

The present work is part of a long-term research project devoted to studying the influence of natural vegetation on phytoseiid colonization in vineyards. Observations were carried out in an experimental site including a vineyard and a contiguous secondary hedgerow. Natural hedgerows situated close to the site were also involved, but the analysis of their effects will be considered in a subsequent paper. The objectives of this study were:
- to evaluate pollen diversity and abundance on the foliage of grapevines and of plants included in the secondary hedgerow;
- to study mite colonization in the vineyard and on different plant species with special emphasis on phytoseiid-pollen relationships.

Materials and methods

Experimental site features

The experimental site included a 15 year-old vineyard with a surface of approximately 1 hectare and rectangular in shape, and a contiguous hedgerow (fig. 1). The vineyard was composed of two varieties (Chardonnay and Cabernet Sauvignon), characterized respectively by low to moderate pubescent leaf undersurfaces. The secondary hedgerow, a permanent meadow and a street surrounded the vineyard. Grass mulching was performed. Fungicide treatments to control diseases, i.e. powdery mildew (wettable sulphur) and downy mildew (mainly dithiocarbamates, cymoxanil, copper oxychloride) were normally applied. A single insecticide treatment (fenitrothion) was performed in mid July to control leafhoppers.

The secondary hedgerow, planted in 1993, was structured in modules (pure) containing 12-16 plants of a single species (*Acer campestre* L., *Sambucus nigra* L., *Cornus sanguinea* L., *Carpinus betulus* L., *Corylus avellana* L.) and modules (mixed) containing 60 plants of mixed species. Two replications per module were considered. The hedgerow was potted in accordance with actual management concepts (Mezzalira 1990).

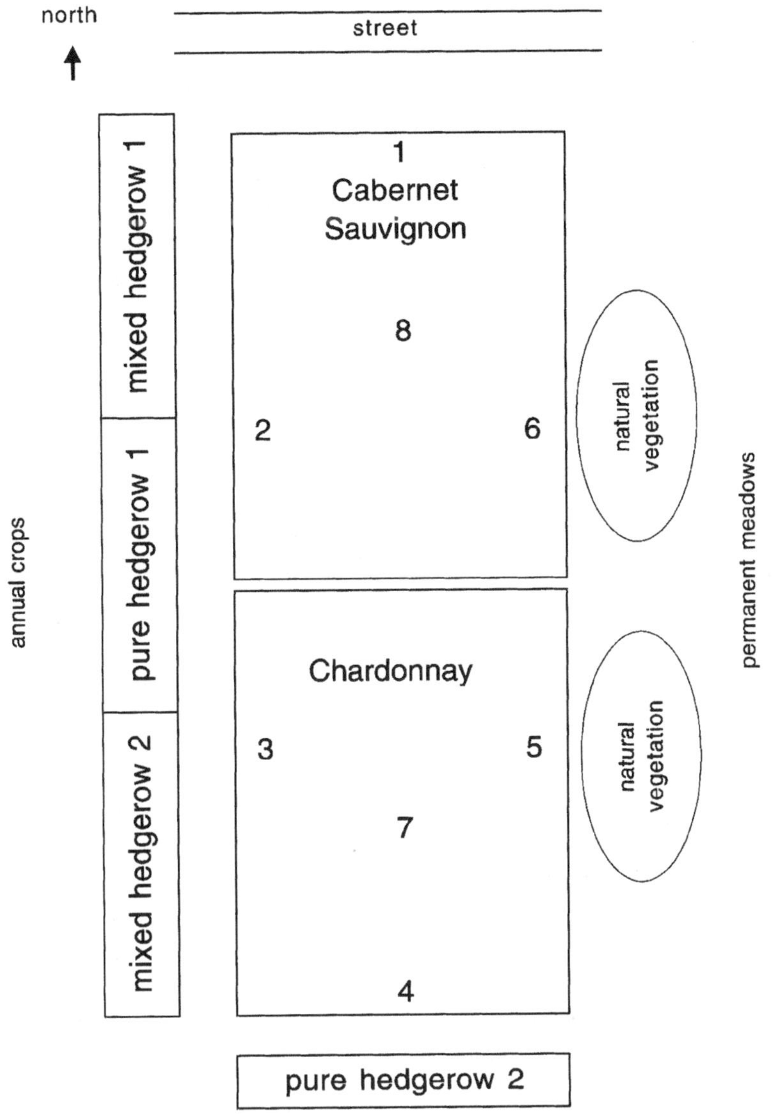

Fig. 1 - The experimental site includes a vineyard consisting of two varieties (Chardonnay and Cabernet Sauvignon, plots 1-8) and a contiguous hedgerow structured in modules (pure or mixed, 2 replications) containing the following species: *Acer campestre* L., *Sambucus nigra* L., *Cornus sanguinea* L., *Carpinus betulus* L., *Corylus avellana* L.

Minor components of this agro-ecosystem were isolated plants of *Salix alba* L., *Populus nigra* L., *P. alba* L., *Acer negundo* L., *Morus* sp. pl. and *Rubus* sp. pl. occurring especially on the east side of the vineyard. The most common weeds belonged to Apiaceae, Asteraceae, Chenopodiaceae, Convolvulaceae, Leguminosae, Papaveraceae, Plantaginaceae, Poaceae, Ranunculaceae, Rosaceae, Solanaceae and Urticaceae.

Sampling methods

Sampling was performed every two weeks from April to October 1997. Eight plots distributed in different areas of the vineyard were considered: plot 1 was placed in front of the street, plots 2, 3 and 4 were in front of the secondary hedgerow, plots 5 and 6 in front of the meadow, plots 7 and 8 in the middle of the vineyard (fig. 1). Ten leaves per plot were collected in each sampling.

In the hedgerow (pure and mixed modules), twenty leaves per plant species (10 leaves per replication, 20 leaves per module) were collected in each sampling. The leaves were transported to the laboratory where densities of Phytoseiidae and of the most common mite families (Tetranychidae, Tenuipalpidae, Eriophyidae, Tydeidae, Tarsonemidae, Winterschmidtiidae, etc.) were evaluated. Mites were identified using a phase-contrast microscope. Among insects, potential prey for phytoseiids (e.g. thrips, whiteflies, etc.) or macropredators (e.g. anthocorids, etc.), were also monitored.

Pollen analysis

The same leaves used for monitoring mite populations were then analysed to assess pollen diversity and abundance. A circle of 0.78 cm^2, located near the petiole area was removed from each leaf. The material was treated using the acetolysis method (Erdtman 1943). A small amount of treated samples was used for microscope analysis (e.g. 2-3 gouts per slide). In several samples more than one slide was observed before the minimum pollen grain number, i.e. 100, was reached.

Pollen taxa were included in two groups: arboreal pollen (AP) and non-arboreal pollen (NAP). Their identification was performed using current keys (Moore *et al.* 1991). Frequencies of the most common taxa were inserted in a diagram together with pollen grain densities per slide.

Results

Pollen diversity and abundance

Pollen production, as well as its retention by the foliage, is affected by climate (Faegri and Iversen 1964). During the vegetative season of 1997, temperatures reached normal values except in May and June when sudden decreases were recorded. Light rains were constantly recorded from April to July. The wind direction was mainly north during May and September, north-west in the remaining summer months.

Considering the total amount of samples, 29 taxa belonging to AP and 39 taxa to NAP were found (tab. 1). In the samples collected in the vineyard, pollen diversity decreased: 19 taxa belonged to AP and 20 to NAP.

The abundance of different taxa over the season was influenced by their flowering periods but also by additional depositions. AP taxa were represented more in spring in accordance with their flowering, while NAP (e.g. Poaceae) were recorded continuously. As far as pollen densities are concerned, relatively high values were recorded from late April to late June for both AP and NAP. Pollen densities declined from July onwards. This trend was probably associated to conditions of high temperature, low humidity and precipitations which affect flowering, especially of grasses.

Pollen and mite colonization in the vineyard

In the grapevine samples, AP taxa (mainly *Salix* sp. pl., *S. nigra*, *Carpinus* sp. pl., *Pinus* sp. pl., *Quercus* sp. pl.) were recorded at low levels, an exception was *Salix* sp. pl. on plots 5 and 6, that were contiguous to isolated plants of white willow. The amount of pollen produced by plant species included in the hedgerow (in particular elder and red dogwood) in the vineyard was negligible even on plots 2, 3 and 4, that were located in front of the hedgerow. *Vitis vinifera* pollen was detected at low levels in most cases. Exceptions were in plots 2 and 5 where grapevine pollen attained respectively 20% and 59% of total grains at the peak of pollen abundance. NAP dominated over AP in most samplings. Among the former, Poaceae were constantly recorded throughout the season, being clearly dominant over Chenopodiaceae, Plantaginaceae and other taxa. Pollen diagrams concerning Cabernet Sauvignon and Chardonnay samples are reported (figs 2-3).

Pollen densities reached the highest levels from mid May to late June but peaks were commonly observed from 4 to 18 June. When

Tab. 1 - Pollen taxa recorded on the leaf samples of grapevine (v), pure (p) and mixed (m) hedgerow modules.

AP	p	m	g	NAP	p	m	g
Larix sp. pl.			x	Alismataceae			x
Abies sp. pl.	x	x		*Ambrosia* sp. pl.		x	
Cedrus sp. pl.	x	x		Apiaceae	x	x	x
Picea sp. pl.	x	x	x	*Artemisia* sp. pl.	x	x	x
Pinus sp. pl.	x	x	x	Asteroideae	x	x	x
Podocarpus sp.		x		Cichorioideae	x	x	x
Acacia sp. pl.	x			*Calystegia* sp. pl.		x	
Acer sp. pl.	x	x	x	Cannabaceae			x
Alnus sp. pl.	x	x	x	Caryophyllaceae	x	x	x
Betula sp. pl.	x	x	x	Cerealia: *Triticum* sp.	x	x	x
Carpinus betulus	x	x	x	*Hordeum* sp.	x	x	
Castanea sativa	x	x	x	*Zea mays*	x	x	
Cornus mas	x	x	x	*Secale* sp.		x	
Cornus sanguinea	x	x	x	Chenopodiaceae	x	x	x
Corylus avellana	x	x		*Centaurea* sp. pl.		x	
Frangula alnus	x	x	x	Dipsacaceae		x	
Fraxinus sp. pl.	x	x		*Epilobium* sp. pl.	x		
Juglans regia	x	x	x	Ericaceae	x		
Ligustrum sp. pl.	x	x		*Fagopyrum esculentum*			x
Magnolia sp. pl.	x	x		Cruciferae	x	x	
Morus sp. pl.		x		*Galium* sp. pl.	x	x	x
Ostrya carpinifolia	x	x	x	*Hottonia palustris*	x	x	
Populus sp. pl.		x		*Hypericum* sp. pl.	x		
Quercus deciduous type	x	x	x	Labiatae			x
Salix sp. pl.	x	x	x	Leguminosae	x	x	
Sambucus nigra	x	x	x	Liliaceae		x	
Tilia cordata	x	x	x	Plantaginaceae	x	x	x
Ulmus sp. pl.	x		x	Poaceae	x	x	x
Vitis vinifera	x	x	x	*Potamogeton* sp. pl.		x	
				Ranunculaceae	x	x	x
				Rosaceae	x	x	x
				Rumex sp. pl.	x	x	x
				Scrophulariaceae		x	
				Solanaceae	x	x	x
				Sparganium sp. pl.		x	
				Typha latifolia	x		x
				Trifolium sp. pl.		x	
				Humulus lupulus		x	
				Urticaceae	x		x

pollen densities peaked, Poaceae were the most important component in all plots, with the exception of plot 5. Pollen abundance trends observed on plots 2 and 6 (Cabernet), 3 and 5 (Chardonnay) are reported as representative of the two varieties (figs 4-7).

In most cases tetranychid mites occurred at negligible levels. The eriophyid *Colomerus vitis* Pagenstecher was seldom recorded. Tydeids (mainly *Tydeus caudatus* Dugés) increased in importance in August and September. Various phytoseiid species were observed. In plots 1 and 2 (fig. 4), *T. pyri* reached moderate densities at sprouting, but the populations declined in spring probably because of the repeated use of dithiocarbamate fungicides. In late summer, predaceous mite numbers (*T. pyri*, *K. aberrans* and *A. andersoni*) increased, especially in plot 1. In this phase, several leaves were infected by downy mildew, a potential alternative food for phytoseiids (Duso *et al.*, *unpubl. data*). In plots 3 (fig. 5) and 4, *A. andersoni* was dominant over *K. aberrans* and *T. pyri*. Densities of the former species reached moderate levels after sprouting and in late summer. In all these plots, pollen densities fell to low levels during the season. In contrast, pollen was relatively abundant in plot 5 (fig. 6), where, *A. andersoni* populations reached high densities in early July. Predaceous mite numbers declined in the second half of July after the use of fenitrothion (15 July). *Kampimodromus aberrans*, *Euseius finlandicus* (Oudemans) and *Paraseiulus talbii* (Athias-Henriot) were also recorded especially in late season. In plot 6 (fig. 7), where pollen was more abundant than in other plots, none of the above-mentioned species became dominant. Populations fluctuated at low levels until mid summer increasing in late season when *E. finlandicus* was also reported. In plot 7, *A. andersoni* was more frequent than *T. pyri* while the reverse was observed in plot 8. In the latter, a peak of *T. pyri* was observed in July despite the fact that the availability of pollen or alternative foods in that period appeared to be low. The predator density decreased after the use of fenitrothion. In plots 6, 7 and 8, *K. aberrans* increased in numbers in late season.

Pollen and mite colonization on the hedgerow

Concerning pollen diversity and abundance on the secondary hedgerow, there were slight differences between pure and mixed modules. The importance of AP taxa was relatively high in late April (mainly *Salix* sp. pl. and *O. carpinifolia*), mid May (first flowering of *S. nigra*) and early July (second flowering of *S. nigra*). In these periods the ratio between AP and NAP was favourable to the former on maple and hornbeam. The same trend was observed on elder from mid May to early June, due to the abundance of the respective pollen. The pollen produced by red dogwood was also frequently detected in the respective samples. On hazel and red dogwood, the ratio between AP and NAP was always favourable to the latter. Other AP taxa commonly found on samples belonged to the genera *Picea*, *Pinus* and *Quercus*. Their abundance, as well as that of *V. vinifera*, was usually low. Concerning NAP taxa, Poaceae were found throughout the season at moderate levels while Chenopodiaceae and Plantaginaceae increased in importance in late summer. The remaining taxa detected on leaves belonged mostly to Apiaceae, Asteroideae, Cichorioideae, Ranunculaceae and Rosaceae.

On maple, hornbeam, elder and red dogwood, the maximum amount of pollen grains was recorded between mid May and early June while on hazel, pollen densities peaked in late April and early July. Concerning cumulative pollen densities, calculated over the season, the highest values were recorded on elder. Red dogwood showed high densities in mid May but only moderate in most of the samplings. Hazel was characterized by the lowest values while hornbeam and maple were intermediate.

In the experimental season, the occurrence of tetranychids on hedgerow plants was negli-

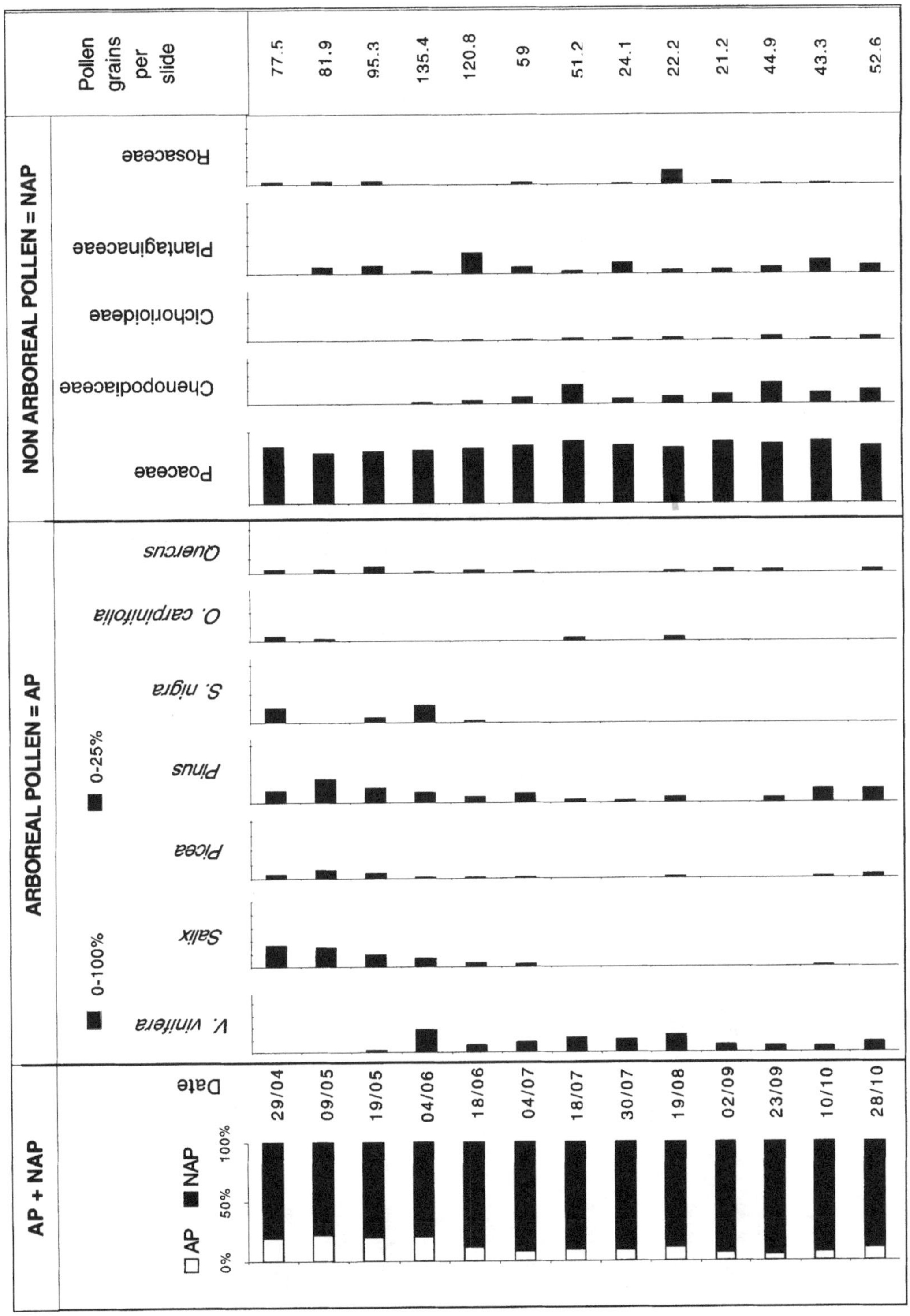

Fig. 2 - Pollen diversity and abundance in Cabernet Sauvignon samples. Pollen taxa were classified into non-arboreal pollen (NAP) and arboreal pollen (AP).

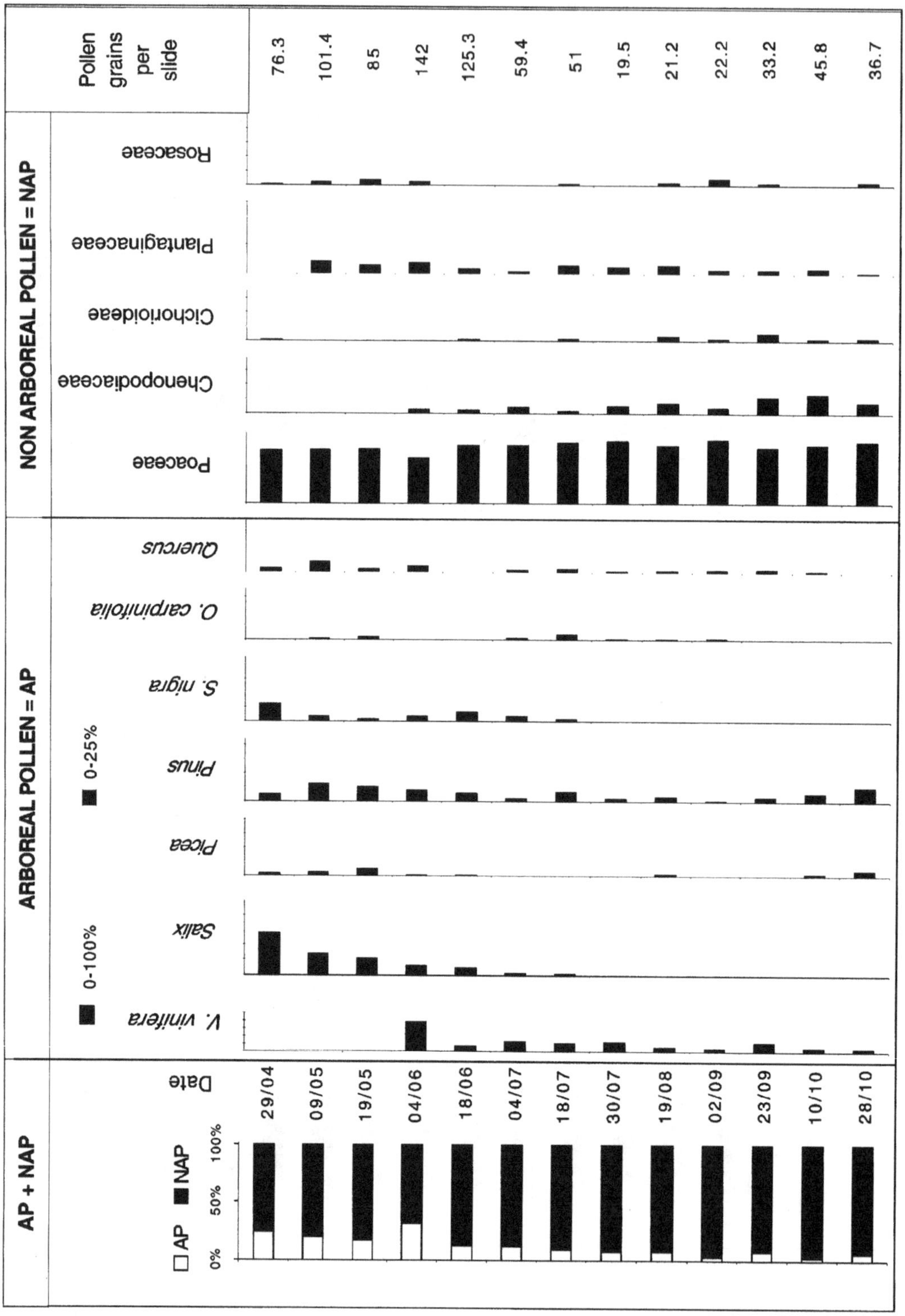

Fig. 3 - Pollen diversity and abundance in Chardonnay samples. Pollen taxa were classified into non-arborcal pollen (NAP) and arboreal pollen (AP).

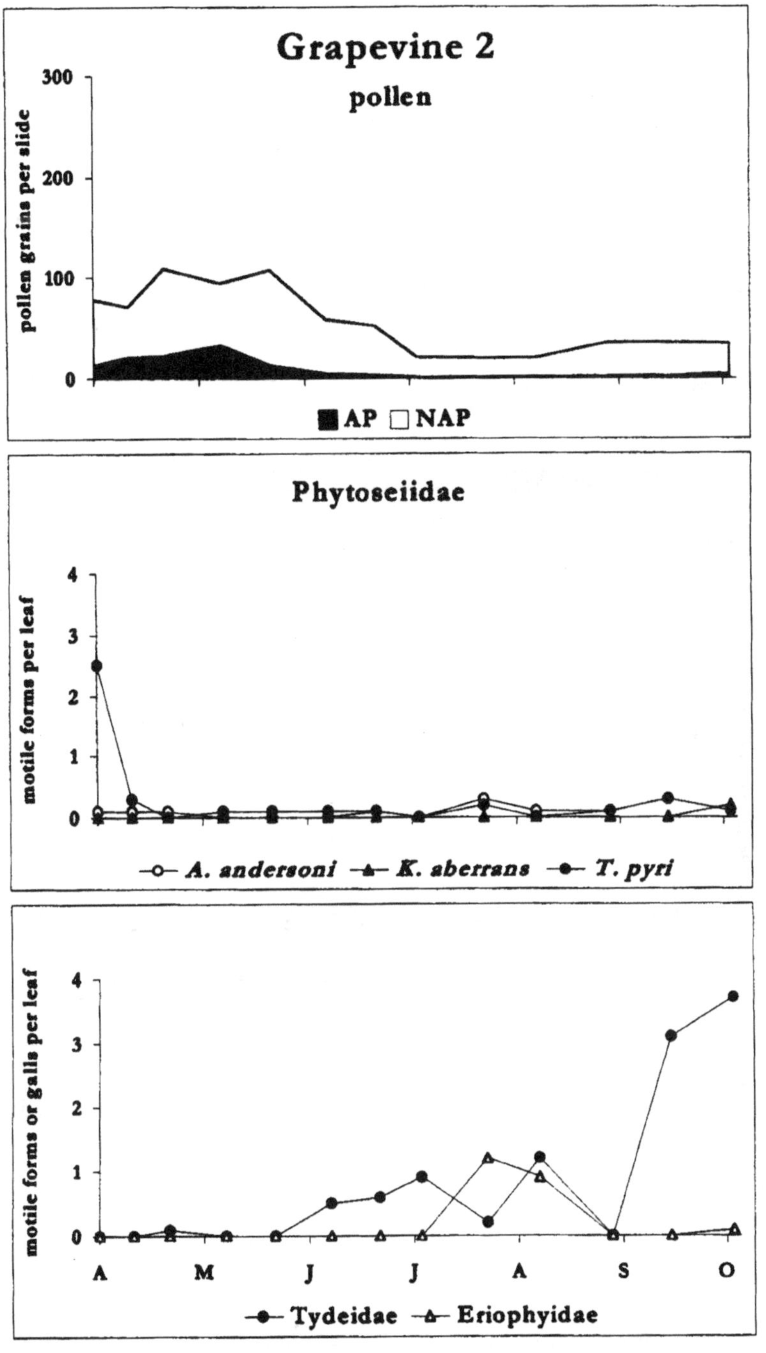

Fig. 4 - Pollen abundance and mite dynamics in the grapevine plot 2.

gible. Eriophyids were constantly recorded on maple, and tenuipalpids reached moderate densities on elder in late summer. Tydeid populations increased in August or September in most plant species while winterschmidtiids increased in late season on maple. In several cases phytoseiids reached low to moderate population levels. *Euseius finlandicus* dominated

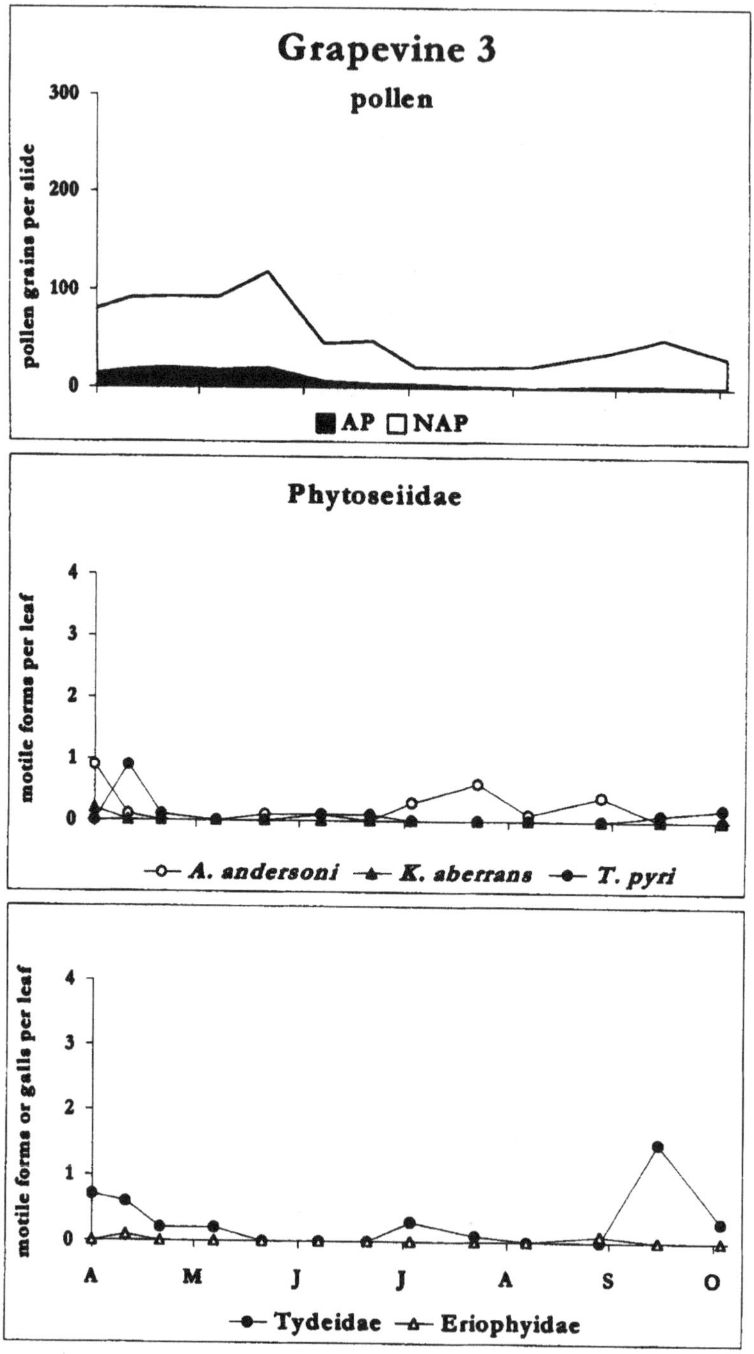

Fig. 5 - Pollen abundance and mite dynamics in the grapevine plot 3.

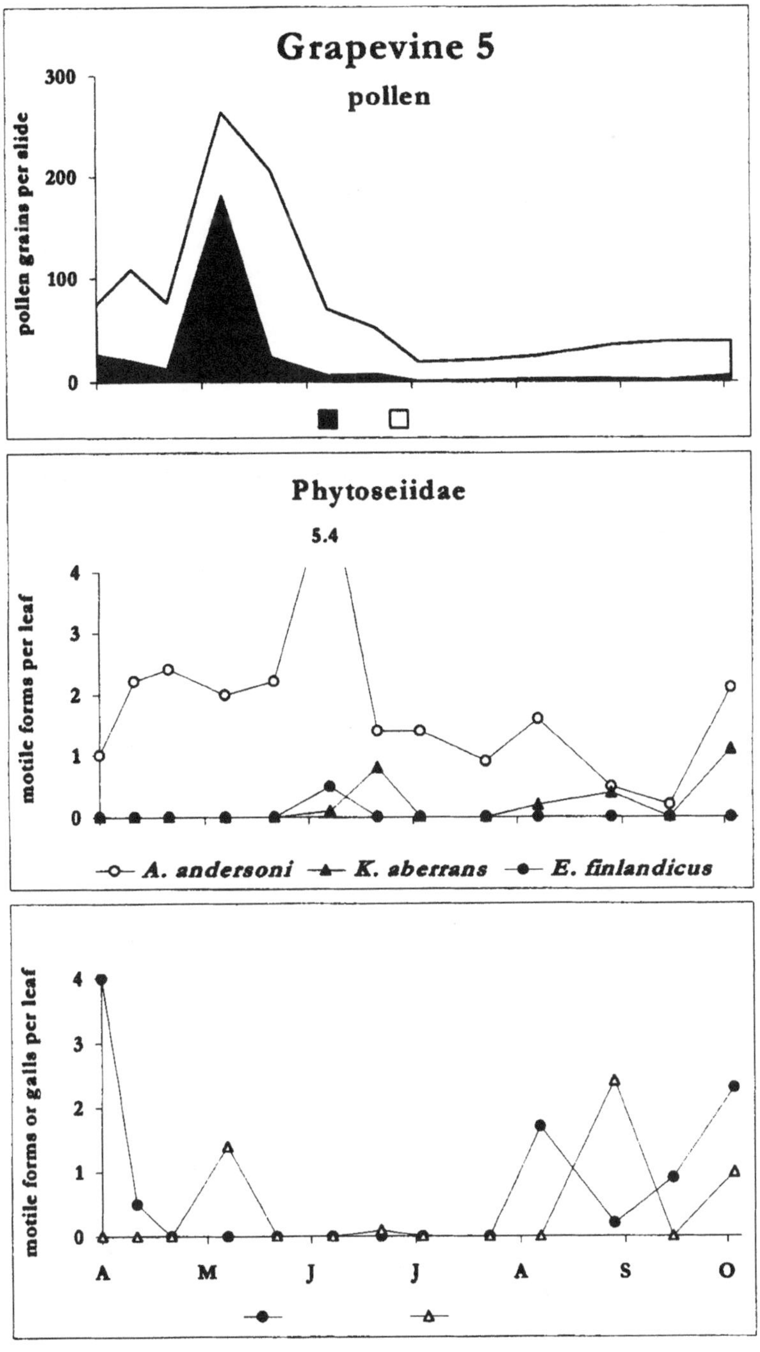

Fig. 6 - Pollen abundance and mite dynamics in the grapevine plot 5.

on hornbeam, elder and red dogwood and was also recorded on maple and hazel. *Neoseiulella tiliarum* (Oudemans) was dominant on maple and hazel and detected on elder and red dogwood. Among phytoseiid species common on the grapevine, *A. andersoni* occurred on all plant species but was relatively abundant on red dogwood only. *Typhlodromus pyri* was seldom found on maple, hornbeam, elder and red dogwood. A few specimens of *K. aberrans* and *P.*

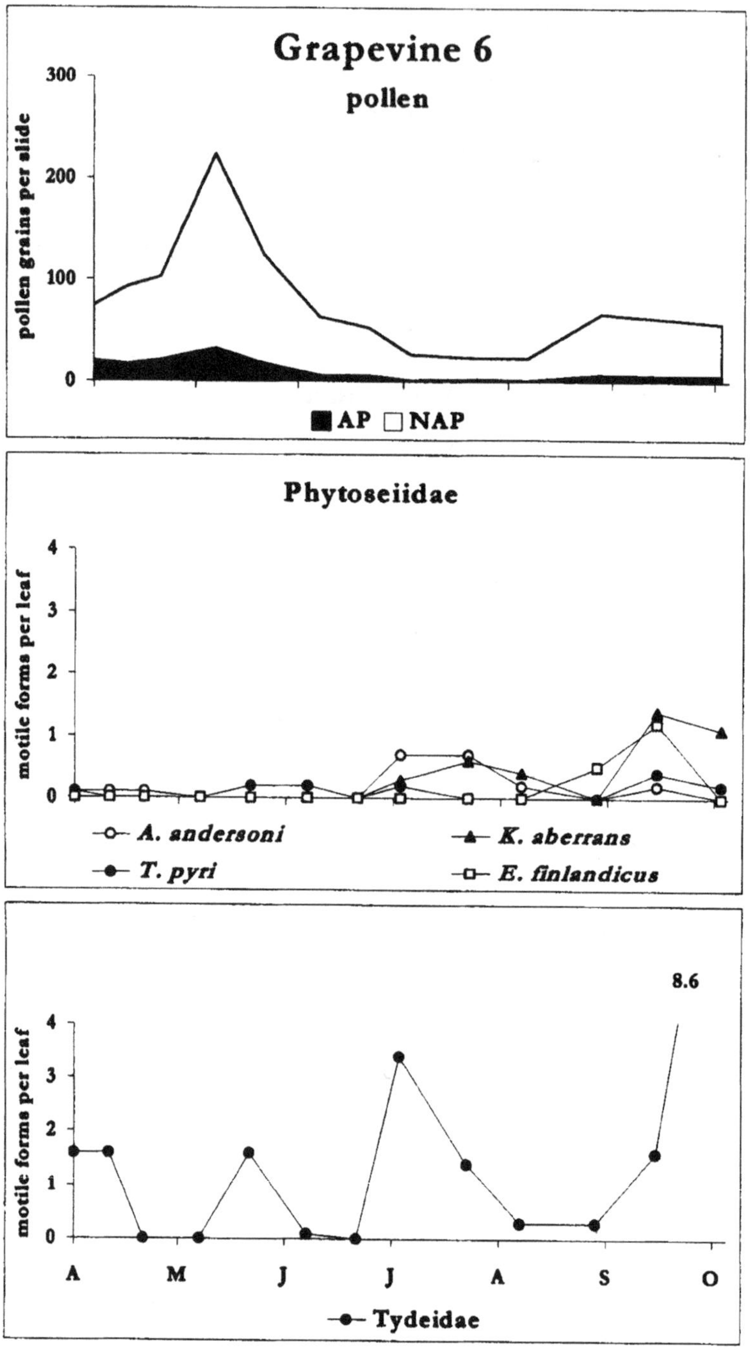

Fig. 7 - Pollen abundance and mite dynamics in the grapevine plot 6.

talbii were collected respectively on red dogwood and on elder. Three other species were recorded: *Neoseiulella aceri* (Collyer) on maple, *Kampimodromus ericinus* Ragusa & Tsolakis on hazel, *Phytoseius plumifer* (Canestrini & Fanzago) on red dogwood.

Discussion and conclusions

Grapevine leaves as a trap for pollen

The vineyard was surrounded by the secondary hedgerow and by natural vegetation. The importance of AP pollen in the vineyard was low. The influence of some isolated plants of white willow, contiguous to plots 5 and 6, appeared to be more important than that of the secondary hedgerow, close to plots 2, 3 and 4. Among NAP, the dominance of Poaceae confirmed a situation commonly found in this region (Duso *et al.* 1997). The role of grasses appears to be crucial and their management in the vineyards is considered to be of increasing importance for phytoseiid persistence (Girolami *et al.* 2000).

Pollen densities recorded in most plots of the vineyard appeared to be lower than those recorded in previous research (Duso *et al.* 1997). The influence of variety may be relevant, as pollen abundance on Chardonnay and Cabernet Sauvignon samples was already seen to be low (Malagnini, *unpubl. data*). Pollen distribution among the grapevine plots was not homogeneous: pollen (mainly Poaceae) densities appeared to be higher on plots 5 and 6, probably because the latter were contiguous to permanent meadows. *Vitis vinifera* pollen was relatively abundant in plot 5 only. In some plots a relevant number of plants were affected by Flavescence dorée that dramatically reduces grapevine flower vitality.

Pollen abundance on hedgerow plants

Pollen diversity on the secondary hedgerow was higher than that recorded in the vineyard. Weeds were more widespread near hedgerow modules than into the vineyard where grass mulching was performed. This situation probably affected NAP diversity. At the same time, the contribution of arboreal plants included in the hedgerow or contiguous to the latter increased AP importance in the respective samples. As well as the diversity and abundance of the most frequent pollen taxa are concerned, differences were more important at plant species level than between pure and mixed modules.

These results probably reflect the fact that the effectiveness in capturing or retaining pollen grains of the various plants considered is different.

Leaf morphological features of different plant species were not clearly associated with pollen retention. The relation between hairiness and pollen retention (Adam *et al.* 1967) was not confirmed for hazel, characterized by tuft domatia and hairy veins. Moreover, elder, characterized by glabrous leaves and low hairy densities along veins, appeared to be one of the most attractive plants. Therefore, more factors (e.g. leaf stickiness, electrostatic forces) are involved (Chamberlain and Chadwick 1972), suggesting a wider approach to this study.

Does phytoseiid abundance reflect pollen availability?

Low phytoseiid densities were found in most plots in the vineyard, probably due to the low levels of prey and the use of pesticides. Concerning the latter factor, dithiocarbamates seemed to affect the decline of *T. pyri* in spring when their use was more frequent. Densities of *A. andersoni* did not appear to be reduced by these fungicides but declined after the use of fenitrothion. The use of non-selective pesticides masked the potential effect of windborne pollen on phytoseiid population dynamics. In any case, the low pollen densities on grapevine foliage, found in most plots, were also associated to low phytoseiid numbers. On the other hand, large populations of *A. andersoni* corresponding to relatively abundant pollen densities colonized plot 5. *Amblyseius andersoni* populations clearly increased after pollen peaks in conditions of prey scarcity. Relatively high pollen densities also characterized plot 6 but, in contrast with the previous situation, negligible phytoseiid populations from sprouting onwards colonized it.

In some plots, phytoseiid numbers increased in late season when pollen abundance on the grapevine foliage was low. In this phase, downy mildew was developing in most plots. Recently, a positive relationship between downy mildew and phytoseiid abundance has been found (Duso *et al.*, *unpubl. data*).

In conclusion, most factors influenced phytoseiid dynamics but the lack of controlled studies of the variables does not allow the clear effect of windborne pollen on phytoseiid populations to be demonstrated. However, the results obtained in the present work for *A. andersoni* are partially consistent with previous findings concerning *K. aberrans* and *T. pyri* in vineyards (Duso *et al.* 1997; Malagnini *et al.*, *unpubl. data*). In the laboratory *A. andersoni* can develop faster on pollen than on tetranychids (Duso and Camporese 1991).

The occurrence of *E. finlandicus* in plot 6 (late season) was somewhat surprising since this species is susceptible to common pesticides. This data probably reflects the proximity of these plants to natural vegetation where the predator was dominant (Duso *et al.* 1993) and the availability of downy mildew on grapevine leaves, since the predator can feed actively on fungi (Kropczynska-Linkiewicz 1971).

Pollen densities recorded on secondary hedgerow were comparable to those found on grapevine plots in most cases. Relationships between pollen availability and phytoseiid abundance are difficult to establish due to the low numbers of predators.

Among factors affecting phytoseiid densities, hedgerow contiguity with the vineyard (e.g. pesticide drift) and its management (e.g. pruning) could be relevant. Moreover, since they were planted (1994) the availability of potential prey has been low (Duso *et al.*, *unpubl. data*). During the experimental season (1997), phytophagous mites were confirmed as being a minor component of the mite fauna. Tydeid populations increased in August irrespective of the plant species. Tydeids are optimal prey for some phytoseiid species (e.g. *P. talbii*) but only a suboptimal food for phytoseiid species more frequently found in this agro-ecosystem.

Euseius finlandicus and *N. tiliarum* were more commonly recorded on hedgerow plants. The former was dominant on hornbeam, red dogwood and elder irrespective of their management. The role of leaf morphology in affecting predatory mite-host plant relationships is well known. Hornbeam, red dogwood and elder are characterized by glabrous leaf undersurfaces, a condition that favours *E. finlandicus* (Collyer 1956). Moreover, population trends on elder and hornbeam were similar, suggesting that the factors involved were moving in the same direction.

In contrast with previous situations, a relationship between pollen availability and *E. finlandicus* abundance was found on natural hedgerows (elder and hornbeam) close to those considered in this experiment. Pollen treatments increased the fecundity and, to a lesser extent, the density of *E. finlandicus* (Duso *et al.*, *unpubl. data*).

Acknowledgements

The work is included in a research project partially financed by MURST (60%) and the Veneto region (Direzione fitosanitaria regionale). We thank Veneto Agricoltura and the Province of Vicenza for their support. We are grateful to G. Galbero and A. Drago for their assistance.

References

Adam D.P., Ferguson C.W., Lamark V.C., 1967 - Enclosed bark as a pollen trap. *Science,* 157: 1067.

Altieri M.A., Letourneau D.K., 1982 - Vegetation management and biological control in agrosystems. *Crop protection,* 1: 405-430.

Boller E.F., Frey B., 1990 - Bluhende Rebberge in der Ostschweiz. I. Zur Bedeutung des Pollens für die Raubmilben. *Schweiz Zeitschrift für Obst- und Weinbau,* 122: 401-405.

Boller E.F., Remund U., Candolfi M.P., 1988 - Hedges as potential sources of *Typhlodromus pyri*, the most important predatory mite in vineyards of northern Switzerland. *Entomophaga*, 33(2): 249-255.

Chamberlain A.C., Chadwick R.C., 1972 - Deposition of spores and particles on vegetation and soil. *Annals of Applied Biology*, 71: 141-158.

Collyer E., 1956 - Notes on the biology of some predacious mites on fruit trees in South-Eastern England. *Bulletin of Entomological Research*, 47: 205-214.

Duso C., 1989 - Role of the predatory mites *Amblyseius aberrans* (Oudemans), *Typhlodromus pyri* Scheuten and *Amblyseius andersoni* (Chant) (Acari: Phytoseiidae) in vineyards. I. The effects of single or mixed phytoseiid population on spider mites densities (Acari: Tetranychidae). *Journal of Applied Entomology*, 107: 474-494.

Duso C., Camporese P., 1991 - Developmental times and ovoposition rates of predatory mites *Typhlodromus pyri* Scheuten and *Amblyseius andersoni* (Chant) (Acari: Phytoseiidae) reared on different foods. *Experimental and Applied Acarology*, 13: 117-128.

Duso C., Fontana P., 1996 - Mite communities on wild plants surrounding vineyards in North-Eastern Italy with special emphasis on phytoseiids (Acari: Phytoseiidae). In: Mitchell R.D., Horn D.J., Needham G.R., Welbourn W.C. (eds), Acarology IX, Ohio Biological Survey, Columbus. Vol. 1: 261-264.

Duso C., Malagnini V., Paganelli A., 1997 - Indagini preliminari sul rapporto tra polline e *Kampimodromus aberrans* (Acari: Phytoseiidae) su *Vitis vinifera* L. *Allionia*, 35: 229-239.

Duso C., Torresan N., Vettorazzo E., 1993 - La vegetazione spontanea come riserva di ausiliari: considerazioni sulla diffusione degli Acari Fitoseidi (Acari: Phytoseiidae) in un vigneto e sulle piante spontanee contigue. *Bollettino di Zoologia agraria e di Bachicoltura,* Serie II, 25(2): 183-203.

Eichhorn K.W., Hoos D., 1990 - Investigations in population dynamics of *Typhlodromus pyri* in vineyards of Palatina, F.R. Germany. *IOBC Bulletin*, 1990/XIII/7: 120-123.

Engel R., Ohnesorge B., 1994 - Die Rolle von Ersatznahrung und Mikroklima im System *Typhlodromus pyri* Scheuten (Acari, Phytoseidae) - *Panonychus ulmi* Koch (Acari, Tetranychidae) auf Weinreben I. Untersuchungen im Labor. *Journal of Applied Entomology*, 118: 129-150.

Erdtman G., 1943 - An introduction to pollen analysis. Kronica Botanica Company, Walthman Massachusetts, USA: 329 pp.

Faegri K., Iversen J., 1964 - Textbook of Pollen Analysis. 3rd ed., Hafner publishing Co., New York, USA: 328 pp.

Girolami V., Borrella E., Di Bernardo A., Malagnini V., 2000 - Influenza positiva sui fitoseidi della fioritura del cotico erboso. *L'informatore agrario*, 21: 71-73.

Grout T.G., Richards R.I., 1992 - The dietary effect of windbreak pollens on longevity and fecundity of a predacious mite *Euseius addoensis addoensis* (Acari: Phytoseiidae) found in citrus orchards in South Africa. *Bulletin of Entomological Research*, 82: 317-320.

Kropczynska-Lienkiewicz D., 1971 - Studies on the feeding of four species of phytoseiid mites (Acarina: Phytoseiidae). In: Daniel M., Rosicky B. (eds.), Proceedings 3rd International Congress of Acarology, Prague: 225-227.

Maixner M., 1990 - Untersuchungen zur Insektizidresistenz der Raubmilbe *Typhlodromus pyri* Scheuten (Acari: Phytoseiidae) an Reben des Weinbaugebites Mosel-Saar-Ruwer. *Mitteilungen aus der Biologischen Bundesanstalt für Land- und Forstwirtschaft*, 257: 1-118.

McMurtry J.A., 1992 - Dynamics and potential impact of generalist phytoseiids in agrosystems and possibilities for establishment of exotic species. *Experimental and Applied Acarology*, 14: 371-382.

Mezzalira G., 1990 - Piantare delle siepi. *Le Foreste*, 5-6: 12-26.

Moore P.D., Webb J.A., Collinson M.E., 1991 - Pollen analysis. Blackwell Scientific Publication, Oxford: 216 pp.

Remund U., Boller E., 1992 - Bluhende Rebberge in der Ostschweiz 3. Erganzende pollenversuche mit raubmilben. *Schweiz Zeitschrift für Obst- und Weinbau*, 128: 237-240.

Schausberger P., 1992 - Vergleichende Untersuchungen über den Einfluss unterschiedlicher Nahrung auf die Praimaginalentwichlung und die Reproduktion von *Amblyseius aberrans* Oud. und *Amblyseius finlandicus* Oud. (Acarina: Phytoseiidae). *Journal of Applied Entomology*, 113: 476-486.

Tixier M.S., Kreiter S., Auger P., Weber M., 1998 - Colonization of Languedoc vineyards by phytoseiid mites (Acari: Phytoseiidae): influence of wind and crop environment. *Experimental and Applied Acarology*, 22: 523-542.

Wiedmer U., Boller E., 1990 - Bluhende Rebberge in der Ostschweiz: 2. Zum Pollenangebot auf den Rebenblattern. *Schweiz Zeitschrift für Obst- und Weinbau,* 126: 426-431.

Influence of carmine spider mite *Tetranychus cinnabarinus* Boisd. (Acarida: Tetranychidae) feeding on ethylene production and the activity of oxidative enzymes in damaged tomato plants

Malgorzata Kielkiewicz
Department of Applied Entomology, Faculty of Horticulture and Landscape Architecture, Warsaw Agricultural University, PL-02-787 Warsaw, Nowoursynowska 166, Poland

Keywords: tomato, *Lycopersicon esculentum* Mill., carmine spider mite, *Tetranychus cinnabarinus* Boisd., ethylene, peroxidase, polyphenol oxidase, ascorbate oxidase, oxidative defence.

Abstract: Ethylene emission and the activity of oxidative enzymes were studied in tomato leaves after injury by carmine spider mite (CSM) (*Tetranychus cinnabarinus* Boisduval) feeding. Based on gas chromatographic data it was found that the feeding of 50 CSM for 3 days on young plants of tomato (*Lycopersicon esculentum* Mill., cv. Slonka) induced an increase of ethylene production by 40%. Elevated ethylene emission by damaged leaves was accompanied by the enhanced activity of oxidative proteins. Activity of total peroxidase and ascorbate oxidase in damaged leaves was about twice as high as compared to undamaged control leaves. Induction of polyphenol oxidase was less pronounced but highly significant. The results obtained suggest that in injured tomato plant's early defence responses against CSM are characterized by the action of ethylene on oxidative enzymes.

Introduction

Ethylene (C_2H_4) is one of the three phytohormones (abscisic acid, ethylene and jasmonate) involved in plant responses to a variety of biotic and abiotic stress factors (Boller 1991; Jones 1994; Ecker 1995; Schneider *et al.* 1996; Pare and Tumlinson 1996; Gawronska and Kielkiewicz 1999). Enhanced emission of this hormone by stress-affected plants may lead to various responses such as stunting of tissue elongation growth, advanced senescence and leaf fall, intensified synthesis of proteins rich in hydroxyproline, activation of phenylalanine ammonia-lyase, which is a key enzyme of phenylpropanoid pathways, advanced necrosis and lignification of tissues (Yang and Pratt 1978; Yang and Hoffman 1984; Dutta and Briggs 1991). It has been proved that most of these responses are vital factors for inducing the defence system of a stress-affected plant (Dutta and Briggs 1991; Jones 1994; Ecker 1995; O'Donnell *et al.* 1996; Schneider *et al.* 1996).

Ethylene is known to induce the activity of a variety of enzymes including peroxidases (Abeles *et al.* 1988; Mehlhorn 1990). Oxidative plant enzymes (peroxidase, ascorbate oxidase, polyphenol oxidase, lipoxygenase) are considered as important plant proteins involved in the defence against insect and mite herbivores (Hildebrand *et al.* 1986; Duffey and Felton 1991; Kielkiewicz 1991; Felton *et al.* 1992; Felton and Summers 1993; Felton *et al.* 1994; Bi and Felton 1995; Duffey and Stout 1996; Kielkiewicz 1996; Stout *et al.* 1996).

In the literature there are substantial data concerning the modified production of ethylene due to such stress factors as drought, salinity, low or high temperatures, pathogens or mechanical injury, but reports regarding the effect of pest feeding on ethylene emission are rare and contro-

F. Bernini, R. Nannelli, G. Nuzzaci, E. de Lillo (eds.), Acarid Phylogeny and Evolution. Adaptations in mites and ticks, 389-392.

versial. (Miller *et al.* 1994; Anderson and Peters 1994 and 1995; Rieske and Raffa 1995). Tetranychid mite feeding is a stress generating both mechanical injury and injection of saliva (Mothes and Seitz 1981; Van der Geest 1985; Kielkiewicz 1988). The aim of this study was to explain if carmine spider mite feeding may affect the rate of ethylene formation by injured tomato leaves and whether this phenomenon is related to the activity of oxidative enzymes.

Materials and methods

Tomato plants (*Lycopersicon esculentum* Mill., Slonka) were grown in pots with peat substrate in glasshouse conditions. Eight week old plants were infested using 50 females of the carmine spider mite (CSM) (*Tetranychus cinnabarinus* Boisd.) per each of the 7th leaf, counting from the bottom of the plant. Leaflets from 3 day infested and uninfested plants were placed separately into 40 ml glass vials and these were closed tightly. After 2 hours, samples of head space gas (1 ml) were withdrawn with a syringe for ethylene estimation. Its quantitative determination was carried out using a model 5890 Hewlett-Packard gas chromatograph (Hewlett-Packard, Orlando, USA) equipped with a flame ionizator and glass column (180 x 0.2 cm) filled with Alumina F-1. Ionizator and detector temperatures were 100 and 180°C, respectively. Argon was used as a flow gas at a rate of 40 ml per min^{-1}.

The activity of total peroxidase, polyphenol oxidase and ascorbate oxidase was estimated according to the methods by English-Loeb *et al.* (1997) and Felton and Summers (1993), respectively. Foliage soluble protein was measured by Bradford's procedure (Bradford, 1976).

Experiments with ethylene were repeated three times using four plants for replication. The measurements of activity of enzymes and content of protein from individual plants were conducted in 5 replications. All data were analyzed by one-way ANOVA. Treatment means were compared using the Fisher's least significant difference (LSD) procedure. The Statgraphics Plus (1994-1997 by Statistical Graphics Corp.) package was used for all analyses.

Results

After 3 days of CSM feeding on young tomato leaves, evident visible injury appeared and at the same time almost a 40% increase in ethylene emission was observed (tab. 1). Elevated production of this gas by affected leaves was accompanied by a higher activity of oxidative enzymes (tab. 1). Activity of total peroxidase (POX) and ascorbic acid oxidase (AOX) was about twice as high in injured leaves as in control leaves. Induction of polyphenol oxidase (PPO) was less pronounced but highly significant. Levels of total proteins in the infested leaves did not significantly differ from the control (tab. 1).

Discussion and conclusions

The data presented indicate that a short period of CSM feeding initiates an increase of ethylene emission from injured leaf tissues and this is accompanied by a higher activity of oxidative enzymes. The elevation of ethylene emission by a deciduous tree, *Tilia americana*, in response to the feeding by the basswood thrips (*Thrips calcarantus*) was reported by Rieske and Raffa (1995). Aphids, *Diuraphis noxia* significantly induced production of this hormone in a susceptible cultivar of barley (Morex), but induced a little ethylene production in PI 366450, a resistant barley cultivar (Miller *et al.* 1994). Other data showing that ethylene formation by wheat seedlings was stimulated by a 6 hour infestation with aphids (*Schizaphis graminum*) (Anderson and Peters 1994) and the rate of its emission dropped 2 days later (Anderson and Peters 1995). The authors concluded that ethylene appeared to be a symptom, not a mediator of *S. graminum* - induced injury.

Peroxidases are regulated by ethylene (Abeles *et al.* 1988). Experiments by Mehlhorn (1990) showed that ethylene production in plants

Tab. 1 - The effect of *T. cinnabarinus* infestation on ethylene production and activity of oxidative enzymes in injured tomato leaf. Abbrev.: SE, standard error of the means; asterisks (*) indicate that differences between means are significantly different at P=0.05 (ANOVA; LSD).

Leaf compound	Undamaged leaf Mean value±SE	Damaged leaf Mean value±SE	F-ratio	P-value of the F-test
Ethylene ($nl \times g^{-1} \times h^{-1}$)	0.8467 ± 0.03	1.1725 ± 0.05	27.94*	P<0.0001
POX ($\Delta A_{430} \times min^{-1} \times g\ f\ wt^{-1}$)	12.35 ± 0.35	25.47 ± 1.37	85.39*	P<0.0001
AOX ($\Delta A_{265} \times 10\ sek^{-1} \times g\ f\ wt^{-1}$)	0.287 ± 0.02	0.598 ± 0.05	29.66*	P=0.0006
PPO ($\Delta A_{470} \times min^{-1} \times g\ f\ wt^{-1}$)	3.42 ± 0.17	4.40 ± 0.13	20.71*	P=0.0019
Soluble protein ($mg \times g\ f\ wt^{-1}$)	8.48 ± 0.19	8.54 ± 0.70	0.01	P=0.98

exposed to stress factor is induced in order to enhance ascorbate peroxidase activity to protect cells against a harmful concentration of hydroperoxides. The induction of oxidative enzymes activity in this study is thought to have negative effects on CSM. It is known that peroxidases and polyphenol oxidases are involved in the defence of a plant against herbivorous insects and mites by the production and polymerization of phenolics, lignification, and hypersensitive responses of injured tissues (Hildebrand *et al.* 1986; Kielkiewicz 1991; Felton and Summers 1993; Felton *et al.* 1994; Bi and Felton 1995; Kielkiewicz 1996; Stout *et al.* 1996). Peroxidases and polyphenol oxidases also have negative effects on food digestibility and protein availability to some herbivorous insects (Felton *et al.* 1989; Duffey and Felton 1991; Duffey and Stout 1996). In plant tissues, ascorbate oxidases oxidize L-ascorbic acid to dehydro-L-ascorbic acid, which may be detrimental to pest growth and survival (Felton and Duffey, 1992; Felton and Summers 1993). Furthermore, ascorbate oxidase may accelerate the hypersensitive reaction to pest attack in plants and enhance the toxic action of polyphenol oxidases, peroxidases and lipoxygenases (Felton and Duffey 1992). Data obtained in the present research provide preliminary evidence that in injured tomato plants early defence responses against CSM can be characterized by the action of ethylene on oxidative enzymes. Further detailed studies are needed to understand if ethylene is an inducer of plant defensive system or is involved in modulating the expression of resistance as suggested by O'Donnell *et al.* (1996).

References

Abeles F.B., Dunn L.J., Callahan A., Dinterman R.E., Schmidt J., 1988 - Induction of a 33-kD and 60-kD peroxidases during ethylene-induced senescence of cucumber cotyledones. *Plant Physiology*, 87: 609-615.

Anderson J.A., Peters D.C., 1994 - Ethylene production from seedlings infested with biotypes of *Schizaphis graminum* (Homoptera: Aphididae). *Environmental Entomology*, 23(4): 992-998.

Anderson J.A., Peters D.C., 1995 - Inhibitors of ethylene biosynthesis and action do not prevent injury to wheat seedlings with *Schizaphis graminum* (Homoptera: Aphididae). *Environmental Entomology*, 24(6): 1644-1649.

Bi J.L., Felton G.W., 1995 - Foliar oxidative stress and insect herbivory: primary compounds, secondary metabolities, and reactive oxygen species as components of induced resistance. *Journal of Chemical Ecology*, 21: 1511-1530.

Bradford M.M., 1976 - A rapid and sensitive method for the quantitation of microgram quantities of protein utilizing the principle of protein dye binding. *Annals of Biochemistry*, 72: 248-254.

Boller T., 1991 - Ethylene in pathogenesis and disease resistance. In: Mattoo A.K., Suttle J.C. (eds.), The plant hormone ethylene. CRC Press, Boca Raton, Florida: 293-313.

Duffey S.S., Felton G.W., 1991 - Enzymatic antinutritive defenses of the tomato plant against insects. In: Hedin P.A. (ed.), Naturally Occurring Pest Bioregulators. American Chemical Society, Washington, DC: 166-197.

Duffey S.S., Stout M.J., 1996 - Antinutritive and toxic components of plant defense against insects. *Archives of Insect Biochemistry and Physiology*, 32(1): 3-37.

Dutta S., Briggs R.H., 1991 - Regulation of ethylene biosynthesis in citrus leaves infected with *Xanthomonas campestris* pv. *citri*. *Physiologia Plantarum*, 82: 225-230.

Ecker J.R., 1995 - The ethylene signal transduction pathway in plants. *Science*, 268 (5211): 667-675.

English-Loeb G., Stout M. J., Duffey S.S., 1997 - Drought stress in tomatoes: changes in plant chemistry and potential nonlinear consequences for insect herbivores. *Oikos*, 79: 456-468.

Felton G.W., Donato K., Del Vecchio R.J., Duffey S.S., 1989 - Activation of plant foliar oxidases by insect feeding reduces nutritive quality of foliage for noctuid herbivores. *Journal of Chemical Ecology*, 15: 2667-2695.

Felton G.W., Duffey S.S., 1992 - Ascorbate oxidation reduction in *Helicoverpa zea* as a scavenging system against dietary oxidants. *Archives of Insect Biochemistry and Physiology*, 19: 27-37.

Felton G.W., Donato K., Brodway R.M., Duffey S.S., 1992 - Impact of oxidazed plant phenolics on the nutritional quality of dietary protein to a noctuid herbivore. *Journal of Insect Physiology*, 38: 277-285.

Felton G.W., Summers C.B., 1993 - Potential role of ascorbate oxidase as a plant defense protein against insect herbivory. *Journal of Chemical Ecology*, 19: 1553-1568.

Felton, G.W., Summers C.B., Mueller A.J., 1994 - Oxidative responses in soybean foliage to herbivory by bean leaf beetle and three-cornered alfaalfa hopper. *Journal Chemical Ecology*, 20: 639-650.

Gawronska H., Kielkiewicz M., 1999 - Effect of the carmine spider mite (Acarida: Tetranychidae) infestation and mechanical injury on the level of ABA in tomato plants. *Acta Physiologiae Plantarum*, 21(3): 297-303.

Hildebrand D.F., Rodriguez J.G., Brown G.C., Luu K.T., Volden C.S., 1986 - Peroxidative responses of leaves in two soybean genotypes injured by two-spotted spider mite (Acari: Tetranychidae). *Journal of Economic Entomology*, 79(6): 1459-1465.

Jones A.M., 1994 - Surprising signals in plant cells. *Science*, 263: 183-184.

Kielkiewicz M., 1988 - Susceptibility of previously damaged strawberry plants to mite attack. *Entomologia Experimentalis et Applicata*, 47: 201-203.

Kielkiewicz M., 1991 - Phenolics in tomato leaves infested by carmine spider mite (*Tetranychus cinnabarinus* Boisduval). In: Dusbabek F., Bukva V. (eds.) Modern Acarology. Academia, Prague and SPB Academic Publishing bv, The Hague, 1: 603-606.

Kielkiewicz M., 1996 - Hypersensitive response of tomato leaf tissues towards *Tetranychus cinnabarinus* Boisd. (Tetranychidae) feeding. In: Mitchell R., Horn D.J., Needham G.R. (eds.), Acarology IX. The Ohio Biological Survey, Columbus, vol. 1: 47-49.

Mehlhorn H., 1990 - Ethylene-promoted ascorbate peroxidase activity protects plants against hydrogen peroxide, ozone and paraquat. *Plant, Cell and Environment*, 13: 971-976.

Miller H. L., Neese P.A., Ketring D.L., Dillwith J.W., 1994 - Involvement of ethylene in aphid infestation of barley. *Journal of Plant Growth Regulation*, 13(4): 167-171.

Mothes U., Seitz K.A., 1981 - Functional microscopic anatomy of the digestive system of *Tetranychus urticae* (Acari, Tetranychidae). *Acarologia*, 22: 257-270.

O'Donnell P. J., Calvert C., Atzorn R., Wasternack C., Leyser H.M.O., Bowles D.J., 1996 - Ethylene as a signal mediating the wound response of tomato plants. *Science*, 274 (5294): 1914-1917.

Pare P.W., Tumlinson J.H., 1996 - Plant volatile signals in response to herbivore feeding. *Florida Entomologist*, 79(2): 93-102.

Rieske L.K., Raffa K.F., 1995 - Ethylene emission by a deciduous tree, *Tilia americana*, in response to feeding by introduced basswood thrips, *Thrips calcarantus*. *Journal of Chemical Ecology*, 21(2): 187-197.

Schneider P., Schweizer P., Meuwly P., Metraux P., 1996 - Systemic Acquired Resistance in plants. *International Review of a Survey of Cytology Cell Biology*, 168: 303-339.

Stout M.J., Workman K.V., Duffey S.S., 1996 - Identity, spatial distribution, and variability of induced chemical responses in tomato plants. *Entomologia Experimentalis et Applicata*, 79(3): 255-271.

Van der Geest L.P.S., 1985 - Physiology and genetics. Aspects of physiology. In: Helle W., Sabelis M.W. (eds) Spider Mites. Their Biology, Natural Enemies and Control. Elsevier, vol. 1A: 171-175.

Yang S.F., Pratt H.K, 1978 - The physiology of ethylene in wounded tissue. In: Kahl G. (ed.), Biochemistry of Wounded Plant Tissue. De Gruyter, Berlin: 596-622.

Yang S.F., Hoffman N.E., 1984 - Ethylene biosynthesis and its regulation in higher plants. *Annual Review of Plant Physiology*, 35: 155-189.

Changes in the content of soluble sugars in leaves of cucumber plant infested with *Tetranychus urticae* Koch and treated with plant growth promoting rhizobacteria (PGPR)

Anna Tomczyk
Department of Applied Entomology, Warsaw Agricultural University, Nowoursynowska 166, PL-02-787 Warsaw, Poland

Key words: *Tetranychus urticae*, cucumber, plant growth promoting rhizobacteria (PGPR), sugar content.

Abstract: Plants of two cucumber cultivars - Corona and Aramis - were grown under glasshouse conditions with or without plant growth promoting rhizobacteria (PGPR). Young plants were infested with *Tetranychus urticae* Koch females. After six weeks of spider mite feeding leaf samples (middle and top leaves) were collected for analysis of sugars. In infested leaves of cv Corona (very susceptible to spider mites) a decrease in soluble sugar content, of about 13% was observed in non-bacterized plants. In the young (top) leaves of infested plants the concentration of soluble sugars increased.
In infested leaves of cv Aramis (less susceptible to spider mites) the minor changes in soluble sugar content were observed) in both bacterized and non-bacterized plants.

Introduction

The results of experiments on the influence of the concentration of sugars in plant tissues on spider mites indicate that sugars belong to basic organic compounds which are used as food for mites. Some sugars, such as glucose or sucrose, act as feeding stimulants (Dabrowski and Rodriguez 1971; Tulisalo 1972; De Angelis *et al.* 1983; Tomczyk 1989). It has been shown earlier that the chemical composition of a host plant has an influence on the bionomy of a pest species (Gerson and Arnovitz 1980).

Changes in the concentration of plant organic substances occurring in plant tissues during herbivore feeding are important for the plant - pest interaction. They can lead to a disturbances in the feeding process and may reduce the pest population on the host plant. For example, a decrease in the concentration of reducing sugars in leaves of some host plants (chrysanthemum, peppermint) as the result of spider mite feeding, limits female fecundity and causes a reduction in the mite population (De Angelis *et al.* 1983; Wermelinger *et al.* 1985; Tomczyk and Kropczynska 1985; Tomczyk 1989).

The induction of plant defence is linked with changes in the ratio of plant nutrients and in the concentration of secondary metabolites of the host plant (Tomczyk and Kropczynska 1985; Tomczyk 1989 and 1996). It was found that the presence of plant growth promoting rhizobacteria (PGPR) in the root system of cucumber can increase its resistance to the cucumber beetle by changes in the concentration of cucurbitacins in attacked plants (Zehnder *et al.* 1997).

The purpose of this study was to investigate the influence of plant growth promoting rhizobacteria on the concentration of soluble sugars in

F. Bernini, R. Nannelli, G. Nuzzaci, E. de Lillo (eds.), Acarid Phylogeny and Evolution. Adaptations in mites and ticks, 393-398.

leaves of two cucumber cultivars infested and uninfested with spider mites and its effect on spider mite - host plant interaction.

Material and methods

The experiment was conducted on two cultivars of cucumber - Corona and Aramis - grown under glasshouse conditions. Half of their seeds were inoculated during 20 minutes with the plant growth promoting rhizobacterium *Pseudomonas* P-112, originating from the roots of cucumber seedlings. Other seeds were treated with water. The inoculum contained 3 x 10^9 bacterial cells in 1 cm^3. Seeds were sown into pots and young plants were subsequently transferred with soil to their final place in a glasshouse. Plants of both cultivars obtained from inoculated and noninoculated seeds were used in the experiment. Twenty plants of each cultivar were infested with females of *Tetranychus urticae* Koch. The initial mite population was 5 individuals per leaf. In effect 4 experimental groups of plants of each cultivar were obtained:

Control (no mites, nonbacterized)
Control + Bacteria (no mites, bacterized)
Infested (with mites, nonbacterized)
Infested + Bacteria (with mites, bacterized)

Six weeks after infestation with spider mites leaf samples were collected for sugar analyses. Leaves of different age were taken from the middle and top of the plant. Middle (mature) leaves collected from infested plants were damaged by spider mites but top (young) were mite free. The collected plant material was lyophilized. The sample weight was about 100 mg. Both total sugars and reducing sugars were estimated according to Nelson (1944) and Yemm and Wilis (1954).

Results

Figure 1 presents the number of spider mites in the different experimental groups of plants, 6 weeks after infestation with the mites. Simultaneously, samples were collected for sugar analysis.

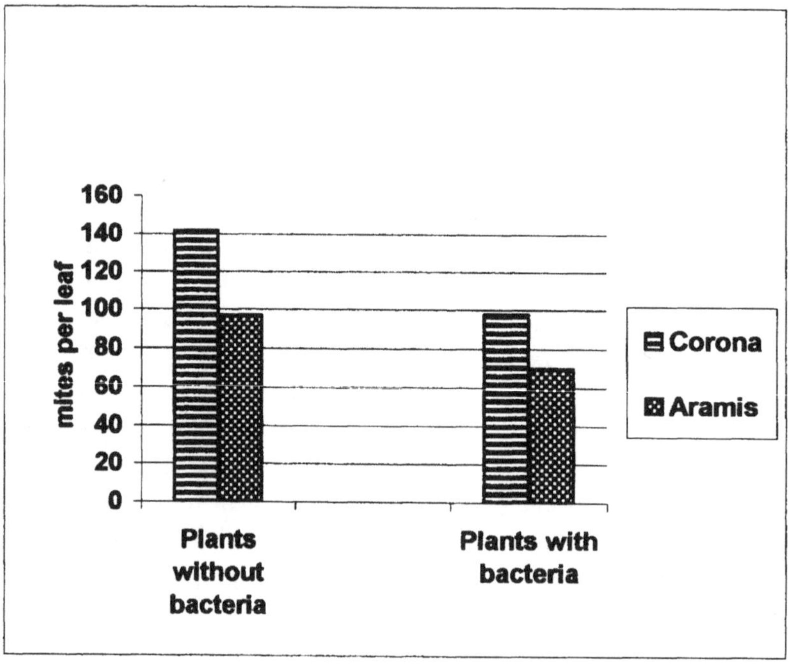

Fig. 1 - *Tetranychus urticae* population on two cucumber cultivars after 6 weeks of feeding on bacterized and nonbacterized plants. SE are presented

Plants of both cucumber cultivars treated with rhizobacteria had smaller mite populations than untreated. This was in particular evident in the case of cv Corona.

Figures 2, 3, 4 and 5 illustrate the content of both total soluble and reducing sugars in leaves of the cultivars tested.

In mite infested leaves of the susceptible cv Corona, grown without bacteria a lower soluble sugar concentration of about 13% was observed as compared to untreated plants. The presence of bacteria in the root system of infested plants inhibited this decline, although uninfested plants grown with these bacteria (control + B) had a lower sugar concentration in their leaves than control. At the same time a high increase (26-35%) in total sugars was observed in the top leaves of all experimental plants compared to control plants. Content of reducing sugars in mite infested leaves of cv Corona was 19% lower than in control plants. Also in this case the presence of rhizobacteria in infested plants reduced the effect of mites on the sugar concentration in their leaves.

In the top leaves of all plant groups studied the content of reducing sugars decreased approximately 10% in comparison with the control plants. The percentage of reducing sugars in the total soluble sugar content in the top leaves of infested plants without bacteria and in those of uninfested but PGPR-treated (control + B) was much lower than in leaves of the control plants (tab. 1). A decline in the proportion of reducing sugars to the total content of soluble sugars as an effect of mite feeding or presence of rhizobacteria was slightly less evident when the mites were feeding on bacterized plants.

In leaves of mite infested and bacterized plants (infested + B) of the less susceptible cultivar Aramis, minor changes in the concentration of soluble sugars were observed as compared to the control. Total content of soluble sugars was about 10% higher in injured mature leaves and 6% in the top (uninjured) leaves of plants treated with rhizobacteria (infested + B) than in respective leaves of control plants. Only in the top leaves of bacterized uninfested plants (control + B) an

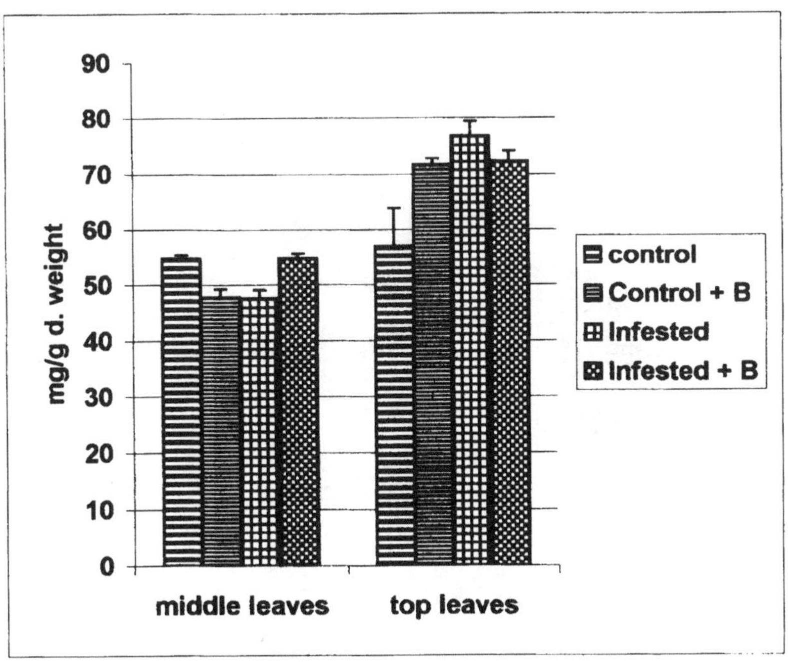

Fig. 2 - Total soluble sugars in leaves of cucumber cv Corona. SE are presented.

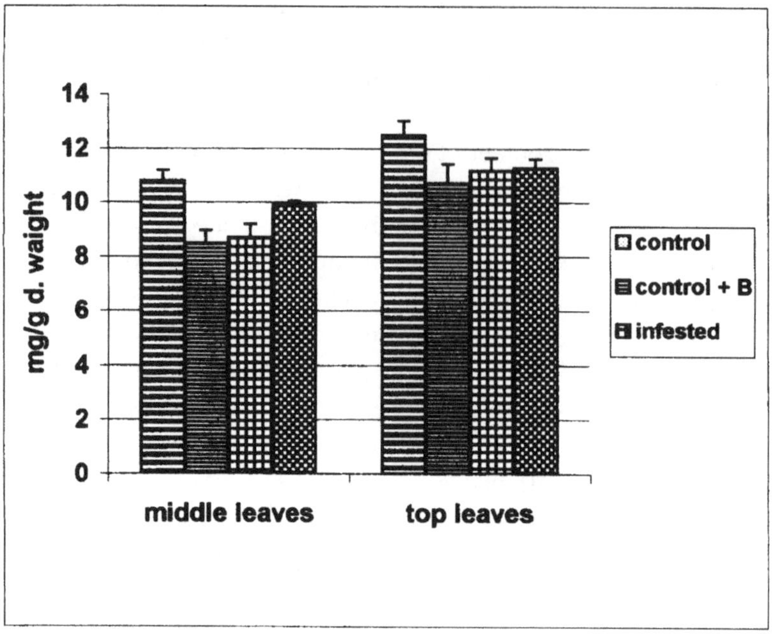

Fig. 3 - Content of reducing sugars in leaves of cucumber cv Corona. SE are presented.

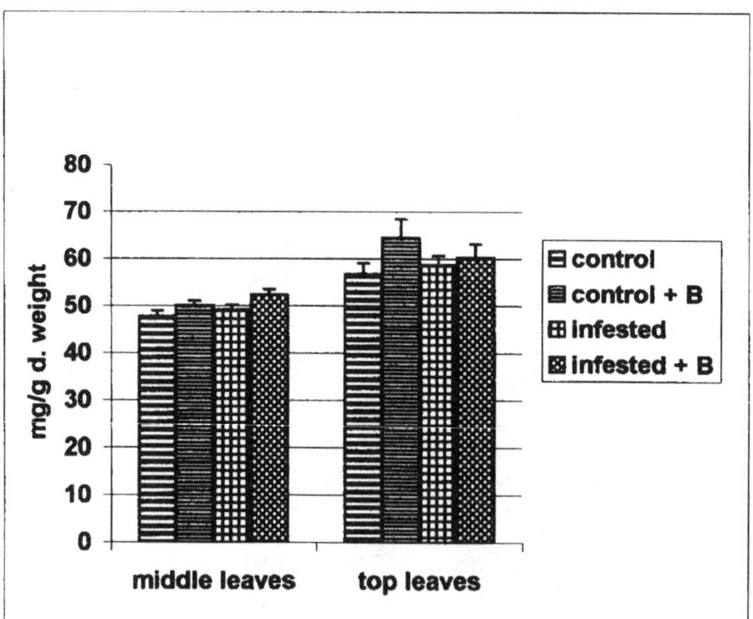

Fig. 4 - Total soluble sugars in leaves of cucumber cv Aramis. SE are presented.

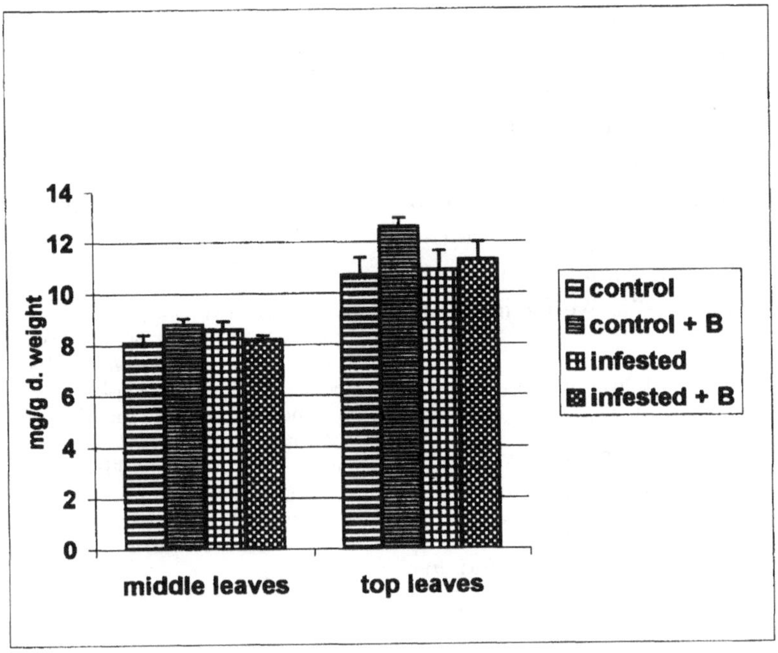

Fig. 5 - Content of reducing sugars in leaves of cucumber cv Aramis. SE are presented.

Tab. 1 - Proportion of reducing sugars in the total soluble sugar content (% ± SE).

Exp. Variant	Corona		Aramis	
	Middle leaves	Top leaves	Middle leaves	Top leaves
Control	19.7 ± 0.95	22.0 ± 2.02	17.0 ± 1.07	18.8 ± 1.27
Control + B	17.8 ± 1.15	14.9 ± 0.92	17.6 ± 0.52	19.5 ± 1.35
Infested	18.3 ± 1.52	14.6 ± 069	17.5 ± 0.54	18.6 ± 0.68
Infested + B	18.1 ± 0.17	15.7 ± 0.58	15.7 ± 0.57	18.8 ± 0.81

enhanced concentration (18%) of reducing sugars was observed. The relative concentration (%) of reducing sugars in the total content of soluble sugars in the top leaves was, however, very similar for all plant groups of this cultivar.

Discussion and conclusions

The injuries caused by mite feeding are responsible for the decrease in the concentration of soluble sugars in middle leaves (with mites) of susceptible Corona cucumber plants and for the increase in top (mite free) leaves of injured plants. The presence of rhizobacteria inhibits sugar decline in mite infested leaves but does not change the high level of sugars in top leaves of injured plants.

Changes in the concentration of sugars in leaves of Corona cucumber, as an effect of spider mite feeding, affect the value of the plant as a host for these pests and can be responsible for a disturbance in the development of mite populations. Similar changes in sugar content can be induced by the presence of rhizobacteria in the root system of mite free plants. Gerson and Arnovitz (1980) found that the pest bionomy is strongly related to the chemical composition of a host plant. Spider mites fecundity and population density was positively correlated with sugar content in apple, chrysanthemum, bean and cucumber leaves (Tulisalo 1972; Wermelinger *et al.* 1985; Tomczyk 1989).

The modified ratio of reducing sugars to total sugars indicates the accelerated changes of simple sugars to oligosaccharides which can be transported to other parts of an infested or bacterized plant of Corona cucumber. Accelerated "flow" of simple sugars to oligosaccharides was found in mite infested chrysanthemum leaves (Tomczyk and Kropczynska 1985; Tomczyk 1996).

Small changes in the concentration of soluble sugars in leaves of Aramis cucumber as an effect of mite feeding or presence of bacteria in the root system is probably related to its lower (as compared to cv Corona) susceptibility to spider mites. It is possible that plants of this cultivar did not react to injuries caused by 6 weeks of mite feeding (density of pest population was lower than on cv Corona). The presence of bacteria in the root system of Aramis cucumber could have an influence on other plant compounds, rather, than on sugar concentration or transportation in the plant. Zehnder *et al.* (1997) observed reduction in the concentration of cucurbitacins in leaves of cucumber treated with PGPR and a negative effect of this phenomenon on the development of cucumber beetle population on these plants.

References

Dabrowski Z.T., Rodriguez J., 1971 - Studies on resistance of strawberries to mites. 3. Preference and nonpreference responses of *Tetranychus urticae* and *T. turkestani* to essential oils of foliage. *Journal of Economic Entomology*, 64: 387-391.

De Angelis J.D., Berry R.E., Krantz G.W., 1983 - Photosynthesis, leaf conductance and leaf chlorophyll content in spider mite (Acari: Tetranychidae) injured peppermint leaves. *Environmental Entomology*, 12: 345-349.

Gerson U., Arnovitz A., 1980 - Feeding of the carmine spider mites on seven host plant species. *Entomologia Experimentalis et Applicata*, 28: 109-115.

Nelson N., 1944 - A photometric adaptation of the Samogyi method for determination of glucose . *Journal of Biological Chemistry*, 153: 375-380.

Tomczyk A., 1989 - Physiological and biochemical responses of different host plants to infestation by spider mites (Acarina: Tetranychidae). Treatises and Monographs, Warsaw Agricultural University Press: 112 pp.

Tomczyk A., 1996 - Mechanisms of physiological and biochemical interactions between spider mites (Tetranychidae) and their host plants. In: Mitchell R., Horn D.J., Needham G.R., Welbourn W.C. (eds.), Acarology IX Proceedings. The Ohio Biological Survey Columbus, Ohio, vol. 1: 25-28.

Tomczyk A., Kropczynska D., 1985 - Effects on the host plants In: Helle W., Sabelis M.W. (eds.), Spider mites, their biology, natural enemies and control. Elsevier, vol. 1A: 317-327.

Tulisalo U., 1972 - The effect of variations in the carbohydrate level of the host plant on the fecundity of the two-spotted spider mite *Tetranychus urticae* Koch. *Annales Entomologici Fennici*, 38: 179-182.

Wermelinger B., Oertli J.J., Delucchi V., 1985 - Effect of host plant nitrogen fertilisation on the biology of the two-spotted spider mite, *Tetranychus urticae* Koch. *Entomologia Experimentalis et Applicata*, 38: 23-28.

Yemm E.N., Wilis A.J., 1954 - The estimation of carbohydrates in plant extracts by anthrone. *Biochemical Journal*, 57: 508-514.

Zehnder G., Klopper J., Yao C., Wei G., 1997 - Induction of systemic resistance in cucumber against cucumber beetles by plant growth-promoting rhizobacteria. *Journal of Economic Entomology*, 90: 391-396.

Adaptation of mites and ticks to parasitism. Medical and veterinary aspects

Frantisek Dusbabek
Institute of Parasitology, Academy of Sciences of the Czech Republic, Branisovska 31, CZ-370 05 Ceske Budejovice, Czech Republic

Key words: Acari, mites, ticks, adaptations, parasitism.

Abstract: Acari are the only group among Chelicerata which includes plant and animal parasites. The ancestors of zooparasites probably were nest inhabiting and free living predators and keratinophages, or phoretic species. They developed (1) adaptations to host location and identification, (2) adaptations to attachment on the host body, (3) adaptations to blood feeding and digestion, and (4) adaptations to host defense mechanisms.

Medical and veterinary consequences of mite and tick parasitism on hosts manifest themselves in the direct damage to hosts and in the adaptations of parasites to transmission of pathogens. Most infectious agents were probably symbionts, commensals or pathogens of different animals before they were transmitted by arthropods. Some arthropod-borne pathogens, however, are considered to be originally parasites or symbionts of arthropods (e.g. *Borrelia* or *Theileria*). Recent studies indicate that some substances produced by salivary glands of ticks during feeding, such as SAT factor (saliva activated transmission factor) or calreticulin, may provide conditions at the feeding site potentiating pathogen transmission.

1. Origin of parasitism in mites and ticks

Parasitism as a biological phenomenon originated independently and repeatedly in different animal groups on different levels of their organization and at different phases of their evolution. The host served more or less as a biotope for parasites and the evolution of the hosts considerably affected the evolution of parasites. Many parasites are strictly host specific and a group of related species or genera of parasites is often restricted to related host species. In 1957 Mayr suggested, that "the association between parasite and host goes far back into geological history, the evolution of the parasite being as closely correlated with that of the host as if it were an organ of the host". Each step in adaptive radiation of hosts was followed by the adequate step in adaptive changes in parasites. Such parallel host-parasite evolution has been confirmed by different authors in many families of parasitic mites, e.g. Spinturnicidae, Macronyssidae, Myobiidae, Chirodiscidae, Atopomelidae, etc. and their specificity is frequently designated as a phylogenetic one (fig. 1). In some of these ectoparasites the restriction to different body regions or specialized setae (e.g. vibrisses) developed.

Many parasites, however, attack a large range of hosts not phylogenetically related, but inhabiting similar biotopes. These parasites which incorporate an *r*-reproductive strategy are adapted rather to the host biotope and their evolution was affected both by the parasitic way of life and by changes in environmental con-

F. Bernini, R. Nannelli, G. Nuzzaci, E. de Lillo (eds.), Acarid Phylogeny and Evolution. Adaptations in mites and ticks, 399-418.

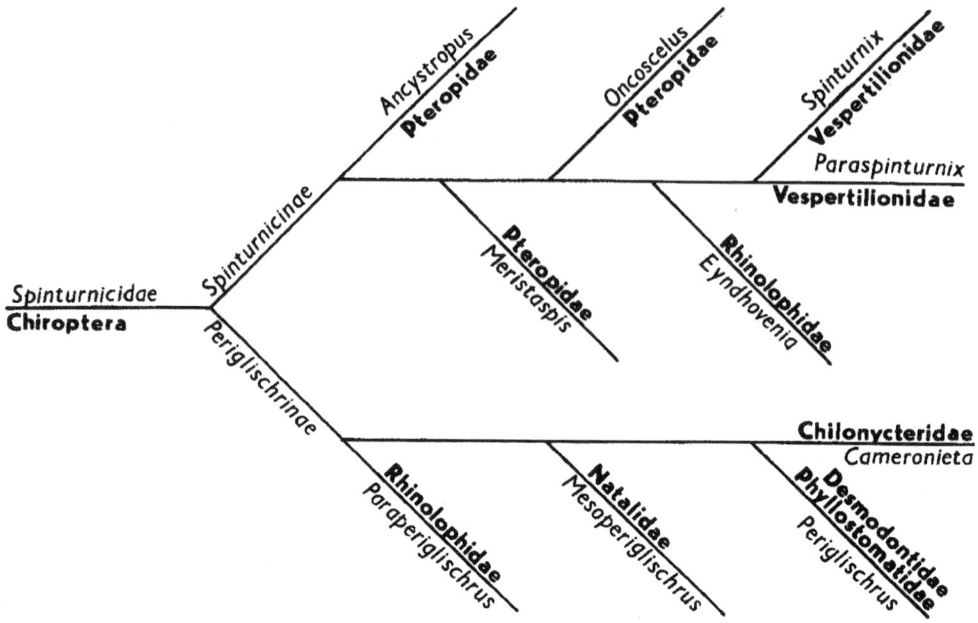

Fig. 1 - Scheme of parallel host-parasite evolution in the mite family Spinturnicidae and the host order Chiroptera. Mites developed in two phylogenetic branches (now subfamilies): Spinturnicinae evolved in parallel with megachiropteran family Pteropidae and Old World families Rhinolophidae and Vespertilionidae. Evolution of the subfamily Periglischrinae was connected with Rhinolophidae and with the New World bat families (Natalidae, Desmodontidae, Phyllostomatidae and Chilonycteridae).

ditions. Their specificity can be designated as an ecological one. This type of specificity occurs in some Ixodoidea, Mesostigmata (e.g. Laelapidae) and Trombiculidae.

Mites are the only group among Chelicerata which include plant and animal parasites. The ancestors of zooparasites were probably forms having contacts with potential hosts in their nests, shelters and lairs on the one hand, or breeding-, resting- and hunting-grounds on the other. They used the potential host and its products in the nest predominantly as a source of food (coprophagic, keratinophagic and predatory forms), but also as a habitat with a suitable microclimate (Beklemishev 1970). Initially, the association of future haematophages may have been one of commensalism with the mite feeding on small arthropods and blood exuding from the abrasion on the body of the host. Such generalized type of feeding is still characteristic in certain Laelapidae, while in the nest-inhabiting *Haemogamasus* a remarkable range of feeding habits occurs, including general predators, polyphagous species with parasitic tendencies and obligate haematophages (Radovsky 1968, 1985 and 1994; Evans 1992). Such evolution to parasitism occurred probably in most of Mesostigmata, such as Spinturnicidae, Dermanyssidae, Macronyssidae, Laelapidae, Spelaeorhynchidae and others. Some of them, such as Rhinonyssidae, Ptilonyssidae, Ixodorhynchidae and Halarachnidae penetrated into the host body cavities and became endoparasites.

Ticks are believed to have evolved from a single ancestral species or species group of the Parasitiformes during the Palaeozoic era. The earliest hosts of ancestral metastriate ticks probably were amphibians and reptiles (Hoogstraal 1985; Oliver 1989; Sonenshine 1991) while primitive Monotremata and Marsupialia are considered to be the earliest hosts of prostriate ticks (Filippova 1977). Klompen *et al.* (1996)

suggest, that biogeography and ecological specificity should be accorded a far more prominent role in further tick evolution than tick-host association. The life cycle of the earliest tick species was probably multi-nymphal, with each stage taking small blood meals, similar to the multi-host argasids present today. Some tick species were transferred secondarily to the nests of hosts and became nidicoles.

Similar extra-nidicolous origin of parasitism can be supposed in some Prostigmata (Trombiculidae, Trombidiidae, Erythreidae, Calyptostomidae, etc.), larvae of which feed on mammals, birds and insects (Beklemishev 1970). Remarkable is the adaptation of some water mites to parasitism in shellfishes (e.g. *Unionicola*) or on insects (larvae of some Hydrachnellae).

Most parasitic Actinotrichida originate probably from nest-inhabiting keratinophages, most of which maintained keratinophagy up to date. They are parasites of birds (feather mites Analgoidea, Pterolichoidea, quill mites Syringophilidae), and mammals (fur mites Listrophoroidea). Other keratinophages became facultative lymph- or blood-feeders (Myobiidae, Psoroptidae, Pterygosomidae) and some of them penetrated into the skin and are now endoparasites (Sarcoptidae, Psorergatidae). Other parasitic families became endoparasitic mucophages (Gastronyssidae, Speleognathidae) or fed on the host cells (Demodecidae). However, also among Actinotrichida the predacious groups occur, e.g. Cheyletidae, from which some parasites developed (e.g. *Cheyletiella, Eucheyletiella, Ornithocheyletia*, etc.).

Some mites parasitizing insects were originally phoretic and their phoresy changed into parasitism (e.g. Tarsonemini - Beklemishev 1970). A special case is Varroidae which inhabit bee nests and hives and feed on the brood.

More detailed data on evolution of distinct groups of Acari were recently summarized in monographs of Houck (1994) and Bruin *et al.* (1999).

2. Adaptation of mites and ticks to parasitism

Adaptations important for a parasitic way of life were well documented in the text-book of Evans (1992). They generally consist of adaptations to host localization and identification, to attachment to the host body, to feeding on the host, and to control of host defense mechanisms. All these adaptations include changes in external morphology, anatomy, physiology, immunology, behaviour and reproduction.

Adaptations to host localization and identification

Adaptation in behavior

Ticks display appetence either by hunting for a host or by preparing an ambush. Some tick species (e.g. *Hyalomma,* some *Amblyomma* and most argasid ticks) behave as hunters, actively looking for hosts. Such hard ticks use eyes for long distance orientation, and host allomones, (odour, the CO_2 concentration and temperature) for short distance orientation (Leonovich 1986). The latter stimuli serve as the main appetence stimuli for soft ticks Argasidae too, although there are some *Ornithodoros* species with eyes present. Other hard tick species prefer to ambush, waiting for the host on vegetation in a questing position. They are able to distinguish the presence of a host according to vibrations, shades and odours, and enhanced concentration of CO_2 (Steulet and Guerin 1992 and 1994). Larvae of *Boophilus microplus* are able to distinguish attractive components (1-octen-3-ol and 2-nitrophenol) of bovine odour in a blend of host volatiles (Osterkamp *et al.* 1999). They are also attracted by squalene, a major lipid component present on the skin and in the blood of many mammals (Yoder *et al.* 1998). Close range stimuli such as warmth and touch or breath components are also important in engagement. The questing behaviour of ticks is connected with the vertical migration of

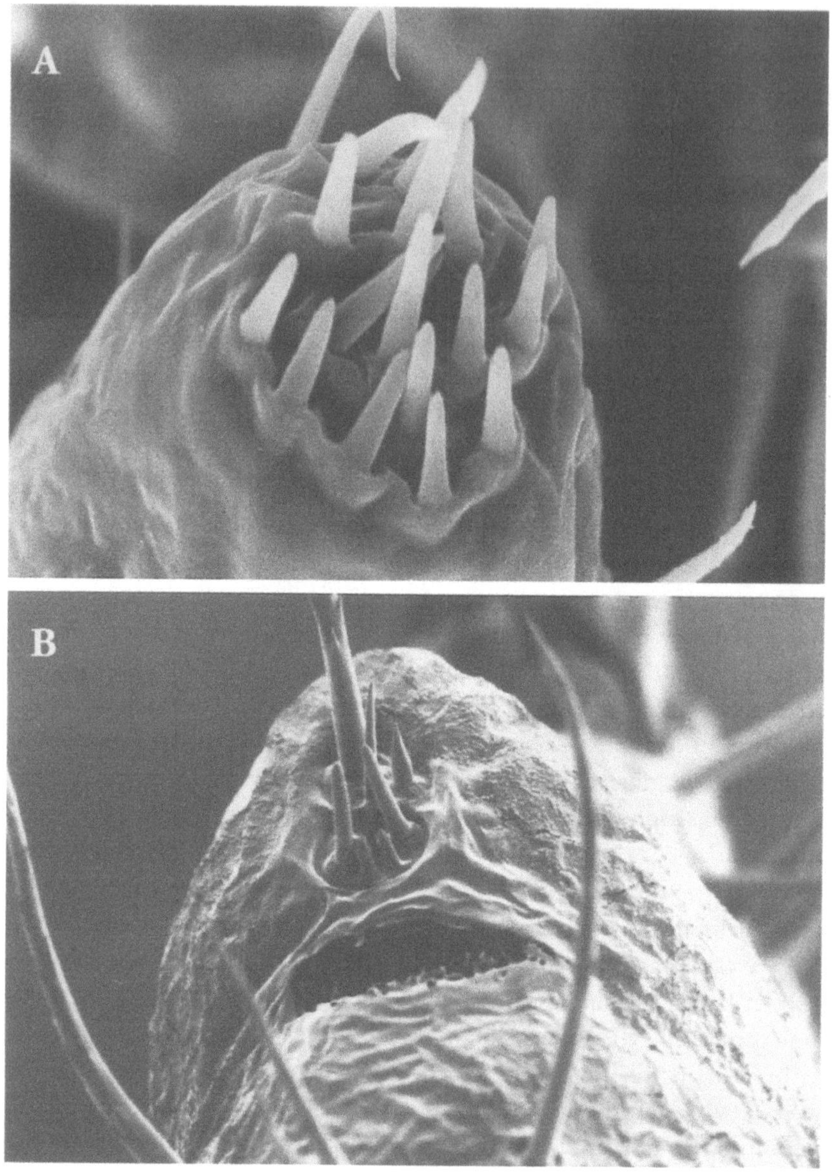

Fig. 2 - A, *Argas persicus* Oken, palpal complex of sensillae. B, *Argas polonicus* Siuda et al., Haller's organ on tarsus I, consisting of the inner capsula and anterior pit setae.

a tick on vegetation, caused primarily by changes of light intensity and air humidity.

Adaptations of sensory system

A special sensory system has developed in ticks and other ectoparasitic mites for detection of the host's presence and its identification. This sensory system consists of specialized sensillae situated on the tarsus of leg I (Haller's organ in ticks or tarsal organ in Mesostigmata) or on other leg segments (genual, tibial and tarsal solenidia) and pulvilli, on palpi (palpal complex), chelicerae (cheliceral digit sensilla), mouthparts and body surface (figs 2A-B, 3). These organs consist of chemosensitive (olfactory and gustatory), thermosensitive, hygrosensitive, photosensitive, mechanosensitive and multifunctional sensilla. The photosensitive cells or eyes are also developed in most tick species.

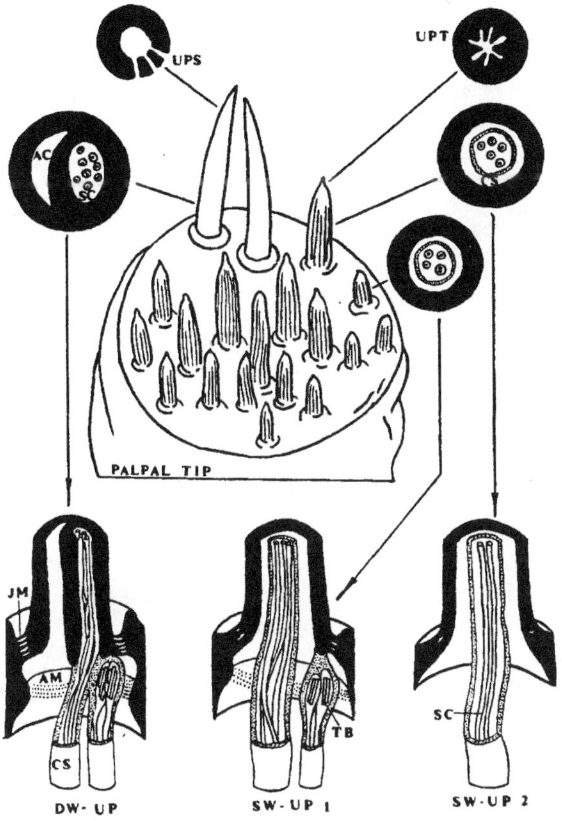

Fig. 3 - *Argas persicus* Oken, fine structural organization of palpal sensory organ. The double-walled upper-pore sensilla (DW-UP) and the single-walled upper pore sensilla (SW-UP 1) are gustatory and mechanoreceptive, the SW-UP 2 are chemoreceptive, probably with pheromone-receptive function.

Adaptations to attachment on the host

Morphological attachment adaptations are apparent on the mouthparts, legs and the idiosoma.

Adaptations of the mouthparts

These adaptations occur mostly in species which feed on the host for a long time and need to be well attached during an entire feeding time. In Ixodoidea, a special denticulated organ, the hypostome, is developed which together with chelicerae enable long-lasting fixation of ticks on the host skin (fig. 4A). In some tick species the cement matter is secreted into the wound from several cell types in the salivary glands serving for better fixation of ticks to the host. In Trombiculidae, a hook-like movable digit of chelicerae helps in the attachment of larvae to host skin; a similar function is seen in the palpal hooks in Myobiidae and some other parasitic Cheyletoidea (fig. 7B).

Adaptations of legs and idiosoma

In many ectoparasitic Mesostigmata, the legs are strong and long and armed with strong claws and pulvilli and the idiosoma is dorso-ventrally compressed. Extremely strong adaptations have developed in Spinturnicidae living on wing and tail membranes (patagium) of bats. In the subfamily Spinturnicinae, both sexes are armed with extremely strong and long legs with strong claws and pulvilli, while in females of the subfamily Periglischrinae the opisthosoma is broadly enlarged and supported by sclerotized endoskeleton (fig. 5), serving as a sucker. In some *Periglischrus,* females are attached to the host wing membrane by a waxy matter, probably the product of opisthosomal glands. In females of the genus *Cameronieta* the waxy matter covers the whole body surface similar to that seen in Coccoidea, insect parasites of plants. Numerous fine spines around the sternal region also aid in better fixation to the host wing membrane (fig. 5A-B). In many endoparasitic Mesostigmata (Rhinonyssidae, Ptilonyssidae, Halarachnidae, Entonyssidae), strong reduction of body setae occurs (fig. 6).

Special adaptations of the legs and idiosoma have developed in some prostigmatic and astigmatic fur mites. In Myobiidae the first pair of legs is strongly reduced, with a great hook-like clasping tubercle on genu I (fig. 7A). In Chirodiscidae the body is laterally compressed and the first two pairs of legs are modified to clasping organs serving for attachment to one host hair (fig. 8). In Atopomelidae two striated

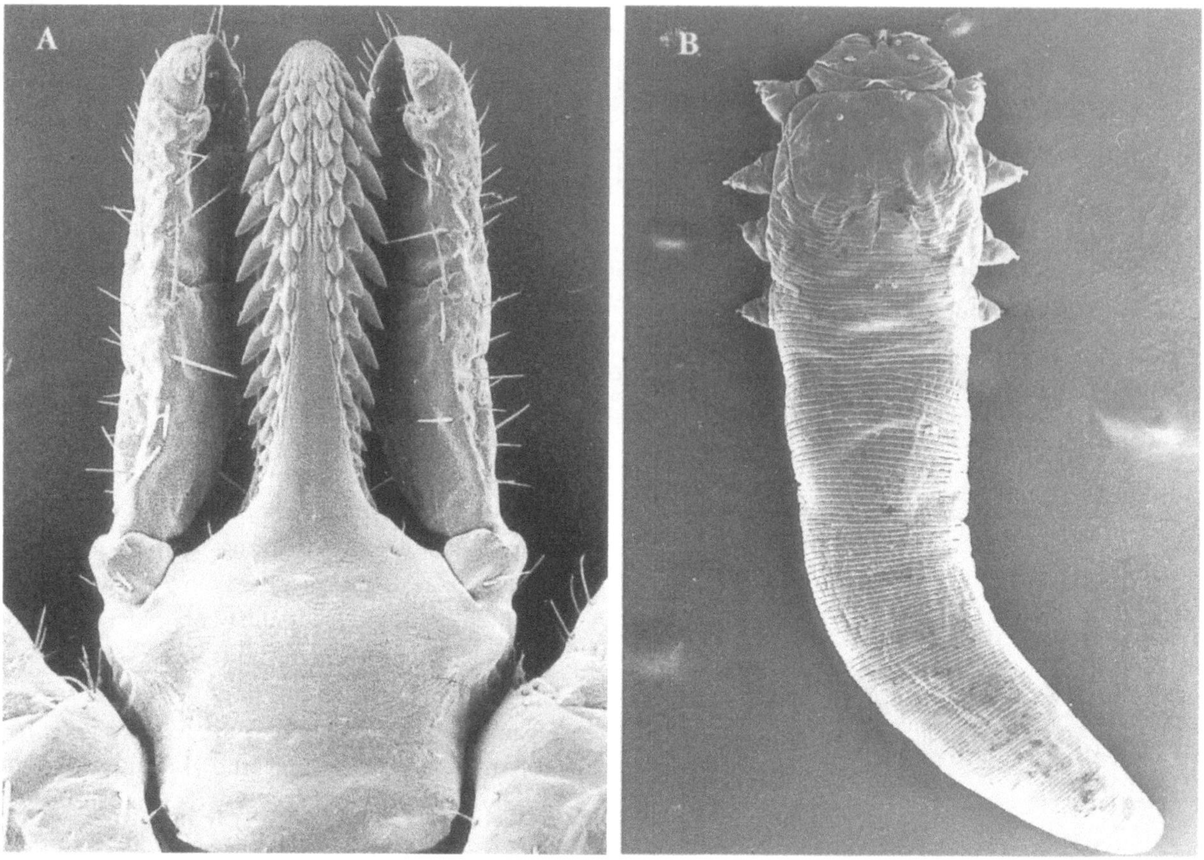

Fig. 4 - A, *Ixodes ricinus* (L.), female capitulum with prominent hypostome and palps. B, *Demodex folliculorum* (Simon), female, with characteristic vermiform body (by V. Bukva).

membranous flaps have developed, arising from the coxal fields of legs I and II, and acting as a clasper (fig. 9). In endoparasitic families the idiosoma is frequently globular (Sarcoptidae, Knemidocoptidae, Psorergatidae) or vermiform (Demodecidae) (figs 4B, 10).

Adaptations in reproduction

Most parasitic Mesostigmata and some parasitic Prostigmata and Astigmata are *k*-strategists in their reproduction, while ticks are *r*-strategists, both with direct sperm transference (Evans 1992). In obligatory blood-feeders, blood proteins are necessary for egg formation and ovulation. Mesostigmata are regularly oviparous, however, viviparity also occurs (e.g. Spinturnicidae, the permanent ectoparasites of bats). In oviparous Mesostigmata, eggs develop one or two at a time and are deposited out of the host body in its shelter or nest, rarely directly on the host body (e.g. *Ornithonyssus*). The larvae and deutonymphs in Macronyssidae molt without feeding while the protonymphs regularly feed on the host. The prostigmatic fur mites (Myobiidae, Chirodiscidae, Listrophoridae) and astigmatic feather mites (Analgoidea, Pterolichoidea) glue their eggs to the host's hair and feathers, while sarcoptids deposit their eggs in burrows in the cornified epithelium of the host skin. Several developmental stages may be eliminated from the developmental cycle and males can copulate with immature female tritonymphs (e.g. Chirodiscidae).

Ixodoidea, the typical *r*-strategists, deposit hundreds or thousands of eggs on the soil surface

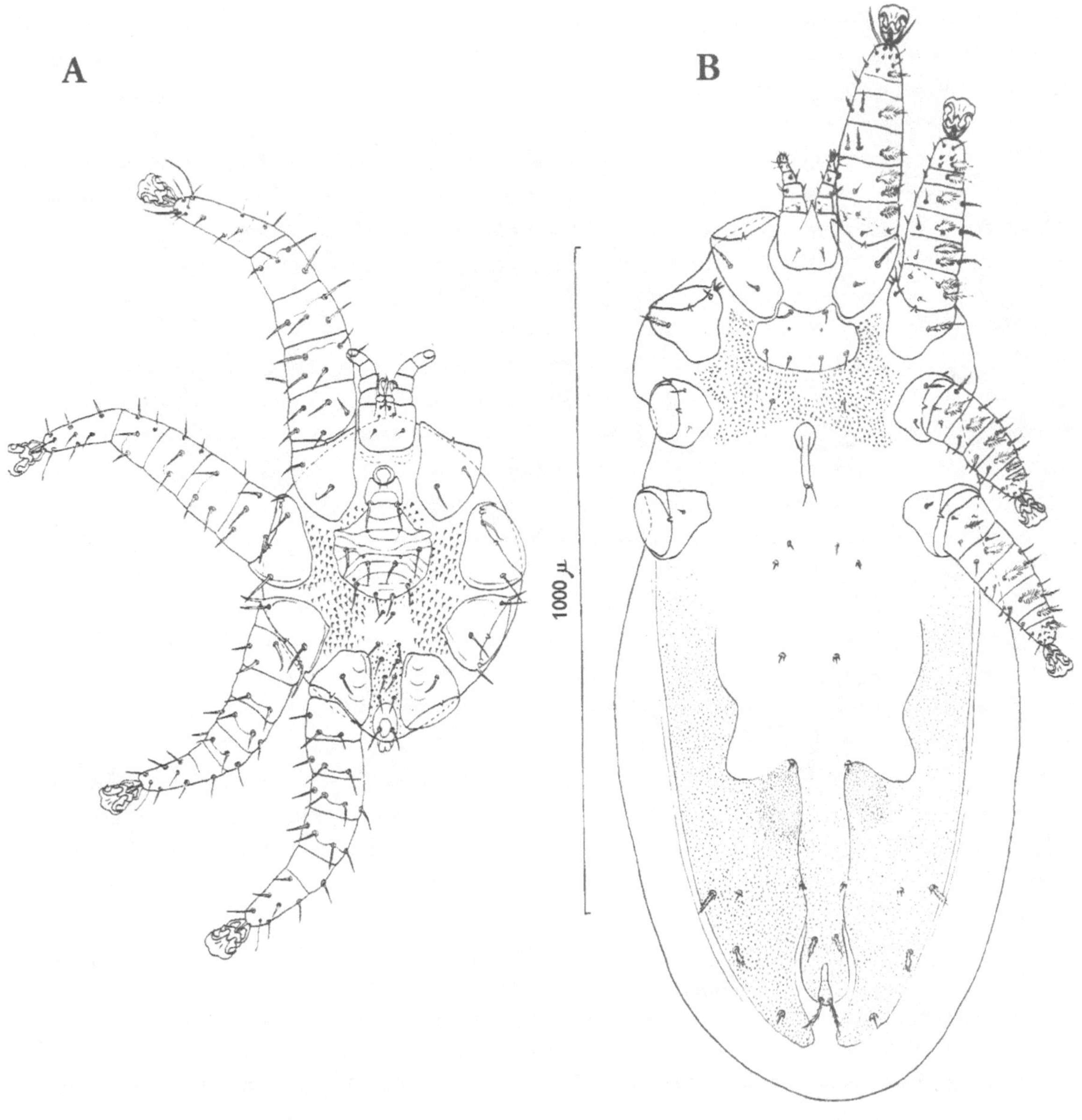

Fig. 5 - *Cameronieta torrei* Dusbabek, male (left) and female (right), ventral view. Enlarged female opisthosoma with chitinized endoskeleton and fine spines in the sternal region help for better fixation on the host wing membrane.

after dropping off the host. Most of developmental stages feed on blood on the hosts, however some exceptions exist. Only larvae are blood-feeders in Antricolinae. Larvae or nymphs of some Ornithodorinae (e.g. *O. moubata, O. savignyi, O. amblus, O. muesebecki*) do not feed at all. In some bat-parasitizing argasids (*Chiropterargas, Paranthricola*), females keep the egg batches and newly hatched larvae on their back which facilitates contact of larvae with roosting bats (Hoogstraal 1985).

Adaptations to feeding

Adaptations of mouthparts

In Mesostigmata there is a distinct trend for elongation of the articles of the cheliceral shaft and narrowing of the cheliceral digits, enabling

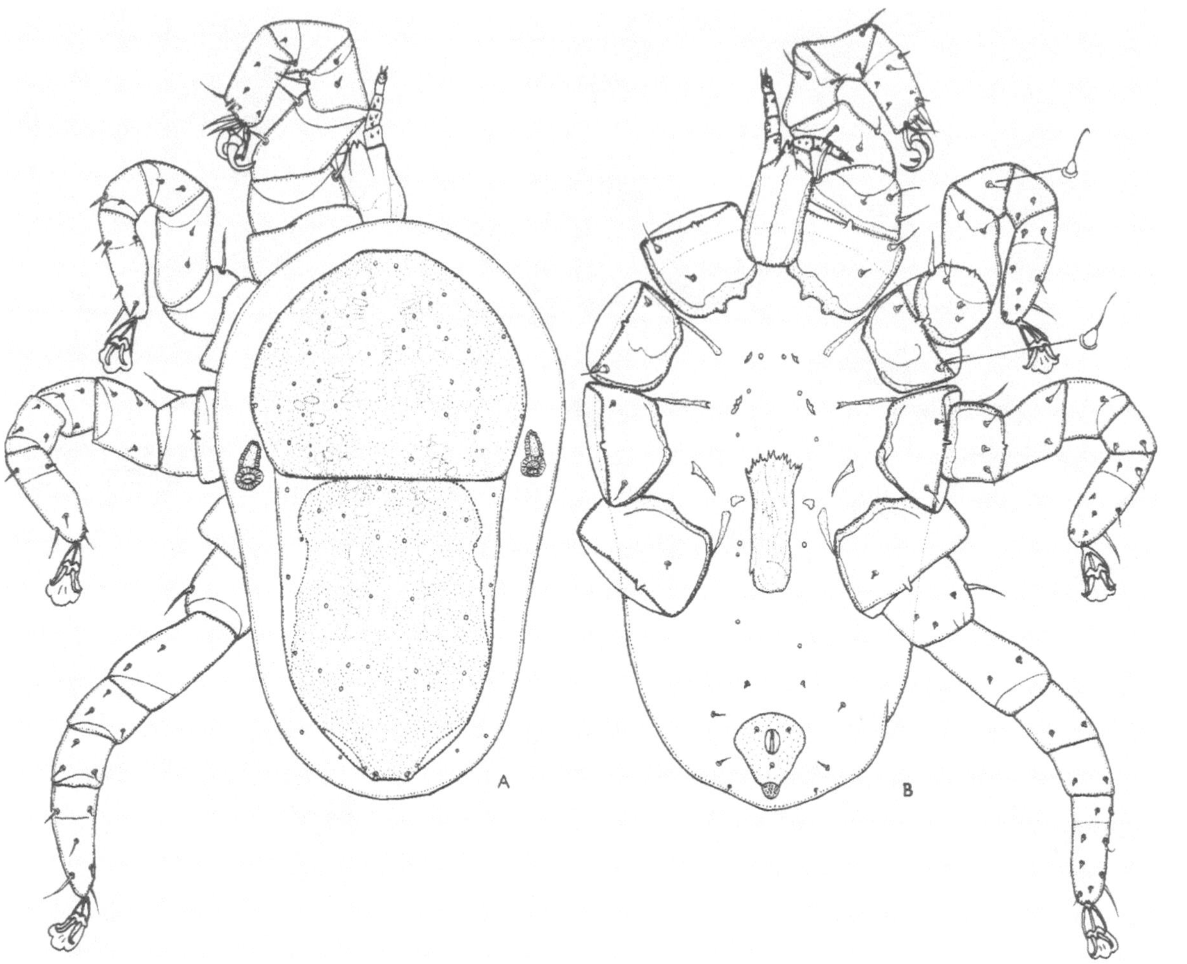

Fig. 6 - *Mesonyssus geotrygoni* Dusbabek, dorsal (A) and ventral (B) view of female, demonstrating the reduction of body and leg setae in endoparasitic rhinonyssid mites.

penetration into host skin and formation of a channel for blood feeding. In Ixodoidea the hypostome is modified into a finger-like denticulated organ (fig. 4A) with a deep groove on its dorsal surface, the food canal, through which blood drawn from the host passes to the mouth and pharynx. Long cheliceral shafts with a small hook-like digit and covered with cheliceral sheaths lie medial to the palps. In some Prostigmata, e.g. Myobiidae, chelicerae are modified into long stilet-like organs for piercing the host skin. The larvae of trombiculid mites have a strong hook-like movable digit of chelicerae, and for feeding they use a special feeding tube (stylostome or histiosiphon) formed by lysis of host tissue with mite lytic enzymes. The feeding tube in some larval Trombidiidae parasitizing insects extends into the hemocoel and is often extensively branched.

Adaptation of cuticle

Engorging a great amount of food (blood or lymph) during host feeding is facilitated by a special structure of the cuticle, enabling a rapid distension of the idiosoma, as apparent mostly in ixodid and argasid ticks. The cuticle forms

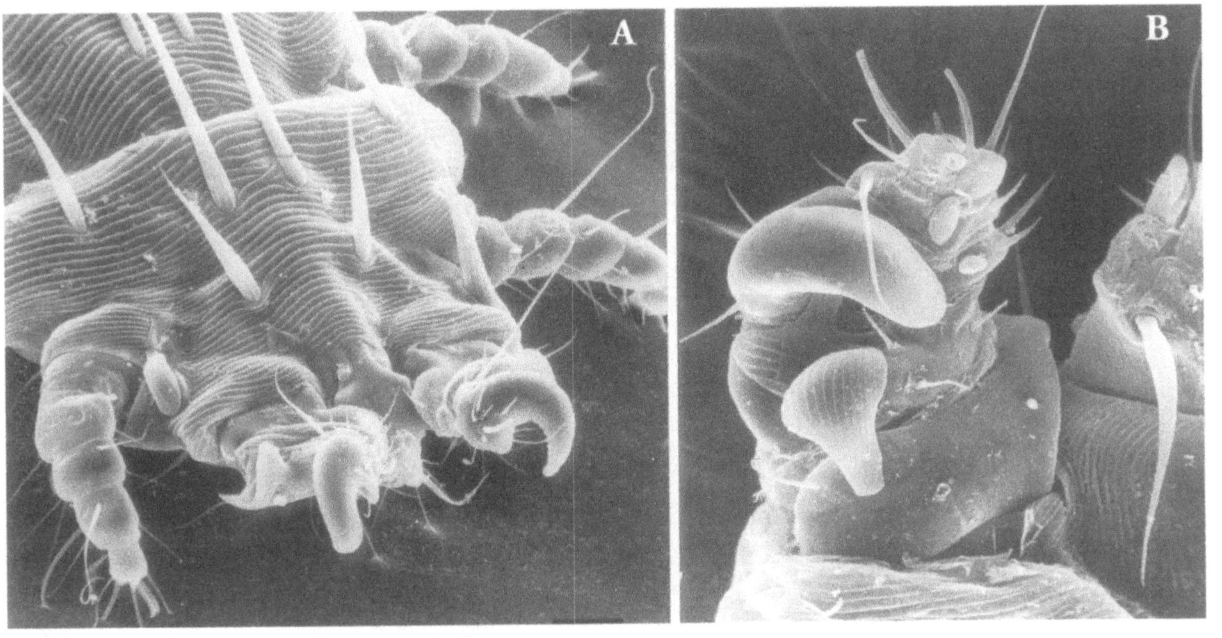

Fig. 7 - *Radfordia lemnina* (C.L. Koch), anterior part of female body, dorsally (left) and *Protomyobia claparedei* (Poppe), leg I ventrally (right). The first pair of legs in both species is strongly reduced and special hook-like clasping tubercles are developed on genu I. Palpal hooks are apparent in *P. claparedei*.

Fig. 8 - *Lawrenceocarpus micropilus* Dusbabek and Cruz, female, lateral view. Body in both sexes is laterally compressed and first two pairs of legs are modified into a clasping organ (after Fain 1973).

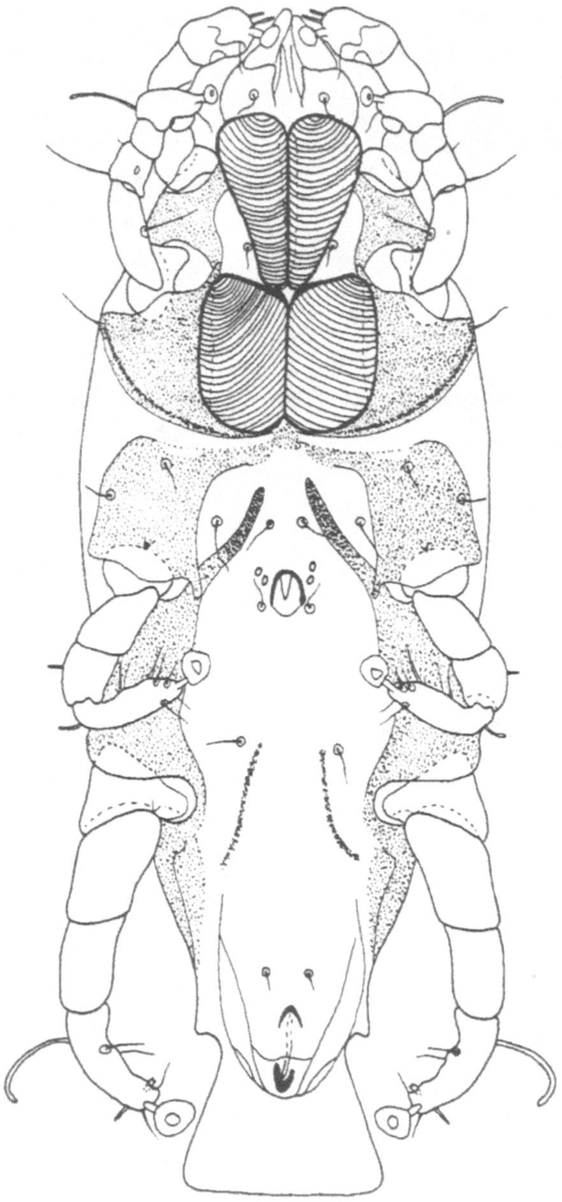

Fig. 9 - *Listrophoroides caudatus* Fain, male, ventral view. Striated membraneous flaps in coxal region I and II act as claspers (after Fain 1972).

parallel microfolds in ixodids and a special star-like structure in argasids, which is able to distend during feeding (fig. 11A-B). The reduction of body sclerotization, mostly of plates or shields, and body setae is also frequent in parasitic forms.

Females of some Heterostigmata, that are parasites or parasitoids (e. g. *Pyemotes),* are capable of physogastry, the broad expansion of opisthosomal integument during development of eggs. This is, however, adaptation to reproduction (Evans 1992).

Adaptation of osmoregulation

In blood and tissue feeding parasitic species (Ixodoidea and others) special salivary glands are developed. The saliva contains many substances enabling blood feeding on the host. Ticks of the family Ixodidae base their osmoregulation on these glands and use them for elimination of the surplus water ingested with blood from the body back into the host tissue. The coxal glands of Argasidae serve the same osmoregulatory function. Moreover, unengorged and engorged ixodid ticks are able to take up water vapor from the subsaturated atmosphere during the long period of starvation and desiccation period (Kahl and Knülle 1988). They are also protected against water loss by a special hydrophobic layer on the exterior surface of the cuticle, which limits transpiration of body water, particularly effective in Argasidae.

Adaptation of digestion

In those species which ingest large quantities of host's tissue fluids and blood, such as some parasitic Mesostigmata and Ixodida, the ventriculus is provided with numerous pairs of caecae, which frequently extend into the leg cavities (e.g. in Ixodoidea, Dermanyssoidea or Spinturnicidae). Both extracellular and intra-cellular digestion occur in Acari but the intracellular digestion prevails in blood-feeders with hematin and guanine as the primary final waste products. In argasid ticks a special adaptation has developed in response to a long period of starvation. The blood-meal digestion consists of three different phases, from which the third one (approximately 10 days after feeding) is a phase of slow digestion. Very few active

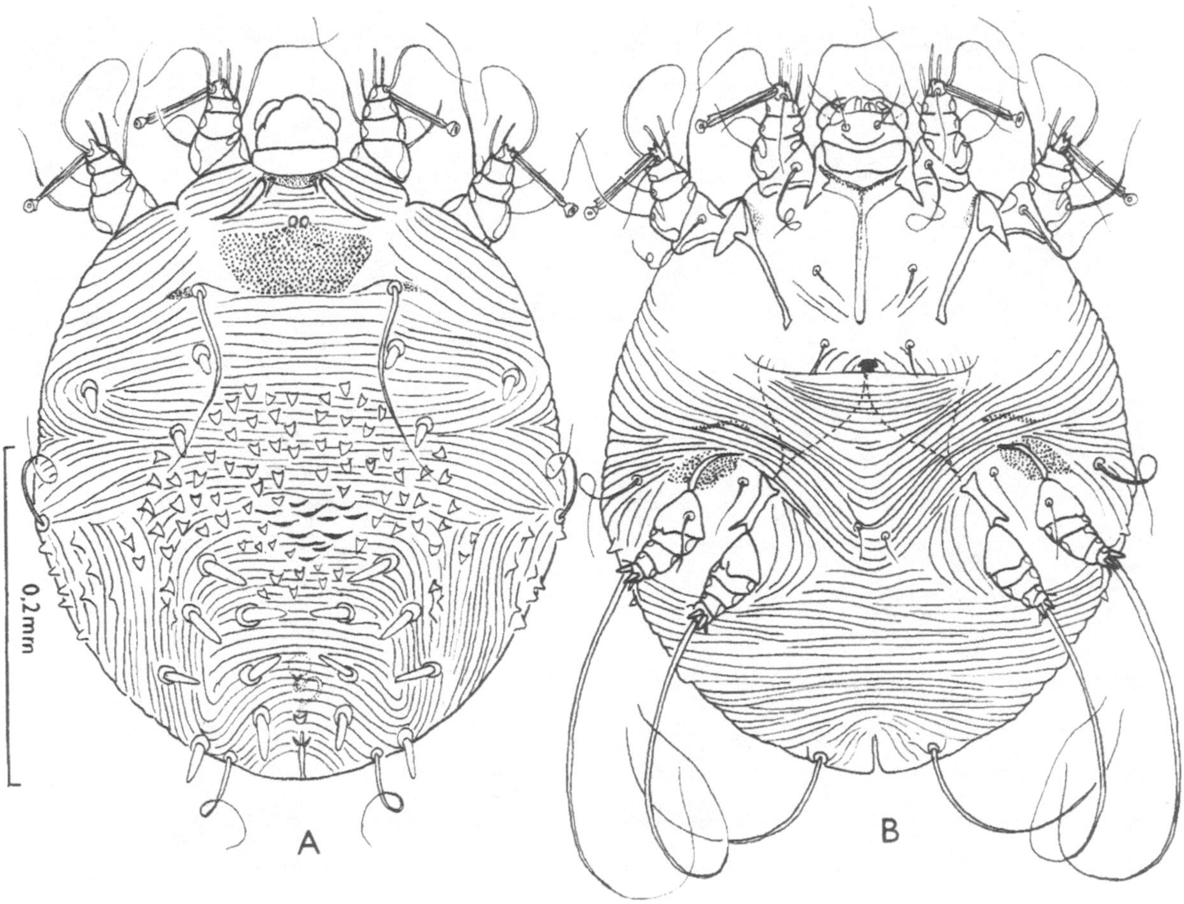

Fig. 10 - *Sarcoptes scabiei* (DeGeer), female from dorsal (A) and ventral (B) view. Globular body form is a characteristic adaptation to endoparasitic way of life.

digestive cells occur in the intestinal epithelium during this phase and digestion practically stops. This enables the ticks to survive periods of starvation for months to years without a blood meal (Tatchell 1964).

Adaptations to host defense mechanisms

Adaptations against haemostasy

Vertebrates prevent blood loss with a variety of interacting haemostatic mechanisms, resulting in blood platelet aggregation mediated by ADP, and blood coagulation. Therefore, blood-feeding arthropods need to have some capacity to inhibit these haemostatic responses. Mechanisms of haemostasy inhibition were recently well summarized by Champagne and Valenzuela (1996) and Bowman *et al.* (1997). Saliva of blood-feeding arthropods contains anticoagulants, anti-platelet and vasodilatory molecules, all of which inhibit host haemostasy and keep blood flowing at the bite site (Ribeiro 1995). Many soft and hard ticks are able to inhibit the blood coagulation cascade of the host at an initial or final phase, inhibiting formation of different factors. Different anticoagulants or different isoforms may work synergistically to ensure that the parasite can secure a blood-meal under different conditions. The main inhibitor of platelet aggregation produced by ticks is *apyrase,* an enzyme that hydrolyses ADP, and *prostaglandins,* which play a role as a vasodilator and inhibitor of production

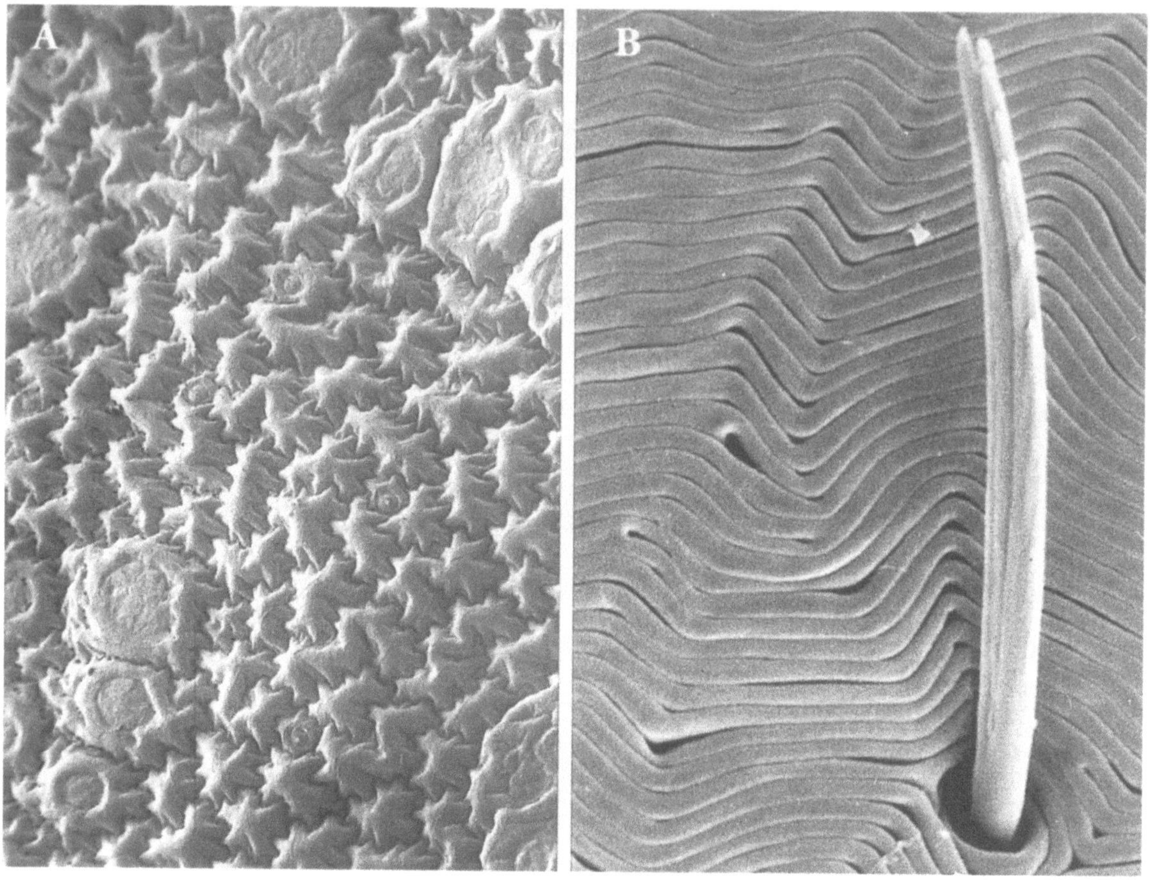

Fig. 11 - Surface structure of epidermis in unengorged females of soft and hard ticks, *Argas polonicus* Siuda et al. (left) and *Ixodes ricinus* (L.) (right). Special star-like structure or parallel microfolds distend during feeding, enabling ingestion of a great amount of blood.

of some cytokines by T_H1 cells (Ribeiro *et al.* 1992). They were recorded in some soft and hard ticks (e.g. *Ornithodoros moubata, Ixodes scapularis, I. holocyclus*, etc.) and other blood-feeding arthropods. Another platelet aggregation inhibitor, *variabilin*, has been reported from saliva of *Dermacentor variabilis* by Wang *et al.* (1996).

A specific disintegrin-like molecule, named *disagregin*, was isolated from saliva of the tick *Ornithodoros moubata* by Karzcewski *et al.* (1994). This peptide was shown to inhibit the aggregation response of a number of agonists including fibrinogen, collagen, platelet activating factor, and thrombin. The formation of thrombin has been reported to be inhibited also by *ixin*, a thrombin inhibitor of *Ixodes ricinus*, by Hoffmann *et al.* (1991). Recently, a specific inhibitor of thrombin, *americanin*, was isolated from salivary glands of *Amblyomma americanum* by Zhu *et al.* (1997). Another peptide, *moubatin*, which is more specific for collagen-mediated aggregation, has been isolated from *Ornithodoros moubata* (Waxman and Connolly 1993).

Adaptations to host immune response

Immune response of vertebrate hosts to feeding of parasitic Acari was studied predominantly in ticks and some skin mites. Tick feeding stimulates host immune regulatory and effector responses involving antigen-presenting cells, cytokines, T lymphocytes, complement, homocy-

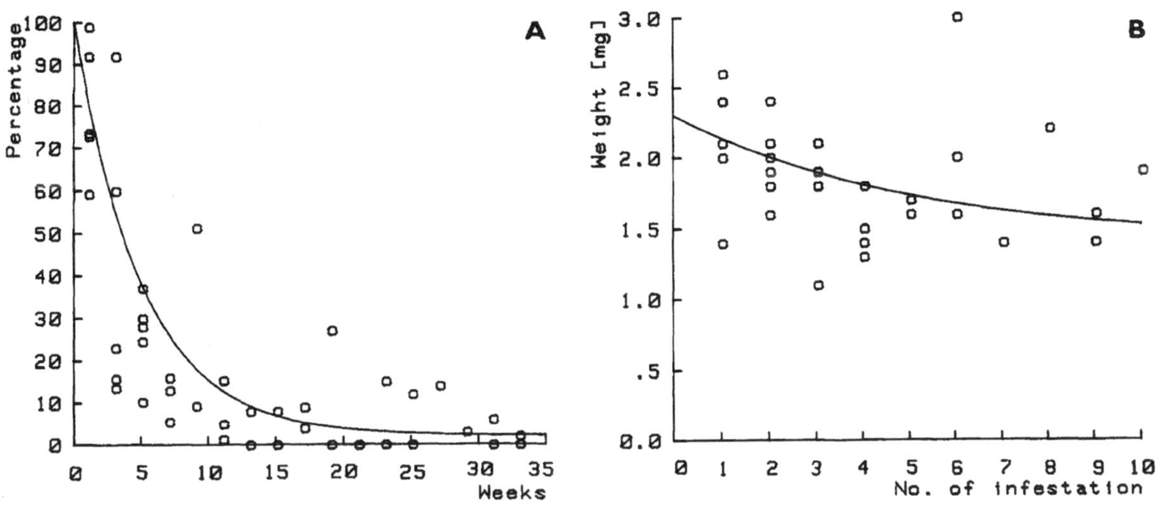

Fig. 12 - Effect of acquired resistance by pigeons to *Argas polonicus* Siuda et al. larval feeding. Apparent is reduced percentage of engorged larvae (A) and reduced engorgement weight (B) following each subsequent infestation.

totropic antibodies and circulating immunoglobulins (Wikel *et al.* 1996). Acquired resistance to ticks is typically expressed as reduced tick engorgement weight, prolonged feeding, impaired production of ova, fewer viable ova, inhibition of molting and death of the feeding ticks (fig. 12).

Ticks can modulate and suppress both innate and adaptive immune responses of hosts. Tick saliva contains inhibitors of the alternative pathway of complement activation and of natural killer cells. *Prostaglandin* E_2 present in tick saliva inhibits neutrophil aggregation (Ribeiro *et al.* 1985). A suppressive effect of salivary gland extract, derived from partially engorged *Ixodes ricinus* females, on important mechanisms of innate immunity, natural killer (NK) cells, interferon and nitric oxide production, was demonstrated by Kopecky *et al.* (1998).

Tick saliva also reduces macrophage cytokine IL 1 and TNF-α elaboration, which impair the earliest steps in development of anti-tick immunity by altering signals to T and B lymphocytes. Furthermore, tick saliva stimulates the production of IL-10, elaborated by T_H2-lymphocytes (Kopecky *et al.* 1999). This contributes to the inhibition of elaboration of pro-inflammatory T_H1-lymphocyte cytokines, e.g. IFNγ and IL-2, which would diminish delayed type hypersensitivity to tick immunogens, and cellular influx at the attachment site. Reduced IFN-γ would also impair macrophage activation (Wikel 1996). Immunosuppressive activity of tick saliva facilitates blood-meal acquisition and in some cases results in an increase of tick feeding success (Dusbabek *et al.* 1995). This may also facilitate the establishment of tick-transmitted pathogens in the host (Brossard and Wikel 1997).

Ixodid ticks produce in haemolymph and salivary glands a unique family of proteins (IGBPs - immunoglobulin-binding proteins), which bind the host immunoglobulins taken up in the blood-meal. These IGBPs act as a "self-defense" system against ingested host immunoglobulins. Moreover, the male-specific IGBPs in *Rhipicephalus appendiculatus* affect positively the feeding site of co-feeding female ticks and facilitate their feeding (Wang *et al.* 1998).

Suppression of circulating IgA in patients infested with *Sarcoptes scabiei* has been reported by Hancock *et al.* (1974). However, the role of this immunoglobulin in scabies mite infestation remains to be defined.

Infestation of canines with the follicle mite, *Demodex canis*, resulted in reduced *in vitro* responsiveness of T lymphocytes to the phytohaemagglutinin mitogen (Wikel *et al.* 1996). Suppressed T lymphocyte responsiveness to Concanavalin A and phytohaemagglutinin mitogens *in vitro* and antibody production were also observed in rabbits infested with *Psoroptes cuniculi* by Uhlir (1991). Suppressed T lymphocyte function can affect both antibody production and cell-mediated immunity.

3. Medical and veterinary aspects of mite and tick parasitism

Medical and veterinary importance of mites and ticks lies in their direct damage to the host by feeding on its body and in transmission of infectious agents. Economic losses caused by mites and ticks on domestic animals are tremendous, estimated in billions of US $. They result from weight loss, reduced weight gain, reduced milk production, reduction in meat, hide, wool and egg production, fetal abortion, and death. Also the indirect losses caused by disease-producing organisms transmitted by arthropod parasites are considerable. Steelman (1976) indicated that in 1965 the losses in livestock caused by ticks were estimated at 65,000,000 US $ annually in cattle production and 4,700,000 US $ annually in sheep production in the USA, and 20,000,000 US $ in cattle production in Australia. The same author referred to the estimation of losses caused only by the cattle scabies mite at 3,000,000 US $ and the sheep scabies mite at 2,000,000 US $ in 1965 in the USA. They were caused by mortality rates, destruction of fleece, loss of body condition caused by stress, and lowered milk and body weight production. These data do not include the expenses connected with prevention, treatment and control of parasitic mites and ticks and tick borne diseases, which raise the expenses for livestock production several times.

Concerning medical importance, Oliver (1996) discussed the statistical data of the Centers for Disease Control and Prevention. According to these data, Lyme disease transmitted by ticks alone forms more than 90% of all reported human vector-borne diseases in the USA. Between 1992 and 1994, 8-11.5 thousand cases of diseases were recorded per year. In the Czech Republic the mean morbidity per 100,000 inhabitants between 1993-1999 was 35.5 for Lyme borreliosis and 5.4 for tick borne encephalitis.

Damage to the host

Bites of a majority of ectoparasitic mites on human skin cause dermatitis with eruptions resembling excoriated papular urticaria, accompanied by pruritus. Sometimes secondary infection with pustule formation occurs. Repeated irritation of skin during infestation with scabies causes pustular and impetiginised lesions of pyoderma. In persons with immunological disorders diffuse erythroderma with hyper/parakeratosis, known as Scabies Norwegica, may occur (Alexander 1984; Nutting 1984).

In mammals, sarcoptids burrowing and feeding in the skin cause irritation and pruritus with consequential skin scratching which leads to inflammation and exudations that form crusts on the skin. The skin wrinkles and thickens with proliferation of connective tissue, which is followed with depilation. Heavy infestation with *Sarcoptes scabiei* or *Notoedres* species may be fatal for the host. Psoroptids puncture the epidermis of the host and feed on lymph and tissue fluid and cause the formation of scabs. As the lesion increases a dry scab forms in the center and is surrounded successively by zones of moist crust and moist, reddened skin. Heavy infestation of cats and dogs by *Otodectes*, penetrating into the ears, may cause convulsions. Two types of cutaneous lesions are known in demodectic mange, squamous and pustular ones. Signs of infestation with *Cheyletiella* often include mild hair loss, scaling, hyperemia, pruritus, serous exudation, and thickening of the skin. A variety of health disorders, including bronchial cough, excessive nasal secretion, lacrimation, listlessness, inappetence, sudden loss of consciousness, and orbital cellulitis occur during infestation of nasal cavities and lungs of monkeys and dogs with lung mites of the genus *Pneumonyssus* and *Pneumonyssoides*.

In birds, matted and discolored feathers and thickened, scabby skin, particularly in the vent region, occur during infestation with *Ornithonyssus* and *Dermanyssus* species. Heavy infestation has been stated to cause anemia, decreased weight gain and egg production, and sometimes death (Yunker 1973). Some Harpyrhynchidae and Hypoderidae form a small nodule in the subcutaneous tissue that calcifies after the mite dies. Members of the family Syringophilidae feed on the calmus inside wing feathers or on the subcutaneous tissue of the wing, leading to feather deformation. Heavy infestation with endoparasitic Cytoditidae, Turbinoptidae, Rhinonyssidae, Ptilonyssidae, parasitizing in the respiratory tract, cause respiratory disorders. Some *Knemidokoptes* burrow beneath the epidermal scales of the legs and feet or invade the feather follicles, skin folds, and smooth epidermis of the face and cere, causing irritation, inflammation, vesicle formation, serous exudation, and crust formation.

Host intoxication by toxins from parasite saliva is known in some hard and soft ticks. Paralysis in people, affecting the lower limbs and ascending within a few hours to the torso, upper limbs and head, may occurs during infestation with *Ixodes holocyclus* in Australia and with *Amblyomma americanum, Dermacentor andersoni,* and *D. variabilis* in North America. In Africa tick paralysis occurs in livestock, sheep and goats after feeding of *Ixodes rubicundus* and *Rhipicephalus evertsi* (Buczek *et al.* 2000). Human toxicosis caused by the tick *Ixodes redikorzevi* has recently been described from Israel (Kassis *et al.* 1997). Feeding by large numbers of larvae of the argasid subgenus *Persicargas* on hens causes paralysis of leg and wing muscles. Heavy infestation with paralysis of respiratory muscles is fatal.

Repeated feeding of mites and ticks on the host causes a hypersensitivity reaction of delayed type and may cause also an anaphylactic shock (Miadonna *et al.* 1982). Allen (1973) characterised the response of sensitized guinea pigs to feeding by *Dermacentor andersoni* as the cutaneous basophil hypersensitivity reaction. This reaction is characterized by basophil and eosinophil infiltration and degranulation connected with histamine deliberation. The Langerhans cells, as antigen presenting cells, and the mast cells participate in this hypersensitivity reaction, mediated by the T_H1 helper lymphocyte subpopulation (Wikel 1996) (fig. 13).

The origin of the arthropod/infectious agent association

Arthropod-borne diseases are caused predominantly by viruses, rickettsiae, bacteria, protozoans and helminths. It is widely believed that most of these infectious agents were already pathogens, commensals or symbionts of vertebrates, before they became transmitted by arthropods (Bushvine 1993). Only after they had penetrated to the host tissue fluid and blood-system, a rich source of nutriment, might they have been transferred to another host by arthropods sucking blood or tissue fluid. One simple way would be via traces of blood on the arthropod's mouthparts and later, penetration of agents into the arthropod body and multiplying there as a secondary parasite. Arthropods began to serve first as simple amplifiers, as many argasid ticks still do. However, more elaborate means of transmission were to develop when salivary glands of the arthropod became infected and thus able to transfer parasites with the next blood meal. Most of the microorganisms have under-

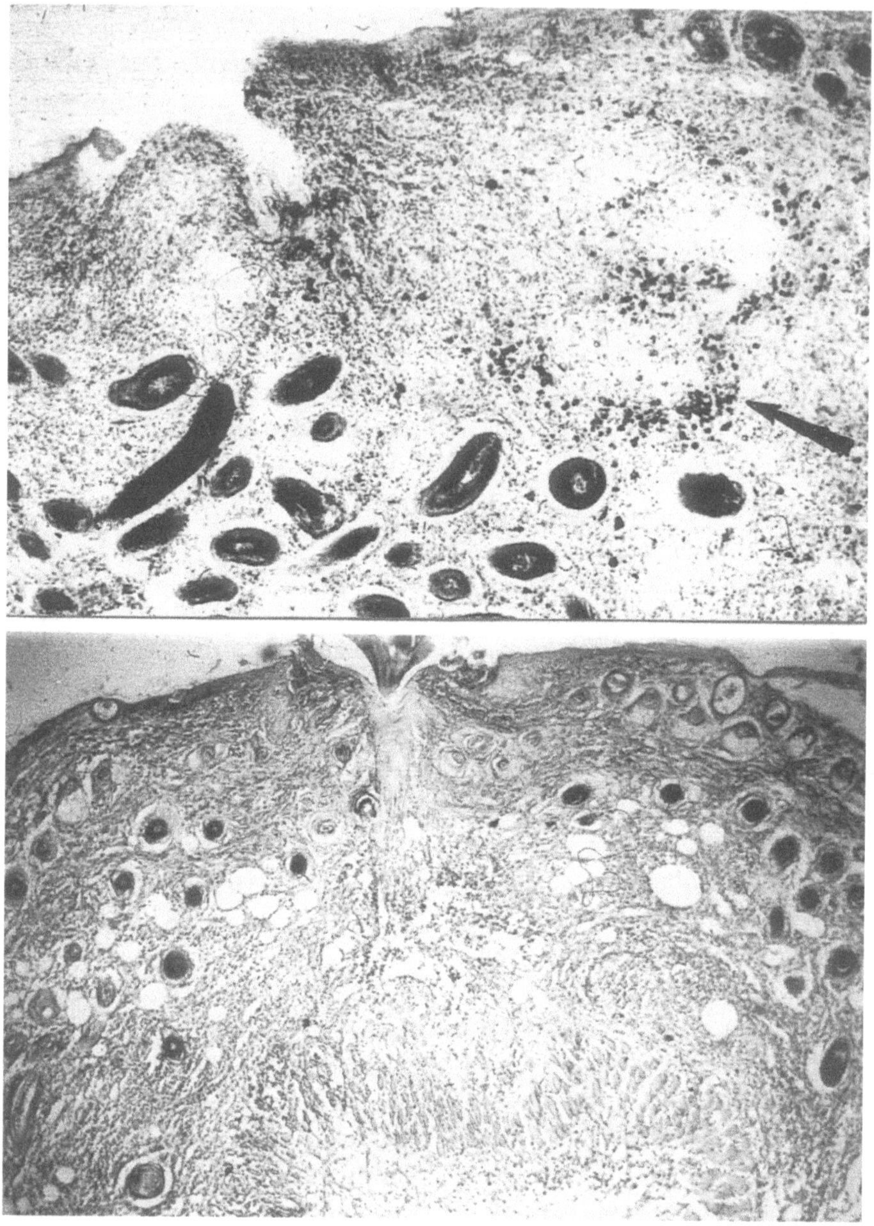

Fig. 13 - Skin section of BALB/c mice at attachment site of *Ixodes ricinus* (L.) nymphs. Secondary infestation, 12h after tick attachment (upper figure). Inflammatory infiltrate is marked and surrounded by numerous mast cells (arrows). After 96h (lower figure) the inflammatory infiltration penetrates throughout skin muscle into subcutaneous tissues. The infiltrate consists mostly of lymphocytes, mast cells are less numerous. Both figures x 40.

gone remarkable transformations and adaptations in arthropods, mostly in their developmental cycle (Bushvine 1993). It is most remarkable in viruses and other parasites transmitted transovarially or transstadially. They must reach the ovaria or some specific organ which is not autolysed during molting, or they have to be resistant to tick digestive enzymes when ingested, or to histolytic enzymes during apolysis and autolysis to survive tick molting (Nuttall *et al.* 1994).

Some arthropod-borne pathogens could originally be parasites or symbionts of arthropods. According to the concept of Hoogstraal (1979), *Borrelia* developed primarily as symbionts of ticks (chiefly Argasidae) and evolved into different strains with the various tick species. Mammals and birds are merely accidental hosts of borreliae, serving as their reservoir and amplifiers following contamination or bites by infected ticks. According to Sonenshine (1993) and Fukunaga *et al.* (1996) there are about 22 known *Borrelia* species causing health problems transmitted by at least 12 species of soft ticks and 6 species of hard ticks. Moreover, the *Borrelia burgdorferi* complex is divided into 8 further genospecies or DNA groups (Oschmann and Kraiczy 1998). Despite their remarkable diversity and specificity for many different tick vector species, DNA homologies of the different borreliae suggest that they may be very closely related (Schmid 1989; Steere 1989). In most of these tick vectors the transovarial or transstadial passage of borreliae to the progeny of infected ticks has repeatedly been confirmed (Burgdorfer and Varma 1967). In soft ticks that emit coxal fluid while engorging, transmission may occur when borrelia-infected coxal fluid contaminates the wounds made by these fast-feeding ticks. However, transmission may also occur via the bite, when tick salivary glands are infected with borreliae, as is the case mostly in hard ticks. Lice, transmitting *Borrelia recurrentis*, are considered to be recent hosts of borreliae as compared to ticks. This insect serves as a culture medium for spirochete development, but transovarial or transstadial transmission does not occur in this vector. The insect remains infected throughout its entire life, but is not capable of transmitting spirochetes either by biting or through its excrements. Transmission to man is of mechanical nature and occurs through contamination of abraded or even normal skin with infectious haemolymph of accidentally crushed lice (Burgdorfer and Varma 1967).

Adaptations within the tick/host/infectious agent system

Ticks have several defense mechanisms against parasites. The first is the secretion of acid phosphatase (*cathepsin*), a digestive proteolytic enzyme released during the preparatory phase of feeding into the gut lumen. Other enzymes, such as lisozymes, phospholipases, asparagine and acid proteases, play a similar role in the gut lumen. The parasites taken up with the blood-meal (e.g. piroplasm or gametes of *Theileria*) have a poor chance of survival in the tick during this phase of digestion. However, the activity of proteolytic enzymes decreases after 72 hours of digestion and the gut content, containing glycogen, lipids and many proteins, becomes a suitable medium for many pathogens. The phagocytic cells of the gut as the second defense mechanism also participate in reduction of the number of parasites in the gut of ticks (Walker 1990). Tick haemocytes, as the third barrier in haemocoel, exhibit coiling phagocytosis of some parasites (e.g. borreliae) (Kuhn *et al.* 1994). Lectins, functional analogues of immunoglobulins in organisms that lack specific immunity, are probably produced and/or stored in tick haemocytes. They contribute to the immune system of ticks by recognition of foreign substances and mediate the coiling phagocytosis in *Ixodes ricinus*. Lectins were also identified and described in soft ticks *Ornithodoros tartakovskyi*, *O. tholozani* and *Argas polonicus* (Grubhoffer and Jindrák 1998). According to Johns *et al.* (1998) the hemolymph of *Dermacentor variabilis* contains different antimicrobial proteins, one of which tentatively identified as tick lysozyme.

It has been suggested that repeatedly infested hosts, which acquired a resistance to ticks, impaired subsequent pathogen transmission. This has been well documented in experiments using Thogoto virus, tick-borne encephalitis virus, *Francisella tularensis, Borrelia burgdorferi, Babesia bovis, B. canis, Anaplasma marginale, Theileria parva, Theileria annulata* and *Trypanosoma congolense,* (Dusbabek *et al.* 1994; Brossard and Wikel 1997). Host resistance to tick feeding may result in an unfavorable environment for pathogen development at the tick-attachment site, and/or host immune effector elements, such as antibodies, may neutralize tick immunosuppressive factors that facilitate pathogen transmission (Wikel 1996).

On the other hand, studies on certain tick-borne viruses have clearly shown that syringe inoculation to a naive host does not reflect the transmission efficiency of the infected arthropod vector. The ability of ticks to effectively transmit a variety of pathogens to the naive primary host, which did not acquire the anti-tick resistance, might result from their ability to modulate the primary host defenses (Wikel *et al.* 1999). Jones *et al.* (1991 and 1992) supposed the presence in tick saliva of one or more proteins, synthesized by salivary glands during feeding, which modify the skin site of attachment to facilitate blood-feeding and enable virus transmission. Such protein(s) is presented as a saliva-activated transmission factor (SAT factor).

Recent studies indicate that immunomodulatory effects induced by tick salivary gland extracts may provide conditions potentiating virus transmission. The SAT factor probably suppresses the antiviral activity of natural killer cells (NK cells), possibly via inhibition of cytokine production (Fuchsberger *et al.* 1999). Jaworski *et al.* (1999) recently identified the calcium-binding protein, *calreticulin,* in the saliva of two tick species, *Amblyomma americanum* and *Dermacentor varabilis,* and they supposed that this protein may influence the transmission of pathogens via the blood-meal. The *calreticulin* bound by antibody in the host can dramatically change the feeding site and the authors suggest some relation between *calreticulin* and the SAT factor.

References

Alexander J.O'D., 1984 - Arthropods and Human Skin. Springer-Verlag, Berlin, Heidelberg, New York, Tokyo: 422 pp.

Allen J.R., 1973 - Tick resistance: Basophils in skin reaction of resistant guinea pigs. *International Journal of Parasitology*, 3: 195-200.

Beklemishev V.N., 1970 - Biological Basis of Comparative Parasitology. Nauka Publishing House, Moscow: 502 pp. (In Russian).

Bowman A.S., Coons L.B., Needham G.R., Sauer J.R., 1997 - Ticks saliva: recent advances and implications for vector competence. *Medical and Veterinary Entomology*, 11: 277-285.

Brossard M., Wikel S.K., 1997 - Immunology of interactions between ticks and hosts. *Medical and Veterinary Entomology*, 11: 270-276.

Bruin J., van der Geest L.P.S., Sabelis M.W. (eds.), 1999 - Ecology and evolution of the Acari. Kluwer Academic Publishers, Dordrecht, Boston and London: 677 pp.

Buczek A., Sodowska H., Magdon T., 2000 - Tick paralysis - cause, symptomatic picture and prevention. In: Buczek A., Blaszak C. (eds.), Stawonogi pasozytnicze i alergogenne. KGM Lublin: 183-190 (in Polish with English summary).

Burgdorfer W., Varma M.G.R., 1967 - Trans-stadial and transovarial development of disease agents in arthropods. *Annual Revue of Entomology*, 12: 347-379.

Bushvine J.R., 1993 - Disease Transmission by Insects. Springer-Verlag, Berlin, Heidelberg, New York: 361 pp.

Champangne D.E., Valenzuela J.G., 1996 - Pharmacology of haematophagous arthropod saliva. In: Wikel S.K. (ed.), The Immunology of Host-Ectoparasitic Arthropod Relationships. CAB International, Wallingford: 85-106.

Dusbabek F., Borsky I., Jelinek F., Uhlir J., 1995 - Immunosuppression and feeding success on *Ixodes ricinus* nymphs on BALB/c mice. *Medical and Veterinary Entomology*, 9: 133-140.

Dusbabek F., Kopecky J., Uhlir J., 1994 - Effect of repeated infestation of BALB/c mice with *Ixodes ricinus* nymphs of tick-borne encephalitis virus infection. *Folia Parasitologica*, 41: 312-316.

Evans G.O., 1992 - Principles of Acarology. C.A.B. International, Wallingford: 563 pp.

Fain A., 1972 - Les listrophorides en Afrique au sud du Sahara (Acarina: Sarcoptiformes) III. Famille Atopomelidae. *Musee Royal de l'Afrique Centrale* - Tervuren, Belgique. Annales-Serie IN-8 - Sciences zoologiques, 197: 1-200.

Fain A., 1973 - Les listrophorides en Amerique Neotropicale (Acarina: Sarcoptiformes) I. Familles Listrophoridae et Chirodiscidae. *Bulletin de l'Institute royal des Sciences naturelles de Belgique*, Bruxelles. Entomologie, 49(6): 1-149.

Filippova N.A., 1977 - Ixodid ticks of the subfamily Ixodinae. Fauna of the USSR, Arachnida. *Nauka Publishing House*, Leningrad, vol. 4(4): 396 pp.

Fuchsberger N., Imanishi J., Labuda M., Kubes M., Kita M., Hajnicka V., Zuffova E., Nuttall P.A., 1999 - Role of tick salivary gland products in virus transmission: immunomodulatory mechanisms. In: Needham G.R., Mitchell R., Horn D.J., Welbourn W.C. (eds.), Acarology IX, Symposia. The Ohio Biological Survey, Columbus, vol. 2: 433-437.

Fukunaga M., Okada K., Nakao M., Konishi T., Sato Y., 1996 - Phylogenetic analysis of *Borrelia* species based on flagelin gene sequences and its application for molecular typing of Lyme disease borreliae. *International Journal of Systematic Bacteriology*, 46: 898-905.

Grubhoffer L., Jindrak L., 1998 - Lectins and tick-pathogen interactions: a minireview. *Folia Parasitologica*, 45: 9-13.

Hancock B.W., Ward A.M., Path M.R.C., 1974 - Serum immunoglobulin in scabies. *Journal of Investigative Dermatology*, 63: 482-484.

Hoffmann A., Walsmann P., Reisener G., Paints M., Markwardt F., 1991 - Isolation and characterization of a thrombin inhibitor from the tick *Ixodes ricinus*. *Pharmazie*, 46: 209-212.

Hoogstraal H., 1979 - Ticks and spirochetes. *Acta tropica*, 36: 133-136.

Hoogstraal H., 1985 - Argasid and nuttalliellid ticks as parasites and vectors. *Advances in Parasitology*, 24: 135-238.

Houck M.A. (ed.), 1994 - Mites: ecological and evolutionary analyses of life history patterns. Chapman & Hall, New York and London: 357 pp.

Jaworski D.C., LaDu A.C., Luo D., Muller M.T., Needham G.R., 1999 - The role of calreticulin in parasitic associations. In: Needham G.R., Mitchell R., Horn D.J., Welbourn W.C., (eds.), Acarology IX, Symposia. The Ohio Biological Survey, Columbus, vol. 2: 445-450.

Johns R., Sonenshine D.E., Hynes E.L., 1998 - Control of bacterial infection in the hard tick *Dermacentor variabilis* (Acari: Ixodidae): Evidence for the existence of antimicrobial protein in tick hemolymph. *Journal of Medical Entomology*, 35: 458-464.

Jones L.D., Hodgson E., Nuttall P.A., 1991 - Characterization of tick salivary gland factor(s) that enhance Thogoto virus transmission. *Archives of Virology*, Supplement, 1: 227-234.

Jones L.D., Hodgson E., Williams T., Higgs S., Nuttall P.A., 1992 - Saliva-activated transmission (SAT) of Thogoto virus: relationship with vector potential of different hematophagous arthropods. *Medical and Veterinary Entomology*, 6: 261-265.

Kahl O., Knülle W., 1988 - Water vapour uptake from subsaturated atmosphere by engorged immature ixodid ticks. *Experimental and Applied Acarology*, 4: 73-83.

Karczewski J., Endris R., Connolly T.M., 1994 - Disagregin is a fibrinogen receptor antagonist lacking the Arg-Gly-Asp sequence from the tick, *Ornithodoros moubata*. *Journal of Biological Chemistry*, 269: 6702-6708.

Kassis I., Ioffe-Uspensky I., Uspensky I., Mumcuoglu K.Y., 1997 - Human toxicosis caused by the tick *Ixodes redikorzevi* in Israel. *Israel Journal of Medical Sciences*, 33: 760-761.

Klompen J.S.H., Black IV W.C., Keirans J.E., Oliver J.H.Jr., 1996 - Evolution of ticks. *Anual Review of Entomology*, 41: 141-161.

Kopecky J., Kuthejlova M., 1998 - Suppressive effect of *Ixodes ricinus* salivary gland extract on mechanisms of natural immunity *in vitro*. *Parasite Immunology*, 20: 169-174.

Kopecky J., Kuthejlova M., Pechova J., 1999 - Salivary gland extract from *Ixodes ricinus* ticks inhibits production of interferon-γ by the upregulation of interleukin-10. *Parasite Immunology*, 21: 351-356.

Kuhn K.-H., Rittig M., Haupl T., Burmester G.R., 1994 - Haemocytes of the hard tick *Ixodes ricinus* express coiling phagocytosis of *Borrelia burgdorferi*. *Developmental and Comparative Immunology*, 18: 15.

Leonovich S.A., 1986 - The orientational behavior of the tick *Hyalomma asiaticum* in a desert. *Parazitologiya*, 20: 431-440 (in Russian).

Mayr E., 1957 - Evolutionary aspects of host specificity among parasites and vertebrates. In: Premier Symposium sur la spécificité parasitaire des parasites de vertébrés. *International Union of Biological Sciences*, series B, 32: 7-14.

Miadonna A., Tedeschi A., Leggieri E., Falagiani P., Nazzari M., Manzoni M., Zanussi C., 1982 - Anaphylactic shock caused by allergy to the venom of *Argas reflexus*. *Annals of Allergy*, 49: 293-294.

Nuttall P.A., Jones L.D., Labuda M., Kaufman W.R., 1994 - Adaptation of arboviruses to ticks. *Journal of Medical Entomology*, 31: 1-9.

Nutting W.B., 1984 - Mammalian Diseases and Arachnids. Vol. I + II, CRC Press Inc., Boca Raton, Florida: 277 pp + 280 pp.

Oliver J.H.Jr., 1989 - Biology and systematics of ticks (Acari: Ixodida). *Annual Review of Entomology and Systematics*, 20: 397-430.

Oliver J.H.Jr., 1996 - Importance of systematics to public health: ticks, microbes, and disease. *Annales of the Missouri Botanical Garden*, 83: 37-46.

Oschmann P., Kraiczy P. (eds.), 1998 - Lyme-Borreliose und Frühsommer-Meningoenzephalitis. Uni-Med Verlag AG, Bremen: 136 pp.

Osterkamp J., Wahl U., Schmalfuss G., Haas W.E., 1999 - Host-odour recognition in two tick species is coded in a blend of vertebrate volatiles. *Journal of Comparative Physiology* A, 185: 59-67.

Radovsky F.J., 1968 - Evolution and adaptive radiation of Gamasina on vertebrates (Acarina: Mesostigmata). *Parazitologiya* (Leningrad), 2: 124-136 (in Russian).

Radovsky F.J., 1985 - Evolution of mammalian mesostigmate mites. In: Ke Chung Kim (ed.), Coevolution of parasitic arthropods and mammals. John Wiley & Sons, Inc., New York: 441-504.

Radovsky F.J., 1994 - The evolution of parasitism and the distribution of some dermanyssoid mites (Mesostigmata) on vertebrate hosts. In: Houck M.A.

(ed.), Mites: ecological and evolutionary analyses of life history patterns. Chapman & Hall, New York and London: 186-217.

Ribeiro J.M.C., 1995 - Blood-feeding arthropods: live syringes or invertebrate pharmacologists? *Infectious Agents Diseases*, 4: 143-152.

Ribeiro J.M.C., Evans P.M., MacAwain J.L., Sauer J., 1992 - *Amblyomma americanum:* Characterization of salivary prostaglandins E_2 and F_2 by RP-HPLC/bioassay and gas chromatography-mass spectrometry. *Experimental Parasitology*, 74: 112-116.

Ribeiro J.M.C., Makoul G.T., Levine J., Robinson D.R., Spielman A., 1985 - Antihaemostatic, antiinflammatory and immunosuppressive properties of the saliva of the tick *Ixodes dammini*. *Journal of Experimental Medicine*, 161: 332-344.

Schmid G.P., 1989 - Epidemiology and clinical similarities of human spirochetal diseases. *Revue of Infectious Disseases*, 2, (Suppl. 6): S1460-S1468.

Sonenshine D.E., 1991 - Biology of Ticks. Oxford University Press, New York, Vol. 1: 447 pp.

Sonenshine D.E., 1993 - Biology of Ticks. Oxford University Press, New York, Vol. 2: 465 pp.

Steelman D.C., 1976 - Effects of external and internal arthropod parasites on domestic livestock production. *Annual Review of Entomology*, 21: 155-178.

Steere A., 1989 - Medical Progress. Lyme Disease. *New England Journal of Medicine,* 321: 586-596.

Steulet P. Guerin P.M., 1992 - Perception of breath components by the tropical bont tick, *Amblyomma variegatum* Fabricius (Ixodidae). 1. CO_2-excited and CO_2-inhibited receptors. *Journal of Comparative Physiology* A, 170: 665-676.

Steulet P., Guerin P.M., 1994 - Identification of vertebrate volatiles stimulating olfactory receptors on tarsus I of the tick *Amblyomma variegatum* Fabricius (Ixodidae). Receptors within the Haller's organ capsule. *Journal of Comparative Physiology* A, 174: 27-38.

Tatchell R., 1964 - Digestion in the tick *Argas persicus* Oken. *Parasitology*, 54: 423-440.

Uhlir J., 1991 - Humoral and cellular immune response of rabbits to *Psoroptes cuniculi*, the rabbit scab mite. *Veterinary Parasitology*, 40: 325-334.

Walker A.R., 1990 - Parasitic adaptations in the transmission of *Theileria* by ticks. A review. *Tropical Animal Health and Production*, 22: 23-33.

Wang H., Coons L.B., Taylor D.B., Stevens S.E., Jr., Gartner T.K., 1996 - Variabilin, a novel RGD-containing antagonist of GPIIb-IIIa and platelet aggregation inhibitor from the hard tick *Dermacentor variabilis*. *Journal of Biological Chemistry*, 271: 17785-17790.

Wang H., Paesen G.C., Nuttall P.A., 1998 - Male ticks help their mates to feed. *Science*, 391: 753-754.

Waxman L., Connolly T.M., 1993 - Isolation of an inhibitor selective for collagen-stimulated platelet aggregation from the soft tick *Ornithodoros moubata*. *Journal of Biological Chemistry*, 268: 5445-5449.

Wikel S.K., 1996 - Host immunity to ticks. *Annual Review of Entomology*, 41: 1-22.

Wikel S.K., Ramachandra R.N., Bergman D.K., 1996 - Arthropod modulation of host immune responses. In: Wikel S.K. (ed.), The Immunology of Host-Ectoparasitic Arthropod Relationships. CAB International, Wallingford: 107-130.

Wikel S.K., Ramachandra R.N., Bergman D.K., 1999 - Tick induced immunomodulation of the host. In: Needham G.R., Mitchell R., Horn D.J., Welbourn W.C. (eds.), Acarology IX, Symposia. The Ohio Biological Survey, Columbus, vol. 2: 423-426.

Yoder J.A., Atwood A.D., Steven B.W., 1998 - Attraction to squalene by ticks (Acari: Ixodidae): First demonstration of a host-derived attractant. *International Journal of Acarology*, 24: 143-147.

Yunker C.E., 1973 - Parasites of endothermal laboratory animals. In: Flynn R.J. (ed.), Parasites of Laboratory Animals. Iowa State University Press, Ames: 425-492.

Zhu K., Bowman A.S., Brigham D.L., Essenberg R.C., Dillwith J.W., Sauer J.R., 1997 - Isolation and characterization of americanin, a specific inhibitor of thrombin, from the salivary glands of the lone star tick *Amblyomma americanum* (L.). *Experimental Parasitology*, 87: 30-38.

Comparison of rabbit skin changes after feeding of *Ixodes ricinus* (L) and *Dermacentor reticulatus* (Fabr.)

A. Buczek[1], A. Kusmierz[1], K. Olszewski[1], L. Buczek[1], K. Czerny[2], M. Lancut[2]
[1] *Department of Biology and Parasitology, Medical University, Lublin, Radziwillowska 11, PL-20-080 Lublin, Poland, parasit@panaceum.am.lublin.pl*
[2] *Department of Histology and Embryology, Medical University, Lublin, Radziwillowska 11, PL-20-080 Lublin, Poland*

Key words: *Dermacentor reticulatus*, *Ixodes ricinus*, skin, histology.

Abstract: The study presents rabbit skin changes caused by feeding of two species of ixodid tick, *Ixodes ricinus* (Linnaeus, 1758) and *Dermacentor reticulatus* (Fabricius, 1794). The skin lesions were examined by use of light and transmission electron microscope. The results indicate different type of skin reactions and more extensive lesions after feeding by *I. ricinus*.

Introduction

Ixodes ricinus (Linnaeus, 1758) and *Dermacentor reticulatus* (Fabricius, 1794), belonging to the Ixodidae (hard ticks) are the most dangerous, and prevalent ticks in Central Europe. They have a wide host range and attack many species of wild and domestic animals, and humans; they serve as vectors of many viruses, rickettsiae, bacteria and protozoa. Besides, a bite of the ixodid tick may cause general reactions (tick paralysis) and local skin changes. The aim of this study was to compare the skin changes caused by *I. ricinus* and *D. reticulatus* by means of light and transmission electron microscope examination.

Materials and methods

Ticks and rabbits

Ixodes ricinus and *D. reticulatus* adults were collected by the flagging method near Lublin (South-Eastern Poland). They were maintained in a laboratory for some days in rearing chambers at ambient temperature of 25°C and at ca. 90-100% R.H., e. g. under similar conditions occurring in habitats of these tick species. Female New Zealand white rabbits from a colony maintained at the Zootechnical Experimental Institute in Chorzelów were used as hosts of both tick species. Rabbits were 4 months old and weighed 2.5-3 kg at the

F. Bernini, R. Nannelli, G. Nuzzaci, E. de Lillo (eds.), Acarid Phylogeny and Evolution. Adaptations in mites and ticks, 419-424.

beginning of the experiment. They exhibited conventional microbial quality and had no previous contact with ticks. During the course of the experiment rabbits were kept under conventional laboratory conditions with full access to water and food.

Experimental design

Each experimental group of rabbits comprised 3 animals. Rabbits from both groups were infested with 10 adult females and 3 adult males of *I. ricinus* or *D. reticulatus*. During infestation, the ticks were confined in a cloth bag placed on the back of each rabbit. The duration of feeding was between 10-12 days for both *I. ricinus* and *D. reticulatus*. Ticks were allowed to feed until detachment.

Histology

Immediately after tick detachment pieces of skin of feeding sites measuring 5 x 5 mm were biopsied. For light microscope examination the skin flanks were fixed in 10% buffered formalin and processed by conventional techniques to paraffin sections that were finally stained with haematoxylin and eosin (H+E). The skin samples (1.5 x 1.5 mm size) for electron microscope analysis were fixed with buffered 2.5% glutaraldehyde and buffered 1% OsO_4, dehydrated in graded concentrations of ethyl alcohol and embedded in Durcupan. The ultrathin sections (60-80 nm thick) prepared by a Reichert Ultracut were stained with uranyl acetate and lead citrate and examined with a transmission electron microscope Tesla BS-500.

Results

After the detachment of *D. reticulatus* female, histopathological changes have been observed in the epidermis and dermis. In the light microscope examination, at the site of tick attachment thinner epidermis with reduction of cell layers (fig. 1) and epidermal focal erosion were observed. Disorders in germinal stratum consist of intracellular oedema,

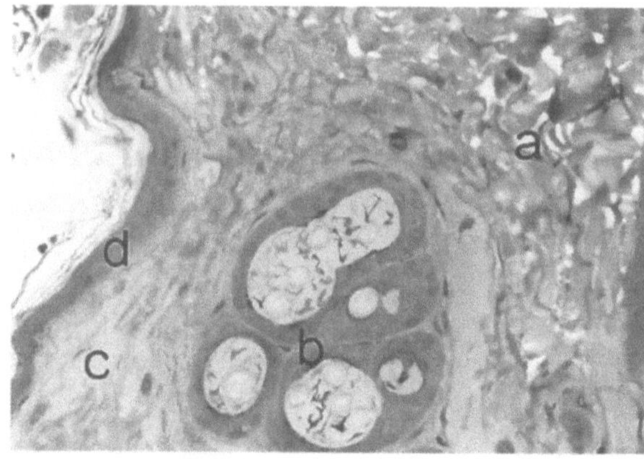

Fig. 1 - Light micrograph: changes in the epidermis and dermis after feeding of *Dermacentor reticulatus*: a - enlargement of intercellular spaces, b - hair follicles with lack of intercellular borders, c - oedema, d - thin epidermis with slightly distinguished cellular borders (400 x).

irregular shape and arrangement of nuclei. Observations in the dermis included: enlargements of intercellular spaces (fig. 1), oedema, and focal necrosis with accompanying inflammatory infiltration containing numerous eosinophils and lipid accumulation. Besides, changes in collagen fibres characterized by thickening, fragmentation and irregularity were exhibited. Moreover, walls of hair follicles were wide and their internal structure was disturbed (changes in nuclei and loss of cellular borders) (fig. 1). There were also wide basement membranes of the sheath of hair follicles.

Electron microscope examination showed epidermis lying directly on clearly distinguished, continuous basement membrane characterized by numerous semi-desmosome junctions. Germinal stratum cells had large, oval nuclei with evenly distributed chromatin. The nuclear membrane contained a lot of stomata (hollows) and open nuclear pores. Within the cytoplasm dilated endoplasmic reticulum were observed as oval vacuoles in the vicinity of the nucleus. Spine cells had oval, flat nuclei with the nuclear membrane forming numerous stomata and large cisternae of endoplasmic reticulum in cytoplasm. These cells, focally exhibited the presence of diluted cyto-

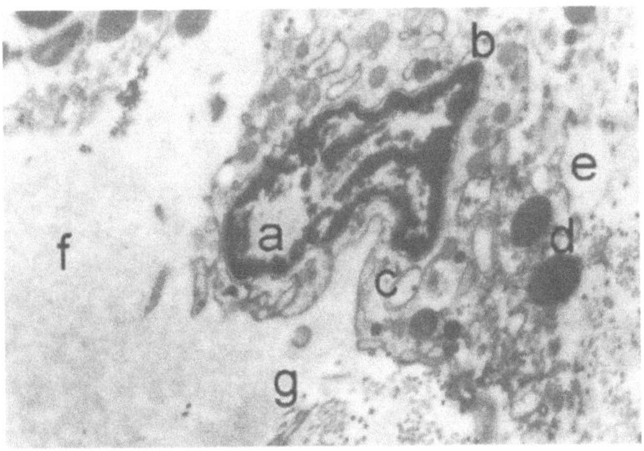

Fig. 2 - Transmission electron micrograph: changes in the dermis after feeding of *Dermacentor reticulatus*: a, fibrocyte nucleus containing dense granules of chromatin; b, enlarged space between two layers of nuclear membrane; c, swollen mitochondrion; d, lysosome; e, wide intercellular space; f, intercellular oedema; g, small collagen fibers (8000 x).

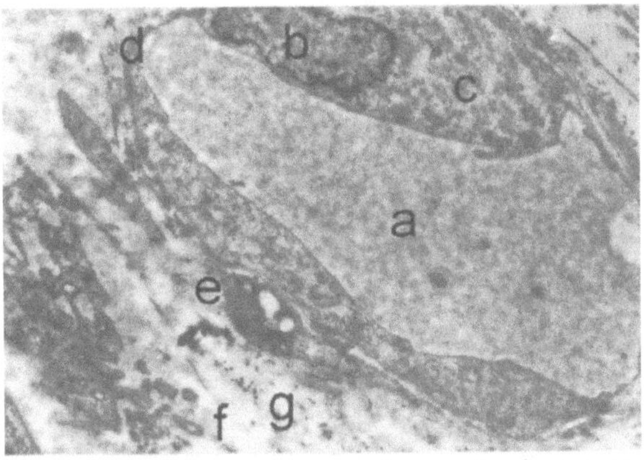

Fig. 3 - Transmission electron micrograph: changes in blood capillaries of the dermis after feeding of *Dermacentor reticulatus*: a, lumen of capillary; b, nucleus of endothelial cell; c, vacuolated cytoplasm of endothelial cell; d, basement membrane; e, pericyte; f, pericapillar oedema; g, few small collagen fibers (6000 x).

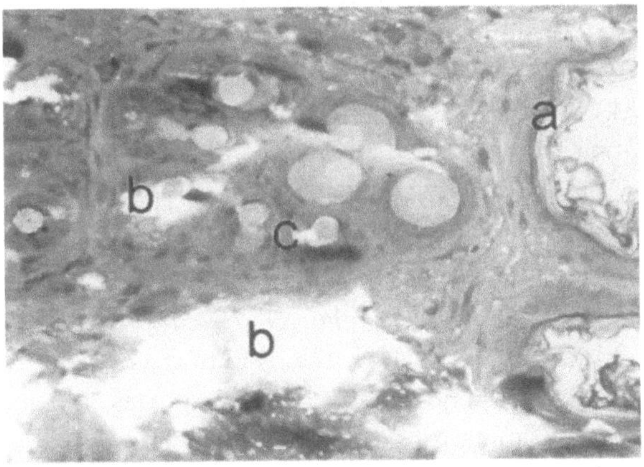

Fig. 4 - Light micrograph: changes in the epidermis and dermis after feeding of *Ixodes ricinus*: a, thin epidermis lying on smooth basement membrane; b, enlargement of intercellular spaces; c, hair follicle with wide, fragmentated basement membrane and disintegrated internal structure (400 x).

plasm characterized by an electronically bright structure. Within intercellular spaces there were small enlargements. Strongly flattened cells of stratum granulosum did not have any nuclei, and their cytoplasm exhibited numerous bundles of tonofilaments and keratohyaline granules. The dermis below the basement membrane of the epidermis contained some collagen fibres and homogenous protein effusion. There were observed changes in fibrocyte structures (fig. 2). Capillaries of the dermis were filled with homogenous protein substance (fig. 3). Endothelial cells, exhibiting normal structure, contained oval nuclei and single mitochondrion. The basement membrane of the endothelium ran continuously. There were a few collagen fibres underneath. Perivascular perfusion as electronically bright protein substance was observed. In the upper part of the dermis abundant accumulation of electronically bright protein substance containing numerous mononuclear cells (infiltration of eosinophils) was seen.

Similarly to *D. reticulatus,* feeding of *I. ricinus* females caused lesions in the epidermis and the dermis. However, the histopathological changes in rabbit skin after feeding of *I. ricinus* had a greater range of distribution.

Light microscope analysis of *I. ricinus* feeding sites indicated a lower thickness of epidermis with slightly distinguished cell membranes and cellular structures, and also desquamation. The basement membrane was smooth (fig. 4). Enlargement of intercellular spaces and nonspecific inflammatory infiltration were observed in the dermis (fig. 4). Purulent changes, with the

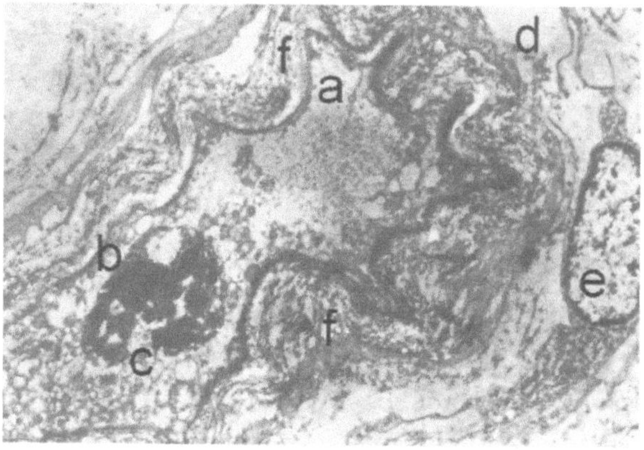

Fig. 5 - Transmission electron micrograph: changes in the epidermis and dermis after feeding of *Ixodes ricinus*: a, intracellular oedema; b, epidermal cell nucleus with dense, granulated chromatin; c, damaged nuclear membrane; d, intercellular oedema; e, fibroblast; f, numerous small collagen fibres (4000 x).

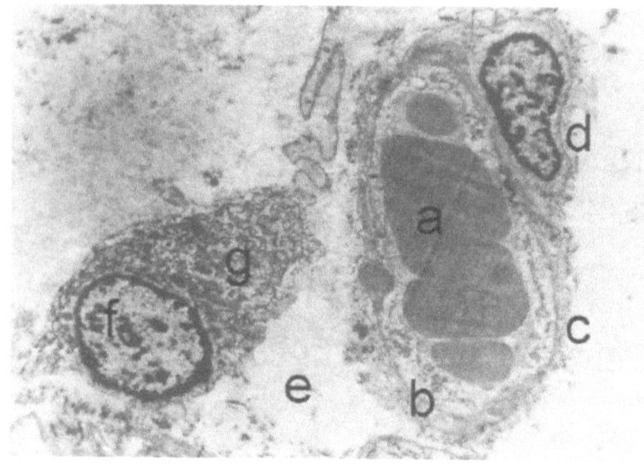

Fig. 6 - Transmission electron micrograph: changes in the dermis after feeding of *Ixodes ricinus*: a, - erythrocyte in the lumen of blood vessel; b, damaged endothelial cell; c, basement membrane; d, pericyte; e, perivascular oedema; f, nucleus of plasmocyte; g, cytoplasm with abundant endoplasmic reticulum (4000 x).

formation of abscesses, dominated in histological pictures. The dominant cells on peripheries of these skin changes were eosinophils. The inflammatory infiltrate also consisted of macrophages loaded with haemosiderine, lymphocytes and rarely basophils. Moreover, there were lipid accumulations and disorders in architecture of connective tissue associated with the scarring process. In the necrotic tissue irregular hair follicles were characterized by wide, partly fragmented basement membranes and disintegration of their internal structure (fig. 4). Blood vessels were widened and filled with blood. They had thin damaged walls.

Electron microscope examination showed oedema containing numerous mononuclear cells, necrosis, and bundles of collagen fibres in the dermis (figs 5-6). There were also exhibited disorders in the structure of blood vessels that were filled with a homogenous substance. There were observed necrotic changes of endothelial cells, disintegration of nuclei, irregular pattern of basement membrane, which forms hollows in the vessel lumen. Under the basement membrane there were large accumulations of collagen fibres providing to vessel fibrosis (fig. 6).

Discussion and Conclusions

The inflammatory reactions in host tissue develop in response to tick bites (Tatchel and Moorhause 1968; Allen 1973; Brossard and Fivaz 1982; Brown 1982; Brown *et al.* 1984; Gill and Walker 1985; Dusbabek *et al.* 1995). The results of light and transmission electron microscope showed differences in skin reactions after feeding by two tick species, i.e. *D. reticulatus* and *I. ricinus* on rabbits. The lesions in host skin caused by ticks depend on the infested species and its instar, duration of tick feeding, species and susceptibility of the host (Tatchel and Moorhause 1968; Brossard and Fivaz 1982; Brown 1982; Brown *et al.* 1983; Nithiuthai and Allen 1984; Gill 1986; Steeves and Allen 1990; Latif *et al.* 1990 and 1991; Schorderet and Brossard 1993; Mbow *et al.* 1994; Dusbabek *et al.* 1995).

The mouthparts of *I. ricinus* are longer and penetrate deeper into the dermis than *D. reticulatus*. At the attachment sites of both species intraepidermal oedema and enlargements of intercellular spaces below basement membrane were formed. In the previous study Latif *et al.* (1991) compared cattle skin changes caused by ticks with different lengths of hypostome:

Amblyomma variegatum with longer, and *Rhipicephalus appendiculatus* with shorter one. Both species caused intraepidermal pustules filled with infiltrating cells of all the types seen in the dermis. At the attachment sites of *A. variegatum*, the number of infiltrating cells (neutrophils, eosinophils, mononucleocytes, and mast cells), the amount of collagen, and degree of necrosis were higher than those seen in *R. appendiculatus* feeding sites. These findings prove our results which indicate that skin changes caused by the long hypostome tick *I. ricinus* are more severe than those of *D. reticulatus*. During feeding of *Dermacentor variabilis* larvae on mast cell-deficient mice basophil-rich skin reactions were observed (Steeves and Allen 1990).

According to Mbow *et al.* (1994) and Dusbabek *et al.* (1995), *I. ricinus* nymphs also caused changes in BALB/c mice skin. As in females the nymph's rostrum penetrated deeply into the dermis. Twelve hours after tick attachment, both the epidermis and dermis looked normal. The amount of intact mast cells and degranulated mast cells did not change during infestation. The number of infiltrating eosinophils remained very low at all times. Relatively low numbers of neutrophils infiltrating the dermis during the first 24 hours increased 72 and 96 hours after initiation of this infestation. The number of mononuclear cells detected 24 hours post-tick attachment remained unchanged. Histological examination of mice skin showed a slight cellular response during primary infestation that became more pronounced at reinfestations of nymphs (Mbow *et al.* 1994).

The histopathology of the attachment sites of *I. ricinus* females on rabbits is generally similar to other ticks of the genus *Ixodes*. *Ixodes holocyclus* larvae caused dermal cellular response dominated by mononuclear cells and relative absence of eosinophils. Accumulation of basophils preceded the development of macroscopic erythema. Lesions consisted of few granulocytes and relatively low cell density. Examination of biopsies revealed obvious intraepidermal vesicles packed with basophilic cells overlying a small but significant basophil response in the dermis. Tick feeding cavities were small and well defined, and most basophils appeared intact and not degranulated. Free red blood cells were apparent in the lower reticular layer of the dermis indicating vascular damage, a typical consequence of tick feeding (Brown *et al.* 1984).

The changes in hair follicles after tick feeding were described for the first time. Desintegration of their structure may lead to hair loss.

References

Allen J.R., 1973 - Tick resistance: basophils in skin reactions of resistant guinea-pigs. *International Journal of Parasitology*, 3: 195-200.

Brossard M., Fivaz V., 1982 - *Ixodes ricinus* L.: mast cells, basophils, and eosinophils in the sequence of cellular events in the skin of infested or reinfested rabbits. *Parasitology*, 85: 583-592.

Brown S.J., 1982 - Antibody- and cell-mediated immune resistance by guinea pigs to adult *Amblyomma americanum* ticks. *American Journal of Tropical Medicine and Hygiene*, 31: 1285-1290.

Brown S.J., Bagnall B.G., Askenase P.W., 1984 - *Ixodes holocyclus*: Kinetics of cutaneous basophil responses in naive and actively and passively sensitized guinea pigs. *Experimental Parasitology*, 57: 40-47.

Brown S.J., Worms M.J., Askenase, P.W., 1983 - *Rhipicephalus appendiculatus*: Larval feeding sites in guinea pig actively sensitized and receiving immune serum. *Experimental Parasitology*, 55: 111-120.

Dusbabek F., Borsky I., Jelinek F., Uhlir J., 1995 - Immunosuppression and feeding success of Ixodes ricinus nymphs on BALB/c mice. *Medical and Veterinary Entomology*, 9: 133-140.

Gill H.S., 1986 - Kinetics of mast cells, basophil, and eosinophil population at *Hyalomma anatolicum anatolicum* feeding sites on cattle and the acquisition of resistance. *Parasitology*, 91: 591-607.

Gill H.S., Walker A.R., 1985 - Different cellular responses at *Hyalomma anatolicum anatolicum* feeding sites on susceptible and tick-resistant rabbits. *Parasitology*, 91: 591-607.

Latif A.A., Maina J.N., Dhadiala T.S., Nokoe S., 1990 - Histological reactions to bites of *Amblyoma variegatum* and *Rhipicephalus appendiculatus* fed simultaneously on naive or sensitized rabbits. *Journal of Medical Entomology*, 27: 316-323.

Latif A.A., Punyua D.K., Capstic P.B., Nokoe S., Walker A.R., Fletcher J.D., 1991 - Histopatology of attachment sites of *Amblyoma variegatum* and *Rhipicephalus appendiculatus* on Zebu cattle of varying resistance of ticks. *Veterinary Parasitology*, 38: 205-213.

Mbow M.L., Christe M., Rutti B., Brossard M., 1994 - Abscence of acquired resistance to nymphal *Ixodes ricinus* tick in BALB/c mice developing cutaneous reactions. *Journal of Parasitology*, 80: 81-87.

Nithiuthai S., Allen J.R., 1984 - Effects of ultraviolet irradiation on the acquisition and expression of tick resistance in guinea pigs. *Immunology*, 51: 153-159.

Schorderet S., Brossard M., 1993 - Changes in immunity to *Ixodes ricinus* by rabbits infested at different level. *Medical and Veterinary Entomology*, 7: 186-192.

Steeves E.B., Allen J.R., 1990 - Basophils in skin reactions of mast cell-deficient mice infested with *Dermacentor variabilis*. *International Journal of Parasitology*, 20: 655-667.

Tatchell R.J., Moorhouse D.E., 1968 - The feeding processes of the cattle tick *Boophilus microplus* (Canestrini). Part II. The sequence of host-tissue changes. *Parasitology*, 58: 441-459.

House-dust mite infestation in temporary prefabricated homes in Central Italy

A. Mansi[1], L. Frusteri[2], M. Maroli[3], A. Salerno[1], A.M. Marcelloni[1], G. Muzi[4], M.P. Accattoli[4], G. Abbritti[4]

[1]*Department of Hygiene, ISPESL, Rome, Italy;*

[2]*CON.T.A.R.P. INAIL, Naples, Italy;*

[3]*Department of Parasitology, ISS, Rome, Italy;*

[4]*Occupational Medicine Institute, University of Perugia, Italy*

Key words: ecology, mucosa and skin symptoms, dust mite distribution.

Abstract: Two years after an earthquake in Central Italy an acarological study was undertaken in prefabricated houses located between 470-500 m a.s.l; brick houses were used as control. House dust samples were collected from mattresses, carpets and sofas using a vacuum cleaner. Mite specimens were identified. Data on indoor activities, house characteristics, heating systems, furniture, maintenance and on irritative symptoms in inhabitants were recorded from a questionnaire. The prefabricated homes were more infested than control houses: 69.6% of mites could be identified. *Dermatophagoides pteronyssinus* (59.3%) was the prevalent species followed by *Dermatophagoides farinae* (18.3%), *Lepidoglyphus destructor* (13.2%), *Euroglyphus maynei* (7.1%) and *Cheyletus tenuipilis* (2.1%). Mite density ranged between 0 and 700 mites/g of dust. Inhabitants of prefabricated homes referred mucosal irritative symptoms more often than people living in control houses.

Introduction

In recent years, several case-control studies have shown house-dust mites in dwellings are associated with increased frequency of allergic diseases. Mite allergens have been identified as the etiologic agent of sensitization and asthma triggering in children (Lau-Schadendorf *et al.* 1989; Sporic *et al.* 1990; Arshad and Hide 1992). The risk levels for mite sensitization were proposed for the first time in 1989 at the First International Workshop on Indoor Allergens and Asthma (Platts-Mills and de Weck 1989). Since the increasing exposure and sensitization to indoor mite allergens in the home and many public places constitute a rising risk factor, assessing environmental mite-concentrations is essential to prevent or reduce asthma symptoms. As few reports are available on mobile homes or temporary houses, we carried out a survey on house-dust mite infestation, in winter and summer, in prefabricated houses installed about 2 years ago, in Nocera Umbra (Central Italy), after an earthquake. This study aimed at: 1) identifying and quantifying house-dust mites; 2) assessing summer and winter prevalence in different types of houses; 3) evaluating the influence of biotic (fungi) and abiotic factors (such as house characteristics, temperature, humidity and ventilation) on mite infestation; 4) determining whether mite infestation was associated with irritative symptoms of the mucosa and skin of inhabitants.

F. Bernini, R. Nannelli, G. Nuzzaci, E. de Lillo (eds.), Acarid Phylogeny and Evolution. Adaptations in mites and ticks, 425-430.

Materials and methods

Sampling

The house dust samples were collected in February and July 1999 from a total of 29 randomly selected houses in the study area located between 470-500 m a.s.l., in Nocera Umbra (Perugia, Italy). Nine were brick houses and were selected as control, twenty were prefabricated houses, 36 m^2 in area, divided into a sitting-room area a single and a double bedroom, toilet and shower unit. All were electrically heated and air-conditioned and equipped with electric cookers. Ten had previously been used (type A) and the other 10 were new (type B). In the 10 old prefabricated houses external walls and roof were made of sheet metal, internal walls and ceiling were painted wood with rock fibre felt in the cavity. In the 10 new prefabricated houses walls, roof and ceiling were made of a double plastic lamina with polyurethane foam in between. Dust samples were collected on a clean cotton filter inserted in a cylindrical trap filter between the dusting attachment and the end of the hose of a portable vacuum cleaner. Dust samples were taken from two mattresses in each dwelling. If only one mattress was present, samples were collected from carpets and/or sofas. In 5 prefabricated houses, it was possible to collect dust samples only from one mattress. One square meter of each sampling site was vacuumed for approximately 2 minutes, as proposed by Fletcher *et al.* (1996). Once the dust was collected, the collection filter was tightly folded and stored in a closed sterile tube to confine the mites, placed in a portable freezing (+4°C), transported to the laboratory and stored refrigerated from 2 to 15 days until microscopic analysis.

Mite extraction and identification

For mite analysis, 100 mg aliquots of dust from each sample were suspended into a beaker containing 30 ml of saturated sodium chloride with a few drops of a detergent (Softi). The beaker was placed in an ultrasonic cleaner. The suspension was decanted into a vacuum pump on three counting paper filters and stained by adding a few drops of 1% crystal violet. Filters were analyzed under a stereomicroscope (magnification 25x10) and all living and dead mites were collected and placed on a microscope slide containing 0.3 ml of water to be frozen (Bigliocchi and Maroli 1995). The specimens were then mounted in Hoyer's medium. Adult specimens were identified by dichotomic and pictorial keys (Fain *et al.* 1990; Coloff and Spieksma 1992). Once the mites were counted, it was possible to estimate the mite density (number of mites per gram of dust).

Climatic measurements and air sampling

Indoor and outdoor climatic conditions were monitored by a thermo-hygrometer (Babuc M, LSI Industrial). Air samples were obtained using a SAS sampler (Surface Air System) positioned on a horizontal surface in the houses. The culturable fraction of fungi was determined by incubation of microrganisms collected directly on contact plates containing Sabouraud Dextrose Agar. Results were expressed as Colony Forming Units per cubic metre of air (UFC/m^3) (Dales *et al.* 1997).

Questionnaire

In winter and the following summer the inhabitants of the 29 dwellings replied to a questionnarie designed to elicit information on the number of inhabitants in each dwelling, indoor activities, house characteristics, building materials and furnishings, heating, ventilation and air-conditioning systems, etc. Symptoms related to staying in the house (irritative symptoms of eyes, nose, throat, skin and lower respiratory tract) were recorded.

Statistical analysis

Statistical analysis were performed using SPSS 9.0 software. The Mann-Whitney test was used to compare mite density. A 'personal symptom score' for each inhabitant (presence of 1 symptom=1; 2 symptoms=2; 3 symptoms=3; etc.) provided the 'house symptom index' (mean of personal symptoms scores in each house). The relationship between mite density and 'house symptom index' was evaluated using the Spearman correlation.

Results

Species

A total of 53 dust samples from mattresses, carpets and sofas were collected. The heterogeneity of samples was necessary because the small area of the containers. House-dust mites were present in all prefabricated houses and controls. Overall 424 live and dead mites were collected (107 in winter and 316 in summer): 69.6% were adults, the rest were in larval (1.9%) and nymphal stages (17.9%) or non-identifiable mites (10.6%). All mite species were present in prefabricated and control houses except *Dermatophagoides farinae* that was not found in control. Figure 1 illustrates the isolated species and their distribution in the three types of houses. *Dermatophagoides pteronyssinus* was the most common species in all houses, while *D. farinae* was isolated predominantly in type B prefabricated houses. Fewer genus *Cheyletus* mites were found in prefabricated than control houses.

Mite density

Different degrees of mite density (number of mites/g dust) were observed in the control and prefabricated houses with values ≥500 only in the prefabricated houses (tab. 1). Mite density varied from a minimum of 0 mites/g dust in control houses during winter to a maximum of 700 mites/g in a type B prefabricated house in summer. In winter, mite density values were not significantly different in control (median 35, range 20-160) and prefabricated (median 30, range 0-360) houses. In summer, the mean mite density was higher in prefabricated houses (median 380, range 10-700) than controls (median 30, range 0-140). The differences were not statistically significant (Mann-Whitney test).

Abiotic and biotic factors

The median value of external winter temperature was 16°C (5.0 SD) and relative humidity (RH) was 65% (0.0 SD). In winter, median indoor values of temperature and relative humidity were 18°C and 42.5% in control houses, 19°C and 45.0% in prefabricated houses. In summer, median indoor were 23°C and 43.0% in control houses, 28°C and 45.0% in prefabricated houses.

Fungal concentrations increased in summer to a median value of 300 UFC/m^3 (range 232-375) in control houses and to 436 UFC/m^3 (range 262-773) in prefabricated houses. These differences in fungal concentrations were not statistically significant (Mann-Whitney test).

The questionnaire was administered to 63 subjects. No inhabitants in control houses referred symptoms related to staying in the house, 36.4% of inhabitants in prefabricated houses referred at least one irritative symptom of eyes, nose, throat or lower respiratory tract. The most common

Tab. 1 - Percentage of houses showing different degrees of estimated mite density (mites/g dust).

Mite density degrees	Control houses	Prefabricated houses	
		Type A	Type B
0–99	66.7	66.7	54.5
100–499	33.3	22.2	27.3
500	0.0	11.1	18.2

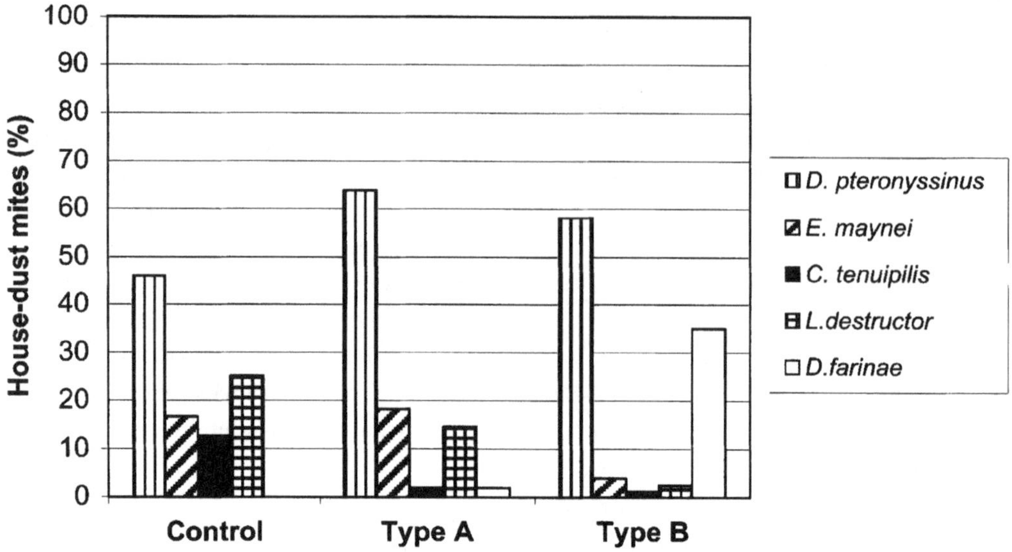

Fig. 1 - Species prevalence in control and prefabricated houses.

symptoms were blocked nose (12.5%), sneezing (12.5%), sore throat (12.5%), running nose (8.3%), burning eyes (8.3%). Personal symptoms scores ranged from 0 to 6. No correlation emerged between house-dust mite density and 'house symptom index' (Spearman correlation coefficient =0.18).

Discussion and conclusions

For the first time in Italy, a survey on house-dust mite infestation in temporary prefabricated houses was carried out. The most common species in prefabricated and control houses was *D. pteronyssinus* (59.3%), followed by *D. farinae* (18.3%) which was mainly present in type B prefabricated houses. Our results are in accordance with other surveys carried out in Italy where local and regional climatic differences explain the different distribution of these house-dust mite species. Several authors reported that *D. pteronyssinus* is predominant in Northern Italy and Sardinia (Ottoboni *et al.* 1979; Castagnoli *et al.* 1983). *Dermatophagoides farinae*, seems to be prevalent in Central and Southern Italy (Noferi *et al.* 1974; Bigliocchi *et al.* 1996), and has adapted to live under very dry conditions, surviving for many months at low humidity levels. *Dermatophagoides pteronyssinus* seems to be prevalent in cold, damp climates. The presence of *L. destructor* can be considered anomalous because it was recovered in the mattresses; nevertheless, it is important to consider the life-style and the small surface inside the containers. Several studies (Harving *et al.* 1994; Korsgaard 1998) have shown the outdoor humidity level, which depends on the climatic region and altitude, is associated with the distribution of house-dust mites. Consequently, the pronounced differences in the number of mites in different dwellings in the same area at the same time of year can be associated with differences in indoor air humidity (Arlian 1977; Korsgaard 1983) in the dwellings. In our summer survey, outdoor conditions were the same for all houses, so they were not responsible for mite overgrowth in prefabricated homes. In winter, mite density values were not different in two types of houses, while higher values of RH were found in prefabricated houses. It is known (Korsgaard 1983) that the level of humidity in winter is a critical factor for the occurrence of house-dust mites. When mites are eradicated during the dry winter season, the more humid summer period is too short for the mites to establish themselves. So, if the absolute humidity is kept below 7.0 g water vapor per kg dry air during the dry winter season (corresponding to 45% relative humidity at 20-23°C), the dwellings will have low mite counts throughout the year, in spite of the humid summer (Korsgaard 1998).

In brick houses, the median value of RH was lower (42.5%) than in prefabricated (45%), so probably low concentrations of some species of house-dust mites were background contamination. As at RH values below 45% house-dust mites do not proliferate, higher indoor humidity in winter could explain the major mite concentration in prefabricated houses, in summer. Although indoor air humidity is a good indicator of the growth conditions for house dust mites, their prevalence is influenced by the micro-ecological conditions in the different parts of the house. These are determined by local water vapour generations, as may take place around condensation foci at cold outer walls and floors or, most importantly, in the bed. Local air humidity in sheltered habitats may greatly exceed average indoor humidity. High humidity might also be responsible for fungal growth and indeed in our prefabricated homes, fungal concentrations were high, increasing during summer to 436 UFC/m^3. Fungi are common in house-dust and are hypothesized to predigest lipids and provide a good source of vitamins and sterols for mites. Moreover, they appear to be important for processing the skin scales mites feed on. Some mite species (*L. destructor*) feed on fungi (Lovik *et al.* 1998). It is difficult to explain the predominance of *D. farinae* in type B prefabricated houses as no significant differences emerged in temperature and humidity levels when type A and B were compared. Furthermore, furnishings, maintenance and overcrowding were similar in all prefabricated houses. The differences in structure of the two types of prefabricated houses is apparently the only factor that could have played a role. Irritative symptoms were more frequent in people living in prefabricated houses. These people were the victims of a devastating earthquake and had been forced to live in a small space for a prolonged time, conditions which could well have caused severe psychological trauma which made them report complaints and symptoms more openly. However several authors (Lau-Schadendorf *et al.* 1989, Sporic *et al.* 1990, Biornsson *et al.* 1995) reported that irritative symptoms and even rhinitis and asthma are linked to indoor mite infestation. In our study irritative symptoms and mite concentrations tended to correlate but the correlation did not reach statistical significance; perhaps because of the low number of participants. The finding that 11.1% of type A prefabricated houses and 18.2% of type B of prefabricated houses showed densities of 500 mites/g dust should be taken into consideration since this level of density is regarded as a high risk factor for sensitization and development of asthma (WHO 1988, Korsgaard 1998).

References

Arlian L.G., 1977 - Humidity as a factor regulating feeding and water balance of the house dust mites. *D. farinae* and *D. pteronyssinus* (Acari: Pyroglyphidae). *Journal of Medical Entomology*, 14: 484-488.

Arshad S.H., Hide D.W., 1992 - Effect of environmental factors on the development of allergic disorders in infancy. *Journal of Allergy and Clinical Immunology*, 90: 235-241.

Bigliocchi F., Maroli M., 1995 - Distribution and abundance of house-dust mites (*Acarina*: *Pyroglyphidae*) in Rome, Italy. *Aerobiologia*,11: 35-40.

Bigliocchi F., Frusteri L., Carrieri M.P., Maroli M., 1996 - Distribution and density of house dust mites *Dermathophagoides spp.* (*Acarina: Pyroglyphidae*) in the matressess of two areas of Rome, Italy. *Parassitologica*, 38: 543-546.

Biornsson E., Norback D., Janson C., Widstrom J., Palmgren U., Strom G., Boman G., 1995 - Asthmatic symptoms and indoor levels of micro-organisms and house-dust mites. *Clinical and Experimental Allergy*, 25: 423-431.

Castagnoli M., Liguori M., Nannelli R., 1983 - Gli acari della polvere delle case in Italia. Atti XII Congresso Nazionale Italiano di Entomologia, Sestriere, 27 Giugno - 1 Luglio 1983: 577-582.

Colloff M.J., Spieksma F.T.H.M., 1992 - Pictorial keys for the identification of domestic mites. *Clinical and Experimental Allergy*, 22: 823-830.

Dales R.E., Miller J.D., McMullan E., 1997 - Indoor air quality and health: validity and determinants of reported home dampness and molds. *International Journal of Epidemiology*, 26: 120-125.

Fain A., Guerin B., Hart B.J., 1990 - In: Mites and Allergic Disease, Guerin B. (ed.), Varennes en Argonne: 190.

Fletcher A.M., Pickering C.A.C., Custovic A., Simpson J., Kennaugh J., Woodcock A., 1996 - Reduction in humidity as a method of controlling mites and mite allergens: the use of mechanical ventilation in British domestic dwellings. *Clinical and Experimental Allergy*, 26: 1051-1056.

Harving H., Korsgaard J., Dahl R., 1994 - Concentration of house-dust mites in dwellings: relation to humidity and ventilation. Ugeskrift for Laeger: Udgivet af den al almindeling dansk. Laeger-forening, 156: 1141-1144.

Korsgaard J., 1983 - House-dust mites and absolute indoor humidity. *Allergy*, 38: 85-92.

Korsgaard J., 1998 - Epidemiology of house-dust mites. *Allergy*, 53(Suppl. 48): 36-40.

Lau-Schadendorf S., Falkenhorst G., Weber A., Werthmann I., Lind P., Buettner-Goetz P., Wahn U., 1989 - High mite-allergen exposure increases the risk of sensitization in atopic children and young adults. *Journal of Allergy and Clinical Immunology*, 84: 718-725.

Lovik M., Gaarder P.I., Mehl R., 1998 - The house-dust mite: its biology and role in allergy. A synopsis. *Allergy*, 53(Suppl. 48): 121-135.

Noferi A., Serino W., Sacerdoti G., Blythe M.E., 1974 - Survey on concentration and type of house dust mites (HDM) in dusts of specifically sensitized people in Naples. Preliminary data. *Folia Allergologica Immunologica Clinica*, 21: 466-469.

Ottoboni F., Falagiani P., Morlacchi C., Centanni G., 1979 - Indagine sugli acari della polvere di casa. *Folia Allergologica Immunologica Clinica*, 26: 427-428.

Platts-Mills T.A.E., de Weck A.L., 1989 - Report of the meeting. Mite allergy-a world-wide problem. Workshop, Bad Kreuznach. *Journal of Allergy and Clinical Immunology*, 83: 416-427.

Sporic R., Holgate S.T., Platts-Mills T.A.E., Cogswell J.J., 1990 - Exposure to house dust mite allergen (Der p1) and the development of asthma in childhood. *New England Journal of Medicine*, 323: 502-507.

WHO. 1988 - Dust mite allergens and asthma: a wordlwide problem. *Bulletin W.H.O.*, 66: 769-780.

Observations on the spread of *Pyemotes ventricosus* (Prostigmata: Pyemotidae) in houses in Umbria, Central Italy

Mario Principato
Facoltà di Medicina Veterinaria, Dipartimento di Scienze Biopatologiche Veterinarie, Via S.Costanzo 4, I-06100 Perugia, Italy

Key words: dermatitis, ecology, population dynamics.

Abstract: The spread of *Pyemotes ventricosus* (Newport, 1850) was observed during three years in 460 houses where dermatoses of environmental origin were reported. The period of highest mite spread was from May to August. This parasitic infestation was among the most important causes of human dermatitis in summer.

Introduction

Pyemotes ventricosus (Newport, 1850) is a prostigmatic mite. It may also be accidentally parasitic on man (Hewitt *et al.* 1976; Rycroft and Kennedy 1981; Betz *et al.* 1982; Kunkle and Greiner 1982; Hanks *et al.* 1992; Principato and Polidori 1993; Letchford *et al.* 1994; Principato 1998; Grob *et al.* 1998). Although the systematics of this mite are known and its presence is frequently reported in food stores, with repeated episodes of human dermatites, very little is known about its frequency and indoor spread and the possible risks for people and domestic animals. The aim of this survey has been, therefore, to determine the period of the maximum development of this pest, related to the outbreak of serious dermatoses, and to establish the main places of its development.

Materials and Methods

The direct examination of indoor dust from 460 Umbrian human dwellings, from which an intense pruritic papular eruption was reported, was continued during three years. Dust was collected by sweeping the floor of the houses carefully. The mites were extracted through flotation with NaCl, observed through a stereo-microscope, isolated with a thin needle and mounted on a slide in Berlese solution. Mites were then counted and identified to species.

Results

P. ventricosus was the fourth most frequent mite in dwellings after *Dermatophagoides pteronyssinus*, *D. farinae* and *Glycyphagus domesticus,* and the second most important acarine cause of human dermatitis after *G. domesticus* (fig. 1).

F. Bernini, R. Nannelli, G. Nuzzaci, E. de Lillo (eds.), Acarid Phylogeny and Evolution. Adaptations in mites and ticks, 431-434.

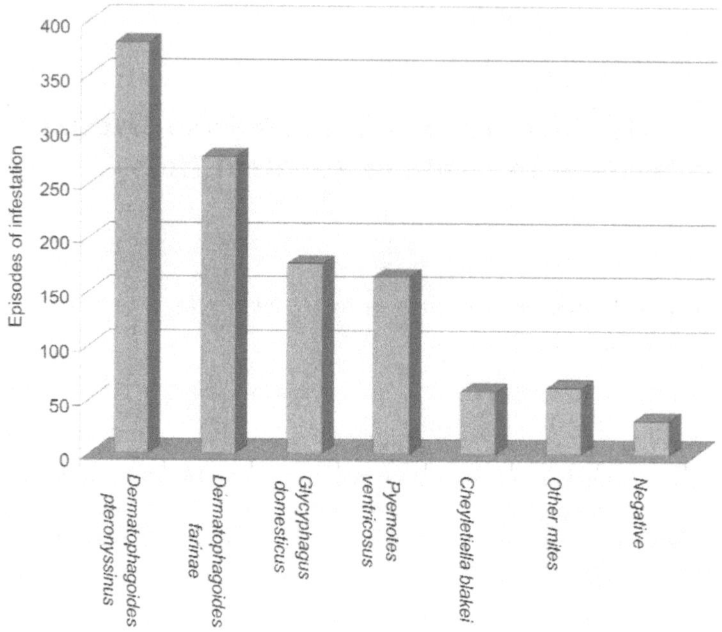

Fig. 1 - Frequency of infestations in the 460 dwellings examined.

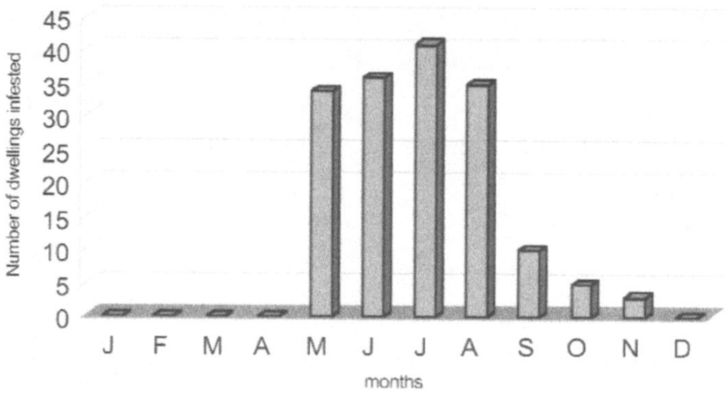

Fig. 2 - Occurrence of *Pyemotes ventricosus* during three years in 460 dwellings.

Higher population density of *P. ventricosus* was detected from May to August; from September to November, the density decreased to zero (fig. 2).

The gravid females of this species were frequently observed in June and by September some males, which were usually rare, were also found (fig. 3).

Even when the number of mites isolated through the direct examination of indoor dust was low, it was always associated with an episode of human dermatitis. The people affected soon recovered after they were given a specific environmental treatment. Furthermore, *P. ventricosus* did not appear to depend on particular temperatures or humidities in the house, since it was isolated, on its own or together with other mites that need a high RH rate or with mites found at 45-55% RH. The main

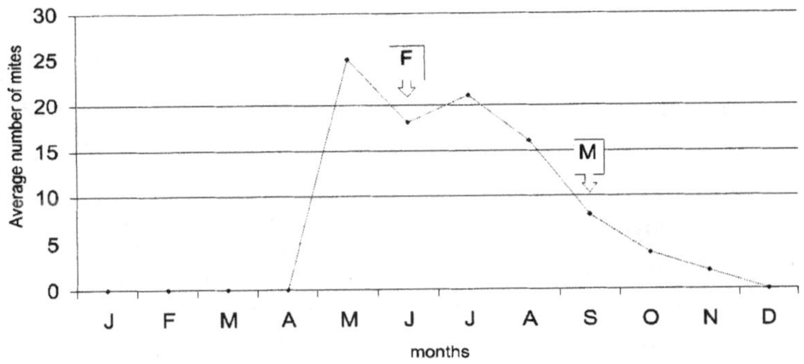

Fig. 3 - *Pyemotes ventricosus*: average number of mites removed by the direct examination of indoor dusts. F, Females; M, Males.

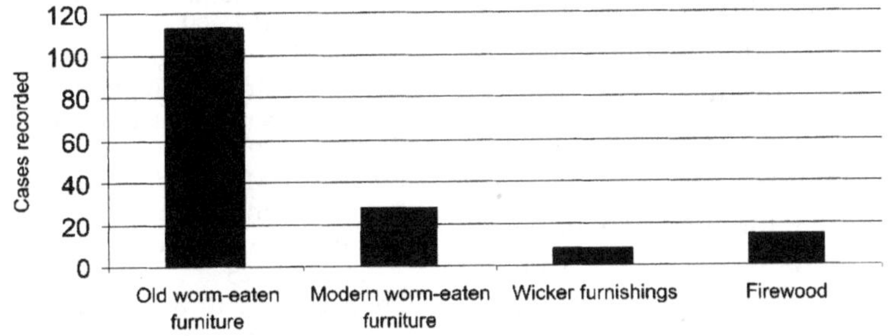

Fig. 4 - *Pyemotes ventricosus*: origin of infestation.

origin of infestations was furniture, both new and old, such as wicker objects infested by the beetle *Anobium punctatum* and also of firewood, which in many cases had remained in the room since the previous winter (fig. 4).

The dermatitis by *P. ventricosus* was reported mainly on man and there were only 3.6% cases when it was observed in dogs living in those houses, showing cutaneous eruptions on their abdomen and the insides of their thighs. The resultant distribution of lesions on humans was characteristic, since they occurred not only on the trunk and the arms, but also in the neck and in the areas tightened by the elastics of clothes. Furthermore, some lesions appeared in small groups and were intensely pruritic.

Discussion and conclusions

We found that human infestations by *P. ventricosus* are very frequent in the spring-summer period in Umbrian houses. Mites showed a clear periodic occurrence, with gravid stages occurring mainly in June-July. In September most of the mites isolated were dead and the infestation decreased in October-November, as did the cases of dermatitis. The most serious episodes were observed in the reproductive period of the mites and in particular when the parasites came from furniture eaten by larvae of hylophagous beetles, but only when they were present. The anamnesis seemed extremely confused, since the patients attested to the outbreak of lesions in quite different conditions, even outdoors, in their cars or while they were having a shower. Actually, this anamnestic diversity leads to the suspicion of infestation by *P. ventricosus,*, for this mite causes lesions some hours after its bite and when people are engaged in different activities. We never found any correlation between the presence of *P. ventricosus* and *Plodia interpunctella* (Lep. Phycitidae) in the houses, a moth that frequently infests farinaceous foods in larders or cupboards. However we always reported the presence of this mite in association with woodworms, in order to isolate the parasite a few times even under the beetle's elytra (Principato and Polidori 1993). As for the lesions observed on domestic animals, we think they could be caused by *P. ventricosus*; Kunkle and Greiner (1982) reported papular eruption due to *Pyemotes* sp. in 12 horses, thus recording cases of dermatitis in animals. Infestations were eliminated and complete recovery was obtained after the removal of the worm-eaten furniture, wherein the mites had developed. Good results were obtained but only temporarily after moth control that was associated with another treatment of the whole flat or house by fumigation. In consideration of the high spread of this mite in Umbrian houses and its pathogenic action on man, we think it is important to deepen the study of its biology and its adaptability with further research.

References

Betz T.G., Davis B.L., Fournier P:V., Rawlings J.A., Elliot L.B., Baggett D.A, 1982 - Occupational dermatitis associated with straw itch mites (*Pyemotes ventricosus*). *Journal of American Medical Association*, 247(20): 2821-2823.

Grob M., Dorn K., Lautenschlager S., 1998 - Grain mites. A small epidemic caused by *Pyemotes* species. *Hautarzt*, 49(11): 838-843.

Hanks L.M., McCelfresh J.S., Millar J.G., Paine T.D., 1992 - Control of the straw itch mite (Acari: Pyemotidae) with sulfur in an insect rearing facility. *Journal of Economical Entomology*, 85(3): 683-686.

Hewitt M., Barrow G.I., Miller D.C., Turk S.M., 1976 - A case of *Pyemotes* dermatitis. With a note on the role of these mites in skin disease.*British Journal of Dermatology*, 94(4): 423-430.

Kunkle G.A., Greiner E.C., 1982 - Dermatitis in horses and man caused by the straw itch mite. *Journal of the American Veterinary Medical Association*, 181(5): 467-469.

Letchford J., Strungs I., Farrell D., 1994 - *Pyemotes* species strongly implicated in an outbreak of dermatitis in a Queensland country hospital. *Pathology*, 26(3): 330-332.

Principato M., Polidori G.A., 1993 - *Pyemotes ventricosus*: un acaro associato al "tarlo del legno", parassita occasionale dell'uomo. *Praxis Veterinaria*, 14(4): 23-25.

Principato M., 1998 - Artropodi patogeni rilevabili nelle abitazioni con l'esame diretto delle polveri ambientali. *Annali Italiani di Dermatologia Clinica e Sperimentale*, 52: 60-72.

Rycroft R.J., Kennedy C., 1981 - *Pyemotes* dermatitis in display artists. *Clinical Experimental Dermatology*, 6 (6): 629-634.

Linden-leaf morphology and the host-plant susceptibility to *Eotetranychus tiliarium* (Hermann) (Acarida: Tetranychidae)

Barbara Czajkowska, Malgorzata Kielkiewicz
Department of Applied Entomology, Faculty of Horticulture and Landscape Architecture, Warsaw Agricultural University, PL-02-787 Warsaw, Nowoursynowska 166, Poland

Key words: *Tilia platyphyllos, Tilia cordata, Tilia tomentosa,* leaf hairs, linden leaf anatomy, *Eotetranychus tiliarium,* linden susceptibility.

Abstract: The population dynamics of *Eotetranychus tiliarium* on three linden species (*Tilia platyphyllos, T. cordata, T. tomentosa*) was studied under urban conditions. At the highest density of the mite population linden leaves were sampled for further morphological and cytological analyses. The study conducted during two growing seasons (1997, 1998) showed that the linden species tested differed distinctly in their respons to *E. tiliarium* feeding. *Tilia platyphyllos* was definitely more infested by mites than the other two species. The extent of leaf damage in the species studied was evidently related to the type and density of hairs on the leaf underside and to the internal structure of the linden leaves.

Introduction

The linden spider mite [*Eotetranychus tiliarium* (Herm.)] is the most important pest of linden trees in urban environments. Results of a previous study have shown that its occurrence and reproduction differ on various linden species (Bozai 1975; Gotoh 1988; Czajkowska 1998; Diwo *et al.* 1998). It was also proved that the density of linden spider mite (LSM) population can be affected in different ways by other factors such as microclimate, natural enemies, leaf quality, dust with heavy metals and polycyclic aromatic hydrocarbons (Hughes 1988; Kropczynska and Czajkowska 1995; Fostad and Pedersen 1997; Tomczyk *et al.* 1998). No literature known reports suggest that the susceptibility of linden trees to *E. tiliarium* might be related to the morpho-anatomical structure of their leaves. Therefore, the objective of the present study was to examine whether and how linden leaf micro-morphology may explain differences in the population dynamics of LSM on various linden species.

Materials and methods

The following two aspects were investigated during this study:

1) the population dynamics of *E. tiliarium* (Herm.) on three linden species, *Tilia platyphyllos* Scop., *T. cordata* Mill. and *T. tomentosa* Moench, during two consecutive growing seasons (1997, 1998). The trees were situated along Zwirki and Wigury street, one of the main Warsaw traffic roads. The abundance of active stages of the mite on each linden species was estimated using 150 and 300 leaf samples during 1997 and 1998, respectively;

2) the micro-morphology and anatomy of

F. Bernini, R. Nannelli, G. Nuzzaci, E. de Lillo (eds.), Acarid Phylogeny and Evolution. Adaptations in mites and ticks, 435-440.

damaged and undamaged leaves from three species of linden trees. Leaves were collected at the peak of pest abundance. Type of leaf hairs and the number of hairs per 2 mm^2 on leaf underside surface were examined using a stereomicroscope SZH 10 (Olympus, Japan). Observations were conducted in 100 replications for each species. At the same time, tissue samples were taken from LSM-damaged and undamaged linden leaves for histological examination. Leaf samples (10 x 10 mm) were fixed in CrAF mixture (chromium trioxide, acetic acid, formaldehyde) and embedded in paraffin. Cross-sections (8-12 μm) were stained with safranin and Fast Green before examination with a light microscope BX40 (Olympus, Japan). Approximately 100 sections from 20 leaves of each linden species were used to characterize internal leaf damage.

Results and discussion

In the present work, we found that the density of LSM population was significantly higher on *T. platyphyllos* than on *T. cordata* in both growing seasons (fig. 1), while the smallest population was recorded on *T. tomentosa*. Weather conditions in the first half of the 1997 growing season were unfavourable for the development of *E. tiliarium*. Data presented in figure 1 indicate a low level of pest infestation in June. In July an increase was noted on all linden species. High temperatures and dry conditions in August caused a rapid increase of the mite populations. In 1998 the weather evidently also affected the level of the pest infestation (fig. 1). In each year the highest mite population was recorded during a dry and warm period in late June/early July, whereas in August (wet and cold)

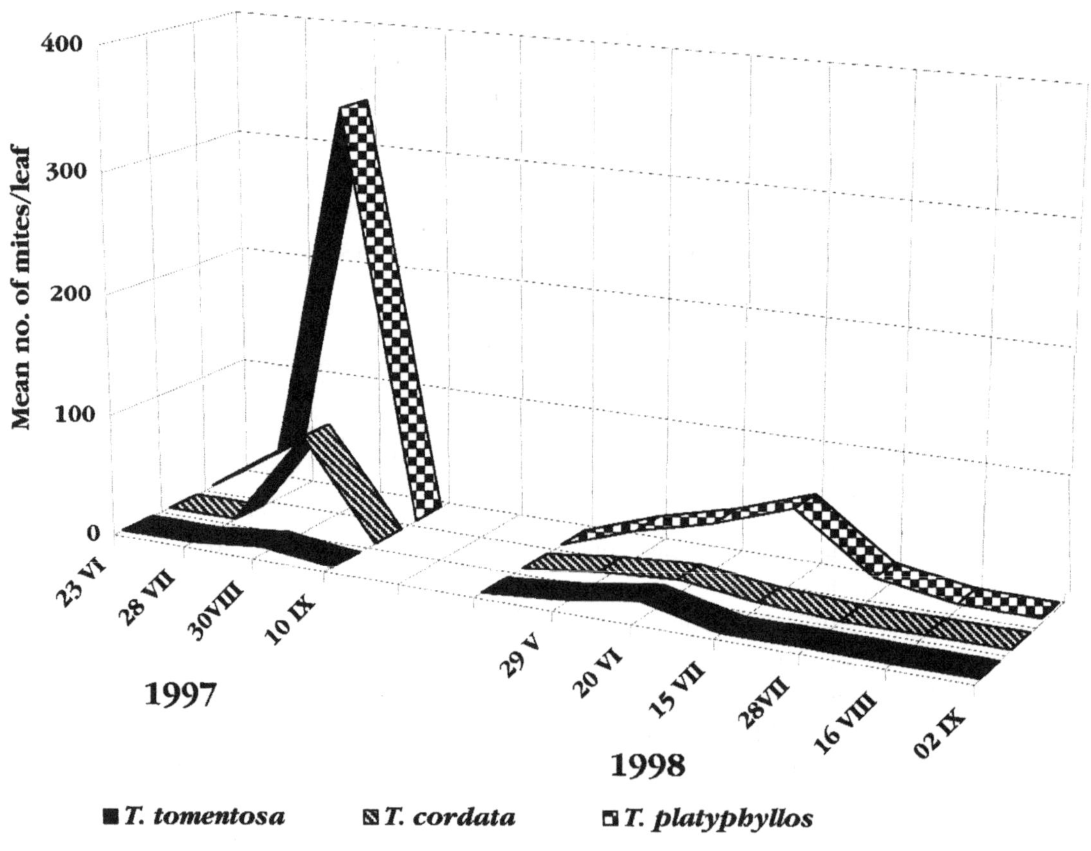

Fig. 1 - Seasonal dynamics of *E. tiliarium* on three species of linden trees (1997, 1998).

Tab. 1 - Density of hairs on the underside of linden leaves. Means followed by the same letters within the column are not significantly different at P=0.05 (Tukey HSD Test).

Linden species	Mean no. of hairs per 2 mm² of leaf blade	Mean no. of hairs per 2 mm² of leaf vein
Tilia cordata	0.9 a (n=100)	50.5 a (n=100)
Tilia platyphyllos	29.9 b (n=100)	153.9 b (n=100)
Tilia tomentosa	125.1 c (n=100)	247.1 c (n=100)
	F=1979.4	F=1700.7
	P< 0.0001	P< 0.0001

Tab. 2a - Mean density of *E. tiliarium* population on linden tree (1997). Means followed by the same letters within the column are not significantly different at P=0.05 (Tukey HSD Test).

Linden species	Mean no. of mites per leaf	No. of replicates	Value of F test
Tilia tomentosa	2.26 a	150	5.43 (P=0.0047)
Tilia cordata	26.8 a	150	
Tilia platyphyllos	127.2 b	150	

Tab. 2b - Mean density of *E. tiliarium* population on linden trees (1998). Means followed by the same letters within the column are not significantly different at P=0.05 (Tukey HSD Test).

Linden species	Mean no. of mites per leaf	No. of replicates	Value of F test
Tilia tomentosa	3.63 a	300	61.35 (P<0.0001)
Tilia cordata	4.06 a	300	
Tilia platyphyllos	23.44 b	300	

it significantly declined. As shown by the present study and confirmed by other authors (Boudreaux 1963; Sabelis 1981), low temperatures and intensive precipitation considerably limit mite populations on linden leaves. Leaf blades of the three studied linden species differed in the type and number of hairs (plate 1a-c, tab. 1). *Tilia platyphyllos* leaves are provided with straight hairs and single glandular trichomes. On the other two species the hairs are more or less branched (plate 1a-c). The most numerous hairs, both on veins and between them, are found on the leaf blades of *T. tomentosa*, whereas the lowest number occur on *T. cordata* foliage (tab. 1). On the underside of *T. tomentosa* leaves the branched hairs form a thick felty structure which probably prevent the movement of mites and limited their access to the leaf tissues (plate 1a-c). In both growing seasons (1997, 1998) the foliage of this species showed the lowest mean density of *E. tiliarium* populations (tab. 1). Against expectations, mite infestation of *T. cordata* leaves was not the highest (tab. 2 a-b) in spite of their light coverage with hairs: 120 and 30 times lower than on *T. tomentosa* and *T. platyphyllos*, respectively (tab. 1). There was no significant difference in the

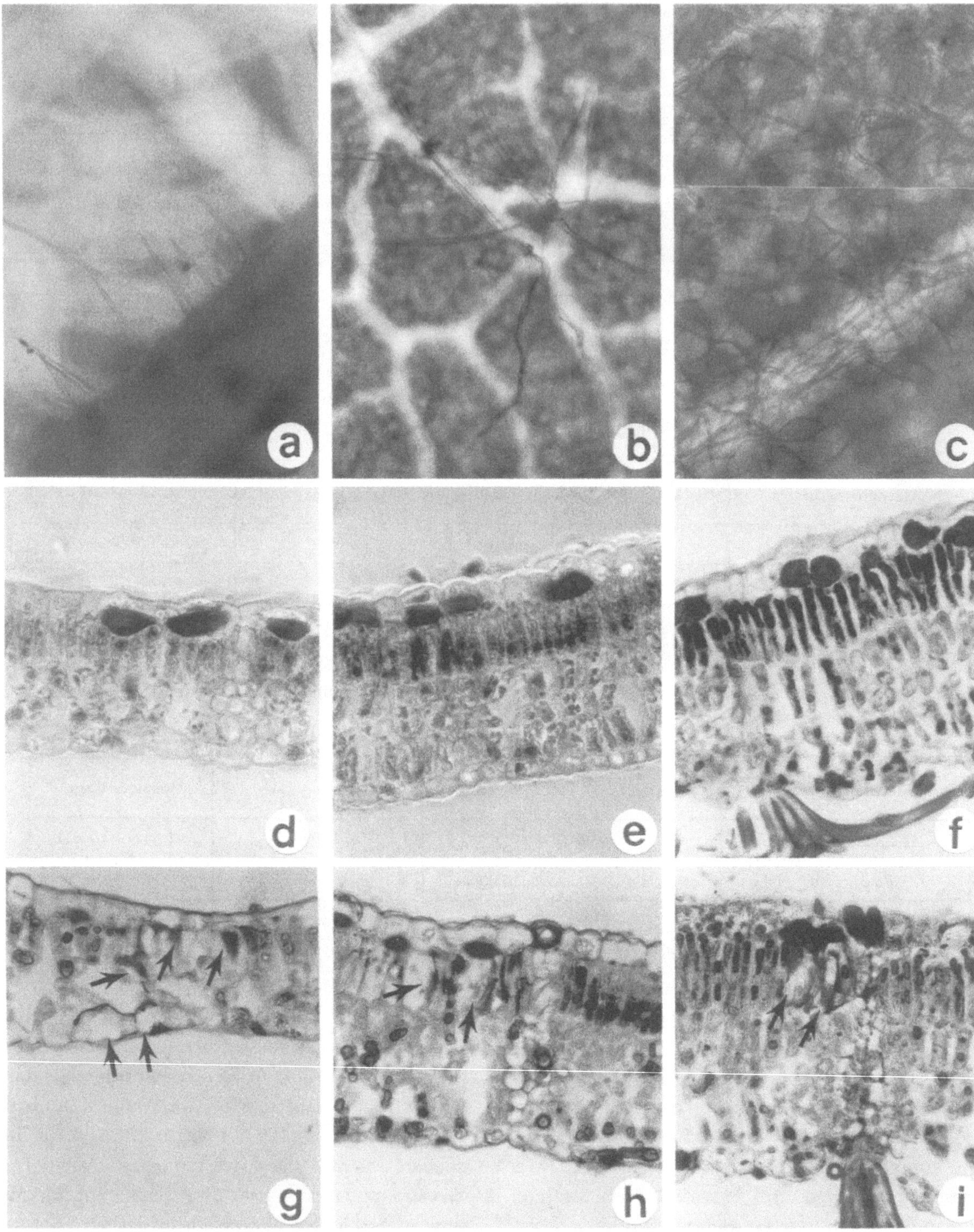

Plate 1 - Hairs on the leaf underside of *T. platyphyllos* (a), *T. cordata* (b) and *T. tomentosa* (c) visible in microscope transmission light (magnification: 25x). Cross-sections of undamaged control leaves of *T. platyphyllos* (d), *T. cordata* (e) and *T. tomentosa* (f) (magnification: 360x). Material stained with safranin visible as black colour. Cross-sections of mite-damaged leaves of *T. platyphyllos* (g), *T. cordata* (h) and *T. tomentosa* (i) showing the appearance of mite injuries (arrows) (magnification: 360x). Arrows indicate partly destroyed cells of palisade and spongy mesophyll.

mean density of mite populations on the foliage of *T. cordata* and *T. tomentosa* (tab. 2a-b). It seems that the straight hairs on *T. platyphyllos* leaves, even more numerous than in *T. cordata*, had a lower preventive effect on the pest than branched hairs (plate 1a-c). Comparison of the cross-sections indicates that the leaves of *T. platyphyllos* were almost twice as thin as those of the two other species (plate 1d-f). It appeared that *T. platyphyllos* leaves consisted of singular layers of palisade mesophyll and spongy mesophyll with small cells loosely distributed and with very large intercellular spaces (plate 1d). Leaf cross-sections of the other two species also revealed a singular layer of palisade mesophyll, whereas spongy mesophyll was much thicker, consisting of large cells tightly fitted together (plate 1e-f). Some mesophyll cells were intensively red-stained with safranin. Deep red colour indicates the presence of phenolic compounds. The symptoms of LSM feeding were well visible in cross-sections of *T. platyphyllos* leaves (plate 1g). Both spongy and palisade mesophyll cells were partly destroyed. Some damaged cells appeared empty the others are filled with red-stained materials or collapsed. The affected tissue area became depressed (plate 1g). In contrary, the symptoms of LSM feeding and the extent of the damage were barely visible in leaves of *T. cordata* and *T. tomentosa* (plate 1h-i). Feeding damage occurred rather on the upper sides of leaves of *T. cordata* and *T. tomentosa* than on the undersides (plate 1h-i). Such a preference might be caused by the presence of abundant branched hairs on the leaf underside and a thick layer of spongy mesophyll, both forming a natural obstacle for the pest. As some authors (Avery and Briggs 1968; Sances *et al.* 1979; Tanigoshi and Davis 1978; Kielkiewicz and Van de Vrie 1983; Mothes and Seitz 1982; Kielkiewicz 1991) observed similar changes in leaves infested with other mite species [*Panonychus ulmi* (Koch), *Tetranychus urticae* (Koch), *Tetranychus cinnabarinus* (Boisd.)] it seems that the type of damage to linden leaves was not a response specific only to *E. tiliarium* feeding. It should be stressed, however, that the extent of linden injury was evidently related to the type and quantity of hairs on the leaf underside and also to the internal structure of leaves of the studied species.

Conclusions

1) Each of the linden species tested (*T. platyphyllos, T. cordata, T. tomentosa*) was infested and damaged by *E. tiliarium* to various degrees.

2) Three linden species distinctly differed in their susceptibility to *E. tiliarium* feeding.

3) *Tilia platyphyllos* was evidently more heavily infested by mites than the other two species.

4) Micro-morphological leaf structure seems to be one of the factors determining the density of *E. tiliarium* population on linden trees.

References

Avery D.J., Briggs J.B., 1968 - Damage to leaves caused by fruit tree red spider mite, *Panonychus ulmi* (Koch). *Journal of Horticulture Science*, 43: 463-473.

Bozai J., 1975 - *Schizotetranychus tiliarium*: the most important foliage pest of *Tilia spp.* grown as forest and ornamental trees in Hungary. *Novenyvedeles*, 11(6): 262-266.

Boudreaux H.B., 1963 - Biological aspects of some phytophagous mites. *Annual Review of Entomology*, 8: 137-154.

Czajkowska B., 1998 - Occurrence and reproduction of linden spider mites (*Eotetranychus tiliarium* (Herm.) on various species of linden growing under urban conditions. Zeszyty Naukowe - Ochrona srodowiska 2, Bydgoszcz, Poland, 214: 17-24.

Diwo S., Miniere T., Kreiter S., Auger J., Chaudoy P., 1998 - The mites of lime trees in the park of Villandry Castle (37): preliminary study on a biological control programme. First transnational workshop on biological, integrated and reasoned control: status and perspectives with regard to regional and European experiences, Lille, France 1998: 97-98.

Fostad O., Pedersen P.A., 1997 - Vitality, variation, and causes of decline of trees in Oslo center (Norway). *Journal of Arboriculture*, 23: 4, 155-165.

Gotoh T., 1988 - Annual life cycles of two populations of *Eotetranychus tiliarium* (Hermann) (Acarina: Tetranychidae) on different alder trees. *Applied Entomology and Zoology*, 23 (3): 304-312.

Hughes P.R., 1988 - Insect populations on host plants subjected to air pollution In: Heinrichs E.A. (ed.),

Plant Stress-Insect Interactions. John Wiley & Sons. Inc., N.Y., 8: 249-319.

Kielkiewicz M., 1991 - Phenolic in tomato leaves infested by the carmine spider mite *Tetranychus cinnabarinus* Boisd. In: Dusbabek F., Bukva V. (eds.), Modern Acarology, Academia, Prague, and SPB Academic Publishing bv, The Hague, The Netherlands, vol. 1: 603-606.

Kielkiewicz M., Van de Vrie M., 1983 - Histological studies on strawberry leaves damaged by the two-spotted spider mite (*Tetranychus urticae* Koch): some aspects of plant self defense. *Mededelingen Faculteit Landbouwwetenschappen Rijksuniversiteit Gent*, 48(2): 235-245.

Kropczynska D., Czajkowska. B., 1995 - Population dynamics of mites inhabiting broadleaved lime in an urban environment. In: Boczek J., Ignatowicz S. (eds.), Proceedings of the Symposium on "Advances of Acarology in Poland" Siedlce, September 26-27, 1995: 56-64.

Mothes U., Seitz K.A., 1982 - Fine structural aberrations of bean plant leaves by feeding injury of *Tetranychus urticae* Koch (Acari, Tetranychidae). *Acarology*, 23(2): 149-157.

Sabelis M.W., 1981 - Biological control of two-spotted spider mites using phytoseiid predators. Part I. Agricultural Research Reports 910. Pudoc Centre for Agriculture Publishing and Documentation, Wageningen: 242 pp.

Sances F.V., Wyman J.A., Ting I.P., 1979 - Morphological responses of strawberry leaves to infestations of two-spotted mite. *Journal of Economic Entomology*, 72(5): 710-713.

Tanigoshi L.K., Davis R.W., 1978 - An ultrastructural study of *Tetranychus mcdanieli* feeding injury to the leaves of "Red Delicious" apple (Acari: Tetranychidae). *International Journal of Acarology*, 4: 47-56.

Tomczyk A., Kielkiewicz M., Czajkowska B., 1998 - Chemical composition of linden tree leaves and spider mite populations in polluted urban conditions. In: Konopnicka D., Goldsworthy G., Nachman R.J., Nawrot J., Orchard J., Rosinski G. (eds.), Chemical, Physiological and Environmental Aspects. Proceedings of the 2nd International Conference on Insects, Stefan Kopec Memorial Conference, September 14-19, 1997, Ladek Zdrój, Poland: 228-232.

In vitro effects of ivermectin on adult and developmental stages of two species of blood feeding arthropods

S.R. Allam[1], H.A.H. Sadaka[1], M.M. Eissa[1], H.A.Rezk[2]
[1]*Parasitology Dept., Faculty of Medicine, Alexandria University, Alexandria, Egypt*
[2]*Economic Entomology Dept., Faculty of Agriculture, Alexandria University, Alexandria, Egypt, hussienrezk@yahoo.com*

Key words: ivermectin, hard ticks, head lice, mortality, hatchability, ovicide.

Abstract: The efficacy of the parasiticide, ivermectin, at different concentrations, on adult and the developmental stages of the hard tick, Boophilus annulatus, and the head louse, Pediculus humanus capitis, was studied in vitro. Mortality, hatchability and egg morphology were included. The results indicated that the control adult lice were alive at 1200±2.5 min and no lice survived over 5 min at 50 ng/ml or higher concentration of ivermectin. Control adult ticks were alive at 57600±1008 min while at 100ng/ml, they survived for only 132±7.8 min. The hatchability of control eggs was 93% and 91% while, at 100 ng/ml of ivermectin it was 1.3% and 7.3% for head lice and ticks, respectively. Using light microscopy, the treated lice nits showed detachment and shrinkage of the embryos inside decolourized operculated shells when treated with ivermectin at 100 ng/ml. While, the treated tick eggs appeared dark with disfigured embryos and detached envelopes. Disfigurement increased with higher ivermectin concentration. On scanning electron micrographs, the treated lice nits were shriveled, disfigured or indented with the presence of operculum. While, the treated tick egg surface appeared nearly smooth or the shell was completely disfigured.

Introduction

Both lice and ticks are blood sucking arthropods. The head lice, *Pediculus humanus capitis* (Anoplura: Pediculidae) are among the most common permanent ectoparasites of humans, though they are not vectors of serious diseases (Meinking *et al.* 1986). Pediculosis (Vagabond's disease) is sometimes accompanied by secondary infection with cervical lymphadenopathy (Goldsmith and Heyneman 1989). Although chemical compounds for treatments are available, the prevalence of head lice remains high and epidemics occur regularly. Ticks are non-permanent obligate ectoparasites that feed on mammals, birds, reptiles and amphibians (Klompen *et al.* 1996). All species are exclusively hematophagous in all feeding stages (Keirans 1992; Klompen *et al.* 1996). They play important roles as pests and as vectors of a wide variety of pathogenic organisms affecting man and animals, including fungi, viruses, rickettsiae, bacteria and protozoa (Wikel 1996). In addition, tick saliva is a cause of human allergic and toxic reactions (Sauer *et al.* 1995). The hard tick, *Boophilus annulatus*, is widely recognized as one of the important ectoparasites causing major economic losses in the cattle industry in Egypt (Hoogstraal 1956). So, the control of these two hematophagous arthropods is a major issue in community practices. Ivermectin is a derivative of the macrocyclic lactone

F. Bernini, R. Nannelli, G. Nuzzaci, E. de Lillo (eds.), Acarid Phylogeny and Evolution. Adaptations in mites and ticks, 441-447.

avermecin B_1. The compound is a natural fermentation product isolated from the soil actinomycete microorganism, *Streptomyces avermitilis* (Campbell 1985; Strong and Brown 1987; Clark *et al.* 1995). It is currently the drug of choice for the treatment of onchocerciasis (*Onchocerca volvulus)* and other lymphatic filariasis in humans (Campbell *et al.* 1983; Greene *et al.* 1985; Taylor and Greene 1989). Recent studies suggest its usefulness in human strongyloidiasis, ascariasis, trichuriasis and entrobiasis in both humans and livestock animals (Campbell 1985). Ivermectin is registered against some human biting arthropods including black flies, fleas and some species of mosquitoes (Miller *et al.* 1986; Tesh and Guzman 1990; Mumcuoglu *et al.* 1990). Other studies have indicated its effects on ixodid ticks, *Amblyomma* and *Dermacentor* (Wilson *et al.* 1991) and some species of mites (Yousses *et al.* 1995).

Because of the wide spread use of ivermectin for treatment of parasitic infections in domestic animals and humans and its demonstrated toxicity for insects. The purpose of the present *in vitro* study was to determine what effect the drug would have when used at different concentrations on adult and developmental stages of two species of blood feeding arthropods, the head louse (*Pediculus humanus capitis)* and the hard tick *(Boophilus annulatus)*.

Materials and Methods

Arthropods

Two species of blood feeding Arthropods were the subject of the present work: Head lice, *Pediculis humanus capitis*, nymph and adult stages were collected from the head of infested children after obtaining an informed consent from their parents. Nits were snipped from the hair. Hard Ticks, *Boophilus annulatus*, were field collected from cattle farms. Cattles have been examined for ticks in two areas in North Tahrir, Alexandria at early Summer (May to July). For identification, different stages of ticks were mounted in a Hoyers medium. Fully fed nymphs and adult ticks were kept in 70% alcohol. After clearing they were identified using a binocular microscope (Zeiss, Germany) (Keirans 1992). Eggs were obtained from engorged females.

Experimental studies: ivermectin

Merck, Sharp and Dohme (Cairo, Egypt) supplied Ivermectin as a liquid solution 0.05% weight to volume. The scientific chemical name of this compound is 5-0 dimethyl-22, 23- dihydro-avermectin A1 (Campbell 1985). After preliminary experiments, six ivermectin concentrations of 7.5, 15, 30, 50, 75 and 100 ng/ml water and a control were selected. Experiments were replicated at least four times for each concentration.

Experimental studies: arthropod mortality time

Five-cm diameter discs were cut from cotton-polyester sterilized ordinary kitchen towels. At the time of the experiment, a disc was placed in a sterile Petri dish and 1 ml of each concentration was evenly distributed over it and leaves to air dry for 10 min. As for the control, 1 ml distilled water was used. All developmental stages of each species tested were divided into two groups:

Experimental group: stages were subjected to the concentrations.

Control group stages were subjected to water contact.

Arthropods were tested within three hours after collection. Adult and immature stages of experimental arthropods were observed under a binocular microscope until death (Meinking *et al.* 1986). The mean mortality time plus the standard error (Mean±SE) were calculated.

Experimental studies: ovicidal activity studies

Sets of lice nits and tick eggs were prepared. Every set of 10 eggs was immersed in each concentration for 15 minutes. The control sets were immersed in water for the same period of time. After several washes in distilled water, eggs were transferred to clean sterile glass vials that were incubated at 30-34°C in the dark for two weeks. Ovicidal activity was expressed as the percentage of egg-hatchability after 14 days (Meinking *et al.* 1986). All egg groups were examined under light and scanning electron microscopy (SEM).

Results

Arthropod Mortality Time (tab. 1)

Control head lice, adults and nymphs, were alive and active at 1200±25 min and 978±26 min, respectively. There were considerable mortality variations between all concentrations of ivermectin. The higher concentration (100 ng/ml) showed the quickest knock down and mortality. In the adult and nymphal stages no lice survived longer than 9 min and 3 min respectively after initial contact with the ivermectin concentration of 50 ng/ml or higher. Dead lice appeared swollen and lethargic. As regard ticks, the control group of adults and immatures were alive and active at 57600±1008 min and 36000±432 min, respectively when removed from the host and given no opportunity to feed. When treated with 100 ng/ml, adults survived for 132±7.8 min and immatures for 15.5±0.7 min. The adult mortality at 50 ng/ml appeared after 216±13.1 min and in immature stages after 75.9±18 min.

Ovicidal Activity (tab. 2)

The control hatchability was 93% and 91% for head lice and *Boophilus* ticks, respectively. The

Tab. 1 - Arthropod mortality time in minutes after the in vitro exposure to different concentrations of ivermectin. *Values are mean numbers ±SE. Number of used arthropod = 10.

Concentration	LICE		TICKS	
(ng/ml)	Immatures	Adult	Immatures	Adult
0	978.0±26*	1200.0±25	36000±432	57600±1008
7.5	17.0±2.1	22.0±7.9	300.8±62	587±44.6
15	13.0±1.4	18.6±6.0	180.6±32	528±21.3
30	8.9±3.7	10.5±2.1	105.5±21	240±15.2
50	2.1±1.1	4.4±0.2	75.9±18	216±13.1
75	2.3±0.9	4.4±0.1	45.4±5.5	186±9.7
100	1.1±0.6	3.7±0.2	15.5±0.7	132±7.8

Tab. 2 - Hatchability of arthropod eggs exposed to different concentrations of ivermectin after 14 days of incubation. N*: Number of used egg.

Concentration	LICE EGGS (NITS)		TICK EGGS	
(ng/ml)	N*	Hatchability (%)	N*	Hatchability (%)
0	100	93.0	100	91.0
7.5	108	24.0	119	40.3
15	101	27.7	127	30.7
30	125	10.4	141	29.7
50	124	8.8	130	25.4
75	127	5.5	144	20.1
100	150	1.3	149	7.3

lower concentration of ivermectin (7.5 ng/ml) was not completely ovicidal since the hatchability was 24% and 40.3% for lice and ticks, respectively. In contrast, the higher concentration acted as an effective ovicide, after exposure to 100 ng/ml of ivermectin, the egg hatchability of the experimental group were 1.3% for lice and 7.3% for ticks.

Egg Morphology (figs 1 and 2)

Using light microscopy, viable control head lice nits appeared plump, shiny and were tan to coffee in colour with an intact operculum (fig. 1a). In contrast, treated lice nits showed detachment and shrinkage of the embryo inside a decolourized shell (fig. 1b-c) when treated with the higher

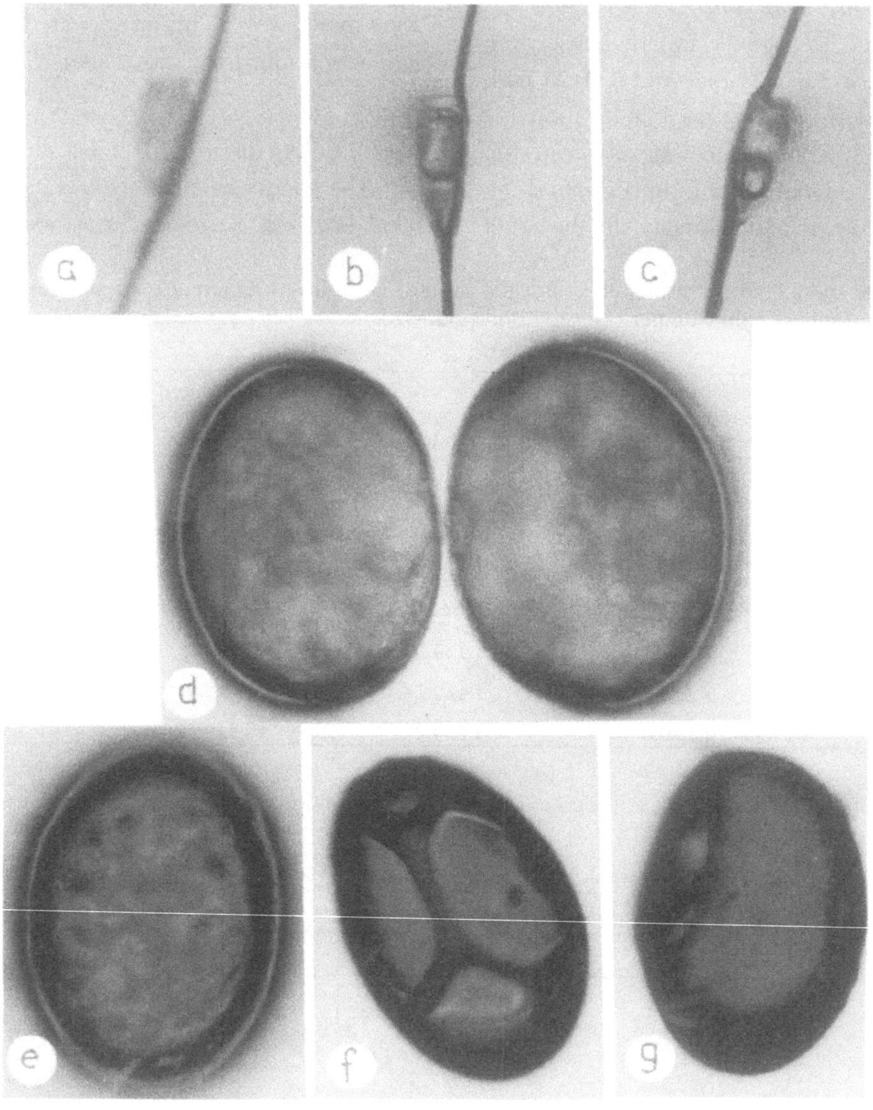

Fig. 1 - Light microscopic photographs of arthropod eggs (x100): a, normal viable head lice, Pediculus capitis nits attached to a human hair; b and c, head lice, Pediculus capitis nits containing shrunken embryo's due to contact with ivermectin at (b) lower concentration and (c) higher concentration; d, normal viable egg of hard ticks; e, f and g, disfigured egg of ticks after exposure to various concentrations of ivermectin.

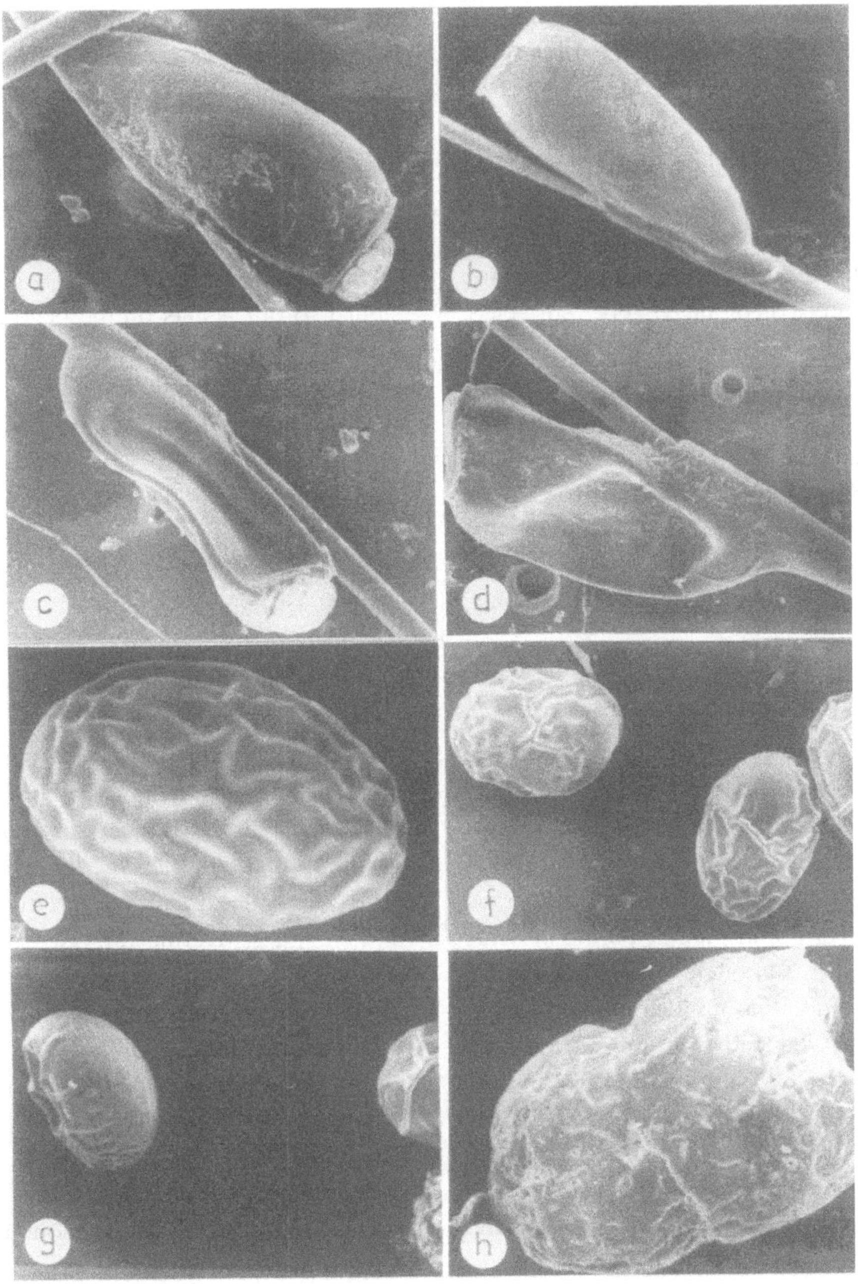

Fig. 2 - Scanning electron micrographs of arthropod eggs: a and b, control viable head lice nits, (a) with intact pored operculum (b) without operculum denoting hatching (x 100); c and d, elongated, collapsed and operculated head lice nits after exposure to various concentrations of ivermectin (x 100); e, control viable egg of ticks showing evident rug on the external surface of the shell (x 200); f and g, tick eggs without variation in the surface pattern after contact with ivermectin at different concentrations (x 100); h, disfigured egg of ticks after exposure to high concentrations of ivermectin (x 200).

ivermectin concentration (100 ng/ml). Control tick eggs appeared light brown in colour, oval with a regular contour containing intact embryo surrounded by enveloping membranes lining the inner wall of the shell (fig. 1d). The treated tick eggs appeared dark with detached internal membranes and disfigured embryo. Disfigurement increased with the higher concentration of ivermectin (fig. 1e-g). In scanning electron micrographs, the control head lice nits laid on a human hair are usually held by cement like coating. Pores or spiracle like structures of the operculum were obvious (fig. 2a). The already hatched nits showed absence of the operculum (fig. 2b). Treated lice nits were shriveled, disfigured or indented with the presence of an operculum (fig. 2c-d). control tick eggs show many rugae covering the external surface of the egg chorion (fig. 2e). The disappearance of the rugae is directly proportional to the ivermectin concentration. At higher concentration, the egg surface appeared smooth (fig. 2f-g), or the eggshell was shrunken and completely disfigured (fig. 2h).

Discussion

Using ivermectin as a parasiticide has proven effective against the infestation by hard ticks and head lice. This effect was apparent on the adult and immature stages. In addition, ivermectin acted as an effective ovicide on both arthropod species. The avermectins are lethal to a wide range of arthropods, although susceptibility to the chemicals and the time of death vary widely among species. Reduced egg production and decreased hatching have also been observed in a variety of insect pests treated with these chemicals (Miller *et al.* 1986). In the present study, three effects (mortality, hatchability and egg morphology) were observed in both lice and ticks after initial contact with ivermectin at six different concentrations. The results indicated that immature stages were more sensitive than the adults. Also, the higher concentration of ivermectin showed the quickest knockdown and mortality. At 100 ng/ml of ivermectin, no lice survived longer than 5 min in adults or 3 min in nymphs. Meinking *et al.* (1986) reported that the normal longevity of adult head lice is 360-900 min when removed from the host and given no opportunity to feed. Mumcuoglu *et al.* (1990) studied the systemic activity of ivermectin on the human body louse, *Pediculus humanus humanus* and found that the effect of ivermectin was very high during the first two to three days. Immature stages (nymphs) were more sensitive than females and the percentage that hatched from those eggs were lower than in controls. This is similar to our results on the head lice adults and nymphs since we found that the mortality was higher after few minutes at higher concentrations. Also, we found a reduction in the hatchability of the head lice when treated with 100 ng/ml concentration of ivermectin. Meinking *et al.* (1986) suggested several drugs for management of pediculosis, 1% lindane shampoo and 0.5% malathion lotion are highly effective. On other insects, Langley and Roe (1984) observed that testes flies that ingested a blood meal containing ivermectin became lethargic, showed reduced flight activity, progressive paralysis of locomotory muscles and died within 48h. Also, Kerlin and East (1992) studied that the survival and fecundity of the buffalo fly, *Haematobia irritans exigua* after the treatment of cattle with three anthelmintics, ivermectin, closantel and levamisole. They found that the blood from animals given closantel or levamisole had no significant effect on the mortality of buffalo flies in a *in vitro* assay. In contrast, blood from animals given ivermectin showed a dose-dependent effect on the mortality of buffalo flies. These results indicated that ivermectin is also useful in the control of buffalo fly populations in the field. As regard ticks, recent studies have indicated its effects on the lone star tick, *Amblyomma americanum* and the American dog tick, *Dermacentor variables* (Wilson *et al.* 1991). Ivermectin has also been administered to domestic livestock to control both internal and external parasites (Wratten *et al.* 1993; Ottesen and Campbell 1994). These results agree with ours when ivermectin is used against the buffalo tick, *Boophilus annulatus*. When given subcutaneosly at 0.2 mg/kg.

Van Starten and Jongejan (1993) found that the ivermectin was not effective against *Hyalomma* tick infestation in camels. This disagrees with our findings since we used the compound directly not subcutaneously.

Lunke and Kaufman (1992) studied the effect of ivermectin on vitellogenesis and reproduction in the ixodid tick, *Amblyomma hebraeum*. They found that the egg laying was almost completely inhibited at 100 micrograms/kg body weight and further more inhibited the process of vitellogensis.

In conclusion, ivermectin, as naturally derived parasiticide, was effective against the two arthropod species infesting both humans and livestock. The compound showed a dose-dependent effect on the mortality of adult and immature stages. Ivermectin was also proven effective as an ovicide on eggs of both species as indicated by the hatchability, egg morphology and viability studies.

References

Campbell W.C., 1985 - Ivermectin: An Update. *Parasitology Today*, 1(1): 10-16.

Campbell W.C., Fisher M.H., Stapley E.O., Schonberg A.G., Jacop T.A., 1983 - Ivermectin: a potent new antiparasitic agent. *Science*, 221: 823-828.

Clark J.M., Scott J.M., Campos F., Bloomquist J.R., 1995 - Resistance to avermectins: extent, mecanisms and management implications. *Annual Review of Entomology*, 40: 1-30.

Goldsmith R., Heyneman D., 1989 - Pharmacology of antiparasitic drugs. Tropical Medicine and Parasitology. London: Perntic-Hall international Inc.: 850-852.

Greene B.M., Taylor H.R., Cupp E.W., 1985 - Comparison on ivermectin and diethylcarbamazine in the treatment of onchocerciasis. *New England Journal of Medicine*, 313: 133-138.

Hoogstraal H., 1956 - African Ixodoidea: Ticks of the Sudan (with special reference to equatorial province and with preliminary reviews of the genera *Boophilus*, *Margaropus* and *Hyalomma*.) U.S. (vol. 1) Naval medical research unit no. 3. Res. Rep. NM 005 050.29.07.

Keirans J.E., 1992 - Systematics of the Ixodida (Argasidae, Ixodidae, Nuttallielidae): an overview and some problems. In: Fivas B.H., Petney T.N., Horak I.G. (eds.), Tick Vector Biology. Medical and Veterinary Aspect, Springer-Verlag, Berlin: 1-21.

Kerlin R.L., East I.L., 1992 - The survival and fecundity of buffalo flies (*Haematobia irritans*) after treatment of cattle with three anthelmintics. *Australian Veterinary*, 69(11): 283-285.

Klompen J.S., Black W.C., Keirans J.E., Oliver J.H.Jr., 1996 - Evolution of Ticks. *Annual Review of Entomology*, 41: 141-161.

Langley P.A., Roe J.M., 1984 - Ivermectin as a possible control agent for the testes fly, *Glossina morsitans*. *Entomology and Experimental Applied*, 36: 137-143.

Lunke M.D., Kaufman W.R., 1992 - Effects of the avermectin analogue MK-243 on vitellogenesis and reproduction in the ixodid tick, *Amblyomma hebraeum*. *Experimental Applied of Acarology*, 13(4): 249-259.

Meinking T.L., Taplin D., Kalter D.C., Eberle M.W., 1986 - Comparative efficacy of treatments for *Pediculosis capitis* infestations. *Archives of Dermatology*, 122: 267-271.

Miller J.H., Oehler D., Siebenaler A.J., Kunz S.E., 1986 - Effect of Ivermectin on survival and fecundity of horn flies and stable flies (Diptera: Muscidae). *Journal of Economic Entomology*, 79: 1564-1569.

Mumcuoglu K.Y., Miller J., Rosen L.J., Galun R., 1990 - Systemic activity of ivermectin on the human body louse (Anoplura: Pediculidae). *Journal of Medical Entomology*, 27(1): 27-75.

Ottesen E.A., Campbell W.C., 1994 - Ivermectin in human medicine. *Journal of Antimicrobial and Chemotherapeutic*, 34(2): 195-203.

Sauer J.R., McSwain J.L., Bowain A.S., Essenberg R.C., 1995 - Tick salivary gland physiology. *Annual Review of Entomology*, 40: 245-267.

Strong L., Brown T.A., 1987 - Avermectins in insect control and biology: a review. *Bulletin of Entomological Research*, 77: 357-389.

Taylor H.R., Greene B.G., 1989 - The status of ivermectin in the treatment of human onchocerciasis. *American Journal of Tropical Medicine and Hygiene*, 41: 460-466.

Tesh R.B., Guzman H., 1990 - Mortality and infertility in adult mosquitoes after ingestion of blood containing ivermectin. *American Journal of Tropical Medicine and Hygiene*, 43: 229-233.

Van Straten M., Jongjan F., 1993 - Ticks (Acari: Lxodidae) infesting the Arabian camel (*Camelus dromedarius*) in the Sinai, Egypt with a note on the acaricidal efficacy of ivermectin. *Experimental and Applied Acarology*, 17(8): 605-616.

Wikel S.k., 1996 - Host immunity ot ticks. *Annual Review of Entomology*, 41: 1-22.

Wilson K.J., Hair J.A., Sauer J.R., Weeks D.L., 1991 - Effect of ivermectin on the volume of blood ingested by two species of ticks (Acari: lxodidae) feeding on cattle. *Journal of Medical Entomology*, 28(3): 465-468.

Wratten S.D., Mead Briggs M., Gettinby G., Ericsson G., Baggott D.G., 1993 - An evaluation of the effects of ivermectin on the decomposition of cattle dung pats. *Veterinary Recored*, 133(15): 365-371.

Youssef M.Y.M., Sadaka H.A.H., Eissa M.M., El-Ariny A.F., 1995 - Topical application of ivermectin for human extoparasites. *American Journal of Tropical Medicine and Hygiene*, 53(6): 652-653.

Index

Page numbers refer to the first page of a chapter in which an index entry is dealt with.